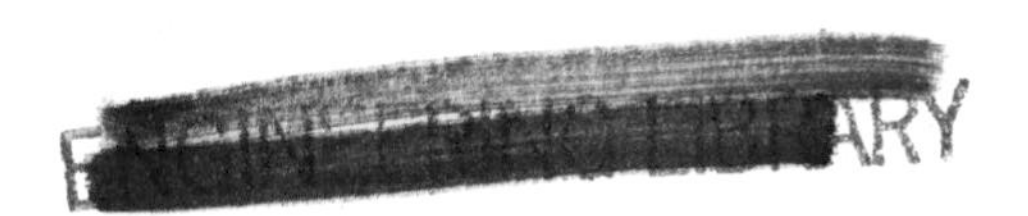

D0880998

Fundamentals of Communications Systems

About the Author

Michael P. Fitz has experience both as a teacher of electrical communications and as a designer of electrical communication systems. He has been a professor of electrical and computer engineering at the University of California Los Angeles (UCLA), the Ohio State University, and Purdue University. Professor Fitz won the D. D. Ewing Undergraduate Teaching Award at Purdue University in 1995. Prof. Fitz was a member of the editorial board of the IEEE Transactions on Communications. Dr. Fitz has worked as a digital communication system engineer for a variety of companies and currently is a senior communication systems engineer at Northrop Grumman. In these roles he has designed, built, and tested modems for land mobile and satellite communications applications. He received the 2001 IEEE Communications Society Leonard G. Abraham Prize Paper Award in the field of communications systems for his contributions to space-time modem technology.

Fundamentals of Communications Systems

Michael P. Fitz

New York Chicago San Francisco
Lisbon London Madrid Mexico City
Milan New Delhi San Juan Seoul
Singapore Sydney Toronto

The McGraw-Hill Companies

Library of Congress Cataloging-in-Publication Data

Fitz, Michael.
Fundamentals of communications systems / Michael Fitz.
p. cm.
ISBN 0-07-148280-6 (alk. paper)
1. Telecommunication systems. I. Title.
TK5101.F54 2007
621.382—dc22
2007017818

McGraw-Hill books are available at special quantity discounts to use as premiums and sales promotions, or for use in corporate training programs. For more information, please write to the Director of Special Sales, Professional Publishing, McGraw-Hill, Two Penn Plaza, New York, NY 10121-2298. Or contact your local bookstore.

Fundamentals of Communications Systems

1234567890 DOC DOC 01987

ISBN-13: 978-0-07-148280-6
ISBN-10: 0-07-148280-6

Sponsoring Editor
Steve Chapman

Editorial Supervisor
Janet Walden

Project Manager
Vasundhara Sawhney

Acquisitions Coordinator
Katie Andersen

Copy Editor
Surendra Nath Shivam

Proofreader
Megha Ghai

Indexer
Broccoli Information Management

Production Supervisor
Jean Bodeaux

Composition
International Typesetting and Composition

Illustration
International Typesetting and Composition

Art Director, Cover
Jeff Weeks

Cover Designer
Pattie Lee

This book is dedicated to

- *the memory of my parents. If my children become half the human beings that my siblings are, I will consider myself a success in life.*
- *Mary. She has given up much to let me pursue my follies. I hope I can now give back to her in an equal amount!*

Contents

PART 2 ANALOG COMMUNICATION

Preface

My goal in teaching communications (and in authoring this text) is to provide students with

- an exposition of the theory required to build modern communication systems.
- an insight into the required trade-offs between spectral efficiency of transmission, fidelity of message reconstruction, and complexity of system implementation that are required for a modern communication system design.
- a demonstration of the utility and applicability of the theory in the homework problems and projects.
- a logical progression in thinking about communication theory.

Consequently, this textbook will be more mathematical than most and does not discuss examples of communication systems except as a way to illustrate how important communication theory concepts solve real engineering problems. My experience has been that my approach works well in an elective class where students are interested in communication careers or as a self-study guide to communications. My approach does not work as well when the class is a required course for all electrical engineering students as students are less likely to see the advantage of developing tools they will not be using in their career. Matlab is used extensively to illustrate the concepts of communication theory as it is a great visualization tool and probably the most prevalent system engineering tool used in practice today. To me the beauty of communication theory is the logical flow of ideas. I have tried to capture this progression in this text. Only you the reader will be able to decide how I have done in this quest.

Teaching from This Text

This book is written for the modern communications curriculum. The course objectives for an undergraduate communication course that can be taught from this text are (along with their ABET criteria)

- Students learn the bandpass representation for carrier modulated signals. (Criterion 3(a))

- Students engage in engineering design of communications system components. (Criteria 3(c),(k))
- Students learn to analyze the performance, spectral efficiency and complexity of the various options for transmitting analog and digital message signals. (Criteria 3(e),(k))
- Students learn to characterize noise in communication systems. (Criterion 3(a))

Prerequisites to this course are probability and random variables and a signal and systems course.

I have taught out of this book material in several ways. I have been lucky to teach at three universities (Purdue University, the Ohio State University, and the University of California Los Angeles) and each of these experiences has profoundly impacted my writing of this book. The material in this book has been used to teach three classes

- Undergraduate analog communications and noise (30 lecture hours)
- Undergraduate digital communications (30 lecture hours)
- Undergradute communications (40–45 lecture hours)

The course outline for the 30 lecture hours of analog communication is

1. Chapters 1 & 2 — 1 hour. This lecture was a review of a previous class.
2. Chapter 4 — 2 hours. These lectures build heavily on signal and system theory and specifically the frequency translation theorem of the Fourier transform.
3. Chapter 5 — 1 hour. This lecture introduces the concept of analog modulation and the performance metrics that engineers use in designing analog communication systems.
4. Chapter 6 — 5 hours. These lectures introduce amplitude modulation and demodulation algorithms. Since this course was an analog only course, I spend more time on the practical demodulation structures for DSB-AM and VSB-AM.
5. Chapter 7 — 5 hours. These lectures introduce angle modulation and demodulation algorithms. I like to emphasize that the understanding of the spectrum of angle modulations is best facilitated by the use of the Fourier series.
6. Chapter 8 — 2 hours. These lectures introduce multiplexing and the phase-locked loop. Multiplexing is an easy concept and yet students enjoy it because of the practical examples that can be developed.
7. Chapters 3 & 9 — 6 hours. These lectures introduce random variables and random processes. This, from the student's perspective, is the most difficult

part of the class as the concept of noise and random processes are mathematically abstract. Undergraduates are not used to abstract concepts being important in practice.

8. Chapter 10 — 3 hours. These lectures introduce bandpass random processes. This goes fairly well once Chapter 9 has been swallowed.
9. Chapter 11 — 3 hours. These lectures introduce fidelity analysis and the resulting SNR for each of the modulation types that have been introduced. The payoff for all the hard work to understand random processes.
10. Test — 1 hour.

The 30 hour digital only communication course followed the analog course so it could build on the material in the previous course. This course often contained some graduate students from outside the communications field that sat in on the course so some review was necessary. The course outline for 30 lecture hours of digital communications is

1. Chapter 1 & 4 — 2 hours. These lectures introduce communications and bandpass signals.
2. Chapter 9 & 10 — 2 hours. These lectures introduce noise and noise in communication systems.
3. Chapter 12 — 1 hour. This lecture introduces the concept of digital modulation and the performance metrics that engineers use in designing systems. This lecture also introduces Shannon's limits in digital communications.
4. Chapter 13 — 6 hours. These lectures emphasize the five-step design process inherent in digital communications. In the end these lectures show how far single-bit transmission is from Shannon's limit.
5. Chapter 14 — 4 hours. These lectures show how to extend the single bit concepts to M-ary modulation. These lectures show how to achieve different performance-spectral efficiency trade-offs and how to approach Shannon's limit.
6. Chapter 15 — 8 hours. These lectures introduce most of the modulation formats used in engineering practice by examing the complexity associated with demodulation.
7. Chapter 16 — 3 hours. These lectures introduce bandwidth efficient transmission and tools used to test digital communication systems.
8. Chapter 17 — 2 hours. These lectures introduce coded modulations as a way to reach Shannon's bounds.
9. Test — 1 hour.

The 40–45 hour analog and digital communication course I taught traditionally had a much more aggressive schedule. The course outline for 40 lecture hours of communications is

1. Chapters 1 & 2 — 1 hour. The signal and systems topics were a review of a previous class.
2. Chapter 4 — 2 hours. These lectures build heavily on Fourier transform theory and the frequency translation theorem of the Fourier transform.
3. Chapter 5 — 1 hour. This lecture introduces the concept of analog modulation and the performance metrics that engineers use in designing systems.
4. Chapter 6 — 4 hours. These lectures introduce amplitude modulation and demodulation algorithms. The focus of the presentation was limited to coherent demodulators and the envelope detector.
5. Chapter 7 — 4.5 hours. These lectures introduce angle modulation and demodulation algorithms. I like to emphasize that the understanding of the spectrum of angle modulations is best facilitated by the use of the Fourier series.
6. Chapter 8 — 0.5 hour. Only covered multiplexing.
7. Chapters 3 & 9 — 5 hours. This is the toughest part of the class as the concept of noise and random processes are mathematically abstract. Undergraduates are not used to abstract concepts being important in practice.
8. Chapter 10 — 2 hours. This goes fairly well once Chapter 9 has been swallowed.
9. Chapter 11 — 2 hours. The payoff for all the hard work to understand random processes.
10. Chapter 12 — 1 hour. This lecture introduces the concept of digital modulation and the performance metrics that engineers use in designing systems. This lecture also introduces Shannon's limits in digital communications.
11. Chapter 13 — 6 hours. These lectures emphasize the five-step design process inherent in digital communications. In the end these lectures show how far single-bit transmission is from Shannon's limit.
12. Chapter 14 — 4 hours. These lectures show how to extend the single bit concepts to M-ary modulation. These lectures show how to achieve different performance-spectral efficiency trade-offs and how to approach Shannon's limit.
13. Chapter 15 — 3 hours. These lectures introduce most of the modulation formats used in engineering practice by examing the complexity associated with demodulation.
14. Chapter 16 — 1 hours. These lectures introduce bandwidth efficient transmission and tools used to test digital communication systems.
15. Chapter 17 — 1 hours. These lectures introduce coded modulations as a way to reach Shannon's bounds.
16. Test — 1 hour.

The 45 hour course added more details in the digital portion of the course.

Style Issues

This book takes a stylistic approach that is different than the typical communication text. A few comments are worth making to motivate this style.

Property–Proof

One stylistic technique that I adopted in many of the sections, especially where tools for communication theory are developed, was the use of a property statement followed by a proof. There are two reasons why I choose this approach

1. The major result is highlighted clearly in the property statement. Students, in a first pass, can understand the flow of the development without getting bogged down in the details. I have found that this flow is consistent with student (and my) learning patterns.
2. Undergraduate students are increasingly not well trained in logical thinking in regard to engineering concepts. The proofs give them some flavor for the process of logical thinking in engineering systems.

General Concepts Followed by Practical Examples

My approach is to teach general concepts and then follow up with specific examples. To me the most important result from a class taught from this book is the learning of fundamental tools. I emphasize these tools by making them the focus of the book. Students entering the later stages of their engineering education want to see that the hard work they have put into an engineering education has practical benefits. The course taught from this book is really fun for the students as old tools (signals and systems and probability) and newly developed tools are needed to understand electronic communication.

Two Types of Homework Problems

This book contains two types of homework problems: (1) direct application problems and (2) extension problems. The application problems try to define a problem that is a straightforward application of the material developed in the text. The extension problem requires the student to think "outside the box" and extend the theory learned in class to cover other important topics or cover practical applications. *As a warning to students and professors*: Often times the direct application problems will appear ridiculously simple if you carefully read the text and the extension problems, as they are often realistic engineering problems, appear to be much too extensive for a homework problem. I have found that both types of problems are important for undergraduate education. Direct application problems allow you to practice the theory but are usually not indicative of the types of problems an engineer sees in practice. Alternatively students often desire realistic problems as they want a feel for "real" engineering but often get overwhelmed with the details needed in realistic problems. All direct application leads to a boring sterile course and all extension problems discourage all but the exceptionally smart and motivated. Having a book with both types of problems allows the student to both exercise and extend their learning in the proper balance.

Examples and Example Solutions

My learning style is one where a very succinct presentation of the issues works best. I originally wrote this book where the material was presented in a condensed version with little or no examples. I then followed each chapter with a set of example solutions to homework problems. During the many revisions of the book I came to realize that many students learn best with examples along with a presentation of the theory. I added a significant number of in-text examples to meet this learning need. I still like a succinct presentation so I did not move all example solutions to in chapter examples. Consequently, each chapter has a set of in-chapter examples and a set of worked solutions. This method was viewed as a compromise between a succinct presentation (my learning style and hopefully a few others) and lots of examples (many students' learning style).

Miniprojects

Both for myself and the students I have taught, learning is consumated in "doing." I include "Miniprojects" in the book to give the students a chance to implement the theory. The project solutions are appropriate for oral presentation and this gives the students experience that will be a valuable part of an engineering career. The format of the project is such that it is most easily done in Matlab as that is the most common computer tool used in communication engineering systems. To aid students who are not familiar with Matlab programming I have included the code for all the Matlab generated figures in the text on the book web page. This also allows students to see how the theory can be implemented in practice.

Writing of the Book

The big question that has to be answered in this preface is "Why should anyone write another communication theory book?" The short answer is "There is no good reason for the book and a rational person would not have written the book." The book resulted from a variety of random decisions and my general enjoyment of communication engineering. A further understanding of this book and my decision to write it can be obtained by understanding the stages I perceived in looking back on the writing this book. This documentation is done in some sense for those who will follow in my folly of attempting to write a book to give them a sense of the journey.

1. **Captured.** As a child, a high school student and a college student I was always drawn to math and science, to problem solving, and to challenges. Quickly my career path steered toward engineering, toward electrical engineering, and finally toward communication engineering. I took a job as a communication engineer while pursuing a graduate education. In the first

four years of working I used all of my graduate classes in solving communication problems and was a key engineer in a team that built and field tested a sophisticated wireless modem. I was hooked by communications engineering: It is a field that has a constant source of problems, a well-defined set of metrics to be used in problem solving, and a clear upperbound (due to Claude Shannon) to which each communication system could aspire.

2. **Arrogance.** I started my academic career, after being reasonably successful while working in industry, with the feeling that I knew a great deal. I was convinced that my way of looking at communications was the best and I started teaching as such. I found the textbooks available at the time to have inadequate coverage of the complex envelope representation of bandpass signals and bandpass noise and other modern topics. Hence my writing career started by preparing handouts for my classes on these topics. I quickly got up to 100 pages of material.
3. **Humbled but Learning.** It was not too long after starting to teach and direct graduate student research that I came to the realization that the field of communications was a mighty river and I had explored only a few fairly minor tributaries. I came to realize I did not know much and still needed to learn much. This realization began to be reflected in my teaching as well. I branched out and learned other fields and reflected my new understanding in my teaching methods and approaches. Much to my students' chagrin, I often used teaching as a method to explore the boundaries of my own learning. This resulted in many poorly constructed homework problems and lectures that were rough around the edges. As I am not very bright, when I synthesized material I always had to put it into my own notation to keep things clear in my own mind. After these bouts with new material that were very confusing for my students, I often felt guilty and wrote up notes to clarify my ramblings. Soon I was up to 200 pages of material on digital communications. As I would discover later these notes, while technically correct, were agonizingly brief for students and lacked sufficient examples to aid in learning.
4. **Cruise Control.** I soon got to the point where I had reasonable notes and homework and my family and professional committments had grown to the point where I needed not to focus so much on my teaching and let things run a bit in cruise control. During this time I added a lot of homework and test problems and continued to write up and edit material that was confusing to students. My research always seems filled with interesting side issues that make great homework problems. I started the practice of keeping a note book of issues that have come up during research and then tried to morph these issues into useful homework problems. Some problems were successes and some were not. In 2002, I was up to about 300 pages.
5. **Well I have 300 Pages...** At the point of 300 pages I felt like I turned a corner and had a book almost done and started shopping this book around to publishers. My feeling was that there was not much left to complete and once I signed a contract the book would appear in 6 months. This writing

would magically take place while my professional and family life flourished. I eagerly wrote more material and was up to 400 pages.

6. **...And Then Depression Set In.** When I looked at my project with the critical eye produced by signing a contract, it quickly became apparent that lots of pages do not necessarily equate to a book of any quality. At this time I also got my first reviews back on the book and quickly came to realize why all textbooks on communications look the same to me. If you took the union of the reviews you would end up with a book close to all the books on the market (that I obviously did not fully appreciate). During this time I struggled mightily at trying to smooth out the rough edges of the book and address reviewers' concerns, while trying to keep what I thought was my personal perspective on communications. This was a significant struggle for me, as I learned the difficult lesson that each person is unique in how they perceive the world and consequently in what they want in a textbook but I stubbornly soldiered through to completion.

In summary the book resulted not from a well thought out plan but from two disjointed themes: (1) my passionate enjoyment of communication engineering, (2) my constant naive thinking as a professional. The book is done and it is much different than I first imagined it. It is unique in its perspective but not necessarily markedly different than the other books out there. It is time to release my creation.

Heresies in the Book

The two things I learned in writing this book is that engineering professionals do not see the field the same and engineering professors do not like change. In trying to tell my version of the communication story I thought long and hard about how to make the most consistent and compelling story of communication engineering for my students. In spite of what I considered a carefully constructed pedagogy, I have been accused (among other things) of making up nomenclature and confusing the student needlessly. Since I now realize that I am guilty of several potential heresies to the field of communication education, I decided to state these heresies clearly in the preface so all (especially those that teach from this book) know my positions (and structure their classes appropriately).

1. **System Engineering Approach.** I have worked as a communication engineer in industry and academia. I do not have a detailed knowledge of circuit theory and yet from all outward appearances I have thrived in my profession by only being an expert in system modeling and analysis. This text will not give circuits to build communication systems as circuits will change over time but will discuss mathematical concepts as these are consistent over time. I am a firm believer that communications engineering is perhaps unique in how theory directly gets implemented in practice.

2. **Fidelity, Complexity, and Spectral Efficiency.** Everything in electronic communications engineering comes down to a trade-off between fidelity of message reconstruction, complexity/cost of the electronic systems used to implement this communication, and the spectral efficiency of the transmission. I have decided to adopt this approach in my teaching. This approach clearly deviates from past practice and has not proven to be universally popular.
3. **Stationary versus Wide-Sense Stationary.** An interesting characteristic in my education was that a big deal was made out of the difference between stationary and wide-sense stationary random processes – and most likely for all communication engineers of my generation. As I went through my life as a communication engineering professional I came to the realization that this additional level of abstraction was only needed because the concept of stationarity was arbitrarily introduced before the concept of Gaussianity. For my book, I introduce Gaussian processes first and then the idea of wide-sense stationarity is never needed. My view is clearly not appreciated by all[1] but my book has only stationary Gaussian processes and random variables. I felt the less new concepts in random processes that are introduced in teaching students how to analyze the fidelity of message reconstruction, the better the student learning experience would be.
4. **Information Theory Bounds.** My view is that digital communication is an exciting field to work in because there are some bounds to motivate what we do. Claude Shannon introduced a bound on the achievable fidelity and spectral efficiency in the 1940s [Sha48]. Communication engineers have been pursuing how to achieve these bounds in a reasonable complexity ever since. Many people feel strongly that Shannon's bounds cannot be introduced to undergraduates and I disagree with that notion! It is arguable that Claude Shannon has a bigger impact on modern life than does Albert Einstein yet name recognition among engineering and science students is not high for Claude Shannon. Hopefully introducing Shannon and his bounds to undergraduates can give him part of his due.
5. **Erfc(•) versus Q(•).** The tail probability of a Gaussian random variable comes up frequently in digital communications. The tail probability of a Gaussian random variable is not given by a simple expression but instead must be evaluated numerically. Past authors have used three different transcendental functions to specify tail probabilities: Erfc(•), Φ(•), and the Q(•). Historically, communication engineers have gravitated to the use of Q(•) as its definition matches more closely how the usage comes up in digital communications. I have chosen to buck this trend because of one simple fact: Matlab is the most common tool used in modern communications engineering and Matlab uses Erfc(•). Most people who read this text after having

[1]Some reviewers went so far as to suggest I needed to review my random processes background to get it right!

used other communication texts do not appreciate the usage of Erfc(•) but frankly this book is written for students first learning communications. This notation serves these students much better (even while it irritates reviewers) as it gives a more consistent view betweeen the text and the common communication tools.

6. **Signal Space Representations.** A major deficiency in my approach, according to some reviewers, is that I do not include signal space representations of digital signals. While I understand the advantages and insights offered by this approach I think signal space representations lead the students off course. Specifically, I do not know of a single communication system that uses the signal space concepts in designing a demodulator or a modulator other than to exploit orthogonality to send bits independently (see orthogonal modulations below). The best example of a high dimension signaling scheme is a direct sequence code division multiple access (DS-CDMA) system. To the best of my knowledge no DS-CDMA system does "chip" level filtering and then combining as would be suggested by a signal space approach but directly implements each spreading waveform or each spreading waveform matched filter. All demodulators I am familiar with are based on the concept of the matched filter. I feel taking the matched filter approach leads to a more consistent discussion, while many of my colleagues feel a signal space approach is necessary for their students to comprehend digital communications.
7. **Noncoherent and Differentially Coherent Detection in Digital Communications.** These subjects never enter this introductory treatment of communication theory as they are really secondary topics in modern communications theory. When I started my career there were three situations where tradition ruled that noncoherent or differentially coherent techniques were mandatory for high performance communications: (1) in the presence of jamming, (2) with short packets, and (3) in land mobile wireless communications. In the past 15 years I have worked on these types of systems in both an academic environment and as part of commercial engineering teams and not once were noncoherent or differentially coherent techiques used[2] in modern communication systems. I decided that rather than confuse the student with a brief section on these topics that I would just not present them in this book and let students pick up this material, if needed, in graduate school or with experience.
8. **Cyclostationarity and Spectrum of Digital Modulations.** This is a sensitive subject for many of my professional colleagues. I am strongly of the opinion that spectral efficiency is a key component of all digital communication discussions. Consequently, all digital transmissions must have an associated bandwidth. Interestingly nowhere in the previous teaching

[2]Except to support legacy systems.

texts was there a method to compute the spectral content of finite length transmissions even though this must be done in engineering practice. To be mathematically consistent with the standard practice for defining the power spectrum of random processes, I settled on the concept of an average energy spectrum. Here the average is over the random data sequences that are transmitted. For students to understand this concept they only need to understand the concept of expectation over a random experiment. This average energy spectrum is defined for any modulation format and for finite or infinite length transmissions. In contrast, many professors who teach communications are wed to the idea of computing the power spectrum of digital transmissions by

(a) Assuming an infinite length transmission
(b) Defining cyclostationarity
(c) Averaging over the period of the correlation function of a cyclostationary process to get a one parameter correlation function
(d) Taking the Fourier transform of this one parameter correlation function to get a power spectrum

This procedure has four drawbacks: (1) it introduces a completely new type of random process (to undergraduates who struggle with random processes more than anything else), (2) it introduces a time averaging for no apparent logical reason (this really confused me as a student and as a young engineer), (3) it only is precise for infinite length transmissions (no stationarity argument can be used on a finite length transmission), and (4) these operations are not consistent with the theory of operation of a spectrum analyzer that will be used in practice. Hopefully it is apparent why this traditional approach seems less logical than computing the average energy spectrum. In addition, the approach used in this book gives the same answers in the cases when cyclostationarity can be used (without the strange concept of cyclostationarity) and gives answers in cases where cyclostationarity cannot be used, and is consistent with spectral analyzer operations. Unfortunately, I have learned (perhaps too late in life) that when you are a heretic, logic does not help your case against true believers of the status quo.

9. **Orthogonal Modulations.** My professional career has led me on many interesting rides in terms of understanding of communication theory. Early in my career the communciation field was roiled by a debate of narrowband modulation versus wideband modulation sparked by Qualcomm's introduction of IS-95. At the time I felt wideband modulation was a special case of a general modulation theory. I remember at the time (roughly 1990) someone making the comment during a discussion that narrowband modulation was a special case of wideband modulation[3] and at the time I was dismissive of

[3] I believe this discussion was with Wayne Stark or Jim Lehnert.

this attitude. These concepts stayed in the back of my mind and fermented. My research focus around 1994 went to multiple antenna modems and in this field one of the most powerful and interesting ideas proposed was the Alamouti signaling scheme [Ala98]. Alamouti signaling was a method to multiplex data across antennas in an orthogonal fashion. Unfortunately, to further upset my thinking, I started doing research on wireless orthogonal frequency division multiplexing (OFDM) systems (2000) and at first the unification of all of these concepts was not apparent to my simple mind. Then I struck upon the idea that Nyquist's criteria, Alamouti signaling, OFDM, and orthogonal spreading waveforms in CDMA systems were all just variations on a theme of orthogonality. This orthogonality allows bits to be sent and decoded independently in a very simple way. Once the orthogonality thread was put in place in my teaching then the relationship among all these apparent disparate systems, with wideband signaling being the most general modulation fell into place nicely. Finally, I hated the idea of calling "normal" modulation "orthogonal time division multiplexing" and wanted a shorter nomenclature. Since the idea was that bits were sent one after another in time and that students were comfortable with the idea of streaming video that comes with internet usage, I adopted the notation of stream modulation. Many of my colleagues and book reviewers really were frustrated by my unified view and by my making up this new terminology[4]. Also the methodology of teaching a general concept (orthogonality) applicable to multiple situations did not resonate with many professors' teaching styles.

10. **Pulse Shapes.** My discussion of spectrally efficient digital modulations again does not follow standard practice. The key ideas in pulse shaping result from orthogonality. I develope these ideas and introduce a cosine pulse shape and a squared cosine pulse shape that can be used either in the time or frequency domain. This approach gives a better understanding of why the pulse shapes evolved but does not use the standard notation in the literature (e.g., spectral square root raised cosine). Hence I have introduced new notation but this notation is only used to make the material more clear conceptually.
11. **OMWM.** Communication engineers have approached Shannon's limits by adding structured redundancy into transmitted waveforms. I try to capture this idea in this undergraduate book and keep a consistent theme by introducing the concept of orthogonal modulations with memory (OMWM). OMWM as a paradigm does not limit coding to time or frequency domain signaling but enables a general approach. This general approach is different from what has been done in the past and hence is not universally accepted as the correct way to teach this subject. A new notation was introduced as a way to highlight the important ideas of modern communications without writing a toothless chapter on capacity approaching signaling.

[4]The first time I taught stream modulation a student Googled the term with zero hits!

Key Teachable Moments in the Book

I thought it useful to enumerate my view of the key teachable moments in an undergraduate communications course. My view will likely give the users of the book a better idea of how to get the most out of the book.

1. **Communication Signals Are Two Dimensional.** Bandpass communication signals are inherently two dimensional. Electronic communication is about embedding information in these two dimensions and retrieving information from the two dimensions of a corrupted received bandpass signal in a manner that best achieves a set of engineering trade-offs. Using the complex envelope representation throughout the book emphasizes this idea of communications signals being two dimensional in a continuous manner.
2. **Fidelity, Complexity, and Spectral Efficiency.** Communication engineering is a vibrant field because there is no one best solution to all communication problems but a wide variety of solutions that define a different operating point within a three dimensional trade-off space. The dimensions of the trade-off space for all communication systems are: (1) the fidelity of message reconstruction, (2) the cost and complexity of the electronics to implement the system, and (3) the amount of bandwidth used to accomodate the electronic transmission of information. An important concept that must be understood is that this trade-off is a constantly changing trade-off. The reason this trade-off evolves with time is that the cost of electronics is decreasing for a fixed complexity with time or equivalently the complexity is increasing for a fixed cost with time. It is important for the readers of this book to understand engineering decisions made in the last century should not be the same as the engineering decisions made in the coming century. Consequently, communications engineering requires lifelong learning.
3. **Filter Design to Improve Spectral Efficiency.** When I was taught about single-sideband and vestigial-sideband amplitude modulation (AM), I came away with the impression that the modulation resulted from a particular judicious choice of a bandpass filter. This gave me no insight into the problem and how I might apply the solution to different problems. I was directly faced with the shortcoming of my education when I had a chance to work on digital television broadcast where vestigial sideband transmission is one option and I realized there was no straightforward way to generalize my education to apply to digital transmissions. My approach in teaching these concepts is to note that a degree of freedom is not used in traditional AM (the quadrature channel) and to turn a desire to achieve better spectral efficiency into a design problem with an intuitive answer. This approach gives a much better flavor for what communication engineering tasks are like as well as giving a nontrivial answer to why single-sideband and vestigial-sideband amplitude modulations are used in communication systems.
4. **Fourier Series to Analyze the Bandwidth of Angle Modulation.** Finding the spectrum of an angle modulated signal is the first problem

encountered by a communication engineer that cannot be solved by application of signal and system theory (as angle modulation is a nonlinear transformation). This is a great case study of how engineers gain insight into complex problems by examining simple special cases, here a simple periodic message signal. This analysis is important for both the insight it provides and the demonstration of a critical engineering approach that works in a wide variety of problems. Additionally, there is no better example of how the Fourier series can help solve practical communication problems.

5. **Pre-emphasis/De-emphasis in FM.** My favorite example of how engineering understanding can lead to significant performance gains is the idea of pre-emphasis and de-emphasis in angle modulations. Pre-emphasis and de-emphasis is used in broadcast radio because engineers realized that in demodulation of frequency modulation (FM) that the noise spectrum at the demodulator output was shaped in the frequency domain (a colored noise). This colored noise and the constant envelope characteristic of FM led to an improved signal processing technique that increased the output demodulation fidelity. It is a great example of how true understanding leads to a 10-dB gain in demodulation fidelity with little increase in complexity.
6. **Single Bit Modulation and Demodulation.** I find the single bit transmission and demodulation process very illustrative of the tasks communication engineering professionals must complete. The process of identifying a need (sending a bit of information), building a mathematical model of the processing (detection theory), and then completing a series of design problems (best threshold, best filter, best signals) to optimize performance of the system is very typical in a communication engineering career. The issue related to single bit detection are all developed in detail to build a comprehensive understanding of the communication engineering process.
7. **Digital Communications and Shannon's Bound.** The generalization of the modulation and demodulation of multiple bits is a straightforward extension of single bit ideas. The interesting part occurs when examples are examined and the realization is made that choices in modulation directly translate to different points in the fidelity versus spectral efficiency performance space parameterized by Shannon's bounds. Showing this relationship between simple modulation ideas and the bound proposed by Shannon demonstrates the power of Shannon's theory.
8. **Two Dimensional Digital Signaling.** The idea of linear modulation is a simple and insightful one. The mapping of bits into symbols in the complex plane is simple to understand. The demodulation by computing the distances between the received signal and all the constellation points is intuitive. What really makes this a teachable moment is that probably more than 90 percent of the digitial communication systems use linear modulations in some form.

9. **Orthogonality to Reduce Complexity.** Almost all modern digital communication systems use the concept of orthogonality but sometimes in disparate forms. The traditional way of introducing these topics is for each type of orthogonality to be a special advanced topic. For example, mobile telephone service is often used to introduce the idea of code division multiplexing. Unfortunately, what is lost in this discussion is that the orthogonality is designed into modems primarily to reduce the complexity of the demodulator. Modern communication engineers should be able to see the big picture and consequently this book teaches a general idea (orthogonality) and introduces code division multiplexing, frequency division multiplexing, and Nyquist theory as the important examples of this general idea. Modern engineers need modern training to understand the trade-offs these systems offer.

10. **Spectral Shaping in Communications.** No modern communication system can afford to ignore the spectral impacts of modulation as that is one of the three dimensions that define the operating point of a modern communication system. Since spectral efficiency has such importance in modern communications, techniques to achieve the practical spectral characteristics are introduced in a separate chapter. The nontrivial idea of pulse shaping in communications is one of the first steps a communication engineer must take to really understand the practical aspects of communication engineering. As an interesting contrast of practice versus academia, academics tend to dismiss this spectral shaping as relatively unimportant and straightforward while most engineering teams in practice agonize over the practical aspects of spectral shaping and meeting the prescribed spectral mask.

11. **Adding Memory to Improve Performance.** While this is a first course in communications, there is a final chapter that shows how far modern communications has progressed and how close the profession of communication engineering is to the bounds Shannon identified in certain situations. This chapter highlights these ideas by showing the fundamental signal design techniques from a communication theory perspective, i.e., signal design to address the Euclidean distance spectrum and average energy spectrum. This is a powerful final perspective on modern communication engineering for an engineer/student to take away from the book.

Acknowledgements

The professional acknowledgements are numerous. First I should thank the vast network of intelligent communication engineers that came before me professionally. Their great work inspired me with the passion to write this book (I will not name names as I will forget a well-deserving person). I must thank Dale Feikema and Dan Jones who trusted a wet behind the ears engineer (me) with a way cool communication system engineering job. Man that was fun, too bad I can never tell anyone about it! William Lindsey my advisor at the University of Southern California was a great role model for hard work and technical competence that I tried to take into my professional life. My time at TRW and Northrop Grumman and the exposure to a wide range of technical expertise taught me a breadth of system engineering expertise that is not often found in a textbook writer. I must thank the industrial sponsors of my research for in working on so-called "practical" and important problems I was able to formulate a modern way of viewing communication theory and derive many interesting homework problems. Hopefully I have brought the sum total of that experience into this text.

I shared my academic life with three superb faculties and these colleagues greatly influenced my growth as an engineer/educator. Professors whose interaction influenced this book include Peter Doerschuk (PD), Saul Gelfand, James Krogmeier, Randy Moses, Oscar Takeshita, Lee Potter, Phil Schniter, Urbashi Mitra (UM), and Rick Wesel (RW). Other people (that I know of) who provided reviews that changed the manuscript for the better include Jerome Gansman (JG) and Youjian Liu. Daniel Costello gave me some well heeded professional advice and provide his talk on the "Genesis of Coding" which I was able to use to help make comparisons of the best modern signaling schemes to Shannon's bounds. Homework problems that I liked and were authored by other people and included in the book are annotated by the author's initials. I thank them for sharing. Oscar Takeshita and Jerome Gansman were helpful in sharing typeset homework solutions.

Financial support by the National Science Foundation (NSF) of the United States of America should be acknowledged. The constant support through my academic career by NSF made this book possible without me having to starve my children.

The greatest professional acknowledgement must go to my students. Nothing teaches you more than working on research with a bunch of very bright individuals. I look back on the people I had the great fortune to advise as graduate students and I realize now how much they caused me to grow and how good of people they were. I thank them publicly now for their enthusiasm for learning and their patience with me as their professor and advisor. I certainly hope that you enjoyed the collaboration as much as I did. The students who sat through my classes deserve a special recognition as I experimented with teaching pedagogy much too often and they paid the price. I hope these people accept my apology for my treatment of them and their time and that the production of this work will in some small way justify their pain.

The final acknowledgement goes to my nuclear family; Mary, Kistern, Julie, Erin, and Anne. Thank you for all you have put up with (more than I want to admit publicly!) during the production of this book. While I certainly have shown you the weakness of my character during this effort, I never stopped loving you.

Chapter

1

Introduction

The students who read this book have grown up with pervasive communications. A vast majority have listened to broadcast radio and television, used a mobile phone, surfed the World Wide Web, and played a compact or video disk recording. Hence there is no need for this book to motivate the student about the utility of communication technology. They use it everyday. This chapter will consequently be focused on the engineering aspects of communication technology that are not apparent from a user perspective.

1.1 Historical Perspectives on Communication Theory

The subject of this book is the transportation of information from point A to point B using electricity or magnetism. This field was born in the mid-1800s with the telegraph and continues today in a vast number of applications. Humans have needed communications since prehistoric times for capitalistic endeavors and the waging of war. These social forces with the aid, at various points in time, of government-sponsored monopolies have continuously pushed forward the performance of communications. It is perhaps interesting to note that the first electronic communications (telegraphy) were sending digital data (words were turned into a series of electronic dashes and dots). As the invention of the telephone took hold (1870s), communication became more focused on analog communication as voice was the information source of most interest to convey. The First World War led to great advances in wireless technology and television and radio broadcast soon followed. Again the transmitted information sources were analog. The digital revolution was spawned by the need for the telephone network to multiplex and automatically switch a variety of phone calls. A further technology boost was given during the Second World War in wireless communications and system theory. The Cold War led to rapid advances in satellite communications and system theory as the race for space gripped the world's major technology innovation centers. The invention of the semiconductor transistor and the impact of Moore's law have spurred the march of innovation since the early 1980s. The evolving power of the microprocessor, the embedded computer, and the signal processor has enabled algorithms, that were considered preposterous at their formulation, to see cost-effective implementation. Distilling this 150 plus years of innovation into a small part of an engineering curriculum is a challenge but one this book arrogantly attempts.

The relative growth rate of electronic communications is phenomenal. Consider, for example, transatlantic transmission of information using undersea cables. This system has gone from roughly 10 bits/s in 1866 to roughly 10^{12} bits/s in the year 2000 [Huu03]. The world community has gone in a very short period of time from accepting message delivery delays of weeks down to seconds. The period from 1850–1900 was one filled with remarkable advances in technology. It is noteworthy that the advances in communications prior to 1900 can almost all be attributed to a single individual or invention. This started to change as technology became more complex in the 1900s. Large corporations and research labs began to be formed to support the large and complex systems that were evolving. The evolution of these technologies and the personalities involved in their development are simply fascinating. Several books that are worth some reading if you are interested in the history of the field are [Huu03, Bur04, SW49, Bra95, Les69]. It is a rare invention that has an uncontested claim to ownership. These intellectual property disputes have existed from the telegraph up until modern times, but the tide of human innovation seems to be ever rising in spite of who gets credit for all the advances.

The ability to communicate has been markedly pushed by advances in technology but this book is not about technology. From the invention of the microphone, to the electric motor, to the electronic tube, to the transistor, and to the laser, engineers and physicists have made great technological leaps forward. These technological leaps have made great advances in communications possible. As technology has advanced, the job of an engineer has become multifaceted and specialized over time. What once was a field where nonexperts could contribute[1] prior to 1900 became a field where great specialization was needed in the post 1900 era. Two areas of specialization formed through the 1900s: the devices engineer and the systems engineer. The devices engineer is focused on designing technology to complete certain tasks. Devices engineers, for example, build antennas, amplifiers, and/or oscillators are heavily involved with current technology. Systems engineers try to put devices together in a way that will work as a system to achieve an overall goal. System engineers try to form mathematical models for how systems operate and use these models to design and specify systems. This text is written with a **systems engineering** perspective. In fact, as a reflection of this focus, this book has exactly one circuit diagram. This is not an academic shortcoming of this book as the author,

[1]For example, Samuel Morse of Morse code and telegraph systems fame in the United States was a professor in the liberal arts.

for example, has worked for 20 years and for five companies designing communication systems with only a rudimentary knowledge of circuit theory. Systems engineering in communications did not really come to be a formalized field until the early 1900s, hence few of the references in this book were published before 1900. Some interesting historical system engineering references are [Car26, Nyq28a, Arm35, Har28, Ric45, Sha48, Wie49]. This systems level perspective is very useful for education in that; while technology will change greatly during an engineer's career, the theory will be reasonably stable.

1.2 Goals for This Text

What this text is attempting to do is to show the mathematical and engineering underpinings of communication systems and systems engineering. While most students have used communication technology, few realize that the technology is built upon a strong core of engineering principles and over 100 years of hard work by a large group of talented people. Without a talented engineering workforce who understood the fundamental theory and put this theory into practice, humans would not have been able to deploy the pervasive communications society experiences. The goal for this text is to have some small part in the education of the workforce that will implement the next 50 years of progress. To reach this goal this text will focus on teaching the fundamentals of communication theory by:

- demonstrating that the mathematical tools the students have learned in their undergraduate education are useful in engineering practice.
- showing that with modern integrated circuits the theory is directly reflected in engineering practice.
- detailing how engineering trade-offs in a communication system are ever evolving and that these trade-offs involve fidelity of message reconstruction, bandwidth efficiency, and complexity of the implementation.

Hopefully in addition to these professional goals, the reader of this text will come away with:

- a historical perspective on the hard work that has led to the current state of the art.
- a sense of how fundamental engineering tools have real impact on system design.
- a realization that fundamental engineering tools have changed little even as the technology to implement designs has evolved at a withering pace.
- an understanding that communications engineering is a growing and evolving entity and that continued education will be an important part of a career as a communication engineer.

1.3 Modern Communication System Engineering

Modern communication systems are very complex systems and no one engineer can be an expert in all the areas of the system. The initial communication systems were very simple point-to-point communication systems (telegraphy) or broadcast systems (commercial radio). As these systems were simple, the engineering expertise could be common. As systems started to get more sophisticated (public telephony), a bifurcation of the needed expertise to address problems became apparent. There was a need to have an engineer who understood the details of the physical channel and how the information was transmitted and decoded. In addition there was a need to have an engineer who abstracted the problem at a higher level. This "higher level" engineer needed to think about switching architectures, supporting multiple users, scalability of networks, fault tolerance, and supporting applications. As the amount of information, system design options, and technology to implement these options grew, further subdisciplines arose within the communications engineering field.

Modern communication systems are typically designed in layers to compartmentalize the different expertise and ease the interfacing of these multitude of expertises. In a modern system, the communication system has a high-level network architecture specification. This high-level architecture is typically broken down into layers for implementation. The advantage of the layered architecture in the design process is that in designing a system for a particular layer the next lower layer can be dealt with as an abstract entity and the higher layer functions do not impact the design. Another advantage to the layered design is that components can be reused at each layer. This allows services and systems to be developed much more quickly in that designs can reuse layers from previous designs when appropriate. This layered design eliminated monolithic communication systems and allowed incremental changes much more readily.

An example of this layered architecture is the *open systems interconnection* (OSI) model. The OSI model was developed by the International Organization for Standardization (ISO) and has found significant utilization in practice. The OSI reference model is shown in Figure 1.1. Each layer of abstraction communicates logically with entities at the same layer but produces this communication by calling the next lower layer in the stack. Using this model, for instance, it is possible to develop different applications (e.g., e-mail vs. web browsing) on the same base architecture (e.g., public phone system) as well as provide a method to insert new technology at any layer of the stack without impacting the rest of the system performance (e.g., replace a telephone modem with a cable modem). This concept of a layered architecture has allowed communications to take great advantage of prior advances and leap-frog technology along at a phenomenal pace.

This text is entirely focused on what is known as **physical layer** communications. The physical layer of communications refers to the direct transfer of physical messages (analog waveforms or digital data bits) over a communications channel. The model for a physical layer communication abstraction is shown in Figure 1.2. Examples of physical communication channels include

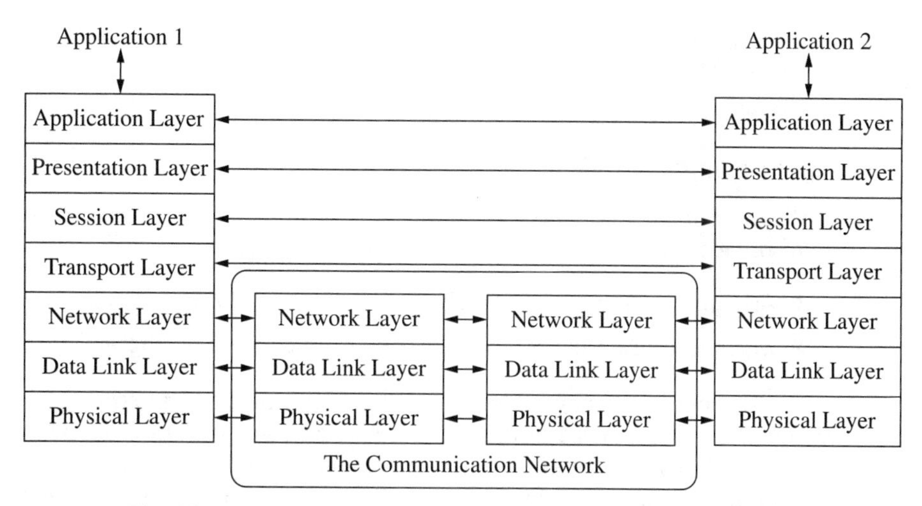

Figure 1.1 The OSI reference model.

copper wire pairs (telephony), coaxial cables, radio channels (mobile telephony), or optical fibers. It is interesting to note that wireless computer modems for local area networking are typically referred to as **WiPhy** in popular culture, which is an acronym for wireless physical layer communications. The engineering tools, the technology, and design paradigms are significantly different at the physical layer than at the higher layers in the stack. Consequently, systems engineering expertise in practice tends to have the greatest divide at the boundary to the physical layer. Engineering education has followed that trend and typically course work in telecommunications at both the undergraduate level and the graduate level tends to be bifurcated along these lines. To reflect the trend in both education and in industrial practice, this book will only try to educate in the area of physical layer communication systems. To reflect this abstraction the perspective in this text will be focused on point-to-point communications. Certainly multiple-access communications is very important in practice but it will not be considered in this text to maintain a consistent focus. Students who

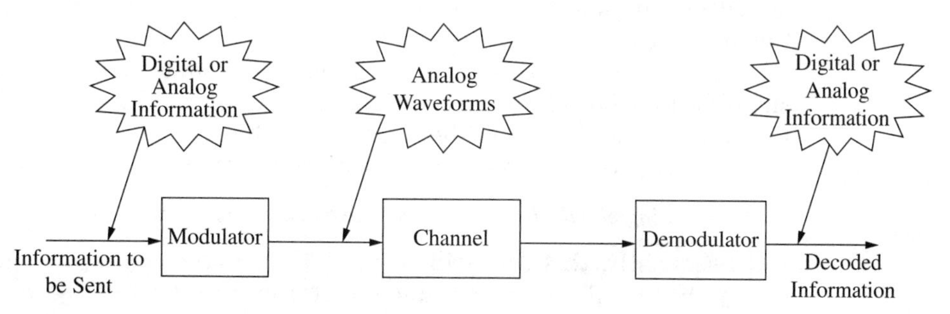

Figure 1.2 The physical layer model.

want a focus more on the upper layers of the communication stack should refer to [KR04, LGW00, Tan02] as examples of this higher layer perspective.

1.4 Technology's Impact on This Text

This text has been heavily influenced by the relatively recent trends in the field of physical layer communications system engineering:

- Advanced communication theory finding utility in practice
- Baseband processing power increasing at a rate predicted by Moore's law

Physical layer communications (with the perhaps partial exception of fiber optics) is filled with examples of sophisticated communication **theory** being directly placed into **practice**. Examples include wireless digital communications and high-speed cable communications. A communication engineer should truly feel lucky to live in a time when theory and practice are linked so closely. It allows people to work on very complex and sophisticated algorithms and have the algorithms almost immediately be put into practice. Because of this reason it is not surprising that many of the prominent communication theorists have also been very successful entrepreneurs (Andrew Viterbi [VO79] and Irwin Jacobs [WJ65] being two obvious examples). In this text, we will attempt to feature the underlying theory as this theory is so important in practical systems from mobile phones to television receivers.

The reason that theory is migrating to practice so quickly is the rapid advance of **baseband processing power**. Moore's law is now almost outstripping the ability of communication theory to use the available processing. In fact, a great paradigm shift occurred in the industry (in my humble opinion) when Qualcomm, in championing the cellular standard IS-95, started the design philosophy of designing a system that was too complicated for the current technology with the knowledge that Moore's law would soon enable the design to be implemented in a cost-effective fashion. Because of this shift in the design philosophy, future engineers are going to be exploring ways to better utilize this ever increasingly cost-efficient processing power. Since the future engineers will be using baseband processing power to implement their algorithms, this text is written with a focus on the baseband signal processing. To reflect this focus, this text starts immediately with the complex envelope representation of carrier signals and uses this representation throughout the entire text. This is a significant deviation from most of the prior teaching texts but directly in line with the notation used in research and in industry.

This book is constructed to align with the quote by Albert Einstein:

Everything should be made as simple as possible, but not one bit simpler.

Consequently, this text will be void of advanced topics in communication theory that I did not see as fundamental in an introductory communication theory book. Examples of material left out of the treatment in this text include, for example, details of noncoherent detection of digital signals, information theory and source coding. While these topics are critical to the training of a communications engineer, it is not necessary to the understanding of analog and digital information transmission. The goal is essentially not to lose the proverbial forest for the trees. Many interesting advanced issues and systems are pursued in the homework problems and projects. The text is written to build up a tool set in students that allow them to flourish in their profession over a full career. Readers looking for a buzzword-level treatment of communications will not find the text satisfactory. Since the focus of this text is the tools that will be important in the future, many ideas are not discussed in detail that traditionally were prominent in communication texts (e.g., pulse modulations). While a communication text can often take the form of an encyclopedia I have purposely avoided this format for a more focused tool-oriented version. Writing this paragraph I feel a little like my mom telling me to "eat my vegtables" but as I grow in age (and hopefully wisdom), I more fully appreciate the wisdom of my parents and of learning fundamental tools in physical layer communication engineering.

1.5 Book Overview

The book consists of four parts:

I Mathematical Foundations

II Analog Communication

III Noise in Communications Systems

IV Fundamentals of Digital Communications

This organization allows a slow logical buildup from a base knowledge in Fourier transforms, linear systems, and probability to an understanding of the fundamental concepts in communication theory. A significant effort has been made to make the development logical and to cover the important concepts.

1.5.1 Mathematical Foundations

This part of the book consists of three chapters that provide the mathematical foundations of communication theory. There are three pieces of test equipment that are critical for a communication engineer to be able to use to understand and troubleshoot communication systems: the oscilloscope, the spectrum analyzer, and the vector signal analyzer. Most communications laboratories contain this equipment and examples of this equipment are:

1. Digital oscilloscope and logic analyzer – Agilent 54622D
2. Spectrum analyzer – Agilent E4402B
3. Vector signal analyzer – Agilent 89600

The first chapter of this part of the book covers material that is included in the core undergraduate electrical engineering curriculum that provides the theory for the operation for the oscilloscope (time domain characterizations of signals) and the theory for the operation of the spectrum analyzer (frequency domain characterization of signals). The fundamental difference between a communication engineer and a technician is that engineers will consider noise to provide a complete characterization of the system trade-offs. The tools used to characterize noise are based on probability theory. The second chapter reviews this material. Courses covering the material on signal and systems and probability theory are taken before a course in communication theory. These two chapters are included strictly as review and to establish notation for the remainder of the book. The chapter on bandpass signals and the complex envelope notation is where the student steps into modern communication systems theory. This chapter will teach students how to characterize bandpass signals in the time domain and in the frequency domain. This chapter also introduces the vector diagram, the theoretical basis of the vector signal analyzer. This tool is frequently used by the modern communication engineer. These mathematical foundations will provide the basis for communications engineering.

1.5.2 Analog Communication

This part of the book consists of four chapters that introduce the theory of bandpass analog communication. The approach taken here is to introduce analog communications before the concepts of random processes. Consequently, the message signal is treated as a known deterministic waveform in the discussion of analog communications. The downside of this approach is that many of the powerful results on the power spectrum of analog communication waveforms cannot be introduced. The advantage of this approach is that since the message signals are known and deterministic, the tools of Fourier series, Fourier transforms, and signals and systems can be applied to the understanding of analog modulation systems. This approach allows students to ease into the world of communications by building upon their prior knowledge in deterministic signal and system analysis. By tying these tools from early undergraduate courses into the process of assessing the spectral efficiency and complexity of analog communications, the student will see the efficacy of an education in the fundamentals of electrical engineering. The first chapter of this part of the book presents the performance metrics in communications: performance, complexity, and spectral efficiency. This three-level trade-off is a recurring theme in the book. The following chapters introduce the classic methods of communicating analog information; amplitude and angle modulation. Amplitude modulation is a modulation format where the message signal is impressed upon the amplitude of the bandpass signal. Similarly, angle modulation is a modulation format where the message signal is impressed in some way in the phase of the bandpass signal. Finally, the important ideas in analog communications of multiplexing and the phase-locked loop are presented.

Many of my professional colleagues have made the suggestion that analog modulation concepts should be removed from the modern undergraduate curriculum. Comments such as "We do not teach about vacuum tubes so why should we teach about analog modulations?" are frequently heard. I heartily disagree with this opinion but not because I have a fondness for analog modulation but because analog modulation concepts are so important in modern communication systems. The theory and notation for analog signals learned in this text is a solid foundation for further explorations into modern communication systems because modern digital communications use analog waveforms. For example, in the testing of modern communication systems and subsystems analog modulation and demodulation concepts are used extensively. In fact, most of my good problems for the analog communication chapters have come as a result of my work in experimental wireless communications even though my research work has always been focused on digital communication systems! Another example of the utility of analog communications is that I am unaware of a synthesized signal generator that does not have an option to produce amplitude modulated (AM) and frequency modulated (FM) test signals. While modern communication engineers do not often design analog communication systems, the theory is still a useful tool. Consequently, this part of the book focuses on analog communications but using a modern perspective that will provide students the tools to flourish in their careers.

1.5.3 Noise in Communication Systems

Understanding the effects of noise and interference on communication systems what makes a communication system engineer uniquely trained. I have always been struck by the fact that engineering technicians have a training in time domain analysis, Fourier analysis, modulation techniques, and demodulation techniques. The main thing a technician does not understand is how to characterize how noise impacts the trade-offs that must be made in system design. On the other hand, the understanding of noise is often a frustrating subject for students as the level of mathematics and abstraction can often seem not worth the gains in useful skills. The approach taken in this text is to introduce the minimal amount of abstraction necessary to get useful results for engineering practice.

There are four topics/chapters that are presented in this section to introduce the techniques to analyze the impact of noise and interference. The first chapter focuses on the characterization of Gaussian stationary random processes and how linear filters impact this characterization. This material builds heavily on probability and random variable concepts. This text offers little new insights than has been available since the 1950s [DR87, Pap84] other than a reordering of topic presentation. The next chapter generalizes the concepts of random processes to the case of noise in bandpass communication receivers. The impact of filters in the receiver on the noise characteristics is explored. After these preliminary tools are in place, a revisiting of all the forms of analog communications in the presence of noise is completed. At this point an understanding of the trade-offs associated with analog communications can be completed and interpreted.

1.5.4 Fundamentals of Digital Communication

This part of the book consists of six chapters that introduce the theory of digital communication. The first chapter of this part of the book again presents the performance metrics in communications: fidelity, complexity, and spectral efficiency and reinterprets these metrics for digital communications. This re-emphasizes the trade-off first introduced with analog modulation. Interestingly there is a more powerful statement, due to Claude Shannon, that can be made about the achievable fidelity and spectral efficiency that will frame this introductory presentation of digital communication presented in this book. The following chapter introduces the classic methods of communicating 1 bit of information. This is developed by going through a five-step design process. The digital communication design problem is generalized for the transmission of multiple bits of information in the following chapter. The unfortunate situation exists with multiple bit transmission that the complexity of the optimum decoder grows exponentially with the number of bits to be transmitted. To address this exponential growth in complexity, signal structures that offer reduced complexity optimum demodulation structures are introduced. Several examples of these reduced complexity modulations that are used in engineering practice are introduced and these examples are considered in detail. The penultimate chapter of the book considers techniques to greatly improve the spectral efficiency of digital communications. A section of this part of the book considers the tools used to test digital communication systems in practice. The ideas and uses of the vector diagram and the eye diagram in the testing of digital communication systems are explained and motivated. The idea of modulations with memory, which provides the structure to approach Shannon's bounds on communication fidelity and spectral efficiency, are briefly introduced in the final chapter.

Now on to the fun stuff!

1.6 Homework Problems

Problem 1.1. Research the following person whose letter matches the first letter of your last name and write a paragraph about why they were important in communication theory.

(a) Jean Baptiste Joseph Fourier
(b) Antoine Parseval
(c) John R. Carson
(d) Friedrich Wilhelm Bessel
(e) John William Strutt
(f) Samuel B. Morse
(g) Emile Baudot
(h) Georg Simon Ohm
(i) Reginald Fessenden
(j) Stephen O. Rice
(k) Claude Shannon
(l) Alexander Graham Bell
(m) Hedy Lamarr
(n) Claude Berrou
(o) Lee de Forest
(p) Harold Nyquist
(q) Andrei Andreyevich Markov
(r) Reverend Thomas Bayes
(s) Gottfried Wilhelm Leibniz
(t) Carl Friedrich Gauss
(u) Pierre Simon Laplace
(v) Christian Doppler
(w) Norbert Wiener
(x) David Hilbert
(y) Gordon Moore
(z) Euclid of Alexandria

Problem 1.2. Research the following person whose letter matches the first letter of your last name and write a paragraph about why they were important in communication theory.

(a) James Maxwell
(b) Ralph Vinton Lyon Hartley
(c) David Sarnoff
(d) William Shockley
(e) Guglielmo Marconi
(f) Hermann Schwarz
(g) Charles Hermite
(h) Philo Farnsworth
(i) Leonhard Euler
(j) Max Planck
(k) Edwin A. Armstrong
(l) Ludwig Boltzmann
(m) Aleksandr Yakovlevich Khinchin
(n) Thomas Edision
(o) Dwight O. North
(p) Andrew J. Viterbi
(q) Edwin A. Armstrong
(r) John Bertrand Johnson
(s) Andre Marie Ampere
(t) Gottfried Ungerboeck
(u) Robert G. Gallagher
(v) Jules Henri Poincare
(w) Alessandro Volta
(x) Heinrich Hertz
(y) Charles Wheatstone
(z) Vladimir Zworykin

Part 1

Mathematical Foundations

Chapter

2

Signals and Systems Review

This chapter provides a brief review of signals and systems theory usually taught in an undergraduate curriculum. The purpose of this review is to introduce the notation that will be used in this text. Most results will be given without proof as these can be found in most undergraduate texts in signals and systems [May84, ZTF89, OW97, KH97].

2.1 Signal Classification

A signal, $x(t)$, is defined to be a function of time ($t \in \mathcal{R}$). Signals in engineering systems are typically described with five different mathematical classifications:

1. Deterministic or random
2. Energy or power
3. Periodic or aperiodic
4. Complex or real
5. Continuous time or discrete time

We will only consider deterministic signals at this point as random signals are a subject of Chapter 9.

2.1.1 Energy versus Power Signals

Definition 2.1 The energy, E_x, of a signal $x(t)$ is

$$E_x = \lim_{T_m \to \infty} \int_{-T_m/2}^{T_m/2} |x(t)|^2 dt \tag{2.1}$$

$x(t)$ is called an energy signal when $E_x < \infty$. Energy signals are normally associated with pulsed or finite duration waveforms (e.g., speech or a finite length information transmission). In contrast, a signal is called a power signal if it does not have finite energy. In reality, all signals are energy signals (infinity is hard to produce in a physical system) but it is often mathematically convenient to model certain signals as power signals. For example, when considering a voice signal it is usually appropriate to consider the signal as an energy signal for voice recognition applications but in radio broadcast applications we often model voice as a power signal.

EXAMPLE 2.1
A pulse is an energy signal:

$$x(t) = \begin{cases} \frac{1}{\sqrt{T_p}} & 0 \leq t \leq T_p \\ 0 & \text{elsewhere} \end{cases} \qquad E_x = 1 \tag{2.2}$$

A common signal that will be used frequently in this text is the sinc function.

Definition 2.2 The sinc function is

$$\text{sinc}(x) = \frac{\sin(\pi x)}{\pi x} \tag{2.3}$$

It should be noted that no standard definition exists among authors for sinc($\bullet$) so care must be exercised in using results from other authors. This text uses the definition of sinc($\bullet$) as adopted by *Matlab*.

EXAMPLE 2.2
Not all energy signals have finite duration:

$$x(t) = 2W \frac{\sin(2\pi Wt)}{2\pi Wt} = 2W \text{sinc}(2Wt) \qquad E_x = 2W \tag{2.4}$$

EXAMPLE 2.3
A voice signal. Figure 2.1 shows the time waveform for a computer-generated voice saying "Bingo." This signal is an obvious energy signal due to its finite time duration.

Definition 2.3 The signal power, P_x, is

$$P_x = \lim_{T_m \to \infty} \frac{1}{T_m} \int_{-T_m/2}^{T_m/2} |x(t)|^2 dt \tag{2.5}$$

Note that if $E_x < \infty$, then $P_x = 0$ and if $P_x > 0$, then $E_x = \infty$.

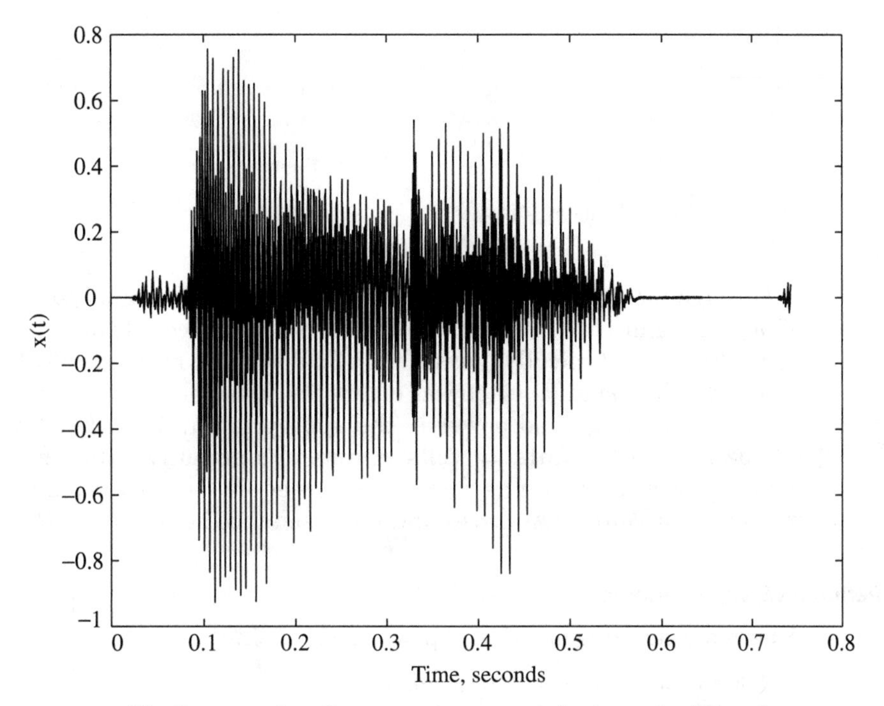

Figure 2.1 The time waveform for a computer-generated voice saying "Bingo."

EXAMPLE 2.4

$$
\begin{aligned}
x(t) &= \cos(2\pi f_c t) \\
P_x &= \lim_{T_m \to \infty} \frac{1}{T_m} \int_{-T_m/2}^{T_m/2} \cos^2(2\pi f_c t) dt \\
&= \lim_{T_m \to \infty} \frac{1}{T_m} \int_{-T_m/2}^{T_m/2} \left(\frac{1}{2} + \frac{1}{2} \cos(4\pi f_c t) \right) dt = \frac{1}{2}
\end{aligned} \tag{2.6}
$$

EXAMPLE 2.5

The waveform in Figure 2.2 is an analytical representation of the transmitted signal in a simple radar or sonar system. A signal pulse is transmitted for τ seconds and the receiver then listens for T seconds for returns from airplanes, ships, stormfronts, or other targets. After T seconds another pulse is transmitted. For this signal

$$
P_x = \lim_{T_m \to \infty} \frac{1}{T_m} \int_{-T_m/2}^{T_m/2} |x(t)|^2 dt = \frac{1}{T} \int_0^{\tau} dt = \frac{\tau}{T} \tag{2.7}
$$

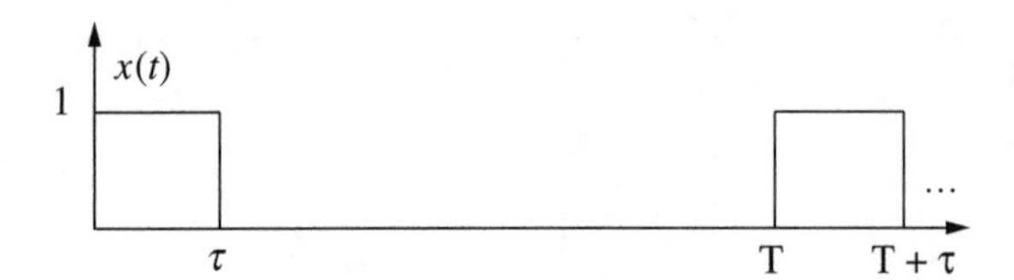

Figure 2.2 A periodic pulse train.

Throughout this text, signals will be considered and analyzed independent of physical units but the concept of units is worth a couple of comments at this point. Energy is typically measured in Joules and power is typically measured in Watts. Most signals electrical engineers consider are voltages or currents, so to obtain energy and power in the appropriate units a resistance needs to be specified (e.g., Watts $=$ Volts^2/Ohms). To simplify notation, we will just define energy and power as earlier which is equivalent to having the signal $x(t)$ measured in Volts or Amperes and the resistance being unity ($R = 1\Omega$).

2.1.2 Periodic versus Aperiodic

A periodic signal is one that repeats itself in time.

Definition 2.4 $x(t)$ is a periodic signal when

$$x(t) = x(t + T_0) \qquad \forall t \quad \text{and for some} \quad T_0 \neq 0 \tag{2.8}$$

Definition 2.5 The signal period is

$$T = \min(|T_0|) \tag{2.9}$$

The fundamental frequency is then

$$f_T = \frac{1}{T} \tag{2.10}$$

EXAMPLE 2.6

A simple example of a periodic signal is

$$x(t) = \cos(2\pi f_m t) \qquad T_0 = \frac{n}{f_m} \qquad T = \frac{1}{f_m} \tag{2.11}$$

Most periodic signals are power signals (note if the energy in one period is nonzero, then the periodic signal is a power signal) and again periodicity is a mathematical convenience that is not rigorously true for any real signal. We use the model of periodicity when the signal has approximately the property in Eq. (2.8) over the time range of interest. An aperiodic signal is defined to be a signal that is not periodic.

EXAMPLE 2.7

Examples of aperiodic signals are

$$x(t) = e^{-\frac{t}{\tau}} \qquad x(t) = \frac{1}{t} \tag{2.12}$$

2.1.3 Real versus Complex Signals

Complex signals arise often in communication systems analysis and design. The most common example is Fourier transforms which is discussed later in this chapter. Another important communication application of complex signals is in the representation of bandpass signals and Chapter 4 discusses this in more detail. For this review, we will consider some simple characteristics of complex signals. Define a complex signal and a complex exponential to be

$$z(t) = x(t) + jy(t) \qquad e^{j\theta} = \cos(\theta) + j\sin(\theta) \tag{2.13}$$

where $x(t)$ and $y(t)$ are both real signals. Note this definition is often known as Euler's rule. A magnitude ($\alpha(t)$) and phase ($\theta(t)$) representation of a complex signal is also commonly used, i.e.,

$$z(t) = \alpha(t)e^{j\theta(t)} \tag{2.14}$$

where

$$\alpha(t) = |z(t)| = \sqrt{x^2(t) + y^2(t)} \qquad \theta(t) = \arg(z(t)) = \tan^{-1}(y(t), x(t)) \tag{2.15}$$

It should be noted that the inverse tangent function in Eq. (2.15) uses two arguments and returns a result in $[0, 2\pi]$. The complex conjugate operation is defined as

$$z^*(t) = x(t) - jy(t) = \alpha(t)e^{-j\theta(t)} \tag{2.16}$$

Some important formulas for analyzing complex signals are

$$\begin{aligned} |z(t)|^2 &= \alpha(t)^2 = z(t)z^*(t) = x^2(t) + y^2(t) & \cos(\theta)^2 + \sin(\theta)^2 &= 1 \\ \Re[z(t)] &= x(t) = \alpha(t)\cos(\theta(t)) = \tfrac{1}{2}[z(t) + z^*(t)] & \cos(\theta) &= \tfrac{1}{2}[e^{j\theta} + e^{-j\theta}] \\ \Im[z(t)] &= y(t) = \alpha(t)\sin(\theta(t)) = \tfrac{1}{2j}[z(t) - z^*(t)] & \sin(\theta) &= \tfrac{1}{2j}[e^{j\theta} - e^{-j\theta}] \end{aligned} \tag{2.17}$$

EXAMPLE 2.8

The most common complex signal in communication engineering is the complex exponential, i.e.,

$$\exp[j2\pi f_m t] = \cos(2\pi f_m t) + j\sin(2\pi f_m t)$$

This signal will be the basis of the frequency domain understanding of signals.

2.1.4 Continuous Time Signals versus Discrete Time Signals

A signal, $x(t)$, is defined to be a continuous time signal if the domain of the function defining the signal contains intervals of the real line. A signal, $x(t)$, is defined to be a discrete time signal if the domain of the signal is a countable subset of the real line. Often a discrete signal is denoted by $x(k)$, where k is an integer and a discrete signal often arises from (uniform) sampling of a continuous time signal, e.g., $x(k) = x(kT_s)$, where T_s is the sampling period. Discrete signals and systems are of increasing importance because of the widespread use of computers and digital signal processors, but in communication systems the vast majority of the transmitted and received signals are continuous time signals. Consequently since this is a course in transmitter and receiver design (physical layer communications), we will be primarily concerned with continuous time signals and systems. Alternately, the digital computer is a great tool for visualization and discrete valued signals are processed in the computer. Section 2.4 will discuss in more detail how the computer and specifically the software package *Matlab* can be used to examine continuous time signal models.

2.2 Frequency Domain Characterization of Signals

Signal analysis can be completed in either the time or frequency domains. This section briefly overviews some simple results for frequency domain analysis. We first review the Fourier series representation for periodic signal, then discuss the Fourier transform for energy signals and finally relate the two concepts.

2.2.1 Fourier Series

If $x(t)$ is periodic with period T, then $x(t)$ can be represented as

$$x(t) = \sum_{n=-\infty}^{\infty} x_n \exp\left[\frac{j2\pi nt}{T}\right] = \sum_{n=-\infty}^{\infty} x_n \exp[j2\pi f_T nt] \tag{2.18}$$

where $f_T = 1/T$ and

$$x_n = \frac{1}{T}\int_0^T x(t)\exp[-j2\pi f_T nt]dt \tag{2.19}$$

This is known as the complex exponential Fourier series. Note a sine-cosine Fourier series is also possible with equivalence due to Euler's rule. Note in general the x_n are complex numbers. In words: *A periodic signal, $x(t)$, with period T can be decomposed into a weighted sum of complex sinusoids with frequencies that are an integer multiple of the fundamental frequency ($f_T = 1/T$).*

EXAMPLE 2.9

Consider again the signal in Example 2.4 where

$$x(t) = \cos(2\pi f_m t) = \frac{1}{2}\exp[j2\pi f_m t] + \frac{1}{2}\exp[-j2\pi f_m t]$$

$$f_T = f_m \qquad x_1 = x_{-1} = \frac{1}{2}, \qquad x_n = 0 \qquad \text{for all other } n \tag{2.20}$$

EXAMPLE 2.10

Consider again the signal in Example 2.8 where

$$x(t) = \exp(j2\pi f_m t)$$

$$f_T = f_m \qquad x_1 = 1, \qquad x_n = 0 \qquad \text{for all other } n \tag{2.21}$$

EXAMPLE 2.11

Consider again the signal in Example 2.5. The Fourier series for this example is

$$\begin{aligned} x_n &= \frac{1}{T}\int_0^{\tau} \exp\left(-j\frac{2\pi nt}{T}\right) dt \\ &= \frac{1}{T}\left.\frac{\exp(\frac{-j2\pi nt}{T})}{\frac{-j2\pi n}{T}}\right|_0^{\tau} \\ &= \frac{1}{T}\frac{1-\exp(\frac{-j2\pi n\tau}{T})}{\frac{j2\pi n}{T}} \end{aligned} \tag{2.22}$$

Using $\sin(\theta) = \frac{e^{j\theta}-e^{-j\theta}}{j2}$ gives

$$x_n = \frac{\tau}{T}\exp\left[-j\frac{\pi n\tau}{T}\right]\frac{\sin(\frac{\pi n\tau}{T})}{\frac{\pi n\tau}{T}} = \frac{\tau}{T}\exp\left[-j\frac{\pi n\tau}{T}\right]\operatorname{sinc}\left(\frac{n\tau}{T}\right) \tag{2.23}$$

A number of things should be noted about this example

1. τ and the bandwidth of the waveform are inversely proportional, i.e., a smaller τ produces a larger bandwidth signal.
2. τ and the signal power are directly proportional.
3. If T/τ = integer, some terms will vanish (i.e., $\sin(m\pi) = 0$).
4. To produce a rectangular pulse requires an infinite number of harmonics.

Properties of the Fourier Series

Property 2.1 If $x(t)$ is real, then $x_n = x^*_{-n}$. This property is known as Hermitian symmetry. Consequently the Fourier series of a real signal is a Hermitian symmetric function of frequency. This implies that the magnitude of the Fourier series is an even function of frequency, i.e.,

$$|x_n| = |x_{-n}| \tag{2.24}$$

and the phase of the Fourier series is an odd function of frequency, i.e.,

$$\arg(x_n) = -\arg(x_{-n}) \tag{2.25}$$

Property 2.2 If $x(t)$ is real and an even function of time, i.e., $x(t) = x(-t)$, then all the coefficients of the Fourier series are real numbers.

EXAMPLE 2.12

Recall Example 2.4, where $x(t) = \cos(2\pi f_m t)$. For this signal $T = 1/f_m$ and the only nonzero Fourier coefficients are $x_1 = 0.5$, $x_{-1} = 0.5$.

Property 2.3 If $x(t)$ is real and odd, i.e., $x(t) = -x(-t)$, then all the coefficients of the Fourier series are imaginary numbers.

EXAMPLE 2.13

For $x(t) = \sin(2\pi f_m t)$, where $T = 1/f_m$ and the only nonzero Fourier coefficients are $x_1 = -j0.5$, $x_{-1} = j0.5$.

Theorem 2.1 (Parseval)

$$P_x = \frac{1}{T}\int_0^T |x(t)|^2 dt = \sum_{n=-\infty}^{\infty} |x_n|^2 \tag{2.26}$$

Parseval's theorem states that the power of a signal can be calculated using either the time or the frequency domain representation of the signal and the two results are identical.

EXAMPLE 2.14

For both $x_1(t) = \cos(2\pi f_m t)$ and $x_2(t) = \sin(2\pi f_m t)$ it is apparent that

$$P_{x_1} = (0.5)^2 + (0.5)^2 = 0.5 = P_{x_2} \tag{2.27}$$

2.2.2 Fourier Transform

If $x(t)$ is an energy signal, then the Fourier transform is defined as

$$X(f) = \int_{-\infty}^{\infty} x(t)e^{-j2\pi f t}dt = \mathcal{F}\{x(t)\} \tag{2.28}$$

$X(f)$ is in general complex and gives the frequency domain representation of $x(t)$. The inverse Fourier transform is

$$x(t) = \int_{-\infty}^{\infty} X(f)e^{j2\pi f t}df = \mathcal{F}^{-1}\{X(f)\}$$

EXAMPLE 2.15

Example 2.1 (cont.). The Fourier transform of

$$x(t) = \begin{cases} 1 & 0 \leq t \leq T_p \\ 0 & \text{elsewhere} \end{cases} \tag{2.29}$$

is given as

$$X(f) = \int_0^{T_p} e^{-j2\pi f t}dt = \left.\frac{\exp[-j2\pi f t]}{-j2\pi f}\right|_0^{T_p} = T_p \exp[j\pi f T_p]\mathrm{sinc}(f T_p) \tag{2.30}$$

EXAMPLE 2.16

Example 2.2(cont.). The Fourier transform of

$$x(t) = 2W\frac{\sin(2\pi Wt)}{2\pi Wt} = 2W\mathrm{sinc}(2Wt)$$

is given as

$$X(f) = \begin{cases} 1 & |f| \leq W \\ 0 & \text{elsewhere} \end{cases} \tag{2.31}$$

Properties of the Fourier Transform

Property 2.4 If $x(t)$ is real then the Fourier transform is Hermitian symmetric, i.e., $X(f) = X^*(-f)$. This implies

$$|X(f)| = |X(-f)| \qquad \arg(X(f)) = -\arg(X(-f)) \tag{2.32}$$

Property 2.5 If $x(t)$ is real and an even function of time, i.e., $x(t) = x(-t)$, then $X(f)$ is a real valued and even function of frequency.

Property 2.6 If $x(t)$ is real and odd, i.e., $x(t) = -x(-t)$, then $X(f)$ is an imaginary valued and odd function of frequency.

Theorem 2.2 (Rayleigh's Energy)

$$E_x = \int_{-\infty}^{\infty} |x(t)|^2 dt = \int_{-\infty}^{\infty} |X(f)|^2 df \tag{2.33}$$

EXAMPLE 2.17

Example 2.1 (cont.). For the Fourier transform pair of

$$x(t) = \begin{cases} 1 & 0 \le t \le T_p \\ 0 & \text{elsewhere} \end{cases} \qquad X(f) = T_p \exp[j\pi f T_p] \text{sinc}(f T_p) \tag{2.34}$$

the energy is most easily computed in the time domain

$$E_x = \int_0^{T_p} |x(t)|^2 dt = T_p \tag{2.35}$$

EXAMPLE 2.18

Example 2.2 (cont.). For the Fourier transform pair of

$$x(t) = 2W \text{sinc}(2Wt) \qquad X(f) = \begin{cases} 1 & |f| \le W \\ 0 & \text{elsewhere} \end{cases} \tag{2.36}$$

the energy is most easily computed in the frequency domain

$$E_x = \int_{-W}^{W} |X(f)|^2 df = 2W \tag{2.37}$$

Theorem 2.3 (Convolution) The convolution of two time functions, $x(t)$ and $h(t)$, is defined as

$$y(t) = x(t) * h(t) = \int_{-\infty}^{\infty} x(\lambda) h(t - \lambda) d\lambda \tag{2.38}$$

The Fourier transform of $y(t)$ is given as

$$Y(f) = \mathcal{F}\{y(t)\} = H(f)X(f) \tag{2.39}$$

Theorem 2.4 (Duality) If $X(f) = \mathcal{F}\{x(t)\}$, then

$$x(f) = \mathcal{F}\{X(-t)\} \qquad x(-f) = \mathcal{F}\{X(t)\} \tag{2.40}$$

Theorem 2.5 Translation and Dilation If $y(t) = x(at + b)$, then

$$Y(f) = \frac{1}{|a|} X\left(\frac{f}{a}\right) \exp\left[j2\pi \frac{b}{a} f\right] \tag{2.41}$$

Theorem 2.6 Frequency Translation Multiplying any signal by a sinusoidal signal results in a frequency translation of the Fourier transforms, i.e.,

$$x_c(t) = x(t)\cos(2\pi f_c t) \Rightarrow X_c(f) = \frac{1}{2}X(f - f_c) + \frac{1}{2}X(f + f_c) \tag{2.42}$$

$$x_c(t) = x(t)\sin(2\pi f_c t) \Rightarrow X_c(f) = \frac{1}{j2}X(f - f_c) - \frac{1}{j2}X(f + f_c) \tag{2.43}$$

Definition 2.6 The correlation function of a signal $x(t)$ is

$$V_x(\tau) = \int_{-\infty}^{\infty} x(t)x^*(t - \tau)dt \tag{2.44}$$

Three important characteristics of the correlation funtion are

1. $V_x(0) = \int_{-\infty}^{\infty} |x(t)|^2 dt = E_x$
2. $V_x(\tau) = V_x^*(-\tau)$
3. $|V_x(\tau)| < V_x(0)$

EXAMPLE 2.19

Example 2.1 (cont.). For the pulse

$$x(t) = \begin{cases} 1 & 0 \le t \le T_p \\ 0 & \text{elsewhere} \end{cases} \tag{2.45}$$

the correlation function is

$$V_x(\tau) = \begin{cases} T_p\left(1 - \dfrac{|\tau|}{T_p}\right) & |\tau| \le T_p \\ 0 & \text{elsewhere} \end{cases} \tag{2.46}$$

Definition 2.7 The energy spectrum of a signal $x(t)$ is

$$G_x(f) = X(f)X^*(f) = |X(f)|^2 \tag{2.47}$$

The energy spectral density is the Fourier transform of the correlation function, i.e.,

$$G_x(f) = \mathcal{F}\{V_x(\tau)\} \tag{2.48}$$

The energy spectrum is a functional description of how the energy in the signal $x(t)$ is distributed as a function of frequency. The units on an energy spectrum are J/Hz. Because of this characteristic, two important properties, of the energy spectral density are

$$G_x(f) \geq 0 \qquad \forall f \qquad \text{(Energy in a signal cannot be negative valued)}$$

$$E_x = \int_{-\infty}^{\infty} |x(t)|^2 dt = \int_{-\infty}^{\infty} G_x(f) df \tag{2.49}$$

This last result is a restatement of Rayleigh's energy theorem and the analogy to Parseval's theorem should be noted.

EXAMPLE 2.20

Example 2.1(cont.). For the Fourier transform pair of

$$x(t) = \begin{cases} 1 & 0 \leq t \leq T_p \\ 0 & \text{elsewhere} \end{cases} \qquad X(f) = T_p \exp[-j\pi f T_p] \text{sinc}(f T_p) \tag{2.50}$$

the energy spectrum is

$$G_x(f) = T_p^{\,2} (\text{sinc}(f T_p))^2 \tag{2.51}$$

The energy spectrum of the pulse is shown in Figure 2.3 (a).

EXAMPLE 2.21

Example 2.2(cont.). For the Fourier transform pair of

$$x(t) = 2W \text{sinc}(2Wt) \qquad X(f) = \begin{cases} 1 & |f| \leq W \\ 0 & \text{elsewhere} \end{cases} \tag{2.52}$$

the energy spectrum is

$$G_x(f) = \begin{cases} 1 & |f| \leq W \\ 0 & \text{elsewhere} \end{cases} \tag{2.53}$$

EXAMPLE 2.22

Example 2.3(cont.). The energy spectrum of the computer-generated voice signal is shown in Figure 2.3 (b). The two characteristics that stand out in examining this spectrum are that the energy in the signal starts to significantly drop off after about 2.5 kHz and that the DC content of this voice signal is small.

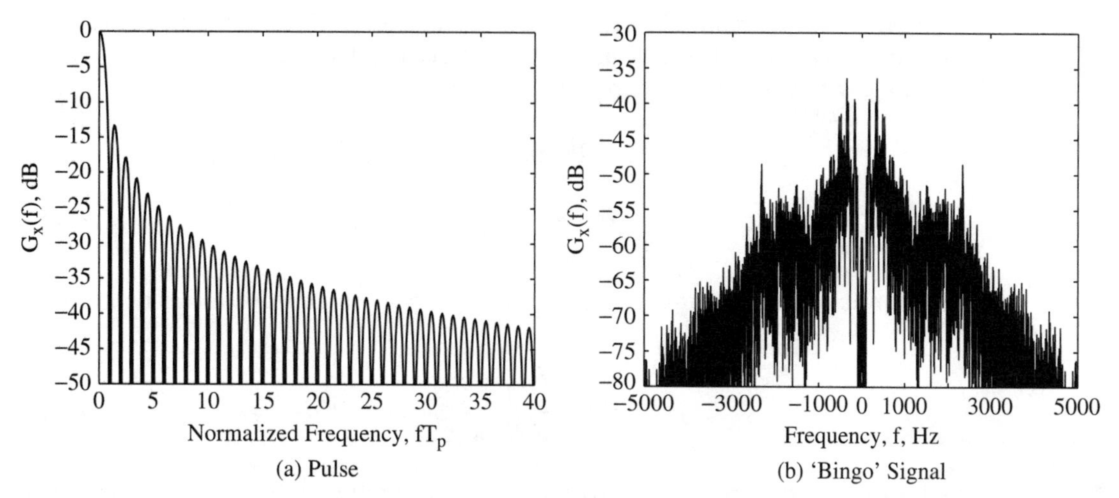

Figure 2.3 (a) The energy spectrum of a pulse from Example 2.1 and (b) the computer-generated voice signal from Example 2.3.

Theory of Operation: Signal Analyzers

Electronic test equipment (such as power meters and spectrum analyzers) are tools that attempt to provide the important characteristics of electronic signals to the practicing engineer. The main differences between test equipment and the theoretical equations like those discussed in this chapter are that test equipment only observes the signal over a finite interval. For instance a power meter outputs a reading, which is essentially a finite time average power.

Definition 2.8 The average power over a measurement time, T_m, is

$$P_x(T_m) = \frac{1}{T_m} \int_{-T_m/2}^{T_m/2} |x(t)|^2 dt \quad \text{W} \tag{2.54}$$

Definition 2.9 The Fourier transform of a signal truncated to a time length of T_m is

$$X_{T_m}(f) = \int_{-T_m/2}^{T_m/2} x(t) e^{-j2\pi f t} dt \tag{2.55}$$

An ideal spectrum analyzer produces the following measurement for a span of frequencies

$$S_x(f, T_m) = \frac{1}{T_m} |X_{T_m}(f)|^2 \quad \text{W/Hz} \tag{2.56}$$

The function in Eq. (2.56) is often termed the *sampled power spectral density* and it is a direct analog to the energy spectral density of Eq. (2.47). The sampled power spectrum is a functional description of how the power in the truncated signal $x(t)$ is distributed as a function of frequency. The units on a power spectrum are Watts/Hz. Because of this characteristic, two important characteristics of the power spectral density are

$$S_x(f, T_m) \geq 0 \qquad \forall f \qquad \text{(Power in a signal cannot be negative valued)}$$

$$P_x(T_m) = \frac{1}{T_m}\int_{-T_m/2}^{T_m/2} |x(t)|^2 dt = \int_{-\infty}^{\infty} S_x(f, T_m) df \tag{2.57}$$

In practice, the actual values output by a spectrum analyzer test instrument are closer to

$$\tilde{S}_x(f, T_m) = \frac{1}{T_m}\int_{f-B_r/2}^{f+B_r/2} |X_{T_m}(\lambda)|^2 d\lambda \quad \text{W} \tag{2.58}$$

where B_r is denoted as the resolution bandwidth (see [Pet89] for a good discussion of spectrum analyzers). Signal analyzers are designed to provide the same information for an engineer as signal and systems theory does within the constraints of electronic measurement techniques.

Also when measuring electronic signals practicing engineers commonly use the decibel terminology. Decibel notation can express the ratio of two quantities, i.e.,

$$P = 10\log\left[\frac{P_1}{P_2}\right] \text{ dB} \tag{2.59}$$

In communication systems, we often use this notation to compare the power of two signals. The decibel notation is used because it compresses the measurement scale ($10^6 = 60$ dB). Also multiplies turn into adds, which is often convenient in the discussion of electronic systems. The decibel notation can also be used for absolute measurements if P_2 becomes a reference level. The most commonly used absolute decibel scales are dBm ($P_2 = 1$ mw) and dBW ($P_2 = 1$ W). In communication systems, the ratio between the signal power and the noise power is an important quantity and this ratio is typically discussed with decibel notation.

2.2.3 Bandwidth of Signals

Engineers often like to quantify complicated entities with simple parameterizations. This often helps in developing intuition and simplifies discussion among different members of an engineering team. Since the frequency domain description of signals is not often easily parameterized, engineers often like to describe signals in terms of a single number. Most often in communications engineering, this single parameter is a description of the bandwidth. Bandwidth in communication engineering most often refers to the amount of **positive** frequency spectrum that a signal occupies. Unfortunately, the bandwidth of a signal does not have a common definition across all engineering disciplines. The common definitions of bandwidth used in engineering practice can be enumerated as

1. X dB relative bandwidth, B_X Hz
2. $P\%$ energy (power) integral bandwidth, B_P Hz

For lowpass energy signals we have the following two definitions

Definition 2.10 If a signal $x(t)$ has an energy spectrum $G_x(f)$, then B_X is determined as

$$10\log(\max_f G_x(f)) = X + 10\log(G_x(B_X)) \tag{2.60}$$

where $G_x(B_X) > G_x(f)$ for $|f| > B_X$

In words, a signal has a relative bandwidth B_X, if the energy spectrum is at least XdB down from the peak at all frequencies at or above B_X Hz. Often used values for X in engineering practice are the 3-dB bandwidth and the 40-dB bandwidth.

Definition 2.11 If a signal $x(t)$ has an energy spectrum $G_x(f)$, then B_P is determined as

$$P = \frac{\int_{-B_P}^{B_P} G_x(f)df}{E_x} \tag{2.61}$$

In words, a signal has an integral bandwidth B_P if the percent of the total energy in the interval $[-B_P, B_P]$ is equal to P%. Often used values for P in engineering practice are 98% and 99%.

EXAMPLE 2.23

Consider the rectangular pulse which is given as

$$x(t) = \begin{cases} 1 & 0 \le t \le T_p \\ 0 & \text{elsewhere} \end{cases} \tag{2.62}$$

The energy spectrum of this signal is given as

$$G_x(f) = |X(f)|^2 = T_p{}^2(\text{sinc}(f T_p))^2 \tag{2.63}$$

Figure 2.3 shows a normalized plot of this energy spectrum. Examining this plot carefully produces the 3-dB bandwidth of $B_3 = 0.442/T_p$ and the 40-dB bandwidth of $B_{40} = 31.54/T_p$. Integrating the power spectrum in Eq. (2.63) gives a 98% energy bandwidth of $B_{98} = 5.25/T_p$. These bandwidths parameterizations demonstrate that a rectangular pulse is not very effective at distributing the energy in a compact spectrum.

EXAMPLE 2.24
Example 2.2 (cont.). For the Fourier transform pair of

$$x(t) = 2W\,\mathrm{sinc}(2Wt) \qquad G_x(f) = \begin{cases} 1 & |f| \leq W \\ 0 & \text{elsewhere} \end{cases} \tag{2.64}$$

the 3-dB bandwidth, $B_3 = W$, and the 40-dB bandwidth, $B_{40} = W$, are identical. Integrating the power spectrum gives a 98% energy bandwidth of $B_{98} = 0.98W$. These bandwidths parameterization demonstrate that a sinc pulse is effective at distributing the energy in a compact spectrum.

EXAMPLE 2.25
Example 2.3 (cont.). For the computer generated voice signal the 3-dB bandwidth is $B_3 = 360$ Hz and the 40-dB bandwidth is $B_{40} = 4962$ Hz. Integrating the power spectrum gives a 98% energy bandwidth of $B_{98} = 2354$ Hz.

It is clear from the preceding examples that one parameter such as bandwidth does not do a particularly good job of characterizing a signal's spectrum. There are many definitions of bandwidth and these numbers while giving insight into the signal characteristics, do not fully characterize a signal.

For lowpass power signals similar ideas hold with $G_x(f)$ being replaced with $S_x(f, T_m)$.

Definition 2.12 If a signal $x(t)$ has a sampled power spectral density $S_x(f, T_m)$, then B_X is determined as

$$10\log(\max_f S_x(f, T_m)) = X + 10\log(S_x(B_X, T_m)) \tag{2.65}$$

where $S_x(B_X, T_m) > S_x(f, T_m)$ for $|f| > B_X$.

Definition 2.13 If a signal $x(t)$ has a sampled power spectral density $S_x(f, T_m)$, then B_P is determined as

$$P = \frac{\int_{-B_P}^{B_P} S_x(f, T_m)df}{P_x(T_m)} \tag{2.66}$$

2.2.4 Fourier Transform Representation of Periodic Signals

The Fourier transform for power signals is not rigorously defined and yet we often want to use frequency representations of power signals. Typically the power signals we will be using in this text are periodic signals that have a Fourier series representation. A Fourier series can be represented in the frequency domain with the help of the following result

$$\delta(f - f_1) = \int_{-\infty}^{\infty} \exp[j2\pi f_1 t]\exp[-j2\pi f t]dt \tag{2.67}$$

In other words, a complex exponential of frequency f_1 can be represented in the frequency domain with an impulse at f_1. Consequently the Fourier transform of a periodic signal is represented as

$$X(f) = \int_{-\infty}^{\infty} \sum_{n=-\infty}^{\infty} x_n \exp\left[j\frac{2\pi nt}{T}\right] \exp[-j2\pi ft]dt = \sum_{n=-\infty}^{\infty} x_n \delta\left(f - \frac{n}{T}\right) \quad (2.68)$$

Throughout the remainder of the text the spectrum of a periodic signal will be plotted using lines with arrows at the top to represent the delta functions. The bandwidth of periodic signals can be defined now by using the results of Section 2.2.3.

EXAMPLE 2.26

Consider again the the cosine signal introduced in Example 2.5.

$$\mathcal{F}(\cos(2\pi f_m t)) = \frac{1}{2}\delta(f - f_m) + \frac{1}{2}\delta(f + f_m)$$

By all definitions the bandwidth of this signal is $B = f_m$.

EXAMPLE 2.27

Consider again the the periodic pulse signal introduced in Example 2.5. The plot of the resultant energy spectrum is given in Figure 2.4 for $\tau = 0.2T$. For $\tau = 0.2T$, $B_3 = 3/T$,

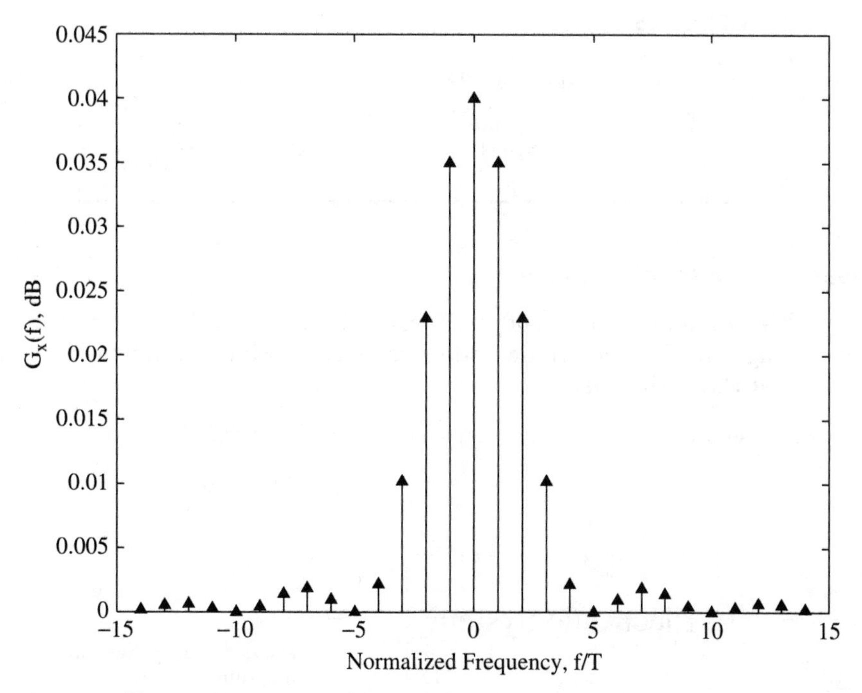

Figure 2.4 The energy spectrum of the periodic pulse. $\tau = 0.2T$.

$B_{40} = 152/T$, and $B_{98} = 26/T$. Note the computation of B_{98} required a Matlab program to be written to sum the magnitude square of the Fourier series coefficients. B_{40} can only be identified by expanding the x-axis in Figure 2.4 as the spectrum has not yet gone 40 dB down in this figure.

2.2.5 Laplace Transforms

This text will use the one-sided Laplace transform for the analysis of transient signals in communications systems. The one–sided Laplace transform of a signal $x(t)$ is

$$X(s) = \mathcal{L}\{x(t)\} = \int_0^\infty x(t)\exp[-st]dt \tag{2.69}$$

The use of the one-sided Laplace transform implies that the signal is zero for negative time. The inverse Laplace transform is given as

$$x(t) = \frac{1}{2\pi j}\oint X(s)\exp[st]ds \tag{2.70}$$

The evaluation of the general inverse Laplace transform requires the evaluation of contour integrals in the complex plane. For most transforms of interest the results are available in tables.

EXAMPLE 2.28

$$x(t) = \sin(2\pi f_m t) \qquad X(s) = \frac{2\pi f_m}{s^2 + (2\pi f_m)^2}$$

$$x(t) = \cos(2\pi f_m t) \qquad X(s) = \frac{s}{s^2 + (2\pi f_m)^2}$$

2.3 Linear Time-Invariant Systems

Electronic systems are often characterized by the input/output relations. A block diagram of an electronic system is given in Figure 2.5, where $x(t)$ is the input and $y(t)$ is the output.

Definition 2.14 A linear system is one in which superposition holds, i.e.,

$$ax_1(t) + bx_2(t) \rightarrow ay_1(t) + by_2(t) \tag{2.71}$$

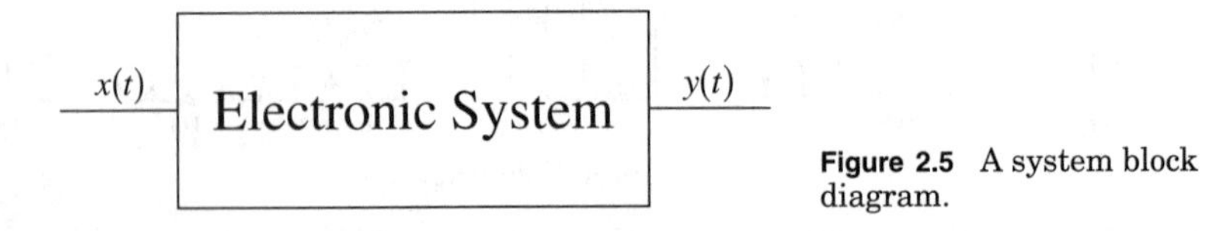

Figure 2.5 A system block diagram.

Definition 2.15 A time-invariant system is one in which a time shift in the input only changes the output by a time shift, i.e.,

$$x(t-\tau) \rightarrow y(t-\tau) \tag{2.72}$$

A linear time-invariant (LTI) system is described completely by an impulse response, $h(t)$. The output of the linear system is the convolution of the input signal with the impulse response, i.e.,

$$y(t) = \int_{-\infty}^{\infty} x(\lambda)h(t-\lambda)d\lambda = x(t) * h(t) \tag{2.73}$$

The Fourier transform (Laplace transform) of the impulse response is denoted the frequency response (transfer function), $H(f) = \mathcal{F}\{h(t)\}(H(s) = \mathcal{L}\{h(t)\})$, and by the convolution theorem for energy signals we have

$$Y(f) = H(f)X(f) \qquad (Y(s) = H(s)X(s)) \tag{2.74}$$

Likewise the linear system output energy spectrum has a simple form

$$G_y(f) = H(f)H^*(f)G_x(f) = |H(f)|^2 G_x(f) = G_h(f)G_x(f) \tag{2.75}$$

EXAMPLE 2.29

If a linear time-invariant filter had an impulse response given in Example 2.2, i.e.,

$$h(t) = 2W\frac{\sin(2\pi Wt)}{2\pi Wt} = 2W \operatorname{sinc}(2Wt)$$

it would result in an ideal lowpass filter. A filter of this form has two problems for an implementation: (1) the filter is anticausal and (2) the filter has an infinite impulse response. The obvious solution to having an infinite duration impulse response is to truncate the filter impulse response to a finite time duration. The way to make an anticausal filter causal is simply to time shift the filter response. Figure 2.6 (a) shows the resulting impulse response when the ideal lowpass filter had a bandwidth of $W =$ 2.5 kHz and the truncation of the impulse response is 93 ms and the time shift of 46.5 ms. The truncation will change the filter transfer function slightly while the delay will only add a phase shift. The resulting magnitude for the filter transfer function is shown in Figure 2.6 (b). If the voice signal of Example 2.3 is passed through this filter, then due to Eq. (2.75), the output energy spectrum should be filtered heavily outside of 2.5 kHz and roughly the same within 2.5 kHz, as shown in Figure 2.7. Figure 2.7 shows the measured energy spectrum of the input and output and these measured values match exactly that predicted by theory. This is an example of how communications engineers might use linear system theory to predict system performance. This signal will be used as a message signal throughout the remainder of this text to illustrate the ideas of analog communication.

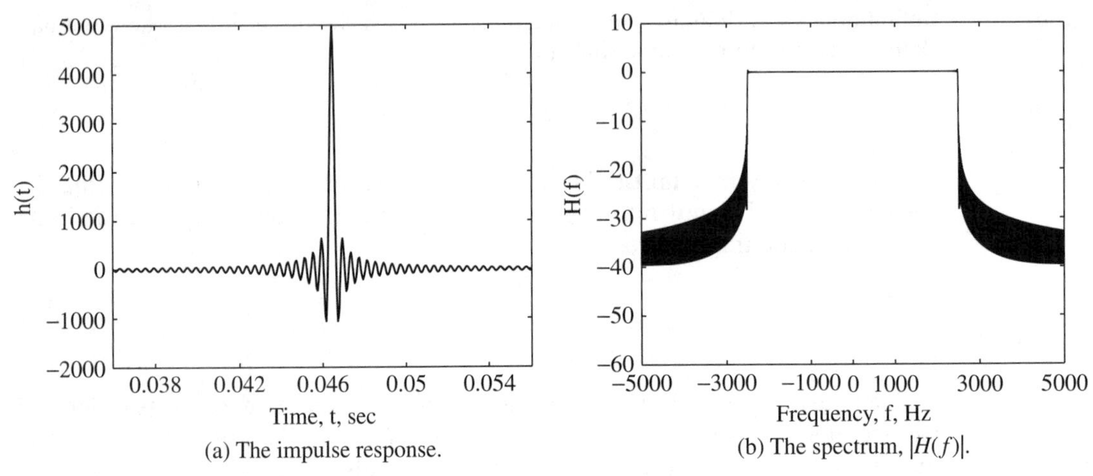

Figure 2.6 A lowpass filter impulse response and the spectrum. (a) The impulse response, $h(t)$, of a filter obtained by truncating (93 ms) and time-shifting (46.5 ms) (b) an ideal lowpass ($W = 2.5$ kHz) filter impulse response.

EXAMPLE 2.30

In this example the signal from Example 2.1 is to be filtered by the filter from the previous example. For this example we assume that $T_p = 907\mu$s. Figure 2.8 shows the measured energy spectrum of the input and output, and Figure 2.9 shows the input and output time waveforms for this example. The impact of filtering in the frequency domain is clearly evident. The elimination of the high-frequency components of the pulse signal eliminates the sharp transitions and produces some ringing effects.

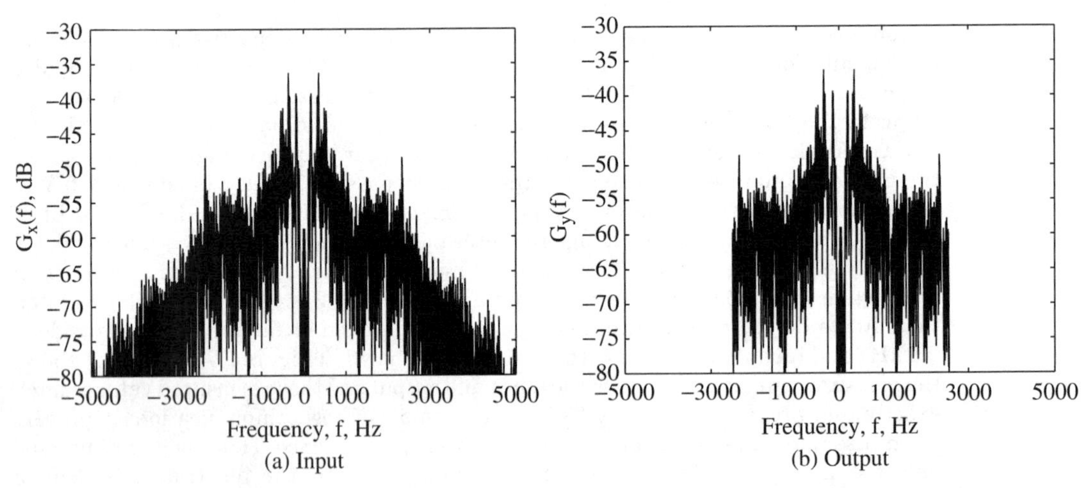

Figure 2.7 (a) The energy spectrum of the voice signal before and (b) after filtering by a lowpass filter of bandwidth 2.5 kHz.

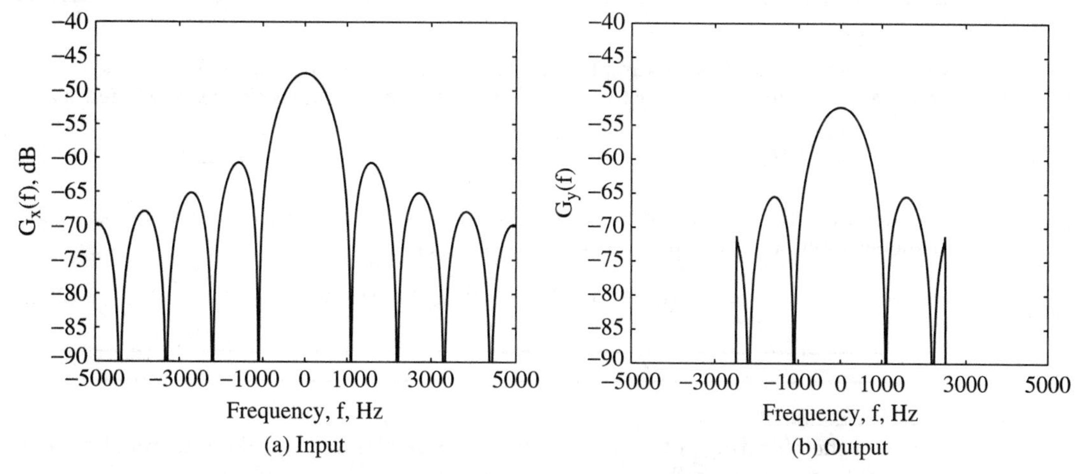

Figure 2.8 (a) The energy spectrum of the pulse signal before and (b) after filtering by a lowpass filter of bandwidth 2.5 kHz.

For a periodic input signal the output of the LTI system will be a periodic signal of the same period and have a Fourier series representation of

$$y(t) = \sum_{n=-\infty}^{\infty} H\left(\frac{n}{T}\right) x_n \exp\left[\frac{j2\pi nt}{T}\right] = \sum_{n=-\infty}^{\infty} y_n \exp\left[\frac{j2\pi nt}{T}\right] \tag{2.76}$$

where $y_n = H(\frac{n}{T})x_n$. In other words the output of an LTI system with a periodic input will also be periodic. The Fourier series coefficients are the product of input signal's coefficients and the transfer function evaluated at the harmonic frequencies.

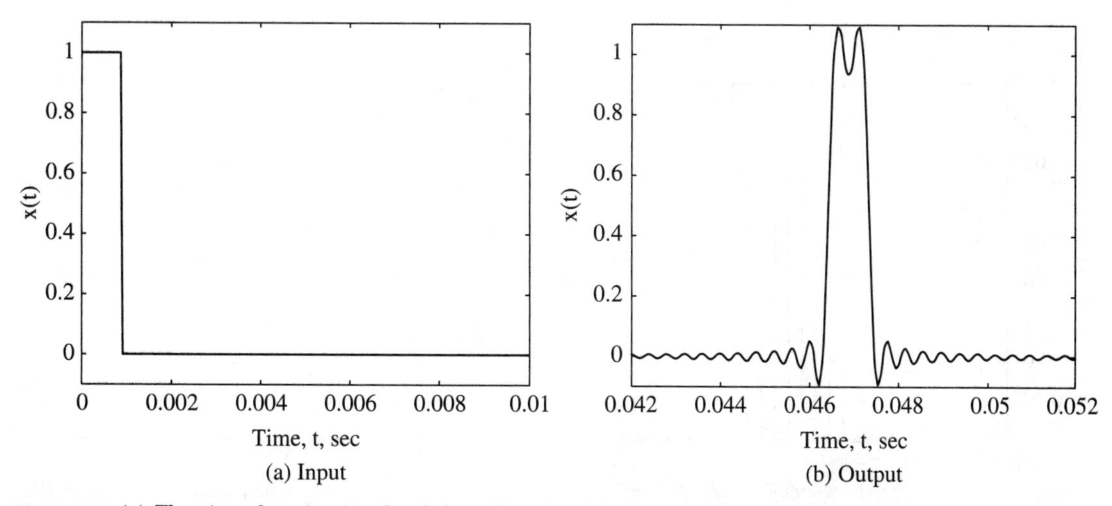

Figure 2.9 (a) The time domain signals of the pulse signal before and (b) after filtering by a lowpass filter of bandwidth 2.5 kHz.

EXAMPLE 2.31

Consider the signal from Example 2.1, $x(t) = \cos(2\pi f_m t)$ input into an arbitrary filter characterized with a transfer function $H(f)$. The output Fourier series coefficients are

$$f_T = f_m \qquad y_1 = \frac{H(f_m)}{2} \qquad y_{-1} = \frac{H(-f_m)}{2} \qquad x_n = 0 \tag{2.77}$$

and denoting the amplitude of the transfer funtion with $H_A(f)$ and the phase of the transfer function with $H_P(f)$ the output signal is

$$y(t) = H_A(f_m)\cos(2\pi f_m t + H_P(f_m)) \tag{2.78}$$

EXAMPLE 2.32

Consider the repeating pulse signal with $1/T = 500$ Hz and $\tau = 2500$ Hz from Example 2.5, input into a filter with an impulse response given in Example 2.2, i.e.,

$$h(t) = 2W\frac{\sin(2\pi Wt)}{2\pi Wt} = 2W\,\mathrm{sinc}(2Wt)$$

where $W = 2500$ Hz. The input signal spectrum is given in Figure 2.4. The output Fourier series coefficients are the same over the passband of the filter and are zero outside the passband. The output spectrum and output time plot are shown in Figure 2.10.

An important linear system for the study of frequency modulated (FM) signals is the differentiator. The differentiator is described with the following transfer function

$$y(t) = \frac{d^n x(t)}{dt^n} \qquad \Leftrightarrow \qquad Y(f) = (j2\pi f)^n X(f) \tag{2.79}$$

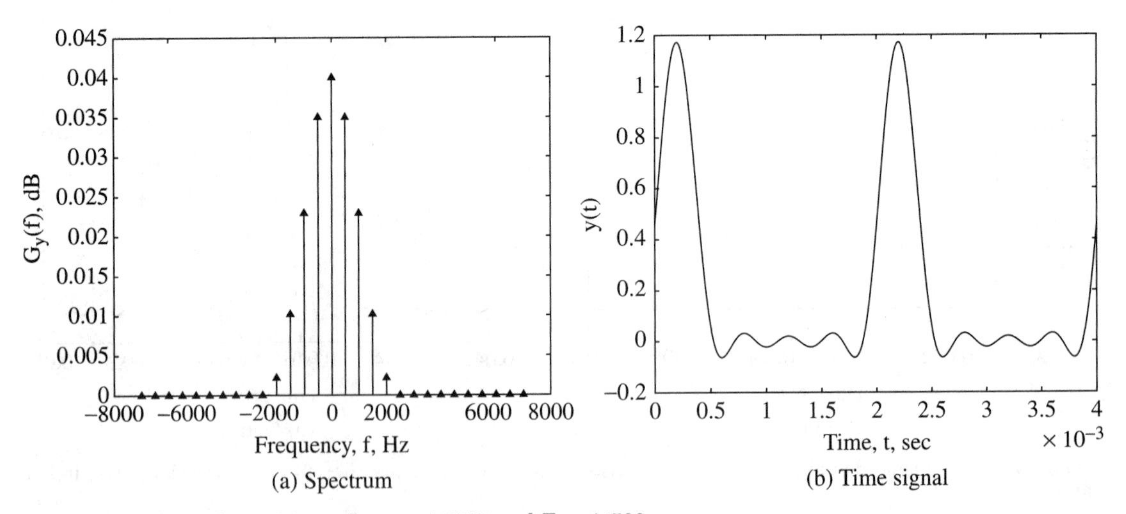

Figure 2.10 A filtered repetitive pulse, $\tau = 1/2500$ and $T = 1/500$.

2.4 Utilizing Matlab

The signals that are discussed in a course on communications are typically defined over a continuous time variable, e.g., $x(t)$. Matlab is an excellent package for visualization and learning in communications engineering and will be used liberally throughout this text. Unfortunately Matlab uses signals that are defined over a discrete time variable, $x(k)$. Discrete time signal processing basics are covered in the prerequisite signals and linear systems and comprehensive treatments are given in [Mit98, OS99, PM88, Por97]. This section provides a brief discussion of how to transition between the continuous time functions (communication theory) and the discrete time functions (Matlab). The examples considered in Matlab will reflect the types of signals you might measure when testing signals in the lab.

2.4.1 Sampling

The simplest way to convert a continuous time signal, $x(t)$, into a discrete time signal, $x(k)$, is to sample the continuous time signal, i.e.,

$$x(k) = x(kT_s + \epsilon)$$

where T_s is the time between samples. The sample rate is denoted $f_s = \frac{1}{T_s}$. This conversion is an important part of analog–to–digital conversion (ADC) and is a common operation in practical communication system implementations. The discrete time version of the signal is a faithful representation of the continuous time signal if the sampling rate is high enough.

To see that this is true it is useful to introduce some definitions for discrete time signals.

Definition 2.16 For a discrete time signal $x(k)$, the discrete time Fourier transform (DTFT) is

$$X(e^{j2\pi f}) = \sum_{k=-\infty}^{\infty} x(k)e^{j2\pi fk} \tag{2.80}$$

For clarity when the frequency domain representation of a continuous time signal is discussed it will be denoted as a function of f, e.g., $X(f)$ and when the frequency domain representation of a discrete time signal is discussed it will be denoted as a function of $e^{j2\pi f}$, e.g., $X(e^{j2\pi f})$. This DTFT is a continuous function of frequency and since no time index is associated with $x(k)$, the range of where the function can potentially take unique values is $f \in [-0.5, 0.5]$. Matlab has built-in functions to compute the discrete Fourier transform (DFT), which is simply the DTFT evaluated at uniformly spaced points.

For a sampled signal, the DTFT is related to the Fourier transform of the continuous time signal via [Mit98, OS99, PM88, Por97].

$$X(e^{j2\pi f}) = \frac{1}{T_s}\sum_{n=-\infty}^{\infty} X\left(\frac{f-n}{T_s}\right) \tag{2.81}$$

Examining Eq. (2.81) shows that if the sampling rate is higher than twice the highest significant frequency component of the continuous time signal then

$$X(e^{j2\pi f}) = \frac{1}{T_s} X\left(\frac{f}{T_s}\right) \tag{2.82}$$

and

$$X(f) = T_s X(e^{j2\pi(f T_s)}) \tag{2.83}$$

Note the highest significant frequency component is often quantified through the definition of the bandwidth (see Section 2.2.3). Consequently the rule of thumb for sampling is that the sampling rate should be at least twice the bandwidth of the signal. A sampling rate of exactly twice the bandwidth of the signal is known as *Nyquist's sampling rate*.

2.4.2 Integration

Many characteristics of continuous time signals are defined by an integral. For example, the energy of a signal is given in Eq. (2.1) as an integral. The Fourier transform, the correlation function, convolution, and the power are other examples of signal characteristics defined through integrals. Matlab does not have the ability to evaluate integrals but the values of the integral can be approximated to any level of accuracy desired. The simplest method of computing an approximation solution to an integral is given by the Riemann sum first introduced in calculus.

Definition 2.17 A Riemann sum approximation to an integral is

$$\int_a^b x(t)dt \approx \frac{b-a}{N} \sum_{k=1}^{N} x\left(a - \epsilon + \frac{k(b-a)}{N}\right) = h \sum_{k=1}^{N} x(k) \tag{2.84}$$

where $\epsilon \in [-(b-a)/N, 0]$ and $h = \frac{b-a}{N}$ is the step size for the sum.

A Riemann sum will converge to the true value of the finite integral as the number of points in the sum goes to infinity. Note that the DTFT for sampled signals can actually be viewed as a Riemann sum approximation to the Fourier transform.

2.4.3 Commonly Used Functions

This section details some Matlab functions that can be used to implement the signals and systems that are discussed in this chapter. Help with the details of these functions is available in Matlab.

Standard Stuff

- `cos` — cosine function
- `sin` — sine function
- `sinc` — sinc function

- `sqrt` — square root function
- `log10` — log base ten, this is useful for working in dB
- `max` — maximum of a vector
- `sum` — sum of the elements of a vector

Complex signals

- `abs` — $|\bullet|$
- `angle` — $\arg(\bullet)$
- `real` — $\Re[\bullet]$
- `imag` — $\Im[\bullet]$
- `conj` — $(\bullet)^*$

Input and Output

- `plot` — 2D plotting
- `xlabel, ylabel, axis` — formatting plots
- `sound, soundsc` — play vector as audio
- `load` — load data
- `save` — save data

Frequency Domain and Linear Systems

- `fft` — computes a discrete Fourier transform (DTF)
- `fftshift` — shifts an DFT output so when it is plotted it looks more like a Fourier transform
- `conv` — convolves two vectors

2.5 Homework Problems

Problem 2.1. Let two complex numbers be given as

$$z_1 = x_1 + jy_1 = a_1 \exp(j\theta_1) \qquad z_2 = x_2 + jy_2 = a_2 \exp(j\theta_2) \tag{2.85}$$

Find

(a) $\Re[z_1 + z_2]$

(b) $|z_1 + z_2|$

(c) $\Im[z_1 z_2]$

(d) $\arg[z_1 z_2]$

(e) $|z_1 z_2|$

Problem 2.2. Plot, find the period, and find the Fourier series representation of the following periodic signals

(a) $x(t) = 2\cos(200\pi t) + 5\sin(400\pi t)$

(b) $x(t) = 2\cos(200\pi t) + 5\sin(300\pi t)$

(c) $x(t) = 2\cos(150\pi t) + 5\sin(250\pi t)$

Problem 2.3. Consider the two signals

$$x_1(t) = m(t)\cos(2\pi f_c t) \qquad x_2(t) = m(t)\sin(2\pi f_c t)$$

where the bandwidth of $m(t)$ is much less than f_c. Compute the simplest form for the following four signals

(a) $y_1(t) = x_1(t)\cos(2\pi f_c t)$

(b) $y_2(t) = x_1(t)\sin(2\pi f_c t)$

(c) $y_3(t) = x_2(t)\cos(2\pi f_c t)$

(d) $y_4(t) = x_2(t)\sin(2\pi f_c t)$

Postulate how a communications engineer might use these results to recover a signal, $m(t)$, from $x_1(t)$ or $x_2(t)$.

Problem 2.4. (Design Problem) This problem gives you a little feel for microwave signal processing and the importance of the Fourier series. You have at your disposal

(1) a signal generator that produces ± 1 V amplitude square wave in a 1 Ω system where the fundamental frequency, f_1, is tunable from 1 kHz to 50 MHz

(2) an ideal bandpass filter with a center frequency of 175 MHz and a bandwidth of 30 MHz (±15 MHz).

The design problem is

(a) Select an f_1 such that when the signal generator is cascaded with the filter that the output will be a single tone at 180 MHz. There might be more than one correct answer (that often happens in real life engineering).

(b) Calculate the amplitude of the resulting sinusoid.

Problem 2.5. This problems exercises the signal and system tools. Compute the Fourier transform of

(a)

$$x(t) = \begin{cases} A & 0 \le t \le T_p \\ 0 & \text{elsewhere} \end{cases} \tag{2.86}$$

(b)

$$x(t) = \begin{cases} A\sin\left(\frac{\pi t}{T_p}\right) & 0 \le t \le T_p \\ 0 & \text{elsewhere} \end{cases} \tag{2.87}$$

and give the value of A such that $E_u = 1$. Compute the 40-dB relative bandwidth, B_{40}, of each signal.

Problem 2.6. This problem is an example of a problem which is best solved with the help of a computer. The signal $x(t)$ is passed through an ideal lowpass filter of bandwidth B/T_p Hz. For the signals given in Problem 2.5 with unit energy make a plot of the output energy versus B.

Hint: Recall the trapezoidal rule from calculus to approximately compute this energy.

Problem 2.7. This problem uses signal and system theory to compute the output of a simple memoryless nonlinearity. An amplifier is an often used device in communication systems and is simply modeled as an ideal memoryless system, i.e.,

$$y(t) = a_1 x(t)$$

This model is an excellent model until the signal levels get large then nonlinear things start to happen, which can produce unexpected changes in the output signals. These changes often have a significant impact in a communication system design. As an example of this characteristic consider the system in Figure 2.11 with the following signal model

$$x(t) = b_1 \cos(200000\pi t) + b_2 \cos(202000\pi t)$$

the ideal bandpass filter has a bandwidth of 10 kHz centered at 100 kHz, and the amplifier has the following memoryless model

$$y(t) = a_1 x(t) + a_3 x^3(t)$$

Give the system output, $z(t)$, as a function of a_1, a_3, b_1, and b_3.

Problem 2.8. (PD) A nonlinear device that is often used in communication systems is a quadratic memoryless nonlinearity. For such a device if $x(t)$ is the input the output is given as

$$y(t) = ax(t) + bx^2(t)$$

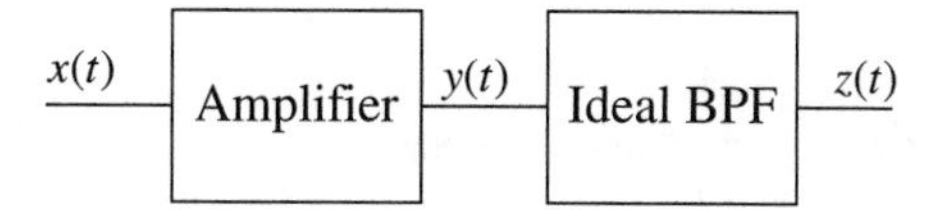

Figure 2.11 The system diagram for Problem 2.7.

(a) If $x(t) = A\cos(2\pi f_m t)$, what is $y(t)$ and $Y(f)$?

(b) If

$$X(f) = \begin{cases} A & ||f| - f_c| \leq f_m \\ 0 & \text{elsewhere} \end{cases} \tag{2.88}$$

what is $y(t)$ and $Y(f)$?

(c) A quadratic nonlinearity is often used in a *frequency doubler*. What component would you need to add in series with this quadratic memoryless nonlinearity such that you could put a sine wave in and get a sine wave out of twice the input frequency?

Problem 2.9. Consider the following signal

$$\begin{aligned} x(t) &= \cos(2\pi f_1 t) + a\sin(2\pi f_1 t) \\ &= X_A(a)\cos(2\pi f_1 t + X_p(a)) \end{aligned} \tag{2.89}$$

(a) Find $X_A(a)$.

(b) Find $X_p(a)$.

(c) What is the power of $x(t)$, P_x?

(d) Is $x(t)$ periodic? If so, what is the period and the Fourier series representation of $x(t)$?

Problem 2.10. Consider a signal and a linear system as depicted in Figure 2.12 where

$$x(t) = A + \cos(2\pi f_1 t)$$

and

$$h(t) = \begin{cases} \dfrac{1}{\sqrt{T_p}} & 0 \leq t \leq T_p \\ 0 & \text{elsewhere} \end{cases} \tag{2.90}$$

Compute the output $y(t)$.

Problem 2.11. For the signal

$$x(t) = 23\frac{\sin(2\pi 147t)}{2\pi 147t} \tag{2.91}$$

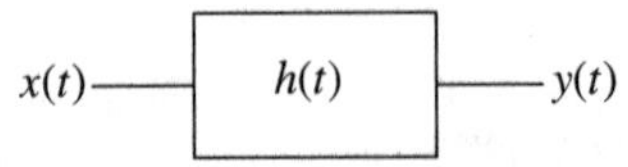

$x(t)$ → [$h(t)$] → $y(t)$

Figure 2.12 The system for Problem 2.10.

(a) Compute $X(f)$.

(b) Compute E_x.

(c) Compute $y(t) = \frac{dx(t)}{dt}$.

(d) Compute $Y(f)$.

Hint: Some of the computations have an easy way and a hard way so think before turning the crank!

Problem 2.12. The three signals seen in Figure 2.13 are going to be used to exercise your signals and systems theory. If no functional form is given for the pulse, assume ideal rectangular transitions to simplify the computations. Compute for each signal

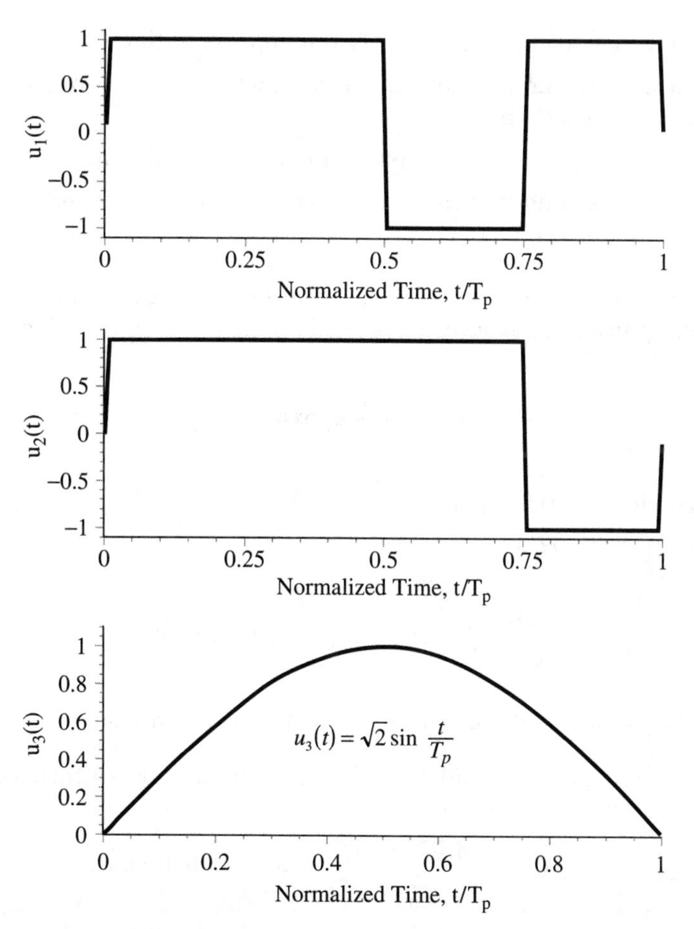

Figure 2.13 Pulse shapes considered for Problem 2.12.

(a) The Fourier transform, $U_i(f)$, $i = 1, 2, 3$.

(b) The energy spectral density, $G_{u_i}(f)$.

(c) The correlation function, $V_{u_i}(\tau)$. What is the energy, E_{u_i}?

Hint: Some of the computations have an easy way and a hard way so think before turning the crank!

Problem 2.13. This problem looks at several simplifying approximations for complex numbers that will be useful in analyzing the characteristics of analog modulations and demodulators. Assume z is a complex number of the form

$$z = 1 + \epsilon_1 + j\epsilon_2 \tag{2.92}$$

where ϵ_i is a real number with $\epsilon_i \ll 1$ for $i = 1, 2$.

(a) Show a logical argument to justify the approximation $|z| \approx 1 + \epsilon_1$.

(b) Assume $\epsilon_1 = \epsilon_2$ and find the values of ϵ_1, where this approximation results in a error of less than 1%.

(c) Show a logical argument to justify the approximation $\arg\{z\} \approx \epsilon_2$.

(d) Assume $\epsilon_1 = \epsilon_2$ and find the values of ϵ_1, where this approximation results in a error of less than 1%.

Problem 2.14. Pulse shapes will be important in digital communication systems. A Gaussian pulse shape arises in several applications and is given as

$$x(t) = \frac{1}{\sqrt{2\pi\sigma^2}} \exp\left[-\frac{t^2}{2\sigma^2}\right] \tag{2.93}$$

(a) Plot $x(t)$ for $\sigma = 0.25, 1, 4$.

(b) Calculate $X(f)$. *Hint:*

$$\frac{1}{\sqrt{2\pi}} \int_{-\infty}^{\infty} \exp\left[-\frac{x^2}{2} + ax\right] dx = \exp\left[\frac{a^2}{2}\right] \tag{2.94}$$

(c) Plot $G_X(f)$ for $\sigma = 0.25, 1, 4$ using a dB scale on the y-axis.

Problem 2.15. A common signal has a Fourier series representation of

$$x(t) = \frac{4}{\pi} \sum_{k=0}^{\infty} \frac{(-1)^k}{2k+1} \cos(10\pi(2k+1)t) \tag{2.95}$$

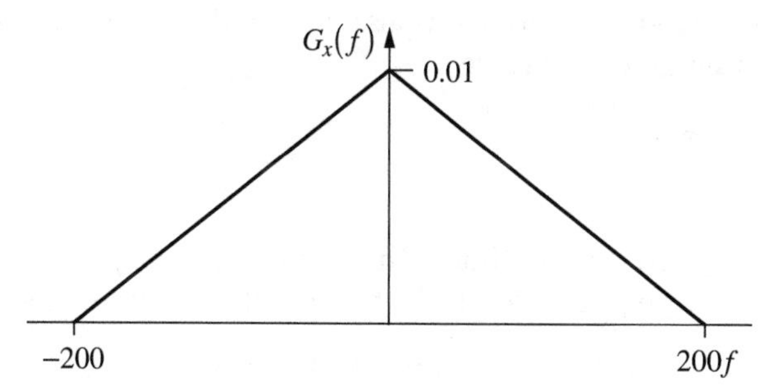

Figure 2.14 An example energy spectrum.

(a) Plot $x(t)$ for one period of the signal.

(b) Find P_x.

(c) What is the 98% power bandwidth, B_{98}, of $x(t)$?

Problem 2.16. Show that

$$\mathcal{F}^{-1}\{X^*(f)\} = x^*(-t) \tag{2.96}$$

Problem 2.17. For the energy spectrum shown in Figure 2.14 find

(a) E_x.

(b) The 3dB bandwidth, B_3.

(c) The 40dB bandwidth, B_{40}.

(d) The 98% energy bandwidth, B_{98}.

Problem 2.18. A signal has a series expansion given as

$$m(t) = \frac{8}{\pi^2}\sum_{n=1}^{\infty}\frac{1}{(2n-1)^2}\cos\left(\frac{2\pi(2n-1)t}{T}\right) \tag{2.97}$$

(a) Characterize the Fourier series.

(b) Find the 3-dB bandwidth, B_3.

(c) Find the 40-dB bandwidth, B_{40}.

(d) Find the 98% power bandwidth, B_{98}.

(e) Find the 99% power bandwidth, B_{99}.

Problem 2.19. Prove that $B_{90} < B_{98}$.

Problem 2.20. An energy spectrum has a peak value of 0.02. Find the value the energy spectrum must go below to be

(a) 40-dB below the peak.

(b) 3-dB below the peak.

Problem 2.21. A concept that sometimes finds utility in engineering practice is the concept of RMS bandwidth. The RMS bandwidth is defined as

$$W_R = \sqrt{\int_{-\infty}^{\infty} f^2 G_x(f) df} \tag{2.98}$$

For the energy spectrum in Figure 2.14 find the RMS bandwidth.

Problem 2.22. A concept that often finds utility in engineering practice is the concept of group delay. For a linear system with a transfer function denoted

$$H(f) = H_A(f) \exp[j H_p(f)] \tag{2.99}$$

the group delay of this linear system is defined as

$$\tau_g(f) = -\frac{1}{2\pi} \frac{d}{df} H_p(f) \tag{2.100}$$

Group delay is often viewed as the delay experienced by a signal at frequency f when passing through the linear system $H(f)$.

(a) An ideal delay element has an impulse response given as $h(t) = \delta(t - \tau_d)$, where $\tau_d > 0$ is the amount of delay. Find the group delay of an ideal delay element.

(b) An example low pass filter has

$$H(f) = \frac{1}{j2\pi f + a} \tag{2.101}$$

Plot the group delay of this filter for $0 \leq f \leq 5a$.

(c) Give a nontrivial $H(f)$ that has $\tau_g(f) = 0$. In your example is the filter causal or anticausal?

Problem 2.23. (RW) An interesting characteristic of the Fourier transform is linearity and linearity can be used to compute the Fourier transform of complicated functions by decomposing these function into sums of simple functions.

(a) Prove if $y(t) = x_1(t) + x_2(t)$, then $Y(f) = X_1(f) + X_2(f)$.

(b) Find the Fourier transform of the signal given in Figure 2.15.

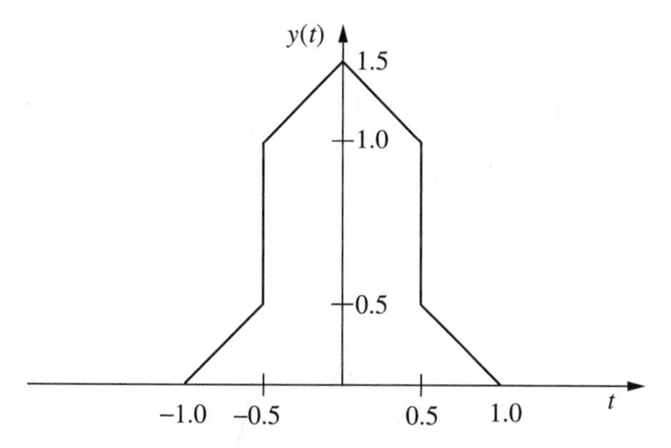

Figure 2.15 The rocket signal.

2.6 Example Solutions

Problem 2.24. Recall that the Fourier series is given as

$$x(t) = \sum_{n=-\infty}^{\infty} x_n \exp\left[j2\pi \frac{n}{T} t\right]$$

Each of these waveforms in this chapter can be put in a Fourier Series representation by the use of Euler's formula or

$$\cos(\theta) = \frac{1}{2}[e^{j\theta} + e^{-j\theta}] \tag{2.102}$$

$$\sin(\theta) = \frac{1}{2j}[e^{j\theta} - e^{-j\theta}] \tag{2.103}$$

(a)

$$\begin{aligned} x(t) &= 2\cos(200\pi t) + 5\sin(400\pi t) \\ &= \exp[j2\pi(100)t] + \exp[j2\pi(-100)t] + \tfrac{5}{j2}\exp[j2\pi(200)t] \\ &\quad - \tfrac{5}{j2}\exp[j2\pi(-200)t] \end{aligned}$$

The waveform is plotted in Figure 2.16.

$$\begin{aligned} &\tfrac{1}{T} = 100 \quad && x_1 = 1 \quad && x_{-1} = 1 \\ &T = 0.01 \quad && x_2 = \tfrac{5}{j2} \quad && x_{-2} = \tfrac{-5}{j2} \\ & && x_n = 0 \quad && \text{otherwise} \end{aligned}$$

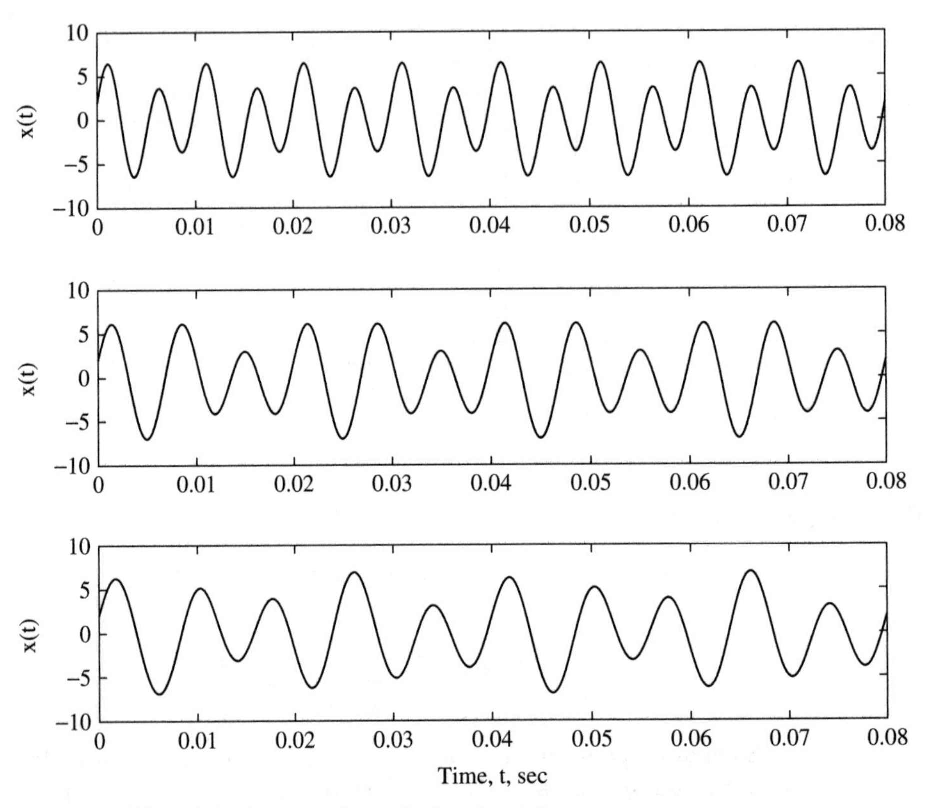

Figure 2.16 Plots of the time waveforms for Problem 2.2.

(b) $x(t) = 2\cos(200\pi t) + 5\sin(300\pi t)$. The waveform is plotted in Figure 2.16.

$$\begin{aligned} \frac{1}{T} &= 50 & x_2 &= 1 & x_{-2} &= 1 \\ T &= 0.02 & x_3 &= \frac{5}{j2} & x_{-3} &= \frac{-5}{j2} \\ && x_n &= 0 & &\text{otherwise} \end{aligned}$$

(c) $x(t) = 2\cos(150\pi t) + 5\sin(250\pi t)$. The waveform is plotted in Figure 2.16.

$$\begin{aligned} \frac{1}{T} &= 25 & x_3 &= 1 & x_{-3} &= 1 \\ T &= 0.04 & x_5 &= \frac{5}{j2} & x_{-5} &= \frac{-5}{j2} \\ && x_n &= 0 & &\text{otherwise} \end{aligned}$$

Problem 2.25. The output of the quadratic nonlinearity is modeled as

$$y(t) = ax(t) + bx^2(t) \tag{2.104}$$

It is useful to recall that multiplication in the time domain results in convolution in the frequency domain, i.e.,

$$z(t) = x(t) \times x(t) \qquad Z(f) = X(f) * X(f) \tag{2.105}$$

(a) $x(t) = A\cos(2\pi f_m t)$, which results in

$$\begin{aligned} y(t) &= aA\cos(2\pi f_m t) + bA^2\cos^2(2\pi f_m t) \\ &= aA\cos(2\pi f_m t) + bA^2\left(\frac{1}{2} + \frac{1}{2}\cos(4\pi f_m t)\right) \\ &= \frac{aA}{2}\exp[j2\pi f_m t] + \frac{aA}{2}\exp[-j2\pi f_m t] + \frac{bA^2}{2} \\ &\quad + \frac{bA^2}{4}\exp[j4\pi f_m t] + \frac{bA^2}{4}\exp[-j4\pi f_m t] \end{aligned} \tag{2.106}$$

The frequency domain representation for the output signal is given as

$$\begin{aligned} Y(f) = \frac{aA}{2}\delta(f - f_m) + \frac{aA}{2}\delta(f + f_m) + \frac{bA^2}{2}\delta(f) + \frac{bA^2}{4}\delta(f - 2f_m) \\ + \frac{bA^2}{4}\delta(f + 2f_m) \end{aligned} \tag{2.107}$$

The first two terms in Eq. (2.107) are due to the linear term in Eq. (2.104) while the last three terms are due to the square law term. It is interesting to note that these last three terms can be viewed as being obtained by convolving the two "delta" function frequency domain representation of a cosine wave, i.e., $2\cos(2\pi f_m t) = \exp[j2\pi f_m t] + \exp[-j2\pi f_m t]$, with itself.

(b) The input signal is

$$X(f) = \begin{cases} A & ||f| - f_c| \le f_m \\ 0 & \text{elsewhere} \end{cases} \tag{2.108}$$

Taking the inverse Fourier transform gives

$$x(t) = 4Af_m \frac{\sin(\pi f_m t)}{\pi f_m t}\cos(2\pi f_c t) \tag{2.109}$$

The output time signal of this quadratic nonlinearity is

$$\begin{aligned} y(t) &= 4aAf_m \frac{\sin(\pi f_m t)}{\pi f_m t}\cos(2\pi f_c t) + b\left(4Af_m \frac{\sin(\pi f_m t)}{\pi f_m t}\cos(2\pi f_c t)\right)^2 \\ &= 4aAf_m \frac{\sin(\pi f_m t)}{\pi f_m t}\cos(2\pi f_c t) + b\left(4Af_m \frac{\sin(\pi f_m t)}{\pi f_m t}\right)^2 \\ &\quad \times \left(\frac{1}{2} + \frac{1}{2}\cos(4\pi f_c t)\right) \end{aligned} \tag{2.110}$$

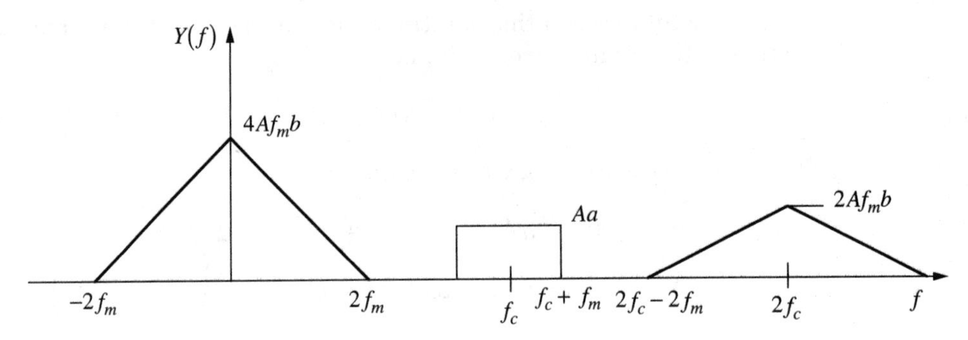

Figure 2.17 Output spectrum of the quadratic nonlinearity for part (b).

The frequency domain representation of $y(t)$ can be computed by taking the Fourier transform or by using $Y(f) = aX(f)+b(X(f)*X(f))$. The resulting Fourier transform is plotted in Figure 2.17.

(c) Looking at part (a) one can see that a tone in will produce a tone out if all frequencies except those at $2f_m$ are eliminated. Consequently a quadratic nonlinearity followed by a BPF will produce a frequency doubler. The block diagram for this frequency doubler is shown in Figure 2.18.

2.7 Miniprojects

Goal: To give exposure

1. To a small scope engineering design problem in communications
2. To the dynamics of working with a team
3. To the importance of engineering communication skills (in this case oral presentations)

Presentation: The forum will be similar to a design review at a company (only much shorter). The presentation will be of 5 minutes in length with an overview of the given problem and solution. The presentation will be followed by questions from the audience (your classmates and the professor). Each team member should be prepared to make the presentation on the due date.

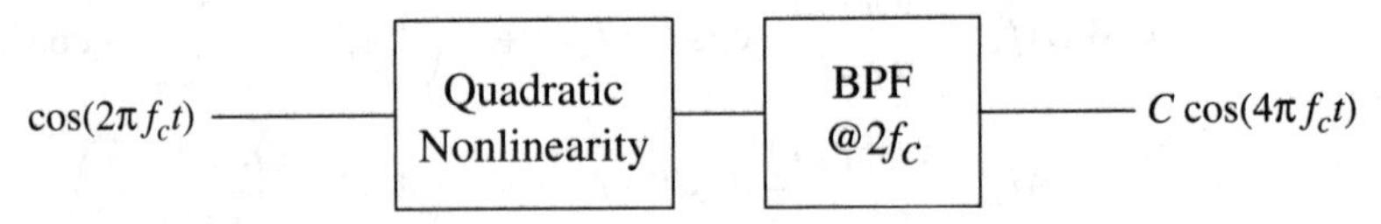

Figure 2.18 A frequency doubler.

2.7.1 Project 1

Project Goals: In designing a communication system, bandwidth efficiency is often a top priority. In analog communications one way to get bandwidth efficiency is to limit the bandwidth of the message signal. The big engineering trade-off is how narrow to make the bandwidth of the message signal. If the bandwidth is made too narrow, the message will be distorted. If the bandwidth is made too wide, spectrum will be wasted. This project examines this trade-off for a realistic signal.

Get the Matlab file `sigsex2.m` from the class web page along with the computer generated voice signal. Use this file to make an estimate how small the bandwidth of a voice signal can be and still enable high-fidelity communications. There is no single or right answer to the problem but engineering judgment must be used. How would your answer be different if you were only concerned with word recognition or if you wanted to maintain speaker recognition as well. Please detail your reasons for your solution. Relate your results to the different notions of bandwidth that were discussed in the text.

2.7.2 Project 2

Project Goals: The computation of a percent energy bandwith, B_P, requires a numerical integration to be performed for most signals that are used in communications. This project will assign you to do this integration for a realistic signal.

Get the computer-generated voice signal from the class web page. Write a program to estimate the the 90% bandwidth of this signal.

Chapter

3

Review of Probability and Random Variables

This chapter is intended to review introductory material on random variables, typically covered in an undergraduate curriculum. The most important use of this chapter will be the introduction of the notation used in this text. More proofs will be provided in this chapter than the last chapter because typically this subject matter is not as synthesized at the start of a senior level communications course. Texts that give a more detailed treatment of the subject of probability and random variables are [DR87, LG89, Hel91, CM86, Dev00, Sti99, YG98].

3.1 Axiomatic Definitions of Probability

Characterization of random events or experiments is critical for communication system design and analysis. A majority of the analyses of random events or experiments are extensions of the simple axioms or definitions presented in this section. A random experiment is characterized by a probability space consisting of a sample space Ω, a field $\mathcal{F}$, and a probability measure $P(\bullet)$. This probability space will be denoted $(\Omega, \mathcal{F}, P)$.

EXAMPLE 3.1

Consider the random experiment of rolling a fair dice

$$\Omega = \{1, 2, 3, 4, 5, 6\} \tag{3.1}$$

The field, $\mathcal{F}$, is the set of all possible combinations of outputs, i.e., consider the following outcomes $A_1 = \{\text{the die shows a 1}\}$, $A_2 = \{\text{the die shows an even number}\}$, $A_3 = \{\text{the die shows a number less than 4}\}$, and $A_4 = \{\text{the die shows an odd number}\}$, which implies

$$A_1, A_2, A_3, A_4 \in \mathcal{F} \tag{3.2}$$

Definition 3.1 Events A and B are mutually exclusive if $A \cap B = \emptyset$

Axioms of Probability. A probability measure, P, for a probability space $(\Omega, \mathcal{F}, P)$ with events $A, B \in \mathcal{F}$ must satisfy the following axioms

- For any event A, $P(A) \geq 0$
- $P(\Omega) = 1$
- If A and B are mutually exclusive events then $P(A \cup B) = P(A) + P(B)$.

These three axioms are the building blocks for probability theory and an understanding of random events that characterize communication system performance.

EXAMPLE 3.2

Example 3.1(cont.). The rolling of a fair die

$$A_1 = \{1\} \quad A_2 = \{2, 4, 6\} \quad A_3 = \{1, 2, 3\} \quad A_4 = \{1, 3, 5\}$$

$$P[\{1\}] = P[\{2\}] = \cdots = P[\{6\}] \tag{3.3}$$

$$P[A_1] = \frac{1}{6} \quad P[A_2] = \frac{1}{2} \quad P[A_3] = \frac{1}{2} \quad P[A_4] = \frac{1}{2}$$

Definition 3.2 (Complement) The complement of a set A, denoted A^C, is the set of all elements of Ω that are not elements of A.

Theorem 3.1 Poincare For N events $A_1, A_2, ...A_N$

$$P[A_1 \cup A_2 \cup .... \cup A_N] = S_1 - S_2 + \cdots + (-1)^{N-1} S_N$$

where

$$S_k = \sum_{i_1 < i_2 < .... < i_k} P[A_{i_1} \cap A_{i_2} \cap .... \cap A_{i_k}] \qquad i_1 \geq 1,\ i_k \leq N$$

Proof: The results are given for $N = 2$ and the proof for other cases is similar (if not more tedious). Note that events A and $B \cap A^C$ are mutually exclusive events. Consequently we have

$$P[A \cup B] = P[(A \cap B^C) \cup B] = P[A \cap B^C] + P[B] \tag{3.4}$$

$$P[A \cup B] = P[(B \cap A^C) \cup A] = P[B \cap A^C] + P[A] \tag{3.5}$$

$$\begin{aligned} P[A \cup B] &= P[(A \cap B^C) \cup (A \cap B) \cup (B \cap A^C)] \\ &= P[A \cap B^C] + P[A \cap B] + P[B \cap A^C] \end{aligned} \tag{3.6}$$

Using Eq. (3.4) + Eq. (3.5) − Eq. (3.6) gives the desired result. □

EXAMPLE 3.3

$$P[A \cup B] = P[A] + P[B] - P[A \cap B]$$

$$P[A \cup B \cup C] = P[A] + P[B] + P[C] - P[A \cap B] - P[B \cap C] - P[A \cap C] + P[A \cap B \cap C] \quad (3.7)$$

EXAMPLE 3.4
Example 3.1(cont.).

$$P[A_2 \cup A_3] = P[A_2] + P[A_3] - P[A_2 \cap A_3] \quad (3.8)$$

$$A_2 \cap A_3 = 2 \quad P[A_2 \cap A_3] = \frac{1}{6} \quad (3.9)$$

$$P[A_2 \cup A_3] = P[1, 2, 3, 4, 6] = \frac{1}{2} + \frac{1}{2} - \frac{1}{6} = \frac{5}{6} \quad (3.10)$$

Definition 3.3 (Conditional Probability) Let $(\Omega, \mathcal{F}, P)$ be a probability space with sets $A, B \in \mathcal{F}$ and $P[B] \neq 0$. The conditional or *a posteriori* probability of event A given an event B, denoted $P[A|B]$, is defined as

$$P[A|B] = \frac{P[A \cap B]}{P[B]}$$

$P[A|B]$ is interpreted as event A's probability after the experiment has been performed and event B is observed.

Definition 3.4 (Independence) Two events $A, B \in \mathcal{F}$ are independent if and only if

$$P[A \cap B] = P[A]P[B]$$

Independence is equivalent to

$$P[A|B] = P[A] \quad (3.11)$$

For independent events A and B, the *a posteriori* or conditional probability $P[A|B]$ is equal to the *a prioir* probability $P[A]$. Consequently, if A and B are independent, then observing B reveals nothing about the relative probability of the occurrence of event A.

EXAMPLE 3.5

Example 3.1(cont.). If the die is rolled and you are told the outcome is even, A_2, how does that change the probability of event A_3?

$$P[A_3|A_2] = \frac{P(A_2 \cap A_3)}{P(A_2)} = \frac{\frac{1}{6}}{\frac{1}{2}} = \frac{1}{3}$$

Since $P[A_3|A_2] \neq P[A_3]$, event A_3 is not independent of event A_2.

Definition 3.5 A collectively exhaustive set of events is one for which

$$A_1 \cup A_2 \cup \ldots . \cup A_N = \Omega$$

Theorem 3.2 Total Probability For N mutually exclusive, collectively exhaustive events $(A_1, A_2, \ldots . . A_N)$ and $B \in \Omega$, then

$$P[B] = \sum_{i=1}^{N} P[B|A_i]P[A_i]$$

Proof: The probability of event B can be written as

$$P[B] = P[B \cap \Omega] = P\left[B \cap \bigcup_{i=1}^{N} A_i\right] = P\left[\bigcup_{i=1}^{N} B \cap A_i\right]$$

The events $B \cap A_i$ are mutually exclusive, so

$$P[B] = P\left[\bigcup_{i=1}^{N} B \cap A_i\right] = \sum_{i=1}^{N} P[B \cap A_i] = \sum_{i=1}^{N} P[B|A_i]P[A_i] \qquad \square$$

EXAMPLE 3.6

Example 3.1(cont.). Note A_2 and A_4 are a set of mutually exclusive collectively exhaustive events with

$$P[A_3|A_4] = \frac{P[A_3 \cap A_4]}{P[A_4]} = \frac{P[\{1, 3\}]}{P[A_4]} = \frac{\frac{2}{6}}{\frac{1}{2}} = \frac{2}{3}$$

This produces

$$P[A_3] = P[A_3|A_2]P(A_2) + P[A_3|A_4]P(A_4) = \frac{1}{3}\left[\frac{1}{2}\right] + \left[\frac{2}{3}\right]\left[\frac{1}{2}\right] = \frac{3}{6} = \frac{1}{2}$$

Theorem 3.3 (Bayes) For N mutually exclusive, collectively exhaustive events $\{A_1, A_2, \ldots A_N\}$ and $B \in \Omega$, then the conditional probability of the event A_j given that event B is observed is

$$P[A_j|B] = \frac{P[B|A_j]P[A_j]}{P[B]} = \frac{P[B|A_j]P[A_j]}{\sum_{i=1}^{N} P[B \cap A_i]}$$

Proof: The definition of conditional probability gives

$$P[A_j \cap B] = P[A_j|B]P[B] = P[B|A_j]P[A_j]$$

Rearrangement and total probability complete the proof. □

EXAMPLE 3.7

(Binary symmetric channel) A facsimile machine divides a document up into small regions (i.e., pixels) and decides whether each pixel is black or white. Reasonable *a priori* statistics for facsimile transmission is

$$P[\text{A pixel is white}] = P[W] = 0.8 \qquad P[\text{A pixel is black}] = P[B] = 0.2$$

This pixel value is transmitted across a telephone line and the receiving fax machine makes a decision about whether a black or white pixel was sent. Figure 3.1 is a simplified representation of this operation in an extremely noisy situation. If a black pixel is decoded (BD) what is the probability a white pixel was sent, $P(W|BD)$? Bayes rule gives a straightforward solution to this problem, i.e.,

$$P(W|BD) = \frac{P[BD|W]P[W]}{P[BD]} \tag{3.12}$$

$$= \frac{P[BD|W]P[W]}{P[BD|W]P[W] + P[BD|B]P[B]}$$

$$= \frac{(0.1)0.8}{(0.1)(0.8) + (0.9)(0.2)} = 0.3077 \tag{3.13}$$

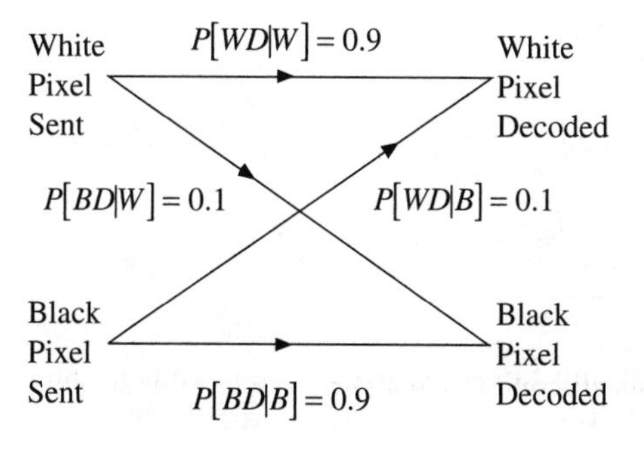

Figure 3.1 The binary symmetric channel.

3.2 Random Variables

In communications, the information transmission is often corrupted by random noise. This random noise manifests itself as a random voltage (a real number) at the output of the electrical circuits in the receiver. The concept of a random variable (RV) links the axiomatic definition of probability with these observed real random numbers. A random variable has an underlying random experiment. The set of all experimental outcomes are denoted Ω and a particular outcome of the random experiment is denoted with ω.

Definition 3.6 Let $(\Omega, \mathcal{F}, P)$ be a probability space. A real random variable $X(\omega)$ is a single-valued function or mapping from Ω to the real line $(\mathcal{R})$.

There are three types of random variables: discrete, continuous, and mixed. A discrete random variable has a finite (or countably infinite) number of possible values. A continuous random variable takes values in some interval of the real line of nonzero length. A mixed random variable is a convex combination of a discrete and a continuous random variable. To simplify the notation when no ambiguity exists, X represents the random variable $X(\omega)$ (the experimental outcome index is dropped) and $x = X(\omega)$ represents a particular realization of this random variable.

Definition 3.7 An observed real number resulting from the random experiment is denoted a **sample** from the random variable.

EXAMPLE 3.8
Matlab has a built-in random number generator. In essence this function when executed performs an experiment and produces a real number output (a sample of the random variable). Each time this function is run, a different real value is returned. Go to Matlab and type `rand(1)` and see what happens. Each time you run this function, it returns a number that is unable to be predicted. The output can be characterized in many ways, but the outputs are not completely predictable.

Random variables are completely characterized by either of two related functions: the cumulative distribution function (CDF) or the probability density function (PDF)[1]. These functions are the subject of the next two sections.

3.2.1 Cumulative Distribution Function

Definition 3.8 For a random variable $X(\omega)$, the CDF is a function $F_X(x)$ defined as

$$F_X(x) = P(\{\omega : X(\omega) \leq x\}) \qquad \forall \quad x \in \mathcal{R}$$

[1]Continuous RVs have PDFs but discrete random variables have probability mass functions (PMF).

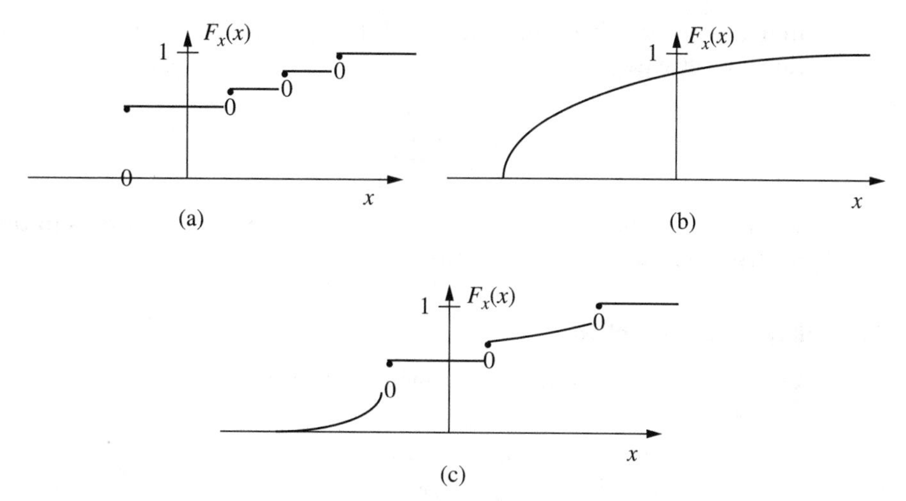

Figure 3.2 Example CDF for (a) discrete RV, (b) continuous RV, and (c) mixed RV.

Again to simplify the notation when no ambiguity exists, the CDF of the random variable X will be written as $F_X(x) = P(X \leq x)$. Figure 3.2 shows example plots of the CDF for discrete, continuous and mixed random variables. The discrete random variable has a CDF with a stairstep form, the steps occur at the points of the possible values of the random variable. The continuous random variable has a CDF, that is a continuous function and the mixed random variable has a CDF containing both intervals where the function is continuous with non-zero derivative and points where the function makes a jumps.

Properties of a CDF

- $F_X(x)$ is a monotonically increasing function[2], i.e.,

$$x_1 < x_2 \qquad \Rightarrow \qquad F_X(x_1) \leq F_X(x_2)$$

- $0 \leq F_X(x) \leq 1$
- $F_X(-\infty) = P(X \leq -\infty) = 0$ and $F_X(\infty) = P(X \leq \infty) = 1$
- A CDF is right continuous, i.e., $\lim_{h \to 0} F_X(x + |h|) = F_X(x)$
- $P(X > x) = 1 - F_X(x)$ and $P(x_1 < X \leq x_2) = F_X(x_2) - F_X(x_1)$
- The probability of the random variable taking a particular value, a, is given as

$$P(X = a) = F_X(a) - \lim_{h \to 0} F_X(x - |h|)$$

Continuous random variables take any value with zero probability since the CDF is continuous, while discrete and mixed random variables take values with

[2]This is sometimes termed a nondecreasing function.

nonzero probabilities since the CDF has jumps. The discrete random variable has a CDF with the form

$$F_X(x) = \sum_{k=1}^{N} P(X = a_k)U(x - a_k) \tag{3.14}$$

where $U()$ is the unit step function, N is the number of jumps in the CDF, and a_k are the locations of the jumps.

3.2.2 Probability Density Function

Definition 3.9 For a continuous random variable $X(\omega)$, the PDF is a function $f_X(x)$ defined as

$$f_X(x) = \frac{d\,P(X(\omega) \leq x)}{dx} = \frac{dF_X(x)}{dx} \quad \forall \quad x \in \mathcal{R} \tag{3.15}$$

Since the derivative in Eq. (3.15) can be rearranged to give

$$\lim_{\Delta \to 0} f_X(x)\Delta = \lim_{\Delta \to 0} P\left(x - \frac{\Delta}{2} < X(\omega) \leq x + \frac{\Delta}{2}\right),$$

the PDF can be thought of as the probability "density" in a very small interval around $X = x$. Discrete random variables do not have a probability "density" spread over an interval but do have probability mass concentrated at points. So the idea of a density function for a discrete random variable is not consistent. The analogous quantity to a PDF for a continuous RV in the case of a discrete RV is the probability mass function (PMF)

$$p_X(x) = P(X = x)$$

The idea of the probability density function can be extended to discrete and mixed random variables by utilizing the notion of the Dirac delta function [CM86].

Properties of a PDF

- $F_X(x) = \int_{-\infty}^{x} f_X(\beta)d\beta$
- $f_X(x) \geq 0$
- $F_X(\infty) = 1 = \int_{-\infty}^{\infty} f_X(\beta)d\beta$
- $P(x_1 < X \leq x_2) = \int_{x_1}^{x_2} f_X(\beta)d\beta$

Again, if there is no ambiguity in the expression, the PDF of the random variable X is written as $f(x)$ and if there is no ambiguity about whether a RV is continuous or discrete, the PDF is written $p(x)$. Knowing the PDF (or equivalently the CDF) allows you to completely describe any random event associated with a random variable.

EXAMPLE 3.9

The `rand(•)` function in Matlab produces a sample from what is commonly termed a uniformly distributed random variable. The PDF for a uniformly distributed random variable is given as

$$f_X(x) = \begin{cases} \dfrac{1}{b-a} & a \le x \le b \\ 0 & \text{elsewhere} \end{cases} \tag{3.16}$$

The function in Matlab has $a = 0$ and $b = 1$. Likewise the CDF is

$$F_X(x) = \begin{cases} 0 & x \le a \\ \dfrac{x-a}{b-a} & a \le x \le b \\ 1 & x \ge b \end{cases} \tag{3.17}$$

Experimenting in Matlab will give you some insight.

3.2.3 Moments and Statistical Averages

A communications engineer often calculates the statistical average of a function of a random variable. The average value or expected value of a function $g(X)$ with respect to a random variable X is

$$E(g(X)) = \int_{-\infty}^{\infty} g(x) p_X(x) dx$$

Average or expected values are numbers, that provide some partial information about the random variable. Average values are one number characterizations of random variables but are not a complete description in themselves like a PDF or CDF. A good example of a statistical average often used to characterize RVs is given by the mean value. The mean value is defined as

$$E(X) = m_X = \int_{-\infty}^{\infty} x p_X(x) dx$$

The mean is the average value of the random variable. The nth moment of a random variable is a generalization of the mean and is defined as

$$E(X^n) = m_{X,n} = \int_{-\infty}^{\infty} x^n p_X(x) dx$$

The mean square value, $E(X^2)$, is frequently used in the analysis of a communication system (e.g., average power). Another function of interest is a central moment (a moment around the mean value) of a random variable. The nth central moment is defined as

$$E((X - m_X)^n) = \sigma_{X,n} = \int_{-\infty}^{\infty} (x - m_X)^n p_X(x) dx$$

The most commonly used second central moment is the variance, $\sigma_{X,2}$, which provides a measure of the spread of a random variable around the mean. The variance has the shorthand notation $\sigma_{X,2} = \sigma_X^2 = \text{var}(X)$. The relation between the variance and the mean square value is given by

$$E(X^2) = m_X^2 + \sigma_X^2$$

The expectation operator is linear, i.e.,

$$E(g_1(x) + g_2(x)) = E(g_1(x)) + E(g_2(x))$$

EXAMPLE 3.10

Consider again the `rand(•)` function in Matlab as an example of a random variable. Since

$$f_X(x) = \begin{cases} \dfrac{1}{b-a} & a \le x \le b \\ 0 & \text{elsewhere} \end{cases} \tag{3.18}$$

the mean is given as

$$E[X] = \frac{1}{b-a}\int_a^b x dx = \frac{b^2 - a^2}{2(b-a)} = \frac{b+a}{2} \tag{3.19}$$

and the variance is

$$\text{var}(X) = \frac{(b-a)^2}{12} \tag{3.20}$$

3.2.4 The Gaussian Random Variable

The most common RV encountered in communications system analysis is the Gaussian random variable. This RV is a very good approximation to many of the physical phenomena that occur in electronic communications (e.g., the noise voltage generated by thermal motion of electrons in conductors). The reason for the common usage of the Gaussian RV is the Central Limit Theorem (see Section 3.3.5) and that analysis with Gaussian RVs is often tractable. This section will review the characteristics of the Gaussian RV. A Gaussian or normal RV, X, of mean m_X and variance σ_X^2 is denoted $N(m_X, \sigma_X^2)$. The Gaussian RV is a continuous random variable taking values over the whole real line and a PDF of

$$f_X(x) = \frac{1}{\sqrt{2\pi\sigma_X^2}} \exp\left[-\frac{(x - m_X)^2}{2\sigma_X^2}\right]$$

The Gaussian RV is characterized completely by its mean and variance. The Gaussian PDF is often referred to as the bell-shaped curve. The mean shifts the centroid of the bell-shaped curve as shown in Figure 3.3(a). The variance is a measure of the spread in the values the random variable realizes. A large variance implies that the random variable is likely to take values far from the mean,

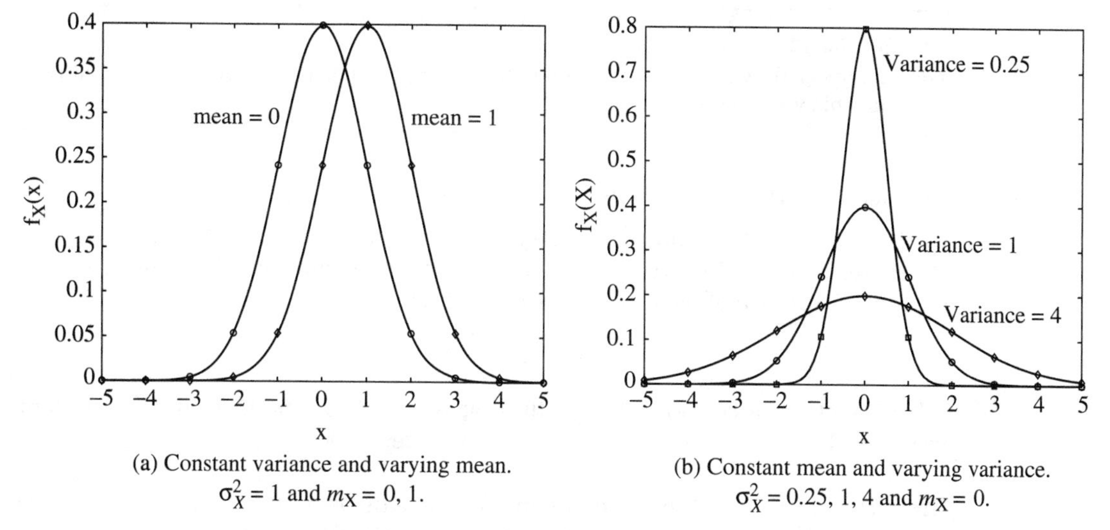

(a) Constant variance and varying mean. $\sigma_X^2 = 1$ and $m_X = 0, 1$.

(b) Constant mean and varying variance. $\sigma_X^2 = 0.25, 1, 4$ and $m_X = 0$.

Figure 3.3 Plots of the PDF of a Gaussian random variable.

whereas a small variance implies that a large majority of the values a random variable takes is near the mean. As an example, Figure 3.3(b) plots the density functions for three zero mean Gaussian RVs with variances of 0.25, 1, and 4.

A closed form expression for the CDF of the Gaussian RV does not exist. The CDF is expressed as

$$F_X(x) = \int_{-\infty}^{x} \frac{1}{\sqrt{2\pi\sigma_X^2}} \exp\left[-\frac{(\alpha - m_X)^2}{2\sigma_X^2}\right] d\alpha$$

The CDF can be expressed in terms of the erf function [Ae72], which is given as

$$\text{erf}(z) = \frac{2}{\sqrt{\pi}} \int_0^z e^{-t^2} dt$$

The CDF of a Gaussian RV is then given as

$$F_X(x) = \frac{1}{2} + \frac{1}{2}\,\text{erf}\left(\frac{x - m_X}{\sqrt{2}\sigma_X}\right)$$

While the CDF of the Gaussian random variable is defined using different functions by various authors, this function is used in this text because it is commonly used in math software packages (e.g., Matlab). Three properties of the erf function important for finding probabilities associated with Gaussian random variables are given as

$$\text{erf}(\infty) = 1$$

$$\text{erfc}(z) = 1 - \text{erf}(z)$$

$$\text{erf}(-z) = -\text{erf}(z)$$

EXAMPLE 3.11
The `randn(•)` function in Matlab produces a sample from a Gaussian distributed random variable with $m_X = 0$ and $\sigma_X = 1$.

3.2.5 A Transformation of a Random Variable

In the analysis of communication system performance it is often necessary to characterize a random variable, which is a transformation of another random variable. This transformation is expressed as

$$Y = g(X)$$

This section is concerned with finding the PDF or the CDF of the random variable, Y. The general technique is a two step process.

Step 1: Find the CDF of Y as a sum of integrals over the random variable X such that $g(X) < y$. This is expressed mathematically as

$$F_Y(y) = \sum_i \int_{R_i(y)} p_X(\beta)d\beta$$

where the $R_i(y)$ are intervals on the real line where X is such that $g(X) < y$.

Step 2: Find the PDF by differentiating the CDF found in Step 1 using Leibniz rule from calculus.

$$p_Y(y) = \frac{dF_Y(y)}{dy} = \sum_i \frac{d}{dy} \int_{R_i(y)} p_X(\beta)d\beta$$

Liebnitz Rule:

$$\frac{d}{dt} \int_{a(t)}^{b(t)} f(x,t)dx = f(b(t),t)\frac{db(t)}{dt} - f(a(t),t)\frac{da(t)}{dt} + \int_{a(t)}^{b(t)} \frac{\partial f(x,t)}{\partial t}dx$$

To illustrate this technique, three examples are given, which are common to communication engineering.

EXAMPLE 3.12
$Y = aX$ (the output of a linear amplifier with gain a having an input random voltage X).

Step 1:

$$F_Y(y) = \begin{cases} \int_{-\infty}^{\frac{y}{a}} p_X(\beta)d\beta & if \ a > 0 \\ U(y) & if \ a = 0 \\ \int_{\frac{y}{a}}^{\infty} p_X(\beta)d\beta & if \ a < 0 \end{cases}$$

where $U(X)$ is the unit step function. Note for $a > 0$ that only one interval of the real line exists, where $g(x) \leq y$ and that is $x \in [-\infty, y/a)$.

Step 2:

$$p_Y(y) = \begin{cases} \frac{1}{a} p_X(\frac{y}{a}) & if\ a > 0 \\ \delta(y) & if\ a = 0 \\ -\frac{1}{a} p_X(\frac{y}{a}) & if\ a < 0. \end{cases}$$

EXAMPLE 3.13

$Y = X^2$ (the output of a square law or power detector with an input random voltage X).

Step 1:

$$F_Y(y) = \begin{cases} \int_{-\sqrt{y}}^{\sqrt{y}} p_X(\beta) d\beta & if,\ y \geq 0 \\ 0 & if\ \ y < 0 \end{cases}$$

Step 2:

$$p_Y(y) = \begin{cases} \frac{1}{2\sqrt{y}} p_X(\sqrt{y}) + \frac{1}{2\sqrt{y}} p_X(-\sqrt{y}) & if\ y \geq 0 \\ 0 & if\ y < 0 \end{cases}$$

EXAMPLE 3.14

$Y = XU(X)$. This example corresponds to the output of an ideal diode with an input random voltage X.

Step 1:

$$F_Y(y) = \begin{cases} \int_{-\infty}^{y} p_X(\beta) d\beta & if\ y \geq 0 \\ 0 & if\ y < 0 \end{cases}$$

$$= U(y) \int_{-\infty}^{y} p_X(\beta) d\beta$$

$$= U(y) F_X(y) \tag{3.21}$$

Step 2:

$$p_Y(y) = U(y) p_X(y) + \delta(y) F_X(0)$$

3.3 Multiple Random Variables

In communication engineering, performance is often determined by more than one random variable. To analytically characterize the performance, a joint description of the random variables is necessary. All of the descriptions of a single random variable (PDF, CDF, moments, etc.) can be extended to the sets of random variables. This section highlights these extensions.

3.3.1 Joint Density and Distribution Functions

Again the random variables are completely characterized by either the joint CDF or the joint PDF. Note, similar simplifications in notation will be used with joint CDFs and PDFs, as was introduced in Section 3.2, when no ambiguity existed.

Definition 3.10 The joint CDF of two random variables is

$$F_{XY}(x, y) = P(\{X \leq x\} \cap \{Y \leq y\}) = P(X \leq x, Y \leq y)$$

Properties of the Joint CDF

- $F_{XY}(x, y)$ is a monotonically nondecreasing function, i.e.,

$$x_1 < x_2 \quad \text{and} \quad y_1 < y_2 \qquad \Rightarrow \qquad F_{XY}(x_1, y_1) \leq F_{XY}(x_2, y_2)$$

- $0 \leq F_{XY}(x, y) \leq 1$
- $F_{XY}(-\infty, -\infty) = P(X \leq -\infty, Y \leq -\infty) = 0$ and $F_{XY}(\infty, \infty) = P(X \leq \infty, Y \leq \infty) = 1$
- $F_{XY}(x, -\infty) = 0$ and $F_{XY}(-\infty, y) = 0$
- $F_X(x) = F_{XY}(x, \infty)$ and $F_Y(y) = F_{XY}(\infty, y)$
- $P(x_1 < X \leq x_2, y_1 < Y \leq y_2) = F_{XY}(x_2, y_2) - F_{XY}(x_1, y_2) - F_{XY}(x_2, y_1) + F_{XY}(x_1, y_1)$

Definition 3.11 For two continuous random variables X and Y, the joint PDF, $f_{XY}(x, y)$, is

$$f_{XY}(x, y) = \frac{\partial^2 P(X \leq x, Y \leq y)}{\partial x \partial y} = \frac{\partial^2 F_{XY}(x, y)}{\partial x \partial y} \quad \forall \quad x, y \in \mathcal{R}$$

Note, jointly distributed discrete random variables have a joint PMF, $p_{XY}(x, y)$, and as the joint PMF and PDF have similar characteristics this text often uses $p_{XY}(x, y)$ for both functions.

Properties of the Joint PDF or PMF

- $F_{XY}(x, y) = \int_{-\infty}^{y} \int_{-\infty}^{x} p_{XY}(\alpha, \beta) d\alpha d\beta$
- $p_{XY}(x, y) \geq 0$
- $p_X(x) = \int_{-\infty}^{\infty} p_{XY}(x, y) dy$ and $p_Y(y) = \int_{-\infty}^{\infty} p_{XY}(x, y) dx$
- $P(x_1 < X \leq x_2, y_1 < Y \leq y_2) = \int_{x_1}^{x_2} \int_{y_1}^{y_2} p_{XY}(\alpha, \beta) d\alpha d\beta$

Definition 3.12 Let X and Y be random variables defined on the probability space $(\Omega, \mathcal{F}, P)$. The conditional or a posteriori PDF of Y given $X = x$, is

$$p_{Y|X}(y|X = x) = \frac{p_{XY}(x, y)}{p_X(x)} \tag{3.22}$$

The conditional PDF, $p_{Y|X}(y|X = x)$, is the PDF of the random variable Y after the random variable X is observed to take the value x.

Definition 3.13 Two random variables x and y are independent if and only if

$$p_{XY}(x, y) = p_X(x)p_Y(y)$$

Independence is equivalent to

$$p_{Y|X}(y|X = x) = p_Y(y)$$

i.e., Y is independent of X if no information is in the RV X about Y in the sense that the conditional PDF is not different than the unconditional PDF.

EXAMPLE 3.15

In Matlab each time `rand(•)` or `randn(•)` is executed it returns an ostensibly independent sample from the corresponding uniform or Gaussian distribution.

Theorem 3.4 Total Probability For two random variables X and Y the marginal density of Y is given by

$$p_Y(y) = \int_{-\infty}^{\infty} p_{Y|X}(y|X = x)p_X(x)dx$$

Proof: The marginal density is given as (Property 3 of PDFs)

$$p_Y(y) = \int_{-\infty}^{\infty} p_{XY}(x, y)dx$$

Rearranging Eq. (3.22) and substituting completes the proof. □

Theorem 3.5 Bayes For two random variables X and Y the conditional density is given by

$$p_{Y|X}(y|X = x) = \frac{p_{X|Y}(x|Y = y)p_Y(y)}{p_X(x)} = \frac{p_{X|Y}(x|Y = y)p_Y(y)}{\int_{-\infty}^{\infty} p_{XY}(x, y)dy}$$

Proof: The definition of conditional probability gives

$$p_{XY}(x, y) = p_{Y|X}(y|X = x)p_X(x) = p_{Y|X}(x|Y = y)p_Y(y)$$

Rearrangement and total probability completes the proof. □

A majority of the results presented in this section correspond to two random variables. The extension of these concepts to three or more random variables is straightforward.

3.3.2 Joint Moments and Statistical Averages

Joint moments and statistical averages are also of interest in communication system engineering. The general statistical average of a function $g(X, Y)$ of two random variables X and Y is given as

$$E[g(X,Y)] = \int_{-\infty}^{\infty}\int_{-\infty}^{\infty} g(x,y)p_{XY}(x,y)dxdy$$

A commonly used joint moment or statistical average is the correlation between two random variables X and Y, defined as

$$E[XY] = \mathrm{corr}(X,Y) = \int_{-\infty}^{\infty}\int_{-\infty}^{\infty} xyp_{XY}(x,y)dxdy$$

A frequently used joint central moment is the covariance between two random variables x and y, defined as

$$E[(X-m_X)(Y-m_Y)] = \mathrm{cov}(X,Y) = \int_{-\infty}^{\infty}\int_{-\infty}^{\infty}(x-m_X)(y-m_Y)p_{XY}(x,y)dxdy$$

Definition 3.14 The correlation coefficient is

$$\rho_{XY} = \frac{\mathrm{cov}(X,Y)}{\sqrt{\mathrm{var}(X)\mathrm{var}(Y)}} = \frac{E[(X-m_X)(Y-m_Y)]}{\sigma_X\sigma_Y}$$

The correlation coefficient is a measure of the statistical similarity of two random variables. If $|\rho_{XY}| = 1$ for random variables X and Y, then X is a scalar multiple of Y. If $\rho_{XY} = 0$ then the random variables are uncorrelated. Values of $|\rho_{XY}|$ between these two extremes provide a measure of the similarity of the two random variables (larger $|\rho_{XY}|$ being more similar).

3.3.3 Two Gaussian Random Variables

Two jointly Gaussian random variables, X and Y, have a joint density function of

$$f_{XY}(x,y) = \frac{1}{2\pi\sigma_X\sigma_Y\sqrt{1-\rho_{XY}^2}}$$
$$\times \exp\left[-\frac{1}{2(1-\rho_{XY}^2)}\left(\frac{(x-m_X)^2}{\sigma_X^2} - \frac{2\rho_{XY}(x-m_X)(y-m_Y)}{\sigma_X\sigma_Y} + \frac{(y-m_Y)^2}{\sigma_Y^2}\right)\right]$$

where $m_X = E(X)$, $m_Y = E(Y)$, $\sigma_X^2 = \mathrm{var}(X)$, $\sigma_Y^2 = \mathrm{var}(Y)$, and ρ_{XY} is the correlation coefficient between the two random variables X and Y. This density function results in a three-dimensional bell-shaped curve which is stretched and

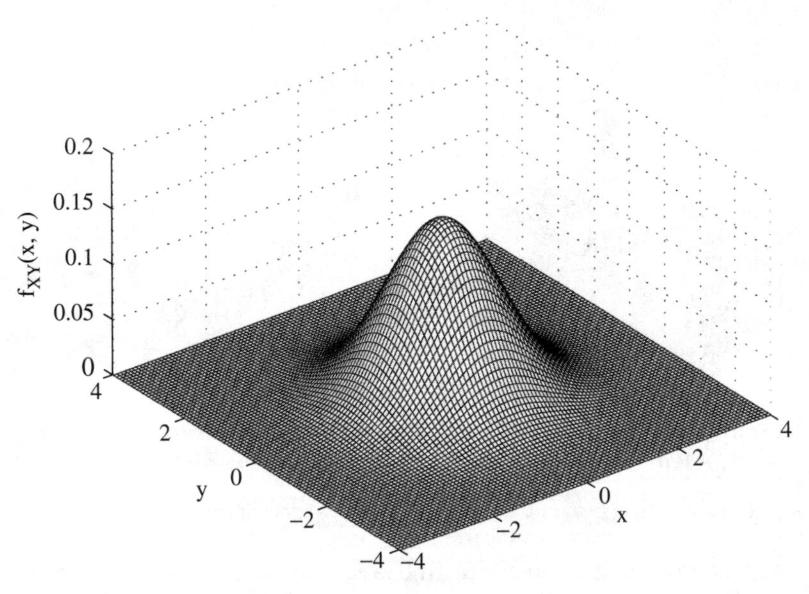

Figure 3.4 Plots of the joint PDF of two zero mean, unit variance, uncorrelated Gaussian random variables.

shaped by the means, variances, and the correlation coefficient. As an example, Figure 3.4 is a plot of the joint density function of two zero mean, unit variance, uncorrelated ($\rho_{XY} = 0$) Gaussian random variables. The uncorrelatedness of the two random variables is what gives this density function its circular symmetry. In fact, for Gaussian random variables uncorrelatedness implies that the two random variables are independent, i.e.,

$$f_{XY}(x, y) = f_X(x) f_Y(y)$$

Changing the mean in the bivariate Gaussian density function again simply changes the center of the bell-shaped curve. Figure 3.5(a) is a plot of a bivariate Gaussian density function with $m_X = 1$, $m_Y = 1$, $\sigma_X = 1$, $\sigma_Y = 1$, and $\rho_{XY} = 0$. The only difference between Figure 3.5(a) and Figure 3.4 (other than the perspective) is a translation of the bell shape to the new mean value (1,1). Changing the variances of joint Gaussian random variables changes the relative shape of the joint density function much like that shown in Figure 3.3(b).

The effect of the correlation coefficient on the shape of the bivariate Gaussian density function is a more interesting concept. Recall that the correlation coefficient is

$$\rho_{XY} = \frac{\text{cov}(X, Y)}{\sqrt{\text{var}(X)\text{var}(Y)}} = \frac{E[(X - m_X)(Y - m_Y)]}{\sigma_X \sigma_Y}$$

If two random variables have a correlation coefficient greater than zero then these two random variables tend probabilistically to take values that are on the

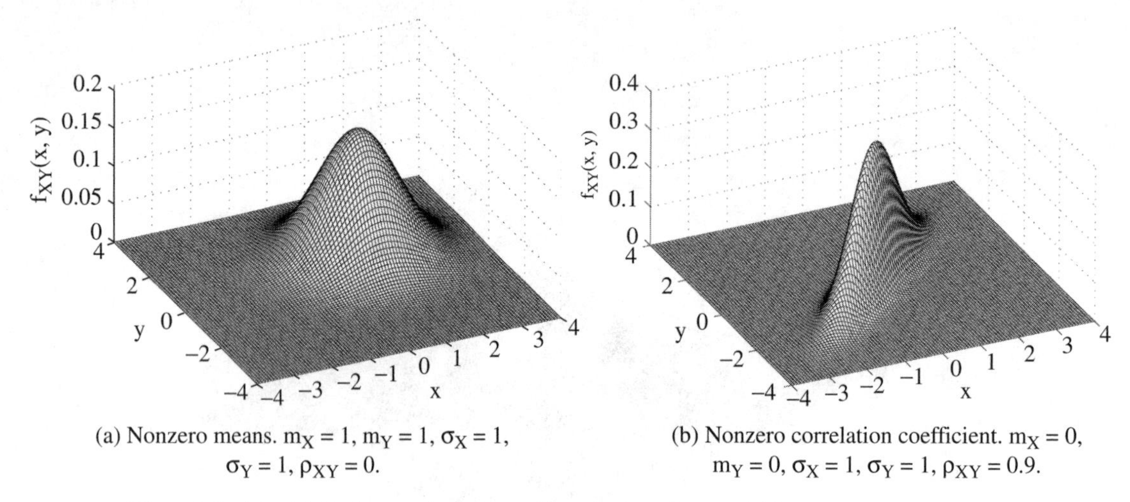

(a) Nonzero means. $m_X = 1$, $m_Y = 1$, $\sigma_X = 1$, $\sigma_Y = 1$, $\rho_{XY} = 0$.

(b) Nonzero correlation coefficient. $m_X = 0$, $m_Y = 0$, $\sigma_X = 1$, $\sigma_Y = 1$, $\rho_{XY} = 0.9$.

Figure 3.5 Plots of the joint PDF of two unit variance Gaussian random variables.

same side of their mean value (if the random variables were zero mean then they would tend to have the same sign). Also if these two random variables have a correlation coefficient close to unity then they probabilistically behave in a very similar fashion. A good example of this characteristic is seen in Figure 3.5(b), which shows a plot of a bivariate Gaussian with $m_X = 0$, $m_Y = 0$, $\sigma_X = 1$, $\sigma_Y = 1$, and $\rho_{XY} = 0.9$.

While it is easy to see that the marginal densities, i.e.,

$$f_X(x) = \int_{-\infty}^{\infty} f_{XY}(x, y) dy$$

from Figure 3.5(b) would be the same as the those obtained in Figure 3.4, the conditional densities would be much different, e.g.,

$$f_{X|Y}(x|y) = \frac{1}{\sigma_X \sqrt{2\pi\left(1 - \rho_{XY}^2\right)}} \times \exp\left[-\frac{1}{2\sigma_X^2\left(1 - \rho_{XY}^2\right)}\left(x - m_X - \frac{\rho_{XY}\sigma_X}{\sigma_Y}(y - m_Y)\right)^2\right] \tag{3.23}$$

It should be noted that Eq. (3.23) says that for a pair of jointly Gaussian random variables, the distribution of one of the Gaussian random variables conditioned on the other Gaussian random variable is also a Gaussian distribution with

$$E[X|Y = y] = m_X + \frac{\rho_{XY}\sigma_X}{\sigma_Y}(y - m_Y) \qquad \text{var}(X|Y = y) = \sigma_X^2\left(1 - \rho_{XY}^2\right) \tag{3.24}$$

Since the random variables whose PDF is shown in Figure 3.5(b) are highly correlated, the resulting values the random variables take are very similar.

This is reflected in the knife edge characteristic that the density function takes along the line $Y = X$.

3.3.4 Transformations of Random Variables

In general, the problem of finding a probabilistic description (joint CDF or PDF) of a joint transformation of random variables is very difficult. To simplify the presentation, two of the most common and practical transformations will be considered: the single function of n random variables and the one-to-one transformation.

Single Function of n Random Variables

A single function transformation of n random variables is expressed as

$$Y = g(X_1, X_2, \ldots, X_n)$$

where $X_1, \ldots, X_n$ are the n original random variables. This section is concerned with finding the PDF or the CDF of the random variable Y. The general technique is the identical two step process used for a single random variable (the regions of integration now become volumes instead of intervals).

Step 1: Find the CDF of Y as a sum of integrals over the random variables $X_1 \ldots X_n$. This is expressed mathematically as

$$F_Y(y) = \sum_i \int \underset{R_i(y)}{\cdots} \int p_{X_1 \cdots X_n}(x_1, x_2 \cdots x_n) dx_1 \cdots dx_n$$

where the $R_i(y)$ are n dimensional volumes where $X_1 \ldots X_n$ are such that $g(X_1, \ldots, X_n) < y$.

Step 2: Find the PDF by differentiating the CDF found in Step 1 using Leibniz rule, i.e.,

$$p_Y(y) = \frac{dF_Y(y)}{dy} = \sum_i \frac{d}{dy} \int \underset{R_i(y)}{\cdots} \int p_{X_1 \cdots X_n}(x_1, x_2, \ldots, x_n) dx_1 \cdots dx_n$$

EXAMPLE 3.16

A transformation which is very important for analyzing the performance of the standard demodulator used with amplitude modulation (AM) is $Y = \sqrt{X_1^2 + X_2^2}$.

Step 1:

$$F_Y(y) = P\left(\sqrt{X_1^2 + X_2^2} \le y\right) = U(y) \int_{-y}^{y} dx_1 \int_{-\sqrt{y^2 - x_1^2}}^{\sqrt{y^2 - x_1^2}} p_{X_1 X_2}(x_1, x_2) dx_2$$

Step 2:

$$p_Y(y) = U(y)\int_{-y}^{y}\frac{y}{\sqrt{y^2-x_1^2}}\left[p_{X_1X_2}\left(x_1,\sqrt{y^2-x_1^2}\right)+p_{X_1X_2}\left(x_1,-\sqrt{y^2-x_1^2}\right)\right]dx_1$$

If in this example X_1 and X_2 are independent ($\rho = 0$) and zero mean Gaussian random variables with equal variances, the joint density is given as

$$f_{X_1X_2}(x_1,x_2) = \frac{1}{2\pi\sigma_X^2}\exp\left(-\frac{x_1^2+x_2^2}{2\sigma_X^2}\right)$$

and $f_Y(y)$ reduces to

$$f_Y(y) = \frac{y}{\sigma_X^2}\exp\left(-\frac{y^2}{2\sigma_X^2}\right)U(y) \tag{3.25}$$

This PDF is known as the Rayleigh density and appears quite frequently in communication system analysis.

One-to-One Transformations

Consider a one-to-one transformation of the form

$$\begin{aligned} Y_1 &= g_1(X_1, X_2, \cdots, X_n) \\ Y_2 &= g_2(X_1, X_2, \cdots, X_n) \\ &\vdots \\ Y_n &= g_n(X_1, X_2, \cdots, X_n) \end{aligned} \tag{3.26}$$

Since the transformation is one-to-one, the inverse functions exist and are given as

$$\begin{aligned} X_1 &= h_1(Y_1, Y_2, ..., Y_n) \\ X_2 &= h_2(Y_1, Y_2, ..., Y_n) \\ &\vdots \\ X_n &= h_n(Y_1, Y_2, ..., Y_n) \end{aligned} \tag{3.27}$$

Since the probability mass in infinitesimal volumes in both the original X coordinate system and the Y coordinate system must be identical, the PDF of the Y's is given as

$$p_Y(y_1, y_2, \ldots y_n) = |J|p_X(h_1(y_1, y_2, \ldots y_n), \ldots, h_n(y_1, y_2, \ldots y_n))$$

where $|J|$ is the Jacobian defined as

$$|J| = \begin{vmatrix} \frac{\partial x_1}{\partial y_1} & \frac{\partial x_2}{\partial y_1} & \cdots & \frac{\partial x_n}{\partial y_1} \\ \frac{\partial x_1}{\partial y_2} & \frac{\partial x_2}{\partial y_2} & \cdots & \frac{\partial x_n}{\partial y_2} \\ \vdots & & \ddots & \\ \frac{\partial x_1}{\partial y_n} & \frac{\partial x_2}{\partial y_n} & \cdots & \frac{\partial x_n}{\partial y_n} \end{vmatrix}$$

EXAMPLE 3.17
A one-to-one transformation which is very important in communications is

$$Y_1 = \sqrt{X_1^2 + X_2^2} \qquad X_1 = Y_1 \sin(Y_2)$$
$$Y_2 = \tan^{-1}\left(\frac{X_1}{X_2}\right) \qquad X_2 = Y_1 \cos(Y_2)$$

The Jacobian of the transformation is given by $|J| = Y_1$ which results in the joint density of the transformed random variables being expressed as

$$p_{Y_1 Y_2}(y_1, y_2) = y_1 p_{X_1 X_2}(y_1 \sin(y_2), y_1 \cos(y_2))$$

3.3.5 Central Limit Theorem

Theorem 3.6 If X_k is a sequence of independent, identically distributed (i.i.d.) random variables with mean m_x and variance σ^2 and Y_n is a random variable defined as

$$Y_n = \frac{1}{\sqrt{n\sigma^2}} \sum_{k=1}^{n} (X_k - m_x)$$

then

$$\lim_{n \to \infty} p_{Y_n}(y) = \frac{1}{\sqrt{2\pi}} \exp\left[-\frac{y^2}{2}\right] = N(0, 1)$$

Proof: The proof uses the characteristic function, a power series expansion, and the uniqueness property of the characteristic function. Details are available in most introductory texts on probability [DR87, LG89, Hel91]. □

The central limit theorem (CLT) implies that the sum of arbitrarily distributed random variables tends to a Gaussian random variable as the number of terms in the sum gets large. Because many physical phenomena in communications systems are due to interactions of large numbers of events (e.g., random electron motion in conductors or large numbers of scatters in wireless propagation), this theorem is one of the major reasons why the Gaussian random variable is so prevalent in communication system analysis.

3.3.6 Multiple Dimensional Gaussian Random Variables

The most common situation encountered in communication system design and analysis is for the corrupting noise samples to be Gaussian random variables. Analytically this is very fortunate since a simple closed-form expression for the joint PDF of L Gaussian random variables is

$$f_{\vec{N}}(\vec{n}) = \frac{1}{\sqrt{(2\pi)^L \det(\mathbf{C}_N)}} \exp\left[-\frac{1}{2}(\vec{n} - \vec{m}_N)^T \mathbf{C}_N^{-1}(\vec{n} - \vec{m}_N)\right] \tag{3.28}$$

where $\vec{N}$ is an $L \times 1$ vector defined as

$$\vec{N} = [N_1, \ldots, N_L]^T$$

$\vec{m}_N$ is the mean vector of $\vec{N}$, and $\mathbf{C}_N$ is the $L \times L$ covariance matrix of the vector $\vec{N}$. The Gaussian assumption is rather powerful. It accurately models noise in most communication systems and the joint PDF, which is a complete probabilistic description of the random process, is given in terms of only the first and second moments.

Definition 3.15 A cross-covariance matrix between two real random vectors is

$$\mathbf{C}_{XY} = E\left[(\vec{X} - \vec{m}_x)(\vec{Y} - \vec{m}_y)^T\right] \tag{3.29}$$

Definition 3.16 If $\vec{X}$ and $\vec{Y}$ are both Gaussian random vectors then $\vec{X}$ conditioned on $\vec{Y} = \vec{y}$ is a Gaussian random vector with

$$E(\vec{X}|\vec{Y} = \vec{y}) = \vec{m}_x + \mathbf{C}_{XY}\mathbf{C}_Y^{-1}(\vec{y} - \vec{m}_y)$$

and

$$\mathbf{C}_{X|Y} = E[(\vec{X} - E(\vec{X}|\vec{Y} = \vec{y}))(\vec{X} - E(\vec{X}|\vec{Y} = \vec{y}))^T] = \mathbf{C}_X - \mathbf{C}_{XY}\mathbf{C}_Y^{-1}\mathbf{C}_{XY}^T$$

Proof: Setting $\vec{X}$ to be a vector of size L and

$$\vec{Z} = \begin{bmatrix} \vec{X} \\ \vec{Y} \end{bmatrix} \qquad \vec{m}_Z = \begin{bmatrix} \vec{m}_X \\ \vec{m}_Y \end{bmatrix} \qquad \mathbf{C}_Z = \begin{bmatrix} \mathbf{C}_X & \mathbf{C}_{XY} \\ \mathbf{C}_{XY}^T & \mathbf{C}_Y \end{bmatrix}$$

and using the definition of a conditional density function gives

$$f_{\vec{X}|\vec{Y}}(\vec{x}|\vec{y}) = \frac{f_{\vec{X}\vec{Y}}(\vec{x},\vec{y})}{f_{\vec{Y}}(\vec{y})} = \frac{f_{\vec{Z}}(\vec{z})}{f_{\vec{Y}}(\vec{y})}$$

$$= \frac{1}{(2\pi)^{L/2}} \frac{\det[\mathbf{C}_Y]}{\det[\mathbf{C}_Z]} \frac{\exp\left[-\frac{1}{2}(\vec{z} - \vec{m}_Z)^T \mathbf{C}_Z^{-1}(\vec{z} - \vec{m}_Z)\right]}{\exp\left[-\frac{1}{2}(\vec{y} - \vec{m}_Y)^T \mathbf{C}_Y^{-1}(\vec{y} - \vec{m}_Y)\right]} \tag{3.30}$$

Equation (3.30) can be simplified using the following two well-known linear algebra identities:

1.

$$\det\begin{bmatrix}\mathbf{A} & \mathbf{B}\\ \mathbf{C} & \mathbf{D}\end{bmatrix} = \det[\mathbf{A} - \mathbf{B}\mathbf{D}^{-1}\mathbf{C}]\det[\mathbf{D}] \tag{3.31}$$

2.

$$\begin{aligned}\begin{bmatrix}\mathbf{A} & \mathbf{B}\\ \mathbf{C} & \mathbf{D}\end{bmatrix}^{-1} &= \begin{bmatrix}\mathbf{A}^{-1} & 0\\ 0 & 0\end{bmatrix} + \begin{bmatrix}-\mathbf{A}^{-1}\mathbf{B}\\ \mathbf{I}\end{bmatrix}[\mathbf{D} - \mathbf{C}\mathbf{A}^{-1}\mathbf{B}]^{-1}\left[-\mathbf{C}\mathbf{A}^{-1}\ \mathbf{I}\right]\\ &= \begin{bmatrix}0 & 0\\ 0 & \mathbf{D}^{-1}\end{bmatrix} + \begin{bmatrix}\mathbf{I}\\ -\mathbf{D}^{-1}\mathbf{C}\end{bmatrix}\left[\mathbf{A} - \mathbf{B}\mathbf{D}^{-1}\mathbf{C}]^{-1}[\mathbf{I}\ -\mathbf{B}\mathbf{D}^{-1}\right]\end{aligned} \tag{3.32}$$

where $\mathbf{A}$, $\mathbf{B}$, $\mathbf{C}$, and $\mathbf{D}$ are arbitrary matrices. Using Eq. (3.31) the ratio of the determinants in Eq. (3.30) is given as

$$\frac{\det[\mathbf{C}_Z]}{\det[\mathbf{C}_Y]} = \det\left[\mathbf{C}_X - \mathbf{C}_{XY}\mathbf{C}_Y^{-1}\mathbf{C}_{YX}\right] = \det[\mathbf{C}_{X|Y}] \tag{3.33}$$

Using Eq. (3.32) to reformulate $\mathbf{C}_Z$ in Eq. (3.30) gives

$$\begin{aligned}\exp\left[-\frac{1}{2}[\vec{z} - \vec{m}_Z]^T\,\mathbf{C}_Z^{-1}[\vec{z} - \vec{m}_Z]\right] &= \exp\left[-\frac{1}{2}[\vec{y} - \vec{m}_Y]^T\,\mathbf{C}_Y^{-1}[\vec{y} - \vec{m}_Y]\right]\\ &\quad\times\exp\left[-\frac{1}{2}\mathbf{F}^T\,\mathbf{G}^{-1}\mathbf{F}\right]\end{aligned} \tag{3.34}$$

where

$$\mathbf{F} = \vec{x} - \vec{m}_X - \mathbf{C}_{XY}\mathbf{C}_Y^{-1}[\vec{y} - \vec{m}_Y] = \vec{x} - E[\vec{X}|\vec{Y} = \vec{y}]$$

and

$$\mathbf{G} = \mathbf{C}_X - \mathbf{C}_{XY}\mathbf{C}_Y^{-1}\mathbf{C}_{XY}^T = \mathbf{C}_{X|Y}$$

Using Eq. (3.33) and Eq. (3.34) in Eq. (3.30) and cancelling the common terms gives the desired result. □

3.4 Homework Problems

Problem 3.1. This problem exercises axiomatic definitions of probability. A communications system used in a home health care system needs to communicate four patient services, two digital and two analog. The two digital services are a 911 request and a doctor appointment request. The two analog services are the transmission of an electrocardiogram (EKG) and the transmission of audio output from a stethoscope. The patient chooses each of the services randomly depending on the doctor's prior advice and the patient's current symptoms. Assume only one service can be requested at a time.

(a) For design purposes it is typical to model a patient's requested service as a random experiment. Define a sample space for the requested services.

A market survey has shown that the probability that a patient will request an EKG is 0.4 and the probability that a patient will request 911 is 0.001. The probability that a digital service will be requested is 0.1.

(b) Compute the probability of requesting a doctor's appointment.

(c) Compute the probability of requesting an audio transmission from a stethoscope.

Problem 3.2. This problem exercises the idea of conditional probability. In the Problem 3.1 if you know the requested service is a digital service what is the probability that a 911 request is initiated?

Problem 3.3. A simple problem to exercise the axioms of probability. Two events A and B are defined by the same random experiment. $P(A) = 0.5$, $P(B) = 0.4$, and $P(A \cap B) = 0.2$

(a) Compute $P(A \cup B)$.

(b) Are events A and B independent?

Problem 3.4. Your roommate challenges you to a game of chance. He proposes the following game. A coin is flipped two times, if heads comes up twice she/he gets a dollar from you and if tails comes up twice you get a dollar from him/her. You know your roommate is a schemer so you know that there is some chance that this game is rigged. To this end you assign the following mathematical framework.

- $P(F)$ = probability that the game is fair $= \frac{5}{6}$
- $P(H|F)$ = probability that a head results if the game is fair $= 0.5$
- $P(H|UF)$ = probability that a head results if the game is unfair $= 0.75$

Assume conditioned on the fairness of the game that each flip of the coin is independent.

(a) What is the probability that the game is unfair, $P(UF)$?

(b) What is the probability that two heads appear given the game is unfair?

(c) What is the probability that two heads appear?

(d) If two heads appear on the first trial of the game, what is the probability that you are playing an unfair game?

Problem 3.5. Certain digital communication schemes use redundancy in the form of an error control code to improve the reliability of communication. The compact disc recording media is one such example. Assume a code can correct two or fewer errors in a block of N coded bits. If each bit detection is independent with a probability of error of $P(E) = 10^{-3}$

1. Plot the probability of correct block detection as a function of N (Note, it is most useful to have log scaling on the y-axis).
2. How small must N be so that the probability the block is detected incorrectly is less than 10^{-6}?

Problem 3.6. The probability of accessing a campus computer system is characterized on a weekday with the following events

1. A ={dialing is between 7:00PM–1:00AM}
2. B = {dialing is between 1:00AM–7:00PM}
3. C = {a connection is made}
4. D = {a connection failed}

The system is characterized with

$$P(\text{Connecting given dialing is between 7:00PM–1:00AM}) = P(C|A) = 0.1$$

$$P(\text{Connecting given dialing is between 1:00AM–7:00PM}) = P(C|B) = 0.75$$

(a) If a person randomly dials with uniform probability during the day what is $P(A)$?

(b) If a person randomly dials with uniform probability during the day what is $P(C)$?

(c) If a person randomly dials with uniform probability during the day what is the $P(A \text{ and } C)$?

(d) Compute $P(A|C)$.

Problem 3.7. In a particular magnetic disk drive bits are written and read individually. The probability of a bit error is $P(E) = 10^{-8}$. Bit errors are independent from bit to bit. In a computer application the bits are grouped into 16-bit words. What is the probability that an application word will be in error (at least one of the 16 bits is in error) when the bits are read from this magnetic disk drive.

Problem 3.8. A simple yet practical problem to exercise the idea of Bayes rule and total probability. Air traffic control radars (ATCR) are used to direct traffic in and around airports. If a plane is detected where it is not suppose to be then the controllers must initiate some evasive actions. Consequently, for the users of a radar system the two most important parameters of a radar system are

- $P(TA|DT)$ = probability of target being absent when a target is detected.
- $P(TP|ND)$ = probability of target being present when no detection is obtained.

Radar designers on the other hand like to quote performance in the following two parameters (because they can easily be measured)

- $P(DT|TA)$ = probability of detecting a target when a target is absent.
- $P(DT|TP)$ = probability of detecting a target when a target is present.

Imagine that you are a high-priced consultant for the Federal Aviation Administration (FAA) and that the FAA has the following requirements for its next generation ATCR

- $P(TA|DT) = 0.01$
- $P(TP|ND) = 0.0001$

A detailed study shows that planes are present only 1% of the time, $P(TP) = 0.01$. A contractor, Huge Aircrash Co., has offered a radar system to the government with the following specifications

- $P(DT|TA) = 0.00005$
- $P(DT|TP) = 0.9$

Would you recommend the government purchase this system and why?

Problem 3.9. The following are some random events

1. The sum of the roll of two dice.
2. The hexadecimal representation of an arbitrary 4 bits in an electronic memory.
3. The top card in a randomly shuffled deck.
4. The voltage at the output of a diode in a radio receiver.

Determine which events are well modeled with a random variable. For each random variable determine if the random variable is continuous, discrete, or mixed. Characterized the sample space and the mapping from the sample space if possible.

Problem 3.10. A random variable has a density function given as

$$\begin{aligned} f_X(x) &= 0 && x < 0 \\ &= K_1 && 0 \le x < 1 \\ &= K_2 && 1 \le x < 2 \\ &= 0 && x \ge 2 \end{aligned}$$

(a) If the mean is 1.2, find K_1 and K_2.

(b) Find the variance using the value of K_1 and K_2 computed in (a).

(c) Find the probability that $X \le 1.5$ using the value of K_1 and K_2 computed in (a).

Problem 3.11. In communications the phase shift induced by propagation between transmitter and receiver, Φ_p, is often modeled as a random variable. A common model is to have

$$\begin{aligned} f_{\Phi_p}(\phi) &= \frac{1}{2\pi} && -\pi \leq \phi \leq \pi \\ &= 0 && \text{elsewhere} \end{aligned}$$

This is commonly referred to as a uniformly distributed random variable.

(a) Find the CDF, $F_{\Phi_p}(\phi)$.

(b) Find the mean and variance of this random phase shift.

(c) It turns out that a communication system will work reasonably well if the coherent phase reference at the receiver, $\hat{\Phi}_p$, is within 30° of the true value of Φ_p. If you implement a receiver with $\hat{\Phi}_p = 0$, what is the probability the communication system will work.

(d) Assuming you can physically move your system and change the propagation delay to obtain an independent phase shift. What is the probability that your system will work at least one out of two times?

(e) How many independent locations would you have to try to ensure a 90% chance of getting your system to work?

Problem 3.12. You have just started work at Yeskia, which is a company that manufactures FM transmitters for use in commercial broadcast. An FCC rule states that the carrier frequency of each station must be within 1 part per million of the assigned center frequency (i.e., if the station is assigned a 100 MHz frequency then the oscillators deployed must be $|f_c - 100\,\text{MHz}| < 100\,\text{Hz}$). The oscillator designers within Yeskia have told you that the output frequency of their oscillators is well modeled as a Gaussian random variable with a mean of the desired channel frequency and a variance of σ_f^2. Assume the lowest assigned center frequency is 88.1 MHz and the highest assigned center frequency is 107.9 MHz.

(a) Which assigned center frequency has the tightest absolute set accuracy constraint?

(b) For the frequency obtained in (a) with $\sigma_f = 40\text{Hz}$, what is the probability that a randomly chosen oscillator will meet the FCC specification?

(c) What value of σ_f should you force the designers to achieve to guarantee that the probability that a randomly chosen oscillator does not meet FCC rules is less than 10^{-7}.

Problem 3.13. X is a Gaussian random variable with a mean of 2 and a variance of 4.

(a) Plot $f_X(x)$.

(b) Plot $F_X(x)$.

(c) Plot the function $g(x) = P(|X - 2| < x)$.

Problem 3.14. For two independent random variables X_1 and X_2 are transformed into a third random variable $Y = X_1 + X_2$. Show

(a) $m_Y = m_{X_1} + m_{X_2}$.

(b) $\sigma_Y^2 = \sigma_{X_1}^2 + \sigma_{X_2}^2$.

Problem 3.15. Consider two random samples, X_1 at the input to a communication system, and X_2, at the output of a communication system and model them as jointly Gaussian random variables. Plot the conditional probability density function, $f_{X_1|X_2}(x_1|x_2)$ for

(a) $m_1 = m_2 = 0$, $\sigma_1 = \sigma_2 = 1$, and $\rho = 0$

(b) $m_1 = m_2 = 0$, $\sigma_1 = 2 \quad \sigma_2 = 1$, and $\rho = 0$

(c) $m_1 = m_2 = 0$, $\sigma_1 = \sigma_2 = 1$, and $\rho = 0.8$

(d) $m_1 = 1 \quad m_2 = 2$, $\sigma_1 = \sigma_2 = 1$, and $\rho = 0$

(e) $m_1 = 1 \quad m_2 = 2$, $\sigma_1 = 2 \quad \sigma_2 = 1$, and $\rho = 0.8$

(f) Rank order the five cases in terms of how much information is available in the output, X_2, about the input X_1.

Problem 3.16. Consider a uniform random variable with the PDF given in Eq. (3.16) with $a = 0$ and $b = 1$.

(a) Compute m_Y.

(b) Compute σ_X.

(c) Compute $P(X \leq 0.4)$.

(d) Use `rand` in Matlab and produce 100 independent samples of a uniform random variable. How many of these 100 samples are less than or equal to 0.4? Does this coincide with the results you obtained in (c)?.

Problem 3.17. The received signal strength in a wireless radio environment is often well modeled as Rayleigh random variable (see Eq. (3.25)). For this problem assume the received signal strength at a particular location for a mobile phone is a random variable, X, with a Rayleigh distribution. A cellular phone using frequency modulation needs to have a signal level of at least $X = 20$mV to operate in an acceptable fashion. Field tests on campus have shown the average signal power is 1mW (in a 1 Ω system).

(a) Using Eq. (3.25) compute the average signal power, $E[X^2]$, as a function of σ_X.

(b) Compute the probability that a cellular phone will work at a randomly chosen location on campus.

(c) Assume you physically move and obtain an independent received signal strength. What is the probability that your cellular phone will work at least one out of two times?

(d) How many independent locations would you have to try to ensure a 90% chance of getting your cellular phone to work?

Problem 3.18. The function `randn` in Matlab produces realizations of a zero mean–unit variance Gaussian random variable each time it is called. This problem leads you to ways to use this function in more general ways. Assume X is a Gaussian random variable with zero mean and unit variance.

(a) Define a random variable $Y = X + b$, find $f_Y(y)$.

(b) Define a random variable $Y = aX$, where $a > 0$, find $f_Y(y)$ (use Example 3.11).

(c) How can you transform X, $Y = g(X)$, so that Y is a Gaussian random variable with $m_Y = b$ and $\sigma_Y = a$?

(d) Test out your answer by invoking `randn` 1000 times and transforming these 1000 samples as you propose in c) with $m_y = 6$ and $\sigma_y = 2$. Plot a histogram of the transformed output.

Problem 3.19. This problem gives both a nice insight into the idea of a correlation coefficient and shows how to generate correlated random variables in simulation. X and W are two zero mean independent random variables where $E[X^2] = 1$ and $E[W^2] = \sigma_W^2$. A third random variable is defined as $Y = \rho X + W$, where ρ is a deterministic constant such that $-1 \leq \rho \leq 1$.

(a) Prove $E[XW] = 0$. In other words prove that independence implies uncorrelatedness.

(b) Choose σ_W^2 such that $\sigma_Y^2 = 1$.

(c) Find ρ_{XY} when σ_W^2 is chosen as in part (b).

Problem 3.20. You are designing a phase-locked loop as an FM demodulator. The requirement for your design is that the loop bandwidth must be greater than 5 kHz and less than 7 kHz. You have computed the loop bandwidth and it is given as

$$B_L = 4R^3 + 2000$$

where R is a resistor in the circuit. It is obvious that choosing $R = 10$ will solve your problem. Unfortunately, resistors are random valued.

(a) If the resistors used in manufacturing your FM demodulator are uniformly distributed between 9 and 11Ω, what is the probability that the design will meet the requirements?

(b) If the resistors used in manufacturing your FM demodulator are Gaussian random variables with a mean of 10 Ω and a standard deviation of 0.5 Ω, what is the probability that the design will meet the requirements?

Problem 3.21. In general, uncorrelatedness is a weaker condition than independence and this problem will demonstrate this characteristic.

(a) If X and Y are zero mean independent random variables, prove $E[XY] = 0$.

(b) Consider the joint discrete random variable X and Y with a joint PMF of

$$p_{XY}(2,0) = \frac{1}{3}$$
$$p_{XY}(-1,-1) = \frac{1}{3}$$
$$p_{XY}(-1,1) = \frac{1}{3}$$
$$p_{XY}(x,y) = 0 \qquad \text{elsewhere} \tag{3.35}$$

What are the marginal PMFs, $p_X(x)$, and $p_Y(y)$? Are X and Y independent random variables? Are X and Y uncorrelated?

(c) Show that two jointly Gaussian random variables which are uncorrelated (i.e., $\rho_{XY} = 0$) are also independent. This makes the Gaussian random variable an exception to the general rule.

Problem 3.22. The probability of having a cellular phone call dropped in the middle of a call while driving on the freeway in the big city is characterized by following events

1. $A = \{\text{call is during rush hour}\}$
2. $B = \{\text{call is not during rush hour}\}$
3. $C = \{\text{the call is dropped}\}$
4. $D = \{\text{the call is normal}\}$

Rush hour is defined to be 7–9:00 AM and 4–6:00 PM. The system is characterized with

$$P(\text{Drop during rush hour}) = P(C|A) = 0.3$$
$$P(\text{Drop during nonrush hour}) = P(C|B) = 0.1.$$

(a) A person's time to make a call, T, might be modeled as a random variable with

$$f_T(t) = \frac{1}{24} \qquad 0 \le t \le 24$$
$$= 0 \qquad \text{otherwise} \tag{3.36}$$

where time is measured using a 24-hour clock (i.e., 4:00 PM $= 16$). Find $P(A)$.

(b) What is $P(C)$?

(c) What is the $P(A \text{ and } C) = P(A \cap C)$?

(d) Compute $P(A|C)$.

Problem 3.23. This problem examines the sum of two random variables and works through the computation of the resulting PDF. Formally $Y = X_1 + X_2$.

(a) Use `rand` in Matlab and produce two vectors, $\vec{X}_1$ and $\vec{X}_2$ of 1000 independent samples of a uniform random variable. The `rand` function produce random variable uniformly distributed on $[0, 1]$. Add these two vectors together and plot a histogram of the resulting vector.

(b) Find the CDF of Y in the form

$$F_Y(y) = \sum_i \int \cdots \int_{R_i(y)} p_{X_1,X_2}(x_1, x_2) dx_1 dx_2 \tag{3.37}$$

as given in Section 3.3.4. Identify all regions in the $x_1 x_2$ plane, $R_i(y)$, where $x_1 + x_2 < y$ where y is a constant.

(c) Find the PDF by taking the derivative of the CDF in (b) and simplify as much as possible.

(d) Show that if X_1 and X_2 are independent random variables then the resultant PDF of Y is given as the convolution of the PDF of X_1 and the PDF X_2.

(e) Compute the PDF of Y if X_1 and X_2 are independent random variable uniformly distributed on $[0, 1]$. Looking back at the histogram produced in (a), does the resulting answer make sense? Verify the result further by considering 10,000 length vectors and repeating (a).

Problem 3.24. A commonly used expected value in communication system analysis is the characteristic function given as

$$\phi_X(t) = E[\exp(jXt)] = \int_{-\infty}^{\infty} \exp(jxt) p_X(x) dx \tag{3.38}$$

(a) Show that

$$E[X] = -j \left. \frac{d\phi_X(t)}{dt} \right|_{t=0} \tag{3.39}$$

(b) Show that

$$E[X^n] = (-j)^n \left. \frac{d^n \phi_X(t)}{dt^n} \right|_{t=0} \tag{3.40}$$

The results in parts (a) and (b) are known as the moment generating property of the characteristic function.

(c) Show that if X is a Gaussian random variable with mean, m_X, and variance, σ_X^2, that $\phi_X(t) = \exp(jtm_X - t^2\sigma_X^2/2)$.

Hint: For any b

$$\int_{-\infty}^{\infty} \frac{1}{\sqrt{2\pi\sigma_X^2}} \exp\left(-\frac{(x-b)^2}{2\sigma_X^2}\right) dx = 1. \tag{3.41}$$

(d) Show that if X_1 and X_2 are independent random variables then the characteristic function of $Y = X_1 + X_2$ is $\phi_Y(t) = \phi_{X_1}(t)\phi_{X_2}(t)$.

(e) Since the characteristic function is unique, deduce from the results in part (d) that the sum of two Gaussian random variable is another Gaussian random variable. Identify m_Y and σ_Y.

Problem 3.25. The probability of having a cellular phone call dropped in the middle of a call while driving on the freeway in a "big" city is characterized by following events

1. $A =$ {call is during rush hour}
2. $B =$ {call is not during rush hour}
3. $C =$ {the call is dropped}
4. $D =$ {the call is normal}

Rush hour is defined to be 7–9:00 AM and 4–6:00 PM. The system is characterized with

$$P(\text{Drop during rush hour}) = P(C|A) = 0.3$$

$$P(\text{Drop during nonrush hour}) = P(C|B) = 0.1$$

Since a call is three time more likely to occur in the daytime than at night, a person's time to make a call, T, might be modeled as a random variable with

$$\begin{aligned} f_T(t) &= K && 0 \le t \le 7 \\ &= 3K && 7 \le t \le 21 \\ &= K && 21 \le t \le 24 \\ &= 0 && \text{otherwise} \end{aligned} \tag{3.42}$$

where time is measured using a 24-hour clock (i.e., 4:00PM = 16).

(a) Find K.

(b) Find $P(A)$.

(c) What is $P(C)$?

(d) What is the $P(A \text{ and } C) = P(A \cap C)$?

3.5 Example Solutions

Problem 3.26. Using 24 hour time keeping to produce a mapping from time to a real number results in a uniform dialing model for the placement of calls having a PDF given as

$$f_T(t) = \frac{1}{24} \quad 0 \le t \le 24$$
$$= 0 \quad \text{elsewhere} \tag{3.43}$$

(a) $P(A) = P(0 \le t \le 1 \cup 19 \le t \le 24) = \int_0^1 f_T(t)dt + \int_{19}^{24} f_T(t)dt = 1/24 + 5/24 = 1/4$.

(b) Here we use total probability over events A and B.

$$P(C) = P(C|A)P(A) + P(C|B)P(B) = 0.1 \times 0.25 + 0.75 \times 0.75 = 0.5875. \tag{3.44}$$

Note here use was made of $P(B) = 1 - P(A) = 0.75$.

(c) Using the definition of conditional probability gives $P(A \cap C) = P(A, C) = P(C|A)$

$P(A) = 0.25 \times 0.1 = 0.025$.

(d) Here we use the definition of conditional probability.

$$P(A|C) = \frac{P(A, C)}{P(C)} = \frac{0.025}{0.5875} = 0.04255 \tag{3.45}$$

Note, if we had not completed parts (b) and (c), then we could have used Bayes rule

$$P(A|C) = \frac{P(C|A)P(A)}{P(C)} = \frac{P(C|A)P(A)}{P(C|A)P(A) + P(C|B)P(B)} \tag{3.46}$$

Problem 3.27.

(a) $F_{\Phi_p}(\phi) = P(\Phi_p \le \phi) = \int_{-\infty}^{\phi} f_{\Phi_p}(x)dx$. Given that

$$f_{\Phi_p}(x) = \frac{1}{2\pi} \quad -\pi \le x \le \pi$$
$$= 0 \quad \text{elsewhere} \tag{3.47}$$

gives

$$F_{\Phi_p}(\phi) = 0 \quad \phi \le -\pi$$
$$= \frac{\phi + \pi}{2\pi} \quad -\pi \le \phi \le \pi$$
$$= 1 \quad \phi \ge \pi \tag{3.48}$$

(b)

$$E[\Phi_p] = \int_{-\pi}^{\pi} x f_{\Phi}(x) dx = 0 \quad (3.49)$$

Since Φ_p is a zero mean random variable

$$\mathrm{var}(\Phi_p) = \int_{-\pi}^{\pi} x^2 f_{\Phi}(x) dx = \frac{1}{2\pi} \int_{-\pi}^{\pi} x^2 dx = \frac{2}{3} \frac{\pi^3}{2\pi} = \frac{\pi^2}{3} \quad (3.50)$$

(c)

$$P(\text{Work}) = P(-30^\circ \leq \Phi_p \leq 30^\circ) = F_{\Phi_p}(30^\circ) - F_{\Phi_p}(-30^\circ) = \frac{1}{6} \quad (3.51)$$

(d) Poincaire's theorem gives

$$P(\text{Work}_1 \text{ or Work}_2) = P\left(W_1 \cup W_2\right) = P(W_1) + P(W_2) - P(W_1 \cap W_2) \quad (3.52)$$

Because of independence of the two trials

$$P(W_1 \cap W_2) = P(W_1) P(W_2) \quad (3.53)$$

consequently

$$P(W_1 \cup W_2) = P(W_1) + P(W_2) - P(W_1) P(W_2) = \frac{1}{6} + \frac{1}{6} - \frac{1}{36} = \frac{11}{36} \quad (3.54)$$

(e) Getting your receiver to work once in N trials is the complement to the event not getting your receiver to work any of N trials, i.e.,

$$P\left(\bigcup_{i=1}^{N} W_i\right) = 1 - P\left(\bigcap_{i=1}^{N} W_i^C\right) \quad (3.55)$$

Again by independence

$$P\left(\bigcap_{i=1}^{N} W_i^C\right) = \prod_{i=1}^{N} P\left(W_i^C\right) = (1 - P(W_i))^N \quad (3.56)$$

so that

$$P\left(\bigcup_{i=1}^{N} W_i\right) = 1 - (1 - P(W_i))^N \quad (3.57)$$

A little exploring in Matlab shows that $N \geq 13$ will give better than a 90% probability of working.

Problem 3.28.

(a) If X and Y are independent, we can compute $E[XY]$ as

$$\begin{aligned}E[XY] &= \int_{-\infty}^{\infty}\int_{-\infty}^{\infty} xyf_{XY}(x,y)dxdy \\ &= \int_{-\infty}^{\infty}\int_{-\infty}^{\infty} xwf_X(x)f_Y(y)dxdy \\ &= \int_{-\infty}^{\infty} xf_X(x)dx \int_{-\infty}^{\infty} yf_Y(y)dy \\ &= E[X]E[Y]\end{aligned}$$

In general $\text{cov}(X,Y) = E[XY] - E[X]E[Y] = 0$, so if X and Y are independent, then X and Y are uncorrelated.

(b) $X \in \{-1, 2\}$ $\quad Y \in \{-1, 0, 1\}$
First, let's compute the marginal PMFs

$$P_X(-1) = \sum_{y\in\Omega_y} P_{XY}(-1,y) = \frac{1}{3} + \frac{1}{3} = \frac{2}{3} \qquad P_X(2) = \sum_{y\in\Omega_y} P_{XY}(2,y) = \frac{1}{3}$$

$$P_Y(-1) = \frac{1}{3} \qquad P_Y(0) = \frac{1}{3} \qquad P_Y(1) = \frac{1}{3}$$

X and Y are independent if and only if $P_X(x)\cdot P_Y(y) = P_{XY}(x,y) \quad \forall x,y$
If $x = 2$ and $y = 0$,

$$\begin{aligned}P_X(2)\cdot P_Y(0) &\neq P_{XY}(x,y) \\ \frac{1}{3}\cdot\frac{1}{3} = \frac{1}{9} &\neq \frac{1}{3}\end{aligned} \tag{3.58}$$

X and Y are not independent.

$$\begin{aligned}E[XY] &= \sum_{x,y} x\cdot y\cdot P_{XY}(x,y) \\ &= 2\cdot 0\cdot\frac{1}{3} + (-1)\cdot(-1)\cdot\frac{1}{3} + (-1)\cdot 1\cdot\frac{1}{3} \\ &= 0 + \frac{1}{3} - \frac{1}{3} = 0\end{aligned}$$

$$E[X] = \sum_x x\cdot P_X(x) = 2\cdot\frac{1}{3} - 1\cdot\frac{2}{3} = 0$$

$$E[Y] = \sum_y y\cdot P_Y(y) = 0\cdot\frac{1}{3} - 1\cdot\frac{1}{3} + 1\cdot\frac{1}{3} = 0$$

$$\text{cov}(X,Y) = 0 - 0\cdot 0 = 0$$

X and Y are uncorrelated random variables. So, here we have shown that if X and Y are uncorrelated random variables this does not imply X and Y are independent.

(c) The joint PDF of two Gaussian r.v's X and Y can be written as

$$f_{XY}(x, y) = \frac{1}{2\pi\sigma_X\sigma_Y\sqrt{1-\rho_{XY}^2}} \exp\left[-\frac{1}{2(1-\rho_{XY}^2)}\left(\left(\frac{x-m_X}{\sigma_X}\right)^2 - 2\rho_{XY}\frac{(x-m_X)(y-m_Y)}{\sigma_X\sigma_Y} + \left(\frac{y-m_Y}{\sigma_Y}\right)^2\right)\right]$$

If X and Y are uncorrelated, then $\text{cov}(X, Y) = 0 \Longrightarrow \rho_{XY} = 0$

$$\begin{aligned} f_{XY}(x, y) &= \frac{1}{\sqrt{2\pi}\sigma_X}\frac{1}{\sqrt{2\pi}\sigma_Y} \exp\left[-\frac{1}{2}\left(\left(\frac{x-m_X}{\sigma_X}\right)^2 + \left(\frac{y-m_Y}{\sigma_Y}\right)^2\right)\right] \\ &= \frac{1}{\sqrt{2\pi}\sigma_X} \exp\left[-\frac{1}{2}\left(\frac{x-m_X}{\sigma_X}\right)^2\right] \cdot \frac{1}{\sqrt{2\pi}\sigma_Y} \exp\left[-\frac{1}{2}\left(\frac{y-m_Y}{\sigma_Y}\right)^2\right] \\ &= f_X(x) \cdot f_Y(y) \end{aligned}$$

This implies that X and Y are independent.

3.6 Miniprojects

Goal: To give exposure

- to a small scope engineering design problem in communications.
- to the dynamics of working with a team.
- to the importance of engineering communication skills (in this case oral presentations).

Presentation: The forum will be similar to a design review at a company (only much shorter). The presentation will be of 5 minutes in length with an overview of the given problem and solution. The presentation will be followed by questions from the audience (your classmates and the professor). Each team member should be prepared to make the presentation on the due date.

3.6.1 Project 1

Project Goals: This problem works you through the generation of realizations of random variables by computer.

Consider real random continuous variables.

(a) Note that if a continuous random variable has a distribution function, $F_X(x)$, defined on the real line which is one-to-one and onto the interval

[0, 1] then an inverse function exists, i.e., $x = F_X^{-1}(u)$ for $u \in [0, 1]$. If U is a uniform random variable in [0, 1] then find the distribution of the random variable $X = F_X^{-1}(u)$.

(b) Consider a transformation of random variables, $X = -\ln(U)/\lambda$. Find the output PDF, when U is a random variable uniformly distributed on [0, 1].

(c) Use the result in (a) and the `rand` function in Matlab to generate a realizations of an exponential random variable, i.e., ones that have a density function

$$f_X(x) = \begin{cases} \lambda \exp[-x\lambda] & x \geq 0 \\ 0 & \text{elsewhere} \end{cases} \tag{3.59}$$

Hint: If U is uniform in the interval [0, 1] then $1 - U$ has the same distribution. Generate 1000 realizations with $\lambda = 1$ and plot a histogram of the resulting numbers.

Chapter

4

Complex Baseband Representation of Bandpass Signals

4.1 Introduction

A majority of communication systems operate by modulating an information bearing waveform onto a sinusoidal carrier. As examples, Table 4.1 lists the carrier frequencies of various methods of electronic communication.

One can see by examining Table 4.1 that the carrier frequency of the transmitted signal is not the component which contains the information. Instead it is the signal modulated on the carrier which contains the information. Hence a method of characterizing a communication signal, which is independent of the carrier frequency, is desired. This has led communication system engineers to use a **complex baseband representation** of communication signals to simplify their job. All of the communication systems mentioned in Table 4.1 can be and typically are analyzed with this complex baseband representation. This chapter develops the complex baseband representation for deterministic signals. Other references that develop these topics well are [Pro89, PS94, Hay83, BB99]. One advantage of the complex baseband representation is simplicity. When communication system engineers use the complex baseband notation, all signals are lowpass signals and the fundamental ideas behind modulation and communication signal processing are easily developed. Also digital processing based receivers use the complex baseband representation in describing the baseband processing algorithms. In fact, complex baseband representation is so prevalent in engineering systems that the most widely used tool, *Matlab*, has been configured by default to process all variables in a program as complex signals. Hopefully by the time you are done with this course the utility of this view will be apparent.

TABLE 4.1 Carrier frequency assignments for different methods of information transmission

Type of Transmission	Center Frequency of Transmission
Telephone Modems	1600–1800 Hz
AM radio	530–1600 kHz
CB radio	27 MHz
FM radio	88–108 MHz
VHF TV	178–216 MHz
Cellular radio	850 MHz, 1.8 GHz
Indoor Wireless Networks	2.4 GHz
Commercial Satellite Downlink	3.7–4.2 GHz
Commercial Satellite Uplink	5.9–6.4 GHz
Fiber Optics	2×10^{14} Hz

4.2 Baseband Representation of Bandpass Signals

The first step in the development of a complex baseband representation is to define a bandpass signal.

Definition 4.1 A bandpass signal, $x_c(t)$, is a signal whose one-sided energy spectrum is both: (1) centered at a nonzero frequency, f_C, and (2) does not extend in frequency to zero (DC).

The two-sided transmission bandwidth of a signal is typically denoted by B_T Hertz so that the one-sided spectrum of the bandpass signal is zero except in $[f_C - B_T/2, f_C + B_T/2]$. This implies that a bandpass signal satisfies the following constraint: $B_T/2 < f_C$. Figure 4.1 illustrates a conformant bandpass energy spectrum. Since a bandpass signal, $x_c(t)$, is a physically realizable signal it is real valued and consequently the energy spectrum will always be even symmetric around $f = 0$. The relative sizes of B_T and f_C are not important, only that the spectrum takes negligible values around DC. In telephone modem communications this region of negligible spectral values is only about 300 Hz wide, while in satellite communications it can be many gigahertz.

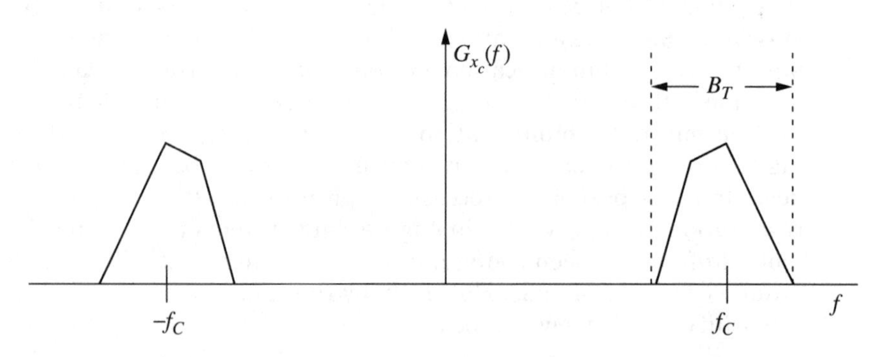

Figure 4.1 Energy spectrum of a bandpass signal.

A bandpass signal has a representation of

$$x_c(t) = x_I(t)\sqrt{2}\cos(2\pi f_c t) - x_Q(t)\sqrt{2}\sin(2\pi f_c t) \tag{4.1}$$

$$= x_A(t)\sqrt{2}\cos(2\pi f_c t + x_P(t)) \tag{4.2}$$

where f_c is denoted the carrier frequency with $f_C - B_T/2 \leq f_c \leq f_C + B_T/2$. The signal $x_I(t)$ in Eq. (4.1) is normally referred to as the **in-phase (I)** component of the signal and the signal $x_Q(t)$ is normally referred to as the **quadrature (Q)** component of the bandpass signal. $x_I(t)$ and $x_Q(t)$ are real valued lowpass signals with a one-sided non–negligible energy spectrum no larger than B_T Hz. Two items should be noted

- The center frequency of the bandpass signal, f_C, (see Figure 4.1) and the carrier frequency, f_c are not always the same. While f_c can theoretically take a continuum of values in Eq. (4.1), in most applications an obvious value of f_c will give the simplest representation[1].
- The $\sqrt{2}$ term is included in the definition of the bandpass signal to ensure that the bandpass signal and the baseband signal have the same power/energy. This will become apparent in Section 4.4.

The carrier signal is normally thought of as the cosine term, hence the I component is in-phase with the carrier. Likewise the sine term is 90° out-of-phase (in quadrature) with the cosine or carrier term, hence the Q component is quadrature to the carrier. Equation (4.1) is known as the canonical form of a bandpass signal. Equation (4.2) is the amplitude and phase form of the bandpass signal, where $x_A(t)$ is the **amplitude** of the signal and $x_P(t)$ is the **phase** of the signal. A bandpass signal has two degrees of freedom. A communication engineer can use either the I/Q representation or the amplitude and phase representation to denote a bandpass signal. The transformations between the two representations are given by

$$x_A(t) = \sqrt{x_I(t)^2 + x_Q(t)^2} \qquad x_P(t) = \tan^{-1}[x_Q(t), x_I(t)] \tag{4.3}$$

and

$$x_I(t) = x_A(t)\cos(x_P(t)) \qquad x_Q(t) = x_A(t)\sin(x_P(t)) \tag{4.4}$$

Note that the inverse tangent function in Eq. (4.3) has a range of $[-\pi, \pi]$ (i.e., both the sign of $x_I(t)$ and $x_Q(t)$ and the ratio of $x_I(t)$ and $x_Q(t)$ are needed to evaluate the function). This inverse tangent function is different than the single argument function that is on most calculators. The particulars of the communication design analysis determine which form for the bandpass signal is most applicable.

[1]This idea will become more obvious in Chapter 6.

A complex valued signal, denoted the **complex envelope**, is defined as

$$x_z(t) = x_I(t) + j x_Q(t) = x_A(t) \exp[j x_P(t)]$$

The original bandpass signal can be obtained from the complex envelope by

$$x_c(t) = \sqrt{2}\Re[x_z(t) \exp[j 2\pi f_c t]]. \tag{4.5}$$

Since the complex exponential only determines the carrier frequency, the complex signal $x_z(t)$ contains all the information in $x_c(t)$. Using this complex baseband representation of bandpass signals greatly simplifies the notation for communication system analysis. As you progress through the text hopefully the additional simplicity provided by the complex envelope representation will become evident.

EXAMPLE 4.1
Consider the bandpass signal

$$x_c(t) = 2\cos(2\pi f_m t)\sqrt{2}\cos(2\pi f_c t) - \sin(2\pi f_m t)\sqrt{2}\sin(2\pi f_c t)$$

where $f_m < f_c$. A plot of this bandpass signal is seen in Figure 4.2 with $f_c = 10 f_m$. Obviously we have

$$x_I(t) = 2\cos(2\pi f_m t) \qquad x_Q(t) = \sin(2\pi f_m t)$$

and

$$x_z(t) = 2\cos(2\pi f_m t) + j\sin(2\pi f_m t)$$

The amplitude and phase can be computed as

$$x_A(t) = \sqrt{1 + 3\cos^2(2\pi f_m t)} \qquad x_P(t) = \tan^{-1}[\sin(2\pi f_m t), 2\cos(2\pi f_m t)].$$

A plot of the amplitude and phase of this signal is seen in Figure 4.3.

As an interesting historical note, the communications field did not always utilize the complex baseband notation. Originally, communication theorist adopted the "analytical signal" denoted

$$x_{an}(t) = x_z(t) \exp[j 2\pi f_c t] \tag{4.6}$$

as a way to understand communication signals. This complex analytical signal was viewed as useful since it was more mathematically tractable and yet capture all the characteristics of communications signals. Slowly over time the field came to realize that the important characteristics of a communication waveform are captured by the complex envelope, $x_z(t)$. An early tutorial paper on the complex analytical signal is given in [Bed62]. Hence the complex envelope

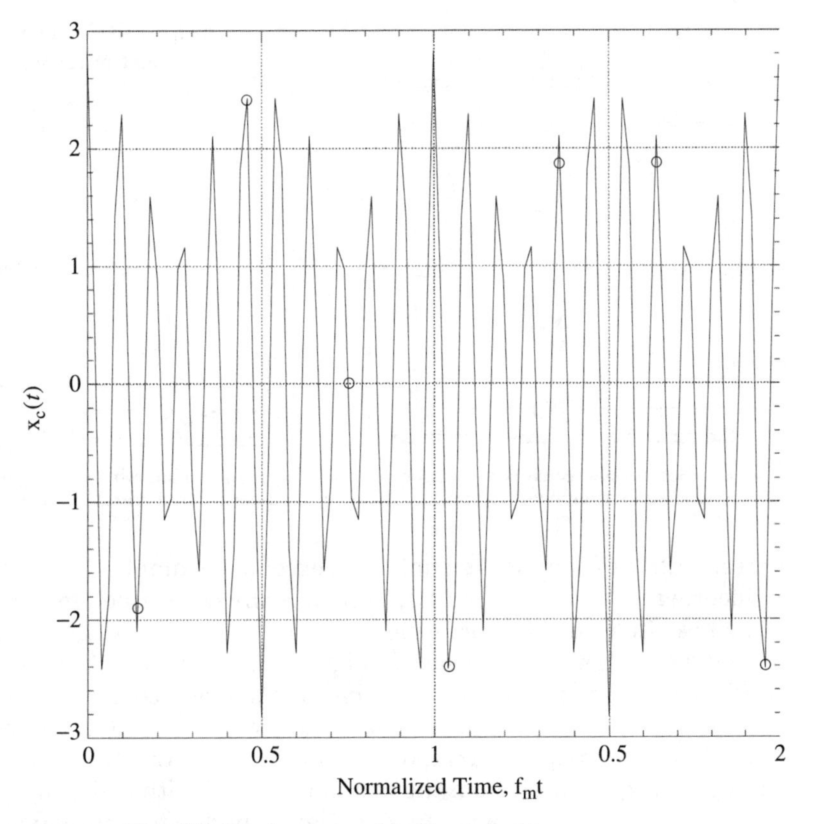

Figure 4.2 Plot of the bandpass signal for Example 4.1.

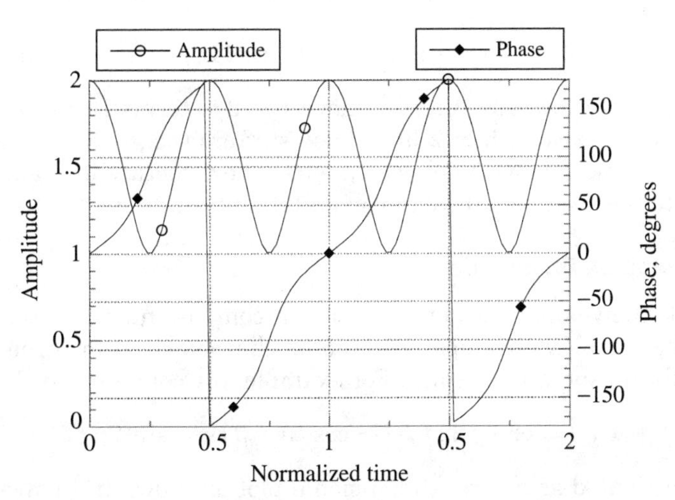

Figure 4.3 Plot of the amplitude and phase for Example 4.1.

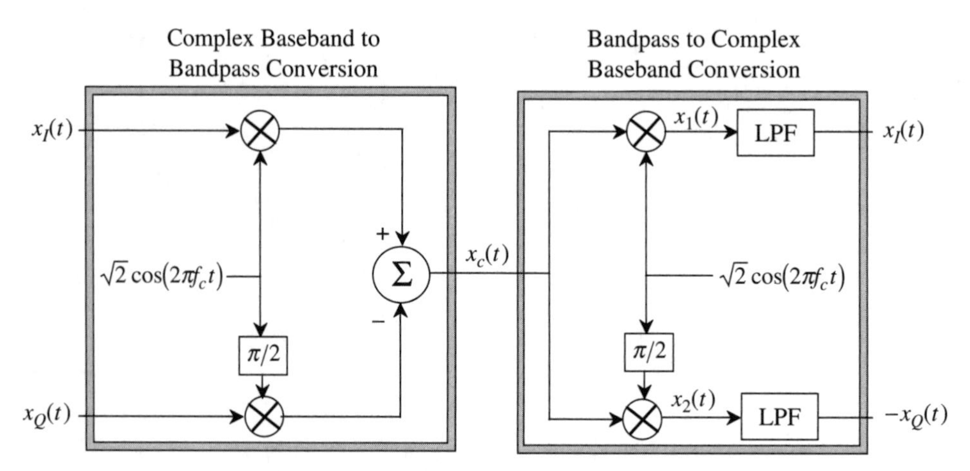

Figure 4.4 Schemes for converting between complex baseband and bandpass representations. Note that the LPF simply removes the double frequency term associated with the down conversion.

representation of bandpass signals is an excellent example of the evolution of an engineering tool where no one person can really be ascribed to the "invention."

The next item to consider is methods to translate between a bandpass signal and a complex envelope signal. Figure 4.4 shows the block diagram of the translation between a complex envelope and a bandpass signal and vice versa. Basically a bandpass signal is generated from its I and Q components in a straightforward fashion corresponding to Eq. (4.1). Likewise a complex envelope signal is generated from the bandpass signal with a similar architecture. Bandpass to baseband downconversion can be understood by using trigonometric identities to give

$$\begin{aligned} x_1(t) &= x_c(t)\sqrt{2}\cos(2\pi f_c t) = x_I(t) + x_I(t)\cos(4\pi f_c t) - x_Q(t)\sin(4\pi f_c t) \\ x_2(t) &= x_c(t)\sqrt{2}\sin(2\pi f_c t) = -x_Q(t) + x_Q(t)\cos(4\pi f_c t) + x_I(t)\sin(4\pi f_c t) \end{aligned} \tag{4.7}$$

In Figure 4.4 the lowpass filters remove the $2f_c$ terms in Eq. (4.7). Note in Figure 4.4 the boxes with $\pi/2$ are phase shifters (i.e., $\cos(\theta - \pi/2) = \sin(\theta)$) typically implemented with delay elements. The structure in Figure 4.4 is fundamental to the study of all carrier modulation techniques.

4.3 Visualization of Complex Envelopes

The complex envelope is a signal that is a complex function of time. Consequently, the complex envelope needs to be characterized in three dimensions (time, in-phase, and quadrature). For example, the complex envelope given as

$$x_z(t) = \exp[j2\pi f_m t] = \cos(2\pi f_m t) + j\sin(2\pi f_m t) \tag{4.8}$$

can be represented as a three dimensional plot as shown in Figure 4.5(a). It is often difficult to comprehend all that is going on in a three-dimensional plot

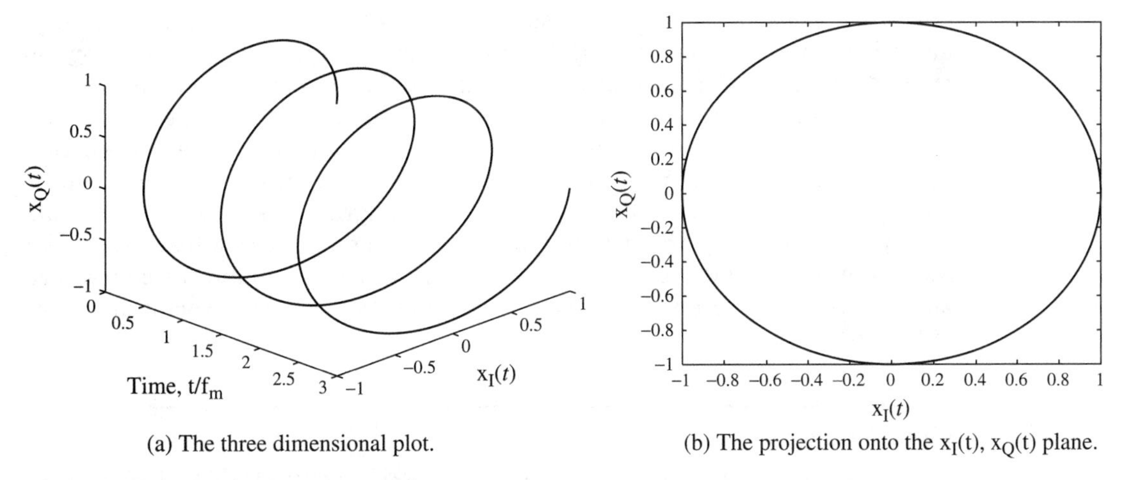

(a) The three dimensional plot.

(b) The projection onto the $x_I(t)$, $x_Q(t)$ plane.

Figure 4.5 Plots of a complex exponential.

when the complex envelope is a typical communication signal hence communication engineers often project these signals into two dimensions. Examples of this two-dimensional projection are shown in Figure 4.5(b) ($x_I(t)$ versus $x_Q(t)$), Figure 4.6(a) (t versus $x_I(t)$), and Figure 4.6(b) (t versus $x_Q(t)$). All of these methods of viewing and visualizing a complex envelope signal are used in engineering practice.

The vector diagram is the two-dimensional projection, where $x_I(t)$ is plotted versus $x_Q(t)$. The vector diagram represents the time trajectory of the complex envelope, $x_z(t)$, in the complex plane. This vector diagram often gives significant insight into the performance or characteristics of a communication system

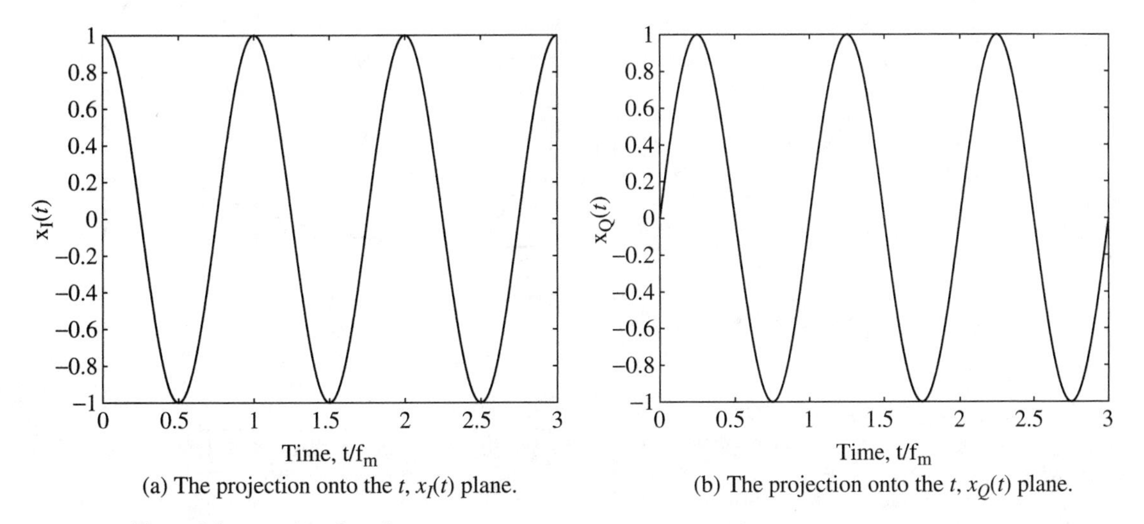

(a) The projection onto the t, $x_I(t)$ plane.

(b) The projection onto the t, $x_Q(t)$ plane.

Figure 4.6 Plots of the two time functions.

and it will be used often in future sections of this text. The vector diagram first gained utility in the days when the standard tool for examining the time domain characteristics of a communication signal was the dual channel analog oscilloscope. To produce a vector diagram with a dual channel oscilloscope, $x_I(t)$ is put into one channel and $x_Q(t)$ is put into the second channel and the scope is configured to plot channel 1 versus channel 2. Since early test instruments could simply generate this visualization, the vector diagram has traditionally found utility in engineering practice. In fact, modern communication test equipment like the Agilent 89600 vector signal analyzer are based around the characterization of the complex envelope and have the ability to measure vector diagrams of bandpass signals.

EXAMPLE 4.2

The vector diagram for the bandpass signal given in Example 4.1 is shown in Figure 4.7. It should be noted that the point $x_z(t) = (2, 0)$ corresponds to time $t = n/f_m$ where n is an

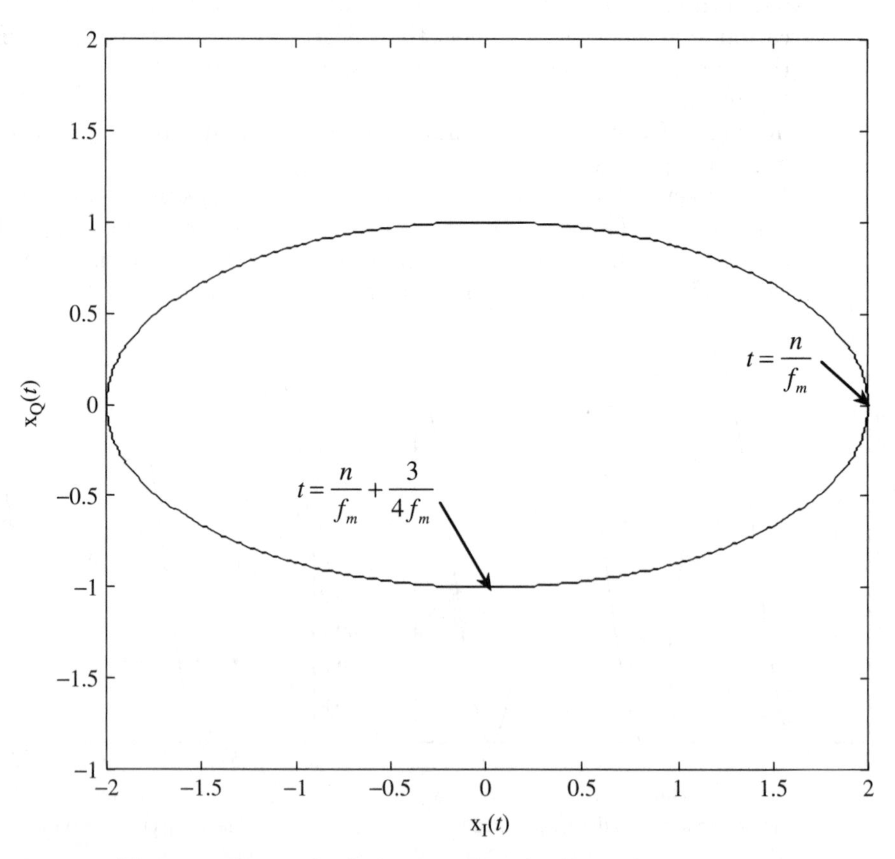

Figure 4.7 The vector diagram for the bandpass signal in Example 4.1.

integer, e.g., $2\cos(2\pi n) = 2$ and $\sin(2\pi n) = 0$. Likewise $t = n/f_m + 1/(4f_m)$ corresponds to the point $x_z(t) = (0, 1)$.

4.4 Spectral Characteristics of the Complex Envelope

4.4.1 Basics

It is of interest to derive the spectral representation of the complex baseband signal, $x_z(t)$, and compare it to the spectral representation of the bandpass signal, $x_c(t)$. Assuming $x_z(t)$ is an energy signal, the Fourier transform of $x_z(t)$ is given by

$$X_z(f) = X_I(f) + jX_Q(f) \tag{4.9}$$

where $X_I(f)$ and $X_Q(f)$ are the Fourier transform of $x_I(t)$ and $x_Q(t)$, respectively, and the energy spectrum is given by

$$G_{x_z}(f) = |X_z(f)|^2 = G_{x_I}(f) + G_{x_Q}(f) + 2\Im[X_I(f)X_Q^*(f)] \tag{4.10}$$

where $G_{x_I}(f)$ and $G_{x_Q}(f)$ are the energy spectrum of $x_I(t)$ and $x_Q(t)$, respectively. The signals $x_I(t)$ and $x_Q(t)$ are lowpass signals with a one-sided bandwidth of less than $B_T/2$ so consequently $X_z(f)$ and $G_{x_z}(f)$ can only take nonzero values for $|f| < B_T/2$.

EXAMPLE 4.3

Consider the case when $x_I(t)$ is set to be the message signal from Example 2.3 (computer generated voice saying “bingo”) and $x_Q(t) = \cos(2000\pi t)$. $X_I(f)$ will be a lowpass spectrum with a bandwidth of 2500 Hz while $X_Q(f)$ will have two impulses located at ± 1000 Hz. Figure 4.8 shows the measured complex envelope energy spectrum for these lowpass signals. The complex envelope energy spectrum has a relation to the voice spectrum and the sinusoidal spectrum exactly as predicted in Eq. (4.10). Note here $B_T = 5000$ Hz.

Equation (4.9) gives a simple way to transform between the lowpass signal spectrums to the complex envelope spectrum. A similar simple formula exists for the opposite transformation. Note that $x_I(t)$ and $x_Q(t)$ are both real signals so that $X_I(f)$ and $X_Q(f)$ are Hermitian symmetric functions of frequency. It is straightforward to show

$$\begin{aligned} X_z(-f) &= X_I^*(f) + jX_Q^*(f) \\ X_z^*(-f) &= X_I(f) - jX_Q(f) \end{aligned} \tag{4.11}$$

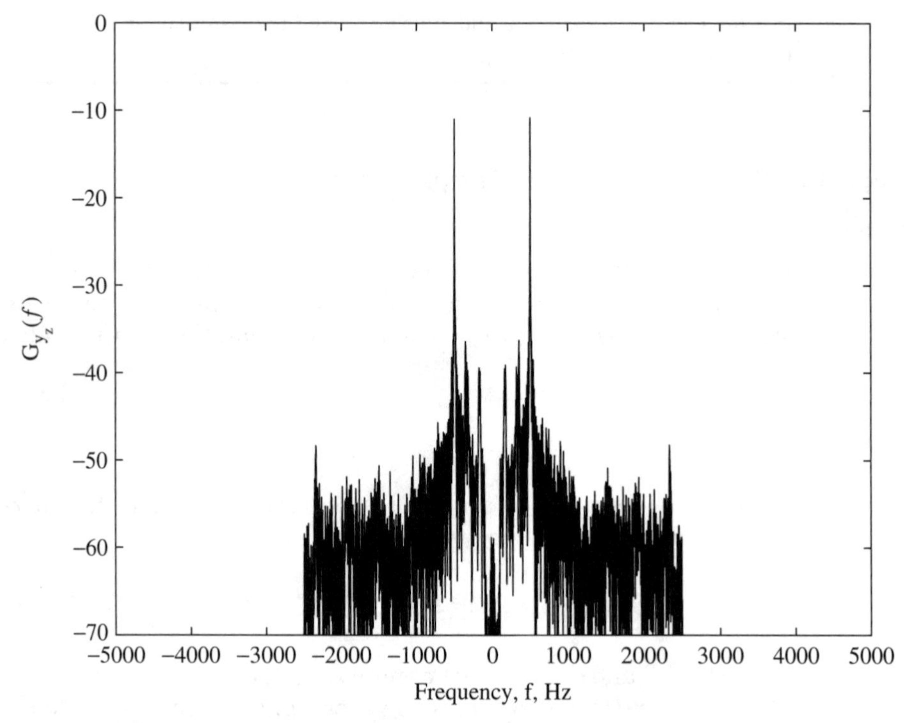

Figure 4.8 The energy spectrum resulting from $x_I(t)$ being a computer generated voice signal and $x_Q(t)$ being a sinusoid.

This leads directly to

$$
\begin{aligned}
X_I(f) &= \frac{X_z(f) + X_z^*(-f)}{2} \\
X_Q(f) &= \frac{X_z(f) - X_z^*(-f)}{j2}
\end{aligned}
\tag{4.12}
$$

Since $x_z(t)$ is a complex signal, in general, the energy spectrum, $G_{x_z}(f)$, has none of the usual properties of real signal spectra. Real signals have a spectral magnitude that is an even function of f and a spectral phase that is an odd function of f. Complex envelopes spectra are not so restricted.

An analogous derivation produces the spectral characteristics of the bandpass signal. Examining Eq. (4.1) and using the Frequency Translation Theorem of the Fourier transform, the Fourier transform of the bandpass signal, $x_c(t)$, is expressed as

$$
\begin{aligned}
X_c(f) = &\left[\frac{1}{\sqrt{2}}X_I(f - f_c) + \frac{1}{\sqrt{2}}X_I(f + f_c)\right] \\
&- \left[\frac{1}{\sqrt{2}j}X_Q(f - f_c) - \frac{1}{\sqrt{2}j}X_Q(f + f_c)\right]
\end{aligned}
$$

This can be rearranged to give

$$X_c(f) = \left[\frac{X_I(f - f_c) + jX_Q(f - f_c)}{\sqrt{2}}\right] + \left[\frac{X_I(f + f_c) - jX_Q(f + f_c)}{\sqrt{2}}\right] \quad (4.13)$$

Using Eq. (4.11) in Eq. (4.13) gives

$$X_c(f) = \frac{1}{\sqrt{2}}X_z(f - f_c) + \frac{1}{\sqrt{2}}X_z^*(-f - f_c) \quad (4.14)$$

This is a very fundamental result. Equation (4.14) states that the Fourier transform of a bandpass signal is simply derived from the spectrum of the complex envelope. For positive values of f, $X_c(f)$ is obtained by translating $X_z(f)$ to f_c and scaling the amplitude by $1/\sqrt{2}$. For negative values of f, $X_c(f)$ is obtained by flipping $X_z(f)$ around the origin, taking the complex conjugate, translating the result to $-f_c$, and scaling the amplitude by $1/\sqrt{2}$. This also demonstrates that if $X_c(f)$ only takes values when the absolute value of f is in $[f_c - B_T/2, f_c + B_T/2]$, then $X_z(f)$ only takes values in $[-B_T/2, B_T/2]$. The energy spectrum of $x_c(t)$ can also be expressed in terms of the energy spectrum of $x_z(t)$ as

$$G_{x_c}(f) = \frac{1}{2}G_{x_z}(f - f_c) + \frac{1}{2}G_{x_z}(-f - f_c) \quad (4.15)$$

Since $E_{x_c} = \int_{-\infty}^{\infty} G_{x_c}(f)\,df = \int_{-\infty}^{\infty} G_{x_z}(f)\,df$, Eq. (4.15) guarantees that the energy of the complex envelope is identical to the energy of the bandpass signal. Additionally, Eq. (4.15) guarantees that the energy spectrum of the bandpass signal is an even function of frequency as it should be for a real signal. Considering these results, the spectrum of the complex envelope of the signal shown in Figure 4.1 will have a form shown in Figure 4.9 when $f_c = f_C$. Other values of f_c would produce a different but equivalent complex envelope representation. This discussion of the spectral characteristics of $x_c(t)$ and $x_z(t)$ should reinforce the idea that the complex envelope contains all the information in a bandpass waveform.

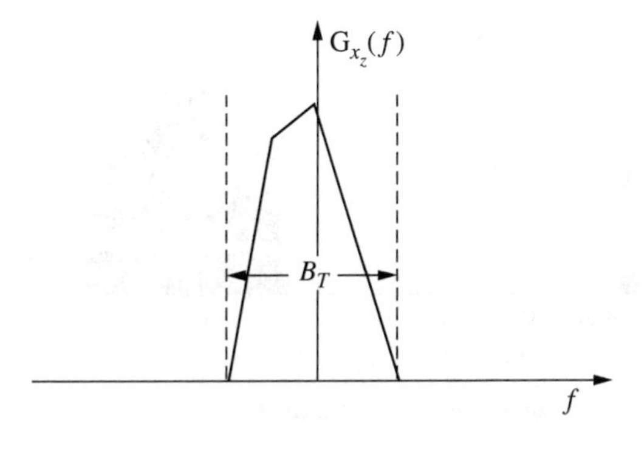

Figure 4.9 The complex envelope energy spectrum of the bandpass signal in Figure 4.1 with $f_c = f_C$.

EXAMPLE 4.4

(Example 4.1 cont.)

$$x_I(t) = 2\cos(2\pi f_m t) \qquad x_Q(t) = \sin(2\pi f_m t)$$

$$X_I(f) = \delta(f - f_m) + \delta(f + f_m) \qquad X_Q(f) = \frac{1}{2j}\delta(f - f_m) - \frac{1}{2j}\delta(f + f_m)$$

$$X_z(f) = X_I(f) + jX_Q(f) = 1.5\delta(f - f_m) + 0.5\delta(f + f_m)$$

and using Eq. (4.14) gives the bandpass signal spectrum as

$$X_c(f) = \frac{1.5}{\sqrt{2}}\delta(f - f_c - f_m) + \frac{1}{2\sqrt{2}}\delta(f - f_c + f_m) + \frac{1.5}{\sqrt{2}}\delta(f + f_c + f_m) + \frac{1}{2\sqrt{2}}\delta(f + f_c - f_m)$$

Note in this example $B_T = 2f_m$

EXAMPLE 4.5

For the complex envelope derived in Example 4.3 the measured bandpass energy spectrum for $f_c = 7000$ Hz is shown in Figure 4.10. Again the measured output is exactly predicted by Eq. (4.15). In this example we have $B_T = 5000$ Hz.

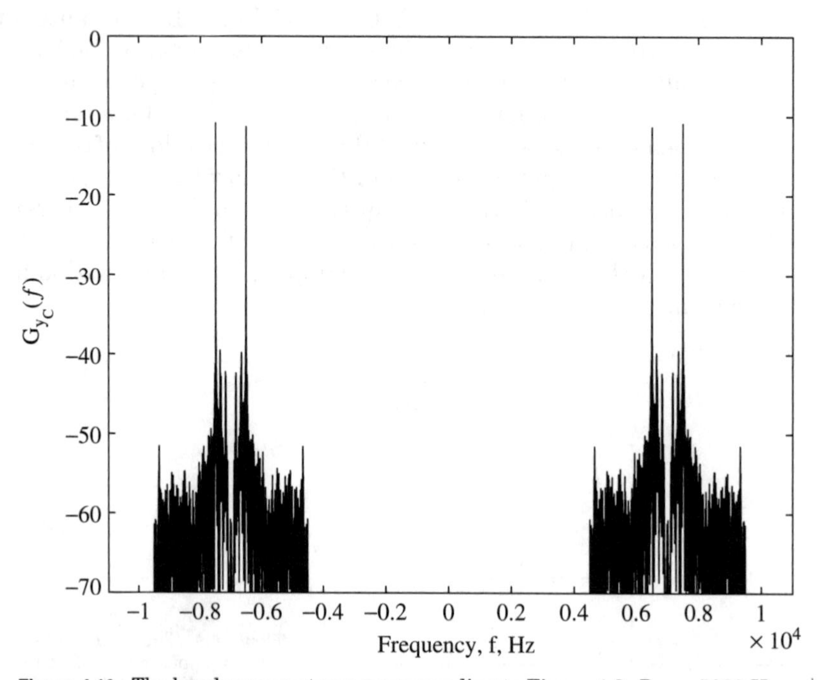

Figure 4.10 The bandpass spectrum corresponding to Figure 4.8. $B_T = 5000$ Hz.

4.4.2 Bandwidth of Bandpass Signals

The ideas of bandwidth of a signal extend in an obvious way to bandpass signals. Recall engineers define bandwidth as being the amount of positive spectrum that a signal occupies. For bandpass energy signals we have the following two definitions that are analogous to the bandwidth definitions in Chapter 2.

Definition 4.2 If a signal $x_c(t)$ has an energy spectrum $G_{x_c}(f)$ then B_X is determined as

$$10\log\left(\max_f G_{x_c}(f)\right) = X + 10\log(G_{x_c}(f_1)) \tag{4.16}$$

where $G_{x_c}(f_1) > G_{x_c}(f)$ for $0 < f < f_1$ and

$$10\log\left(\max_f G_{x_c}(f)\right) = X + 10\log(G_{x_c}(f_2)) \tag{4.17}$$

where $G_{x_c}(f_2) > G_{x_c}(f)$ for $f > f_2$ where $f_2 - f_1 = B_X$.

Definition 4.3 If a signal $x_c(t)$ has an energy spectrum $G_{x_c}(f)$ then $B_P = \min(f_2 - f_1)$ such that

$$P = \frac{2\int_{f_1}^{f_2} G_{x_c}(f)\,df}{E_{x_c}} \tag{4.18}$$

where $f_2 > f_1$.

Note the reason for the factor of 2 in Eq. (4.21) is that half of the energy of the bandpass signal is associated with positive frequencies and half of the energy is associated with negative frequencies.

Again, for bandpass power signals similar ideas hold with $G_{x_c}(f)$ being replaced with $S_{x_c}(f, T)$.

Definition 4.4 If a signal $x_c(t)$ has a sampled power spectral density $S_{x_c}(f, T_m)$ (see Eq. (2.56)) then B_X is determined as

$$10\log\left(\max_f S_{x_c}(f, T_m)\right) = X + 10\log(S_{x_c}(f_1, T_m)) \tag{4.19}$$

where $S_{x_c}(f_1, T_m) > S_{x_c}(f, T_m)$ for $0 < f < f_1$ and

$$10\log\left(\max_f S_{x_c}(f, T_m)\right) = X + 10\log(S_{x_c}(f_2, T_m)) \tag{4.20}$$

where $S_{x_c}(f_2, T_m) > S_{x_c}(f, T_m)$ for $f > f_2$ where $f_2 - f_1 = B_X$.

Definition 4.5 If a signal $x_c(t)$ has an power spectrum $S_{x_c}(f, T_m)$ then $B_P = \min(f_2 - f_1)$ such that

$$P = \frac{2\int_{f_1}^{f_2} S_{x_c}(f, T_m)\,df}{P_{x_c}(T_m)} \tag{4.21}$$

where $f_2 > f_1$ and $P_{x_c}(T_m)$ is defined in Eq. (2.57).

4.5 Power of Carrier Modulated Signals

The output power of a carrier modulated signal is often important in evaluating the trade-offs between analog communication options. For generality of the exposition in this text, power will be computed as a time average power. The time average power is defined as

$$P_{x_c} = \lim_{T_m \to \infty} \frac{1}{T_m} \int_{-T_m/2}^{T_m/2} x_c^2(t)dt \tag{4.22}$$

Using the Rayleigh energy theorem and Eq. (2.56) the time average power of a bandpass modulated signal is

$$P_{x_c} = \lim_{T_m \to \infty} \int_{-\infty}^{\infty} S_{X_c}(f, T_m)\, df \tag{4.23}$$

Using the power spectrum analog to Eq. (4.15) gives

$$P_{x_c} = \lim_{T_m \to \infty} \frac{1}{T_m} \int_{-T_m/2}^{T_m/2} |x_z(t)|^2 dt = P_{x_z} \tag{4.24}$$

Hence we have shown that the power contained in a carrier modulated signal is exactly the power in the complex envelope. This result is the main reason why the notation used in this text uses the $\sqrt{2}$ term in the carrier terms.

4.6 Linear Systems and Bandpass Signals

This section discusses methods for calculating the output of a linear, time-invariant (LTI) filter with a bandpass input signal using complex envelopes. Linear system outputs are characterized by the convolution integral given as

$$y_c(t) = \int_{-\infty}^{\infty} x_c(\tau)h(t-\tau)d\tau \tag{4.25}$$

where $h(t)$ is the impulse response of the filter. Since the input signal is bandpass, the effects of an arbitrary filter, $h(t)$, in Eq. (4.25) can be modeled with an equivalent bandpass filter, $h_c(t)$, with no loss in generality. The bandpass filter, $H_c(f)$, only needs to equal the true filter, $H(f)$ over the frequency support of the bandpass signal and the two filters need not be equal otherwise. Because of this characterization the bandpass filter is often simpler to model (and to simulate).

EXAMPLE 4.6

For example consider the lowpass filter given in Figure 4.11(a). Since the bandpass signal only has a nonzero spectrum in a bandwidth of B_T around the carrier frequency, f_c, the bandpass filter shown in Figure 4.11(b) would be input–output equivalent to the filter in Figure 4.11(a).

This bandpass LTI system also has a canonical representation given as

$$h_c(t) = 2h_I(t)\cos(2\pi f_c t) - 2h_Q(t)\sin(2\pi f_c t) \tag{4.26}$$

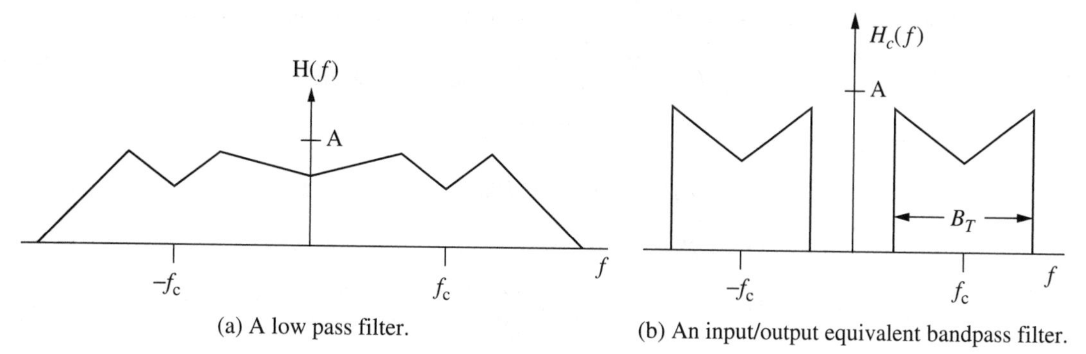

Figure 4.11 An example of a filter and its bandpass equivalent filter.

The complex envelope for this bandpass impulse response and transfer function associated with this complex envelope are given by

$$h_z(t) = h_I(t) + jh_Q(t) \qquad H_z(f) = H_I(f) + jH_Q(f)$$

where the bandpass system impulse response is

$$h_c(t) = 2\Re[h_z(t)\exp[j2\pi f_c t]]$$

The representation of the bandpass system in Eq. (4.26) has a constant factor of $\sqrt{2}$ difference from the bandpass signal representation of Eq. (4.1). This factor results because the system response at baseband and at bandpass should be identical. This notational convenience permits a simpler expression for the system output (as is shown shortly). Using similar techniques as in Section 4.4, the transfer function is expressed as

$$H_c(f) = H_z(f - f_c) + H_z^*(-f - f_c) \tag{4.27}$$

EXAMPLE 4.7

Consider the signal

$$x_c(t) = (\cos(2\pi f_m t) + \cos(6\pi f_m t))\sqrt{2}\cos(2\pi f_c t) - (\sin(2\pi f_m t) + \sin(6\pi f_m t))\sqrt{2}\sin(2\pi f_c t) \tag{4.28}$$

that is input into a bandpass filter with a transfer function of

$$H_c(f) = \begin{cases} 2 & f_c - 2f_m \leq |f| \leq f_c + 2f_m \\ 0 & \text{elsewhere} \end{cases} \tag{4.29}$$

Since the frequency domain representation of $x_c(t)$ is

$$X_c(f) = \frac{1}{\sqrt{2}}[\delta(f - (f_c + f_m)) + \delta(f - (f_c + 3f_m) + \delta(f + (f_c + f_m)) + \delta(f + (f_c + 3f_m)] \tag{4.30}$$

the output bandpass signal will have the frequency domain representation of

$$Y_c(f) = \frac{2}{\sqrt{2}}[\delta(f - (f_c + f_m) + \delta(f + (f_c + f_m)] \quad (4.31)$$

The complex envelopes of the input and output signals are

$$x_z(t) = \exp[j2\pi f_m t] + \exp[j6\pi f_m t] \quad y_z(t) = 2\exp[j2\pi f_m t] \quad (4.32)$$

consequently it makes sense to have

$$H_z(f) = \begin{cases} 2 & |f| \leq 2f_m \\ 0 & \text{elsewhere} \end{cases} \quad (4.33)$$

and $H_c(f) = H_z(f - f_c) + H_z^*(-f - f_c)$.

Equations (4.27) and (4.14) combined with the convolution theorem of the Fourier transform produce an expression for the Fourier transform of $y_c(t)$ given as

$$Y_c(f) = X_c(f)H_c(f) = \frac{1}{\sqrt{2}}[X_z(f - f_c) + X_z^*(-f - f_c)][H_z(f - f_c) + H_z^*(-f - f_c)]$$

Since both $X_z(f)$ and $H_z(f)$ only take nonzero values in $[-B_T/2, B_T/2]$, the cross terms in this expression will be zero and $Y_c(f)$ is given by

$$Y_c(f) = \frac{1}{\sqrt{2}}[X_z(f - f_c)H_z(f - f_c) + X_z^*(-f - f_c)H_z^*(-f - f_c)] \quad (4.34)$$

Since $y_c(t)$ will also be a bandpass signal, it will also have a complex baseband representation. A comparison of Eq. (4.34) with Eq. (4.14) demonstrates the Fourier transform of the complex envelope of $y_c(t)$, $y_z(t)$, is given as

$$Y_z(f) = X_z(f)H_z(f)$$

Linear system theory produces the desired form

$$y_z(t) = \int_{-\infty}^{\infty} x_z(\tau)h_z(t - \tau)d\tau = x_z(t) * h_z(t) \quad (4.35)$$

EXAMPLE 4.8

(Example 4.1 cont.) The input signal and Fourier transform are

$$x_z(t) = 2\cos(2\pi f_m t) + j\sin(2\pi f_m t) \qquad X_z(f) = 1.5\delta(f - f_m) + 0.5\delta(f + f_m)$$

Assume a bandpass filter with

$$H_I(f) = \begin{cases} 2 & -2f_m \leq f \leq 2f_m \\ 0 & \text{elsewhere} \end{cases} \qquad H_Q(f) = \begin{cases} j\frac{f}{f_m} & -f_m \leq f \leq f_m \\ j & f_m \leq f \leq 2f_m \\ -j & -2f_m \leq f \leq -f_m \\ 0 & \text{elsewhere} \end{cases} \quad (4.36)$$

This produces

$$H_z(f) = \begin{cases} 2 - \frac{f}{f_m} & -f_m \le f \le f_m \\ 1 & f_m \le f \le 2f_m \\ 3 & -2f_m \le f \le -f_m \\ 0 & \text{elsewhere} \end{cases} \tag{4.37}$$

and now the Fourier transform of the complex envelope of the filter output is

$$Y_z(f) = H_z(f)X_z(f) = 1.5\delta(f - f_m) + 1.5\delta(f + f_m) \tag{4.38}$$

The complex envelope and bandpass signal are given as

$$y_z(t) = 3\cos(2\pi f_m t) \qquad y_c(t) = 3\cos(2\pi f_m t)\sqrt{2}\cos(2\pi f_c t) \tag{4.39}$$

In other words, convolving the complex envelope of the input signal with the complex envelope of the filter response produces the complex envelope of the output signal. The different scale factor was introduced in Eq. (4.26) so that Eq. (4.35) would have a familiar form. This result is significant since $y_c(t)$ can be derived by computing a convolution of baseband (complex) signals, which is generally much simpler than computing the bandpass convolution. Since $x_z(t)$ and $h_z(t)$ are complex, $y_z(t)$ is given in terms of the I/Q components as

$$\begin{aligned} y_z(t) = y_I(t) + jy_Q(t) &= [x_I(t) * h_I(t) - x_Q(t) * h_Q(t)] \\ &+ j\,[x_I(t) * h_Q(t) + x_Q(t) * h_I(t)] \end{aligned}$$

Figure 4.12 shows the lowpass equivalent model of a bandpass system. The two biggest advantages of using the complex baseband representation are that it simplifies the analysis of communication systems and permits accurate digital computer simulation of filters and the effects on communication systems performance.

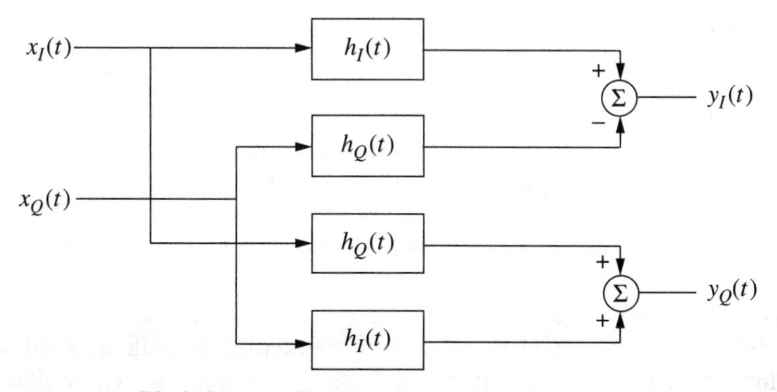

Figure 4.12 Block diagram illustrating the relation between the input and output complex envelope of bandpass signals for a linear time invariant system.

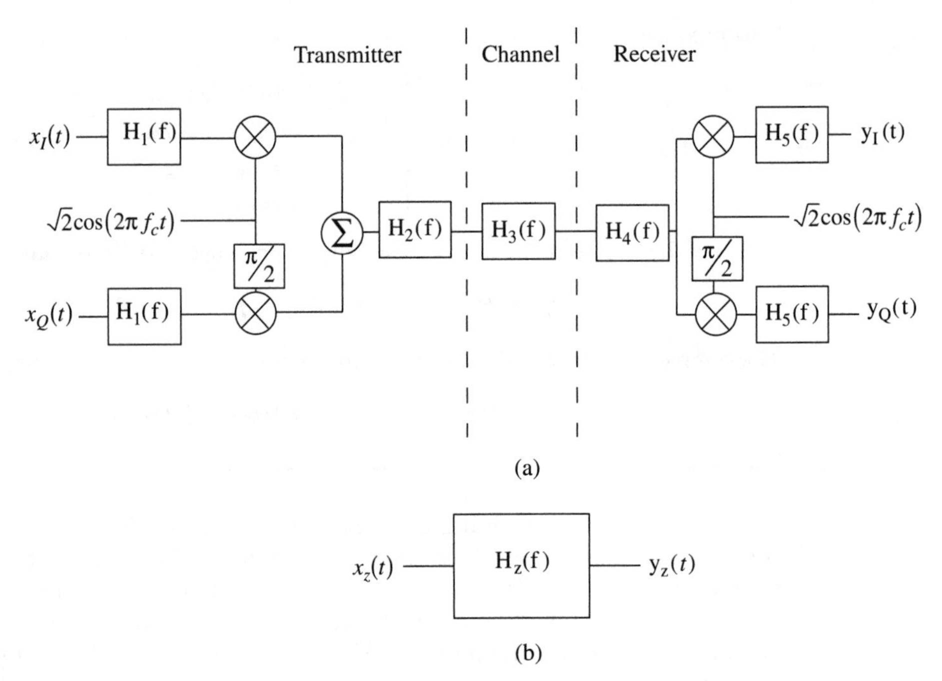

Figure 4.13 A comparison between (a) the actual communication system model and (b) the complex baseband equivalent model.

4.7 Conclusions

The complex baseband representation of bandpass signals permits accurate characterization and analysis of communication signals independent of the carrier frequency. This greatly simplifies the job of the communication systems engineer. A linear system is often an accurate model for a communication system, even with the associated transmitter filtering, channel distortion, and receiver filtering. As demonstrated in Figure 4.13, the complex baseband methodology truly simplifies the models for a communication system performance analysis.

4.8 Homework Problems

Problem 4.1. Many integrated circuit implementations of the quadrature upconverters produce a bandpass signal having a form

$$x_c(t) = x_I(t)\sqrt{2}\cos(2\pi f_c t) + x_Q(t)\sqrt{2}\sin(2\pi f_c t) \tag{4.40}$$

from the lowpass signals $x_I(t)$ and $x_Q(t)$ as opposed to Eq. (4.1). How does this sign difference affect the transmitted spectrum? Specifically for the complex envelope energy spectrum given in Figure 4.9 plot the transmitted bandpass energy spectrum.

Problem 4.2. Find the form of $x_I(t)$ and $x_Q(t)$ for the following $x_c(t)$

(a) $x_c(t) = \sin(2\pi(f_c - f_m)t)$

(b) $x_c(t) = \cos(2\pi(f_c + f_m)t)$

(c) $x_c(t) = \cos(2\pi f_c t + \phi_p)$

Problem 4.3. If the lowpass components for a bandpass signal are of the form

$$x_I(t) = 12\cos(6\pi t) + 3\cos(10\pi t)$$

and

$$x_Q(t) = 2\sin(6\pi t) + 3\sin(10\pi t)$$

(a) Calculate the Fourier series of $x_I(t)$ and $x_Q(t)$.

(b) Calculate the Fourier series of $x_z(t)$.

(c) Assuming $f_c = 40$ Hz calculate the Fourier series of $x_c(t)$.

(d) Calculate and plot $x_A(t)$. Computer might be useful.

(e) Calculate and plot $x_P(t)$. Computer might be useful.

Problem 4.4. A bandpass filter has the following complex envelope representation for the impulse response

$$h_z(t) = \begin{cases} 2\left(\frac{1}{2}\exp\left[-\frac{t}{2}\right]\right) + j2\left(\frac{1}{4}\exp\left[-\frac{t}{4}\right]\right) & t \geq 0 \\ 0 & \text{elsewhere} \end{cases} \tag{4.41}$$

(a) Calculate $H_z(f)$.

Hint: *The transforms you need are in a table somewhere.*

(b) With $x_z(t)$ from Problem 4.3 as the input, calculate the Fourier series for the filter output, $y_z(t)$.

(c) Plot the output amplitude, $y_A(t)$, and phase, $y_P(t)$.

(d) Plot the resulting bandpass signal, $y_c(t)$ using $f_c = 40$ Hz.

Problem 4.5. The picture of a color television set proposed by the National Television System Committee (NTSC) is composed by scanning in a grid pattern across the screen. The scan is made up of three independent beams (red, green, and blue). These independent beams can be combined to make any color at a particular position. In order to make the original color transmission compatible with black and white televisions the three color signals ($x_r(t)$, $x_g(t)$, $x_b(t)$) are transformed into a luminance signal (black and white level), $x_L(t)$, and two independent chrominance signals, $x_I(t)$ and $x_Q(t)$. These chrominance signals are modulated onto a carrier of 3.58 MHz to produce a bandpass signal for transmission. A commonly used tool for video engineers to understand these coloring patterns is the vectorscope representation shown in Figure 4.14.

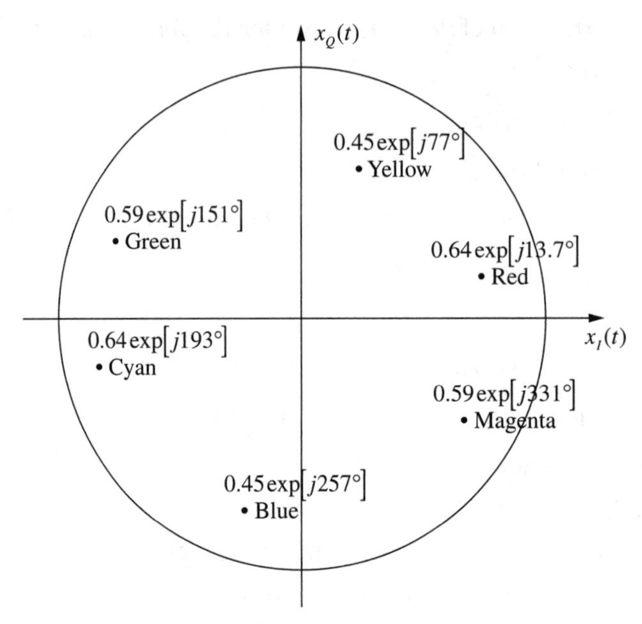

Figure 4.14 Vector scope representation of the complex envelope of the 3.58 MHz chrominance carrier.

(a) If the video picture is making a smooth transition from a blue color (at $t = 0$) to green color (at $t = 1$), make a plot of the waveforms $x_I(t)$ and $x_Q(t)$.

(b) Plot $x_I(t)$ and $x_Q(t)$ that would represent a scan across a red and green striped area. For consistency in the answers assume the red starts at $t = 0$ and extends to $t = 1$, the green starts at $t = 1^+$ and extends to $t = 2, \cdots$.

Problem 4.6. Consider two lowpass spectra, $X_I(f)$ and $X_Q(f)$ in Figure 4.15 and sketch the energy spectrum of the complex envelope, $G_{x_z}(f)$.

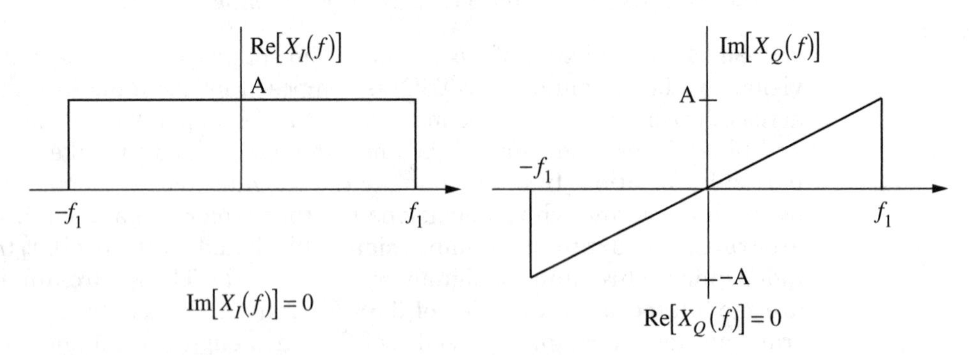

Figure 4.15 Two lowpass Fourier transforms.

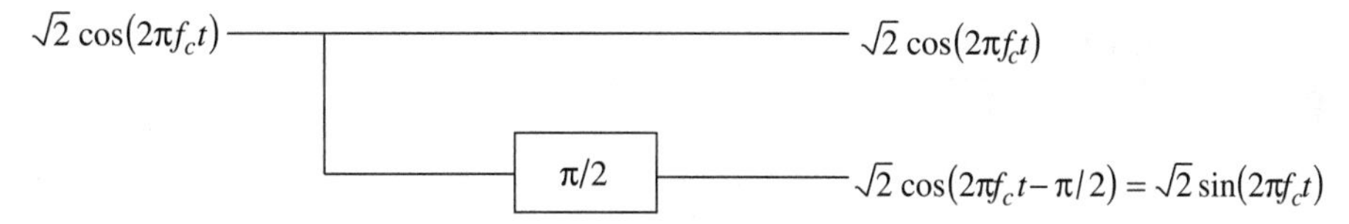

Figure 4.16 Sine and cosine generator.

Problem 4.7. **(Design Problem)** A key component in the quadrature up/down converter is the generator of the sine and cosine functions. This processing is represented in Figure 4.16 as a shift in the phase by 90° of a carrier signal. This function is done in digital processing in a trivial way but if the carrier is generated by an analog source the implementation is more tricky. Show that this phase shift can be generated with a time delay as in Figure 4.17. If the carrier frequency is 100 MHz, find the value of the delay to achieve the 90° shift.

Problem 4.8. The lowpass signals, $x_I(t)$ and $x_Q(t)$, which comprise a bandpass signal are given in Figure 4.18.

(a) Give the form of $x_c(t)$, the bandpass signal with a carrier frequency f_c, using $x_I(t)$ and $x_Q(t)$.

(b) Find the amplitude, $x_A(t)$, and the phase, $x_P(t)$, of the bandpass signal.

(c) Give the simplest form for the bandpass signal over $[2T, 3T]$.

Problem 4.9. The amplitude and phase of a bandpass signal is plotted in Figure 4.19. Plot the in-phase and quadrature signals of this baseband representation of a bandpass signal.

Problem 4.10. The block diagram in Figure 4.20 shows a cascade of a quadrature upconverter and a quadrature downconverter where the phases of the two (transmit and receive) carriers are not the same. Show that $y_z(t) = y_I(t) + jy_Q(t) = x_z(t)\exp[-j\theta(t)]$. Specifically consider the case when the frequencies of the two carriers are not the same and compute the resulting output energy spectrum $G_{Y_z}(f)$.

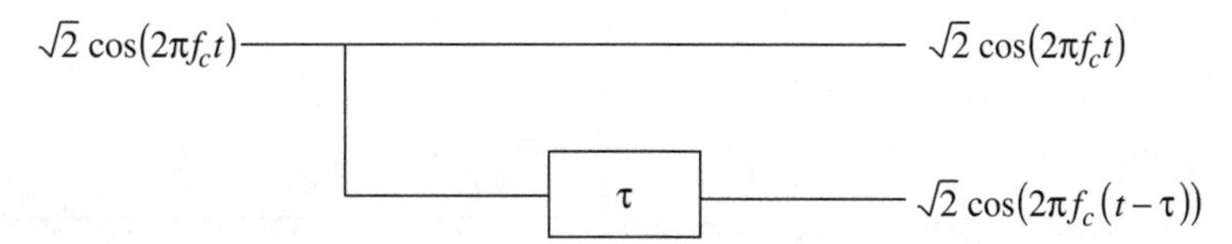

Figure 4.17 Sine and cosine generator implementation for analog signals.

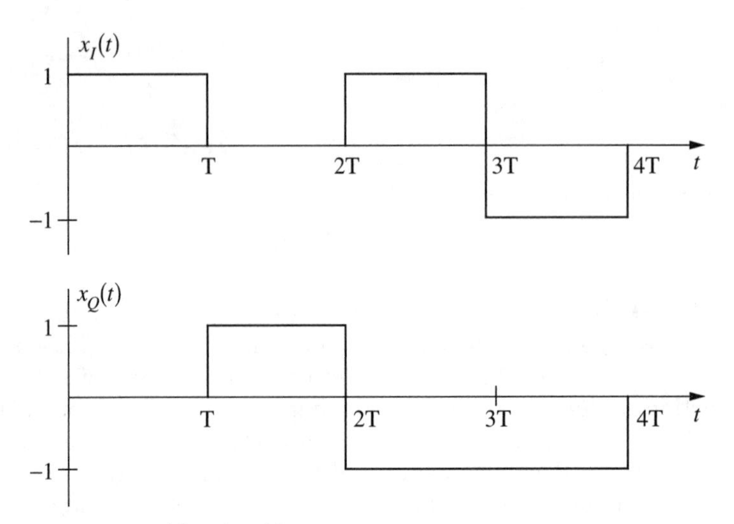

Figure 4.18 $x_I(t)$ and $x_Q(t)$.

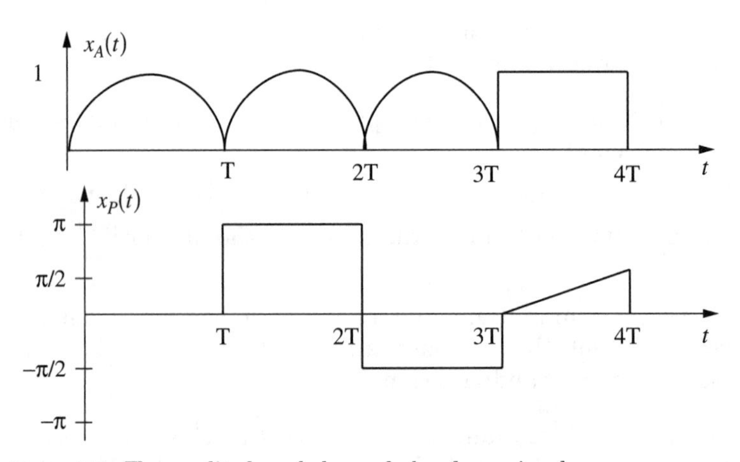

Figure 4.19 The amplitude and phase of a bandpass signal.

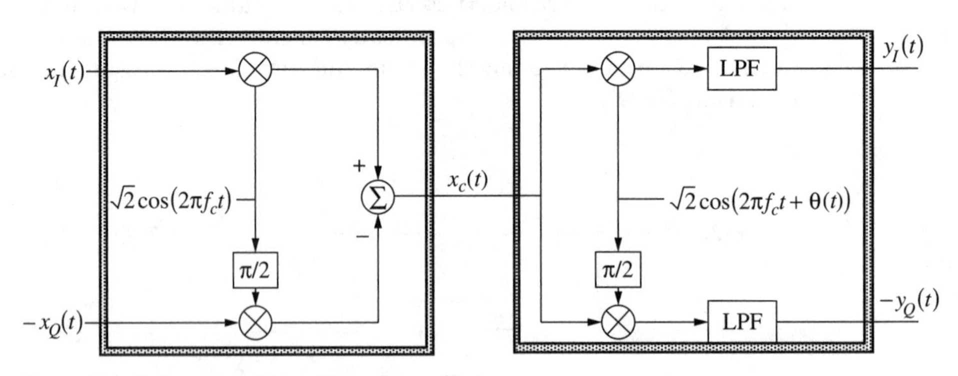

Figure 4.20 A downconverter with a phase offset.

Problem 4.11. A periodic real signal of bandwidth W and period T is $x_I(t)$ and $x_Q(t) = 0$ for a bandpass signal of carrier frequency $f_c > W$.

(a) Can the resulting bandpass signal, $x_c(t)$, be periodic with a period of $T_c < T$? If yes, give an example.

(b) Can the resulting bandpass signal, $x_c(t)$, be periodic with a period of $T_c > T$? If yes, give an example.

(c) Can the resulting bandpass signal, $x_c(t)$, be periodic with a period of $T_c = T$? If yes, give an example.

(d) Can the resulting bandpass signal, $x_c(t)$, be aperiodic? If yes, give an example.

Problem 4.12. In communication systems bandpass signals are often processed in digital processors. To accomplish the processing, the bandpass signal must first be converted from an analog signal to a digital signal. For this problem assume this is done by ideal sampling. Assume the sampling frequency, f_s, is set at four times the carrier frequency.

(a) Under what conditions on the complex envelope will this sampling rate be greater than the Nyquist sampling rate (see Section 2.4.1) for the bandpass signal?

(b) Give the values for the bandpass signal samples for $x_c(0)$, $x_c(\frac{1}{4f_c})$, $x_c(\frac{2}{4f_c})$, $x_c(\frac{3}{4f_c})$, and $x_c(\frac{4}{4f_c})$.

(c) By examining the results in (b) can you postulate a simple way to downconvert the analog signal when $f_s = 4f_c$ and produce $x_I(t)$ and $x_Q(t)$? This simple idea is frequently used in engineering practice and is known as $f_s/4$ downconversion.

Problem 4.13. A common implementation problem that occurs in an I/Q upconverter is that the sine carrier is not exactly 90° out of phase with the cosine carrier. This situation is depicted in Figure 4.21.

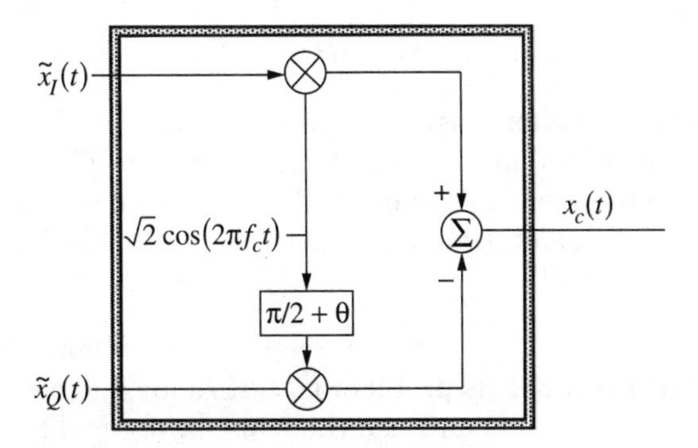

Figure 4.21 The block diagram for Problem 4.13.

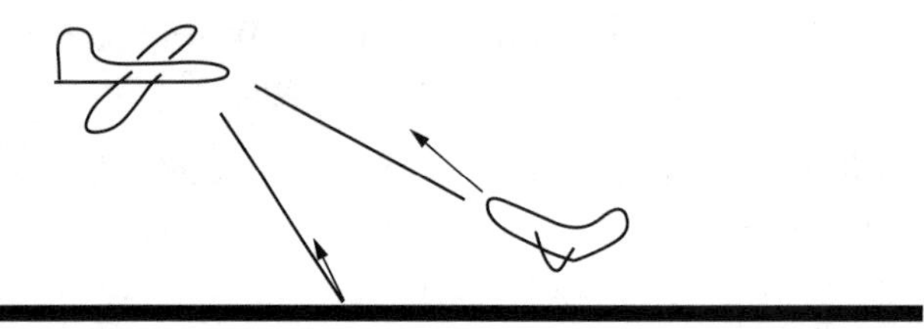

Figure 4.22 An airborne air traffic control radar example.

(a) What is the actual complex envelope, $x_z(t)$, produced by this implementation as a function of $\tilde{x}_I(t)$, $\tilde{x}_Q(t)$, and θ?

(b) Often in communication systems it is possible to correct this implementation error by preprocessing the baseband signals. If the desired output complex envelope was $x_z(t) = x_I(t) + jx_Q(t)$, what should $\tilde{x}_I(t)$ and $\tilde{x}_Q(t)$ be set to as a function of $x_I(t)$, $x_Q(t)$, and θ to achieve the desired complex envelope with this implementation?

Problem 4.14. A commercial airliner is flying 15,000 feet above the ground and pointing its radar down to aid traffic control. A second plane is just leaving the runway as shown in Figure 4.22. The transmitted waveform is just a carrier tone, $x_z(t) = 1$ or $x_c(t) = \sqrt{2}\cos(2\pi f_c t)$.

The received signal return at the radar receiver input has the form

$$y_c(t) = A_P\sqrt{2}\cos(2\pi(f_c + f_P)t + \theta_P) + A_G\sqrt{2}\cos(2\pi(f_c + f_G)t + \theta_G) \quad (4.42)$$

where the P subscript refers to the signal returns from the plane taking off and the G subscript refers to the signal returns from the ground. The frequency shift is due to the Doppler effect you learned about in your physics classes.

(a) Why does the radar signal bouncing off the ground (obviously stationary) produce a Doppler frequency shift?

(b) Give the complex baseband form of this received signal.

(c) Assume the radar receiver has a complex baseband impulse response of

$$h_z(t) = \delta(t) + \beta\delta(t - T) \quad (4.43)$$

where β is a possibly complex constant, find the value of β which eliminates the returns from the ground at the output of the receiver. This system was a common feature in early radar systems and has the common name Moving Target Indicator (MTI) as stationary target responses will be canceled in the filter given in Eq. (4.43).

Modern air traffic control radars are more sophisticated than this problem suggests. An important point of this problem is that radar and communication systems are similar in many ways and use the same analytical techniques for design.

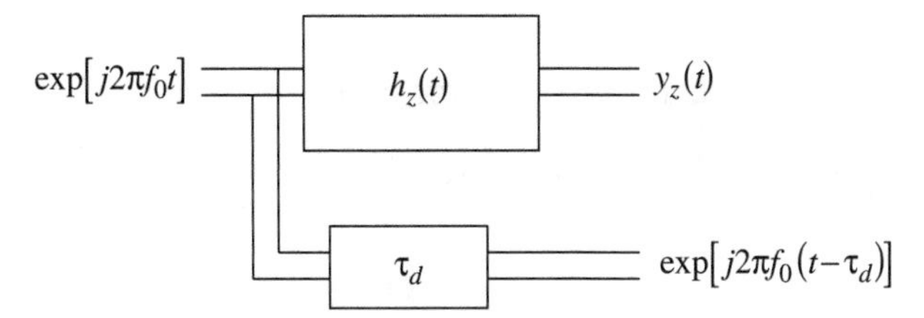

Figure 4.23 The block diagram for Problem 4.15.

Problem 4.15. A baseband signal (a complex exponential) and two linear systems are shown in Figure 4.23. The top linear system in Figure 4.23 has an impulse response of

$$h_z(t) = \begin{cases} \dfrac{1}{\sqrt{T_p}} & 0 \le t \le T_p \\ 0 & \text{elsewhere} \end{cases} \tag{4.44}$$

The bottom linear system in Figure 4.23 is an ideal delay element (i.e., $y_z(t) = x_z(t - \tau_d)$).

(a) Give the bandpass frequency response $H_c(f)$.

(b) What is the input power? Compute $y_z(t)$.

(c) Select a delay, τ_d, in the bottom system in Figure 4.23 such that $\arg[y_z(t)] = 2\pi f_0(t - \tau_d)$ for all f_0.

(d) What is the output power as a function of f_0, $P_{y_z}(f_0)$?

(e) How large can f_0 be before the output power, $P_{y_z}(f_0)$, is reduced by 10 dB compared to the output power when $f_0 = 0$, $P_{y_z}(0)$?

Problem 4.16. The following bandpass filter has been implemented in a communication system that you have been tasked to simulate

$$H_c(f) = \begin{cases} 1 & f_c + 7500 \le |f| \le f_c + 10000 \\ 2 & f_c + 2500 \le |f| < f_c + 7500 \\ \dfrac{4}{3} & f_c \le |f| < f_c + 2500 \\ \dfrac{3}{4} & f_c - 2500 \le |f| < f_c \\ 0 & \text{elsewhere} \end{cases} \tag{4.45}$$

You know because of your great engineering education that it will be much easier to simulate the system using complex envelope representation.

(a) Find $H_z(f)$.

(b) Find $H_I(f)$ and $H_Q(f)$.

(c) If $x_z(t) = \exp(j2\pi f_m t)$ find $x_c(t)$.

(d) If $x_z(t) = \exp(j2\pi f_m t)$ compute $y_z(t)$ for $2000 \leq f_m < 9000$.

Problem 4.17. Consider two bandpass filters

$$h_{z1}(t) = \begin{cases} \frac{1}{\sqrt{0.2}} & 0 \leq t \leq 0.2 \\ 0 & \text{elsewhere} \end{cases} \tag{4.46}$$

$$h_{z2}(t) = \frac{\sin(10\pi t)}{10\pi t} \tag{4.47}$$

Consider the filters and an input signal having a complex envelope of $x_z(t) = \exp(j2\pi f_m t)$.

(a) Find $x_c(t)$.

(b) Find $H_z(f)$.

(c) Find $H_c(f)$.

(d) For $f_m = 0, 7, 14$ Hz find $y_z(t)$.

Problem 4.18. Find the amplitude signal, $x_A(t)$, and phase signal, $x_P(t)$ for

(a) $x_z(t) = z(t)\exp(j\phi)$ where $z(t)$ is a complex valued signal.

(b) $x_z(t) = m(t)\exp(j\phi)$ where $m(t)$ is a real valued signal.

Problem 4.19. A bandpass signal has a complex envelope given as

$$x_z(t) = j\,\exp[-j2\pi f_m t] + 3\exp[j2\pi f_m t] \tag{4.48}$$

where $f_m > 0$.

(a) Find $x_I(t)$ and $x_Q(t)$.

(b) Plot the frequency domain representation of this periodic baseband signal using impulse functions.

(c) Plot the frequency domain representation of the bandpass signal using impulse functions.

(d) What is the bandpass bandwidth of this signal, B_T?

Problem 4.20. The amplitude and phase of a bandpass signal is plotted in Figure 4.24. Plot the in-phase and quadrature signals of this baseband representation of a bandpass signal.

Problem 4.21. A bandpass signal has a complex envelope given as

$$x_z(t) = j\,\exp[-j2\pi f_m t] + 3\exp[j2\pi f_m t] \tag{4.49}$$

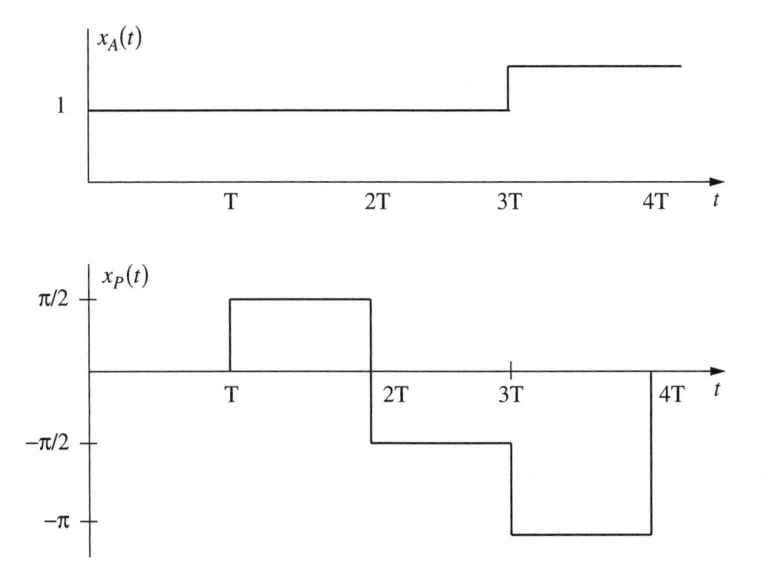

Figure 4.24 The amplitude and phase of a bandpass signal.

where $f_m > 0$. This signal is put into a bandpass filter which has a complex envelope characterized with

$$H_Q(f) = \begin{cases} 1 & |f| \leq 4000 \\ 0 & \text{elsewhere} \end{cases} \qquad H_I(f) = \begin{cases} -j & -4000 \leq f \leq 0 \\ j & 0 \leq f \leq 4000 \\ 0 & \text{elsewhere} \end{cases} \tag{4.50}$$

The output of the filter at bandpass is denoted $y_c(t)$ and at baseband is denoted $y_z(t)$.

(a) What is $H_z(f)$.

(b) Find $y_z(t)$ as a function of f_m.

(c) Plot the frequency domain representation of the output bandpass signal using impulse functions for the case $f_m = 2000$ Hz.

Problem 4.22. **(PD)** In Figure 4.25 are drawings of the $f \geq 0$ portions of the Fourier transforms of two bandpass signals. For each transform plot the magnitude and phase for the entire f axis (i.e., filling in the missing $f < 0$ part). Also for each transform plot the magnitude and phase for the entire f axis for $X_I(f)$ and $X_Q(f)$.

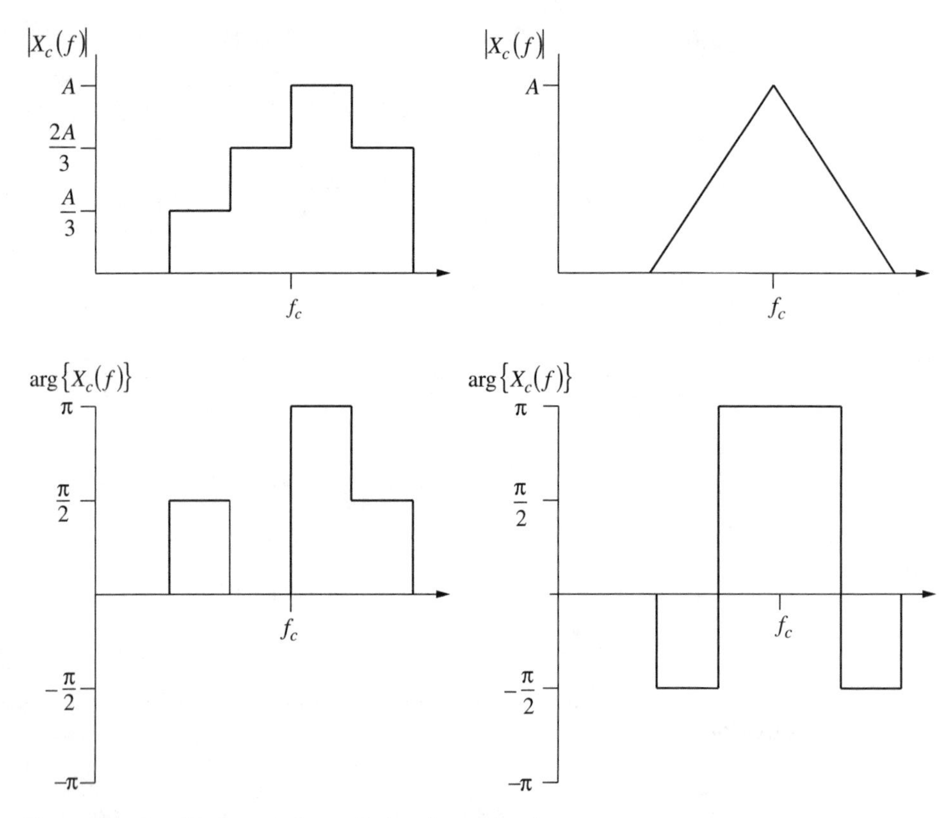

Figure 4.25 Amplitude and phase of a bandpass signal.

Problem 4.23. If $\mathcal{F}\{x_z(t)\} = X_z(f)$ what is $\mathcal{F}\{x_z^*(t)\}$?

Problem 4.24. An often used pulse in radar systems has a complex envelope of

$$x_z(t) = \begin{cases} A_c \exp[j2\pi g_0 t^2] & 0 \le t \le T_p \\ 0 & \text{elsewhere} \end{cases} \tag{4.51}$$

(a) What are $x_A(t)$ and $x_P(t)$.

(b) Plot the bandpass signal, $x_c(t)$ for $f_c = 10$ Hz and $g_0 = 100$ Hz/s, $A_c = 1$, and $T_p = 1$.

(c) What is the E_{x_z}?

(d) Plot $G_{x_z}(f)$ and estimate B_{98} when $g_0 = 100$ Hz/s and $T_p = 1$.

Problem 4.25. A complex baseband signal is given as $x_z(t) = x_I(t) + jx_Q(t)$, where $x_I(t)$ and $x_Q(t)$ are real lowpass signals. Find

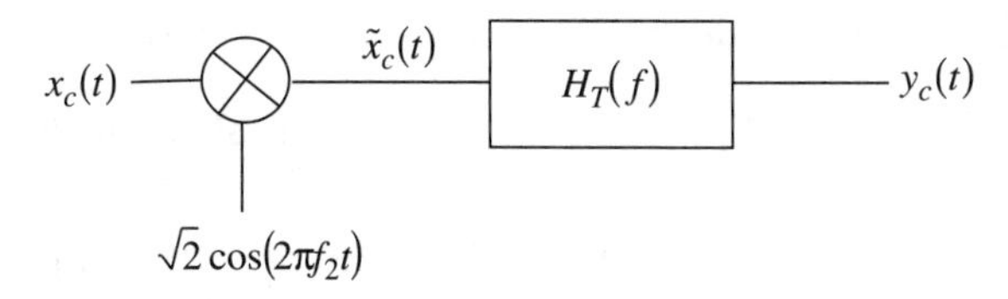

Figure 4.26 A heterodyne upconverter.

(a) The bandpass signal, $x_c(t)$, represented by $x_z(t)$

(b) $X_z(f)$ and $X_z^*(f)$ in terms of $X_I(f)$ and $X_Q(f)$

(c) $X_Q(f)$ in terms of $X_z(f)$

(d) $X_c(f)$ in terms of $X_I(f)$ and $X_Q(f)$

(e) Show that $|X_c(f)|$ is an even function of frequency.

Problem 4.26. Consider a bandpass signal, $x_c(t)$ with $f_c = 10.7$ MHz and a complex envelope given as

$$x_z(t) = 0.5\exp[j2000\pi t] + 1.5\exp[-j2000\pi t] \tag{4.52}$$

in a system with a block diagram given in Figure 4.26 where $f_2 = 110.7$ MHz. This block diagram is often described as a heterodyne upconverter and is frequently used in practice. Further assume that the bandpass filter, $H_T(f)$ is characterized as

$$H_T(f) = \begin{cases} a|f| + b & 99 \text{ MHz} \leq |f| \leq 101 \text{ MHz} \\ 0 & \text{elsewhere} \end{cases} \tag{4.53}$$

where $a = 0.3 \times 10^{-6}$ and $b = -29$.

(a) Plot the spectrum of the bandpass signal, $X_c(f)$.

(b) The output of the multiplier (mixer) is denoted as $\tilde{x}_c(t)$. Plot the spectrum $\tilde{X}_c(f)$.

(c) Plot the transfer function of the bandpass filter, $H_T(f)$.

(d) Plot the bandpass output spectrum $Y_c(f)$.

(e) Give the complex envelope of the output signal, $y_z(t)$.

Problem 4.27. A common implementation issue that arises in circuits that implement the I/Q up and down converters is an amplitude imbalance between the I channel and the Q channel. For example, if the complex envelope is given as $x_z(t) = x_I(t) + jx_Q(t)$ then the complex envelope of the signal that is transmitted or received after imperfect conversion needs to be modeled as

$$\tilde{x}_z(t) = Ax_I(t) + jBx_Q(t) \tag{4.54}$$

The amplitude imbalance is denoted $\gamma = (B/A)^2$ in practice.

(a) Assume that $P_{x_I} = P_{x_Q} = P_{x_z}/2$ and find the values of A and B that achieve a specified γ and $P_{x_z} = P_{\tilde{x}_z}$.

(b) For the transformation detailed in (a) find the output complex envelope, $\tilde{x}_z(t)$, when $x_z(t) = \exp[j2\pi f_m t]$. Plot the signal spectrum before and after the amplitude imbalance.

Problem 4.28. (PD) Let $x_c(t)$ be a bandpass signal with

$$X_c(f) = \begin{cases} X_0 & ||f| - f_a| \leq W \\ 0 & 0 \text{ elsewhere} \end{cases} \tag{4.55}$$

(a) Find E_{x_c}.

(b) Plot $X_z(f)$ for $f_c = f_a$. Is $x_z(t)$ a real valued signal?

(c) Plot $X_z(f)$ for $f_c = f_a + W$. Is $x_z(t)$ a real valued signal?

4.9 Example Solutions

Problem 4.2.

(a) Using $\sin(a - b) = \sin(a)\cos(b) - \cos(a)\sin(b)$ gives

$$x_c(t) = \sin(2\pi f_c t)\cos(2\pi f_m t) - \cos(2\pi f_c t)\sin(2\pi f_m t) \tag{4.56}$$

By inspection we have

$$x_I(t) = \frac{-1}{\sqrt{2}}\sin(2\pi f_m t) \qquad x_Q(t) = \frac{-1}{\sqrt{2}}\cos(2\pi f_m t) \tag{4.57}$$

(b) Recall $x_c(t) = x_A(t)\sqrt{2}\cos(2\pi f_c t + x_P(t))$ so by inspection we have

$$x_z(t) = \frac{1}{\sqrt{2}}\exp(j2\pi f_m t) \quad x_I(t) = \frac{1}{\sqrt{2}}\cos(2\pi f_m t)$$
$$x_Q(t) = \frac{1}{\sqrt{2}}\sin(2\pi f_m t) \tag{4.58}$$

(c) Recall $x_c(t) = x_A(t)\sqrt{2}\cos(2\pi f_c t + x_P(t))$ so by inspection we have

$$x_z(t) = \frac{1}{\sqrt{2}}\exp(j\phi_p) \quad x_I(t) = \frac{1}{\sqrt{2}}\cos(\phi_p) \quad x_Q(t) = \frac{1}{\sqrt{2}}\sin(\phi_p) \tag{4.59}$$

Problem 4.6. We can write

$$X_I(f) = A\,rect\left(\frac{f}{2f_1}\right) \quad \text{and} \quad X_Q(f) = j\frac{Af}{f_1}rect\left(\frac{f}{2f_1}\right)$$

Then

$$
\begin{aligned}
G_{X_I}(f) = |X_I(f)|^2 &= A^2 rect\left(\frac{f}{2f_1}\right) \\
G_{X_Q}(f) = |X_Q(f)|^2 &= \frac{A^2 f^2}{f_1^2} rect\left(\frac{f}{2f_1}\right) \\
X_I(f) X_Q^*(f) &= -\frac{j A^2 f}{f_1} rect\left(\frac{f}{2f_1}\right) \\
\Im\left\{X_I(f) X_Q^*(f)\right\} &= -\frac{A^2 f}{f_1} rect\left(\frac{f}{2f_1}\right)
\end{aligned}
$$

Hence

$$
\begin{aligned}
G_{X_z}(f) &= G_{X_I}(f) + G_{X_Q}(f) + 2\Im X_I(f) X_Q^*(f) \\
&= A^2 \left[rect\left(\frac{f}{2f_1}\right) + \left(\frac{f}{f_1}\right)^2 rect\left(\frac{f}{2f_1}\right) - 2\frac{f}{f_1} rect\left(\frac{f}{2f_1}\right)\right] \\
G_{X_z}(f) &= \left(f - \frac{f}{f_1}\right)^2 rect\left(\frac{f}{2f_1}\right)
\end{aligned}
$$

Problem 4.11. $x_I(t)$ is periodic with period T. Because of this $x_I(t)$ can be represented in a Fourier series expansion

$$
x_I(t) = \sum_{k=-\infty}^{\infty} x_k \exp\left[\frac{j2\pi kt}{T}\right] \tag{4.60}
$$

If the bandwidth of the signal is less than W Hz then the Fourier series will be truncated to a finite summation. Define k_m to be the largest integer such that $k_m/T \leq W$ then

$$
x_I(t) = \sum_{k=-k_m}^{k_m} x_k \exp\left[\frac{j2\pi kt}{T}\right] \tag{4.61}
$$

The bandpass signal will have the form

$$
\begin{aligned}
x_c(t) &= x_I(t)\sqrt{2}\cos(2\pi f_c t) \\
&= x_I(t)\left[\frac{1}{\sqrt{2}}\exp(j2\pi f_c t) + \frac{1}{\sqrt{2}}\exp(-j2\pi f_c t)\right] \qquad (4.62) \\
&= \sum_{k=-k_m}^{k_m} \frac{1}{\sqrt{2}} x_k \left(\exp\left[j2\pi\left(\frac{k}{T} + f_c\right)t\right] + \exp\left[j2\pi\left(\frac{k}{T} - f_c\right)t\right]\right)
\end{aligned}
$$

Since the bandpass signal has a representation as a sum of weighted sinusoids there is a possibility that the bandpass signal will be periodic.

This bandpass signal will only be period if all the frequencies are an integer multiple of a fundamental frequency, $1/T_c$.

(a) The bandpass signal can be periodic with $T_c < T$. For example, choose $x_I(t)$ to be a 2 Hz sinusoid, i.e., $x_I(t) = \cos(2\pi(2)t)$ and the carrier frequency to be $f_c = 6$ Hz. Clearly here $T = 1/2$ seconds. The bandpass signal in this case is

$$x_c(t) = x_I(t)\sqrt{2}\cos(2\pi(6)t) = \frac{\sqrt{2}}{4}(\exp[j2\pi(-8)t] + \exp[j2\pi(-4)t] + \exp[j2\pi(4)t] + \exp[j2\pi(8)t]) \quad (4.63)$$

The bandpass signal is a sum of four sinusoids that have frequencies of $f_1 = -8$, $f_2 = -4$, $f_3 = 4$, $f_4 = 8$. Clearly the fundamental frequency bandpass signal is 4 Hz and $T_c = 0.25 < 0.5$.

(b) The bandpass signal can be periodic with $T_c < T$. For example, choose $x_I(t)$ to be a 3 Hz sinusoid, i.e., $x_I(t) = \cos(2\pi(3)t)$ and the carrier frequency to be $f_c = 5$ Hz. Here, $T = 1/3$ seconds. The bandpass signal in this case is

$$x_c(t) = x_I(t)\sqrt{2}\cos(2\pi(5)t) = \frac{\sqrt{2}}{4}(\exp[j2\pi(-8)t] + \exp[j2\pi(-2)t] + \exp[j2\pi(2)t] + \exp[j2\pi(8)t]) \quad (4.64)$$

The bandpass signal is a sum of four sinusoids that have frequencies of $f_1 = -8$, $f_2 = -2$, $f_3 = 2$, $f_4 = 8$. The fundamental frequency of the bandpass signal is 2 Hz and $T_c = 0.5 > 1/3$.

(c) The bandpass signal can be periodic with $T_c = T$. For example, choose $x_I(t)$ to be a 2 Hz sinusoid, i.e., $x_I(t) = \cos(2\pi(2)t)$ and the carrier frequency to be $f_c = 4$ Hz. Here, $T = 0.5$ seconds. The bandpass signal in this case is

$$x_c(t) = x_I(t)\sqrt{2}\cos(2\pi(4)t) = \frac{\sqrt{2}}{4}(\exp[j2\pi(-6)t] + \exp[j2\pi(-2)t] + \exp[j2\pi(2)t] + \exp[j2\pi(6)t]) \quad (4.65)$$

The bandpass signal is a sum of four sinusoids that have frequencies of $f_1 = -6$, $f_2 = -2$, $f_3 = 2$, $f_4 = 6$. The fundamental frequency of the bandpass signal is 2 Hz and $T_c = 0.5$.

(d) The bandpass signal can be aperiodic. For example, again choose $x_I(t)$ to be a 2 Hz sinusoid, i.e., $x_I(t) = \cos(2\pi(2)t)$. Here, $T = 0.5$ seconds. The bandpass signal in this case is

$$\begin{aligned} x_c(t) &= x_I(t)\sqrt{2}\cos(2\pi(f_c)t) \\ &= \frac{\sqrt{2}}{4}(\exp[j2\pi(-2-f_c)t] + \exp[j2\pi(-f_c+2)t] \\ &\quad + \exp[j2\pi(f_c-2)t] + \exp[j2\pi(f_c+2)t]) \end{aligned} \quad (4.66)$$

The bandpass signal will be aperiodic if the two frequency are not integer multiples of a common frequency. That implies that the two bandpass frequencies must be irrational numbers. Choosing $f_c = \sqrt{2}$ will produce an aperiodic signal.

4.10 Miniprojects

Goal: To give exposure

- to a small scope engineering design problem in communications.
- to the dynamics of working with a team.
- to the importance of engineering communication skills (in this case oral presentations).

Presentation: The forum will be similar to a design review at a company (only much shorter). The presentation will be of 5 minutes in length with an overview of the given problem and solution. The presentation will be followed by questions from the audience (your classmates and the professor). All team members should be prepared to give the presentation.

4.10.1 Project 1

Project Goals: In engineering often in the course of system design or test anomalous performance characteristics often arise. Your job as an engineer is to use your knowledge of the theory to identify the causes of these characteristics and correct them. Here is an example of such a case.

Get the Matlab file `bpex1.m` from the class web page. In this file the carrier frequency was chosen as 7 kHz. If the carrier frequency is chosen as 8 kHz an anomalous output is evident from the quadrature downconverter. This is most easily seen in the output energy spectrum, $G_{y_z}(f)$. Postulate a reason why this behavior occurs.

Hint: It happens at 8 kHz but not at 7 kHz and Matlab is a sampled data system. What problems might one have in sampled data system? Assume that this upconverter and downconverter were products you were designing how would you specify the performance characteristics such that a customer would never see this anomalous behavior?

4.10.2 Project 2

Project Goals: Use the knowledge of the theory of bandpass signals to identify unknown parameters of a bandpass signal.

A signal has a form

$$x_c(t) = x_I(t)\sqrt{2}\cos(2\pi f_c t + \theta) - x_Q(t)\sqrt{2}\sin(2\pi f_c t + \theta) \qquad (4.67)$$

It is known that

$$x_I(t) = \cos(200\pi t) \quad x_Q(t) = \cos(400\pi t) \tag{4.68}$$

The values of θ and f_c are unknown. The signal has been recorded with a sample frequency of 22050 Hz (assume Nyquist criterion was satisfied) and is available in the file `CompenvPrjt2.mat` on the class web page. Using this file make an estimate of θ and f_c. Detail the logic that led to the estimates.

Part 2

Analog Communication

Chapter

5

Analog Communications Basics

Analog communication involves transferring an analog waveform containing information (no digitization at any point) between two users. Typical examples where analog information is transmitted in this fashion are

- Music — broadcast radio
- Voice — citizen band radio, amateur radio, walkie-talkies, cellular radio
- Video — broadcast television

These message signals all have unique characteristics that engineers need to understand to be able to build communication systems that can implement this communication with high fidelity and low cost. The characterization of these message signals build upon the foundation of an undergraduate signals and systems education as reviewed in Chapter 2. Signals are typically characterized in both the time and frequency domain and that is the approach that will be taken in this chapter.

5.1 Message Signal Characterization

Characterizing the message to be transmitted will be important in understanding analog communications. The information bearing analog waveform is denoted $m(t)$ and it is assumed to be an energy signal with a Fourier transform of $M(f)$ and energy spectral density of $G_m(f)$, respectively. The message signal is assumed to be a real valued signal. Figures 5.1 and 5.2 show an example of a short time record of a message signal and an energy spectral density of a message signal. This text will use these examples, time signal and spectrum, to demonstrate modulation concepts. Often a DC value is not an important component in an analog message signal so we will often use the fact that the

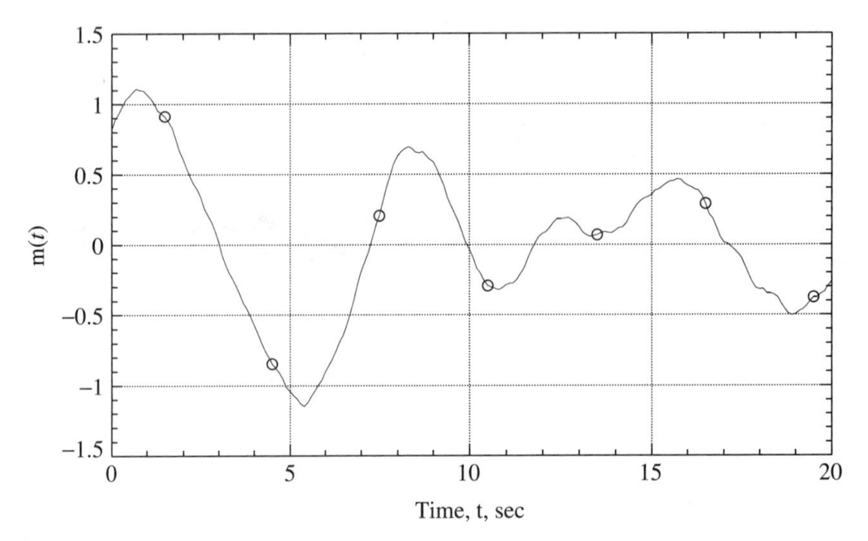

Figure 5.1 An example message signal, $m(t)$.

energy spectrum of the message signal goes to zero for small frequencies and that the message signal will pass through a DC block unchanged. In analog modulations it is often important to discuss the signal bandwidth and this will be denoted W. Note that this W could correspond to either relative or integral bandwidth and will not be specified unless necessary for clarity.

Several parameters will occur frequently to characterize analog communications. The most important characteristic of an analog signal to the communication engineer is the bandwidth of the signal. The bandwidth of analog baseband

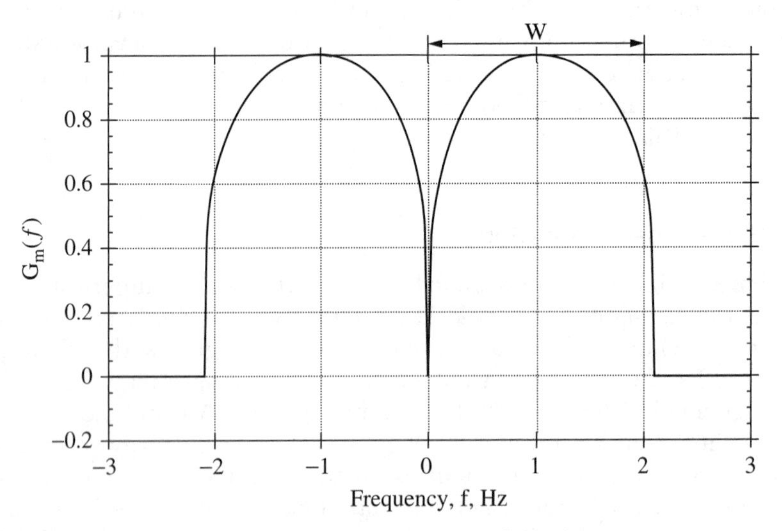

Figure 5.2 An example of a message energy spectral density.

signals have been characterized in Chapter 2 and these results will be used through the remainder of the text. The time average signal power will be an important characteristic of a message signal.

Definition 5.1 The time average message power, P_m, is

$$P_m = \lim_{T_m \to \infty} P_m(T_m) \tag{5.1}$$

where $P_m(T_m)$ is defined in (2.54).

Other characteristics of the message signal that are important include

- The largest value of the message signal, $\max m(t)$
- The smallest value of the message signal, $\min m(t)$
- The maximum rate of change of the message signal, $\max |\frac{d}{dt} m(t)|$

A characteristic often used to characterize communication signals is the peak to average power ratio (PAPR).

Definition 5.2 The peak power to average power ratio of $m(t)$ is

$$PAPR_m = \frac{(\max |m(t)|)^2}{P_m} \tag{5.2}$$

The remainder of this book will consistently relate performance of communication systems back to these parameters that describe the message signal.

EXAMPLE 5.1

Recall the example from Chapter 2 of a filtered computer generated voice saying the word "Bingo" whose time waveform is shown in Figure 2.1 and whose energy spectrum is shown in Figure 2.8. This is pretty typical of a spoken word signal that would be communicated in an analog communication system. A spoken word signal typically has the characteristic that there is little DC value for the signal and this is reflected in the notch in the energy spectrum of this signal at 0 Hz. This characteristic implies that a spoken word signal will pass through a DC block with little distortion. Examining this signal both the 40 dB bandwidth and 98% energy bandwidth are about 2.5 kHz so when this signal is used in future examples it will be assumed that $W = 2.5$ kHz. In general, spoken word signals have been characterized as having a bandwidth of about $W = 3.5$–4 kHz and many commonly employed communication systems are built on this premise (e.g., the public telephone system).

EXAMPLE 5.2

High quality audio signals have different characteristics than spoken word signals. Figure 5.3(a) shows the time signal of a short excerpt from a song by Andrea Bocelli sampled at 44100 Hz. Figure 5.3(b) shows the measured power spectrum from this same song. This is pretty typical of a high fidelity audio signal that would be communicated

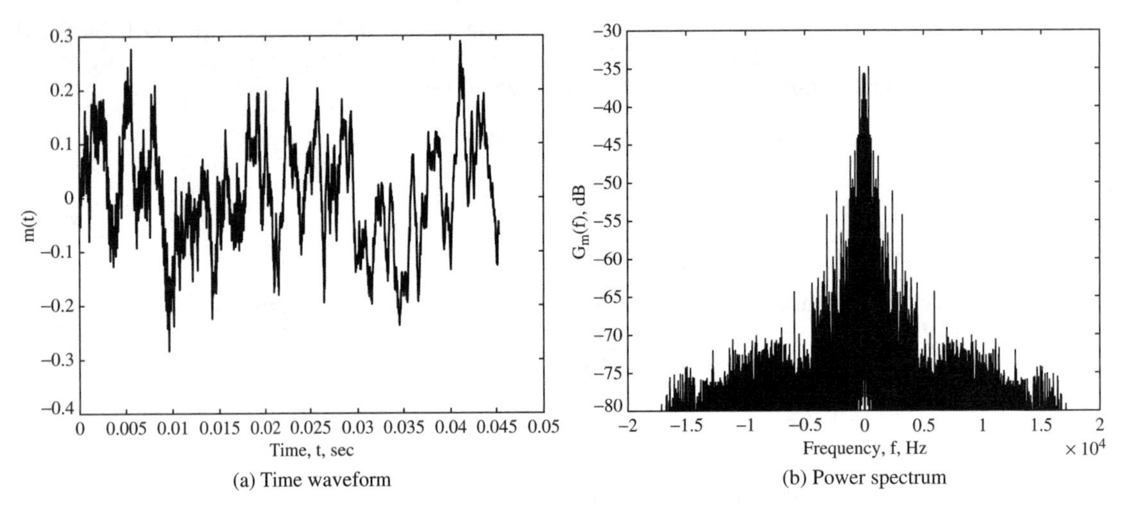

Figure 5.3 A high fidelity audio signal.

in an analog communication system. A high fidelity audio signal typically has the characteristic that there is little DC value for the signal and this is reflected in the notch in the power spectrum of this signal at 0 Hz. This characteristic implies that a high fidelity audio signal will pass through a DC block with little distortion. Examining this signal the 40 dB bandwidth is about $B_{40} = 15.5$ kHz. In general, high fidelity audio signals have been characterized as having a bandwidth of about $W = 15$–20 kHz and many commonly employed communication systems are built on this premise (e.g., audio broadcasting).

EXAMPLE 5.3

Video signals have different characteristics than audio signals. Figure 5.4(a) shows the time signal of a short excerpt from an NTSC video signal. NTSC stands for National Television System Committee, which devised the NTSC television (TV) broadcast system in 1953. All analog television broadcasts in the United States must have a message signal that meet this standard. This message signal format was chosen so that the transformation to a raster scan on a TV tube is a fairly simple circuit. Figure 5.4(b) shows the measured power spectrum from an NTSC video signal. This is pretty typical of an NTSC video signal that would be communicated in an analog communication system. An NTSC video signal typically has the characteristic that the DC value for the signal is important and this is reflected in the fact that there is no notch in the power spectrum of this signal at 0 Hz. DC blocks cannot be used in the processing of an NTSC video signals. Examining this signal the 40 dB bandwidth is about 4.5 MHz. In general, NTSC video signals have been characterized as having a bandwidth of about $W = 4.5$ MHz and many commonly employed communication systems are built on this premise (e.g., TV broadcasting). More details about NTSC video signals can be found in [WB00].

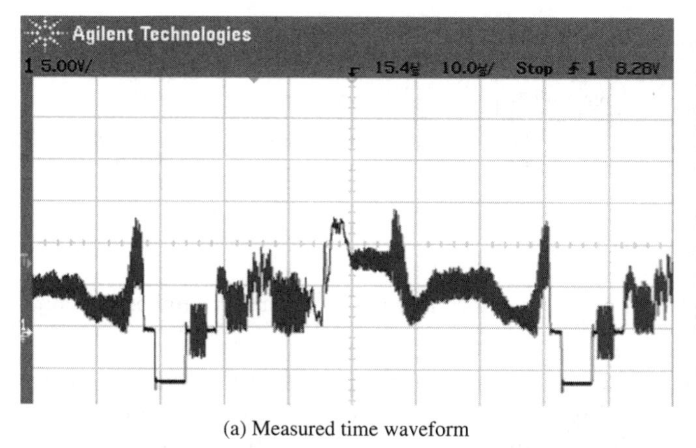

(a) Measured time waveform

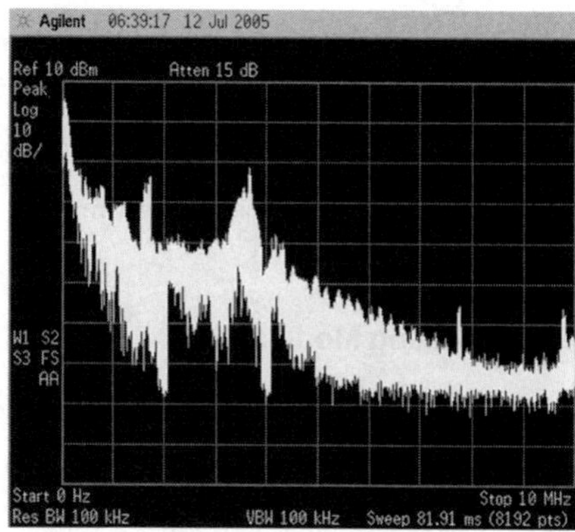

(b) Measured power spectrum

Figure 5.4 An NTSC video signal.

5.2 Analog Transmission

The conventional communication system has a modulator producing a signal that is transmitted over a channel (a cable or radio propagation) and a demodulator which takes this signal and constructs an estimate of the transmitted message signal. Figure 5.5 is a block diagram of this system where $r_c(t)$ is the output of the channel, $Y_c(t)$ is the waveform observed with the receiver[1], and $\hat{m}(t)$ is the estimate of the transmitted message signal. The noise added by a radio receiver is usually a combination of signal distortion and additive interfering noise. It is this noise that makes the job of a communication engineer challenging as this noise makes it impossible to perfectly reconstruct the transmitted signal. Noise in communication systems will be characterized later (see Chapter 11). The job of communication system engineers is to design and optimize the modulators and demodulators in communication systems.

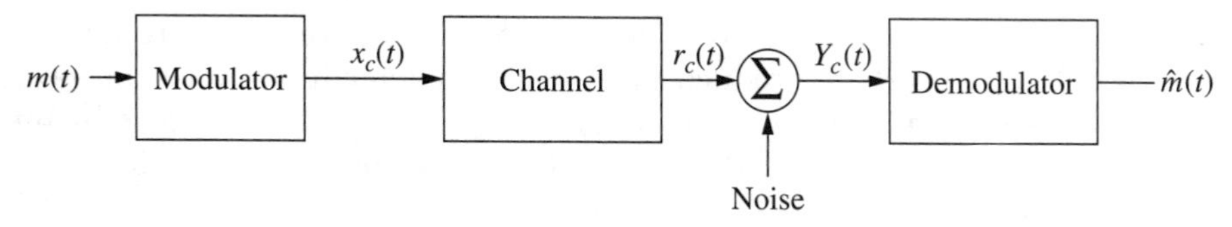

Figure 5.5 An analog communication system block diagram.

[1]For clarity this text will try to consistently represent estimates with a caret, random quantities with capital letters and the deterministic quantities with lower case letters.

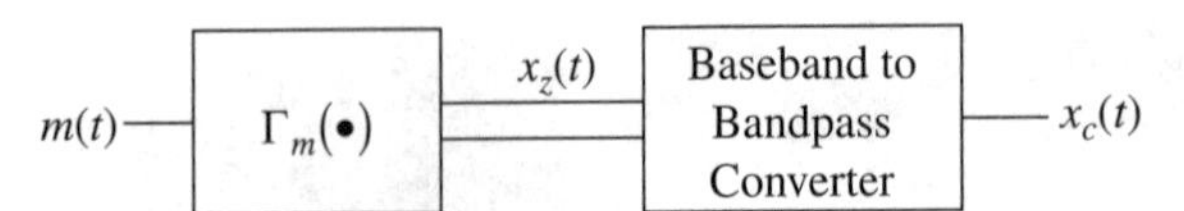

Figure 5.6 The analog modulation process. Note the baseband to bandpass converter is given in Figure 4.4.

5.2.1 Analog Modulation

Definition 5.3 Analog modulation is a transformation of $m(t)$ into a complex envelope, $x_z(t)$.

This transformation is equivalent to transforming $m(t)$ into a bandpass signal, $x_c(t)$. There are conceivably an infinite number of ways to transform a message signal into a complex envelope but only a handful have found utility in practice. The ubiquitous AM (amplitude modulation) and FM (frequency modulation) actually refer to specific transformations of message signals into bandpass signals. Later chapters will discuss the specific characteristics of these two modulations and some other less well-known analog modulations. The analog modulation process, $x_z(t) = \Gamma_m(m(t))$ is represented in Figure 5.6[2]. This transformation mapping can be given as

$$x_I(t) = g_I(m(t)) \qquad x_Q(t) = g_Q(m(t)) \tag{5.3}$$

Historically, analog modulations were invented long before the invention of the transistor (hence large scale integration) so many of the commonly used analog modulations evolved because of the simplicity of the implementation. While the importance of analog modulation is decreasing in an ever increasingly digital world, analog modulation is still used in many important applications and serves as a good introduction to the idea of the modulation process. Hopefully, these comments will become more clear in the remainder of this book.

5.2.2 Analog Demodulation

Definition 5.4 Analog demodulation is a transformation of the received complex envelope, $Y_z(t)$ into an estimate of the message signal, $\hat{m}(t)$.

The system on the receiver side essentially takes the received signal, $Y_c(t)$, downconverts to produce $Y_z(t)$ and then demodulation produces an estimate of the transmitted message signal, $m(t)$. Since this is an introductory treatment of analog communications, the channel output in this book is always assumed to be

$$r_c(t) = L_p x_c(t - \tau_p)$$

---

[2]In figures single lines will denote real signals and double lines will denote complex analytical signals.

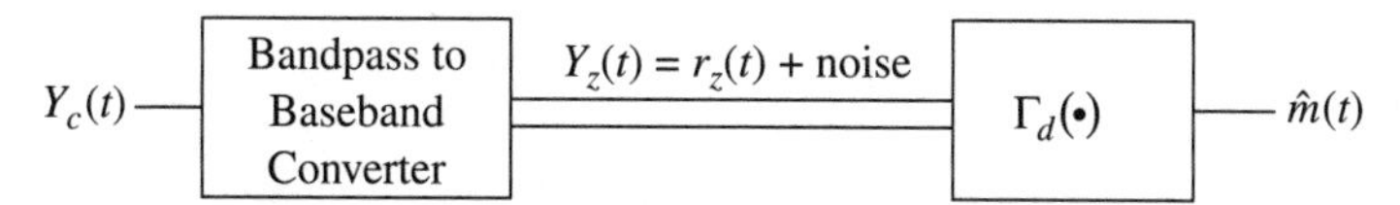

Figure 5.7 The analog demodulation process. Note the bandpass to baseband converter is given in Figure 4.4.

where L_p is the propagation loss and τ_p is the propagation time delay. The propagation loss is due to the inability to perfectly couple the transmitted power to the receiver input and the propagation delay is due to the limited speed that electronic signals can achieve in transmission. This channel with only a propagation loss and delay is an idealized channel but one that captures many of the important challenges in analog communications. Define $\phi_p = -2\pi f_c \tau_p$ so that the channel output is given as

$$\begin{aligned} r_c(t) &= \sqrt{2} L_p x_A(t-\tau_p)\cos(2\pi f_c(t-\tau_p) + x_P(t-\tau_p)) \\ &= \Re[\sqrt{2} L_p x_z(t-\tau_p)\exp[j\phi_p]\exp[j2\pi f_c t]]. \end{aligned} \tag{5.4}$$

It is obvious from Eqs. (5.4) and (4.5) that the received complex envelope is $r_z(t) = L_p x_z(t-\tau_p)\exp[j\phi_p]$. It is important to note that a time delay in a carrier modulated signal will produce a phase offset. Consequently the demodulation process conceptually is a down conversion to baseband and a reconstruction of the transmitted signal from $Y_z(t)$. The block diagram for the demodulation process is seen in Figure 5.7.

EXAMPLE 5.4

Radio broadcast. For a carrier frequency of 100 MHz and a receiver 30 kilometers from the transmitter we have

$$\tau_p = \frac{\text{distance}}{c} = \frac{3\times 10^4}{3\times 10^8} = 100\ \mu\text{s} \qquad \phi_p = -2\pi(10^8)(10^{-4}) = -2\pi\times 10^4 \text{ radians.} \tag{5.5}$$

For the example of radio broadcast a typical channel produces a relatively short (perhaps imperceivable) delay but a very large phase shift.

EXAMPLE 5.5

An example that will be used in the sequel has a carrier frequency of 7 kHz and a propagation delay of $\tau_p = 45.3\ \mu$s gives

$$\phi_p = -2\pi(7000)(0.0000453) = -1.995 \text{ radians} = -114^\circ. \tag{5.6}$$

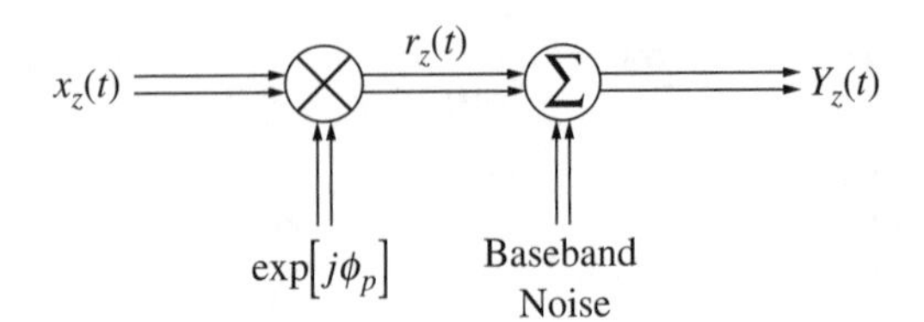

Figure 5.8 The equivalent complex baseband channel model.

The following further simplifications will be made when discussing analog demodulation

- A time delay in the output is unimportant. This is typically true in analog modulation especially since $\tau_p \ll 1$ second.
- An amplitude gain in the output is unimportant. Output amplitude control is a function of the receiver circuitry and will not be discussed in detail.

Consequently, for this book it will be assumed that $r_z(t) = x_z(t)\exp[j\phi_p]$, where ϕ_p is an unknown constant, is the channel output for the remainder of the discussion of analog modulation and demodulation. Figure 5.8 is a diagram of the equivalent complex baseband channel model.

Given this formulation, demodulation can be thought of as the process of producing an $\hat{m}(t)$ from $Y_z(t)$ via a function $\Gamma_d(Y_z(t))$. Some demodulation structures that will be considered will need to have both the received signal, $Y_z(t)$, and the phase shift induced by the channel, ϕ_p. These demodulators will be known as **coherent** demodulators and are generically specified with

$$\hat{m}(t) = g_c(Y_I(t), Y_Q(t), \phi_p) \tag{5.7}$$

Other demodulation structures can produce a message estimate without knowning the phase shift induced by the channel. These demodulators will be known as **noncoherent** demodulators and are given as

$$\hat{m}(t) = g_n(Y_I(t), Y_Q(t)) \tag{5.8}$$

The remainder of the discussion on analog modulations in this text will focus on identifying $\Gamma_m(m(t))$ (modulators) and $\Gamma_d(Y_z(t))$ (demodulators) and assessing the performance of these modulators and demodulators in the presence of noise.

5.3 Performance Metrics for Analog Communication

Engineering design is all about defining performance metrics and building systems that optimize some trade-off between these metrics. To evaluate the efficacy of various communication system designs presented in this book, the performance metrics commonly applied in engineering design must be defined. The most commonly used metrics for analog communications are

- Complexity – This metric almost always translates directly into cost.
- Fidelity – This metric typically measures how accurate the received message estimate is given the amount of transmitted power.
- Spectral Efficiency – This metric measures how much bandwidth a modulation uses to implement the communication.

Complexity is a quantity that requires engineering judgment to estimate. Often the cost of a certain level of complexity changes over time. What were seen as good design choices in terms of complexity in the 1930s when vacuum tubes were prevalent seem somewhat silly today with the prevalance of integrated circuits and complex electronic devices.

The fidelity of the communication system is typically a measure of how well the message is reconstructed at the demodulator output. The message at the output of the demodulator can always be classified as

$$\hat{m}(t) = Am(t) + N_L(t) = m_e(t) + N_L(t) \tag{5.9}$$

where A is an amplitude gain on the signal, $m_e(t)$ is the effective message estimate, and $N_L(t)$ is a combination of the noise and distortion produced in demodulation. Engineers in practice have many ways to characterize the fidelity of the message estimate. In this introductory course we will concentrate on the idea of signal power-to-noise power ratio (SNR). The demodulator output SNR is defined as

$$\text{SNR} = \frac{A^2 P_m}{P_N} = \frac{P_{m_e}}{P_N} \tag{5.10}$$

where P_m is defined in Eq. (5.1) and the noise or distortion power will require the tools in Chapter 9 to define. The SNR is typically a function of the received signal power, the noise power, and the demodulator processing. It should be noted in this text that since the channel model that is being adopted is $r_z(t) = x_z(t)e^{j\phi_p}$ this implies that the received power is equal to the transmitted power ($P_{r_z} = P_{x_z}$). This is obviously not the case in a real communication system but is true in the idealized models assumed in this book.

Communications engineers usually like to compare performance to an idealized standard to get a metric for fidelity produced by a communication system. The idealized metric for performance used in this text is the system where the message is transmitted directly across a baseband channel. The block diagram of this system is shown in Figure 5.9. The resulting output SNR from this system will be denoted SNR_b. The figure of merit in assessing system performance is the ratio of the output SNR to the reference SNR for a common received signal power, P_{r_z}. We denote this quantity as the transmission efficiency, i.e.,

$$E_T = \frac{\text{SNR}}{\text{SNR}_b} \tag{5.11}$$

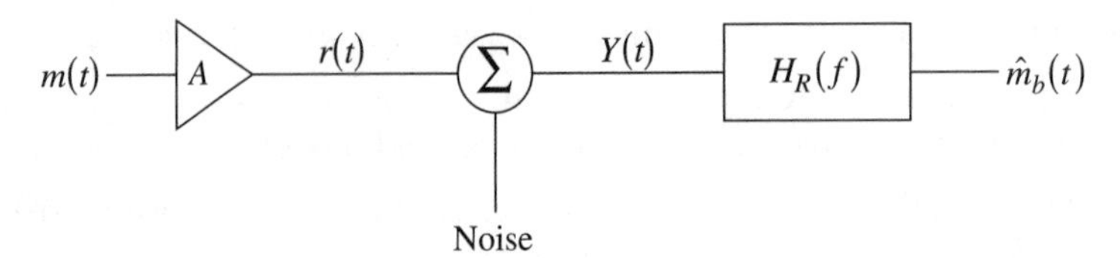

Figure 5.9 The baseline standard for comparison in analog modulation.

This transmission efficiency is a measure of how effectively the modulation and demodulation algorithms process the message signal and the corrupting noise in comparison to the case where no modulation is used.

The spectral efficiency of a communication system is typically a measure of how well the system is using the bandwidth resource. Bandwidth costs money to acquire. Examples are licenses to broadcast radio signals or the installation of copper wires to connect two points. Hence spectral efficiency is very important for people who try to make money selling communication resources. For instance, if one communication system has a spectral efficiency that is twice the spectral efficiency of a second system then the first system can support twice the users on the same bandwidth. Twice the users implies twice the revenue. The measure of spectral efficiency that we will use in this class is called bandwidth efficiency and is defined as

$$E_B = \frac{W}{B_T}$$

where W is the message bandwidth and B_T is the transmission bandwidth. Bandwidth efficiency is a measure of how effectively the modulation uses bandwidth in comparison to the case where no modulation is used in sending the message (as in Figure 5.9).

EXAMPLE 5.6

The most common example of voice transmission in the United States is AM broadcasting. AM broadcasting in the United States usually refers to transmissions confined to a band from 535 kHz to 1,700 kHz. The channels are set up to have center frequencies spaced at 10 kHz spacings and the US Federal Communications Comission (FCC) allows each station to use about 8 kHz of bandwidth ($B_T = 8$ kHz). Since each voice band signal has a bandwidth of around $W = 4$ kHz, AM broadcast in the United States achieves a bandwidth efficiency of

$$E_B = \frac{4}{8} = 50\% \tag{5.12}$$

EXAMPLE 5.7

The most common example of high fidelity audio transmission in the United States is FM broadcasting. FM broadcasting in the United States usually refers to transmissions confined to a band from 88 MHz to 108 MHz. The channels are set up to have center frequencies spaced at 200 kHz spacings and the US Federal Communications Comission (FCC) allows each station to use about 180 kHz of bandwidth ($B_T = 180$ kHz). Since a high fidelity audio has a bandwidth of around $W = 15$ kHz, FM broadcast in the United States achieves a bandwidth efficiency of

$$E_B = \frac{15}{180} = 8.3\% \tag{5.13}$$

EXAMPLE 5.8

The most common example of video transmission in the United States is TV broadcasting. TV broadcasting in the United States has many noncontiguous bands for transmission (e.g., 54–88 MHz for channels 2–6 and 174–220 MHz for channels 7–13). The US Federal Communications Comission (FCC) allows each station to use about 6 MHz of bandwidth ($B_T = 6$ MHz). Since video has a bandwidth of around $W = 4.5$ MHz, TV broadcast in the United States achieves a bandwidth efficiency of

$$E_B = \frac{4.5}{6} = 75\% \tag{5.14}$$

It is clear from the preceding examples that communication engineers have made different choices for system designs for different applications. From the development in Chapter 4 it is clear that the center frequency of the transmission has little impact on system design so the marked differences seen in system designs must be due to the differences in the signal characteristics, the desired fidelity of the communication, and the time varying nature of engineering trade-offs. The time variations in communication system design trade-offs are mostly due to the advances in technology to implement communication systems and the increasing scarcity of bandwidth that can be used to communicate. As the different analog modulation techniques are discussed in this book, constant comparisons will be made to the complexity, the fidelity of the message reconstruction, and the spectral efficiency. This will help make the trade-offs available in the communication system readily apparent.

5.4 Preview of Pedagogy

The next three chapters introduce specific techniques in analog modulation and demodulation. This will be done assuming no noise is present in the waveform observed at the receiver. While no communication system operates without noise, experience has shown that students learn best by immediately starting to discuss the modulation and demodulation process. Consequently, for the next three chapters the observations, $y_c(t)$ or $y_z(t)$, will be represented with lower case letters to represent the fact that they are not random but determinisitc. These chapters will give an understanding of the trade-offs that various analog modulations schemes offer for two of the important performance metrics, complexity, and spectral efficiency.

After introducing the modulation and demodulation process students will be motivated to learn about the trade-offs that various analog modulations schemes offer for the final of the important performance metrics, fidelity. Fidelity essentially reflects how the noise affects the various demodulation processing and distorts the message reconstruction. So at that point the text will return to the characterization of noise in communication systems and eventually the characterization of the noise on analog communication system performance.

5.5 Homework Problems

Problem 5.1. Many characteristics of the message signal appear prominently in the ways we characterize analog modulations. A message signal given as

$$m(t) = -\cos(200\pi t) + \sin(50\pi t)$$

is to be transmitted in an analog modulation. A plot of the signal is given in Figure 5.10.

(a) This signal is periodic. What is the period?

(b) Give the Fourier series coefficients for $m(t)$.

(c) Compute the power of $m(t)$, P_m.

(d) Compute the min $m(t)$.

(e) Compute the max $|m(t)|$.

(f) Compute the max $|\frac{d}{dt}m(t)|$.

Problem 5.2. A message signal of the form shown in Figure 5.11 has a series expansion given as

$$m(t) = \frac{8}{\pi^2}\sum_{n=1}^{\infty}\frac{1}{(2n-1)^2}\cos\left(\frac{2\pi(2n-1)t}{T}\right) \tag{5.15}$$

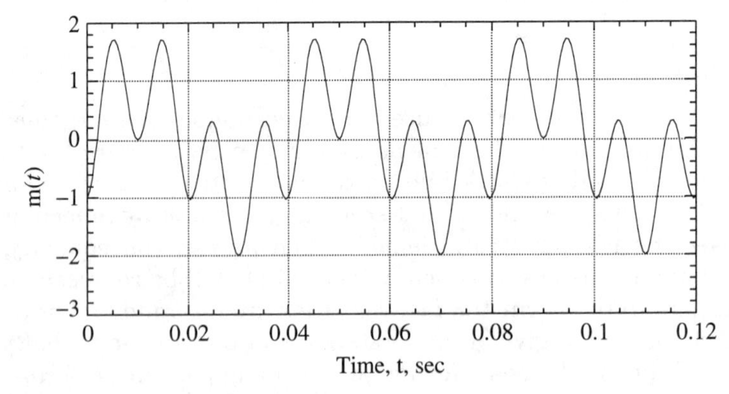

Figure 5.10 An example message signal.

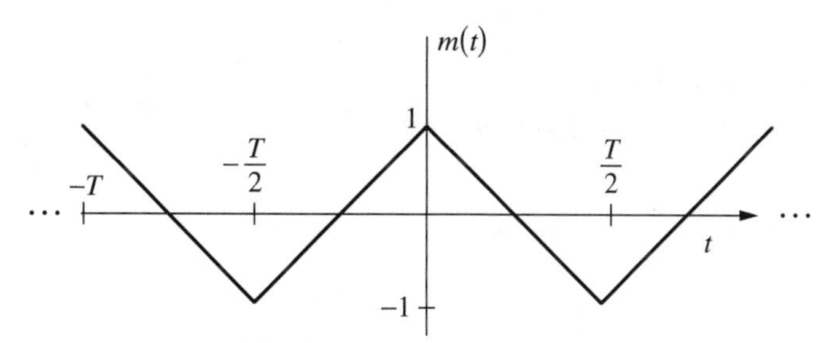

Figure 5.11 A message signal.

(a) Select T such that the fundamental frequency of the Fourier series expansion is 2 Hz.

(b) If m_k represents the Fourier series coefficients of $m(t)$, what is m_0? What is m_1?

(c) Compute the power of $m(t)$, P_m.

(d) For transmission in an analog communication system it is desired to limit the bandwidth to 9 Hz. Using the T from part (a) compute how many terms in the Fourier series will remain after this bandlimiting operation. Give their frequencies and Fourier coefficients.

(e) Compute the resulting message power after bandlimiting.

(f) Plot the message signal after bandlimiting.

Problem 5.3. A message signal is to be transmitted using analog modulation. The message signal Fourier transform has the form

$$M(f) = \begin{cases} A\left|\sin\left(\frac{\pi f}{W}\right)\right| & |f| \leq W \\ 0 & \text{elsewhere} \end{cases} \tag{5.16}$$

(a) Compute the value of A such that E_m is equal to 1 Joule in a 1 Ω system.

(b) Compute the min $m(t)$.

(c) Compute the max $|m(t)|$.

(d) Compute the max $|\frac{d}{dt}m(t)|$.

Problem 5.4. Distortion of the message signal during the communication process is often a major source of SNR degradation. This problem is a simple example of how frequency selectivity can produce distortion and how a communication engineer might characterize it. A message signal given as

$$m(t) = -\cos(200\pi t) + \sin(50\pi t)$$

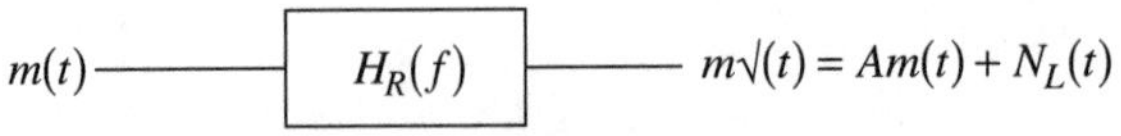

$m(t)$ —— $H_R(f)$ —— $m\surd(t) = Am(t) + N_L(t)$

Figure 5.12 An example of filtering a message signal.

is input into the system shown in Figure 5.12 where

$$H_R(f) = \frac{2\pi 500}{j2\pi f + 2\pi 500} \tag{5.17}$$

(a) Find and plot the output signal.

(b) If $\hat{m}(t) = Am(t) + n_L(t)$ for a chosen value of A find the power of the distortion $n_L(t)$.

(c) Find the value of A that minimizes the distortion power.

(d) Redesign $H_R(f)$ to reduce the effective distortion power.

Problem 5.5. A message signal is of the form

$$m(t) = 2\cos(4\pi t) + \cos(6\pi t)$$

A plot of the message signal is given in Figure 5.13.

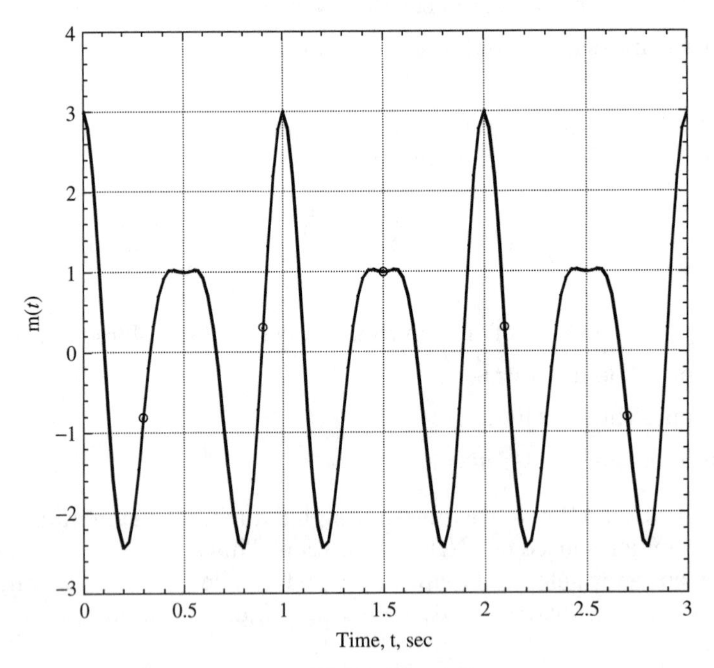

Figure 5.13 The message signal for Problem 5.5.

(a) $m(t)$ is periodic. Identify the period and the Fourier series coefficients.

(b) Sketch using impulse functions the Fourier transform of the periodic signal, $m(t)$.

(c) Calculate the message signal power, P_m.

(d) What is the message signal bandwidth, W?

(e) Compute the $\min m(t)$.

(f) Compute the $\max |m(t)|$.

(g) Compute the $\max |\frac{d}{dt}m(t)|$.

5.6 Example Solutions

Problem 5.2.

(a) Consider the term

$$\cos\left(\frac{2\pi(2n-1)t}{T}\right) = \frac{1}{2}\exp\left(\frac{j2\pi(2n-1)t}{T}\right) + \frac{1}{2}\exp\left(\frac{-j2\pi(2n-1)t}{T}\right) \tag{5.18}$$

For $n = 1$ the frequency of the sinusoid is $f = 1/T$. Likewise for an arbitrary n the frequency of the sinusoid is $f = (2n-1)/T$. Consequently, the smallest frequency is $f = 1/T$ and all freqeuncies are integer multiples of this smallest frequency. Consequently, $f = 1/T$ will be the fundamental frequency of the Fourier series of $m(t)$. Choose $T = 0.5$s will produce a fundamental frequency of 2 Hz.

(b)

$$m(t) = \sum_{n=-\infty}^{\infty} m_n \exp\left(\frac{j2\pi nt}{T}\right) \tag{5.19}$$

$n = 0$ has $m_0 = 0$ consequently there is no DC term. This is expected by examining the plot of the signal as the average value is zero. $n = 1$ has

$$m_1 = \frac{8}{\pi^2}\frac{1}{(2-1)^2}\frac{1}{2} = \frac{4}{\pi^2} \tag{5.20}$$

(c)

$$P_m = \frac{1}{T}\int_0^T |m(t)|^2\,dt = \frac{1}{T}\left[\int_0^{T/2}\left(1-\frac{4t}{T}\right)^2 dt + \int_{T/2}^{T}\left(-3+\frac{4t}{T}\right)^2 dt\right] \tag{5.21}$$

$$= \frac{2}{T}\left[\int_0^{T/2}\left(1-\frac{8t}{T}+\frac{16t^2}{T^2}\right)dt\right] \tag{5.22}$$

$$= \frac{2}{T}\left[\frac{T}{2}-\frac{4T^2}{T\,2^2}+\frac{16T^3}{3T^2 2^3}\right] = \frac{2}{T}\left[\frac{T}{6}\right] = \frac{1}{3} \tag{5.23}$$

(d) $n = 1 \Longrightarrow f = 2$ Hz
$n = 2 \Longrightarrow f = 6$ Hz
$n = 3 \Longrightarrow f = 10$ Hz

Since the bandwidth is limited to 9 Hz, only the first two terms of the summation will be kept and there are 4 nonzero terms in the Fourier Series

$$m_1 = \frac{4}{\pi^2} \cdots f = 2 \text{ Hz} \tag{5.24}$$

$$m_{-1} = \frac{4}{\pi^2} \cdots f = -2 \text{ Hz} \tag{5.25}$$

$$m_3 = \frac{4}{9\pi^2} \cdots f = 6 \text{ Hz} \tag{5.26}$$

$$m_{-3} = \frac{4}{9\pi^2} \cdots f = -6 \text{ Hz} \tag{5.27}$$

$$m_k = 0 \qquad \text{otherwise} \tag{5.28}$$

The message signal is given as

$$m(t) = \frac{8}{\pi^2} \cos(2\pi(2)t) + \frac{8}{9\pi^2} \cos(2\pi(6)t) \tag{5.29}$$

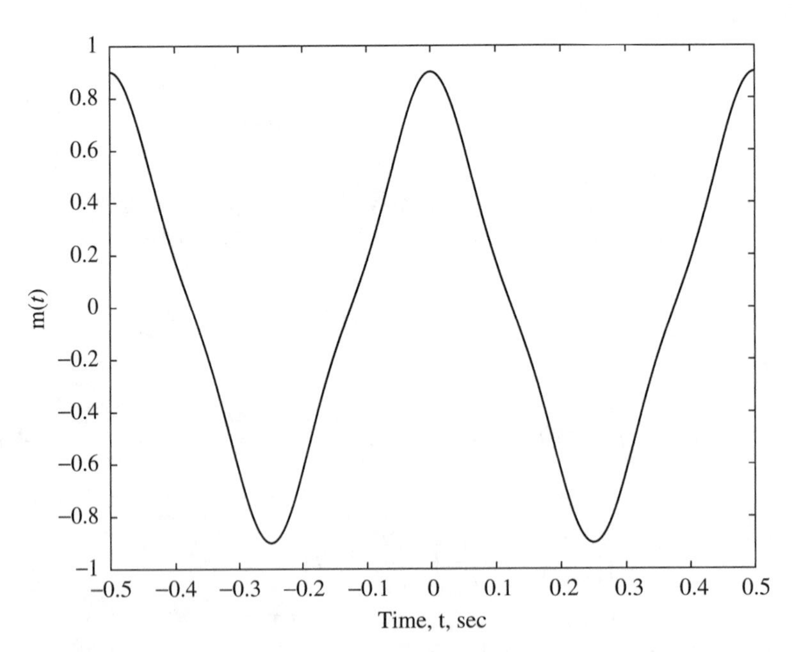

Figure 5.14 The filtered message signal.

(e) Parsevals theorem allows the power to be computed in the frequency domain as

$$P_m = \sum_{n=-\infty}^{\infty} |m_n|^2 = 2\left[\left(\frac{4}{\pi^2}\right)^2 + \left(\frac{4}{9\pi^2}\right)^2\right] = 0.3325 \tag{5.30}$$

(f) The Matlab code

```
%
% M-file for EE501
% Problem 5.2
% Author: M. Fitz
% Last Revision: 01/29/01
%
time=linspace(-0.5,0.5,1000);
%
% e) Computing the power after filtering
%
m_1=4/pi^2;
m_3=4/9/pi^2;
p_m=2*(m_1^2+m_3^2)
%
% f) plotting the output message waveform
%
mess=2*m_1*cos(2*pi*2*time)+2*m_3*cos(2*pi*6*time);
% plotting
figure(1)
plot(time,mess)
xlabel('Time, t, seconds')
ylabel('m(t)')
```

produces the plot shown in Figure 5.14.

Chapter

6

Amplitude Modulation

Amplitude modulation (AM) was historically the first modulation developed and conceptually the easiest to understand. Consequently, AM is developed first in this text.

6.1 Linear Modulation

The simplest analog modulation is to make $\Gamma_m(m(t)) = A_c m(t)$, i.e., a linear function of the message signal. The complex envelope and the spectrum of this modulated signal are given as

$$x_z(t) = A_c m(t) \qquad G_{x_z}(f) = A_c^2 G_m(f)$$

This modulation has $x_I(t) = A_c m(t)$ and $x_Q(t) = 0$, so the imaginary portion of the complex envelope is not used in a linear analog modulation. The resulting bandpass signal and spectrum are given as

$$x_c(t) = \Re\left[\sqrt{2}x_z(t)\exp[j2\pi f_c t]\right] = A_c m(t)\sqrt{2}\cos(2\pi f_c t) \tag{6.1}$$

$$G_{x_c}(f) = \frac{A_c^2}{2}G_m(f - f_c) + \frac{A_c^2}{2}G_m(-f - f_c) = \frac{A_c^2}{2}G_m(f - f_c) + \frac{A_c^2}{2}G_m(f + f_c) \tag{6.2}$$

where the fact that $m(t)$ was real was used to simplify Eq. (6.2). Figure 6.1 shows the complex envelope and an example bandpass signal for the message signal shown in Figure 5.1 with $A_c = 1/\sqrt{2}$. It is quite obvious from Figure 6.1 that the amplitude of the carrier signal is modulated directly proportional to the absolute value of the message signal (hence the name amplitude modulation). Figure 6.2 shows the resulting energy spectrum of the linearly modulated signal for the message signal shown in Figure 5.2. A very important characteristic of this modulation is that if the message signal has a bandwidth of W Hz then the bandpass signal will have a transmission bandwidth of $B_T = 2W$.

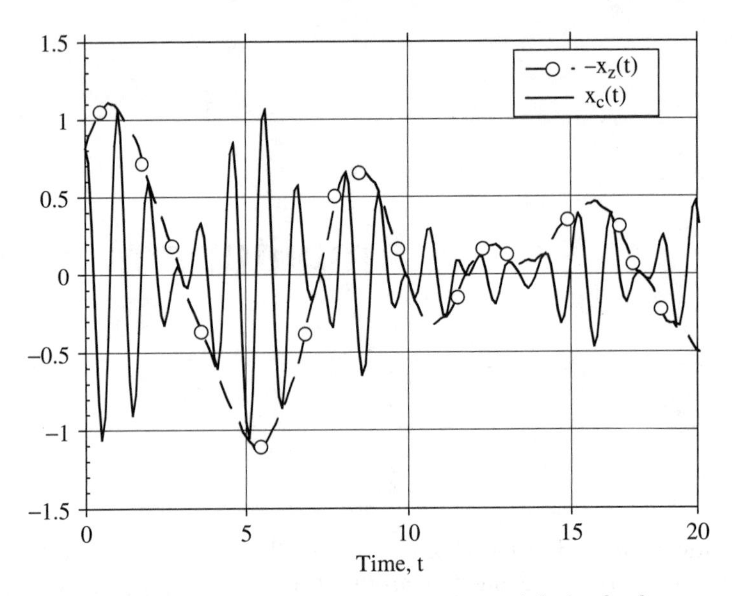

Figure 6.1 Example waveforms for linear analog modulation for the message signal in Figure 5.1.

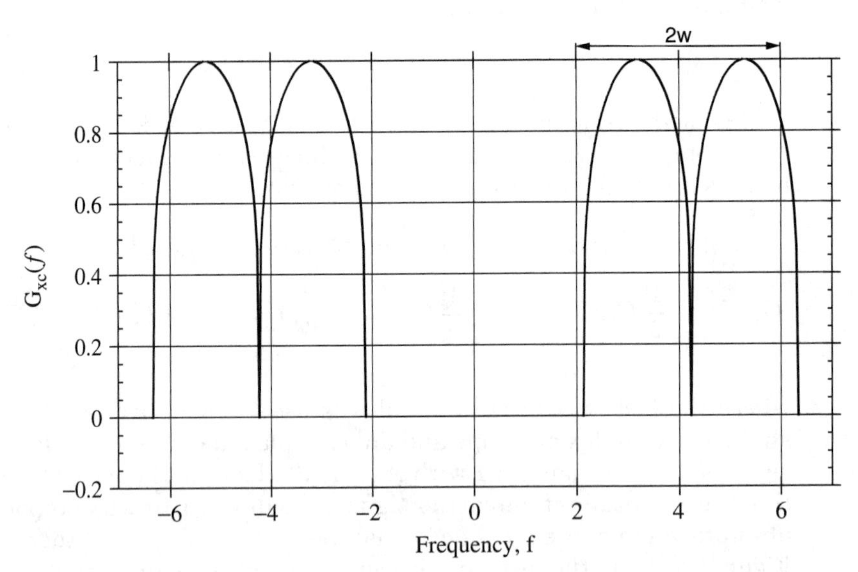

Figure 6.2 The energy spectrum of the linearly modulated signal for the message signal spectrum in Figure 5.2.

This implies $E_B = 50\%$. Because of this characteristic the modulation is often known as double sideband-amplitude modulation (DSB-AM). An efficiency of 50% is wasteful of the precious spectral resources but obviously the simplicity of the modulator is a positive attribute.

The power of a DSB-AM modulation is often of interest (e.g., to specify the characteristics of amplifiers or to calculate the received signal-to-noise ratio). To this end the power is given as

$$P_{x_c} = P_{r_c} = \lim_{T_m \to \infty} \frac{1}{T_m} \int_{-T_m/2}^{T_m/2} x_c^2(t)dt$$

$$= \lim_{T_m \to \infty} \frac{1}{T_m} \int_{-T_m/2}^{T_m/2} A_c^2 m^2(t) 2\cos^2(2\pi f_c t)dt = P_m A_c^2 = P_{x_z} \quad (6.3)$$

For a DSB-AM modulated waveform the output power is usually given as the product of the power associated with the carrier amplitude (A_c^2) and the power in the message signal (P_m).

EXAMPLE 6.1

Linear modulation with

$$m(t) = \beta \sin(2\pi f_m t) \qquad G_m(f) = \frac{\beta^2}{4}\delta(f - f_m) + \frac{\beta^2}{4}\delta(f + f_m)$$

produces

$$x_c(t) = A_c \beta \sin(2\pi f_m t)\sqrt{2}\cos(2\pi f_c t)$$

and

$$G_{x_c}(f) = \frac{A_c^2 \beta^2}{8}(\delta(f - f_m - f_c) + \delta(f + f_m - f_c) + \delta(f - f_m + f_c) + \delta(f + f_m + f_c))$$

The output power is

$$P_{x_c} = \frac{A_c^2 \beta^2}{2}$$

The mesage signal and the output modulated time domain signal are plotted in Figure 6.3(a) for $f_c = 20 f_m$ and $A_c = 1/\sqrt{2}$. Note that both the message signal and the modulated signal are periodic with a period of $T = 1/f_m$. The plot of the energy spectrum of the message and the modulated signals are plotted in Figure 6.3(b) for $f_c = 20 f_m$, $\beta = 1$, and $A_c = 1/\sqrt{2}$. It should be noted that $P_{x_z} = P_{r_z} = 0.25$.

EXAMPLE 6.2

The computer generated voice signal given in Chapter 2 ($W = 2.5$ kHz) is used to DSB-AM modulate a 7-kHz carrier. A short time record of the message signal and the resulting modulated output signal is shown in Figure 6.4(a). The energy spectrum of the signal is shown in Figure 6.4(b). Note the bandwidth of the carrier modulated signal is 5 kHz.

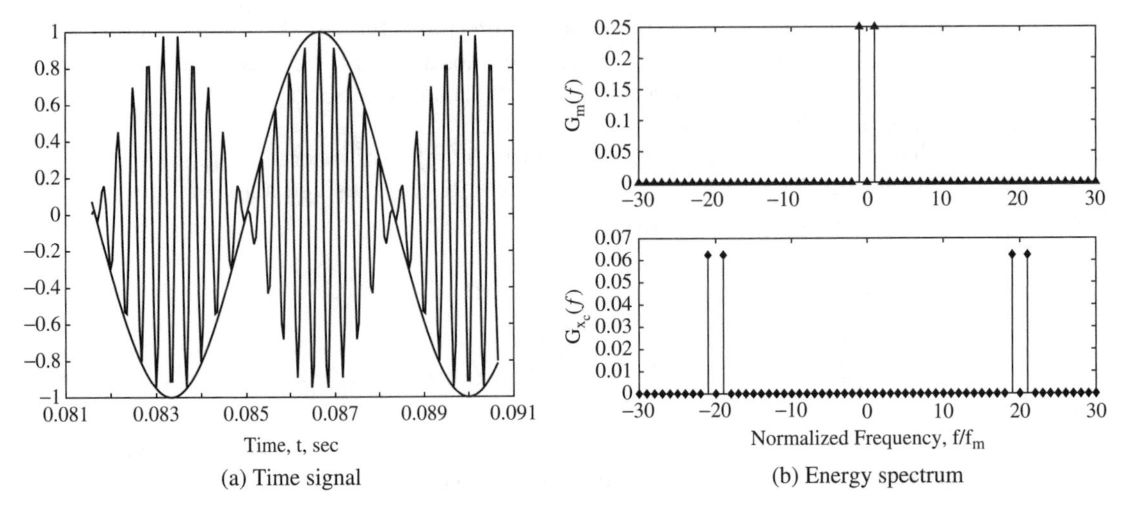

Figure 6.3 DSB-AM with $m(t) = \sin(2\pi f_m t)$.

Historically, there is no one person that can be attributed to the invention of DSB-AM. The early days of wireless electronic communications (around 1900) were filled with a variety of people working on technological gadgets that allowed voice to be transmitted. These gadgets allowed the "modulation" of a voiceband signal on a high-frequency carrier and the recovery of the voiceband signal from the received modulated high-frequency carrier. The mathematical analysis of the signal processing was not as mature as the experimental circuits that were built hence a real understanding of the modulation process was not available until much later. The first person to note that two sidebands arose in modulation was Carl Englund in 1914 [Osw56, SS87]. The credit for

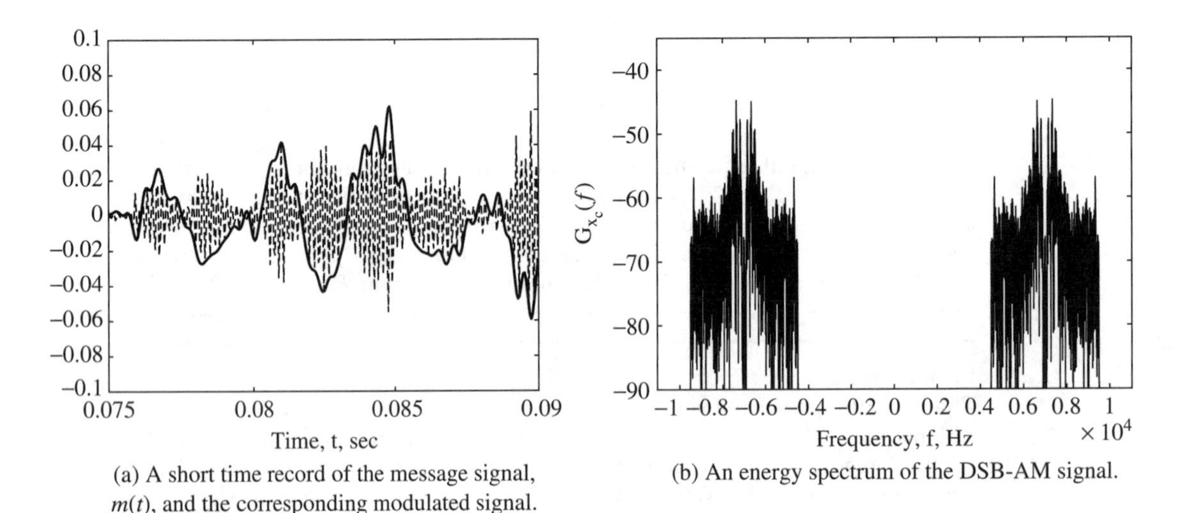

Figure 6.4 Example of DSB-AM with the computer generated voice signal. $f_c = 7$ kHz.

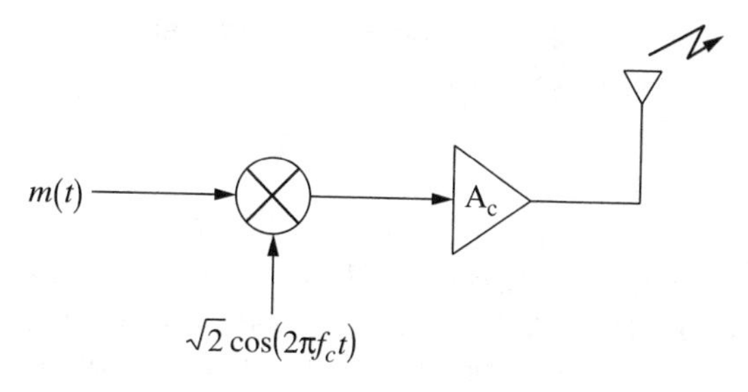

Figure 6.5 A DSB-AM modulator.

formulating a mathematical basis for the modulation and demodulation process is given to J.R. Carson [Car22, Car26].

6.1.1 Modulator and Demodulator

The modulator for a DSB-AM signal is simply the structure in Figure 4.4 except with DSB-AM there is no imaginary part to the complex envelope. Figure 6.5 shows the simplicity of this modulator. Using the terminology of Chapter 5, the modulator is denoted with

$$x_I(t) = g_I(m(t)) = A_c m(t) \qquad x_Q(t) = g_Q(m(t)) = 0 \tag{6.4}$$

Demodulation can be accomplished in a very simple configuration for DSB-AM. Given the channel model in Figure 5.8 a straightforward demodulator is seen in Figure 6.6. This demodulator simply derotates the received complex envelope by the phase induced by the propagation delay and uses the real part of this derotated signal after filtering by a low pass filter as the estimate of the message signal. The low pass filter is added to give noise and interference immunity and the effects of this filter will be discussed later. Note that the output of the demodulator is given as

$$\hat{m}(t) = A_c m(t) + N_I(t) = m_e(t) + N_I(t)$$

Using the terminology of Chapter 5 where the low pass filter impulse response is denoted $h_L(t)$, it can be noted that the DSB-AM demodulator is a **coherent** demodulator with

$$\hat{m}(t) = g_c(y_I(t), y_Q(t), \phi_p) = h_L(t) * (y_I(t)\cos(\phi_p) + y_Q(t)\sin(\phi_p)) \tag{6.5}$$

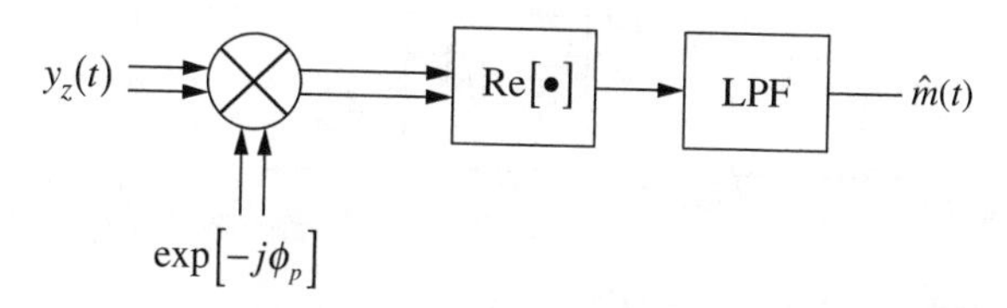

Figure 6.6 The block diagram of a DSB-AM demodulator.

The demodulator is quite simple once the phase induced in the propagation from transmitter to receiver is identified.

6.1.2 Coherent Demodulation

An important function of a DSB-AM demodulator is producing the appropriate value of ϕ_p for good message reconstruction. Demodulators that require an accurate phase reference like DSB-AM requires are often called phase **coherent** demodulators. Often in practice this phase reference is obtained manually with a tunable phase shifter. This is unsatisfactory if one or both ends of the link are moving (hence a drifting phase) or if automatic operation is desired.

Automatic phase tracking can be accomplished in a variety of ways. The techniques available for automatic phase tracking are easily divided into two sets of techniques: a phase reference derived from a transmitted reference and a phase reference derived from the received modulated signal. Note, a transmitted reference technique will reduce the efficiency of the transmission since part of the transmitted power is used in the reference signal and is not available at the output of the demodulator. Though a transmitted reference signal is wasteful of transmitted power it is often necessary for more complex modulation schemes (e.g., see Section 6.3.3). For each of these above mentioned techniques two methodologies are typically followed in deriving a coherent phase reference; open loop estimation and closed loop or phase-locked estimation. Consequently, four possible architectures are available for coherent demodulation in analog communications.

An additional advantage of DSB-AM is that the coherent reference can easily be derived from the received modulated signal. Consequently, in the remainder of this section the focus of the discussion will be on architectures that enable automatic phase tracking from the received modulated signal for DSB-AM. The block diagram of a typical open loop phase estimator for DSB-AM is shown in Figure 6.7. The essential idea in open loop phase estimation for DSB-AM is that any channel induced phase rotation can easily be detected since DSB-AM only uses the real part of the complex envelope. Note that the received DSB-AM signal has the form

$$y_z(t) = x_z(t)\exp[j\phi_p] = A_c m(t)\exp[j\phi_p] \tag{6.6}$$

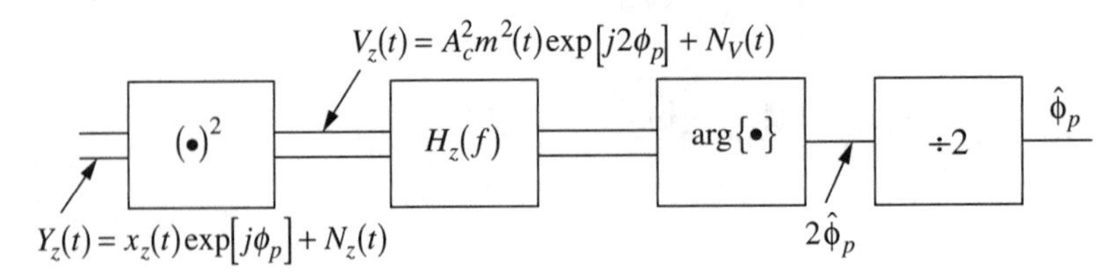

Figure 6.7 An open loop phase estimator for DSB-AM.

The phase of $y_z(t)$ (in the absence of noise) will either take value of ϕ_p (when $m(t) > 0$) or $\phi_p + \pi$ (when $m(t) < 0$). Squaring the signal gets rid of this bi-modal phase characteristic as can be seen by examining the signal

$$v_z(t) = (x_z(t)\exp[j\phi_p])^2 = A_c^2 m^2(t)\exp[j2\phi_p] \tag{6.7}$$

because $A_c^2 m^2(t) > 0$ so the $\arg(v_z(t)) = 2\phi_p$. In Figure 6.7 the filtering, $H_z(f)$, is used to smooth the phase estimate in the presence of noise.

EXAMPLE 6.3

Consider the DSB-AM with the computer generated voice signal given in Example 6.2 with a carrier frequency of 7 kHz and a propagation delay in the channel of 45.6 μs. This results in a $\phi_p = -114°$ (see Example 5.5.) The vector diagram which is a plot of $x_I(t)$ versus $x_Q(t)$ is a useful tool for understanding the operation of the carrier phase recovery system. The vector diagram was first introduced in Chapter 4. The vector diagram of the transmitted signal will be entirely on the x-axis, since $x_Q(t) = 0$. The plot of the vector diagram for the channel output (in the absence of noise) is shown in Figure 6.8(a). The $-114°$ phase shift is evident from this vector diagram. The output vector diagram from the squaring device is shown in Figure 6.8(b). This signal now has only one phase angle ($2 \times -114°$) and the coherent phase reference for demodulation can now be easily obtained.

Other types of coherent demodulators are available for DSB-AM. The historically most famous DSB-AM demodulator uses a phase-locked loop (PLL) for phase tracking and has a block diagram given in Figure 6.9. This structure

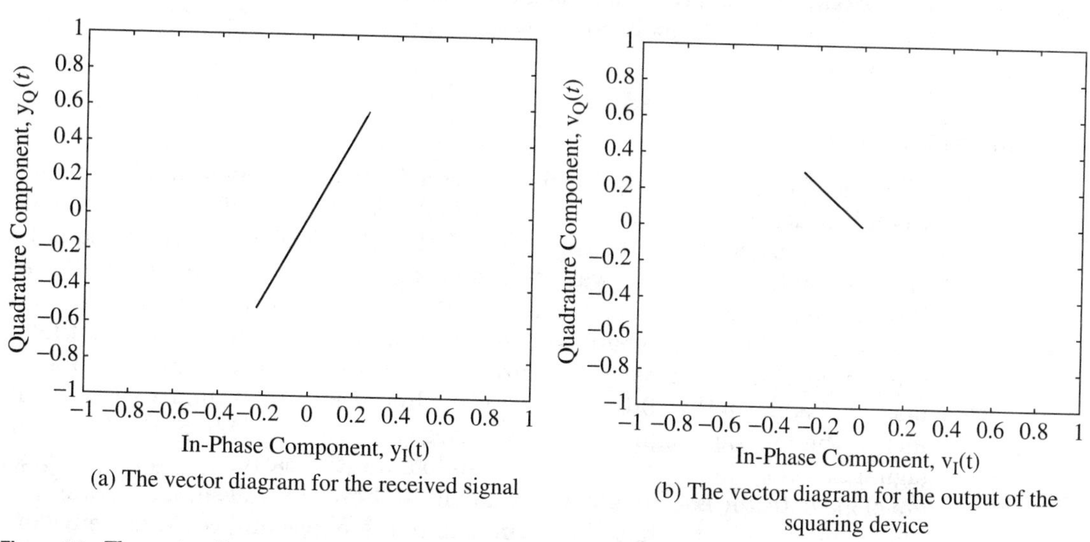

(a) The vector diagram for the received signal

(b) The vector diagram for the output of the squaring device

Figure 6.8 The vector diagram for coherent DSB-AM demodulation. $\phi_p = -114°$.

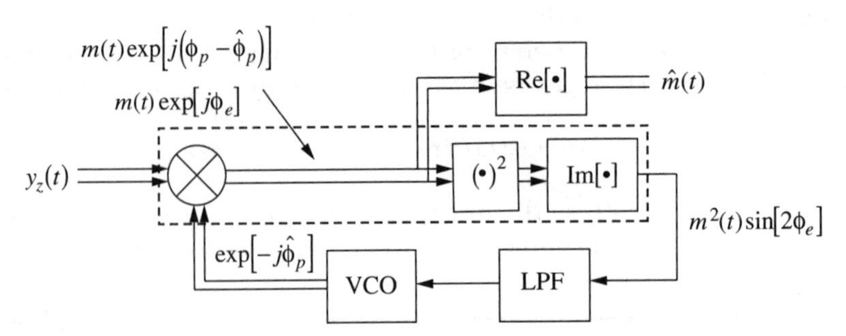

Figure 6.9 The block diagram of a Costas loop for synchronous DSB-AM demodulation.

is known as a Costas loop DSB-AM demodulator [Cos56]. The Costas loop is a feedback system that consists, as do all phase-locked loops, of three components: a voltage controlled oscillator, a phase detector, and a loop filter. The basic idea in a Costas loop demodulator is the phase detector measures the phase difference between a locally generated phase reference and the incoming received signal. The feedback system attempts to drive this phase difference to zero and hence implement carrier phase tracking. A simple theory of operation for the PLL will be explored in Chapter 8. The Costas loop has all the components of a PLL and ideas behind a Costas loop synchronous AM demodulator are explored in the homework (see Problem 6.3).

6.1.3 DSB-AM Conclusions

The advantage of DSB-AM is that DSB-AM is very simple to generate, e.g., see Figure 6.5. The disadvantages of DSB-AM are that phase coherent demodulation is required (relatively complex demodulator) and $E_B = 50\%$ (wasteful of bandwidth).

6.2 Affine Modulation

While now in the age of large-scale integrated circuits it may be hard to fathom, when broadcast radio was being developed DSB-AM was determined to have too complex a receiver to be commercially feasible. The coherent demodulator discussed in Section 6.1.2 was too complex in the days of the vacuum tubes. Early designers of radio broadcast systems noted that the message signal modulates the envelope of the bandpass signal, $x_A(t)$ in a continuous fashion that is proportional to the message amplitude but modulates the phase in a binary fashion, i.e., if $m(t) > 0$ then $x_P(t) = 0$ while if $m(t) < 0$ then $x_P(t) = \pi$. In fact, if the message signal never goes negative the envelope of the bandpass signal and the message are identical up to a multiplicative constant. Since an envelope detector is a simple device to build, these early designers formulated a modulation scheme that did not modulate the phase and could use envelope detectors to reconstruct the message signal at the receiver.

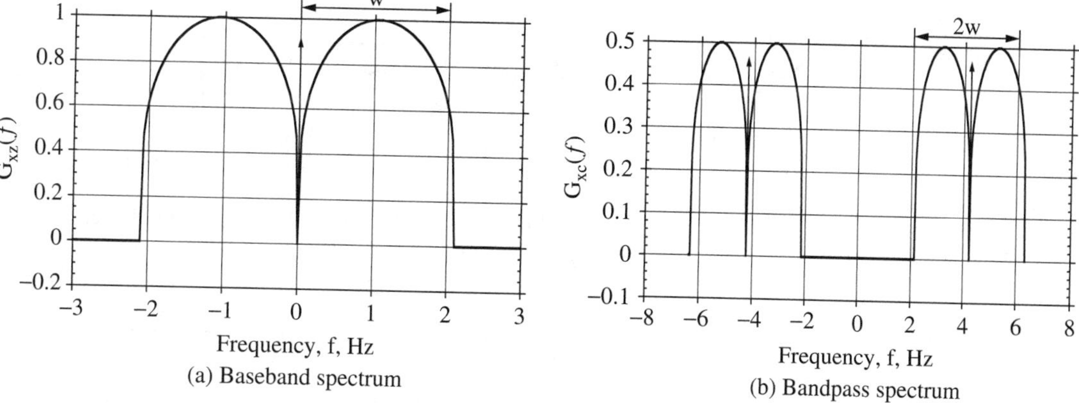

Figure 6.10 An example energy spectrum for LC-AM.

This desired characteristic is obtained if a DC signal is added to the message signal to guarantee that the resulting signal always is positive. This implies the complex envelope is an affine[1] function of the message signal, i.e.,

$$x_z(t) = A_c(1 + am(t)) \qquad G_{x_z}(f) = A_c^2\left[\delta(f) + a^2 G_m(f)\right]$$

where a is a positive number. This modulation has $x_I(t) = A_c + A_c am(t)$ and $x_Q(t) = 0$, so the imaginary portion of the complex envelope is not used again in an affine analog modulation. The resulting bandpass signal and spectrum are given as

$$x_c(t) = \Re[\sqrt{2}x_z(t)\exp[j2\pi f_c t]] = (A_c + A_c am(t))\sqrt{2}\cos(2\pi f_c t) \qquad (6.8)$$

$$G_{x_c}(f) = \frac{A_c^2}{2}\left(\delta(f - f_c) + a^2 G_m(f - f_c)\right) + \frac{A_c^2}{2}\left(\delta(f + f_c) + a^2 G_m(f + f_c)\right) \qquad (6.9)$$

Because of the discrete carrier term in the bandpass signal (see Eq. (6.9)) this modulation is often referred to as large carrier AM (LC-AM). Figure 6.10(a) shows an example message energy spectrum with the DC term added (the impulse at $f = 0$) and Figure 6.10(b) shows the resulting bandpass energy spectrum.

LC-AM has many of the same characteristics as DSB-AM. LC-AM still modulates the amplitude of the carrier and the bandwidth of bandpass signal is still $B_T = 2\,W$ ($E_B = 50\%$). The imaginary part of the complex envelope is also not used in LC-AM. LC-AM differs from DSB-AM in that a DC term is added to the complex envelope. This DC term is chosen such that $x_I(t) > 0$ or equivalently

[1]Affine is a linear term plus a constant term.

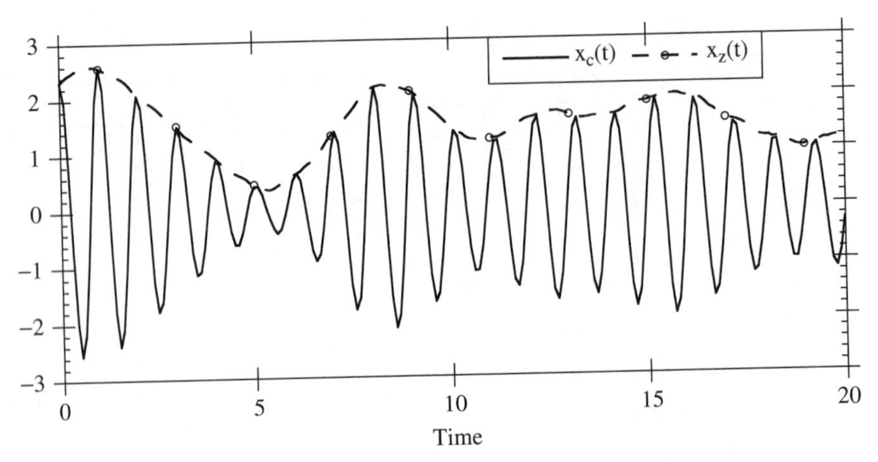

Figure 6.11 A LC-AM time waveform for the example given in Figure 5.1 with $A_c = 1/sqrt2$.

the envelope of the bandpass signal never passes through zero. This implies that $am(t) > -1$ or equivalently

$$a < \frac{-1}{\min m(t)} \tag{6.10}$$

This constant a, here denoted the modulation coefficient, is important in obtaining good performance in a LC-AM system. An example of a LC-AM waveform is shown in Figure 6.11 where the message waveform is given in Figure 5.1 ($A_c = 3/\sqrt{8}$ and $a = 2/3$).

Average power is given by

$$P_{x_c} = A_c^2 \left(\lim_{T_m \to \infty} \frac{1}{T_m} \int_{-T_m/2}^{T_m/2} (1 + am(t))^2 dt \right) \tag{6.11}$$

Since typically the time average of $m(t)$ is zero the average power simplifies to

$$P_{x_c} = P_{r_z} = A_c^2 (1 + a^2 P_m)$$

Note that there are two parts to the transmitted/received power: (1) the power associated with the added carrier transmission, A_c^2, and (2) the power associated with the message signal transmission, $A_c^2 a^2 P_m$. A designer usually wants to maximize the power in the message signal transmission and a factor that characterizes this split in power in LC-AM is denoted the message to carrier power.

Definition 6.1 The message to carrier power ratio for LC-AM is

$$\text{MCPR} = \frac{A_c^2 a^2 P_m}{A_c^2} = a^2 P_m \tag{6.12}$$

Assuming that the negative peaks of the message signal are at about the same level as the positive peaks and using Eq. (6.10) and the definition of PAPR given in Chapter 5, the MCPR can be approximated with

$$\text{MCPR} \approx \frac{1}{\text{PAPR}_m} \tag{6.13}$$

Consequently, the smaller the PAPR of the message signal the easier it is to put a larger percentage of the transmitted power into the message signal. It should be noted that most signals of interest in analog communications have a PAPR $>$ 10, so typically MCPR $<$ 10%. Later after the impacts of noise are evaluated on LC-AM demodulation it will be shown that the MCPR is directly related to the achieved transmission efficiency of LC-AM.

EXAMPLE 6.4

Affine modulation with

$$m(t) = \beta \sin(2\pi f_m t) \qquad G_m(f) = \frac{\beta^2}{4}\delta(f - f_m) + \frac{\beta^2}{4}\delta(f + f_m)$$

produces

$$x_c(t) = A_c(1 + a\beta \sin(2\pi f_m t))\sqrt{2}\cos(2\pi f_c t)$$

and

$$G_{x_c}(f) = \frac{A_c^2 a^2 \beta^2}{8}[\delta(f - f_m - f_c) + \delta(f + f_m - f_c) + \delta(f - f_m + f_c)$$
$$+\delta(f + f_m + f_c)] + \frac{A_c^2}{2}[\delta(f - f_c) + \delta(f + f_c)]$$

The transmitted power is

$$P_{x_c} = A_c^2\left(1 + \frac{a^2\beta^2}{2}\right)$$

To maintain $x_I(t) > 0$ implies that $a < 1/\beta$. The mesage signal and the output modulated time domain signal are plotted in Figure 6.12(a) for $f_c = 20f_m$, $a = 1/\beta$, and $A_c = 1/\sqrt{2}$. Note that both the message signal and the modulated signal are periodic with a period of $T = 1/f_m$. The plot of the energy spectrum of the message and the modulated signals are plotted in Figure 6.12(b) for $f_c = 20f_m$, $a = 1/\beta$, and $A_c = 1/\sqrt{2}$. It should be noted that $P_{x_z} = P_{r_z} = 0.75$ and MCPR $= 50\%$.

EXAMPLE 6.5

The computer-generated voice signal given in Chapter 2 ($W = 2.5$ kHz) is used to LC-AM modulate a 7-kHz carrier. A short time record of the scaled complex envelope and the resulting output modulated signal is shown in Figure 6.13(a). Note, the minimum value of the voice signal is -3.93 so the modulation coefficient was set to $a = 0.25$. Figure 6.13(a)

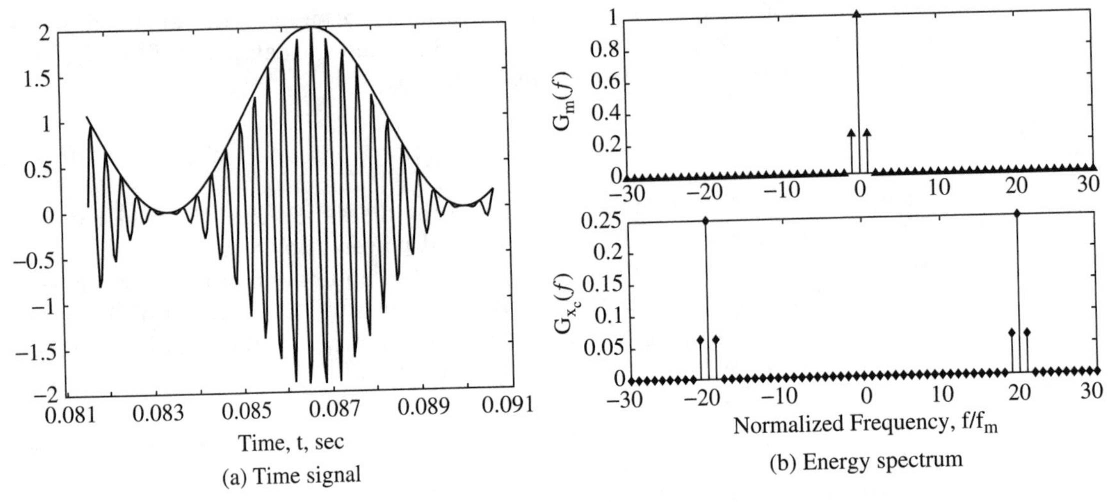

Figure 6.12 LC-AM with $m(t) = \sin(2\pi f_m t)$.

shows the modulated signal for large values $m(t)$ and envelope comes close to zero. The energy spectrum of the signal is shown in Figure 6.13(b). Note, the bandwidth of the carrier modulated signal is 5 kHz and the large carrier is evident in the plot.

6.2.1 Modulator and Demodulator

The modulator for LC-AM is very simple. It is just one arm (the in-phase one) of a quadrature modulator where a DC term has been added to the message signal. Figure 6.14 shows a common implementation of this modulator.

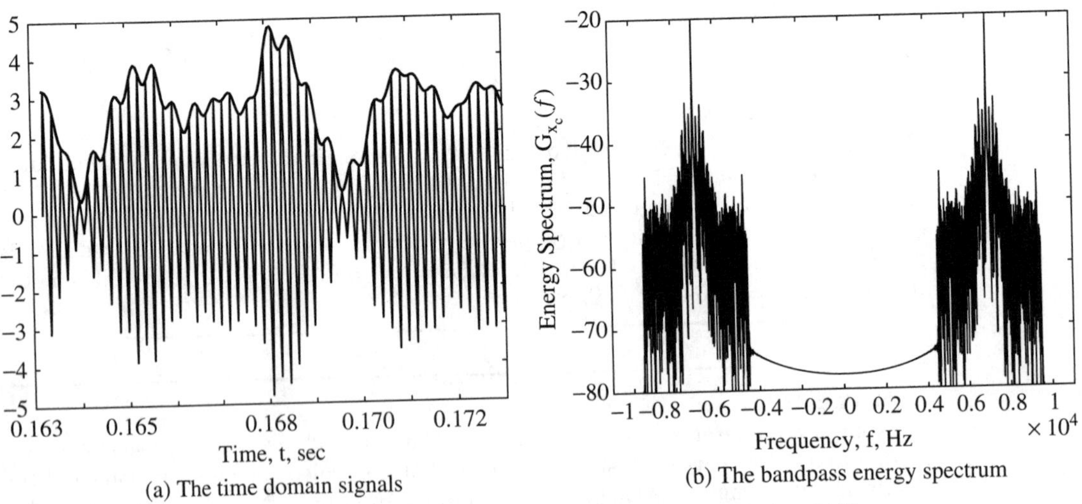

Figure 6.13 The resulting LC-AM signal for a computer generated voice signal. $f_c = 7$ kHz.

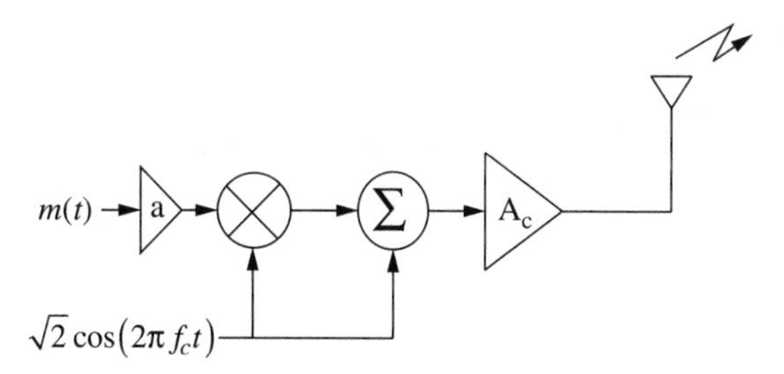

Figure 6.14 The block diagram of a modulator for LC-AM.

Again the simplicity of the modulator is an obvious advantage for LC-AM. Using the terminology of Chapter 5, the modulator is denoted with

$$x_I(t) = g_I(m(t)) = A_c(1 + am(t)) \qquad x_Q(t) = g_Q(m(t)) = 0 \tag{6.14}$$

The demodulator for LC-AM is simply an envelope detector followed by a DC block. Following the developed notation, the output of the envelope detector for $1 + am(t) > 0$ in the absence of noise is

$$|y_z(t)| = \left|A_c(1 + am(t))e^{j\phi_p}\right| = A_c(1 + am(t))$$

The DC block will remove the DC term to give

$$\hat{m}(t) = A_c am(t) = m_e(t)$$

Figure 6.15 shows the block diagram of a LC-AM demodulator. Using the terminology of Chapter 5 where the DC block impulse response is denoted $h_H(t)$, it can be noted that the LC-AM demodulator is a **noncoherent** demodulator with

$$\hat{m}(t) = g_n(y_I(t), y_Q(t)) = h_H(t) * \left(\sqrt{y_I^2(t) + y_Q^2(t)}\right) = h_H(t) * y_A(t) \tag{6.15}$$

It is important to note that the demodulation performance is unaffected by the random phase induced by the propagation delay in transmission, ϕ_p.

Figure 6.16 shows a circuit implementation of the bandpass version of the demodulator. Note that no active devices are contained in this implementation, which was a big advantage in the vacuum tube days. Currently, with large scale integrated circuits so prevalent, the advantage is not so large. The reason that a simple structure like the envelope detector can be used to recover the message signal is that the phase of the transmitted signal is always zero so that the

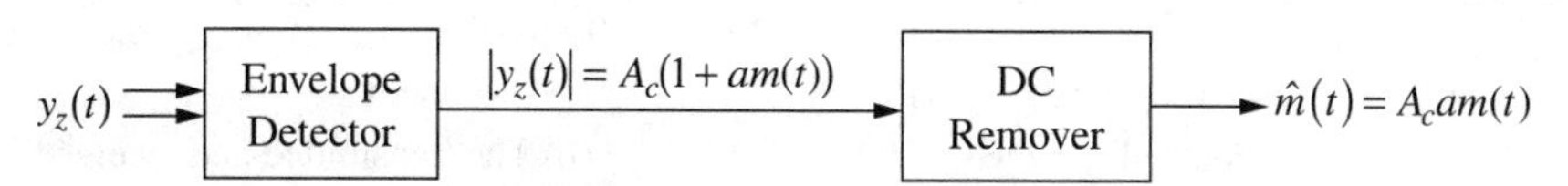

Figure 6.15 The block diagram of a baseband LC-AM demodulator.

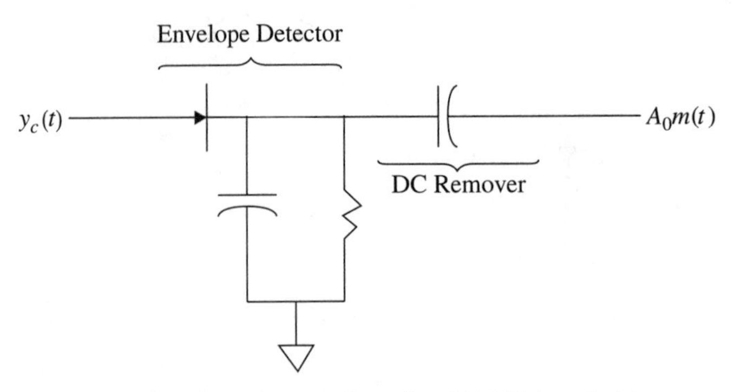

Figure 6.16 Circuit implementation of an LC-AM demodulator.

phase shift induced by the propagation delay in the channel is entirely limited to shifting the phase of the bandpass signal and has no impact on the amplitude of bandpass signal.

EXAMPLE 6.6

Consider again affine modulation with

$$m(t) = \beta \sin(2\pi f_m t)$$

with

$$x_z(t) = A_c(1 + a\beta \sin(2\pi f_m t))$$

The vector diagram of $x_z(t)$ is shown in Figure 6.17(a) and it is obvious from this plot that only the real part of the complex envelope is used in producing a LC-AM signal and

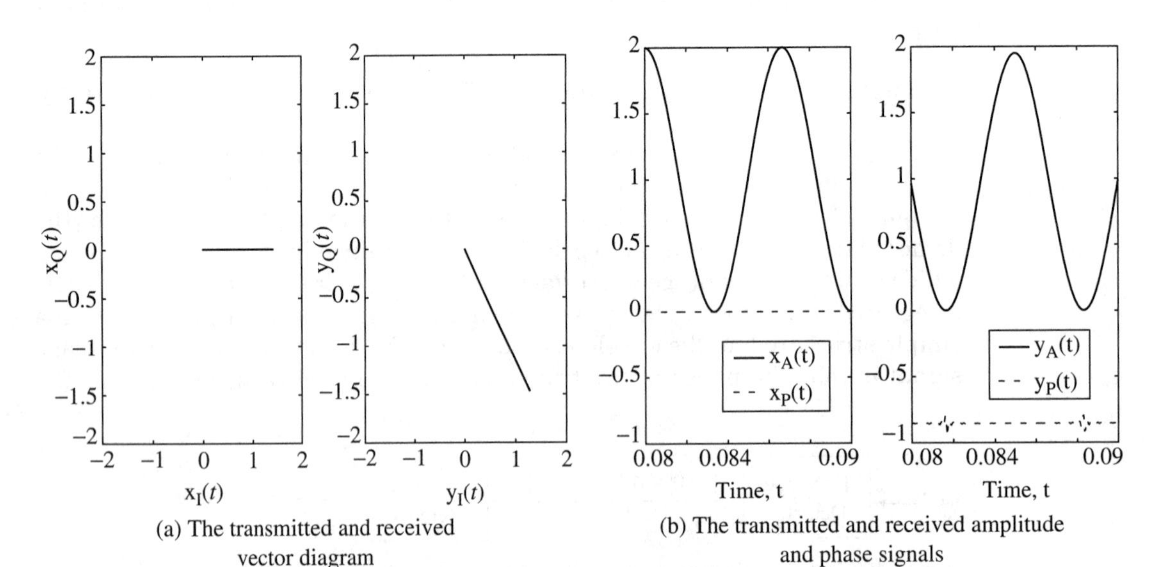

Figure 6.17 The signals for a sinusoidal message signal. $\phi_p = -49°$.

the transmitted signal phase is always zero. The received signal has the form

$$y_z(t) = A_c(1 + a\beta \sin(2\pi f_m(t - \tau_p))) \exp[j\phi_p] \tag{6.16}$$

The vector diagram for the noiseless received signal, $y_z(t)$, is plotted in Figure 6.17(a). By examining this figure it is clear that the phase shift has rotated the complex envelope but not changed the amplitude of the signal. This is evident in the plots of the transmitted and received amplitude and phase (measured in radians) signals shown in Figure 6.17(b). The received amplitude signal and the transmitted amplitude signal are different only by a time delay, τ_p. The received phase is exactly ϕ_p.

EXAMPLE 6.7

Consider the LC-AM computer-generated voice signal given in Example 6.6 with a carrier frequency of 7 kHz and a propagation delay in the channel of 45.6 μs. This results in a $\phi_p = -114°$ (see Example 6.4.) The plot of the envelope of the received signal,

$$y_A(t) = \sqrt{y_I^2(t) + y_Q^2(t)}$$

is shown in Figure 6.18(a) (Note the LPFs in the quadrature downconverter did not have unity gain). $y_A(t)$ has a DC offset due to the large carrier component but the message signal is clearly seen riding on top of this DC offset. The demodulated output after the DC block is shown in Figure 6.18(b). Note, the impulse at the start of this plot is due to the transient response of the filters and the DC block.

Again with LC-AM the working circuit implementation preceded the mathematical understanding of the modulation process. In fact, early on the modulation was thought to be a linear process (the carrier wave was ostensibly

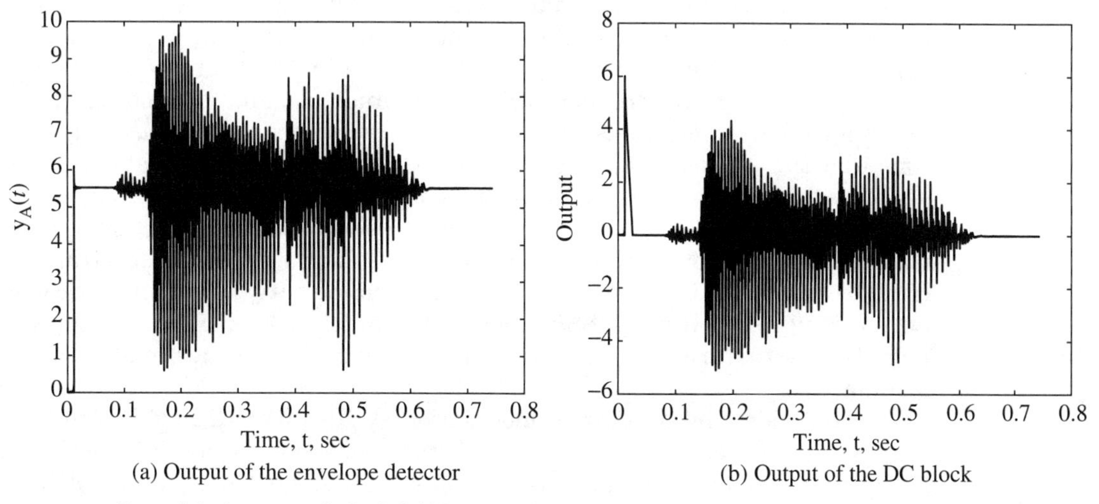

(a) Output of the envelope detector

(b) Output of the DC block

Figure 6.18 Demodulation example for LC-AM.

summed with the voiceband signal). Unbeknownst to the inventors a circuit nonlinearity in this "linear" modulation process was in fact producing LC-AM signal. Envelope detection of the modulated waveforms was actually easy to accomplish with a wide variety of nonlinear devices. In fact, an early LC-AM detector used a crystal (the Dunwoody and Picard crystal detector) in a very simple circuit to implement to the envelope detector. Many amatuer radio hobbyists have used crystals to build radio demodulators without a single active radio frequency device (only an audio amplifier). Again this pre-1910 work was mostly pushed by technologists and the theory was not completely understood until much later.

6.2.2 LC-AM Conclusions

The advantage of LC-AM is again it is easy to generate and it has a simple cheap modulator and demodulator. The disadvantage is that $E_B = 50\%$. Later we will learn that noncoherent detection suffers a greater degradation in the presence of noise than coherent detection.

6.3 Quadrature Modulations

Both DSB-AM and LC-AM are very simple modulations to generate but they have $E_B = 50\%$ so bandwidth sensitive applications need to explore other modulation options. The spectral efficiency of analog modulations can be improved. Note that both DSB-AM and LC-AM only use the real part of the complex envelope. The imaginary component of the complex envelope can be used to shape the spectral characteristics of the analog transmissions. As an example, the bandwidth of analog video signals is approximately 4.5 MHz. DSB-AM and LC-AM modulation would produce a bandpass bandwidth for video signals of 9 MHz. Broadcast analog television (TV) signals have a bandwidth of approximately 6 MHz and this is achieved with a quadrature modulation. Engineers in 1915 first realized that there were two sidebands (positive frequencies and negative frequencies) in amplitude modulation and theorized and realized in experiments that only one of the sidebands was needed to reconstruct the message signal at the demodulator. This mathematical observation led to a great flurry of development in quadrature modulation. Since only one sideband needs to be transmitted, this makes a transmission scheme that has $E_B = 100\%$ possible. This two times improvement in spectral efficiency compared to DSB-AM and LC-AM is what led early telecommunications engineering groups to adopt SSB-AM for efficient multiplexing of voiceband signals. This section tries to summarize the current state of the art in quadrature modulations.

Vestigial sideband amplitude modulation (VSB-AM) is a modulation that improves the spectral efficiency of analog transmission by specially designed linear filters at the modulator or transmitter. The goal with VSB-AM is to

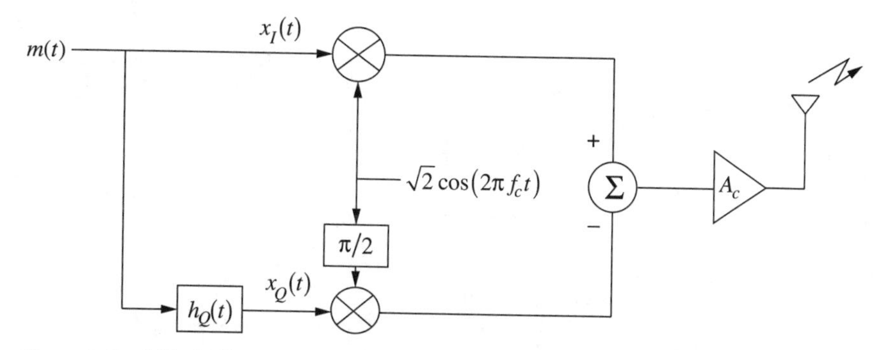

Figure 6.19 A block diagram of a VSB-AM modulator.

achieve $E_B > 50\%$ using the imaginary component of the modulation complex envelope. A VSB-AM signal has a complex envelope of

$$x_z(t) = A_c[m(t) + j(m(t) * h_Q(t))]$$

where $h_Q(t)$ is the impulse response of a real LTI system. Two interpretations of VSB-AM are useful. The first interpretation is that $x_I(t)$ is generated exactly the same as DSB-AM (linear function of the message) and $x_Q(t)$ is generated as a filtered version of the message signal. A block diagram of a baseband VSB-AM modulator using this interpretation is seen in Figure 6.19. Recalling the results in Section 4.6, the second interpretation is that a bandpass VSB-AM signal is generated by putting a bandpass DSB-AM signal through a bandpass filter (asymmetric around f_c) whose complex envelope impulse response and transfer function are

$$h_z(t) = \delta(t) + jh_Q(t) \qquad H_z(f) = 1 + jH_Q(f) \qquad -W \le f \le W \tag{6.17}$$

Note the transmitted power of a VSB-AM signal is going to be higher than a similarly modulated DSB-AM signal since the imaginary portion of the complex envelope is not zero. The actual resulting output power is a function of the filter response, $h_z(t)$, and an example calculation will be pursued in the homework.

6.3.1 VSB Filter Design

The design of the filter, $h_Q(t)$, is critical to achieving improved spectral efficiency. The Fourier transform of a VSB-AM signal (using Eq. (4.9)) is given as

$$X_z(f) = X_I(f) + jX_Q(f) = A_c M(f)[1 + jH_Q(f)] \tag{6.18}$$

Additionally, note $X_z(f) = A_c H_z(f) M(f)$ with $H_z(f) = [1 + jH_Q(f)]$, which is the Fourier transform of the impulse response given in Eq. (6.17). Since the message signal spectrum is nonzero over $-W \le f \le W$, reduction of the bandwidth of the bandpass signal requires that $[1 + jH_Q(f)]$ be zero over

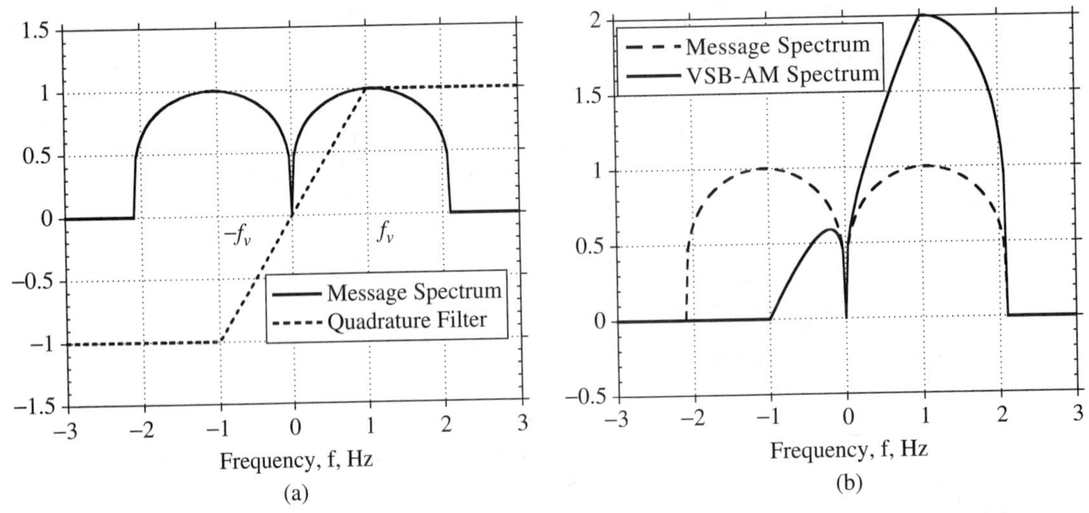

Figure 6.20 An example to show how quadrature modulation can be used to reduce the bandwidth of the transmitted signal. Upper sideband modulation.

some part of $f \in [-W, W]$. The remainder of the discussion will examine the conditions which must hold to produce

$$X_z(f) = 0 \qquad -W \leq f \leq -f_v \leq 0$$

where f_v is commonly called the vestigial frequency. In other words, what conditions on $H_z(f)$ must hold so that a portion of the lower sideband of the bandpass signal is eliminated to achieve

$$E_B = \frac{W}{W + f_v} > 50\%$$

Note that VSB-AM is always more bandwidth efficient than either DSB-AM or LC-AM and the improvement in bandwidth efficiency is a function of f_v. Since the lower portion of the signal spectrum is eliminated, this type of transmission is often termed upper sideband VSB-AM (the upper sideband is transmitted). Similar results as the sequel can be obtained to produce lower sideband VSB-AM transmissions. A set of example spectra for upper sideband VSB-AM is shown in Figure 6.20 to illustrate the idea of quadrature modulation. The figure uses the energy spectrum of the baseband transmitted signal since it is simple to represent the energy spectrum with a two dimensional graph. Note that the remains of the lower sideband in this example illustrates the reason for the name VSB-AM[2] and that the resulting transmission bandwidth is clearly $B_T = W + f_v$.

[2]A dictionary definition of vestigial is *pertaining to a mark, trace, or visible evidence of something that is no longer present or in existence.*

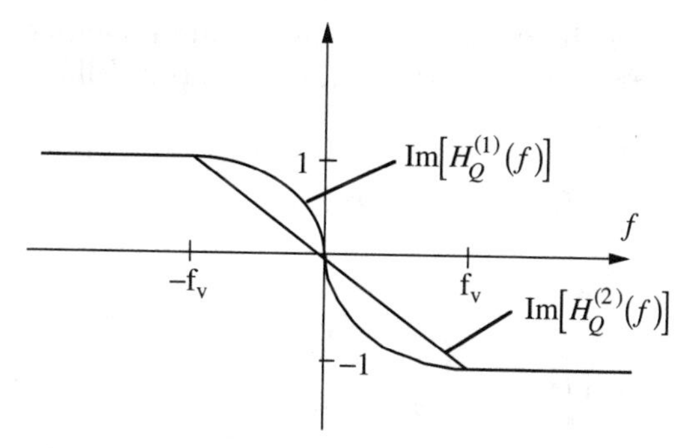

Figure 6.21 Examples of the imaginary part of the quadrature filter for VSB-AM.

We want

$$H_z(f) = [1 + jH_Q(f)] = 0 \qquad -W \leq f \leq -f_v \tag{6.19}$$

Since $H_Q(f) = \Re[H_Q(f)] + j\Im[H_Q(f)]$ it is simple to see that Eq. (6.19) implies that

$$\Im[H_Q(f)] = 1 \qquad \Re[H_Q(f)] = 0 \qquad -W \leq f \leq -f_v \tag{6.20}$$

Note, since $h_Q(t)$ is real then this implies

$$\Im[H_Q(f)] = -1 \qquad \Re[H_Q(f)] = 0 \qquad f_v \leq f \leq W \tag{6.21}$$

but the values for $H_Q(f), -f_v \leq f \leq f_v$ are unconstrained. A plot of two possible realizations of the filter $H_Q(f)$ is shown in Figure 6.21. It should be noted that our discussion focused on eliminating a portion of the lower sideband but similar results hold for eliminating the upper sideband.

EXAMPLE 6.8

Analog television broadcast in the United States uses a video signal with a bandwidth of approximately 4.5 MHz. The transmitted signals are a form of quadrature modulation, where $f_v = 1.25$ MHz for a total channel bandwidth of <6 MHz. Television stations in the United States have 6 MHz spacings. Consequently

$$E_B = \frac{4.5 \text{ MHz}}{6 \text{ MHz}} = 75\%$$

6.3.2 Single Sideband AM

An interesting special case of VSB-AM occurs when $f_v \to 0$. In this case the $E_B \to 100\%$. This is accomplished by eliminating one of the sidebands and hence this modulation is often termed single sideband amplitude modulation

(SSB-AM). The $x_Q(t)$ that results in this case is important enough to get a name; the Hilbert transform [ZTF89]. The transfer function of the Hilbert transformer is

$$H_Q(f) = -j\,\mathrm{sgn}(f) \tag{6.22}$$

where

$$\mathrm{sgn}(f) = \begin{cases} 1 & f > 0 \\ -1 & f < 0 \end{cases} \tag{6.23}$$

Note, because of the sharp transition in the transfer function at DC it is only possible to use SSB-AM with message signals that do not have significant spectral content near DC. It should be noted that in analog video signals the DC value is important in a simple way to synchronize the scanning of the picture so SSB-AM cannot be used with video signals. The transmitted signal for SSB-AM is

$$x_z(t) = A_c(m(t) + j m_h(t)) \tag{6.24}$$

where $m_h(t)$ is the Hilbert transform of $m(t)$.

EXAMPLE 6.9
Single-sideband modulation with a message signal

$$m(t) = \beta \sin(2\pi f_m t)$$

has an in-phase signal of

$$x_I(t) = A_c\beta \sin(2\pi f_m t) \qquad X_I(f) = \frac{A_c\beta}{2j}\delta(f - f_m) - \frac{A_c\beta}{2j}\delta(f + f_m)$$

Applying the Hilbert transform, $H_Q(f) = -j\,\mathrm{sgn}(f)$, to $x_I(t)$ produces a quadrature signal with

$$X_Q(f) = H_Q(f)X_I(f) = \frac{-A_c\beta}{2}\delta(f - f_m) - \frac{A_c\beta}{2}\delta(f + f_m)$$

or

$$x_Q(t) = -A_c\beta \cos(2\pi f_m t)$$

This results in

$$x_z(t) = A_c\beta(\sin(2\pi f_m t) - j\cos(2\pi f_m t)) = -jA_c\beta \exp(j2\pi f_m t)$$

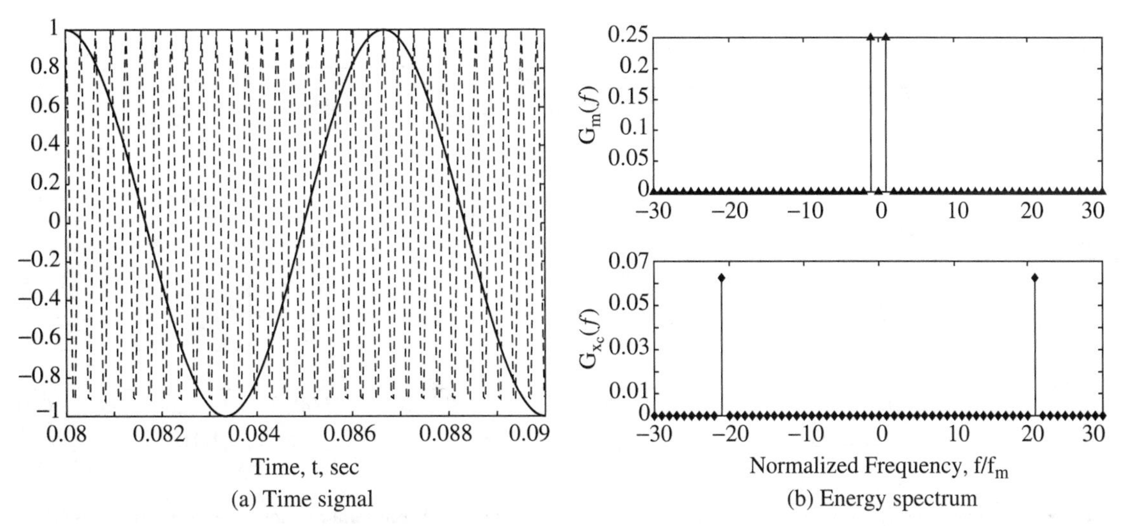

Figure 6.22 SSB-AM with $m(t) = \sin(2\pi f_m t)$.

and

$$G_{x_z}(f) = A_c^2 \beta^2 \delta(f - f_m)$$

The bandpass spectrum is

$$G_{x_c}(f) = \frac{A_c^2 \beta^2}{2} \delta(f - f_m - f_c) + \frac{A_c^2 \beta^2}{2} \delta(f + f_m + f_c)$$

Note that the lower sideband of the message signal has been completely eliminated with SSB-AM. The message signal and the output modulated time domain signal are plotted in Figure 6.22(a) for $f_c = 20 f_m$, $\beta = 1$, and $A_c = 1/\sqrt{2}$. Note that the message signal is periodic with a period of $T = 1/f_m$ and the modulated signal has a much shorter period, $T = 1/(21 f_m)$. The plot of the energy spectrum of the message and the modulated signals are plotted in Figure 6.22(b) for $f_c = 20 f_m$, $\beta = 1$, and $A_c = 1/\sqrt{2}$. It should be noted that $P_{x_z} = P_{r_z} = 0.5$.

EXAMPLE 6.10

The computer generated voice signal given in Chapter 2 ($W = 2.5$ kHz) is used to SSB-AM modulate a 7-kHz carrier. The quadrature filter needs to be a close approximation to a Hilbert transformer given in Eq. (6.22). The magnitude and phase response of a filter designed for a voice signal is shown in Figure 6.23. The resulting complex envelope energy spectrum is given in Figure 6.24(a). This implementation provides greater than 80 dB of sideband rejection over most of the band. The bandpass energy spectrum is shown in Figure 6.24(b). This plot clearly shows that SSB-AM uses half the bandwidth of DSB-AM or LC-AM.

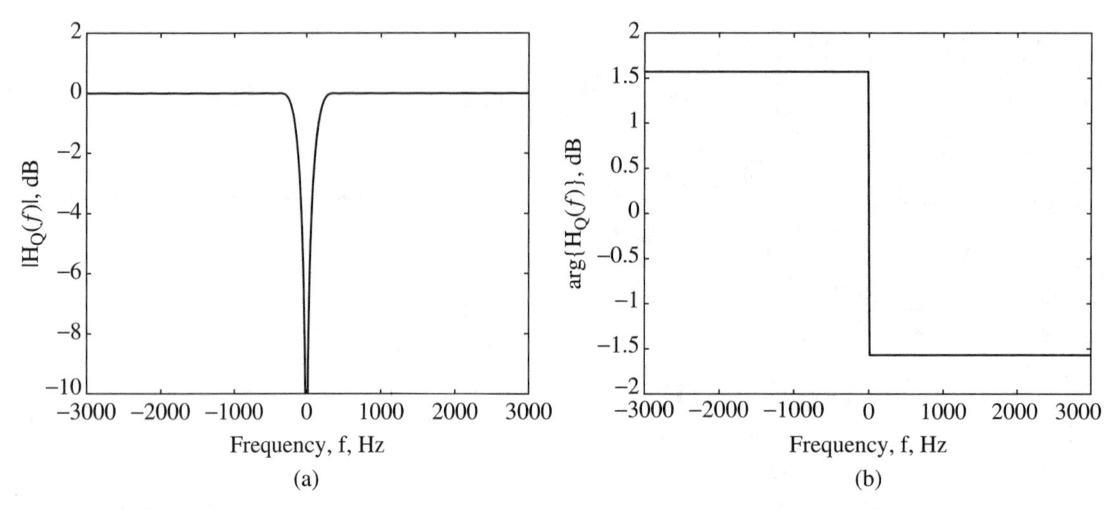

Figure 6.23 The magnitude and phase response of the Hilbert transform implementation used in Example 6.10.

6.3.3 Modulator and Demodulator

The modulator for VSB-AM is given in Figure 6.19. Note also an implementation can be achieved with a bandpass filter that has an asymmetric frequency response around f_c. In practice both the baseband and bandpass implementations are used. An overview of analog processing techniques for SSB-AM transmitters is provided in [Kur76]. The high performance digital processing that has emerged over the recent past has pushed the implementations toward the

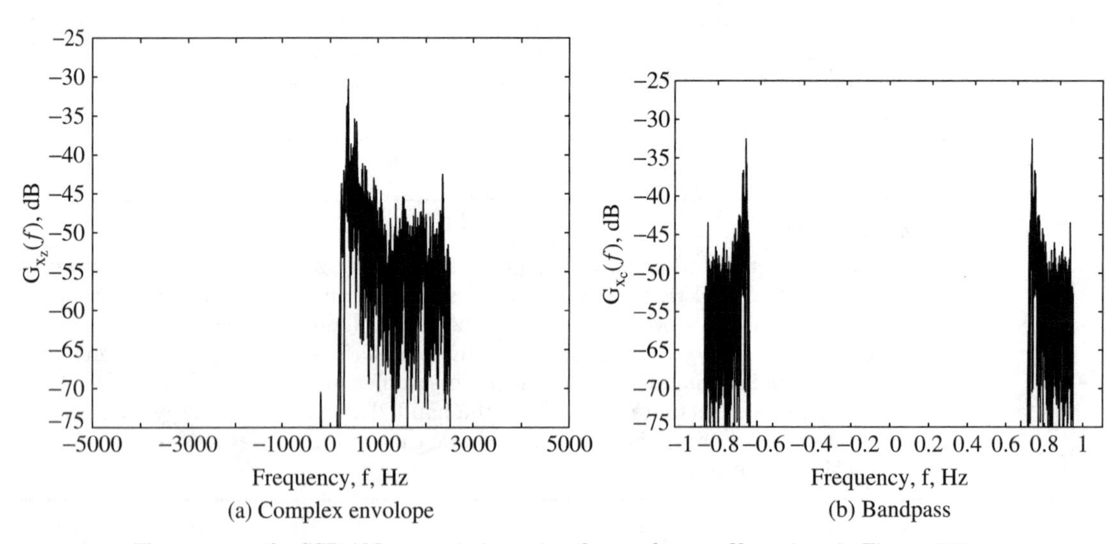

Figure 6.24 The spectrum for SSB-AM transmission using the quadrature filter given in Figure 6.23.

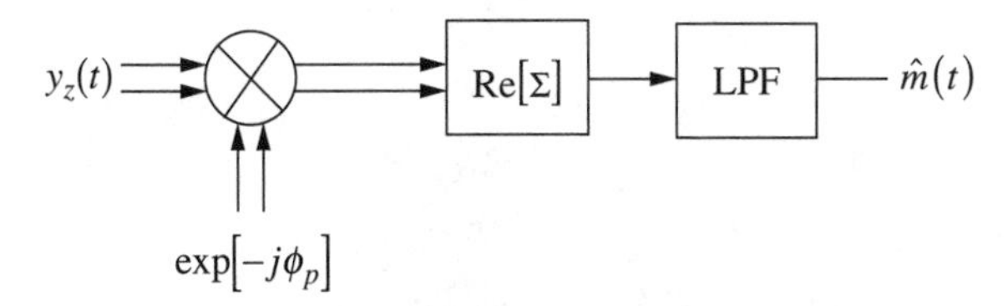

Figure 6.25 A coherent demodulator for VSB-AM. Note it is exactly the same as DSB-AM.

digital baseband realizations due to the precise nature with which filter responses can be controlled in digital circuits.

The demodulator for VSB-AM can be exactly the same as for DSB-AM. Recall that the received signal for VSB-AM is given as

$$y_z(t) = A_c(m(t) + jm_h(t))\exp[j\phi_p] \tag{6.25}$$

Again it is clear that $m(t)$ can be recovered by derotating by ϕ_p and looking at the real part of the resulting signal. The coherent demodulator for VSB-AM is shown in Figure 6.25. It is obvious from examining Figure 6.25 that the VSB-AM demodulator is identical to the coherent demodulator for DSB-AM and the output is

$$\hat{m}(t) = A_c m(t) + N_I(t) = m_e(t) + N_I(t)$$

Using the terminology of Chapter 5 where the low pass filter impulse response is denoted $h_L(t)$, it can be noted that the VSB-AM demodulator is a **coherent** demodulator with

$$\hat{m}(t) = g_c(y_I(t), y_Q(t), \phi_p) = h_L(t) * (y_I(t)\cos(\phi_p) + y_Q(t)\sin(\phi_p)) \tag{6.26}$$

The demodulator is quite simple once the phase induced in the propagation from transmitter to receiver is identified.

The theory of quadrature modulation was developed pretty quickly [Osw56, SS87]. In 1915, H. D. Arnold built an antenna that was tuned only to the upper sideband of a DSB-AM signal and demonstrated that the voiceband signal could be recovered from only one sideband. This result quickly led J. R. Carson to propose a SSB-AM system for multiplexing many voiceband messages in a telephony system. After a contentious and long patent examination a patent for this SSB-AM system was finally issued in 1923 [Car23] even though commercial usage of SSB-AM commenced in 1918.

6.3.4 Transmitted Reference Based Demodulation

A coherent demodulator is also necessary for VSB-AM transmission. Unfortunately, the phase reference cannot be derived from the received signal like in DSB-AM because VSB-AM uses both the real and imaginary components of the complex envelope. This characteristic is best demonstrated by the vector diagram of the complex envelope. Figure 6.26 shows the vector diagram of

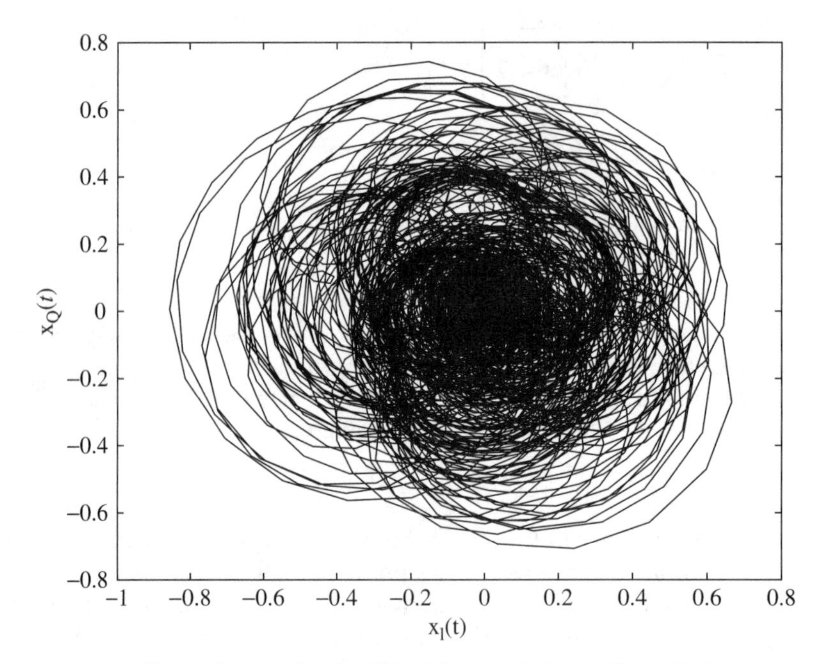

Figure 6.26 Vector diagram for the SSB-AM transmission in Example 6.10.

the SSB-AM transmission of Example 6.10. Clearly if one compares the vector diagram shown in Figure 6.8(a) to the vector diagram in Figure 6.26, it is easy to see why automatic phase recovery from the received signal is tough. So, while VSB-AM provides improved spectral efficiency it does it at the cost of increased demodulator complexity.

If automatic phase tracking is desired with VSB-AM, a transmitted reference signal is often used. We will explore this idea in a SSB-AM application. The block diagram of an example SSB-AM transmitter system that uses a transmitted reference is given in Figure 6.27. The notch filter removes the message signal components from around DC. The in-phase signal consists of this notched-out message and a DC term. This DC term when upconverted will result in a carrier

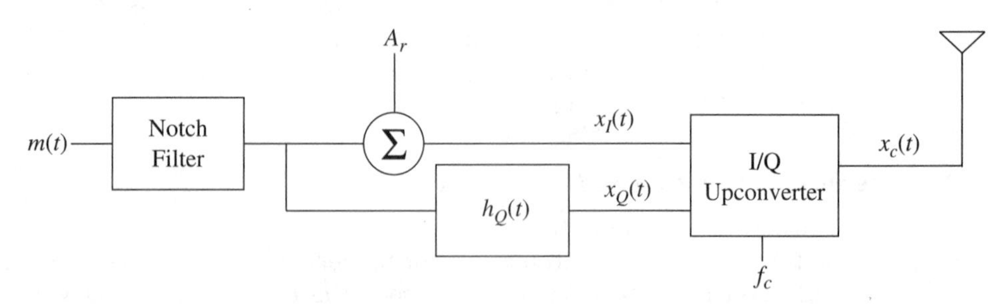

Figure 6.27 A modulator implementation for SSB-AM using a transmitted reference signal.

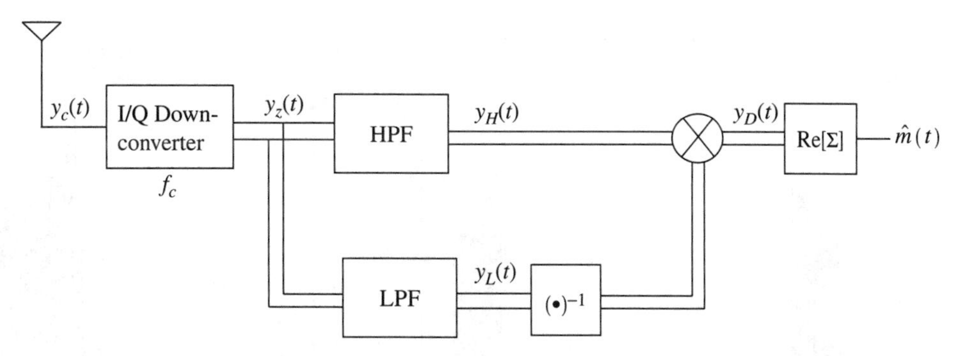

Figure 6.28 A transmitted reference based demodulator implementation for SSB-AM.

signal much like LC-AM. The quadrature signal is a Hilbert transform of the notched-out message signal.

The demodulator uses the transmitted reference as a carrier phase reference to recover the message signal. The demodulator block diagram is shown in Figure 6.28. In essence, at the receiver two filters are used to separate the transmitted modulated signal and the transmitted reference signal into two separate path, i.e.,

$$y_H(t) = A_c[m(t) + j(m(t) * h_Q(t))]\exp[j\phi_p] \tag{6.27}$$

$$y_L(t) = A_c A_r \exp[j\phi_p] \tag{6.28}$$

Since each of these paths experience the same channel distortion and phase shift, the reference can be used to derotate the demodulated signal and recover the message signal. This is easily seen with

$$y_D(t) = \frac{y_H(t)}{y_L(t)} = \frac{1}{A_r}[m(t) + j(m(t) * h_Q(t))] \tag{6.29}$$

This transmitted reference demodulation scheme is very useful especially in systems where automatic operation is desired or if the channel is varying rapidly. Consequently, transmitted reference systems are often used in land mobile radio where multipath and mobility can often cause significant channel variations [Jak74, Lee82]. An important practical design consideration is how large the reference signal power should be in relation to the modulated signal power. This consideration is not important unless noise is considered in the demodulation so this discussion will also be left until Chapter 11.

EXAMPLE 6.11

In this example we consider again the computer-generated voice signal given in Chapter 2 ($W = 2.5$ kHz). This signal is SSB-AM modulated with a transmitted reference at a 7 kHz carrier in a fashion as shown in Figure 6.27. The notch bandwidth

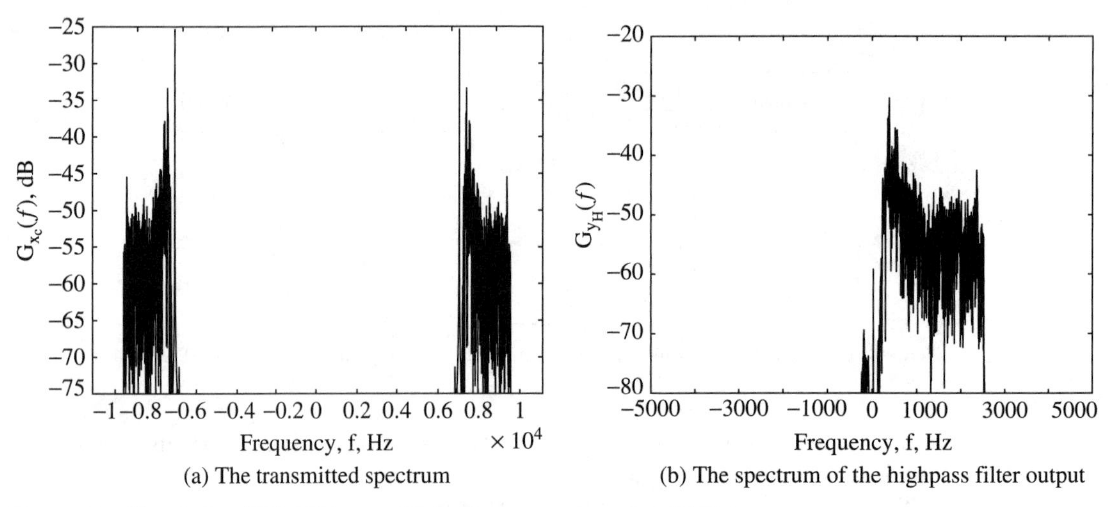

(a) The transmitted spectrum

(b) The spectrum of the highpass filter output

Figure 6.29 Signals in a SSB-AM transmitted reference system. $f_c = 7$ kHz

is chosen to be about 200 Hz as this does not affect the audio quality of the signal. The measured transmitted signal spectrum is shown in Figure 6.29(a). The transmitted reference signal is the tone at the carrier frequency. The signal out of the notch filter in the demodulator of Figure 6.28 is approximately a standard baseband SSB-AM modulation. The measured power spectrum of $y_H(t)$ is shown in Figure 6.29(b) for a notch bandwidth of 100 Hz. The transmitted reference tone has been greatly attenuated but the modulated signal is roughly unaltered. The signal out of the low pass filter in the demodulator of Figure 6.28 is approximately the phase shifted reference tone. The measured power spectrum of $y_L(t)$ is shown in Figure 6.30(a) for a notch bandwidth of

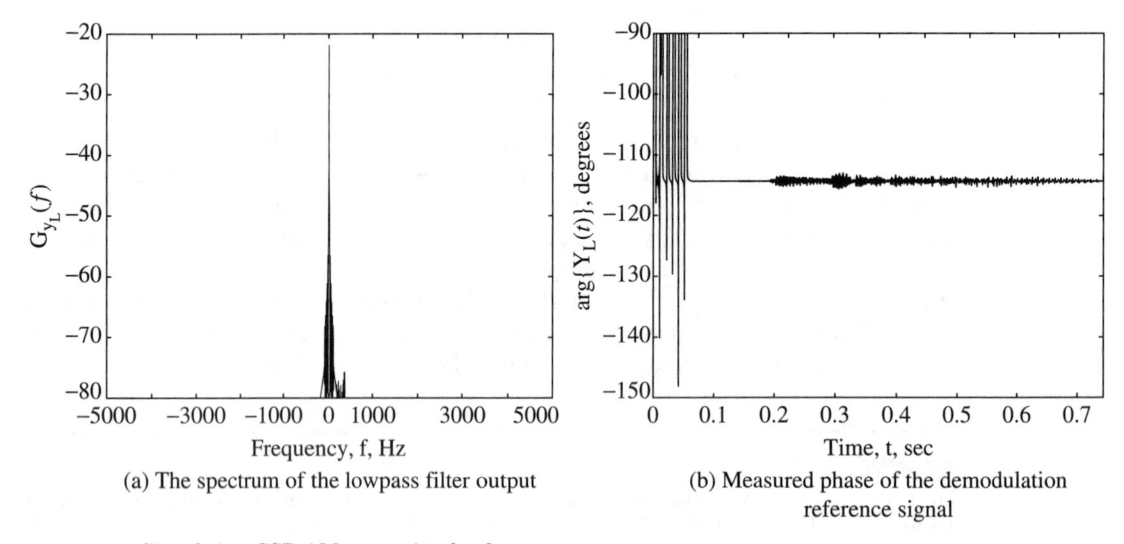

(a) The spectrum of the lowpass filter output

(b) Measured phase of the demodulation reference signal

Figure 6.30 Signals in a SSB-AM transmitted reference system.

100 Hz. The modulated signal has been greatly attenuated in $y_L(t)$ while the tone used in demodulation is clearly evident. For example, in a channel with $\tau_p = 45.3\ \mu$s a phase shift of -114° occurs. The measured phase of the output tone is shown in Figure 6.30(b) and it is close to -114° once the filter transients have passed. The imperfections in the filter can be seen to allow the message signal to cause some phase jitter to be induced on the demodulator reference but the overall system produces a high quality audio output.

Noncoherent demodulation (envelope detection) is also used in conjunction with a large carrier version of VSB-AM in analog television reception in the United States. The details of how this is possible with a quadrature modulation will be addressed in the homework.

6.3.5 Quadrature Modulation Conclusions

Quadrature modulation provides better spectral efficiency than either DSB-AM or LC-AM but with a more complex modulator. Quadrature modulation accomplishes this improved spectral efficiency by using the imaginary portion of the complex envelope. Like LC-AM, VSB-AM wastes part of the transmitted power (the imaginary component is not used in demodulation) but for a useful purpose. LC-AM contains a carrier signal that transmits no information while VSB-AM transmits a filtered signal in the quadrature component of the modulation but never uses this signal in demodulation. VSB-AM is also suitable for use with a large carrier and envelope detection (this is what is used in broadcast TV) and this idea will be explored in the homework.

In conclusion, three types of amplitude modulated signals have been presented in this chapter. These three signaling schemes provide different options in cost/bandwidth efficiency tradeoffs.

6.4 Homework Problems

Problem 6.1. The message signal in a DSB-AM system is of the form

$$m(t) = 12\cos(6\pi t) + 3\cos(10\pi t)$$

(a) Calculate the message power, P_m.

(b) If this message is DSB-AM modulated on a carrier with amplitude A_c, calculate the Fourier series of $x_z(t)$.

(c) Assuming $f_c = 20$ Hz calculate the Fourier series of $x_c(t)$ and plot the resulting time waveform when $A_c = 1$.

(d) Compute the output power of the modulated signal, P_{x_c}.

(e) Calculate and plot $x_P(t)$. Computer might be useful.

Problem 6.2. A message signal of the form

$$m(t) = \cos(2\pi f_m t)$$

is to be transmitted by DSB-AM with a carrier frequency f_c.

(a) Give the baseband and bandpass forms of the modulated signal for this message signal that has a power of 8 W (in a 1 Ω system).

(b) Assume $f_c = 10 f_m$ give the baseband and bandpass spectral representation of this modulation. What is the transmission bandwidth, B_T of such a system?

(c) Plot the resulting bandpass signal for $f_m = 2$ Hz.

(d) Give the simplest form of the modulator and demodulator (assume you know ϕ_p).

Problem 6.3. The received bandpass DSB-AM signal has the form

$$y_c(t) = A_c m(t)\sqrt{2}\cos(2\pi f_c t + \phi_p)$$

and the Costas loop used in phase synchronous demodulation of DSB-AM is shown in Figure 6.31

(a) What is $y_z(t)$?

(b) Find expressions for the signals at points A and B. Assume the LPF are only to remove the double frequency terms.

(c) Define a complex signal $z_1(t) = A - jB$. Show that $z_1(t) = y_z(t)\exp[-j\hat{\theta}(t)]$.

(d) Note that $C = -0.5 \times \Im[z_1(t)^2]$. Find C in terms of the phase error, $\phi_p - \hat{\theta}(t)$.

Problem 6.4. As your first task, your new boss at Fony asks you to design a LC-AM modulator (obviously busy work until she can find something real for you to do

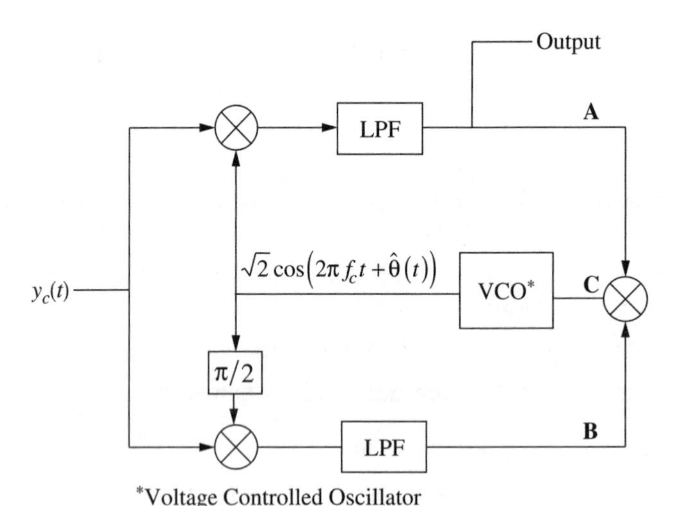

Figure 6.31 A bandpass version of the Costas loop.

since AM is not used much in practice) for a 1 kHz, 3 V peak amplitude test tone, e.g.,

$$m(t) = 3\sin(2000\pi t) \tag{6.30}$$

(a) She wants the transmitter to deliver a 5 W output power across a 1 Ω resistor with MCPR = 8%. Provide the values of A_c and a to achieve this specification.

(b) What is the maximum MCPR that can be achieved in this problem and have distortionless envelope detection still possible? Give the values of A_c and a to achieve this efficiency while maintaining a 5 W output power.

(c) Plot $x_I(t)$ and $x_Q(t)$ for the design in part (b).

Problem 6.5. A message signal of the form

$$m(t) = 2\cos(2\pi f_m t)$$

is to be transmitted by LC-AM with a carrier frequency f_c.

(a) Give the baseband and bandpass forms of the modulated signal for this message signal that has a power of $P_{x_z} = 8$ W (in a 1 Ω system) and MCPR = 1/8.

(b) Assume $f_c = 10 f_m$ give the baseband and bandpass spectral representation of this modulation. What is the transmission bandwidth, B_T of such a system?

(c) Give the simplest form of the modulator and a noncoherent demodulator

(d) How big can a be made in this system and still ensure distortionless envelope detection?

Problem 6.6. A term often used to characterize LC-AM waveforms is the percent modulation defined as

$$P_{LC} = \frac{\text{Maximum value of the envelope} - \text{Minimum value of the envelope}}{\text{Maximum value of the envelope}} \tag{6.31}$$

An example of this usage is seen on most microwave signal generators where the LC-AM signal is characterized by percent modulation. For this problem consider a sinusoidal message signal of unit amplitude and LC-AM. An example LC-AM waveform is shown for $P_{LC} = 0.5$ in Figure 6.32.

(a) What is the range for P_{LC}?

(b) Solve for the value of a to give a particular P_{LC}.

(c) The bandpass frequency domain representation (Fourier series or Fourier transform) of this signal will have three frequencies (the carrier and two modulation sidebands). Plot the relative power ratio (in dB) between the carrier and one of the modulation sidebands as a function of P_{LC}.

(d) Plot the MCPR as a function of P_{LC}.

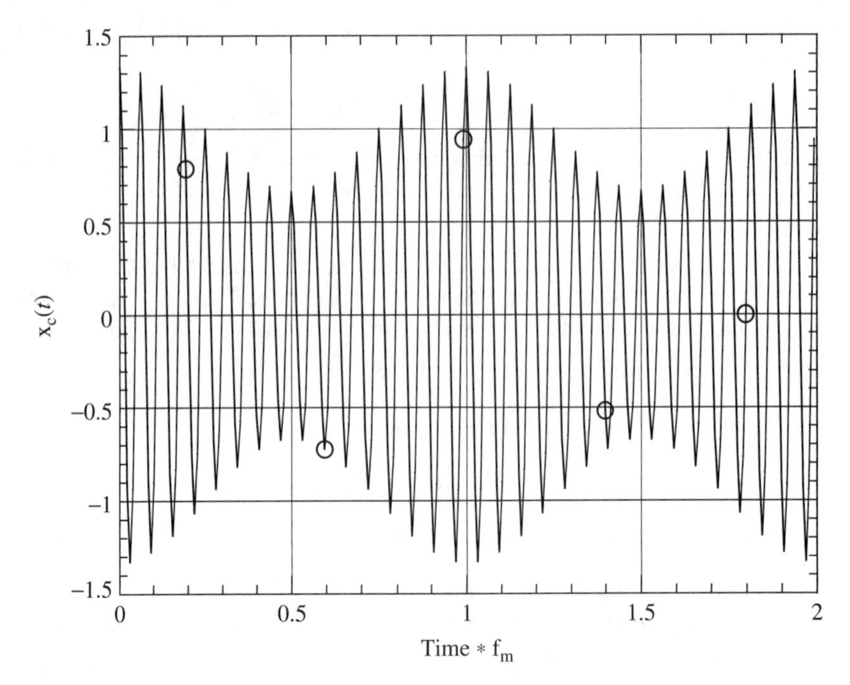

Figure 6.32 A LC-AM signal with $P_{LC} = 0.5$. Sinusoidal signal.

Problem 6.7. The Federal Communications Commission (FCC) makes each station strictly limit their output frequency content. Assume AM stations produce a usable audio bandwidth of 10 kHz. AM stations in the same geographical area are normally spaced at least 30 kHz apart. As an example of why limits on the frequency content are necessary the following problem is posed.

Consider two stations broadcasting in the same geographic area with a receiver tuned to one of them as shown in Figure 6.33. Station A is broadcasting a 1 V peak sinewave of 2 kHz, $m_A(t)$ (a test of the emergency broadcast

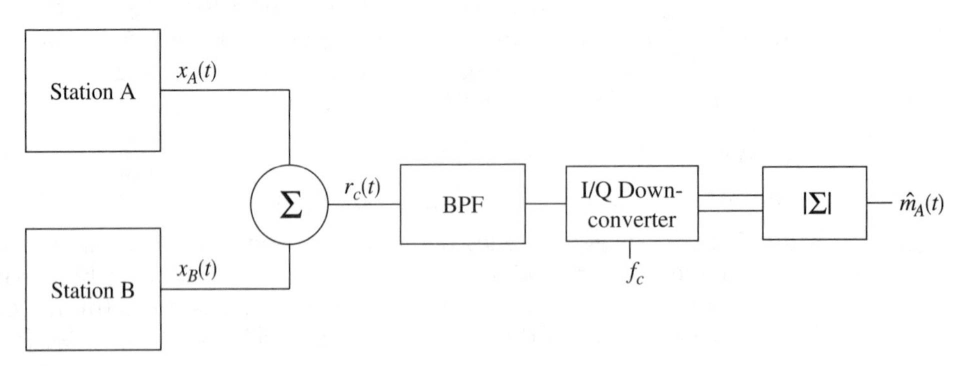

Figure 6.33 A model for two transmitting AM radio stations for Problem 6.7.

system), at a center frequency of f_c Hz. Station B is broadcasting a 1 V peak 3 kHz square wave, $m_B(t)$, at $f_c + 30$ kHz with no filtering at the transmitter. Since each transmitted waveform will have a different propagation loss, the received waveform is given by the form

$$r_c(t) = A_A[(1.5 + m_A(t))\sqrt{2}\cos(2\pi f_c t + \phi_A)]$$
$$+ A_B[(1.5 + m_B(t))\sqrt{2}\cos(2\pi(f_c + 30000)t + \phi_B)]$$

Assume without loss of generality that the phase shift for the station A is zero, i.e., $\phi_A = 0$.

(a) Give the complex envelope of the received signal, $r_z(t)$.

(b) Find the spectrum (Fourier series coefficients) of the complex envelope, $r_z(t)$.

(c) The demodulator has an ideal bandpass filter with a center frequency of f_c and a two-sided bandwidth of $2W = 15$ kHz followed by an envelope detector as shown in Figure 6.33. Plot the output demodulated waveform, $\hat{m}_A(t)$, over 5 ms of time for several values of ϕ_b in the range $[0, 2\pi]$ and $A_B = 0.1, 1, 10$.

The distortion you see in this example is called adjacent channel interference and one of the FCC's functions is to regulate the amount of interference each station produces for people trying to receive another station.

Problem 6.8. Consider the message signal in Figure 6.34 and a bandpass signal of the form

$$x_c(t) = (5 + bm(t))\sqrt{2}\cos(2\pi f_c t)$$

(a) What sort of modulation is this?

(b) Plot the bandpass signal, $x_c(t)$, with $b = 2$ when $f_c \gg \frac{1}{T}$.

(c) How big can b be and still permit envelope detection for distortion-free message signal recovery.

(d) Compute the transmitted power for $b = 3$.

(e) Compute MCPR for $b = 3$.

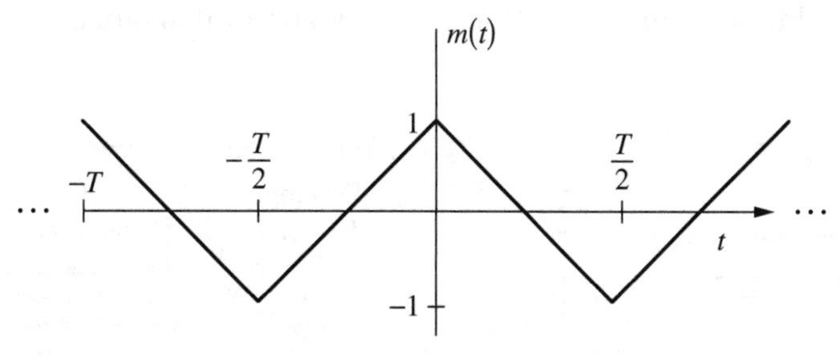

Figure 6.34 A message signal for Problem 6.8.

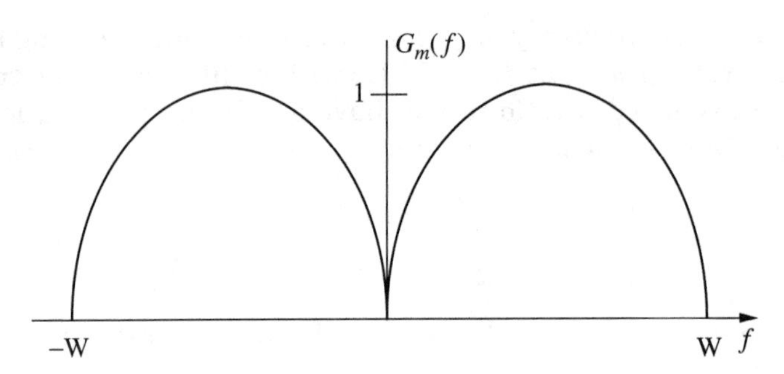

Figure 6.35 A message signal energy spectrum.

Problem 6.9. The complex baseband model for analog communications is given in Figure 5.8. Ignore noise and assume $m(t) = \sin(2\pi f_m t)$. If $y_z(t) = (5 + 3\sin(2\pi f_m t))\exp[j\pi/3]$,

(a) What kind of modulation does this represent? Identify all the important parameters.

(b) Plot the envelope of the output, $y_A(t)$.

(c) Plot the phase of the output, $y_P(t)$.

(d) Show how to process $y_z(t)$ to recover $m(t)$.

Problem 6.10. A message signal with an energy spectrum given in Figure 6.35 is to be transmitted with large carrier amplitude modulation. Additionally, $\min(m(t)) = -2$ and $P_m = 1$.

(a) If the MCPR = 25%, what is the modulation index, a.

(b) Will the modulation index obtained in part (a) be sufficiently small to allow demodulation by envelope detection. *Note if you cannot solve for part (a) just give the necessary conditions such that envelope detection is possible.*

(c) Plot the modulator output spectrum (either bandpass or baseband is fine) and compute the E_B.

Problem 6.11. Commercial TV uses large carrier AM transmission in conjunction with VSB-AM for the intensity signal (black and white levels). A typical TV receiver block diagram for the intensity signal demodulation is shown in Figure 6.36.

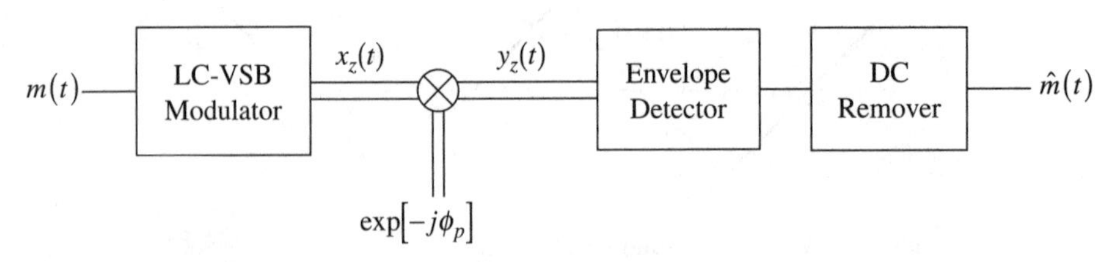

Figure 6.36 Typical TV demodulator.

This problem tries to lead you to an explanation of why the filtering does not significantly affect the envelope detector's performance. Consider the simple case of a sinusoidal message signal

$$m(t) = \cos(2\pi f_m t)$$

with $f_m > 0$ and a VSB modulator that produces a complex envelope signal of the form

$$x_z(t) = \begin{cases} A_c(1 + a\cos(2\pi f_m t) + ja\sin(2\pi f_m t)) & f_m > f_v \\ A_c(1 + a\cos(2\pi f_m t)) & f_m \le f_v \end{cases} \quad (6.32)$$

(a) Sketch the modulator that would produce Eq. (6.32) and derive the transfer function of the quadrature filter, $H_Q(f)$, that is necessary to produce this signal.

(b) Calculate the output envelope, $|y_z(t)|$.

(c) Show that the envelope detector output is the desired signal (the message signal plus a DC offset).

(d) Consider the case where modulator is exactly the same as above and the message signal is the sum of two sinusoids

$$m(t) = A_1\cos(2\pi f_1 t) + A_2\cos(2\pi f_2 t) \quad (6.33)$$

what would the form of $x_z(t)$ be?

(e) Show if a is chosen such that a^2 is small that the envelope detector output is approximately the desired signal (the message signal plus a DC offset).

(f) Compute the MCPR for $f_m < f_v$ and $f_m > f_v$ assuming a^2 is small.

(g) Choose a value of a for which a^2 is small compared to a and the envelope detector can be used with a small resulting distortion. What does this say about the MCPR of typical TV broadcast?

Problem 6.12. RF engineers that design and build quadrature upconverters (see Figure 6.37) need tests to estimate how close to ideal their circuits are performing. The standard test used is known as a single sideband rejection test. This test uses an input of $x_I(t) = \cos(2\pi f_m t)$ and $x_Q(t) = \sin(2\pi f_m t)$ and measures the resulting bandpass power spectrum, $|X_c(f)|^2$ on a spectrum analyzer.

(a) Compute what the output bandpass spectrum, $|X_c(f)|^2$, should be for an ideal quadrature upconverter.

(b) A common design issue in quadrature modulators is that the quadrature carrier has a phase offset compared to the in-phase carrier, i.e.,

$$x_c(t) = x_I(t)\sqrt{2}\cos(2\pi f_c t) - x_Q(t)\sqrt{2}\sin(2\pi f_c t + \theta)$$

For the test signal in a single sideband rejection test what will be the output bandpass spectrum as a function of θ.

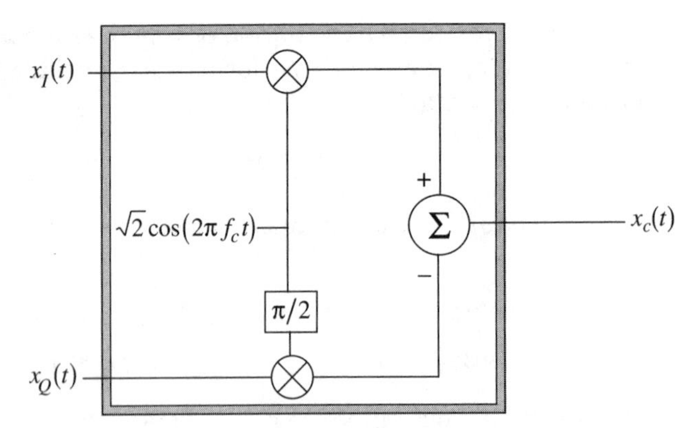

Figure 6.37 A quadrature upconverter.

(c) If $|X_c(f_c + f_m)|^2 = 100|X_c(f_c - f_m)|^2$ what is the value of θ that would produce this spectrum.

(d) Postulate why this test is known as a single sideband rejection test.

Problem 6.13. A message with a Fourier transform of

$$M(f) = \begin{cases} \frac{1}{\sqrt{W}}\left|\sin\left(\frac{\pi f}{W}\right)\right| & -W \leq f \leq W \\ 0 & \text{elsewhere} \end{cases} \tag{6.34}$$

is to be transmitted with a quadrature modulation.

(a) Calculate the message energy, E_m.

(b) For SSB-AM (upper sideband) compute the output transmitted energy, E_{x_z}.

(c) Design a simple quadrature filter for VSB-AM where $f_v = W/4$. Give either $H_Q(f)$ or $H_z(f)$.

(d) Compute the resulting output energy for your design, E_{x_z}, in part c). *Note, depending on your design the solution might be easiest done with the aid of a computer*

Problem 6.14. (UM). An amplitude modulated (AM) signal has the following form

$$x_c(t) = [A + 1.5\cos(10\pi t) - 3.0\cos(20\pi t)]\cos(2\pi f_c t)$$

where the message signal is

$$m(t) = \cos(10\pi t) - 2.0\cos(20\pi t)$$

(a) What type of amplitude modulation is this?

(b) What is the spectrum of $x_c(t)$, $X_c(f)$?

(c) Determine the conditions on A such that envelope detection could be used in the demodulation process with no distortion.

(d) Determine the MCPR of this modulation scheme as a function of A.

Problem 6.15. You need to test a VSB-AM system that your company has purchased. The modulation scheme is DSB-AM followed by a bandpass filter, $H_c(f)$. The demodulator is exactly the same as DSB-AM when the phase offset, ϕ_p, is known. The system does not work quite right and testing has led you to suspect the filter $H_c(f)$. After analysis and testing, you determine that the filter has the following bandpass characteristic

$$H_c(f) = \begin{cases} 1 & f_c + 7500 \leq |f| \leq f_c + 10000 \\ 2 & f_c + 2500 \leq |f| < f_c + 7500 \\ \frac{4}{3} & f_c \leq |f| < f_c + 2500 \\ \frac{3}{4} & f_c - 2500 \leq |f| < f_c \\ 0 & \text{elsewhere} \end{cases}$$

You will be using the system to transmit voice signals with a bandwidth of 5000 Hz.

(a) Compute $H_z(f) = H_I(f) + jH_Q(f)$.

(b) What are the conditions on $H_I(f)$ and $H_Q(f)$ for $f \in [-W, W]$ that will produce a distortionless demodulator.

(c) Does the VSB system produce a distortionless output? That is, does $\hat{m}(t) = Km(t)$, where K is some constant?

(d) The filter above consists of five frequency segments with constant gain. Because of cost restrictions, you can change the gain on only **one** of the segments. Change one segment gain such that the system **will** produce a distortionless output.

(e) What is the bandwidth efficiency of your resulting VSB system and the resulting savings in transmission bandwidth over DSB-AM?

(f) What is the spectrum of the corresponding complex envelope equivalent, $H_z(f)$, of the improved filter?

Problem 6.16. This problem is concerned with double-sided band amplitude modulation (DSB-AM) of the message signal given in Problem 5.1. Assume a carrier frequency of $f_c = 200$ Hz and a carrier amplitude of A_c.

(a) Give the baseband, $x_z(t)$, and bandpass, $x_c(t)$, time waveforms for DSB-AM for this message signal as a function of $m(t)$.

(b) The bandpass signal is also periodic. What is the period? Give the Fourier series representation for the bandpass signal

(c) Give the value of A_c that will produce a $P_{x_z} = 50$ Watt transmitter output power in a 1 Ω system.

(d) Sketch the demodulation process for a received signal $y_c(t)$.

Problem 6.17. Repeat Problem 6.16 with the bandlimited message signal of Problem 5.2.

Problem 6.18. Repeat Problem 6.16 with the message signal of Problem 5.5.

Problem 6.19. This problem is concerned with large carrier amplitude modulation (LC-AM) by the message signal given in Problem 5.1. Assume a carrier frequency of $f_c = 200$ Hz, a carrier amplitude of A_c, and a modulation coefficient of a.

(a) Give the baseband, $x_z(t)$, and bandpass, $x_c(t)$, time waveforms for LC-AM for this message signal as a function of $m(t)$.

(b) Give the Fourier series representation of the bandpass modulated signal as a function of A_c and a.

(c) Give the value of a such that the MCPR is maximized and distortion-free envelope detection is still possible.

(d) What is the MCPR with the value of a computed in (c)?

(e) With the value of a computed in (c) give the value of A_c that will produce a $P_{x_z} = 50$ W transmitter output power in a 1-Ω system.

(f) Sketch the demodulation block diagram for a received signal $y_c(t)$.

Problem 6.20. Repeat Problem 6.19 with the message signal of Problem 5.2.

Problem 6.21. Repeat Problem 6.19 with the message signal of Problem 5.5.

Problem 6.22. This problem is concerned with single sideband amplitude modulation (SSB-AM) by the message signal given in Problem 5.1. Assume a carrier frequency of $f_c = 200$ Hz, and a carrier amplitude of A_c.

(a) Give the baseband, $x_z(t)$, and bandpass, $x_c(t)$, time waveforms for an upper sideband SSB-AM for this message signal.

(b) Give the Fourier series representation of the bandpass modulated signal as a function of A_c.

(c) Give the value of A_c that will produce a 50 W output power in a 1-Ω system.

(d) Sketch the demodulation block diagram for a received signal, $y_c(t)$, assuming that you know ϕ_p.

Problem 6.23. Repeat Problem 6.22 with the message signal of Problem 5.2.

Problem 6.24. Repeat Problem 6.22 with the message signal of Problem 5.5.

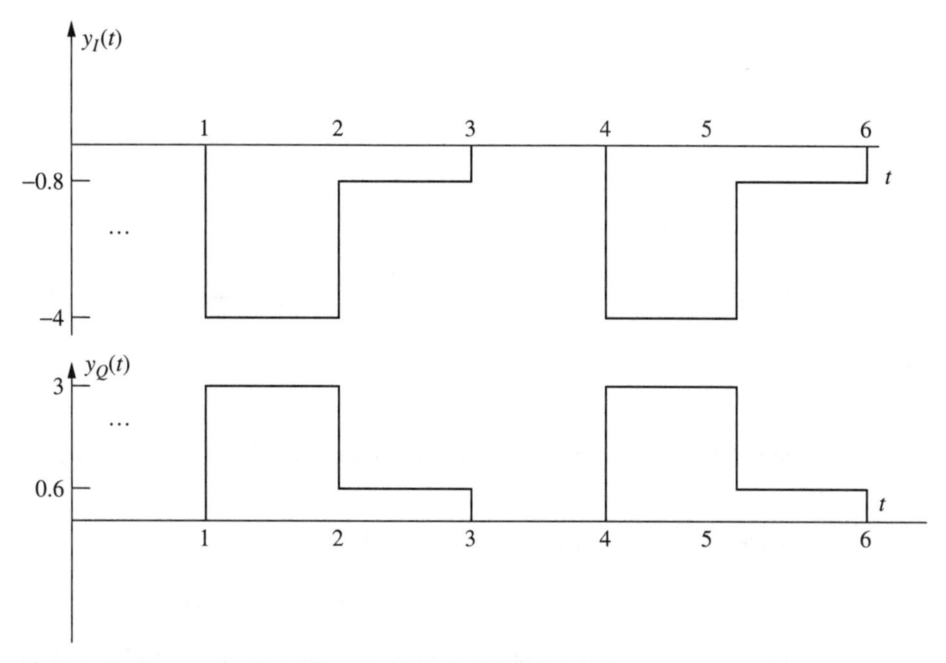

Figure 6.38 The received complex envelope of a LC-AM signal.

Problem 6.25. You have been given the noiseless received complex envelope, $y_z(t) = y_I(t) + jy_Q(t)$, of a periodic message signal transmitted via large carrier–amplitude modulation (LC-AM) as shown in Figure 6.38. Assume that $1 + a \min m(t) > 0$. As is standard practice the message signal has a time average value of zero so that it will pass through a DC block unchanged.

(a) Give the form of the received complex envelope for LC-AM, $y_z(t)$ as a function of A_c, a, $m(t)$, and ϕ_p.

(b) Find the best estimate of the message signal, $\hat{m}(t) = Am(t)$, from the data in Figure 6.38.

(c) Find the transmitted power, P_{x_z} using the data in Figure 6.38.

(d) Find the MCPR using the data in Figure 6.38.

Problem 6.26. Consider the periodic message signal in Figure 6.39 and LC-AM modulation.

(a) Compute the DC value of $m(t)$.

(b) Compute the message signal power, P_m as a function of ρ.

(c) What is the maximum value of a as a function of ρ that can be used and still maintain distortionless envelope detection.

(d) Compute and plot the MCPR as a function of ρ.

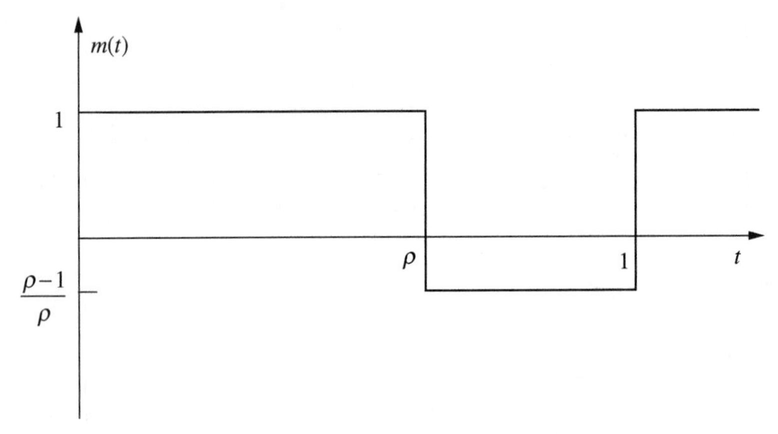

Figure 6.39 A periodic message signal for Problem 6.26.

Problem 6.27. A message signal is of the form

$$m(t) = 2\cos(4\pi t) + \cos(6\pi t)$$

If $x_I(t) = A_c m(t)$ and the modulator output complex envelope spectrum is given by Figure 6.40.

(a) What is W and what is B_T?

(b) What type of modulation is being implemented?

(c) What is the value of A_c?

(d) What should $x_Q(t)$ be to produce the output spectrum?

Problem 6.28. In a single sideband amplitude demodulator the following signal is output from the down converter

$$y_z(t) = \exp[j6\pi t] + j\exp[j8\pi t] \qquad (6.35)$$

The channel phase is known to be $\phi_p = \pi/2$.

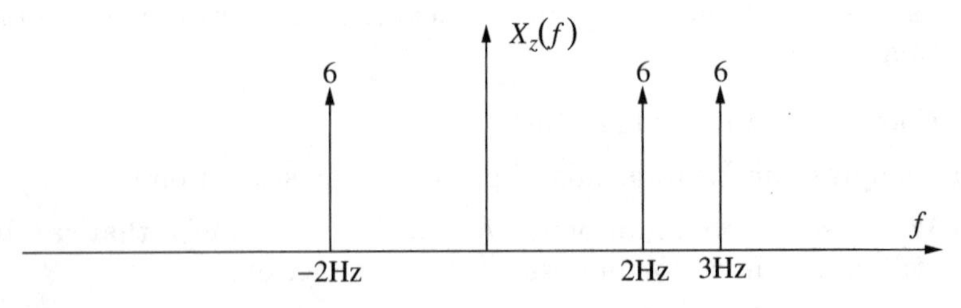

Figure 6.40 The baseband modulator output spectrum.

(a) Is the modulation using the upper sideband or the lower sideband?

(b) Find $\hat{m}(t)$.

Problem 6.29. At the output of a double sideband amplitude modulator (DSB-AM) the signal

$$x_c(t) = 8\cos(200\pi t)\sqrt{2}\cos(2\pi f_c t) \tag{6.36}$$

is observed. You know that the input message signal had a power of 2 W.

(a) Find $x_z(t)$.

(b) Find $x_A(t)$ and $x_P(t)$ over $t \in [0, 0.01]$.

(c) What is the input message signal and what is A_c?

(d) Plot the Fourier transform of the complex envelope, $X_z(f)$ using impulse functions. What the bandpass bandwidth, B_T, that this signal occupies?

Problem 6.30. Consider the DSB-AM signal from the previous problem given as

$$x_c(t) = 8\cos(200\pi t)\sqrt{2}\cos(2\pi f_c t) \tag{6.37}$$

This signal is put into a bandpass filter which has a complex envelope characterized with

$$H_Q(f) = 0 \qquad H_I(f) = \begin{cases} 1 - \dfrac{0.5f}{100} & |f| \leq 200 \\ 0 & \text{elsewhere} \end{cases} \tag{6.38}$$

The output of the filter at bandpass is denoted $y_c(t)$ and at baseband is denoted $y_z(t)$.

(a) Find and plot $H_c(f)$.

(b) Find $y_z(t)$.

(c) Find $y_A(t)$ and $y_P(t)$ over $t \in [0, 0.01]$. Why is the phase varying as a function of time for the DSB-AM signal?

Problem 6.31. A message signal with an energy spectrum given in Figure 6.35 is to be transmitted with single sideband amplitude modulation (SSB-AM). Additionally, $\min(m(t)) = -2$ and $P_m = 10$.

(a) Specify the quadrature filter transfer function to achieve a lower sideband transmission.

(b) Find the transmitted power, P_{x_z}.

(c) Plot the modulator output spectrum (either bandpass or baseband is fine) and compute the E_B.

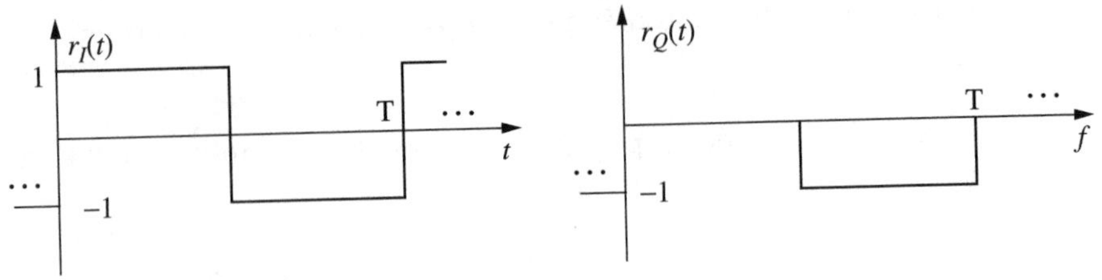

Figure 6.41 A received complex envelope.

Problem 6.32. Recall that for a quadrature modulation that the overall modulator transfer function is

$$H_z(f) = 1 + j H_Q(f) \tag{6.39}$$

Specify conditions on $H_Q(f_o)$ such that the sideband attenuation at frequency f_o is greater than 50 dB.

Problem 6.33. A periodic received signal, $r_z(t)$, from an analog modulation is given in Figure 6.41.

(a) Assume the transmitted signal is given as $x_z(t)$, how have we modeled the form for $r_z(t)$ in terms of $x_z(t)$ for this class?

(b) Given $r_z(t)$ in Figure 6.41, is it possible for the transmitted signal, $x_z(t)$, to be a LC-AM signal that is able to be detected without distortion by an envelope detector? Why?

Problem 6.34. **(JG)** Let $m(t)$ be a message signal and define the bandpass signals $x_c(t)$ and $y_c(t)$ as

$$m(t) = (\mathrm{sinc}(t))^2 \tag{6.40}$$

$$x_c(t) = m(t)\sqrt{2}\cos(2\pi f_c t) - \hat{m}(t)\sqrt{2}\sin(2\pi f_c t) \tag{6.41}$$

$$y_c(t) = m(t)\sqrt{2}\cos(2\pi f_c t) + \hat{m}(t)\sqrt{2}\sin(2\pi f_c t) \tag{6.42}$$

where $\hat{m}(t)$ is the Hilbert transform of $m(t)$.

(a) Find the complex envelopes, $x_z(t)$ and $y_z(t)$, for the bandpass signals $x_c(t)$ and $y_c(t)$.

(b) Determine and sketch the Fourier transform of $x_c(t)$. What is the bandwidth of $X_c(f)$?

(c) Determine and sketch the Fourier transform of $y_c(t)$.

Problem 6.35. **(PD)** Recall the received signal for a SSB-AM system (upper sideband) is given as

$$y_z(t) = A_c(m(t) + j m_h(t)) \exp[j\phi_p] \tag{6.43}$$

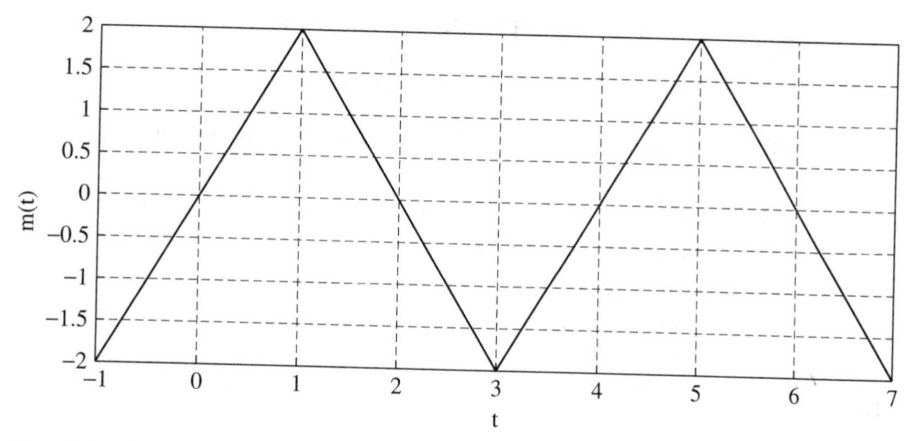

Figure 6.42 A message signal.

where $m_h(t)$ is the Hilbert transform of $m(t)$. An alternative coherent demodulation structure has been proposed as

$$\hat{m}(t) = y_I(t) + \alpha(\phi_p) y_Q(t) \tag{6.44}$$

where $\alpha(\phi_p)$ is a real constant that is a function of ϕ_p.

(a) Give the form of $\hat{m}(t)$ as a function of $m(t)$, $m_h(t)$, ϕ_p, and $\alpha(\phi_p)$.

(b) Find the value of $\alpha(\phi_p)$ where $\hat{m}(t) = A(\phi_p)m(t)$, where $A(\phi_p)$ is not a function of time.

(c) What is the value of $A(\phi_p)$ for the value of $\alpha(\phi_p)$ derived in (b).

(d) What are your perceived advantages and disadvantages of this demodulator versus that presented in the text.

Problem 6.36. **(JG)** Consider the message signal, $m(t)$, shown in Figure 6.42.

(a) If an arbitrary $m(t)$ is modulated using DSB-LC AM, what is the general expression for the complex envelope?

(b) If the message signal shown above is modulated using DSB-LC AM, what is the maximum possible power efficiency?

(c) If the message signal shown above is modulated using DSB-LC AM, sketch the bandpass form of the DSB-LC AM signal which achieves the maximum power efficiency. Clearly label your figure.

(d) If the message signal shown above is modulated using DSB-SC and demodulated using a standard envelope detector, sketch the resulting output signal. Clearly label your figure.

Problem 6.37. Given the LC-AM waveform of the form

$$x_z(t) = A_c(1 + a\cos(2\pi f_m t)) \tag{6.45}$$

(a) For $A_c = 1$ and $a = 0.1$ plot the instantaneous power of $x_z(t)$ in dB scale as a function of time.

(b) Find the value of a such that the difference between the maximum and the minimum power is 1 dB.

(c) Find the value of A_c and a such that $P_{x_z} = -10$ dBm and the difference between the maximum and the minimum power is 1 dB.

6.5 Example Solutions

Problem 6.17. For DSB-AM we have

$$x_z(t) = A_c m(t) \tag{6.46}$$

(a) Since

$$m(t) = \frac{8}{\pi^2}\cos(2\pi(2)t) + \frac{8}{9\pi^2}\cos(2\pi(6)t) \tag{6.47}$$

we have

$$x_z(t) = A_c\left(\frac{8}{\pi^2}\cos(2\pi(2)t) + \frac{8}{9\pi^2}\cos(2\pi(6)t)\right) \tag{6.48}$$

The bandpass signal is

$$x_c(t) = x_I(t)\sqrt{2}\cos(2\pi f_c t) = A_c m(t)\sqrt{2}\cos(2\pi f_c t) \tag{6.49}$$

(b) The baseband signal has a Fourier series representation of

$$x_I(t) = \frac{A_c 4}{\pi^2}\exp[j2\pi(2)t] + \frac{A_c 4}{\pi^2}\exp[-j2\pi(2)t] + \frac{A_c 4}{9\pi^2}\exp[j2\pi(6)t]$$
$$+\frac{A_c 4}{9\pi^2}\exp[-j2\pi(6)t] \tag{6.50}$$

and has a period $T = 0.5$. The modulated signal Fourier series is

$$x_c(t) = \frac{A_c 4}{\sqrt{2}\pi^2}\exp[j2\pi(202)t] + \frac{A_c 4}{\sqrt{2}\pi^2}\exp[j2\pi(198)t]$$
$$+\frac{A_c 4}{\sqrt{2}9\pi^2}\exp[j2\pi(206)t] + \frac{A_c 4}{\sqrt{2}9\pi^2}\exp[j2\pi(194)t]$$
$$+\frac{A_c 4}{\sqrt{2}\pi^2}\exp[-j2\pi(202)t] + \frac{A_c 4}{\sqrt{2}\pi^2}\exp[-j2\pi(198)t]$$
$$+\frac{A_c 4}{\sqrt{2}9\pi^2}\exp[-j2\pi(206)t] + \frac{A_c 4}{\sqrt{2}9\pi^2}\exp[-j2\pi(194)t] \tag{6.51}$$

The frequencies contained in the bandpass signal are $f = 194, 198, 202, 206$ so the bandpass signal will be periodic with period $T = 1/2$. Since

$$x_c(t) = \sum_{k=-\infty}^{\infty} x_k \exp[j2\pi 2kt] \tag{6.52}$$

the Fourier series coefficients are

$$x_{97} = \frac{8A_c}{9\pi^2\sqrt{2}} \qquad x_{-97} = \frac{8A_c}{9\pi^2\sqrt{2}} \qquad x_{99} = \frac{8A_c}{\pi^2\sqrt{2}} \qquad x_{-99} = \frac{8A_c}{\pi^2\sqrt{2}} \tag{6.53}$$

$$x_{101} = \frac{8A_c}{\pi^2\sqrt{2}} \qquad x_{-101} = \frac{8A_c}{\pi^2\sqrt{2}} \qquad x_{103} = \frac{8A_c}{9\pi^2\sqrt{2}} \qquad x_{-103} = \frac{8A_c}{9\pi^2\sqrt{2}} \tag{6.54}$$

(c) The power of DSB-AM is $P_{x_z} = A_c^2 P_m = A_c^2 0.3325$. If a transmitted power of 50 W is to be achieved then $A_c^2 = 50/.3325$ or $A_c = 12.26$.

(d) See Figure 6.6.

Problem 6.23. For SSB-AM we have

$$x_z(t) = x_I(t) + jx_Q(t) = A_c[m(t) + jh_Q(t) * m(t)] \tag{6.55}$$

where $h_Q(t)$ is the Hilbert transformer where $H_Q(f) = -j\,\mathrm{sgn}(f)$.

(a) Since

$$m(t) = \frac{8}{\pi^2}\cos(2\pi(2)t) + \frac{8}{9\pi^2}\cos(2\pi(6)t) \tag{6.56}$$

we have

$$x_I(t) = A_c\left(\frac{8}{\pi^2}\cos(2\pi(2)t) + \frac{8}{9\pi^2}\cos(2\pi(6)t)\right) \tag{6.57}$$

The complex envelope of the transmitted bandpass signal is given as

$$\begin{aligned} x_z(t) &= x_I(t) + jx_Q(t) \\ &= A_c\left(\frac{8}{\pi^2}\cos(2\pi(2)t) + \frac{8}{9\pi^2}\cos(2\pi(6)t) + \frac{j8}{\pi^2}\sin(2\pi(2)t)\right. \\ &\quad \left. + \frac{j8}{9\pi^2}\sin(2\pi(6)t)\right) \\ &= A_c\left(\frac{8}{\pi^2}\exp(j2\pi(2)t) + \frac{8}{9\pi^2}\exp(j2\pi(6)t)\right) \end{aligned} \tag{6.58}$$

The frequency domain representation of the complex envelope is

$$X_z(f) = \frac{A_c 8}{\pi^2}\delta(f-2) + \frac{A_c 8}{9\pi^2}\delta(f-6) \tag{6.59}$$

The complex envelope is periodic with period $T = 1/2$.

(b) $x_c(t) = x_I(t)\sqrt{2}\cos(2\pi f_c t) - x_Q(t)\sqrt{2}\sin(2\pi f_c t)$ and

$$X_c(f) = \frac{1}{\sqrt{2}}X_z(f - f_c) + \frac{1}{\sqrt{2}}X_z^*(-f - f_c) \tag{6.60}$$
$$= \frac{8A_c}{\pi^2\sqrt{2}}\delta(f - 202) + \frac{8A_c}{9\pi^2\sqrt{2}}\delta(f - 206) + \frac{8A_c}{\pi^2\sqrt{2}}\delta(f + 202)$$
$$+ \frac{8A_c}{9\pi^2\sqrt{2}}\delta(f + 206)$$

The frequencies contained in the bandpass signal are $f = 202, 206$ so the bandpass signal will be periodic with period $T = 1/2$. The period of the bandpass signal is the same as the baseband signal. Since

$$x_c(t) = \sum_{k=-\infty}^{\infty} x_k \exp[j2\pi 2kt] \tag{6.61}$$

the Fourier series co-efficients are

$$x_{101} = \frac{A_c 8}{\pi^2\sqrt{2}} \qquad x_{-101} = \frac{A_c 8}{\pi^2\sqrt{2}} \tag{6.62}$$
$$x_{103} = \frac{A_c 8}{9\pi^2\sqrt{2}} \qquad x_{-103} = \frac{A_c 8}{9\pi^2\sqrt{2}} \tag{6.63}$$

(c) Parseval's theorem gives $P_{x_z} = 2A_c^2 P_m$. If a transmitted power of 50 W is to be achieved then $A_c^2 = 25/0.3325$ or $A_c = 8.671$.

(d) See Figure 6.25.

6.6 Miniprojects

Goal: To give exposure

- to a small scope engineering design problem in communications.
- to the dynamics of working with a team.
- to the importance of engineering communication skills (in this case oral presentations).

Presentation: The forum will be similar to a design review at a company (only much shorter) The presentation will be of 5 minutes in length with an overview of the given problem and solution. The presentation will be followed by questions from the audience (your classmates and the professor). All team members should be prepared to give the presentation.

6.6.1 Project 1

Project Goals: Undergraduate classes often only look at idealized problems. For example, one thing that is ignored in a course like this one is that frequency sources at the transmitter and receiver cannot be made to have exactly the same value. This project challenges the student to think about the impact of this practical constraint that is not discussed in the text. This mini-project will investigate the effects of this frequency offset on one demodulation algorithm for DSB-AM.

For a signal received with a frequency offset the complex envelope of the received signal will be rotating due to this frequency offset, i.e.,

$$y_z(t) = x_z(t)\exp[j(\phi_p + 2\pi f_o t)] \tag{6.64}$$

where f_o is the exisiting frequency offset. Get the Matlab files `ampmodex1.m` and `ampmodex2.m` from the class web page. In these files a DSB-AM transmitter and receiver is implemented. If the carrier frequency offset is set to 10 Hz (`deltaf=10 in the Matlab code`) the output of the squaring device will be

$$V_z(t) = A_c^2 m^2(t)\exp[j40\pi t + j2\phi_p] + N_V(t) \tag{6.65}$$

The demodulator implemented in the m-file still works pretty well with this frequency offset. Explain why. Specifically, the filter in the open loop phase estimator, $H_z(f)$, has a specific characteristic that enables the frequency offset to be tracked in such a way as to not cause significant distortion. Note, at higher frequency offsets (e.g., 100 Hz) the performance suffers noticeable distortion. Extra credit will be given if you can figure out a method to eliminate this distortion. Understanding Problem 4.15 will help.

6.6.2 Project 2

Project Goals: The selection of the modulation index can allow some distortion and hence becomes a trade-off between MCPR and distortion. This project challenges the student to examine this trade-off with a typical audio signal.

LC-AM uses a DC offset in $x_I(t)$ to ensure that envelope detection is possible on the received signal. An envelope detector is a very simple demodulator. This DC offset results in wasted transmitted power. Typically the best MCPR that can be achieved with voice signals is around 10% to maintain $x_I(t) > 0$. Better MCPR can be achieved if $x_I(t)$ is allowed to go negative for a small amount of time. This of course will cause distortion in the demodulated signal. Get the Matlab files `ampmodex3.m` and `ampmodex4.m` from the class web page. In these files a LC-AM transmitter and receiver is implemented. Find the maximum value of a such that no distortion envelope detection is possible. Increase a beyond this value and see how large it can be made before the distortion becomes significant. Calculate the maximum MCPR for the no distortion case and calculate the MCPR when the amount of audio distortion is acceptable to your ear.

6.6.3 Project 3

Project Goals: This project examines the impact of phase and frequency error in the SSB-AM demodulator.

SSB-AM uses the Hilbert transform in $x_Q(t)$ to eliminate half of the spectral content of a DSB-AM signal. In class we showed that for demodulation a coherent phase reference signal is needed to recover the $x_I(t)$ after the phase shift in the channel. One might imagine that the an offset in the carrier phase reference signal might result in a distorted demodulated signal. For example, if the coherent phase reference is $\phi_p + \varphi_e$ then the demodulated output given in Figure 6.25 is

$$\hat{m}(t) = m(t)\cos(\varphi_e) + m(t) * h_Q(t)\sin(\varphi_e) \tag{6.66}$$

Get the Matlab file `ampmodex5.m` and `ampmodex6.m` from the class web page. In these files a SSB-AM transmitter and receiver is implemented (note, these systems are much more complicated than you will actually need so prune the code accordingly). Find the maximum value of φ_e such that the amount of perceived distortion in the demodulated audio output is small. Note, the answer might surprise you (it did me the first time). *Extra credit will be given if the team can explain the reason for this surprising behavior*. Repeat the experiment with a small frequency offset instead of a phase offset.

Chapter

7

Analog Angle Modulation

Analog angle modulation embeds the analog message signal in the phase of the carrier or equivalently in the time varying phase angle of the complex envelope (instead of the amplitude as was the case for AM modulation). The general form of angle modulation is

$$x_z(t) = A_c \exp[j\Gamma_a(m(t))]$$

It is important to note that analog angle modulated signals have a constant envelope (i.e., $x_A(t) = A_c$). This is a practical advantage since most high power amplifiers are not linear and amplitude variations produce distortion (e.g. signal compression).

7.1 Angle Modulation

Amplitude modulation was the first modulation type to be considered in analog communication systems. Amplitude modulation has the obvious advantage of being simple and relatively bandwidth efficient. The disadvantages of amplitude modulations are

- The message is embedded in the amplitude of the carrier signal. Consequently, linear amplifiers are very important to obtaining good performance in AM systems. Linear amplifiers are difficult to achieve in applications when either cost or small size are important.
- When the message signal goes through a quiet period in DSB-AM or SSB-AM systems, very small carrier signals are transmitted. This absence of signal tends to accentuate the noise.
- The bandpass bandwidth in AM systems is directly dependent on the message signal bandwidth. There is no opportunity to use a wider bandwidth to achieve better performance.

These enumerated shortcomings of AM can be addressed by using angle modulation. Angle modulations modulate the angle of the complex envelope with the message signal.

The introduction of angle modulation into commercial engineering systems was wrought with controversy. It should be noted that angle modulation was proposed during the 1910s and in a paper that set communication theory back 10 years angle modulation was dismissed as being inferior to amplitude modulation by J. R. Carson [Car22], a proponent and credited inventor of SSB-AM working at Bell Laboratories. Roughly a decade later Edwin H. Armstrong became a proponent of frequency modulation (a type of angle modulation) in spite of the theoretical analysis of Carson. Armstrong was able to build and demonstrate a working angle modulation system that had, for the time period, a remarkable fidelity [Arm35]. While battles raged between commercial interests in radio engineering that were proponents of amplitude and angle modulation [Les69] the engineering community quickly realized the trade-offs, angle modulations offered that were not available with amplitude modulation [CF37]. Over time, angle modulation has become the de facto standard for high quality analog audio broadcast worldwide. This chapter will investigate the theoretical underpinnings of angle modulation.

The first issue to address is what is meant by the phase and frequency of a bandpass signal and how it relates to the complex envelope.

Definition 7.1 When a signal has the form

$$x(t) = \sqrt{2}\cos(\theta(t))$$

then the instantaneous phase is $\theta(t)$ and the instantaneous frequency in Hertz is

$$f_i(t) = \frac{1}{2\pi}\frac{d\theta(t)}{dt} \tag{7.1}$$

EXAMPLE 7.1

When

$$x(t) = \sqrt{2}\cos(2\pi f_m t)$$

then $\theta(t) = 2\pi f_m t$ and $f_i(t) = f_m$.

Definition 7.2 For a bandpass signal having the form

$$x_c(t) = \sqrt{2}x_A(t)\cos(2\pi f_c t + x_P(t))$$

then the instantaneous phase is $2\pi f_c t + x_P(t)$ and the instantaneous frequency in Hertz is

$$f_i(t) = f_c + \frac{1}{2\pi}\frac{dx_P(t)}{dt} \tag{7.2}$$

Definition 7.3 For a bandpass signal the instantaneous frequency deviation in Hertz is

$$f_d(t) = \frac{1}{2\pi}\frac{d x_P(t)}{dt} \tag{7.3}$$

The instantaneous frequency deviation is a measure of how far the instantaneous frequency of the bandpass signal has deviated from the carrier frequency.

There are two prevalent types of analog angle modulated signals; phase modulated (PM) and frequency modulated signals (FM). Analog PM changes the phase angle of the complex envelope in direct proportion to the message signal

$$x_z(t) = \Gamma_m(m(t)) = A_c \exp[jk_p m(t)]$$

where k_p (units of radians/volt) is the phase deviation constant. Similarly, FM changes the instantaneous frequency deviation in direct proportion to the message signal. Note that the instantaneous radian frequency is the derivative of the phase so that the phase of a bandpass signal can be expressed as

$$x_P(t) = \int_{-\infty}^{t} 2\pi f_d(\lambda) d\lambda$$

Consequently, FM signals will have $f_d(t) = f_k m(t)$ and so for this book an FM signal will have the following form

$$x_z(t) = \Gamma_m(m(t)) = A_c \exp\left[j \int_{-\infty}^{t} k_f m(\lambda) d\lambda\right] = A_c \exp\left[j \int_{-\infty}^{t} 2\pi f_k m(\lambda) d\lambda\right]$$

where k_f is the radian frequency deviation constant (units radians/second/volt) and f_k is the frequency deviation constant (units Hertz/volt). It should be noted that the complex envelope of an angle modulated signal is a nonlinear function of the message signal. This nonlinear relationship will complicate the analysis of the fidelity of message reconstruction and spectral efficiency of transmission. The subject of angle modulation is typically where students interested in communications first find the limits of the undergraduate signals and systems education. The nonlinearities in angle modulation and demodulation do not permit closed form analysis and consequently approximate/simplified results need to be examined. The approaches taken to understand the fidelity of message reconstruction and spectral efficiency in angle modulation are useful not only for the results that are derived but also for the methodology used to solve the nonlinear problems.

The bandpass angle modulated signals are constant envelope signals with a phase or a frequency modulation. For PM the modulation is defined as

$$x_I(t) = g_I(m(t)) = A_c \cos(k_p m(t)) \qquad x_Q(t) = g_Q(m(t)) = A_c \sin(k_p m(t))$$

and bandpass signal is

$$x_c(t) = A_c\sqrt{2}\cos(2\pi f_c t + k_p m(t))$$

For FM the modulation is defined as

$$x_I(t) = g_I(m(t)) = A_c \cos\left(k_f \int_{-\infty}^{t} m(\lambda)d\lambda\right)$$

$$x_Q(t) = g_Q(m(t)) = A_c \sin\left(k_f \int_{-\infty}^{t} m(\lambda)d\lambda\right)$$

and the bandpass signal is

$$x_c(t) = A_c\sqrt{2}\cos\left(2\pi f_c t + k_f \int_{-\infty}^{t} m(\lambda)d\lambda\right)$$

An example message signal and the integral of this message signal is given in Figure 7.1. The corresponding bandpass phase modulated signal is shown in Figure 7.2. The corresponding bandpass FM modulated signal is given in Figure 7.3. It is obvious from this plot that the instantaneous frequency of the carrier[1] is being modulated in proportion to the message signal.

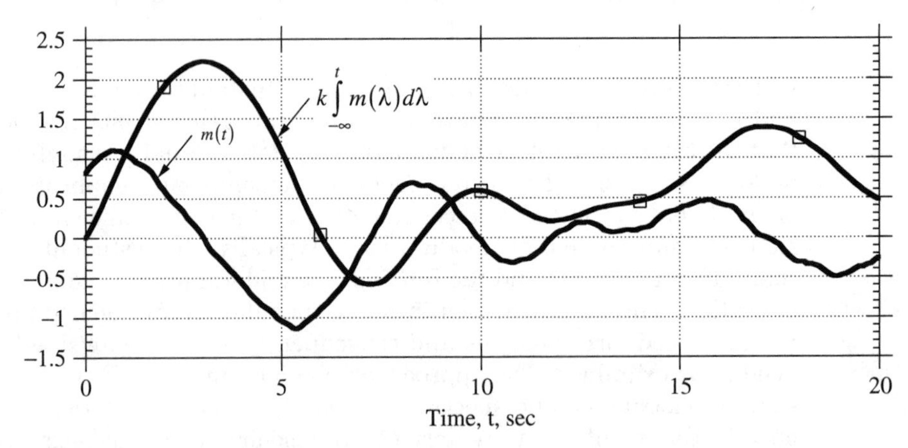

Figure 7.1 An example message signal and message signal integral.

[1]The instantaneous carrier frequency can be estimated by noting the frequency of the peaks of the sinusoid.

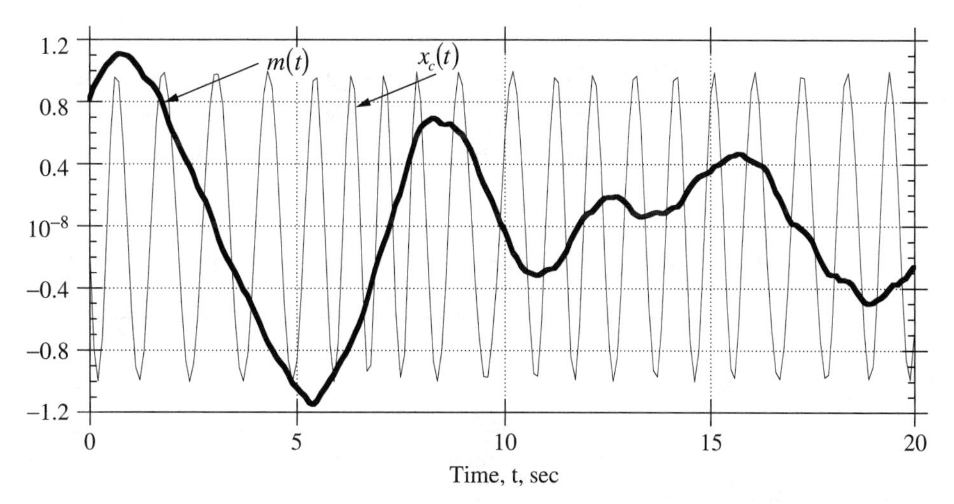

Figure 7.2 The PM modulated bandpass waveform corresponding to the message in Figure 7.1.

EXAMPLE 7.2
Angle modulation with

$$m(t) = A_m \sin(2\pi f_m t)$$

produces a complex envelope of

$$x_z(t) = A_c \exp[j\beta_p \sin(2\pi f_m t)] \text{ (PM)} \qquad x_z(t) = A_c \exp[j(\beta_f \cos(2\pi f_m t) + \theta_f)] \quad \text{(FM)}$$

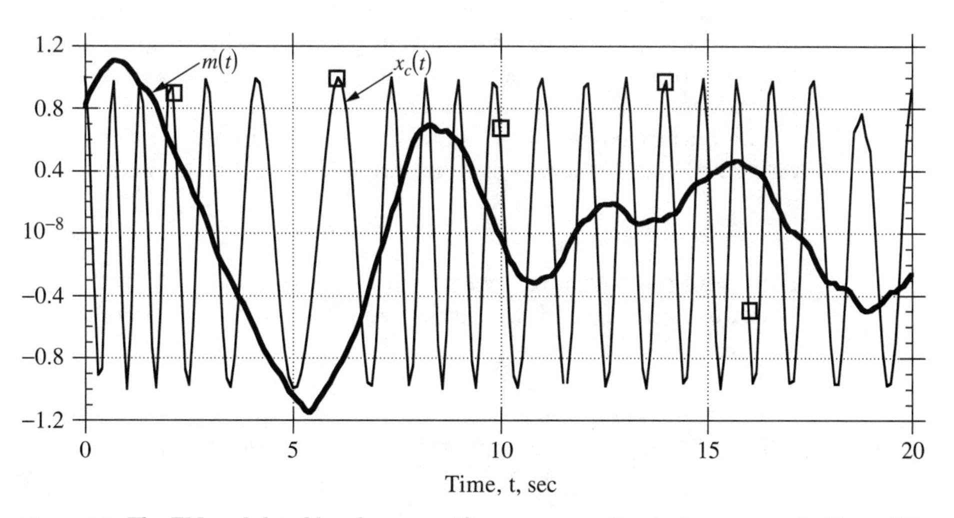

Figure 7.3 The FM modulated bandpass waveform corresponding to the message in Figure 7.1.

where $\beta_p = A_m k_p$, $\beta_f = \frac{-k_f A_m}{2\pi f_m}$ and θ_f is a constant phase angle that depends on the initial conditions of the integration. The in-phase and quadrature signals then have the form

$$x_I(t) = A_c \cos[\beta_p \sin(2\pi f_m t)] \qquad x_Q(t) = A_c \sin[\beta_p \sin(2\pi f_m t)] \qquad \text{(PM)}$$

and

$$x_I(t) = A_c \cos[\beta_f \cos(2\pi f_m t) + \theta_f] \qquad x_Q(t) = A_c \sin[\beta_f \cos(2\pi f_m t) + \theta_f] \qquad \text{(FM)}$$

7.1.1 Angle Modulators

Two methods are typically used to produce analog angle modulation in modern communication systems; the voltage controlled oscillator (VCO) and the direct digital synthesizer. The VCO is a device which directly generates an FM signal and the block diagram is given in Figure 7.4. The VCO varies the frequency of the oscillator in direct proportion to the input signal and it is very useful in a variety of applications besides FM modulation as well. Note, the VCO was first introduced as a component in a Costas loop in Chapter 6.

Direct digital synthesis (DDS) is a powerful technique used in the generation of radio frequency signals for use in a variety of applications in analog and digital communications. The technique has become widespread with the advances being made in integrated circuit technology that allow high performance digital synthesis to be made cheaply. The DDS is detailed in Figure 7.5. The input to a DDS is an analog signal (for angle modulation $m(t)$). This analog signal can be processed in a variety of ways but for angle modulation this processing corresponds to a multiplication (for PM) or an integration (for FM). The output of this processing is then input into two parallel trigonometric functions (cosine and sine). The structure is most often implemented with digital processing and these trigonometric functions are implemented in a digital lookup table. Often the DDS function is integrated with an upconverter so that a bandpass signal is produced at the output of the DDS.

The output power of an angle modulated signal is of interest. To this end using Eq. (4.24), the power in an angle modulated signal is given as

$$P_{x_c} = A_c^2 \tag{7.4}$$

For an angle modulated waveform the output power is only a function of the power associated with the carrier (A_c^2) and not a function of the message signal at all.

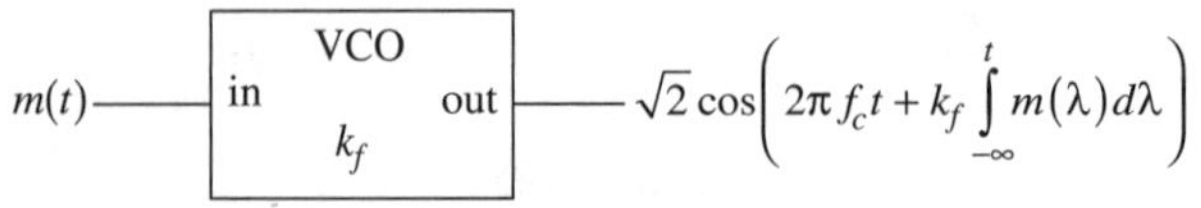

$m(t)$ — in — VCO k_f — out — $\sqrt{2}\cos\left(2\pi f_c t + k_f \int_{-\infty}^{t} m(\lambda)d\lambda\right)$

Figure 7.4 A voltage controlled oscillator.

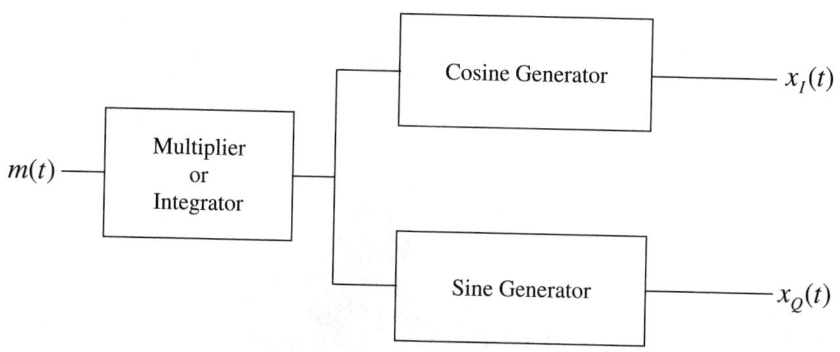

Figure 7.5 A direct digital synthesizer.

7.2 Spectral Characteristics

Unfortunately, a general expression cannot be found for a Fourier transform of the form

$$X_z(f) = \mathcal{F}\{\exp[ju(t)]\}$$

so no general results can be given for angle modulated spectrums.

EXAMPLE 7.3

The computer-generated voice signal given in Chapter 2 ($W = 2.5$ kHz) is used with phase modulation ($k_p = 1$). The power spectrum of the complex envelope of the PM signal is shown in Figure 7.6. Note, there is no apparent relationship between the baseband message spectrum (see Figure 2.8) and the modulated spectrum. The nonlinear nature of angle modulation produces a situation where the prerequisite signals and systems material fails to characterize the resulting modulation bandwidth. It is interesting to note that the apparent tone at DC is due to the periods of silence in the message signal ($m(t) = 0$).

7.2.1 A Sinusoidal Message Signal

As a first step in understanding the spectrum of angle modulated signals, consider the simple case of

$$x_z(t) = A_c \exp[j\beta \sin(2\pi f_m t)] \tag{7.5}$$

It should be noted here that by looking at the signal in Eq. (7.5) that insight can be achieved about the spectrum of both phase and frequency modulation. It should be noted that an achieved β is a function of both the message signal characteristics and the modulator characteristics (k_p in the case of PM and

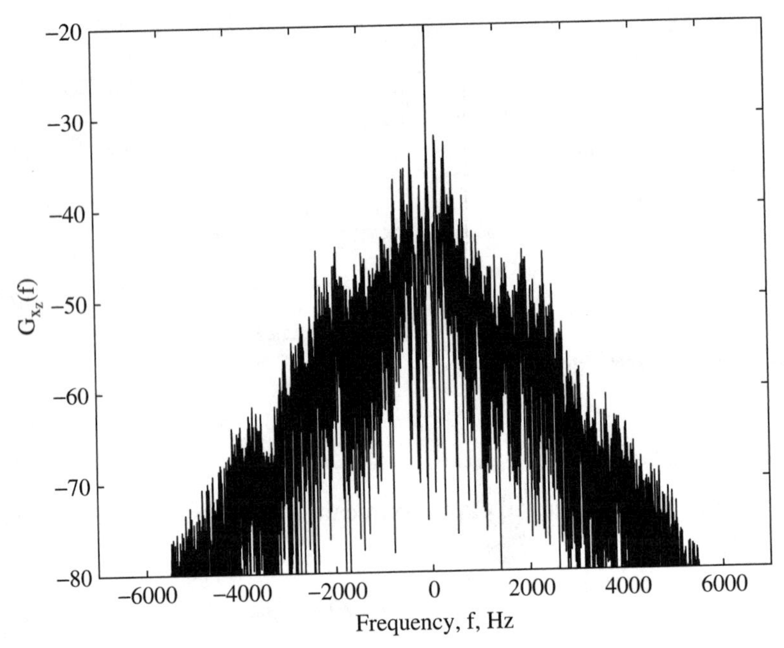

Figure 7.6 The power spectral density of the complex envelope of a phase modulated carrier. The message signal is the computer-generated voice signal from Chapter 2 ($W = 2.5$ kHz). $k_p = 1$.

k_f in the case of FM). The complex envelope in Eq. (7.5) can correspond to a phase modulated signal where

$$m(t) = \frac{\beta}{k_p}\sin(2\pi f_m t) = A_{pm}\sin(2\pi f_m t) \tag{7.6}$$

or a frequency modulated signal where

$$m(t) = \beta\frac{f_m}{f_k}\cos(2\pi f_m t) = A_{fm}\cos(2\pi f_m t) \tag{7.7}$$

Or stated another way, for PM if the message signal has the form $m(t) = A_m \sin(2\pi f_m t)$ then the resulting angle modulated complex envelop will be as in Eq. (7.5) with $\beta = A_m k_p$. Likewise, for FM if the message signal has the form $m(t) = A_m \cos(2\pi f_m t)$ then the resulting angle modulated complex envelop will be as in Eq. (7.5) with $\beta = \frac{A_m f_k}{f_m}$. It is important in this setup to note that for PM signals β is proportional to the message amplitude and the phase deviation constant. In FM β is directly proportional to the message amplitude and the radian frequency deviation constant while inversely proportional to the message frequency (see Example 7.2).

The signal $x_z(t)$ is a periodic signal with period $T = 1/f_m$, i.e., it is obvious that $x_z(t) = x_z(t + 1/f_m)$. Since the signal is periodic, the obvious tool for computing the spectral representation is the Fourier series.

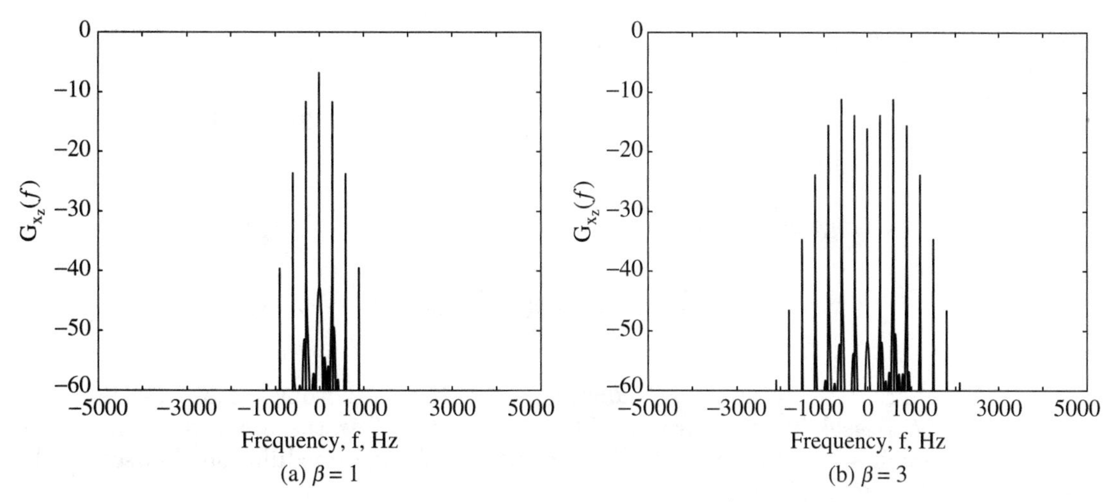

Figure 7.7 The measured power spectrum of an angle modulation with a sinusoidal message signal. $f_m = 300$ Hz.

EXAMPLE 7.4

Consider an angle modulation with a sinusoidal message signal with $f_m = 300$ Hz, $A_c = 1$, and $\beta = 1$, and $\beta = 3$. The measured power spectrum of the angle modulated complex envelope is shown in Figure 7.7. The impulsiveness of the spectrum indicates the signal is periodic and the fundamental frequency is $f_m = 300$ Hz. It can be deduced from Figure 7.7 that a larger value of β produces a larger transmission bandwidth.

The Fourier series is given as

$$x_z(t) = \sum_{n=-\infty}^{\infty} z_n \exp[j2\pi f_m nt] \tag{7.8}$$

where

$$z_n = \frac{1}{T} \int_{-\frac{T}{2}}^{\frac{T}{2}} x_z(t) \exp\left[-j\frac{2\pi nt}{T}\right] dt$$

A first spectral characteristic obvious on examining Eq. (7.8) is that angle modulation can potentially produce a bandwidth expansion compared to AM modulation. Note that a DSB-AM signal with a message signal of

$$m(t) = \sin(2\pi f_m t)$$

will produce a bandpass bandwidth of $2f_m$. If $z_n \neq 0$ for any $|n| \geq 2$ then an angle modulated signal will have a bandwidth greater than that of DSB-AM and $E_B < 50\%$. The actual coefficients of Fourier series can be obtained by noting that

$$z_n = f_m \int_{-\frac{1}{2f_m}}^{\frac{1}{2f_m}} A_c \exp[-j(2\pi f_m nt - \beta \sin(2\pi f_m t))]dt \tag{7.9}$$

The change of variables $\theta = 2\pi f_m t$ simplifies Eq. (7.9) to give

$$z_n = \frac{A_c}{2\pi} \int_{-\pi}^{\pi} \exp[-j(\theta n - \beta \sin\theta)]d\theta = A_c J_n(\beta)$$

where $J_n(\bullet)$ is the Bessel function of the first kind order n. More details about the Bessel function are provided in the sequel as these characteristics give insights into the spectral characteristics of angle modulation. Consequently, we now have

$$x_z(t) = \sum_{n=-\infty}^{\infty} A_c J_n(\beta) \exp[j2\pi f_m nt] \tag{7.10}$$

and a frequency domain representation of

$$X_z(f) = \sum_{n=-\infty}^{\infty} A_c J_n(\beta)\delta(f - nf_m) \tag{7.11}$$

A convenient form for the spectrum of the angle modulation for a sinusoidal message signal is now available once we characterize the Bessel function.

The Bessel Function

The Bessel function of the first kind is a transcendental function much like sine or cosine and has the following definition [Ae72]

$$J_n(x) = \frac{1}{2\pi} \int_{-\pi}^{\pi} \exp[-j(n\theta - x\sin\theta)]d\theta$$

This Bessel function can completely characterize the spectrum of an angle modulated signal with a sinusoidal message signal. Plots of the Bessel function of the first kind for various orders are seen in Figure 7.8. The Bessel function has the following important characteristics

1. $J_n(x)$ is real valued
2. $J_n(x) = J_{-n}(x)$ n even
3. $J_n(x) = -J_{-n}(x)$ n odd
4. $\sum_{n=-\infty}^{\infty} J_n^2(x) = 1$

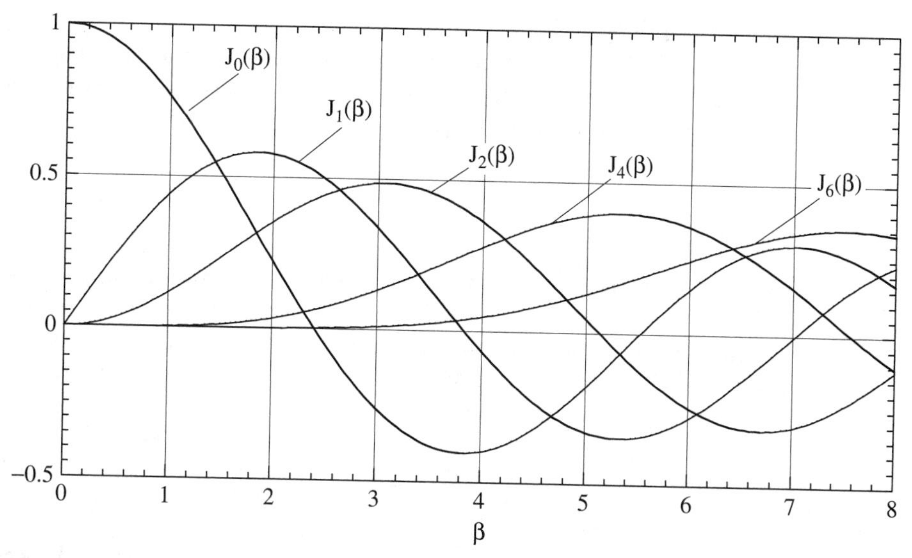

Figure 7.8 The Bessel function of the first kind.

5. $\lim_{n\to\infty} |J_n(x)| = 0$
6. When $\beta \ll 1$
   (a) $J_0(\beta) \approx 1$
   (b) $J_1(\beta) \approx \frac{\beta}{2}$
   (c) $J_n(\beta) \approx 0 \quad \forall \quad |n| > 1$

The first three characteristics of the Bessel function indicate that the power of the angle modulated signal is distributed evenly around zero. Characteristic 4 of the Bessel function is simply a consequence of Parseval's theorem, i.e.,

$$P_{x_z} = \frac{1}{T}\int_0^T |x_z(t)|^2 dt = \sum_{n=-\infty}^{\infty} |z_n|^2 = A_c^2 \sum_{n=-\infty}^{\infty} |J_n(\beta)|^2 = A_c^2$$

The fifth characteristic simply indicates that the angle modulated signal has a finite practical bandwidth (the Fourier coefficients converge to zero). Characteristic 6 of the Bessel function will be important for understanding the case of narrowband angle modulation.

EXAMPLE 7.5

The previous example considered a sinusoidal message signal. Comparing the measured power spectrum in Figure 7.7 to the results one could obtain from Figure 7.8 shows a perfect match between theory and measurement.

Engineering Bandwidth

The bandwidth of angle modulation for a sinusoidal message signal is controlled by β. All terms of the Fourier series expansion of the complex envelope are such that, $z_n = J_n(\beta) \neq 0$, so to transmit this angle modulated waveform without distortion requires an infinite bandwidth. As noted above

$$\lim_{n\to\infty} J_n(\beta) = 0$$

so a bandwidth can be selected such that the distortion that occurs due to not including higher frequencies is appropriately small. A traditional design practice used by communications engineers sets the filtering at the transmitter such that the transmission bandwidth is the 98% power bandwidth. Since the Bessel function has the characteristics

$$\sum_{n=-\infty}^{\infty} J_n^2(x) = 1 \qquad \lim_{n\to\infty} |J_n(x)| = 0 \tag{7.12}$$

it is clear that the power in an angle modulation is finite and the spectral distribution of power of the angle modulated signal dies off eventually with increasing frequency. For angle modulation of a sinusoidal message signal selecting the 98% bandwidth implies selecting how many harmonics of the message signal are transmitted. Consequently, the 98% bandwidth of a angle modulated sinusoid is $B_{98} = 2Kf_m$, where the value of K satisfies

$$\sum_{n=-K}^{K} J_n^2(\beta) > 0.98 \geq \sum_{n=-K+1}^{K-1} J_n^2(\beta) \tag{7.13}$$

While this value can be evaluated for each β, a good approximation for $\beta > 1$ is given as $K = \beta + 1$. The 98% transmission bandwidth is accurately approximated as

$$B_T = 2(\beta + 1) f_m \tag{7.14}$$

and consequently the bandwidth efficiency of angle modulation is approximated by

$$E_B = \frac{W}{B_T} = \frac{1}{2(\beta + 1)} \tag{7.15}$$

Figure 7.9 shows a comparison between the actual 98% transmission bandwidth and the simple approximation given in Eq. (7.14). Consequently, setting β for a particular message frequency, f_m, sets the transmission bandwidth. Commercial FM broadcast can be thought of as having $0.8 < \beta < 5$, but we will discuss this in more detail later.

Early engineers noted that the peak frequency deviation of an angle modulated carrier largely determines the bandwidth occupied by the signal [CF37].

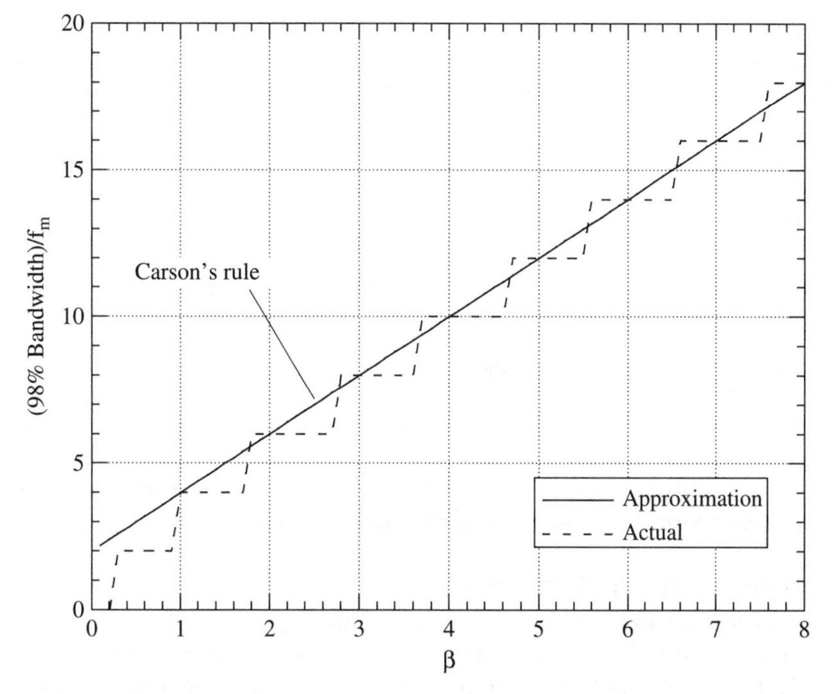

Figure 7.9 A comparison of the actual transmission bandwidth Eq. (7.13) and the rule of thumb approximation Eq. (7.14) for angle modulation with a sinusoidal message signal.

Definition 7.4 The peak frequency deviation is

$$f_p = \max|f_d(t)| \qquad \text{Hz} \tag{7.16}$$

For the sinusoidal message signal the frequency deviation is given as

$$f_d(t) = \frac{1}{2\pi}\frac{dx_P(t)}{dt} = \beta f_m \cos(2\pi f_m t) \tag{7.17}$$

and the peak frequency deviation is given as

$$f_p = \max|f_d(t)| = \beta f_m \tag{7.18}$$

Consequently, the transmission bandwidth is approximated as

$$B_T = 2(f_p + f_m) \tag{7.19}$$

This approximation is valid for other signals besides sinusoidal message signals.

In conclusion, the bandwidth of an angle modulated sinusoidal message signal is completely a function of β. Returning to Example 7.2 shows the value of β for PM is directly proportional to the message amplitude, A_m, and the phase deviation constant, k_p. For FM the value of β is directly proportional to the

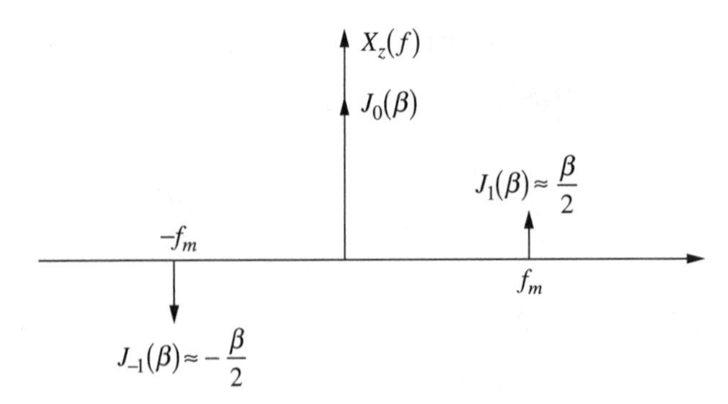

Figure 7.10 The resulting spectrum of a narrowband angle modulation with a sinusoidal message signal.

message amplitude, A_m, and the radian frequency deviation constant, k_f, but inversely proportional to the message frequency, f_m.

Narrowband Angle Modulation

Interesting insights can be obtained by examining $\beta \ll 1$ which is known as narrowband (NB) angle modulation. Using the characteristics of the Bessel function for small arguments, only DC and f_m spectral lines are significant, i.e., $z_0 \approx A_c$ and $z_{\pm 1} = \pm A_c \beta/2$. The resulting spectrum is given in Figure 7.10 and we have

$$\begin{aligned} x_z(t) &\approx A_c(1 + \beta/2 \exp[j 2\pi f_m t] - \beta/2 \exp[-j 2\pi f_m t]) \\ &= A_c(1 + j\beta \sin[2\pi f_m t]) \end{aligned} \tag{7.20}$$

Note this implies that

$$x_I(t) = A_c \qquad x_Q(t) = A_c \beta \sin[2\pi f_m t] \tag{7.21}$$

It is interesting to note that by examining Eq. (7.21) one can see that NB angle modulation is quite similar to LC-AM. Both have a large carrier component but the message signal is modulated on the in-phase component of the complex envelope in LC-AM while the message signal is modulated on the quadrature component in NB angle modulation. Also it should be noted that no bandwidth expansion of an angle modulated signal compared to an AM signal occurs in this case as the bandpass bandwidth in both cases is $2f_m$. This implies that the bandwidth efficiency of narrowband angle modulation is also $E_B = 50\%$.

7.2.2 General Results

We cannot obtain similar analytical results for a general message signal but we can use the results in Section 7.2.1, which was for the sinusoidal message signal, to provide general guidelines for angle modulation characteristics.

Engineering Bandwidth

Again we can use the approximation using f_p found for sinusoidal message signals to get some insight into the occupied bandwidth for general signals. For FM $f_p = f_k \max|m(t)|$ and for PM $f_p = (k_p/2\pi)\max|\frac{d}{dt}m(t)|$. Recall the sinusoidal message signal case detailed in Section 7.2.1 and we have

$$f_p = f_k A_{fm} \quad \text{FM} \qquad\qquad f_p = k_p A_{pm} f_m \quad \text{PM} \tag{7.22}$$

Hence noting $W = f_m$, we can see that Eq. (7.14) reduces to

$$B_T = 2(f_p + W) \tag{7.23}$$

Generalizing Eq. (7.14) can be accomplished by identifying the peak frequency deviation.

Definition 7.5 The bandwidth expansion factor of an angle modulated signal is the ratio of the peak frequency deviation to the message bandwidth, i.e.,

$$D = \frac{f_p}{W} \tag{7.24}$$

For the angle modulations, considered in this chapter, the bandwidth expansion factor is given as

$$D = \begin{cases} \dfrac{f_k \max|m(t)|}{W} & \text{FM} \\ \dfrac{k_p \max\left|\frac{d}{dt}m(t)\right|}{2\pi W} & \text{PM} \end{cases} \tag{7.25}$$

D is roughly the equivalent factor for an arbitrary message signal that β is for a sinusoidal message signal[2]. Consequently, an engineering approximation to the 98% power bandwidth of an angle modulated signal is

$$B_T = 2(D+1)W \tag{7.26}$$

and the bandwidth efficiency is

$$E_B = \frac{1}{2(D+1)} \tag{7.27}$$

This is known as Carson's rule and is a common rule of thumb used for the engineering bandwidth of angle modulated signals. Early work by Carson and Fry [CF37] showed theoretically how the bandwidth varies with f_p and W, and since Carson was an early pioneer of communication theory his name became associated with this powerful rule of thumb.

[2]In fact, if $m(t)$ is a sinusoidal signal it is easy to show $D = \beta$.

Another way to interpret Carson's rule is to note that

$$B_T = 2(f_p + W) \tag{7.28}$$

f_p corresponds to the largest frequency deviation of the complex envelope, $x_z(t)$, and so the total bandwidth required on one side of the carrier is f_p plus the bandwidth necessary to support the message signal (W). Obviously Carson's rule assumes $m(t)$ behaves similiarly for positive values as it does for negative values.

EXAMPLE 7.6

Commercial FM broadcast was originally configured with the following characteristics

1. Channel separation = 200 kHz,
2. Peak frequency deviation = $f_p = f_d\,[\max|m(t)|] \leq 75$ kHz (set by the FCC),
3. Message bandwidth = W = 15 kHz (mono broadcast).

Consequently we have

$$D \leq \frac{75\text{ kHz}}{15\text{ kHz}} = 5$$

and

$$B_T = 2(D+1)\ 15\text{ kHz} = 180\text{ kHz} < 200\text{ kHz}$$

Modern FM broadcast has two audio channels (stereo) and sometimes even a auxiliary channel so $W \geq 15$ kHz. The later discussion of multiplexing analog signals will provide more details.

EXAMPLE 7.7

The computer-generated voice signal given in Chapter 2 (W = 2.5 kHz) is used with phase modulation ($k_p = 3.0$). The peak frequency deviation constant is measured to be about 2608 Hz in this case. Carson's bandwidth rule of thumb predicts a B_T = 10000 Hz (5000 Hz one-sided) would be required. The measured cumulative power as a function of frequency is plotted in Figure 7.11 for the whole time record in the left plot and for a smaller time record which includes the largest values of the message signal and the frequency deviation in the right plot. The measured 98% bandwidth of the whole time record (approximately 3.5 kHz) is much less than that predicted by Carson's rule. This characteristic is due to the message signal having fairly long passages of silence ($m(t) \approx 0$) which tend to concentrate power at low frequencies. On the other hand the measured spectrum of the part of the time record where the signal and the frequency deviation take their largest values, the spectral distribution of power is quite accurately modeled by Carson's rule. Filtering the transmitted signal to a bandwidth less than Carson's bandwidth will distort the bandpass signal when the message signal is large and results in degraded performance. The bandpass PM signal (f_c = 5500 Hz) filtered to Carson's bandwidth has a measured power spectrum shown in Figure 7.12.

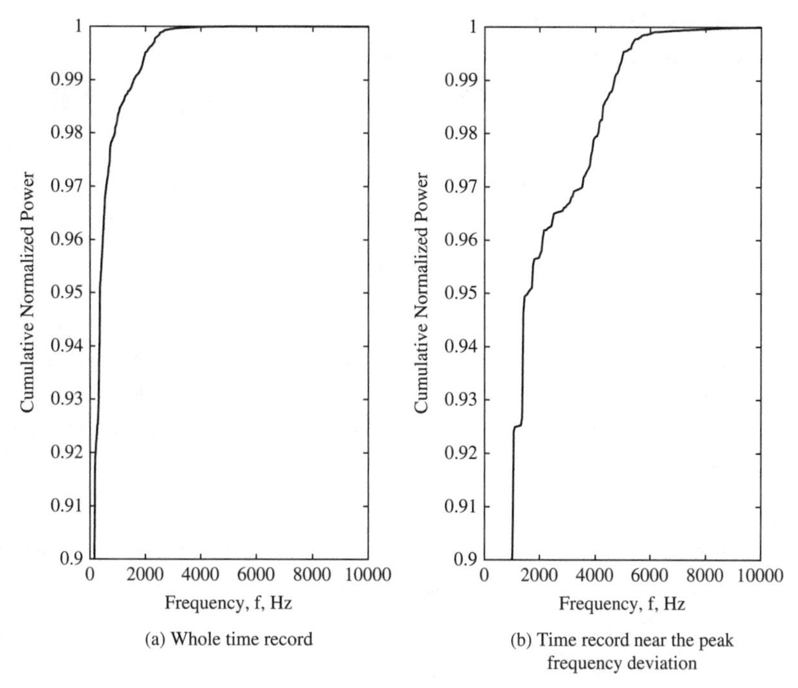

Figure 7.11 The normalized cumulative power of the phase modulated computer generated voice saying "Bingo". $k_p = 3.0$.

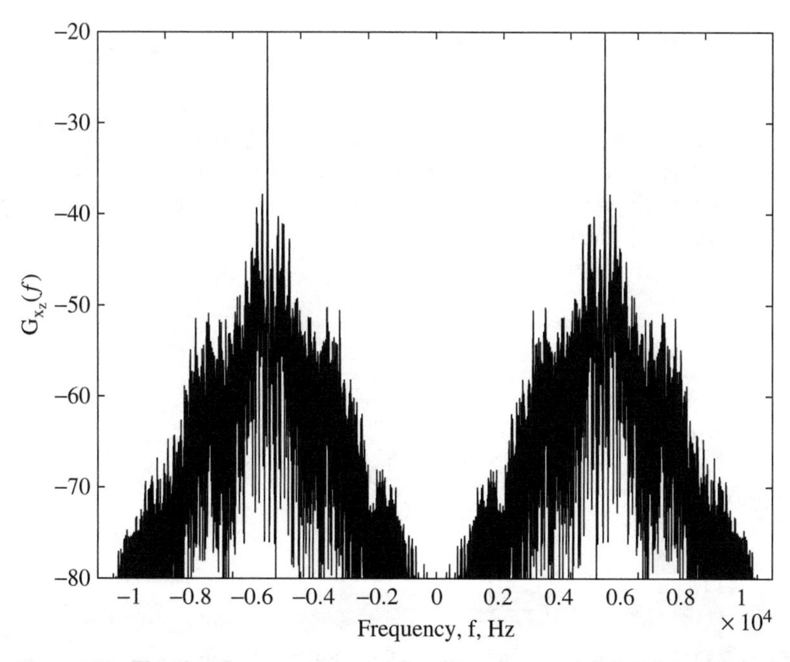

Figure 7.12 The bandpass spectrum for the phase modulated computer-generated voice saying "Bingo" filtered at Carson's bandwidth. $k_p = 0.8$, $f_c = 5500$ Hz.

Narrowband Angle Modulation

Recall that PM has the complex envelope

$$x_z(t) = A_c \exp[jk_p m(t)]$$

and FM has the complex envelope

$$x_z(t) = A_c \exp\left[j \int_{-\infty}^{t} k_f m(\lambda) d\lambda\right]$$

Narrowband angle modulation results when either k_p (PM) or k_f (FM) is small such that the total phase deviation caused by the modulation is small. Consequently, we can use the Taylor series expansion and examine the first order approximation, i.e.,

$$x_z(t) \approx A_c(1 + jk_p m(t)) \quad \text{(PM)} \qquad x_z(t) \approx A_c\left(1 + jk_f \int_{-\infty}^{t} m(\lambda) d\lambda\right) \quad \text{(FM)}$$

This leads to a bandpass signal of the form

$$x_c(t) \approx A_c\sqrt{2}\cos(2\pi f_c t) - A_c k_p m(t)\sqrt{2}\sin(2\pi f_c t) \quad \text{(PM)}$$

$$x_c(t) \approx A_c\sqrt{2}\cos(2\pi f_c t) - A_c k_f \int_{-\infty}^{t} m(\lambda) d\lambda \sqrt{2}\sin(2\pi f_c t) \quad \text{(FM)} \quad (7.29)$$

The bandpass implementation is seen in Figure 7.13 and is very similar to the LC-AM modulator presented in Figure 6.14. This modulator was often used in the early days of broadcast as one stage in a wideband angle modulator. This idea will be explored in the homework.

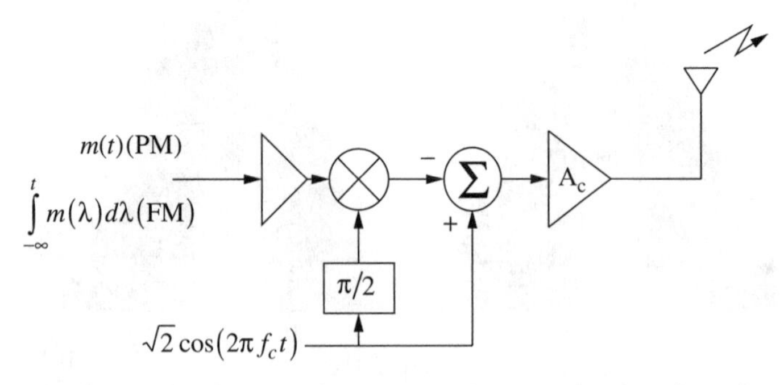

Figure 7.13 A bandpass implementation of a narrowband angle modulator.

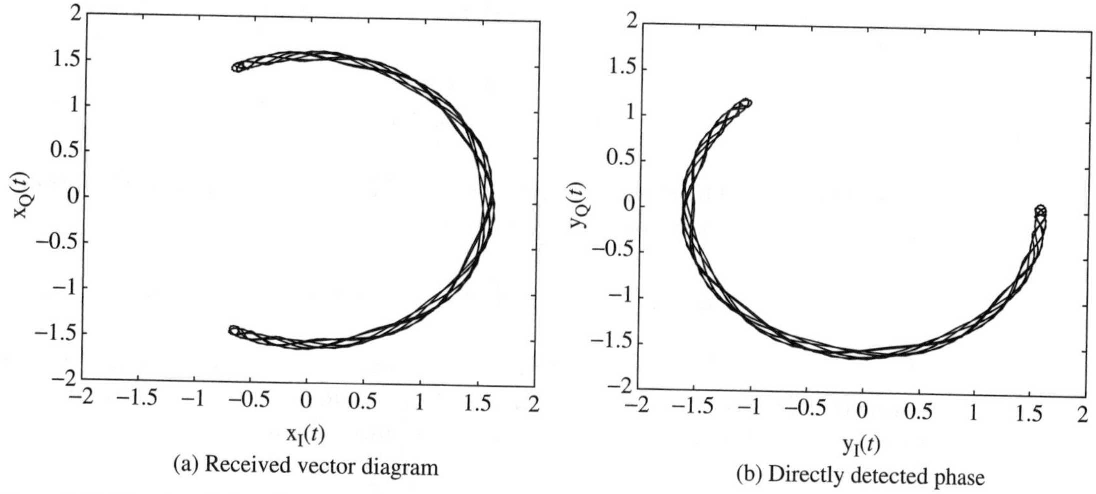

Figure 7.14 The signals in a direct phase detector.

7.3 Demodulation of Angle Modulations

Since angle modulation only affects the angle of the complex envelope, the demodulation algorithms also focus on the phase of the received complex envelope. Recall the received complex envelope has the form

$$y_z(t) = x_z(t) \exp[j\phi_p] + \text{noise}$$

Consequently, the received complex envelope of the angle modulated signal is given as

$$y_z(t) = A_c \exp[j(k_p m(t) + \phi_p)] + \text{Noise} \quad \text{(PM)}$$

$$y_z(t) = A_c \exp\left[j\left(k_f \int_{-\infty}^{t} m(\lambda)d\lambda + \phi_p\right)\right] + \text{Noise} \quad \text{(FM)} \qquad (7.30)$$

EXAMPLE 7.8

Consider phase modulation of a unit amplitude sinusoidal message signal with $k_p = 2$ and $\phi_p = -112.6^\circ = -1.97$ radians. The complex envelope of transmitted signal, $x_z(t)$, plotted in Figure 7.14(a), demonstrates a phase deviation of 2 radians as expected. It should be noted that the slight amplitude variation of the transmitted complex envelope is due to a filtering at Carson's bandwidth. The complex envelope output from the channel has a phase rotation as predicted by Eq. (7.30). The plot of the received complex envelope shown in Figure 7.14(b) clearly demonstrates this phase shift.

The obvious detection approach for PM is to compute the angle of this complex envelope,

$$y_p(t) = \tan^{-1}(y_Q(t), y_I(t)) \tag{7.31}$$

and implement a DC block to eliminate the phase shift induced by the propagation delay in the channel. The demodulator is the form

$$\hat{m}(t) = g_n(y_I(t), y_Q(t)) = h_H(t) * y_p(t) = k_p m(t) + \text{Noise} \tag{7.32}$$

where $h_H(t)$ represents the DC block. To recover the message signal in the FM case this direct phase detector could be implemented and followed by a differentiator so that the demodulator would have the form

$$\hat{m}(t) = g_n(y_I(t), y_Q(t)) = \frac{d}{dt} y_P(t) = \frac{d}{dt}\left(k_f \int_{-\infty}^{t} m(\lambda)d\lambda + \text{Noise}\right)$$
$$= k_f m(t) + \text{Noise} \tag{7.33}$$

Figure 7.15 shows the block diagram for such a detector. This detector is known as a direct phase detector. Note that the DC block is unnecessary in the case of FM because the constant phase shift will be eliminated by the derivative. This demodulation technique does not have to know or estimate ϕ_p hence it is a **noncoherent** demodulator. The resulting fidelity of demodulation of angle modulations will be detailed after noise is introduced in Chapter 11.

One practical characteristic of a direct phase detector that must be accommodated is that phase is typically measured in the range $[-\pi, \pi]$. A phase shift or a large phase deviation will cause the measured phase to wrap around the range $[-\pi, \pi]$. Recovery of the message signal requires that measured phase be unwrapped. The unwrapping ensures that the output phase from the direct phase detector is a smooth function of time. The unwrapping function is implemented in Matlab in the function `unwrap`.

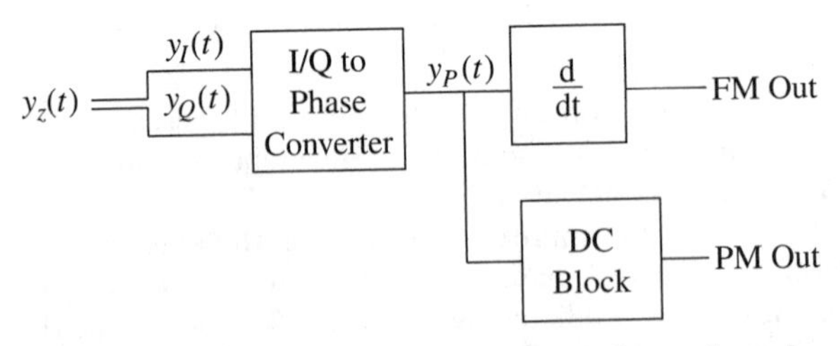

Figure 7.15 The direct phase detector for analog angle modulation.

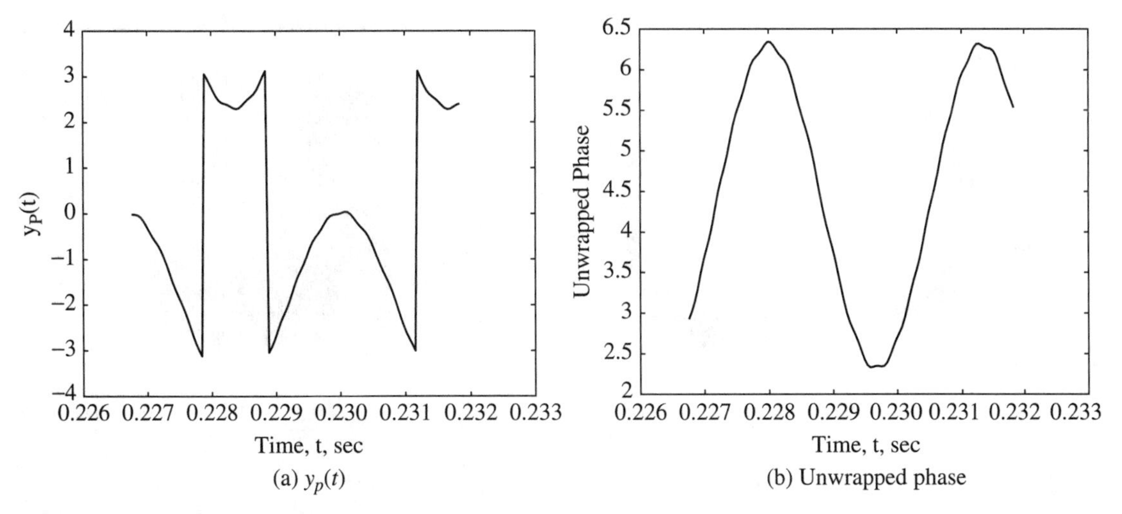

(a) $y_p(t)$

(b) Unwrapped phase

Figure 7.16 An example of phase unwrapping with a sinusoidal message signal.

EXAMPLE 7.9

Consider phase modulation of a unit amplitude sinusoidal message signal with $k_p = 2$ and $\phi_p = -112.6^\circ = -1.97$ radians. The plot of $y_P(t)$ is shown in Figure 7.16(a). It is clear that the direct detected phase has the form of a sinusoid offset roughly -2 radians. The deviation from this sinusoidal characteristic occurs when the value of the direct detected phase goes below $-\pi$, in this case the direct phase detector will return a positive phase angle. These abrupt changes in the detected phase can be reliably identified since the message signal is bandlimited. The unwrapped phase is plotted in Figure 7.16(b). In this case the unwrapping function has moved all the negative values of the phase to positive values of phase by adding 2π.

EXAMPLE 7.10

Consider the PM signal generated in Example 7.7 with a measured transmitted power spectrum given in Figure 7.12. In a channel with a propagation delay $\tau_p = 45.3\mu$s a phase shift of $\phi_p = -89.8^\circ$ is produced in transmission. The vector diagram of the received complex envelope is shown in Figure 7.17. The received signal does not have a constant amplitude due to the filtering to Carson's bandwidth at the transmitter. The phase shift due to the propagation delay in transmission is clearly evident in Figure 7.17(a). It is clear that the phase shift incurred in transmission will require phase unwrapping in a direct phase demodulator as the received signal phase crosses the $[-\pi, \pi]$ boundary frequently. The direct detected phase is given in Figure 7.17(b) and the unwrapped phase is given in Figure 7.18(a). The unwrapped phase has a DC offset due to the propagation delay in the channel. The output of the DC block is also given in Figure 7.18(b). The demodulated output signal has little distortion compared to the original computer-generated message signal introduced in Chapter 2. Note, in this

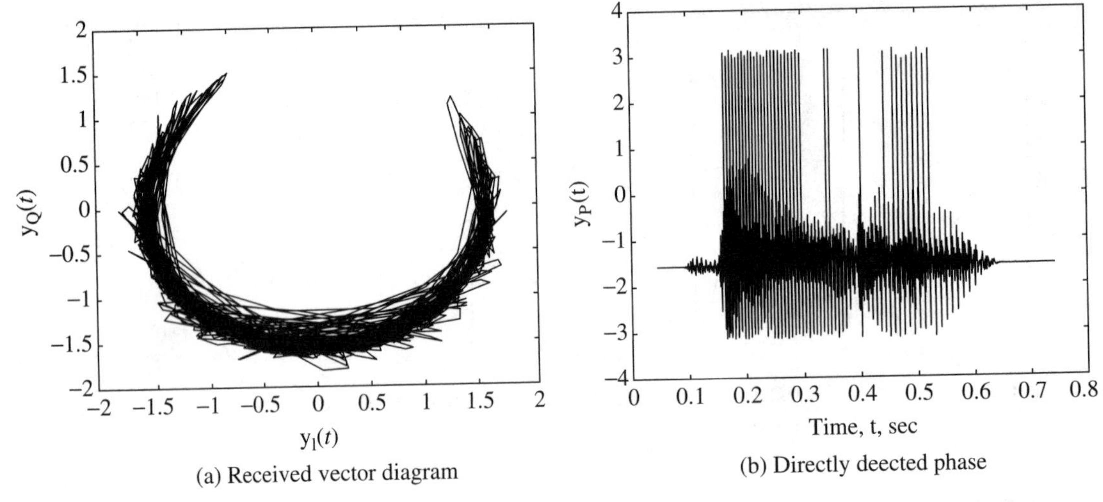

(a) Received vector diagram

(b) Directly deected phase

Figure 7.17 The signals in a direct phase detector.

example there is a little glitch at the beginning of the transmission due to the transient response of the DC block.

A second detector called a discriminator has frequently been used in practice because it can be implemented without an active device. The basis for the operation of the discriminator is found in the relation

$$\frac{d}{dt}e^{a(t)} = \frac{d\ a(t)}{dt}e^{a(t)}$$

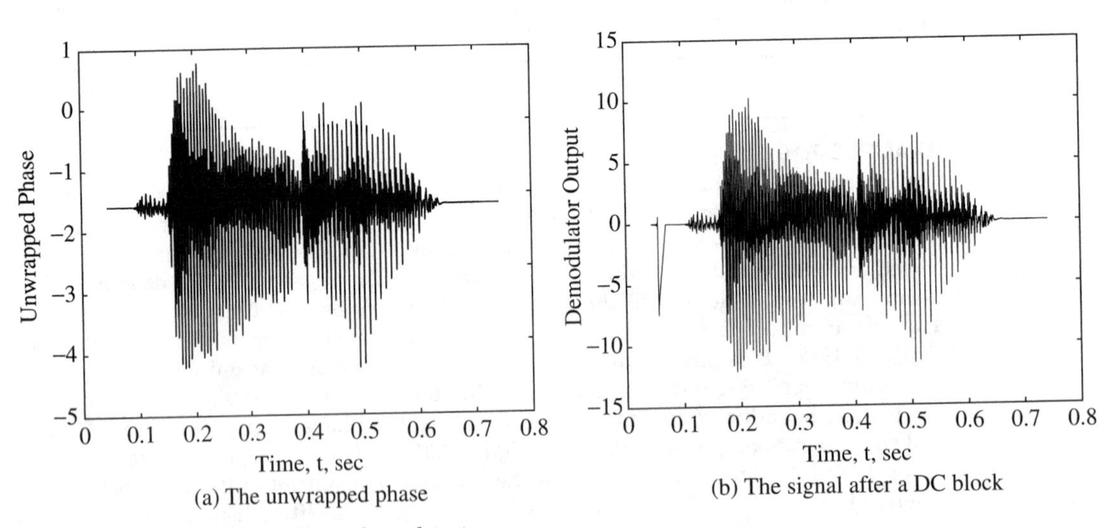

(a) The unwrapped phase

(b) The signal after a DC block

Figure 7.18 The signals in a direct phase detector.

or equivalently

$$\frac{d}{dt}\cos(a(t)) = \frac{d\ a(t)}{dt}\cos(a(t) + \pi/2)$$

For angle modulated signals this implies that the derivative of the complex envelope will have the form

$$\begin{aligned} \frac{d\ y_z(t)}{dt} &= jk_p\frac{d\ m(t)}{dt}A_c\exp[j(k_pm(t) + \phi_p)] \quad \text{(PM)} \\ \frac{d\ y_z(t)}{dt} &= jk_f\, m(t)A_c\exp\left[j\left(k_f\int_{-\infty}^{t} m(\lambda)d\lambda + \phi_p\right)\right] \quad \text{(FM)} \end{aligned} \tag{7.34}$$

Note that taking the derivative of an angle modulated signal produces an amplitude modulation which is a direct function of the message signal. If we can simply demodulate the AM then we have a simple angle demodulator. Recall the simplest AM demodulator is the envelope detector but this only works for LC-AM. Fortunately, similar processing can be implemented to produce a large carrier signal amplitude modulated signal. Recall that the form of the bandpass signal is

$$\begin{aligned} y_c(t) &= A_c\sqrt{2}\cos(2\pi f_ct + k_pm(t) + \phi_p) \quad \text{(PM)} \\ y_c(t) &= A_c\sqrt{2}\cos\left(2\pi f_ct + k_f\int_{-\infty}^{t} m(\lambda)d\lambda + \phi_p\right) \quad \text{(FM)} \end{aligned} \tag{7.35}$$

so taking a derivative at bandpass actually gives a LC-AM signal, i.e.,

$$\begin{aligned} \frac{d\ y_c(t)}{dt} &= A_c\left(k_p\frac{d\ m(t)}{dt} + 2\pi f_c\right)\sqrt{2}\cos(2\pi f_ct + k_pm(t) + \phi_p + \pi/2) \quad \text{(PM)} \\ \frac{d\ y_c(t)}{dt} &= A_c(k_f\, m(t) + 2\pi f_c)\sqrt{2}\cos\left(2\pi f_ct + k_f\int_{-\infty}^{t} m(\lambda)d\lambda + \phi_p + \pi/2\right) \quad \text{(FM)} \end{aligned} \tag{7.36}$$

Note that f_c is usually very large in comparison with the frequency deviation of the angle modulation so that the resultant signal is an LC-AM and the modulation can be recovered by envelope detection. Figure 7.19 shows the block diagram of the discriminator for angle modulations. Implementation of the discriminator can be accomplished without active devices (see Figure 5.39 in [PS94]) so it was popular in the early days of FM broadcast. The discriminator also does not need to know or estimate ϕ_p so it is also a **noncoherent** demodulation scheme.

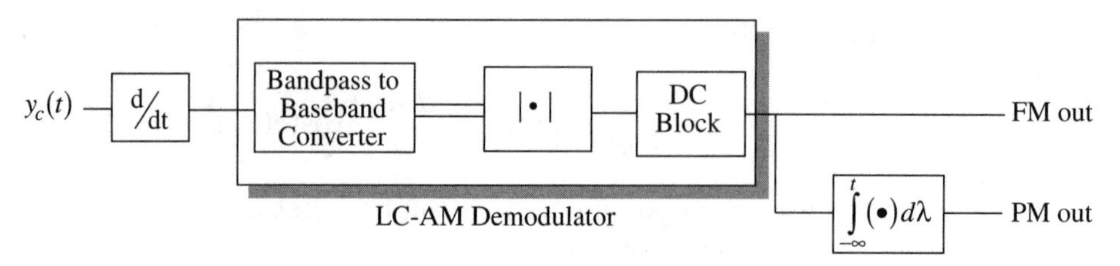

Figure 7.19 Model of a discriminator detector for angle modulation.

7.4 Comparison of Analog Modulation Techniques

A summary of the important characteristics of analog modulations is given in Table 7.1. At the beginning of Chapter 5 the performance metrics for analog communication systems were given as: complexity, fidelity, and spectral efficiency. At this point in our development we are able to understand the trade-offs involved in two of the three metrics (spectral efficiency and complexity). Characterizing fidelity of reconstruction can only be addressed after the tools to analyze the effects of noise and interference in electronic systems are introduced.

When examining the complexity of the implementation, angle modulations offer the best characteristics. Transmitters for all of the analog modulations are reasonably simple to implement. The most complicated transmitters are needed for VSB-AM and SSB-AM. The complexity is needed in VSB-AM and SSB-AM to achieve the improved bandwidth efficiency. The angle modulations have a transmitted signal with a constant envelope. This makes designing the output amplifiers much simpler since the linearity of the amplifier is not a significant issue. Receivers for LC-AM and angle modulation can be implemented in simple noncoherent structures. Receivers for DSB-AM and VSB-AM require the use of a more complicated coherent structure. SSB-AM additionally will require a transmitted reference for automatic coherent demodulation and operation.

When examining the spectral efficiency, SSB-AM offers the best characteristics. SSB-AM is the only system that achieves 100% bandwidth efficiency. DSB-AM and LC-AM achieve a bandwidth efficiency of 50%. VSB-AM offers performance somewhere between SSB-AM and DSB-AM. Angle modulations are the least bandwidth efficient. At best they achieve $E_B = 50\%$, and often much less.

TABLE 7.1 Summary of the important characteristics of analog modulations

Modulation	E_B	E_T	Transmitter Complexity	Receiver Complexity
DSB-AM	50%	?	moderate	moderate
LC-AM	50%	?	moderate	small
VSB-AM	$> 50\%$	?	large	large
SSB-AM	100%	?	large	large
PM	$< 50\%$	?	small	moderate
FM	$< 50\%$	?	small	moderate

This concluding discussion on the characteristics of analog modulations demonstrates the trade-offs typically encountered in engineering practice. Rarely does one option lead in all of the possible performance metric categories. The trade-offs between analog modulations will become even more complicated when the fidelity of message reconstruction is included.

7.5 Homework Problems

Problem 7.1. A message signal is given as

$$m(t) = \begin{cases} -1+2t & 0 \le t < 1 \\ 2-t & 1 \le t < 2 \\ 0 & \text{elsewhere} \end{cases} \tag{7.37}$$

Assume $A_c = 1$.

(a) If $m(t)$ is phase modulated with $k_p = \pi$ radians/volt, plot $x_P(t)$, $x_I(t)$, and $x_Q(t)$.

(b) If $m(t)$ is frequency modulated with $k_f = \pi$ radians/second/volt, plot $x_P(t)$, $x_I(t)$, and $x_Q(t)$.

(c) If $m(t)$ is frequency modulated with $k_f = 10\pi$ radians/second/volt, plot $x_P(t)$, $x_I(t)$, and $x_Q(t)$.

Problem 7.2. As a first task at Swensson, your boss asks you to design a PM modulator (obvious busy work) for a 1 kHz, 2 V peak amplitude test tone.

(a) She wants a 5 W output power across a 1-Ω resistor with a 98% power bandwidth of 20 kHz. Provide the values of A_c and k_p to achieve this specification.

(b) If you chose $m(t) = 2\sin(2\pi f_m t)$, then the Fourier series coefficients are a direct lift from lecture. What if $m(t) = 2\cos(2\pi f_m t)$?

Problem 7.3. A true angle modulated signal has an infinite bandwidth but the Federal Communications Commission (FCC) makes each station strictly limit their output frequency content as shown in Figure 7.20. For the case of sinusoidal angle modulation as discussed in class with $\beta = 3$ plot the output waveforms, $\tilde{x}_c(t)$, when the angle modulation is ideally bandlimited before

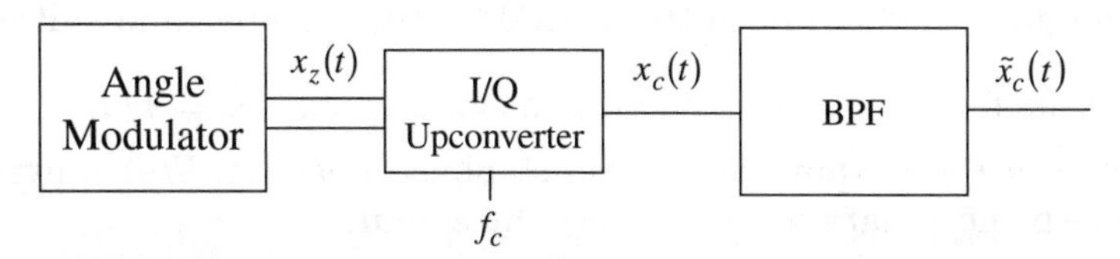

Figure 7.20 Bandlimiting the output of an angle modulator for Problem 7.3.

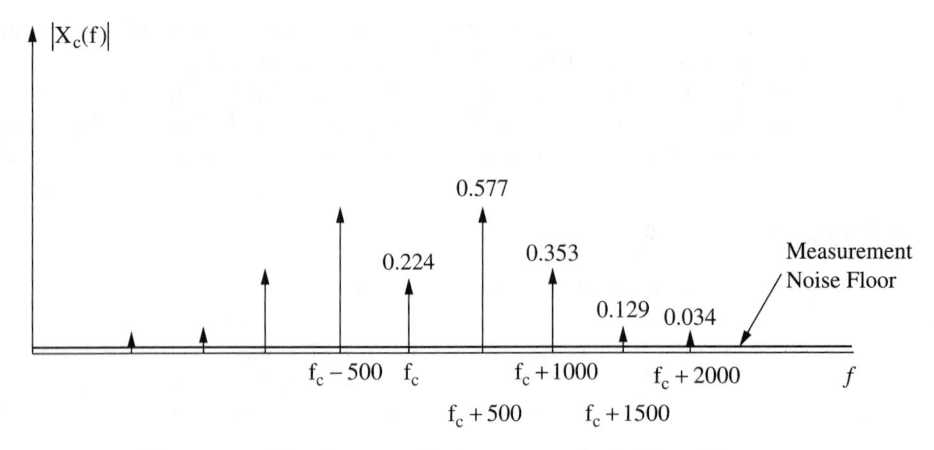

Figure 7.21 The measured bandpass amplitude spectrum for Problem 7.4.

transmission. Consider plots where the bandlimiting is done at 30%, 75%, 100%, and 150% of Carson's bandwidth. For uniformity in the answers assume that $f_m = 1$ kHz and $f_c = 20$ kHz. The plots should have time points no more than 5 μs apart.

Problem 7.4. You work for North Dakota Instruments and they are pathetically behind the competition in building portable PM radio transmitters. Your boss has obtained a transmitter from the competition and has asked you to reverse engineer the product. In the first test you perform, you input a signal into the phase modulator of the form

$$m(t) = \sin(2\pi f_m t)$$

The resulting normalized amplitude spectrum is measured (where $A_c = \sqrt{2}$) and plotted in Figure 7.21.

(a) What is f_m?

(b) What is the phase deviation constant, k_p? *Hint:* It is a whole number for simplicity.

(c) Draw a block diagram for a demodulator for this signal.

Problem 7.5. In Problem 7.3 you were asked to compute the output waveforms when a sinusoidally modulated FM or PM signal is bandlimited at various fractions of Carson's bandwidth. We return to this problem and look at the distortion in demodulation of this signal for the various bandwidths.

(a) Find the functions $y_I(t)$ and $y_Q(t)$ assuming that $\phi_p = \pi/4$.

(b) Compute the output from a direct phase detector. Will unwrapping the phase be necessary to reconstruct the signal?

(c) Show that the signal can be written in the form $\hat{m}(t) = Am(t) + n(t)$ and identify A and $n(t)$. $n(t)$ here represents the distortion due to nonideal processing at the transmitter and the receiver. This distortion is much different than the noise that will be introduced in Chapters 9 and 10 but produces similar effects.

(d) With

$$SNR = \frac{A^2 P_m}{P_n}$$

Find B_T such that the SNR > 30 dB *Hint:* $n(t)$ is periodic with the same period as the message signal so P_n can be computed in a simple way.

Problem 7.6. A test tone of frequency $f_m = 1000$ Hz with an amplitude $A_m = 2$ volts is to be transmitted using both frequency modulation (FM) and phase modulation (PM).

(a) Give values of k_p and k_f to achieve a $\beta = 2$. Using Carson's rule what is the resulting transmission bandwidth.

(b) Keeping A_m, k_p, and k_f the same as in part (a) and setting $f_m = 200$ Hz compute the resulting transmission bandwidth for both modulations.

(c) Keeping A_m, k_p, and k_f the same as in part (a) and setting $f_m = 5000$ Hz compute the resulting transmission bandwidth for both modulations.

(d) What characteristic does FM have that would be an advantage in a practical system?

Problem 7.7. A message signal, $m(t)$, with bandwidth $W = 10$ kHz, $\max |m(t)| = A_m$, and average power P_m is frequency modulated (FM) (the frequency deviation constant is k_f radians/second/volt) onto a carrier (f_c) with amplitude A_c.

(a) Give the form of the transmitted signal (either in bandpass form or complex baseband form).

(b) Choose the signal parameters such that the transmitted power is $P_{x_c} = 25$ W in a 1-Ω system.

(c) Define the deviation ratio, D, for an FM signal and choose the signal parameters such that $D = 4$.

(d) With $D = 4$ give a transmission bandwidth which will contain approximately 98% of the signal power.

Problem 7.8. As a new engineer at Badcomm, you are given the task to reverse engineer a phase modulator made by your competitor.

(a) Given your lab bench contains a signal generator and a spectrum analyzer, describe a test procedure that would completely characterize the phase modulator.

(b) Given your lab bench contains a signal generator, a bandpass filter with center frequency f_c and bandwidth 2000 Hz, a power meter, and an I/Q downconverter describe a test procedure that would completely characterize the phase modulator.

Problem 7.9. This problem is concerned with frequency modulation (FM) by the message signal given in Problem 5.1. Assume a carrier frequency of $f_c = 1000$ Hz, a radian frequency deviation constant of k_f and a carrier amplitude of A_c.

(a) Give the baseband, $x_z(t)$, and bandpass, $x_c(t)$, time waveforms.

(b) Using the results of Problem 5.1 what is the peak frequency deviation?

(c) State Carson's rule for an arbitrary message signal that is angle modulated. Select a value of k_f that will achieve a Carson's rule bandwidth of 600 Hz for FM with a message signal given in Problem 5.1.

(d) For the value of k_f selected in (c) plot the bandpass time waveform, $x_c(t)$ and the vector diagram of $x_z(t)$.

(e) Use Matlab to compute a measured power spectrum. How do the results compare to that predicted by the Carson's rule approximation?

(f) Give the value of A_c that will produce a 50-W output power in a 1-Ω system.

Problem 7.10. Repeat Problem 7.9 with the message signal of Problem 5.5 except with a Carson's rule bandwidth of 20 Hz and a carrier frequency of $f_c = 40$ Hz.

Problem 7.11. This problem is concerned with phase modulation (PM) by the message signal given in Problem 5.1. Assume a carrier frequency of $f_c = 1000$ Hz, a phase deviation constant of k_p and a carrier amplitude of A_c.

(a) Give the baseband, $x_z(t)$, and bandpass, $x_c(t)$, bandpass time waveforms.

(b) Using the results of Problem 5.1 what is the peak frequency deviation?

(c) State Carson's rule for an arbitrarily message signal that is angle modulated. Select a value of k_p that will achieve a Carson's rule bandwidth of 600 Hz for PM with a message signal given in Problem 4.1.

(d) For the value of k_p selected in (c) plot the bandpass time waveform, $x_c(t)$ and the vector diagram of $x_z(t)$.

(e) Use Matlab to compute a measured power spectrum. How do the results compare to that predicted by the Carson's rule approximation?

(f) Give the value of A_c that will produce a 50-W output power in a 1-Ω system.

Problem 7.12. Repeat Problem 7.11 with the message signal of Problem 5.5 except with a Carson's rule bandwidth of 20 Hz and a carrier frequency of $f_c = 40$ Hz.

Problem 7.13. Angle modulation with a sinusoidal message signal has the form

$$x_z(t) = A_c \exp[j\beta \sin(2\pi f_m t)]$$

The vector diagram of the received signal $y_z(t)$ is plotted in Figure 7.22 for $0 \leq t \leq (f_m)^{-1}$.

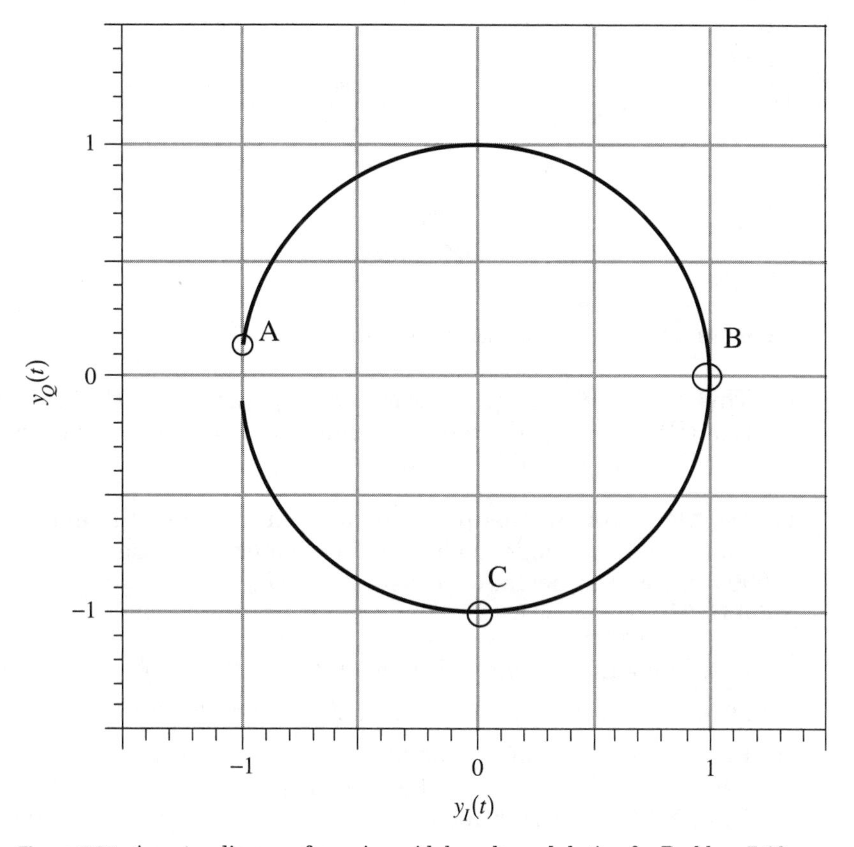

Figure 7.22 A vector diagram for a sinusoidal angle modulation for Problem 7.13.

(a) What are the approximate values of A_c, β, and ϕ_p?

(b) Give the approximate values of time (or times) associated with each labeled point (A, B, & C).

Problem 7.14. For $m(t) = \sin(2\pi f_m t)$ the received complex envelope is

$$y_z(t) = 5\exp[j(0.1\cos(2\pi f_m t - \pi) + \pi/5)]$$

Identify as many parameters of the modulation and channel as possible.

Problem 7.15. The waveform shown in Figure 7.23 is phase modulated on a carrier with $k_p = \pi/3$.

(a) Plot the output bandpass waveform for $f_c = 10 f_m$. For uniformity in the answers assume that $f_m = 1$ kHz and $f_c = 10$ kHz. The plots should have time points no more than 5 μs apart.

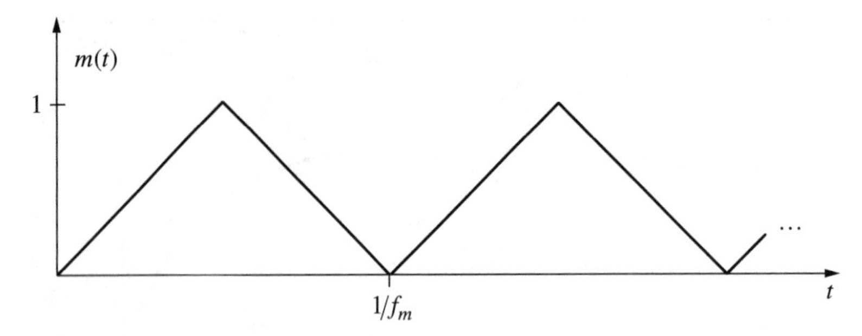

Figure 7.23 Message waveform for Problem 7.15.

(b) This same bandpass waveform can be generated with frequency modulation (FM). Specify the message signal and the f_k that achieves the same waveform as plotted in part (a).

Problem 7.16. (UM) In this problem, you will consider the effects of the message amplitude on angle modulation. Consider a message signal, $m(t) = 2\cos(2000\pi t)$. Let the deviation constants be $k_p = 1.5$ radians/volt and $f_k = 3000$ Hz/V.

(a) Determine the modulation indices: β_p for PM and β_f for FM.

(b) Determine the engineering transmission bandwidths.

(c) Plot the spectrum of the complex envelopes, considering only those components that will lie within the transmission bandwidth determined above. *Hint:* You'll need to think about what the modulated signal will look like.

(d) If the amplitude of $m(t)$ is decreased by a factor of two, how will your previous answers change?

You can use Matlab to compute the Bessel function evaluated for relevant orders and relevant arguments (`besselj(n,x)`).

Problem 7.17. Consider the following complex envelope signal for angle modulation with a sinusoidal message $x_z(t) = \exp[j\beta\cos(2\pi f_m t)]$ and compute the Fourier series. Recall that $\sin\left(\theta + \frac{\pi}{2}\right) = \cos(\theta)$. How does the power spectrum of this complex envelope compare to the signal considered in Eq. (7.5)?

Problem 7.18. You are making a radio transmitter with a carrier frequency f_c that uses an angle modulation. The message signal is a sinusoid with $A_m = 1$ and the complex envelope of the signal output from your voltage controlled oscillator is

$$x_z(t) = 2\exp[j\beta\sin(2\pi 500t)] \tag{7.38}$$

where β is a constant.

(a) Choose β such that $B_T = 6000$ Hz (use Carson's rule).

(b) The output spectrum for this transmitted signal will have a line (or impulsive) spectrum since the baseband signal is periodic. What is the value of the output bandpass spectrum impulse at $f = f_c$ for the value of β given in part (a).

(c) What is the message signal and the value of k_p if the signal is phase modulated (PM)?

(d) What is the message signal and the value of k_f if the signal is frequency modulated (FM)?

(e) For PM if f_m is reduced from 500 Hz to 100 Hz choose a value of k_p such that the output bandwidth is still $B_T = 6000$ Hz according to Carson's rule.

(f) For FM if f_m is reduced from 500 Hz to 100 Hz choose a value of k_f such that the output bandwidth is still $B_T = 6000$ Hz according to Carson's rule.

Problem 7.19. Recall that narrowband phase modulation has a very similar form to LC-AM, i.e.,

$$x_c(t) \approx A_c\sqrt{2}\cos(2\pi f_c t) - A_c k_p m(t)\sqrt{2}\sin(2\pi f_c t) \quad \text{(PM)}$$
$$x_c(t) = A_c\sqrt{2}\cos(2\pi f_c t) + A_c a m(t)\sqrt{2}\cos(2\pi f_c t) \quad \text{(AM)}$$

Because of this similarity one might be tempted to use an envelope detector as a simple demodulator.

(a) Compute the envelope, $y_A(t)$, of the received signal.

(b) Consider a simple example with $m(t) = \cos(2\pi(100)t)$ and $k_p = 0.1$. Plot the vector diagram of the true PM signal and the approximate form in Eq. (7.20).

(c) Plot the envelope for the signal in Eq. (7.20).

(d) Could the envelope detector be used for demodulating narrowband PM?

Problem 7.20. For a sinusoidal angle modulated signal

$$x_z(t) = A_c \exp[j\beta\sin(2\pi 500 t)] \tag{7.39}$$

and a propagation phase rotation of ϕ_p.

(a) For $\phi_p = 0$, how large could β be and not require phase unwrapping?

(b) For $\phi_p = \frac{\pi}{2}$, how large could β be and not require phase unwrapping?

(c) For $\phi_p = -\frac{7\pi}{8}$, how large could β be and not require phase unwrapping?

Problem 7.21. What condition must hold for the message signal, $m(t)$, and the modulation constants, k_p and k_f, such that the signals in Eq. (7.36) are guaranteed to be of the form of a LC-AM signal.

Problem 7.22. In a digital receiver samples of the complex envelope, $y_z(kT_s)$, are often processed to demodulate a message signal. Here T_s is the sample period and k enumerates the sample number. Assume the received signal is FM modulated, i.e.,

$$y_z(t) = A_c \exp\left(jk_f \int_{-\infty}^{t} m(\lambda)d\lambda + \phi_p\right) \tag{7.40}$$

(a) Show that if the sampling frequency is high enough that

$$q(k) = y_z(kT_s)y_z^*((k-1)T_s) \approx A_c^2 \exp(jKm(t)) \tag{7.41}$$

(b) Postulate how small T_s must be such that the result in (a) holds with a good accuracy.

(c) Identify the value of K.

(d) Using the idea in (a) as a basis for an FM demodulator, show a block diagram of the whole FM demodulator.

Problem 7.23. You have been given the received complex envelope, $y_z(t) = y_I(t) + jy_Q(t)$, of a periodic message signal transmitted via phase modulation (PM) as shown in Figure 7.24. The message signal has an average value of 0 so that it will pass through a DC block unchanged and $\max|m(t)| = 1$.

(a) Give the form of the received complex envelope for PM, $y_z(t)$ as a function of the phase deviation constant k_p, A_c, $m(t)$, and ϕ_p.

(b) Find $y_A(t)$ and use that to find A_c.

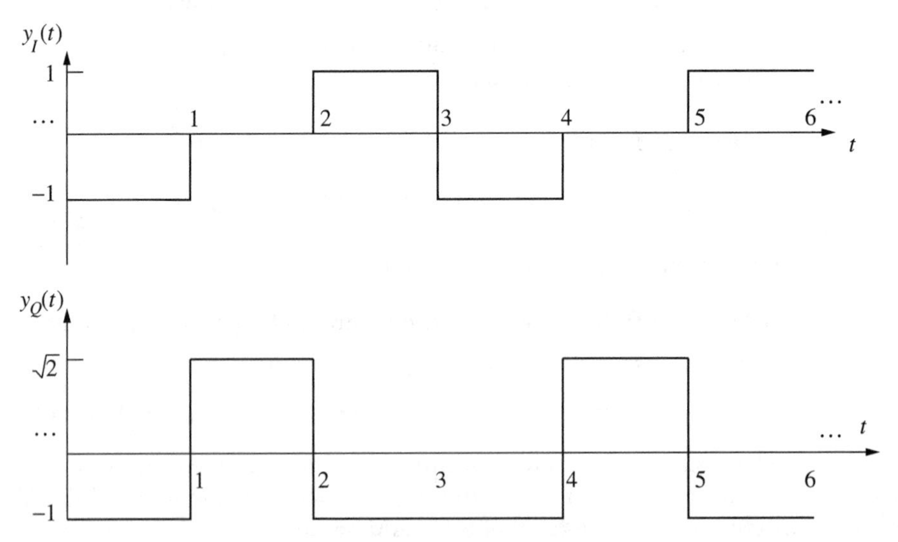

Figure 7.24 The received complex envelope of a PM signal.

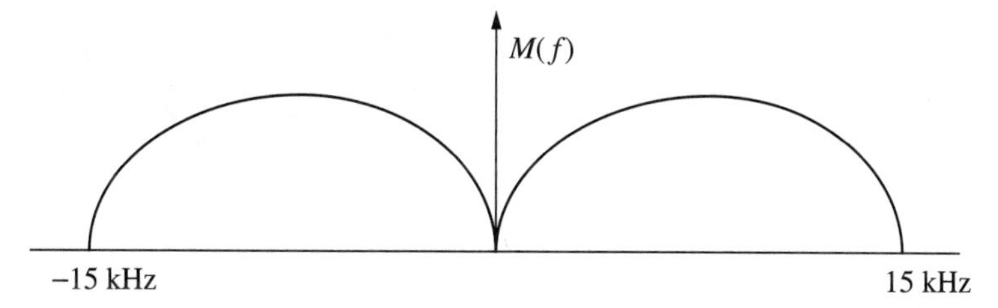

Figure 7.25 An example spectrum for audio broadcast.

(c) Find received phase $y_P(t)$ from the data in Figure 7.24. *Hint: I made it easy enough to do without calculators or Matlab.*

(d) Pass $y_P(t)$ through a PM demodulator and use the result to find a possible message signal and a value of k_p. Will phase unwrapping be necessary?

(e) Find the transmitted power, P_{x_z}.

Problem 7.24. High fidelity audio broadcast typically tries to maintain a 15 kHz of bandwidth in transmitting to listener. An example audio spectrum is given in Figure 7.25.

(a) In the days before stereo became a reality FM broadcasts frequency modulated a message signal of the form seen in Figure 7.25 onto a carrier. The FCC set $f_p \leq 75$ kHz. Using Carson's rule determine the D that would achieve a 98% power bandwidth of 180 kHz.

(b) After these allocations were set in concrete by the FCC stereo audio became a reality. After long discussions by the communication engineers of the day it was decided that stereo FM was possible if the input message signal to the FM modulator had the spectrum seen in Figure 7.26. Essentially the signal resulting from the sum of the left and right channels, $m_S(t) = m_L(t) + m_R(t)$ is combined with an DSB-AM modulated signal ($f_c = 38$ kHz) of the difference of the left and the right channels, $m_D(t) = m_L(t) - m_R(t)$. This way of combining the left and right channel was considered a nice

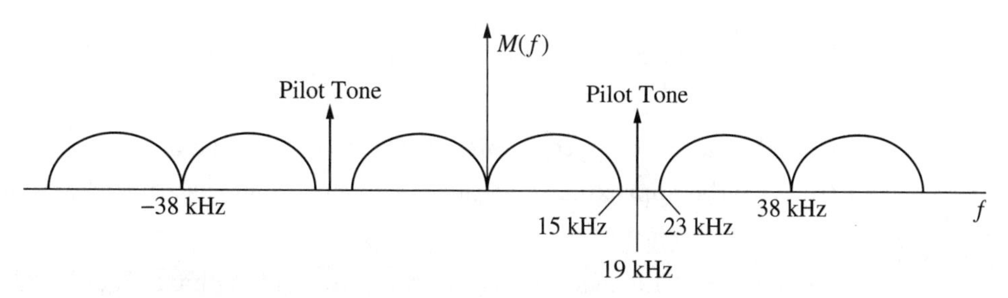

Figure 7.26 Stereo FM message signal.

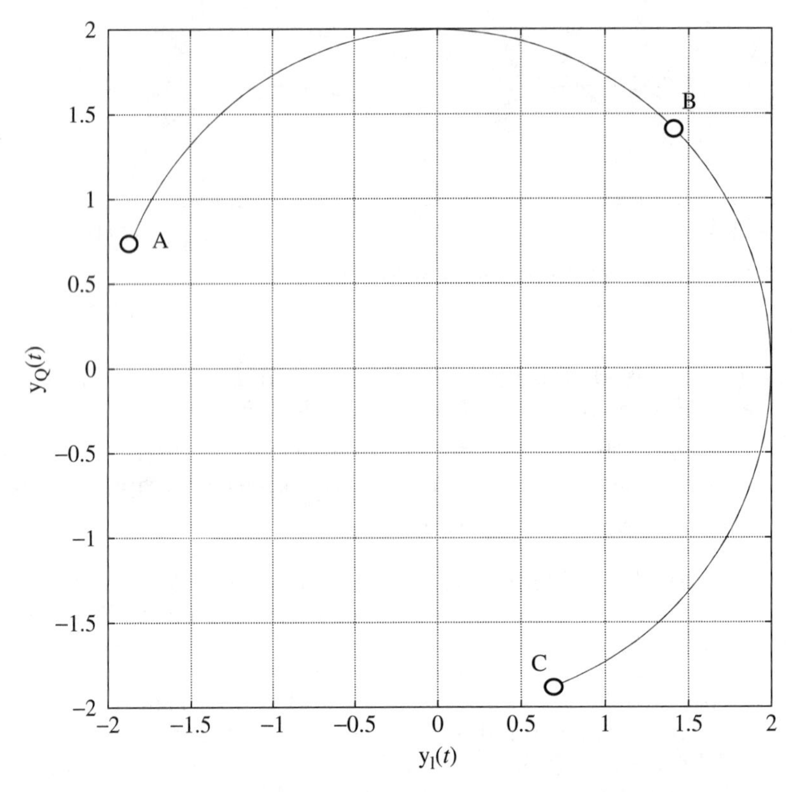

Figure 7.27 Vector diagram of angle modulation with a sinusoidal message signal.

solution since a radio that did not have stereo capability would still decode $m_S(t)$ and not lose any capability. Again using Carson's rule determine the D that would achieve a 98% power bandwidth of 180 kHz.

Problem 7.25. Angle modulation with a sinusoidal message signal has the form

$$x_z(t) = A_c \exp[j\beta \sin(2\pi f_m t)]$$

The vector diagram of the received signal $y_z(t)$ is plotted in Figure 7.27 for $0 \leq t \leq (f_m)^{-1}$.

(a) What are the approximate values of A_c, β, and ϕ_p?

(b) Give the approximate values of time (or times) associated with each labeled point (A, B, & C).

Problem 7.26. The message signal in Figure 5.11 is phase modulated with a $k_p = \pi/2$ and an amplitude $A_c = 2$.

(a) What is the form for $x_z(t)$.

(b) What is the transmitted power?

(c) Plot the instantaneous frequency deviation, $f_d(t)$.

(d) Plot a vector diagram of $x_z(t)$. Clearly label the points corresponding to $t = 0, T/4, T/2, 3T/4, T$.

(e) Is $x_z(t)$ a periodic signal? If so what is the fundamental frequency?

(f) Using Carson's rule estimate the 98% bandpass bandwidth of this phase modulated signal.

(g) The spectrum of the bandpass signal will be impulsive (discrete frequencies). What will be the value of the impulse at $f = f_c$.

Problem 7.27. If $m(t)$ is periodic with period T

(a) What is the largest possible period of a phase modulated signal with phase deviation constant k_p?

(b) Is it possible for the period of $x_z(t)$ to be less than T? If yes, give an example.

(c) State a condition on $m(t)$ such that Fourier series coefficient for DC,

$$z_0 = \int_0^T x_z(t)dt \tag{7.42}$$

is real valued.

Problem 7.28. If $m(t)$ is periodic with period T and the integral of $m(t)$ is

$$m_i(t) = \int_{-\infty}^{t} m(\lambda)d\lambda \tag{7.43}$$

(a) State a condition on $m(t)$ such that $m_i(t)$ is periodic with the same period.

(b) Can $m_i(t)$ be aperiodic but $x_z(t) = \exp[j m_i(t)]$ be periodic? Give an example if the statement is true.

Problem 7.29. A periodic received signal, $r_z(t)$, from an analog modulation is given in Figure 7.28.

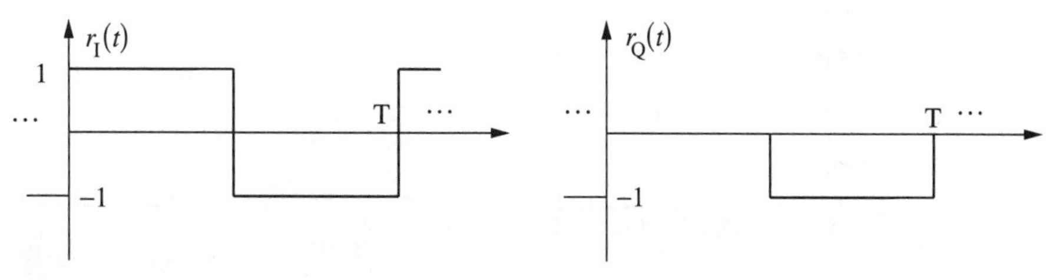

Figure 7.28 A received complex envelope.

(a) Assume the transmitted signal is given as $x_z(t)$, how have we modeled the form for $r_z(t)$ in terms of $x_z(t)$ for this class?

(b) Given $r_z(t)$ in Figure 7.28, is it possible for the transmitted signal to be a PM signal? Why?

Problem 7.30. For $m(t) = \cos(200\pi t)$ a received complex envelope in the absence of any noise is

$$y_z(t) = 5\exp[j(0.8\sin(200\pi t) + \pi/5)]$$

(a) With what modulation has the analog message been transmitted?

(b) Identify as many parameters of the modulation and the channel as possible.

(c) You have been tasked to build a lowpass filter for the signals $y_I(t)$ and $y_Q(t)$ such that you pass about 98% of the power of the signals. With what bandwidth should the lowpass filters be designed?

Problem 7.31. The message signal detailed in Problem 5.5 is used in a phase modulation. Assume the amplitude is A_c and that $k_p = 3$.

(a) Choose A_c such that $P_{x_z} = 16$.

(b) With the A_c as specified in a) plot $x_I(t)$ and $x_Q(t)$.

(c) Find D.

(d) Use D and Carson's rule to estimate B_T. What is E_B?

(e) Assume $\phi_p = \pi/4$ and plot $y_I(t)$, $y_Q(t)$, and $y_p(t)$.

Problem 7.32. **(JG)** If the waveform in Figure 7.29 is input into an angle modulator,

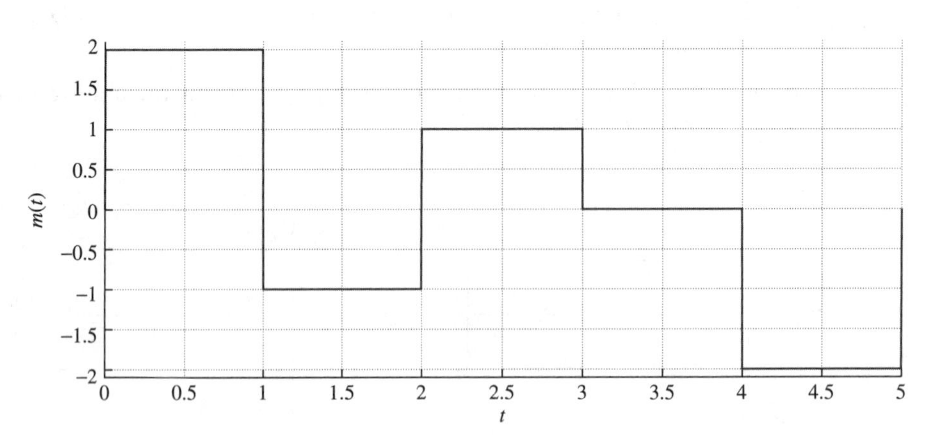

Figure 7.29 A message signal.

(a) Derive an expression for the time domain bandpass waveform if PM is used with $k_p = \pi$ and $A_c = 1/\sqrt{2}$. Plot the PM signal for $f_c = 4$.

(b) Derive an expression for the time domain bandpass waveform if FM is used with $k_f = 2\pi$ and $A_c = 1/\sqrt{2}$. Plot the FM signal for $f_c = 4$.

Problem 7.33. An often used pulse in radar systems has a complex envelope of

$$x_z(t) = \begin{cases} A_c \exp\left[j2\pi g_0 t^2\right] & 0 \leq t \leq T_p \\ 0 & \text{elsewhere} \end{cases} \tag{7.44}$$

(a) Is this an angle modulated signal? Justify your answer.

(b) Plot the bandpass signal, $x_c(t)$ for $f_c = 10$ Hz and $g_0 = 100$ Hz/s, $A_c = 1$, and $T_p = 1$.

(c) Find the instantaneous frequency deviation, $f_d(t)$, and the peak frequency deviation, f_p, for this signal.

(d) Plot $G_{x_z}(f)$ and estimate B_{98} when $g_0 = 100$ Hz/s and $T_p = 1$.

(e) Show that Carson's rule of thumb does not hold for this signal and identify the characteristic of this radar signal that is different than a typical angle modulated message signal that causes this well respected rule of thumb not to hold.

7.6 Example Solutions

Problem 7.7.

(a) The complex baseband signal for FM is given as

$$x_z(t) = A_c \exp\left[jk_f \int_{-\infty}^{t} m(\lambda)d\lambda\right] \tag{7.45}$$

The bandpass signal is given as

$$x_c(t) = A_c\sqrt{2}\cos\left(2\pi f_c t + k_f \int_{-\infty}^{t} m(\lambda)d\lambda\right) \tag{7.46}$$

(b) Since $P_{x_c} = 25 = A_c^2$ it is straightforward to see that $A_c = 5$

(c) The definition of the deviation ratio is

$$D = \frac{f_k \max|m(t)|}{W} \tag{7.47}$$

Rearranging gives

$$\max|m(t)| = \frac{40,000}{f_k} \tag{7.48}$$

(d) Using Carson's rule gives $B_T = 2(D+1)W = 2 \times 5 \times 10,000 = 100,000$ Hz.

Problem 7.18.

(a) Carson's rule gives

$$B_T = 2(\beta + 1) f_m \tag{7.49}$$

$$6000 = 2(\beta + 1)500 \tag{7.50}$$

Solving gives $\beta = 5$.

(b) Using the results in Eq. (6.7) gives

$$x_z(t) = \sum_{n=-\infty}^{\infty} A_c J_n(\beta) \cdot e^{(j2\pi f_m nt)} \tag{7.51}$$

and consequently

$$X_z(f) = \sum_{n=-\infty}^{\infty} A_c J_n(\beta)\delta(f - nf_m) \tag{7.52}$$

Note $A_c = 2$ and $\beta = 5$ and recall the relationship between the baseband and bandpass spectrum is

$$X_c(f) = \frac{1}{\sqrt{2}} X_z(f - f_c) + \frac{1}{\sqrt{2}} X_z^*(-f - f_c) \tag{7.53}$$

$$= \sum_{n=-\infty}^{\infty} \frac{A_c J_n(\beta)}{\sqrt{2}} \delta(f - nf_m - f_c) + \sum_{n=-\infty}^{\infty} \frac{A_c J_n(\beta)}{\sqrt{2}} \delta(f + nf_m + f_c) \tag{7.54}$$

so consequently the value of the spectral impulse at f_c is $\frac{2J_0(5)}{\sqrt{2}} = -0.3678$

(c) Recall

$$k_p \cdot m(t) = 5\sin(2\pi f_m t) \tag{7.55}$$

and since $A_m = 1$ we have $k_p = 5$ and $m(t) = \sin(1000\pi t)$.

(d) Recall

$$k_f \int_{-\infty}^{t} m(\lambda) d\lambda = 5\sin(2\pi f_m t) \tag{7.56}$$

consequently

$$k_f m(t) = 5\frac{d}{dt}\sin(2\pi f_m t) = 5(2\pi f_m)\cos(2\pi f_m t) \tag{7.57}$$

and since $A_m = 1$ we have $k_f = 5(2\pi f_m) = 5000\pi$ rad/s/V and $m(t) = \cos(1000\pi t)$.

(e) Carson's rule gives

$$B_T = 2(\beta + 1) f_m \tag{7.58}$$

$$6000 = 2(\beta + 1)100 \tag{7.59}$$

Solving gives $\beta = k_p = 29$.

(f) Carson's rule again gives $\beta = 29$ and $k_f = \beta(2\pi f_m) = 5800\pi$ rad/s/V. Parts (e) and (f) show one of the interesting differences between FM and PM. A characteristic of FM is that once the frequncy deviation is chosen the transmission bandwidth remains relatively constant regardless of the signal bandwidth. PM on the otherhand has the transmission bandwidth varying in direct proportion to the message signal bandwidth for a fixed phase deviation constant. This is one reason why FM is preferred in practice to PM.

7.7 Miniprojects

Goal: To give exposure

- to a small scope engineering design problem in communications.
- to the dynamics of working with a team.
- to the importance of engineering communication skills (in this case oral presentations).

Presentation: The forum will be similar to a design review at a company (only much shorter). The presentation will be of 5 minutes in length with an overview of the given problem and solution. The presentation will be followed by questions from the audience (your classmates and the professor). Each team member should be prepared to make the presentation on the due date.

7.7.1 Project 1

Project Goals: Undergraduate communication courses often only look at idealized problems. For example, one thing that is ignored in a course is that frequency sources at the transmitter and receiver cannot be made to have exactly the same value. This miniproject will investigate the effects of this frequency offset on one phase demodulation algorithm.

When a frequency offset exists the complex envelope of the received signal will be rotating due to this frequency offset, i.e.,

$$y_z(t) = x_z(t) \exp[j(\phi_p + 2\pi f_o t)] \tag{7.60}$$

where f_o is the existing frequency offset. Get the Matlab file `angmodex1.m` and `angmodex3.m` from the class web page. In these files a PM transmitter (for a sinusoid message signal) and receiver is implemented. If the carrier frequency offset is set to 10 Hz `(in the Matlab code deltaf = 10)` the demodulator implemented in the m-file still works pretty well with this frequency offset. Explain why. Note, at higher frequency offsets (e.g., 100 Hz) the performance suffers noticeable distortion. *Extra credit will be given if you can figure out a method to eliminate this distortion.*

7.7.2 Project 2

Project Goals: Use your knowledge of communication theory to decode an unknown analog modulated signal.

You have been hired by the National Security Agency of the US Federal government as an engineer. The first job you have been given by your new boss is to decode a radio message signal between two members of a South Albuman drug cartel. This signal was intercepted by a spy satellite and the data has been processed and forwarded on to you for decoding.

The technical details are

- The sampled data file contains about 1 second worth of data at a 44,100-Hz sampling rate. (Precisely 44,864 samples). This data is available at the class web site in the file entitled `Intercept.mat`.
- The radio signal is analog modulated. The type of analog modulation is unclear. Human intelligence sources have indicated that the modulation used is either DSB-AM, LC-AM, SSB-AM, or PM.
- The carrier frequency is unknown. Obviously since the data is provided to you with a sample rate of 44,100 Hz the carrier frequency must be between 0 and 22,050 Hz.
- The channel includes an unknown delay that produces an unknown phase shift, φ_p.

Your job is to decode the signal. It should be obvious when you are correct as you can play the demodulated output as a sound. The solution should include

- a discussion of the methodology that led you to identifying the modulation.
- a discussion of the methodology that led you to identifying the carrier frequency.
- a discussion of the demodulator architecture and the method used to estimate the phase shift or to operate in the presence of an unknown phase shift.
- submission of Matlab code for the demodulator.

Chapter

8

More Topics in Analog Communications

This chapter covers two topics which are important for both amplitude and angle modulation methods of analog communications: the phase-locked loop and the multiplexing of message signals. In fact, almost all modern communications systems, analog or digital, contain a phase-locked loop and contain some form of multiplexing.

8.1 Phase-Locked Loops

8.1.1 General Concepts

The phase-locked loop (PLL) is a commonly used component in communication systems. It can be used for tracking the phase and frequency of signals, demodulating angle modulated signals, and frequency source synthesis. The PLL is a tracking system that uses feedback. We have seen one example so far in the Costas loop for synchronous demodulation of DSB-AM. This section will investigate the use of the PLL for FM and PM demodulation (feedback demodulation) but the general characteristics are valid for any application of the PLL. The input complex signal into a PLL has amplitude variation $A(t)$, a phase variation, $\theta(t)$, and a nuisance phase variation, $d(t)$. The goal of a PLL, in most cases, is to track and/or estimate the value of $\theta(t)$.

The typical PLL, seen in Figure 8.1, has three components: the phase detector, the loop filter, and the voltage controlled oscillator (VCO). Recall the VCO has the characteristics shown in Figure 8.2. The VCO has a quiescent frequency of f_c and a gain of $k_f = 2\pi f_d$ radians/s/volt.

EXAMPLE 8.1

For demodulation of PM we have $A(t) = A_c$, $\theta(t) = k_p m(t) + \phi_p$, and $d(t) = 0$.

$A(t)\exp[j(d(t)+\theta(t))]$
Phase Detector
$e_d(t)$
Loop Filter
$y_d(t)$
VCO
$\exp[j\hat{\theta}(t)]$

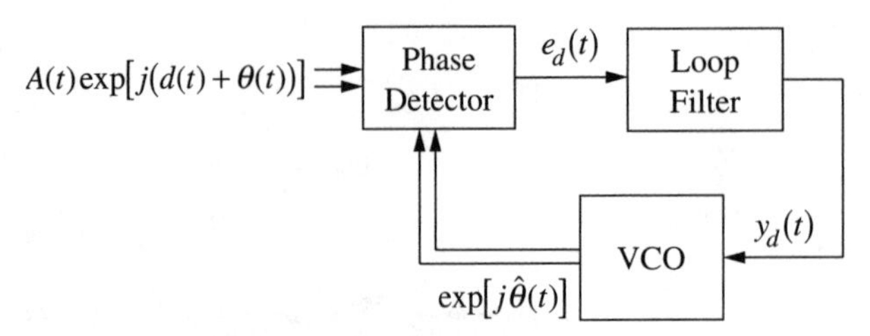

Figure 8.1 Phase-locked loop block diagram.

EXAMPLE 8.2

For the Costas loop used as a synchronous demodulator for DSB-AM we have $A(t) = A_c|m(t)|$, $\theta(t) = \phi_p$, and $d(t) = (1 - m(t))\frac{\pi}{2}$.

The phase detector simply measures the phase difference between the input signal and the locally generated reference. Denoting the input signal to the PLL as $y_z(t)$ the phase detector is a nonlinear function denoted $e_d(t) = g(y_z(t), \hat{\theta}(t))$, which is usually designed to be proportional to the phase error $\varphi(t) = \theta(t) - \hat{\theta}(t)$.

EXAMPLE 8.3

The most common phase detector for an unmodulated sinusoidal input signal ($A(t) = A_c$ and $d(t) = 0$) is the quadrature multiplier and the analytical model is seen in Figure 8.3. This phase detector is given as

$$g(y_z(t), \hat{\theta}(t)) = A_c k_m \Im[y_z(t)\exp[-j\hat{\theta}(t)]]$$

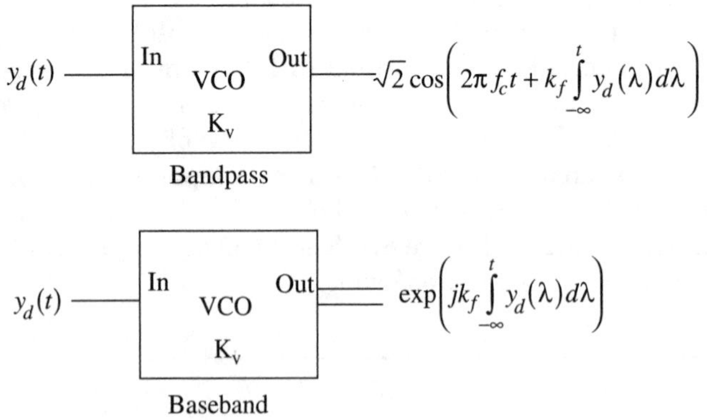

$y_d(t)$
In
VCO
K_v
Out
$\sqrt{2}\cos\left(2\pi f_c t + k_f \int_{-\infty}^{t} y_d(\lambda)d\lambda\right)$
Bandpass
$y_d(t)$
In
VCO
K_v
Out
$\exp\left(jk_f \int_{-\infty}^{t} y_d(\lambda)d\lambda\right)$
Baseband

Figure 8.2 The bandpass and baseband block diagrams for a VCO.

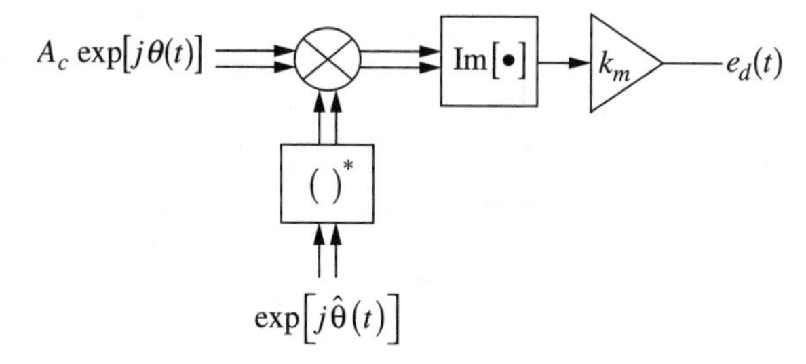

Figure 8.3 The quadrature multiplier phase detector block diagram.

where k_m is the multiplier gain. The phase detector output can be expressed in terms of the phase error as

$$e_d(t) = A_c k_m \sin(\theta(t) - \hat{\theta}(t)) = A_c k_m \sin(\varphi(t))$$

The signal $e_d(t)$ is a function of the phase error and it can be used as a feedback signal in a closed loop tracking system.

EXAMPLE 8.4

Note, Figure 6.9 shows another type of a phase detector which is useful for phase tracking with DSB-AM signals ($A(t) \neq A_c$ and $d(t) \neq 0$). The phase detector is given as

$$\begin{aligned} g(y_z(t), \hat{\theta}(t)) &= k_m \Im\left[(y_z(t)\exp[-j\hat{\theta}(t)])^2\right] \\ &= 2k_m \Im[y_z(t)\exp[-j\hat{\theta}(t)]]\Re[y_z(t)\exp[-j\hat{\theta}(t)]] \end{aligned} \tag{8.1}$$

Expressing the phase detector output in terms of the phase error gives

$$e_d(t) = k_m A_c^2 m^2(t) \sin(2\varphi(t))$$

The loop filter is a linear filter and its design is often critical in getting the desired performance out of a PLL. For this development the loop filter is represented with $F(s)$. Two common types of loops are the first and second order PLL. The first-order PLL loop filter has a transfer function given as

$$F(s) = 1$$

and the second-order PLL loop filter is given as

$$F(s) = \frac{s+a}{s+b}$$

It should be noted that the most common type of second order loop is one with a perfect integrator ($b = 0$). The second-order PLL loop filter is more commonly used in practice because it can both track a phase and a frequency offset.

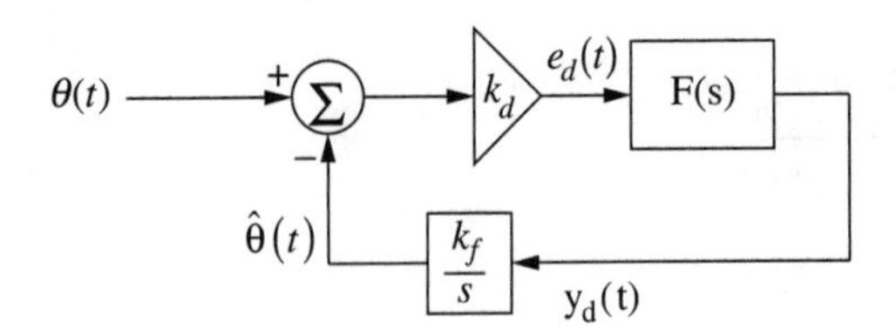

Figure 8.4 The linear model of the phase-locked loop.

8.1.2 PLL Linear Model

The linear analysis of the PLL is valid when the loop is closely tracking the input signal ($\varphi(t) \approx 0$). Close tracking implies that $\varphi(t)$ is small and that the phase detector can be linearized around $\varphi(t) = 0$ so that

$$g(y_z(t), \hat{\theta}(t)) \approx k_d \varphi(t)$$

where k_d is denoted the phase detector gain. This results in a linear model for the PLL that is shown in Figure 8.4. Note, since the phase of the VCO output is proportional to the integral of the input signal the VCO can be modeled by a transfer function with a Laplace transform of k_f/s where k_f is the gain of the VCO.

EXAMPLE 8.5

The quadrature multiplier has $e_d(t) = A_c k_m \sin(\varphi(t))$ so that the linear approximation has $k_d = A_c k_m$ and $\sin(\varphi(t)) \approx \varphi(t)$.

8.2 PLL–Based Angle Demodulation

8.2.1 General Concepts

Demodulation of angle modulations with a PLL is actually quite simple conceptually. This section will concentrate on demodulation of FM signals and the differences for PM will be pointed out when appropriate. Recall a received FM signal has the form

$$y_z(t) = A_c \exp\left[jk_f \int_{-\infty}^{t} m(\lambda) d\lambda + j\phi_p\right] \tag{8.2}$$

Figure 8.5 shows a block diagram of an FM system with a feedback demodulator. A VCO is used to produce the FM signal and the proposed receiver consists of a two input, one output black box and a VCO identical to the one used for modulation. After a transient period if $\theta(t) = \hat{\theta}(t)$ then we know $\hat{m}(t) = m(t)$. The job of the black box in the diagram is to measure the phase difference

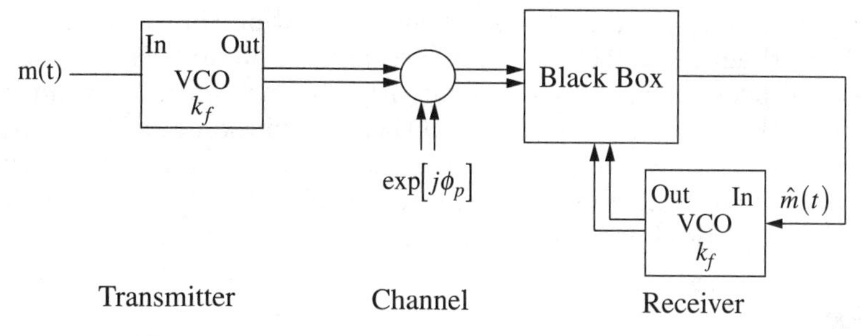

Figure 8.5 The conceptual diagram for FM demodulation.

$\varphi(t) = \theta(t) - \hat{\theta}(t)$ and update $\hat{m}(t)$ in such a way as to drive $\varphi(t)$ to zero. Note that if

$$\varphi(t) = \theta(t) - \hat{\theta}(t) < 0 \quad \Rightarrow \quad \int_{-\infty}^{t} m(\lambda)d\lambda < \int_{-\infty}^{t} \hat{m}(\lambda)d\lambda$$

$$\varphi(t) = \theta(t) - \hat{\theta}(t) > 0 \quad \Rightarrow \quad \int_{-\infty}^{t} m(\lambda)d\lambda > \int_{-\infty}^{t} \hat{m}(\lambda)d\lambda \tag{8.3}$$

Equation (8.3) indicates that $\varphi(t)$ can be driven to zero if $\hat{m}(t)$ is updated in proportion to $\varphi(t)$. Thus the black box in Figure 8.5 is composed of the other two elements of the PLL: the phase detector and the loop filter.

Since the FM signal has a constant amplitude the quadrature multiplier (see Figure 8.3) can be used as the phase detector. Consequently, the overall block diagram is seen in Figure 8.6 and the equation defining the operation is

$$\hat{m}(t) = f(t) * k_d \sin\left(k_f \int_{-\infty}^{t} (m(\lambda) - \hat{m}(\lambda))d\lambda + \phi_p\right) \tag{8.4}$$

where $k_d = A_c k_m$ is the quadrature multiplier phase detector gain. It is important to note that the phase detector gain is a function of the received signal amplitude. While this amplitude is treated as a known constant in these notes in practical systems accounting for this amplitude is an important function in a FM demodulator based on a PLL.

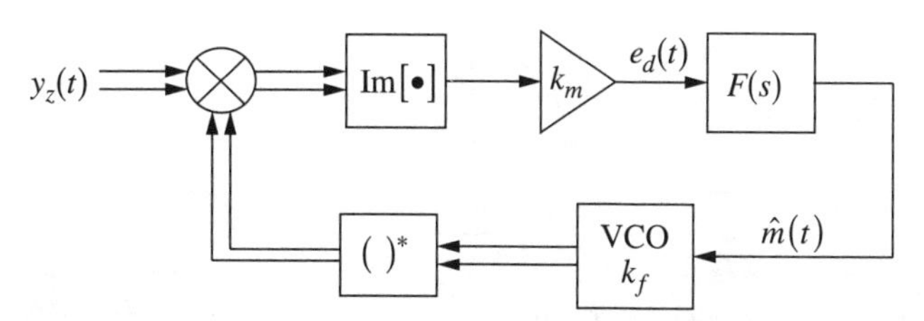

Figure 8.6 The block diagram of a PLL for FM demodulation.

Equation (8.4) is a nonlinear integral equation that can be solved but only with mathematical tools that undergraduate engineering students typically have not been exposed to. Consequently, this presentation examines approximations that simplify the mathematical description of a PLL. This simplified description enables the use of undergraduate engineering tools to be applied to the problem of analyzing the performance of a PLL. These approximations will give some design insights and provide a rough performance characterization.

8.2.2 PLL Linear Model

The linear analysis of the PLL is valid when the loop is closely tracking the input signal. Close tracking implies that $\varphi(t)$ is small and for the quadrature multiplier phase detector

$$e_d(t) = A_c k_m \sin(\varphi(t)) \approx A_c k_m \varphi(t) = k_d \varphi(t)$$

This results in a linear model for a PLL used in FM demodulation that is shown in Figure 8.7. Rearranging and defining $k = k_f k_d$ gives the block diagram shown in Figure 8.8. It is quite obvious from Figure 8.8 that the linear model for the PLL reduces to a simple linear feedback control system commonly studied in an undergraduate program. Taking Laplace transforms yield

$$(M(s) - \hat{M}(s))\frac{kF(s)}{s} = \hat{M}(s)$$

Solving for the transfer function gives

$$H_L(s) = \frac{\hat{M}(s)}{M(s)} = \frac{kF(s)}{s + kF(s)} \tag{8.5}$$

Often the error between the demodulated output and the message signal is of interest. Defining $e(t) = m(t) - y_d(t)$ and using identical techniques a transfer function for the error is given as

$$H_E(s) = \frac{E(s)}{M(s)} = \frac{s}{s + kF(s)}$$

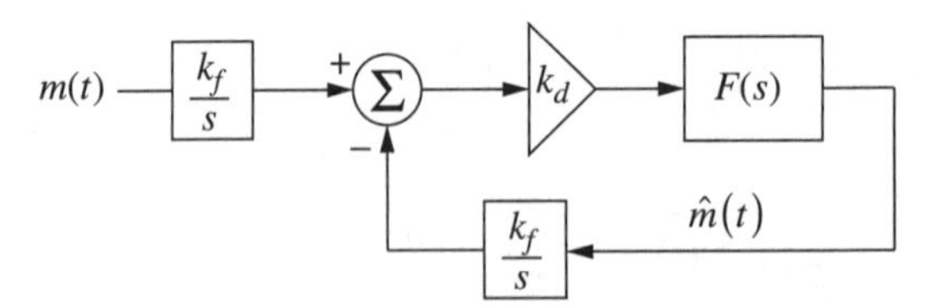

Figure 8.7 The PLL linear model for FM demodulation.

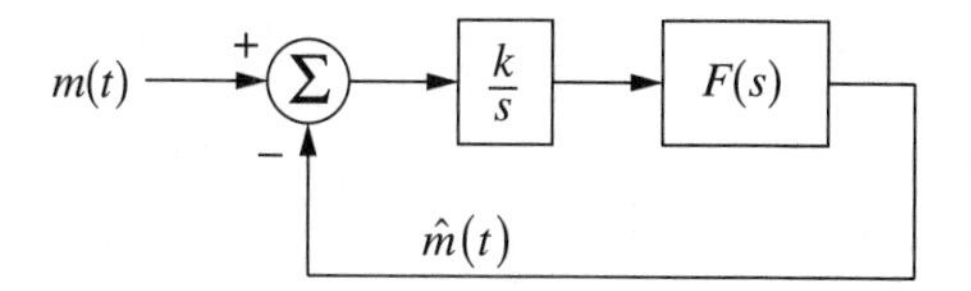

Figure 8.8 The PLL linear model rearranged to have a classic feedback control systems structure.

Frequency Response

The frequency response of the loop is of interest. The frequency response is given as

$$H(f) = H(s)|_{s=j2\pi f}$$

The loop and error transfer functions are plotted Figure 8.9 for a first-order loop with $k = 1000$. Consequently, when the linear model is valid the PLL appears much like a lowpass filter to the message signal. For example, for the case considered in Figure 8.9 ($k = 1000$) the 3 dB-loop bandwidth is about 100 Hz. Consequently, design of PLLs for FM demodulation must carefully consider the bandwidth of the message signal in choosing the bandwidth of the loop.

Acquisition

Let's consider the transient response produced by a first-order loop ($F(s) = 1$) to a sinusoidal input signal, $m(t) = \sin(2\pi f_m t)$. The Laplace transform of this signal is

$$M(s) = \frac{2\pi f_m}{s^2 + (2\pi f_m)^2} \tag{8.6}$$

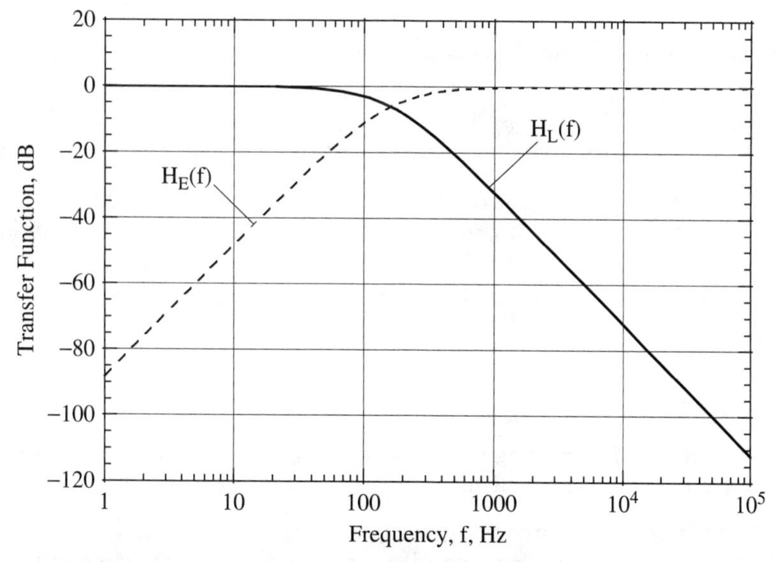

Figure 8.9 The frequency response of a first-order PLL used for FM demodulation. $k = 1000$

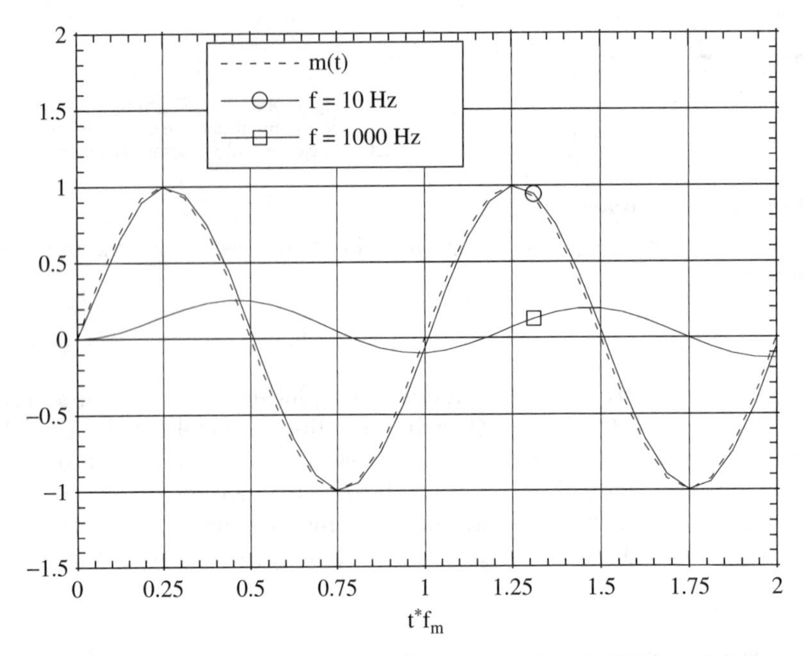

Figure 8.10 The time response of the linear model of a PLL for FM demodulation.

Using Eq. (8.6) in Eq. (8.5) produces a loop output Laplace transform of

$$\hat{M}(s) = \frac{2\pi f_m}{s^2 + (2\pi f_m)^2} \frac{k}{s+k}$$

Inverse transforming this signal gives

$$\hat{m}(t) = \frac{k(2\pi f_m)}{k^2 + (2\pi f_m)^2} e^{-kt} + \frac{k}{\sqrt{k^2 + (2\pi f_m)^2}} \sin\left[2\pi f_m t + \tan^{-1}\left(\frac{2\pi f_m}{k}\right)\right]$$

Figure 8.10 shows a plot of the time response for $k = 1000$ and $f_m = 10$ and $f_m = 1000$. Note that if the signal is inside the passband of the loop, the loop quickly locks onto the signal and tracks it very closely. If the signal is outside the loop bandwidth the signal is not tracked and the FM demodulation performance is poor.

8.3 Multiplexing Analog Signals

Definition 8.1 Multiplexing for analog message signals is the common transmission of multiple message signals using a single communications resource.

The common resource is typically spectrum (e.g., broadcast radio), expensive electronic equipment (e.g., satellites), or both (e.g., analog cellular telephony). For this discussion we will denote the ith message signal as $m_i(t)$ and K as

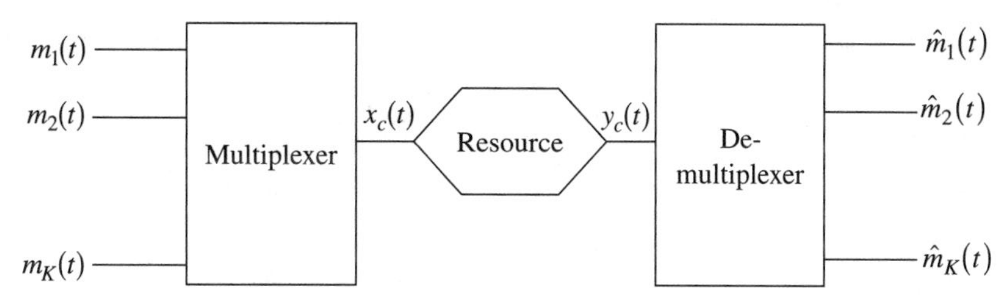

Figure 8.11 The multiplexing concept.

the number of users of the resource. Consequently, the idea of multiplexing is shown in Figure 8.11. As a measure of how efficiently we are using the resource we define the multiplexing efficiency as

$$E_M = \frac{\sum_{i=1}^{K} W_i}{B_T} \tag{8.7}$$

where W_i is the ith message bandwidth and B_T is the transmission bandwidth of the composite signal. Note, the best multiplexing efficiency without significant distortion for analog communications is unity. For analog signals there are two basic types of multiplexing techniques: (1) quadrature carrier multiplexing and (2) frequency division multiplexing.

EXAMPLE 8.6

For those homes that subscribe to cable television services a single cable enters the home with an electronic signal. Typically greater than 100 channels are supported with that one cable.

8.3.1 Quadrature Carrier Multiplexing

Quadrature carrier multiplexing (QCM) uses the two degrees of freedom in a bandpass signal to transmit two message signals on the same carrier. This type of multiplexing is achieved by setting $x_I(t) = A_c m_1(t)$ and $x_Q(t) = A_c m_2(t)$. The block diagram is shown in Figure 8.12. The efficiency of QCM is

$$E_M = \frac{W_1 + W_2}{2 \max W_1, W_2} \tag{8.8}$$

Note, if $W_1 = W_2$ then $E_M = 1$ which is the best that can be accomplished in analog communications. The advantage of QCM is that it is a very efficient method of multiplexing message signals. The disadvantage is that the applicability is limited to only two message signals and that coherent demodulation must be used (typically with a transmitted reference carrier which will reduce the bandwidth efficiency somewhat).

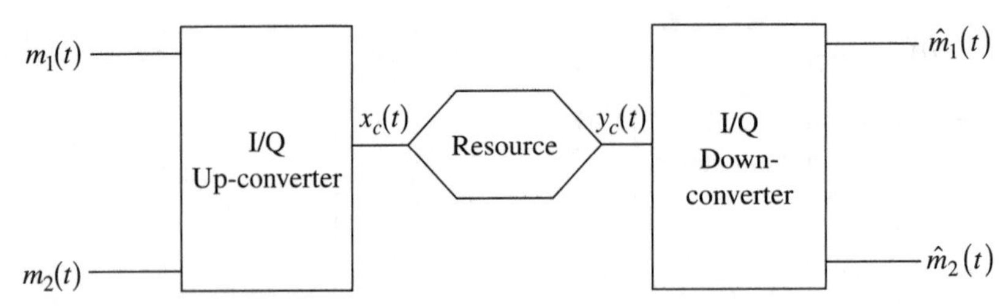

Figure 8.12 Quadrature carrier multiplexing.

8.3.2 Frequency Division Multiplexing

Frequency division multiplexing (FDM) assigns each user a different carrier frequency for transmission. Typically the carrier frequencies for each user are chosen so that the transmissions do not overlap in frequency. Figure 8.13 shows a typical energy spectrum for a FDM system. Typically each user in the system is modulated with an analog modulation and has a bandwidth of B_i. This modulation is then separated from its spectral neighbor by a guard band. The size of this guard band and the modulation type limits the spectral efficiency. The multiplexing for FDM is accomplished by simply summing together appropriately modulated bandpass waveforms as shown in Figure 8.14. Demultiplexing for FDM, seen in Figure 8.15, simply corresponds to bandpass filtering to isolate each user followed by a standard I/Q demodulator. Note, the guard bands between users are usually chosen to be compatible with the bandpass filtering operation (i.e., the guard band is made large enough to ensure each user can be adequately separated at the demultiplexer).

EXAMPLE 8.7

Broadcast communications in the United States use the resource of free space radio wave propagation. The multiplexing is typically implemented with a combination of frequency

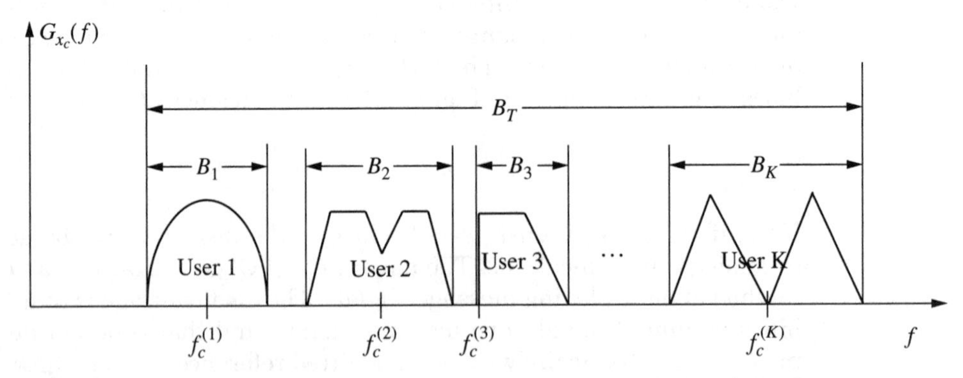

Figure 8.13 A typical energy spectrum for a frequency division multiplexed signal.

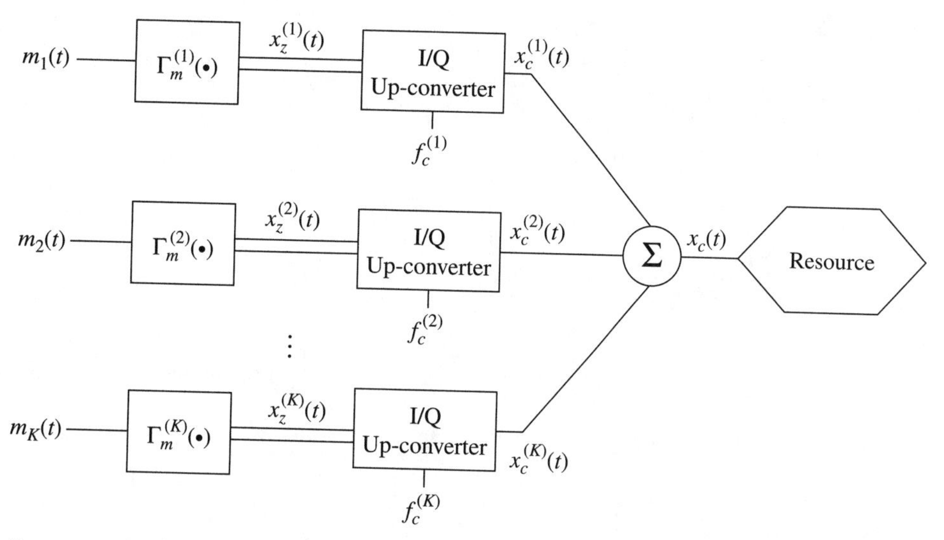

Figure 8.14 A FDM multiplexing system.

multiplexing and location multiplexing (stations far enough away from each other can use the same transmission frequency). AM radio broadcast in the United States has been assigned the spectrum of 535–1605 kHz, $B_T = 1070$ kHz. There are 107 FDM channels spaced about 10 kHz apart ($f_c^{(i)} = 540 + 10i$ kHz). Each AM station provides an audio

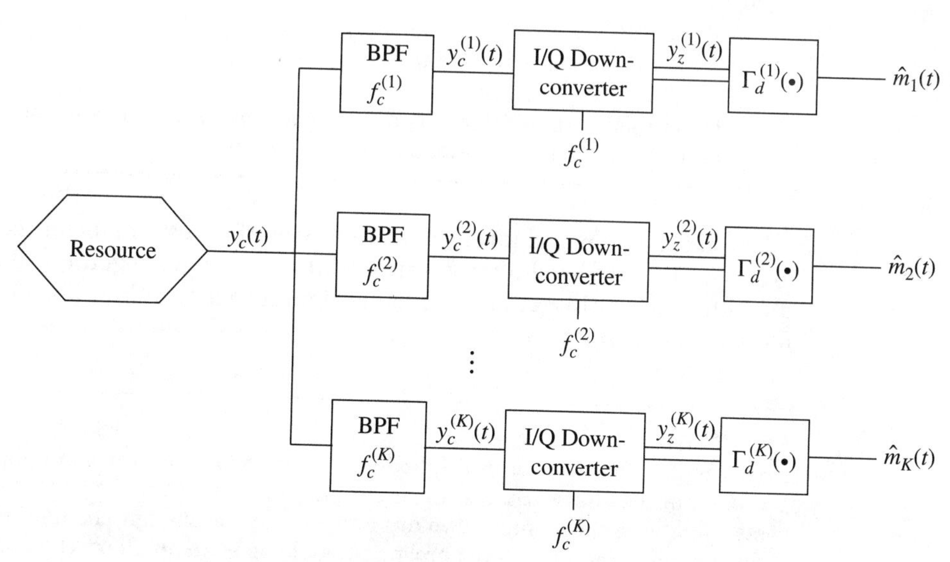

Figure 8.15 A FDM demultiplexing system.

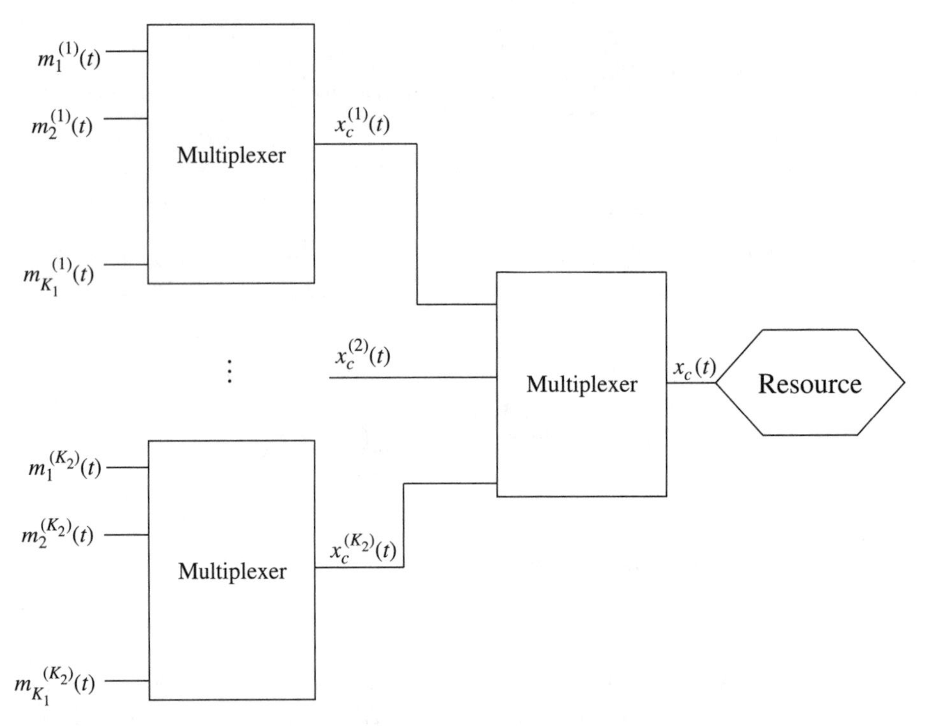

Figure 8.16 A two-level multiplexing scheme.

signal of about 4 kHz bandwidth ($W_i = 4$ kHz). The efficiency of AM broadcast is then given as

$$E_M = \frac{107 \times 4}{1070} = 40\%$$

Note, 20% of the loss in efficiency is due to guard bands and 40% of the loss is due to using spectrally inefficient LC-AM as the modulation.

There can be several levels of multiplexing that are implemented for a given system. The first level often groups related messages together and the subsequent levels multiplex several groups of messages together. The block diagram for a system using two levels of multiplexing is shown in Figure 8.16.

EXAMPLE 8.8

An example of multilevel multiplexing is seen in broadcast communications in the FM band in the United States. Each FM station in the United States typically broadcasts three message signals. Two are audio channels: the left channel, $m_L(t)$ and the right channel, $m_R(t)$ for stereo broadcast, and one is an auxiliary channel, $m_A(t)$ (e.g., Musak, or Radio Data Services). These three message signals are multiplexing onto one composite message signal which is FM modulated and transmitted. For discussion purposes the bandwidth of these message signals can be assumed to be $W_i = 15$ kHz. The first level of multiplexing FDM with $m_1(t) = m_L(t) + m_R(t)$, $m_2(t) = m_L(t) - m_R(t)$, and $m_3(t) = m_A(t)$, $\Gamma_m^{(1)} = \Gamma_m^{(2)} = \Gamma_m^{(3)} =$ DSB-AM, and $f_c^{(1)} = 0$, $f_c^{(2)} = 38$ kHz, and $f_c^{(3)} = 76$ kHz.

8.4 Conclusions

This chapter considered two ideas that have found utility in analog communications; phase locked loops and multiplexing. Phase-locked loops are used in a variety of applications in analog communications. Examples include phase reference estimation and tracking in DSB-AM demodulation and in demodulation of angle modulations. This chapter gave an introductory overview and showed how tools from feedback control are useful in gaining an understanding of the operation of phase-locked loops. Multiplexing is an important aspect of communications. Multiplexing allows multiple information sources to be transmitted on a common resource. The common forms of multiplexing for analog communications include quadrature carrier multiplexing and frequency multiplexing. Again the trade-offs between spectral efficiency and complexity were evident in multiplexing techniques.

8.5 Homework Problems

Problem 8.1. Your boss at Enginola has assigned you to be the systems engineer in charge of an FM demodulator. Unfortunately, this design must be done in one day. The demodulator is required to be a PLL-based system and since you are pressed for time, you will use the linear approximation for the loop. Figure 8.17 is the model for your system.

The transfer function, $H_L(f)$, from the input $(m(t))$ to the output $(\hat{m}(t))$ has the form

$$H_L(s) = \frac{2\zeta\omega_n s + \omega_n^2}{s^2 + 2\zeta\omega_n s + \omega_n^2}$$

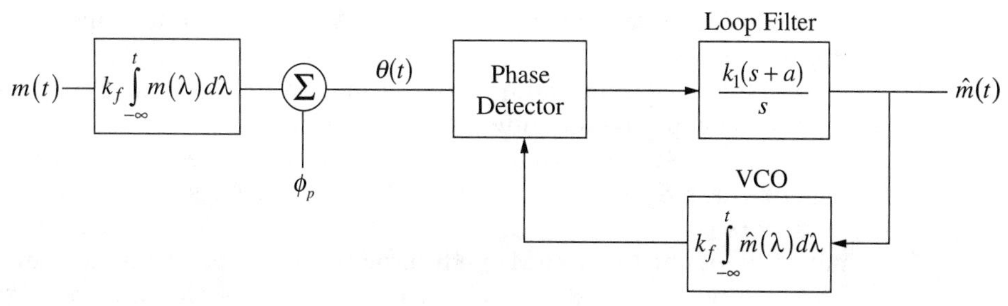

Figure 8.17 Block diagram of a PLL demodulator for FM.

(a) Select values of k_f, k_1, and a such that $\zeta = 1.00$ and $|H_L(12\text{ kHz})| > 0.707$ (i.e., the 3-dB bandwidth is 12 kHz). Note, while this problem can be completed with pencil and paper, a proficiency with a computer would be useful.

(b) If the input signal is $m(t) = \sin(2\pi(10\text{ kHz})t)U(t)$ where $U(t)$ is the unit step function, calculate the output response to this signal and plot the response.

Problem 8.2. As a communication engineer you need to transmit eight voice signals ($W = 4$ kHz) across a common communication resource at a carrier frequency of 1 GHz. For hardware reasons you need to maintain a 1 kHz guard band between frequency multiplexed transmissions.

(a) If you use DSB-AM and FDM, what is the total used bandwidth, B_T, and the multiplexing efficiency, E_M?

(b) If you use SSB-AM and FDM what is the total used bandwidth and the multiplexing efficiency, E_M?

(c) If a two-stage multiplexing scheme is employed such that two voice signals are grouped together with QCM and the resulting four signals are multiplexed using FDM, what is the total bandwidth and the multiplexing efficiency, E_M?

Problem 8.3. In the text the example of a first-order loop was considered for FM demodulation. $k = 1000$ gave a design of a loop with a bandwidth of about 100 Hz.

(a) Redesign the loop gain, k, to give a 3-dB bandwidth of 5 kHz.

(b) Note that the loop gain is a function of the signal amplitude, A_c. Keeping the remainder of the loop the same and if the signal amplitude is cut in half what will be the loop 3 dB bandwidth?

(c) Keeping the remainder of the loop the same and if the signal amplitude is doubled what will be the loop 3 dB bandwidth?

(d) Postulate how PLL-based FM demodulators might deal with amplitude variations in the received signal.

Problem 8.4. High fidelity audio broadcast typically tries to maintain a 15 kHz of bandwidth in transmitting to listener. An example audio spectrum is given in Figure 8.18. In the days before stereo became a reality FM broadcasts frequency modulated a message signal of the form seen in Figure 8.18 onto a carrier. This message signal had a bandwidth of 15 kHz.

After FM had been around a while stereo audio became a reality and a financially desirable service to provide. After long discussions by the communication engineers of the day it was decided that stereo FM was possible if the input message signal to the FM modulator had the spectrum seen in Figure 8.19. Essentially the signal resulting from the sum of the left and right channels,

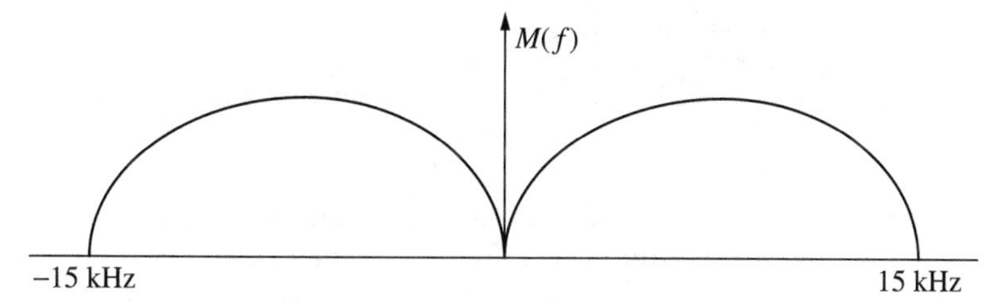

Figure 8.18 An example spectrum for audio broadcast.

$m_S(t) = m_L(t) + m_R(t)$ is combined with an DSB-AM modulated signal ($f_2 =$ 38 kHz) of the difference of the left and the right channels, $m_D(t) = m_L(t) - m_R(t)$. There is a third signal that is a tone at 19 kHz that is used as a reference for the DSB-AM demodulator. The composite message signal has the form

$$m(t) = m_S(t) + m_D(t)\sqrt{2}\cos(2\pi f_2 t) + A_m\sqrt{2}\cos(\pi f_2 t) \tag{8.9}$$

This way of combining the left and right channel was considered a nice solution since a radio that did not have stereo capability would still decode $m_S(t)$ and not lose any capability.

(a) For this method of multiplexing these two signals, $m_L(t)$ and $m_R(t)$, on to one signal, $m(t)$ compute the multiplexing efficiency, E_M.

(b) Draw a block diagram of the multiplexing system.

(c) After demodulation assume the decoded message signal has the form

$$\hat{m}(t) = m_S(t) + m_D(t)\sqrt{2}\cos(2\pi f_2 t) + A_m\sqrt{2}\cos(\pi f_2 t) + N_L(t) \tag{8.10}$$

Assume a bandpass filter is centered at $f = 19$ kHz, what bandwidth must the filter have to not pass anything but the tone at $f = 19$ kHz so that the resulting bandpass filter output is

$$m_r(t) = A_m\sqrt{2}\cos(\pi f_2 t) + N_r(t) \tag{8.11}$$

(d) Show that when the noise is ignored Figure 8.20 will produce the needed carrier signal for a coherent demodulator of the DSB-AM.

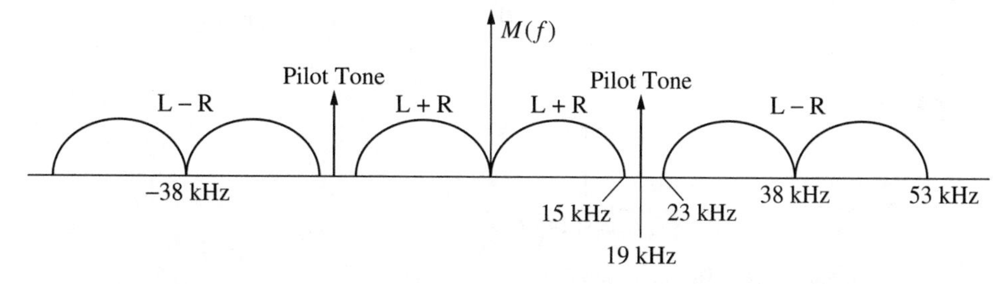

Figure 8.19 Stereo FM message signal.

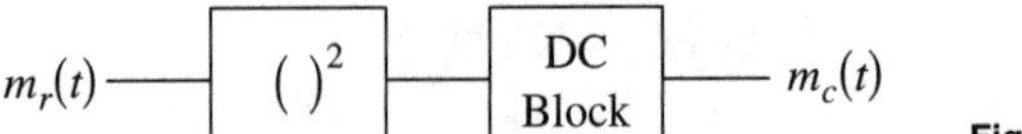

Figure 8.20 Tone processor.

(e) Draw a block diagram of the stereo demultiplexer having an input of $\hat{m}(t)$ and outputs $m_L(t)$ and $m_R(t)$, using the idea presented in (d).

(f) If SSB-AM (lower sideband) replaced DSB-AM in Eq. (8.9), draw the block diagram of the multiplexing system and compute the multiplexing efficiency, E_M.

(g) Draw a block diagram of the stereo demultiplexer for the system using SSB-AM using the idea presented in (d). Why do you think the engineers of the day chose DSB-AM over SSB-AM?

Problem 8.5. In systems where frequency division multiplexing is used the concept of a superheterodyning receiver has found utility in practice. The idea is to take a bandpass signal, which is centered on a carrier frequency, and shift the frequency to some intermediate value, called the intermediate frequency (IF). For practical purposes, the superheterodyne receiver always reduces to the same value of IF. The block diagram of the superheterodyne receiver is shown in Figure 8.21 where the filters are bandpass filters centered at frequency f_c and f_{IF}, respectively.

(a) Assuming a typical bandpass modulated spectrum at the input to the radio what is the output spectrum corresponding to $y_1(t)$?

(b) A common IF frequency is $f_2 = 10.7$ MHz. Find the value(s) of f_1 such that this receiver architecture could be used to demodulate a radio signal at 89.3 MHz.

(c) In a superheterodyne receiver two frequencies arrive at the IF frequency after the mixing with f_1. One signal is the desired and one signal is denoted

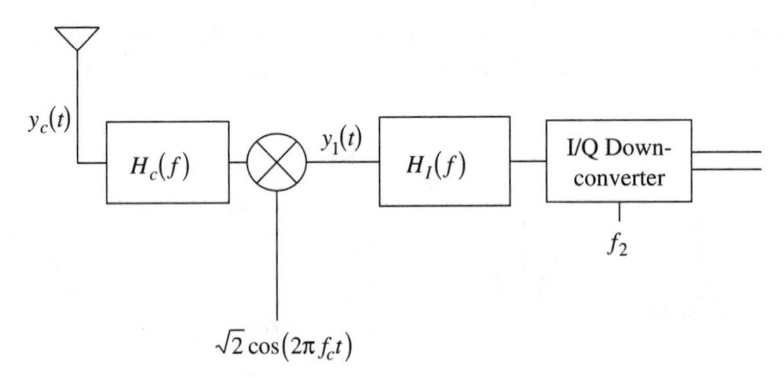

Figure 8.21 The superheterodyning receiver.

the "image" frequency. For the case detailed in (b) what is the image frequency? It should be noted that one of the main functions of $H_c(f)$ in a superheterodyne receiver is to make sure this image frequency has been appropriately suppressed.

(d) The FM radio band is from 88 to 108 MHz. The FM stations are assigned center frequencies at 200 kHz separation starting at 88.1 MHz, for a maximum of 100 stations. Assume a frequency synthesizer that can step f_1 to whatever frequencies that is desired and $f_2 = 10.7$ MHz. Specify the filters $H_c(f)$ and $H_I(f)$ so that it would be possible to perfectly demodulate every FM channel and eliminate interference from any adjacent transmissions.

Problem 8.6. In television (TV) broadcast there are four signals that must be multiplexed into one channel. These signals are

- Black and white video – $W_1 = 4.2$ MHz
- Color 1 – $W_2 = 1.5$ MHz
- Color 2 – $W_3 = 0.5$ MHz
- Stereo audio – $W_4 = 54$ kHz.

Design a multiplexer and a modulator that would fit the transmission of these four signals in a small bandpass bandwidth. Do not feel constrained by past practice in developing a design. Sketch block diagrams for both the modulator/multiplexer and the demodulator/demultiplexer. What do you think the minimum achievable bandwidth would be for a TV channel that would contain these four independent signals?

Problem 8.7. You are working for Glenfyre a leading paging company. Your first assignment is to set up a communications link from just outside the main paging headquarters to the telephone company home office. The FCC has allocated to the Glenfyre 20 kHz of spectrum for this communication at a carrier frequency of 1 GHz. You are told that you have four voice channels containing speech (W = 4 kHz). The guard band between frequency multiplexed systems must be at least 1 kHz.

(a) Design a complete multiplexing scheme that transmits the four voice channels on the 20 kHz allocated to Glenfyre. Specify the center frequencies of any bandpass signals, the guard bands between frequency division multiplexed signals, and sketch the multiplexer.

(b) Plot an example output spectrum and draw the demultiplexer. Compute E_M.

(c) You are asked to support eight voice channels in the same 20 kHz of spectrum. Is this possible? If yes, sketch the multiplexer and demultiplexer.

Problem 8.8. A voiceband signal is typically specified to have frequencies from 200 to 3200 Hz. A telegraph signal is usually specified to have frequencies from

0 to 100 Hz. These signals are easily frequency multiplexed. Draw a possbile multiplexer and a demultiplexer. Compute the multiplexer efficiency, E_M.

Problem 8.9. A voiceband signal is typically specified to have frequencies from 200 to 3200 Hz. In the United States analog voice communications was the standard for telephone communications in the early days of the telephone network. One of the important jobs of the telephone company is to efficiently multiplex many voice signals across the cables that were laid between cities. The first level of multiplexing in analog telephony in the United States (denoted 12VF by American Telephone and Telegraph (AT&T)) used SSB modulation for each message and frequency multiplexed 12 voiceband signals into a $B_T = 48$ kHz bandwidth. The second level of multiplexing (denoted 60VF by AT&T) used SSB modulation for each message and frequency multiplexed 5 12VF signals into a $B_T = 240$ kHz bandwidth.

(a) Compute the overall multiplexing efficiency, E_M, for this two-level multiplexing scheme.

(b) Draw an example block diagram of the two-level multiplexer.

(c) Draw an example block diagram of the two-level demultiplexer.

Problem 8.10. The telecommunication infrastructure in the United States was originally an analog transmission system. Telephone calls were multiple layer multiplexed on cables that were strung across the nation. The lowest levels of multiplexing are detail as

- V12 – 12 voiceband channels grouped together
- V60 – 5 V12 signals multiplexed together
- V600 – 10 V60 signals multiplexed together
- V3600 – 6 V600 signals multiplexed together.

The cost of the multiplexer and demultiplexer grow with the number of voice band signals that are supported. A typical design problem is to identify the size of the multiplexer needed to support a desired quality of service in a telecommunication system. A cable is to be run between Akron, Ohio, and Columbus Ohio to carry phone calls between these two cities and measurements show that the number of calls is well modeled as a Poisson random variable, i.e.,

$$P_N(n) = \frac{\lambda^k \exp[-\lambda]}{k!} \tag{8.12}$$

where λ is the average number of calls. Assuming that between Akron and Columbus the average number of calls is $\lambda = 40$, find the size of multiplexer that should be deployed for this cable such that there is less than a 1% chance that a call will be blocked.

8.6 Example Solutions

Problem 8.2.

(a) If eight signals are frequency multiplexed using DSB-AM the transmitted signal spectrum is shown in Figure 8.22. The transmission bandwidth is given as

$$B_T = 8 \text{ users} \times 2\ W\text{Hz/user} + 7 \text{ guardbands} \times 1 \text{ kHz/guardband} \quad (8.13)$$

Therefore, $B_T = 8 \times 8 + 7 = 71$ kHz and

$$E_M = \frac{8 \times W}{B_T} = \frac{32}{71} = 45.07\% \quad (8.14)$$

(b) If eight signals are frequency multiplexed using SSB-AM the transmission bandwidth is given as

$$B_T = 8 \text{ users} \times W\text{Hz/user} + 7 \text{ guardbands} \times 1 \text{ kHz/guardband} \quad (8.15)$$

Therefore, $B_T = 8 \times 4 + 7 = 39$ kHz and

$$E_M = \frac{8 \times W}{B_T} = \frac{32}{39} = 82\% \quad (8.16)$$

(c) If two signals are quadrature carrier multiplexed then the resulting bandwidth is 8 kHz. FDM four pairs of signals will give a transmission bandwidth of

$$B_T = 4 \text{ pairs} \times 2\ W\text{Hz/pair} + 3 \text{ guardbands} \times 1 \text{ kHz/guardband} \quad (8.17)$$

Therefore, $B_T = 4 \times 8 + 3 = 35$ kHz and

$$E_M = \frac{8 \times W}{B_T} = \frac{32}{35} = 91.4\% \quad (8.18)$$

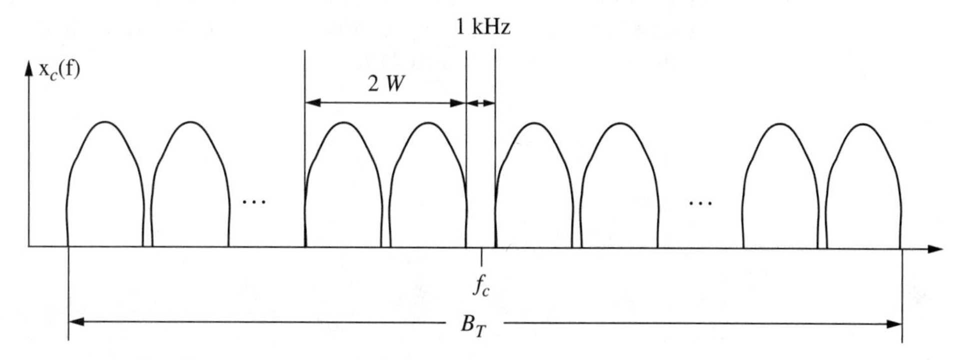

Figure 8.22 Bandpass spectrum of a DSB-AM FDM system.

8.7 Miniprojects

Goal: To give exposure

- to a small scope engineering design problem in communications.
- to the dynamics of working with a team.
- to the importance of engineering communication skills (in this case oral presentations).

Presentation: The forum will be similar to a design review at a company (only much shorter). The presentation will be of 5 minutes in length with an overview of the given problem and solution. The presentation will be followed by questions from the audience (your classmates and the professor). All team members should be prepared to give the presentation.

8.7.1 Project 1

Project Goals: Analyze a signal and figure out a multiplexing and demultiplexing strategy.

Problem 8.4 gives the details of the multiplexing of stereo signals in analog FM broadcast in the United States. To multiplex two channels of 15 kHz audio requires 53 kHz of spectrum resulting in a

$$E_M = \frac{30}{53} = 57\% \tag{8.19}$$

Prof. Fitz would like to propose a more efficient multiplexing strategy that would require no change in the demultiplexer. Get the Matlab file `projfdm1data.m` from the class web page. This file contains a multiplexed sum and difference audio signal sampled at $f_s = 132.3$ kHz using Prof. Fitz's proposed multiplexing strategy.

(a) Examining the signal identify the multiplexing strategy.

(b) Compute the multiplexing efficiency of Prof. Fitz's multiplexing strategy.

(c) Identify a demultiplexer and produce the two audio signals. If this signal processing is done well a song should be able to be identified in the output using the `sound` command in Matlab.

Part

3

Noise in Communications Systems

Chapter

9

Random Processes

An additive noise is characteristic of almost all communication systems. This additive noise typically arises from thermal noise generated by random motion of electrons in the conductors comprising the receiver. In a communication system the thermal noise having the greatest effect on system performance is generated at and before the first stage of amplification. This point in a communication system is where the desired signal takes the lowest power level and consequently the thermal noise has the greatest impact on the performance. This characteristic is discussed in more detail in Chapter 10. This chapter's goal is to introduce the mathematical techniques used by communication system engineers to characterize and predict the performance of communication systems in the presence of this additive noise. The characterization of noise in electrical systems could comprise a course in itself and often does at the graduate level. Textbooks that provide more detailed characterization of noise in electrical systems are [DR87, Hel91, LG89, Pap84, YG98].

A canonical problem formulation needed for the analysis of the performance of a communication systems design is given in Figure 9.1. The thermal noise generated within the receiver is denoted $W(t)$. This noise is then processed in the receiver and will experience some level of filtering, represented with the transfer function $H_R(f)$. The simplest analysis problem is examining a particular point in time, t_s and characterizing the resulting noise sample, $N(t_s)$, to extract a parameter of interest (e.g., average signal–to–noise ratio (SNR)). Additionally, we might want to characterize two or more samples, e.g., $N(t_1)$ and $N(t_2)$, output from this filter.

To accomplish this analysis task this chapter first characterizes the thermal noise, $W(t)$. It turns out that $W(t)$ is accurately characterized as a stationary, Gaussian, and white random process. Consequently, our first task is to define a random process (Section 9.1). The exposition of the characteristics of a Gaussian random process (Section 9.2) and a stationary Gaussian random process (Section 9.3) then will follow. A brief discussion of the characteristics of thermal noise is then followed by an analysis of stationary random processes

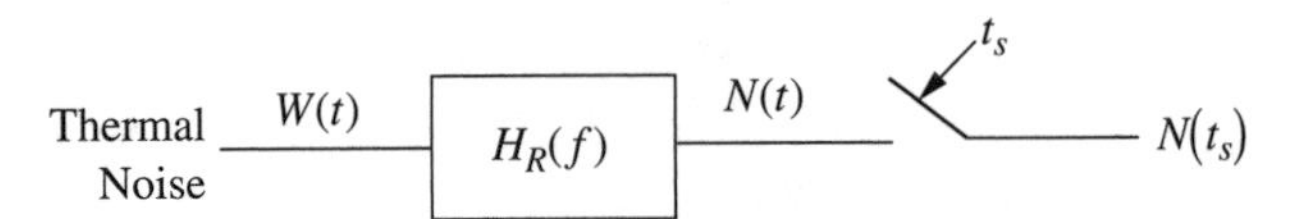

Figure 9.1 The canonical communications noise formulation.

and linear systems. In conclusion, we return and solve the canonical problem posed in Figure 9.1 in Section 9.6.

9.1 Basic Definitions

Understanding random processes is fundamental to communications engineering. Essentially random or stochastic processes are indexed families of random variables where the index is normally associated with different time instances.

Definition 9.1 Let $(\Omega, \mathcal{F}, P)$ be a probability space. A real random process, $N(\omega, t)$, with $\omega \in \Omega$ is a single-valued function or mapping from Ω to real valued functions of an index set variable t.

The index set in this text will always be time but generalizations of the concept of a random process to other index sets is possible (i.e., space in image processing). The idea behind the definition of a random process is shown in Figure 9.2. Essentially there is an underlying random experiment. The outcome of this random experiment is then mapped to a function of time. The example in Figure 9.2 shows the mapping of three of the possible experiment outcomes into three different functions of time.

Definition 9.2 The function $N(\omega_1, t)$ for ω_1 fixed is called a *sample path* of the random process.

Communication engineers observe the sample paths of random processes and the goal of this chapter is to develop the tools to characterize these sample paths.

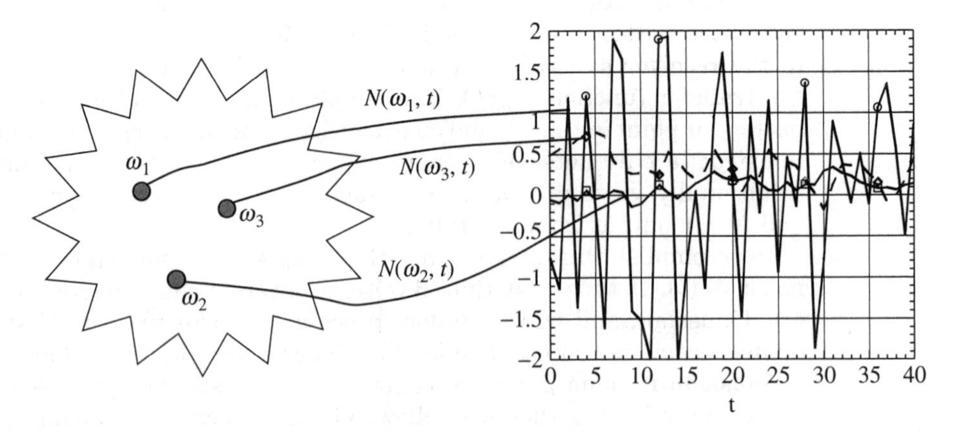

Figure 9.2 A pictorial explanation of the definition of a random process.

From this point forward in the text the experimental outcome index will be dropped and random processes will be represented as $N(t)$.

EXAMPLE 9.1
A particular random process is defined as

$$N(t) = U \exp[-|t|] + V \tag{9.1}$$

where U and V are independent random variables. It is clear that with each sample value of the random variables $U(\omega)$ and $V(\omega)$ there will be a time function $N(t, \omega)$. This example of a random process is not typical of a noise process produced in real communication systems but it is an example process that proves insightful as we develop tools to characterize noise in communications.

EXAMPLE 9.2
A noise generator and a lowpass filter are implemented in Matlab with a sample rate of 22,050 kHz. Recall each time Matlab is run this is equivalent to a different experiment outcome, i.e., a different ω. A sample path of the input noise to the filter, $W(t)$, and a sample path at the output of the filter, $N(t)$, is shown in Figure 9.3 for a filter with a bandwidth of 2.5 kHz. It is clear that filtering drastically alters the characteristics of the noise process. Each time the Matlab program is run a different set of sample paths will be produced.

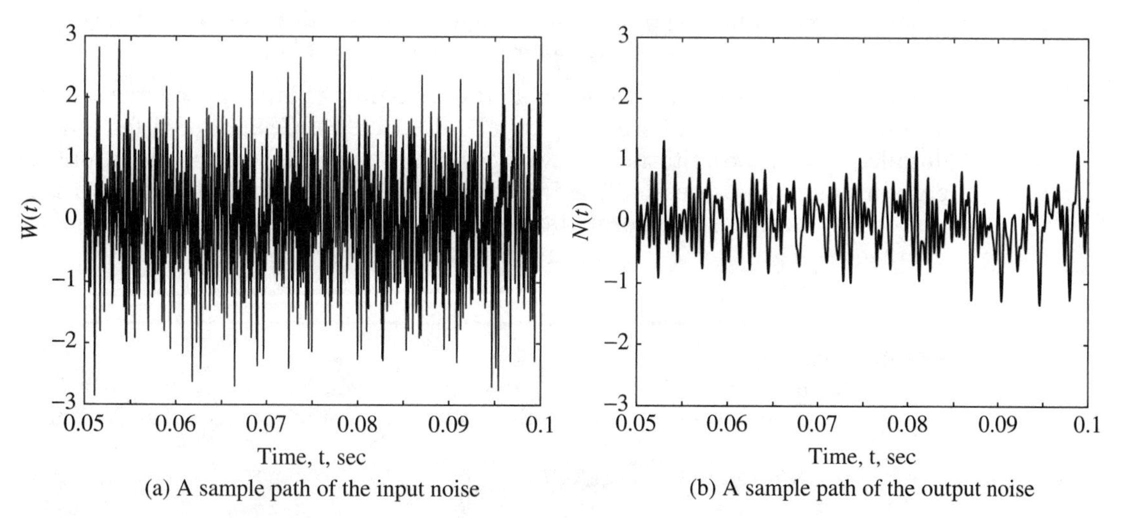

(a) A sample path of the input noise

(b) A sample path of the output noise

Figure 9.3 Input and output noise typical of the canonical problem highlighted in Figure 9.1 when $H_R(f)$ is a lowpass filter of bandwidth 2.5 kHz.

Property 9.1 A sample of a random process is a random variable.

If we focus on a random process at a particular time instance then we have a mapping from a random experiment to a real number. This is exactly the definition of a random variable (see Section 3.2). Consequently, the first important question in the canonical problem stated at the beginning of this chapter has been answered: *the sample $N(t_s)$ of the random process, $N(t)$, is a random variable*. Consequently, to characterize this sample of a random process we use all the tools that characterize a random variable (distribution function, density function, etc.). If random processes are indexed sets of random variables then a complete statistical characterization of the random process would be available if the random variables comprising the random process were completely characterized. In other words, for a random process, $N(t)$, if M samples (where M can be an arbitrary value) of the random process are taken at various times and if the PDF given by

$$f_{N(t_1),N(t_2),\cdots,N(t_M)}(n_1, n_2, \cdots, n_M)$$

is known then the random process is completely characterized. This full characterization, in general, is very difficult to do. Fortunately, the case of the Gaussian random process is the exception to the rule as Gaussian random processes are very often accurate models for noise in communication systems.

> **Point 1:** The canonical problem can be solved when the random variable $N(t_s)$ or a set of M random variables, $N(t_1), \ldots, N(t_M)$, can be statistically characterized with a PDF or a CDF.

Property 9.2 A numeric average of a set of independent samples of a random variable or sample paths of a random process will converge to the statistical average of that random variable or random process as the number of samples grows large.

This property is formally known as the weak law of large numbers (WLLN) and will not be proven here (see e.g., [DR87, Hel91, LG89, Pap84] for a more detailed discussion). Statistical averages play a very important role in understanding random processes. The intersting characteristic of the WLLN for this text is that Matlab can generate large number of sample paths and the numeric averages will give some insight into the statistical averages.

EXAMPLE 9.3

An example of the WLLN is given by the sample mean. For this example it will be assumed that there is a random variable X with finite set of moments. The N sample mean is

$$m_N = \frac{1}{N}\sum_{n=1}^{N} X(\omega_n) \tag{9.2}$$

If $X(\omega_n)$ is independently obtained, then the WLLN implies that

$$E[X] = \lim_{N \to \infty} m_N. \tag{9.3}$$

For example, Matlab can be used to generate N independent samples of a random variable or a random process and insight can be gained into the ensemble average by looking at m_N as N gets large.

9.2 Gaussian Random Processes

Definition 9.3 A Gaussian random process is a random process where any set of samples taken from the random process are jointly Gaussian random variables.

Many important random processes in communication system analysis are well modeled as Gaussian random processes. This is important since Gaussian random variables are simply characterized. A complete characterization of Gaussian random variables (their joint PDF) is obtained by knowing their first- and second-order moments.

EXAMPLE 9.4

Considering the Matlab experiment briefly highlighted in Example 9.2, histograms of values of the sample paths at $t = 45$ ms are shown for the input in Figure 9.4(a) and for the output in Figure 9.4(b). These histograms show that the noise in the canonical problem is well modeled by Gaussian random processes. Also the filtering has obviously changed the distribution of the random variables $W(0.045)$ and $N(0.045)$ since the measured variance of the input is clearly larger than the measured variance of

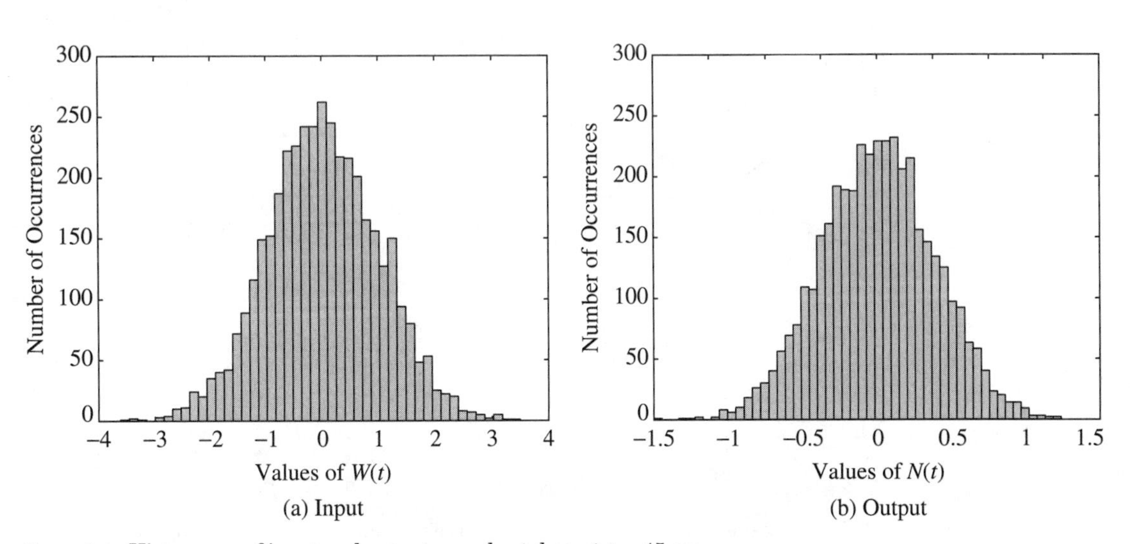

Figure 9.4 Histograms of input and output samples taken at $t = 45$ ms.

the output. The goal in the remainder of this chapter is to develop the needed tools to predict the change in statistical characterization of the noise as a function of the filtering, $H_R(f)$.

EXAMPLE 9.5

Considering random process defined in Example 9.1, if U and V are independent, zero mean, unit variance Gaussian random variables then, for every t, $N(t)$ will be the sum of two Gaussian random variables and hence a Gaussian random variable. Likewise, joint distributions of the samples of $N(t)$ will be jointly Gaussian random variables. Three example sample paths of the Gaussian random process $N(t) = U \exp[-|t|] + V$ is shown in Figure 9.5.

Property 9.3 If $N(t)$ is a Gaussian random process then one sample of this process, $N(t_s)$, is completely characterized with the PDF

$$f_{N(t_s)}(n_s) = \frac{1}{\sqrt{2\pi\sigma_N^2(t_s)}} \exp\left(-\frac{(n_s - m_N(t_s))^2}{2\sigma_N^2(t_s)}\right) \tag{9.4}$$

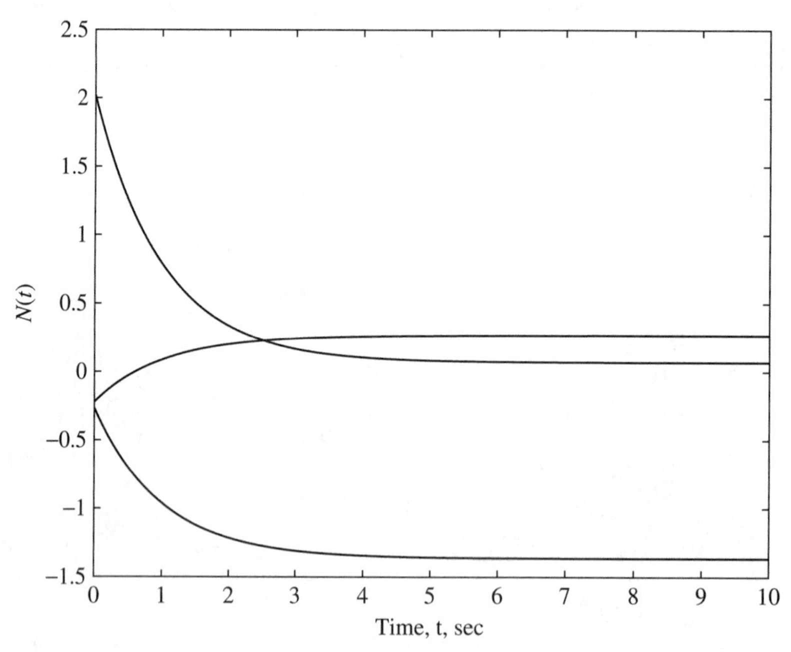

Figure 9.5 Three example sample paths of the Gaussian random process $N(t) = U \exp[-|t|] + V$.

where

$$m_N(t_s) = E[N(t_s)] \qquad \sigma_N^2(t_s) = \mathrm{var}(N(t_s))$$

Consequently, the complete characterization of one sample of a Gaussian random process only requires the mean value and the variance of $N(t_s)$ to be evaluated. Here is the first example of how ensemble averages of random quantities help characterize the random quantities. The Gaussian random variable has great utility because only two ensemble averages completely characterize the randomness. Section 3.2.4 has more discussion on the PDF of a Gaussian random variable.

EXAMPLE 9.6

Consider the random process defined in Example 9.1, where U and V are independent, zero mean, unit variance Gaussian random variables and samples are taken at $t = 0$ and $t = 2$. The appropriate moments needed to characterize the marginal PDFs of these two samples are

$$E[N(0)] = E[U] + E[V] = 0 \qquad E[N(2)] = E[U]e^{-2} + E[V] = 0 \tag{9.5}$$

$$E[N^2(0)] = E[(U+V)^2] = E[U^2] + E[V^2] = 2 \tag{9.6}$$

$$E[N^2(2)] = E[(Ue^{-2}+V)^2] = E[U^2]e^{-4} + E[V^2] = 1 + e^{-4} \tag{9.7}$$

Property 9.4 Most thermally generated noise corrupting a communication system typically has a zero mean.

This is motivated by the physics of thermal noise generation. If the average voltage or current in a conductor is not zero that means that electrons are on average leaving the device. Physically this nonzero mean is not possible without an external force to induce an average voltage or current flow. **All random processes will be assumed to have a zero mean throughout the remainder of this chapter.**

Property 9.5 If $N(t)$ is a Gaussian random process then two samples of this process, $N(t_1)$ and $N(t_2)$, are completely characterized with the PDF

$$f_{N(t_1)N(t_2)}(n_1, n_2) = \frac{1}{2\pi\sqrt{\sigma_N^2(t_1)\sigma_N^2(t_2)\left(1-\rho_N(t_1,t_2)^2\right)}} \times \exp\left[\frac{-1}{1-\rho_N(t_1,t_2)^2}\left(\frac{n_1^2}{2\sigma_N^2(t_1)} - \frac{2\rho_N(t_1,t_2)n_1n_2}{2\sigma_N(t_1)\sigma_N(t_2)} + \frac{n_2^2}{2\sigma_N^2(t_2)}\right)\right]$$

where

$$\sigma_N^2(t_i) = E\left[N^2(t_i)\right] \qquad \rho_N(t_1,t_2) = \frac{E[N(t_1)N(t_2)]}{\sigma_N(t_1)\sigma_N(t_2)}$$

The joint PDF of two Gaussian random variables can be characterized by evaluating the variance of each of the samples and the correlation coefficient between the two samples. Here is a second example of how random quantities are entirely characterized by ensemble averages (variance and correlation). Recall a correlation coefficient describes how similar two random variables behave. The two random variables in this case are samples from the same random process. Section 3.3.3 has more discussion on the PDF of a bivariate Gaussian random variable. Similarly, three or more samples of a Gaussian random process can be characterized by evaluating the variance of each of the samples and the correlation coefficient between each of the samples as was shown in Section 3.3.6.

Property 9.6 The correlation function,

$$R_N(t_1, t_2) = E[N(t_1)N(t_2)] \tag{9.8}$$

contains all the information needed to characterize the joint distribution of any set of samples from a zero mean Gaussian random process.

Proof: Property 9.5 shows that only $\sigma_N^2(t)$ and $\rho_N(t_1, t_2)$ need to be identified to characterize the joint PDF of the Gaussian random variables resulting from samples of the random process. To this end

$$\sigma_N^2(t) = R_N(t, t) = E[N^2(t)] \qquad \rho_N(t_1, t_2) = \frac{R_N(t_1, t_2)}{\sqrt{R_N(t_1, t_1)R_N(t_2, t_2)}}. \qquad \square$$

The correlation function plays a key role in the analysis of Gaussian random processes.

EXAMPLE 9.7
For the random process defined in Example 9.1, the correlation function is given as

$$R_N(t_1, t_2) = E[N(t_1)N(t_2)] = e^{-|t_1|}e^{-|t_2|} + 1 \tag{9.9}$$

The correlation function, $R_N(t_1, t_2)$, is plotted in Figure 9.6(a). Consequently, the correlation coefficient between $N(0)$ and $N(2)$ is

$$\rho_N(0, 2) = \frac{e^{-2} + 1}{\sqrt{2(e^{-4} + 1)}} \tag{9.10}$$

The joint density of these two samples is plotted in Figure 9.6(b).

Point 2: When the noise, $N(t)$, is a Gaussian random process the canonical problem can be solved with knowledge of the correlation function, $R_N(t_1, t_2)$. The correlation function completely characterizes any joint PDF of samples of the process $N(t)$.

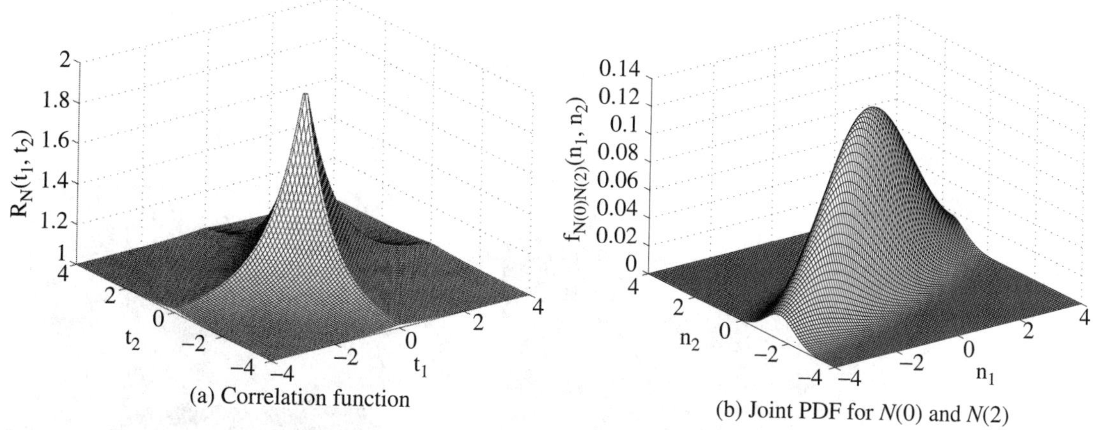

Figure 9.6 The functions that characterize $N(t) = U \exp[-|t|] + V$.

9.3 Stationary Random Processes

If the statistical description of a random process does not change over time, it is said to be a stationary random process.

EXAMPLE 9.8

Figure 9.7(a) shows a sample function of a nonstationary noise, $W_{NS}(t)$, and Figure 9.7(b) shows a sample function of a stationary noise, $W_S(t)$. The random process plotted in Figure 9.7(a) appears to have a variance, $\sigma_N^2(t)$, that is growing with time, while the random process in Figure 9.7(b) appears to have a constant variance.

Examples of random processes which are not stationary are prevalent

- Temperature in Columbus, Ohio
- The level of the Dow-Jone's Industrial Average

In these nonstationary random processes the statistical description of the resulting random variables will change greatly depending on when the process is sampled. For example, the average temperature in Columbus, Ohio is significantly different in July than in January. Nonstationary random processes are much more difficult to characterize than stationary processes. Fortunately, the thermal noise that is often seen in communication systems is well modeled as stationary over the time spans which are of interest in understanding communication system's performance. **For the remainder of this book all noise will be assumed to be stationary.**

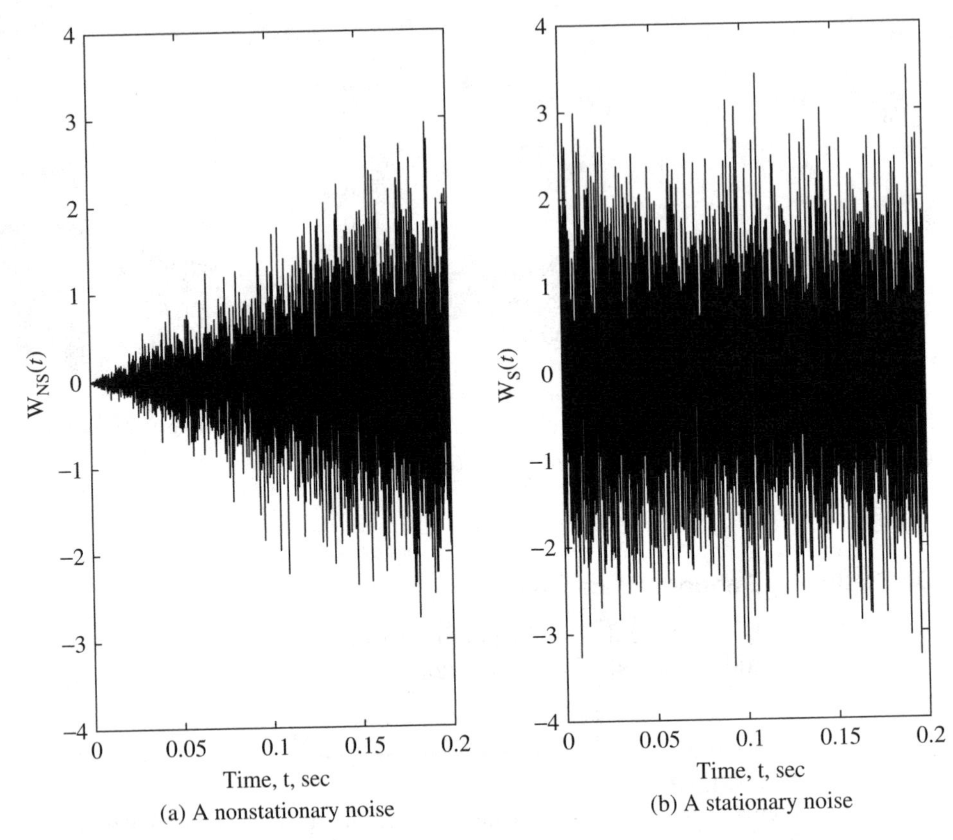

Figure 9.7 Sample paths of random processes.

9.3.1 Basics

Definition 9.4 A random process, $N(t)$, is called stationary if the density function describing M samples of the random process is independent of time shifts in the sample times, i.e.,

$$f_{N(t_1),N(t_2),\cdots,N(t_M)}(n_1, n_2, \cdots, n_M) = f_{N(t_1+t_0),N(t_2+t_0),\cdots,N(t_M+t_0)}(n_1, n_2, \cdots, n_M)$$

for any value of M and t_0.

In particular the random variable obtained by sampling a stationary random processes is statistically identical regardless of the selected sample time, i.e., $f_{N(t)}(n_1) = f_{N(t+t_0)}(n_1)$. Also two samples taken from a stationary random process will have a statistical description that is a function of the time difference between the two time samples but not the absolute location of the time samples, i.e., $f_{N(0),N(t_1)}(n_1, n_2) = f_{N(t_0),N(t_1+t_0)}(n_1, n_2)$.

EXAMPLE 9.9

For the random process defined in Example 9.1, since the second moment is given as

$$E[N^2(t)] = R_N(t,t) = 1 + e^{-2|t|} \tag{9.11}$$

it is obvious that $N(t)$ is not a stationary process.

9.3.2 Gaussian Processes

Stationarity for Gaussian random processes implies a fairly simple complete description for the random process. Recall a Gaussian random process is completely characterized with $R_N(t_1,t_2) = E[N(t_1)N(t_2)]$. Since $f_{N(t)}(n_1) = f_{N(t+t_0)}(n_1)$ this implies that the mean ($E[N(t)] = 0$) and the variance ($\sigma_N^2(t_i) = \sigma_N^2$) are constants. Since $f_{N(0),N(t_1)}(n_1,n_2) = f_{N(t_0),N(t_1+t_0)}(n_1,n_2)$ we know from Property 9.5 that $R_N(0,t_1) = R_N(t_0,t_0+t_1)$ for all t_0. This implies that the correlation function is essentially a function of only one variable, $R_N(t_1,t_2) = R_N(\tau) = E[N(t)N(t-\tau)]$ where $\tau = t_1 - t_2$. For a stationary Gaussian process this correlation function $R_N(\tau)$ completely characterizes any statistical event associated with the noise $N(t)$.

EXAMPLE 9.10

This example examines again the random process defined in Example 9.2 where a wideband noise is lowpass filtered. Recall that the mean function of the input and output is defined as

$$m_W(t) = E[W(t)] \qquad m_N(t) = E[N(t)] \tag{9.12}$$

Insight into the ensemble mean can be obtained by looking at the sample mean defined as

$$\bar{m}_W(t) = \frac{1}{N}\sum_{i=1}^{N} W(t,\omega_i) \qquad \bar{m}_N(t) = \frac{1}{N}\sum_{i=1}^{N} N(t,\omega_i) \tag{9.13}$$

Figure 9.8 shows a plot of the sample mean function for 4000 sample paths. It is pretty clear that both the input and the output random processes from the canonical problem have a zero mean. Similarly, the sample variance functions of the input and output process are computed from 4000 sample paths, e.g.,

$$\bar{\sigma}_W^2(t) = \frac{1}{N}\sum_{i=1}^{N} W^2(t,\omega_i) \qquad \bar{\sigma}_N^2(t) = \frac{1}{N}\sum_{i=1}^{N} N^2(t,\omega_i) \tag{9.14}$$

Figure 9.9 shows the resultant averaged square value of the noise functions. It is clear from Figure 9.9 that the variance of both the input and output noise processes for the canonical problem are a constant function of time. These curves are empirical evidence

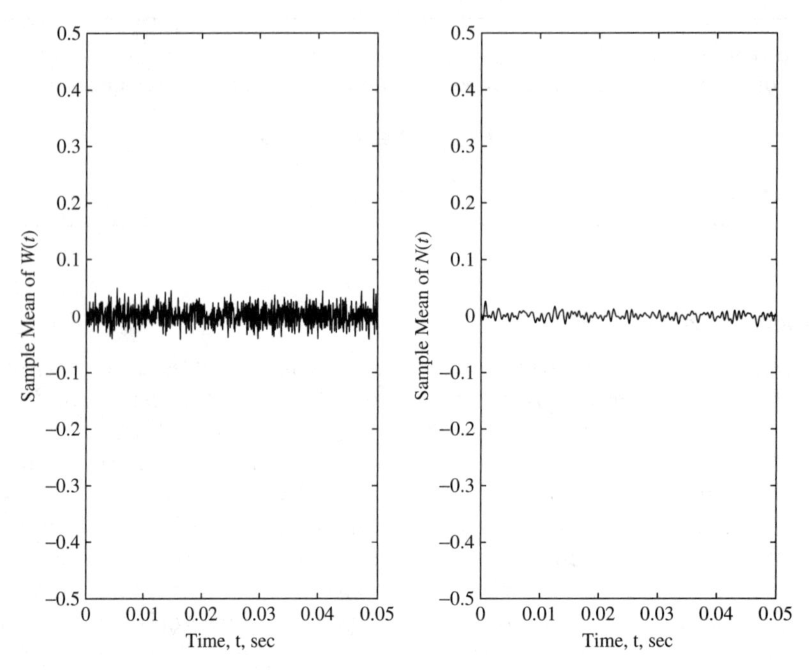

Figure 9.8 The ensemble averages of the process mean function for the input and output of the filter. 4000 trials.

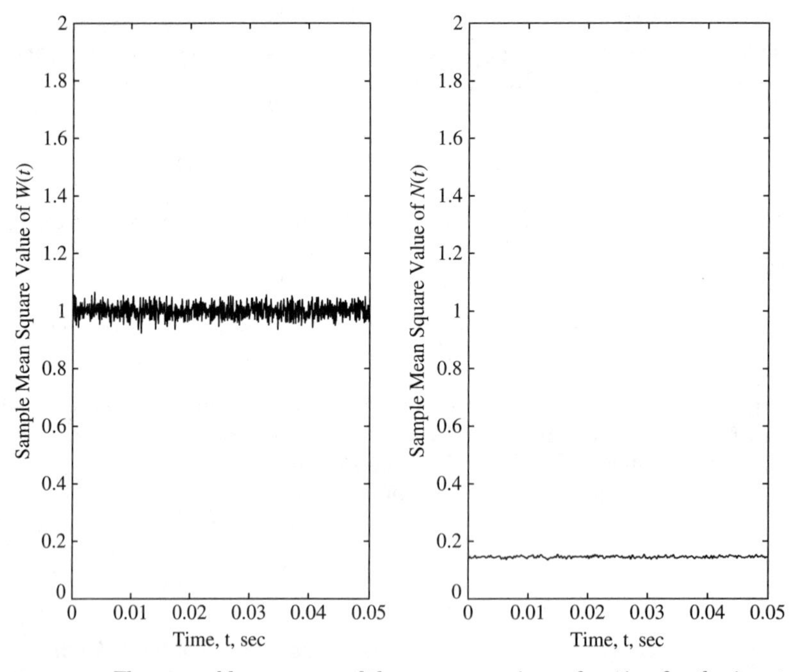

Figure 9.9 The ensemble averages of the process variance function for the input and output of the filter. 4000 trials.

that the processes in the canonical problem are stationary processes, since the mean and variance appear to be constant functions of time. The sample mean and variance of the processes match the histograms in Figure 9.4.

The correlation function gives a description of how rapidly the random process is changing with time. For a stationary random process the correlation function can be expressed as

$$R_N(\tau) = \sigma_N^2 \rho_N(\tau) \tag{9.15}$$

where $\rho_N(\tau)$ is the correlation coefficient between two samples taken τ seconds apart. Recall that if $\rho_N(\tau) \approx 0$ then the two samples will behave statistically in an independent fashion. If $\rho_N(\tau) \approx 1$ then the two samples will be very close in value (statistically dependent). In addition to giving an engineer an idea of how rapidly the random process varies as a function of time, the correlation function also fully specifies the PDF of any number of samples from a Gaussian random process. Recall the quantities needed to specify the PDF of zero mean, jointly Gaussian random variables are the variance which is constant given as

$$\sigma_N^2 = R_N(0) \tag{9.16}$$

and the correlation coefficient between the random variables. The correlation coefficient between the random variables $N(t)$ and $N(t-\tau)$ is given as

$$\rho_N(\tau) = \frac{R_N(\tau)}{\sigma_N^2} \tag{9.17}$$

Consequently for a zero mean, stationary Gaussian process the correlation function completely determines the statistical characteristics of the process.

The idea of a correlation function has been introduced before in the context of deterministic signal analysis. For deterministic signals the correlation function of a signal $x(t)$ was defined as

$$V_x(\tau) = \int_{-\infty}^{\infty} x(t)x^*(t-\tau)dt \tag{9.18}$$

This correlation function also was a measure of how rapidly a signal varied in time just like the correlation function for random processes measures the time variations of a noise. Table 9.1 summarizes the comparison between the correlation functions for random processes and deterministic energy signals.

TABLE 9.1 Summary of the important characteristics of the correlation function

Signal	Definition	Units	Correlation value at $\tau = 0$
Deterministic	$V_x(\tau) = \int_{-\infty}^{\infty} x(t)x^*(t-\tau)dt$	Joules	$V_x(0) = E_x$ (Energy of $x(t)$)
Random	$R_N(\tau) = E[N(t)N(t-\tau)]$	Watts	$R_N(0) = \sigma_N^2$ (Average power of $N(t)$)

EXAMPLE 9.11

Consider a correlation coefficient parameterized as

$$\rho_N(\tau) = \text{sinc}(2W\tau) \tag{9.19}$$

For two samples taken $\tau = 250\ \mu$s apart, a process characterized with $W = 400$ Hz would have these samples behaving very similar ($\rho_N(.00025) = 0.935$). A process characterized with $W = 6400$ Hz would have $\rho_N(.00025) = -0.058$ and essentially the two samples will be independent. Figure 9.10 shows a plot of several sample functions of Gaussian

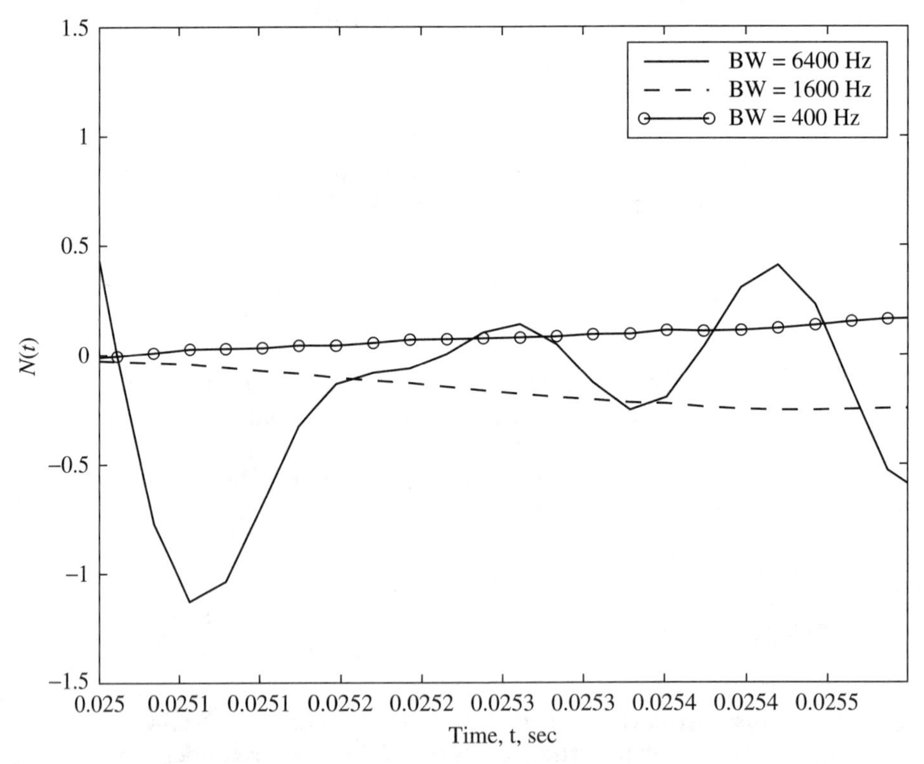

Figure 9.10 Sample paths for several Gaussian random processes parameterized by Eq. (9.19). Note this plot show samples of random processes where the sample rate is $f_s = 44,100$ Hz. The jaggedness and piecewise linear nature of the high bandwidth noise is due to this sampling. A higher sampling rate would result in a smoother appearance.

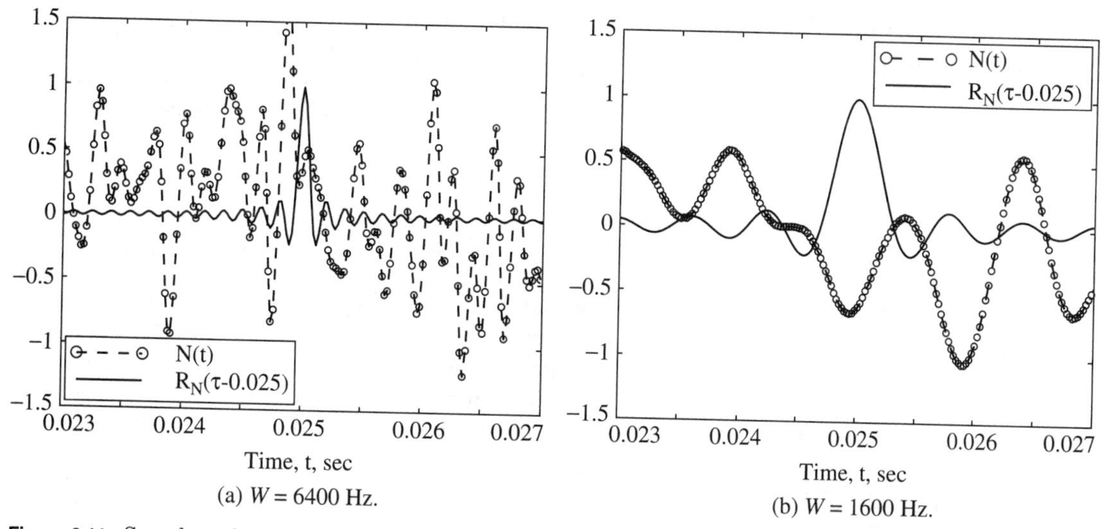

(a) $W = 6400$ Hz.

(b) $W = 1600$ Hz.

Figure 9.11 Sample paths of a stationary Gaussian process along with the associated correlation function.

random processes whose correlation function is given Eq. (9.19) with different values of W. The process parameterized with $W = 6400$ Hz varies rapidly over 250 μs, while the process parameterized with $W = 400$ Hz is nearly constant. Another way to view the relation between the correlation function and the time variability of the random process intuitively is to plot sample paths of the random process along with the correlation function. Examples of this type of plot are shown in Figure 9.11 for $W = 6400$ Hz and $W = 1600$ Hz. This type of plot clearly highlights how the correlation function relates to the variability of a random process; the narrower the correlation function the faster the random process varies in time.

Point 3: When the noise, $N(t)$, is a stationary Gaussian random process the canonical problem can be solved solely by knowing the form of $R_N(\tau) = E[N(t)N(t-\tau)]$ as the PDF of any number of samples taken arbitrarily from the process $N(t)$ can be determined from $R_N(\tau)$.

9.3.3 Frequency Domain Representation

Stationary noise can also be described in the frequency domain. Frequency domain analysis has been very useful for providing an alternate view (compared to time domain) in deterministic signal analysis. This characteristic holds true when analyzing stationary random processes. Recall the definition of a finite time Fourier transform given in Eq. (2.8), i.e.,

$$N_{T_m}(f,\omega) = \int_{-T_m}^{T_m} N(t,\omega)\exp[-j2\pi f t]dt \tag{9.20}$$

The finite time Fourier transform is a complex random function of frequency. Since $N(t)$ is a stationary random process the energy of $N(t)$ grows unbounded as the measurement interval, T_m, gets large. Because of the unbounded energy it proves useful to examine the finite time measured power spectral density (units Watts per Hertz), which is given as

$$S_N(f, \omega, T_m) = \frac{1}{2T_m} |N_{T_m}(f, \omega)|^2 \tag{9.21}$$

It should be noted that the finite time measured power spectral density is now a real random function of frequency. In analogy to the energy spectrum, $G_x(f)$, introduced in Chapter 2, the measured power spectral density gives a distribution of power in the frequency domain for the sample path $N(t, \omega)$. Stationary Gaussian random processes are characterized in the frequency domain by the average power spectral density. The average of the measured power spectral density is given as

$$S_N(f, T_m) = \frac{1}{2T_m} E\left[|N_{T_m}(f, \omega)|^2\right] = E\left[S_N(f, \omega, T_m)\right] \tag{9.22}$$

and the limit as the measurement time gets large of this average is known as the power spectral density of a stationary random process.

Definition 9.5 The power spectral density of a random process $N(t)$ is

$$S_N(f) = \lim_{T_m \to \infty} S_N(f, T_m)$$

Since the power spectral density is obtained by taking the ensemble average and the time average of the energy spectrum of the sample paths, the power spectral density is a measure of how the average power of the process is distributed as a function of frequency. The characteristics of the power spectral density of stationary random signals, $S_N(f)$, will parallel that of the energy spectral density of deterministic signals, $G_X(f)$. Intuitively, random processes that vary rapidly must have power at higher frequencies in a very similar way as deterministic signals. This definition of the power spectral density puts that intuition on a solid mathematical basis.

Property 9.7 The power spectrum of a real valued[1] random process is always a nonnegative and an even function of frequency, i.e., $S_N(f) \geq 0$ and $S_N(f) = S_N(-f)$.

Proof: The positivity of $S_N(f)$ is a direct consequence of Definition 9.5 and Eq. (9.22). The evenness is due to Property 2.4. □

[1]This chapter only considers real valued random processes but the next chapter will extend these ideas to the complex envelope of a bandpass noise, hence the need to distinguish between a real and a complex random process.

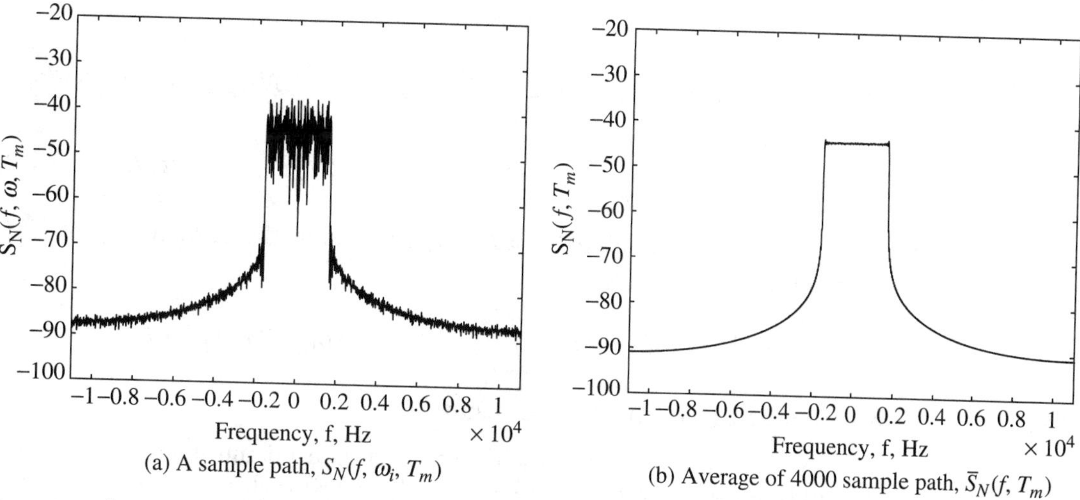

(a) A sample path, $S_N(f, \omega_i, T_m)$

(b) Average of 4000 sample path, $\bar{S}_N(f, T_m)$

Figure 9.12 A sample path and the average of 4000 sample paths.

EXAMPLE 9.12

For the output random process defined in Example 9.2, sample power spectra, $S_N(f, \omega, T_m)$, can be computed from one sample path. Insight into the ensemble average can be computed by looking at the sample mean of this function, i.e.,

$$\bar{S}_N(f, T_m) = \frac{1}{N} \sum_{i=1}^{N} S_N(f, \omega_i, T_m) \tag{9.23}$$

Figure 9.12 shows one sample path, $S_N(f, \omega_i, T_m)$, and the average of the sample paths, $\bar{S}_N(f, T_m)$, for 4000 samples paths. This is empirical evidence that $S_N(f)$ is well defined and exists for this canonical process that is the focus of this chapter.

Spectral analysis of stationary random processes proves extremely valuable because the power spectral density is related to the correlation function as a Fourier transform pair.

Property 9.8 (Wiener-Khinchin) The power spectral density for a stationary random process is given by

$$S_N(f) = \mathcal{F}\{R_N(\tau)\}$$

Proof: The power spectral density (see Definition 9.5) can be rewritten as

$$S_N(f) = \lim_{T_m \to \infty} \frac{1}{2T_m} E\left[|N_{T_m}(f)|^2\right]$$

Using the definition in Eq. (9.20) gives

$$E\left[|N_{T_m}(f)|^2\right] = E\left[\int_{-T_m}^{T_m} N(t_1)\exp[-j2\pi f t_1]dt_1 \int_{-T_m}^{T_m} N(t_2)\exp[j2\pi f t_2]dt_2\right]$$

$$= \int_{-T_m}^{T_m}\int_{-T_m}^{T_m} E[N(t_1)N(t_2)]\exp[-j2\pi f(t_1-t_2)]dt_1dt_2 \qquad (9.24)$$

Making a change of variables $\tau = t_1 - t_2$ and using the stationarity of $N(t)$ reduces Eq. (9.24) to

$$E\left[|N_{T_m}(f)|^2\right] = \int_{-T_m}^{T_m}\int_{-T_m-t_2}^{T_m-t_2} R_N(\tau)\exp[-j2\pi f\tau]d\tau dt_2 \qquad (9.25)$$

Taking the limit gives the desired result. □

Actually the true Wiener-Khinchin theorem is a bit more general but this form is all that is needed for this text. The Wiener-Khinchin theorem has utility in communication system engineering because it implies that $R_N(\tau)$ can be computed once $S_N(f)$ is identified. This correspondence between $R_N(\tau)$ and $S_N(f)$ implies that the statistical description of $N(t)$ is fully characterized with $S_N(f)$.

The power spectral density can be used to find the average power of a random process with a direct analogy to Rayleigh's energy theorem. Recall that

$$E[N(t)N(t-\tau)] = R_N(\tau) = \mathcal{F}^{-1}\{S_N(f)\} = \int_{-\infty}^{\infty} S_N(f)\exp(j2\pi f\tau)df \quad (9.26)$$

so that the total average power of a random process is given as

$$E[N^2(t)] = R_N(0) = \sigma_N^2 = \int_{-\infty}^{\infty} S_N(f)df \qquad (9.27)$$

Equation (9.27) demonstrates that the average power of a random process is the area under the power spectral density curve.

EXAMPLE 9.13

If

$$R_N(\tau) = \sigma_N^2 \,\mathrm{sinc}(2W\tau) \qquad (9.28)$$

then

$$S_N(f) = \begin{cases} \dfrac{\sigma_N^2}{2W} & |f| \le W \\ 0 & \text{elsewhere} \end{cases} \qquad (9.29)$$

A random process having this correlation–spectral density pair has power uniformly distributed over the frequencies from $-W$ to W.

TABLE 9.2 Summary of the duality of spectral densities nonnegativity

Signal	Definition	Units	Correlation Value at $\tau = 0$	
Deterministic	$G_x(f) = \mathcal{F}\{V_x(\tau)\}$	J/Hz	$V_x(0) = E_x = \int_{-\infty}^{\infty} G_x(f)df$	$G_x(f) \geq 0$
Random	$S_N(f) = \mathcal{F}\{R_N(\tau)\}$	W/Hz	$R_N(0) = \sigma_N^2 = \int_{-\infty}^{\infty} S_N(f)df$	$S_N(f) \geq 0$

The duality of the power spectral density of random signals, $S_N(f)$ and the energy spectral density of deterministic energy signals, $G_x(f)$, is also strong. Table 9.2 summarizes the comparison between the spectral densities for random processes and deterministic energy signals.

> **Point 4:** When the noise, $N(t)$, is a stationary Gaussian random process the canonical problem can be solved with knowledge of the power spectral density of the random process, $S_N(f)$. The power spectral density determines the correlation function, $R_N(\tau)$ through an inverse Fourier transform. The correlation function completely characterizes any joint PDF of samples of the process $N(t)$.

9.4 Thermal Noise

An additive noise is characteristic of almost all communication systems. This additive noise typically arises from thermal noise generated by random motion of electrons in conductors. From the kinetic theory of particles, given a resistor R at a physical temperature of T degrees Kelvin, there exists a random additive noise voltage, $W(t)$, across the resistor. This noise is zero mean, stationary[2], and Gaussian distributed (due to the large number of randomly moving electrons and the central limit theorem), with a spectral density given as [Zie86, Joh28, Nyq28b]

$$S_W(f) = \frac{2Rh|f|}{\exp\left[\frac{h|f|}{KT}\right] - 1} \tag{9.30}$$

where $h = 6.62 \times 10^{-34}$ is Planck's constant and $K = 1.3807 \times 10^{-23}$ is Boltzmann's constant. A plot of the power spectral density of thermal noise is shown in Figure 9.13. The major characteristic to note is that this spectral density is constant up to about 10^{12} Hz at a value of $2KTR$. Since Johnson [Joh28] made measurements to characterize the thermally generated noise in conductors and Nyquist [Nyq28b] provided the theory that matched these measurements, thermal noise is often known as Johnson noise or Johnson–Nyquist noise.

[2] As long as the temperature is constant.

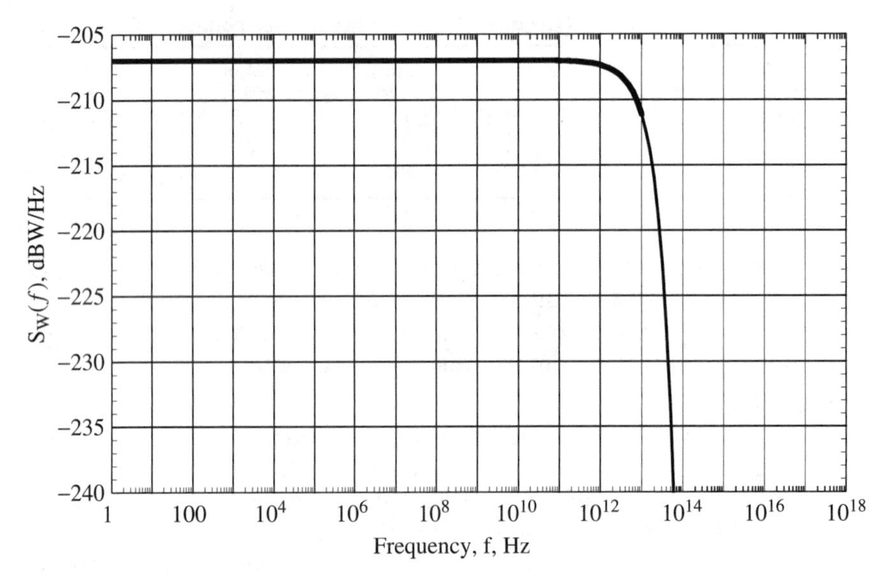

Figure 9.13 The PSD of thermal noise.

Since a majority of communications occurs at frequencies much lower than 10^{12} Hz[3] an accurate and simplifying model of the thermal noise is to assume a flat spectral density (white noise). This spectral density is a function of the receiver temperature. The lower the temperature of the receiver the lower the noise power. Additive white Gaussian noise (AWGN) is a basic assumption made about the corrupting noise in most communication systems performance analyses. White noise has a constant spectral density given as

$$S_W(f) = \frac{N_0}{2} \tag{9.31}$$

where $N_0/2$ is the noise power spectral density. Note that since

$$E[W^2(t)] = \int_{-\infty}^{\infty} S_W(f)df \tag{9.32}$$

this thermal noise model has an infinite average power. This characteristic is only a result of modeling Eq. (9.30) with the simpler form in Eq. (9.31). Traditionally some engineers have been loath to admit to the idea of negative frequencies hence they are not comfortable with the model in Eq. (9.31). These engineers like to think about noise power distributed over positive only frequencies with a spectral density of double the height. In the literature one will often see people denote $N_0/2$ as the two-sided noise spectral density and N_0 as the one-sided noise spectral density.

[3]Fiber optic communication is a notable exception to this statement.

The autocorrelation function of AWGN is given by

$$R_W(\tau) = E[W(t)W(t-\tau)] = \frac{N_0}{2}\delta(\tau) \tag{9.33}$$

AWGN is a stationary random process so a great majority of the analysis in communication systems design can utilize the extensive theory of stationary random processes. White noise has the interesting characteristic that two samples of the process $W(t)$ will always be independent no matter how closely the samples are taken. Consequently, white noise can be thought of as a very rapidly varying noise with no predictability.

A resistor in isolation will produce an approximate white noise with a two-sided spectral density of $2KTR$. If this system is placed in an electronic system, the greatest value of the average power that would be able to be transferred out of the resistor is one fourth of value that would exist if the resistor was shorted (using the maximum power transfer theorem of linear systems, see Problem 9.13). Hence an important number for communication system engineers to remember is $N_0 = KT$. For room temperature ($T = 290°K$) $N_0 = 4 \times 10^{-21}$ W/Hz (-174 dBm/Hz). When signals get amplified by active components in a receiver the value of this noise spectral density will change. We explore some of these ideas in the homework and refer the interested student to the detailed references [Zie86, Cou93, Skl88] on the ideas of noise figure, link budgets, and computing output signal to noise ratio.

This noise spectral density at the input to a communication system is very small. If this thermal noise is to corrupt the communication signal then the signal must also take a small value. Consequently, it is apparent from this discussion that thermal noise is only going to be significant when the signal at the receiver has been attenuated to a relatively small power level.

EXAMPLE 9.14

The noise plotted in Figure 9.7(b) could result from sampling of an wideband noise at a sample rate of $f_s = 22,050$ Hz. The rapidly varying characteristic of the noise samples predicted by Eq. (9.33) is evident from Figure 9.7(b). A sample path of the measured power spectrum and the average of 4000 trials of this experiment (see Eq. (9.22)) for this process is shown in Figure 9.14. This average spectrum clearly shows the white noise model for the PSD is appropriate. Figure 9.15 shows the histogram of the samples of the AWGN. The Gaussian nature of this histogram is also clearly evident. In a vast majority of communication system analysis the model of a AWGN is very accurate and used in practice.

Point 5: The dominant noise in communication systems is often generated by the random motion of electrons in conductors. This noise process is accurately modeled as stationary, Gaussian, and white.

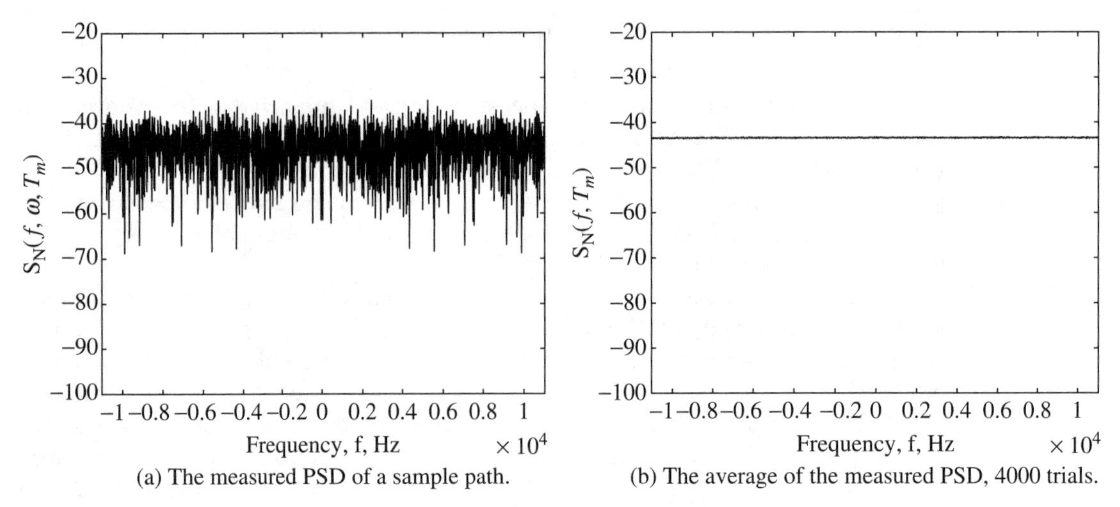

(a) The measured PSD of a sample path.

(b) The average of the measured PSD, 4000 trials.

Figure 9.14 Spectral characteristics of sampled wideband Gaussian noise.

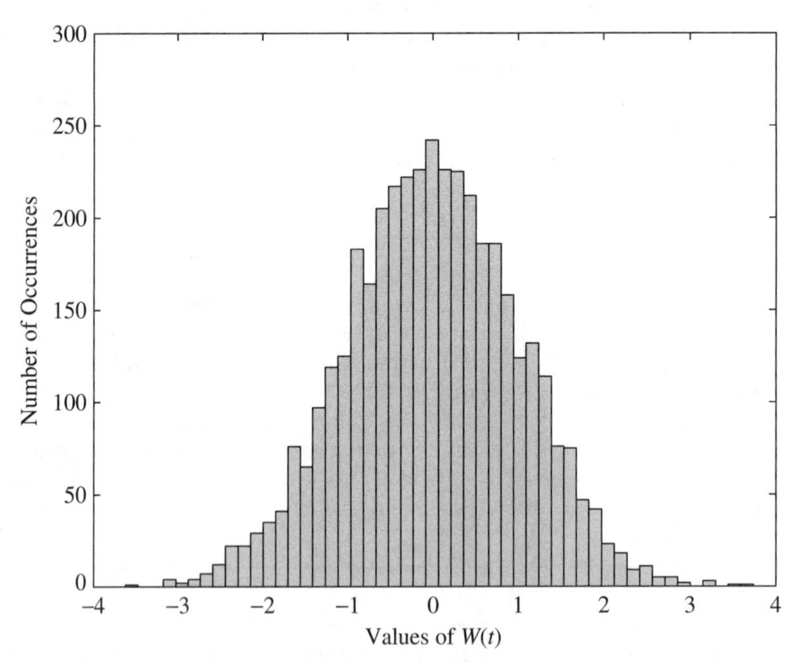

Figure 9.15 The measured histogram of wideband Gaussian noise, 4000 trials.

9.5 Linear Systems and Random Processes

The final step in solving the canonical problem is to characterize the output of a linear time-invariant filter when the input is a stationary, white, Gaussian random process.

Property 9.9 Random processes that result from linear filtering of Gaussian random processes are also Gaussian processes.

Proof: The technical machinery needed to prove this property is beyond the scope of this class but most advanced texts concerning random processes will prove this. For example [DR87]. It should be noted that a majority of the important ideas used in this proof can be understood by examining Problem 3.24.

□

Property 9.10 A random process that results from linear filtering of a zero mean process will also be zero mean.

Proof: Assume the filter input is $W(t)$, the filter impulse response is $h_R(t)$, and the filter output is $N(t)$.

$$E[N(t)] = E\left[\int_{-\infty}^{\infty} h_R(\lambda)W(t-\lambda)d\lambda\right] = \int_{-\infty}^{\infty} h_R(\lambda)E[W(t-\lambda)]d\lambda = 0. \quad (9.34)$$

□

Property 9.11 A Gaussian random process that results from linear time-invariant filtering of another stationary Gaussian random process is also a stationary Gaussian process.

Proof: Assume the filter input is $W(t)$, the filter impulse response is $h_R(t)$, and the filter output is $N(t)$. Since we know the output is Gaussian, the stationarity of $N(t)$ only requires that the correlation function of $N(t)$ be a function of τ only, i.e., $R_N(t_1, t_2) = R_N(\tau)$. The two argument correlation function of $N(t)$ is given as

$$\begin{aligned} R_N(t_1, t_2) &= E[N(t_1)N(t_2)] \\ &= E\left[\int_{-\infty}^{\infty} h_R(\lambda_1)W(t_1-\lambda_1)d\lambda_1 \int_{-\infty}^{\infty} h_R(\lambda_2)W(t_2-\lambda_2)d\lambda_2\right] \end{aligned} \quad (9.35)$$

Rearranging gives

$$\begin{aligned} R_N(t_1, t_2) &= \int_{-\infty}^{\infty} h_R(\lambda_1)\int_{-\infty}^{\infty} h_R(\lambda_2)E[W(t_1-\lambda_1)W(t_2-\lambda_2)]d\lambda_1 d\lambda_2 \\ &= \int_{-\infty}^{\infty} h_R(\lambda_1)\int_{-\infty}^{\infty} h_R(\lambda_2)R_W(t_1-\lambda_1-t_2+\lambda_2)d\lambda_1 d\lambda_2 \\ &= \int_{-\infty}^{\infty} h_R(\lambda_1)\int_{-\infty}^{\infty} h_R(\lambda_2)R_W(\tau-\lambda_1+\lambda_2)d\lambda_1 d\lambda_2 \\ &= g(\tau). \end{aligned} \quad (9.36)$$

□

Point 6: The output noise in the canonical problem, $N(t)$, is accurately modeled as a stationary Gaussian process.

The canonical problem given in Figure 9.1 can now be solved by finding the correlation function or spectral density of the stationary Gaussian random process $N(t)$. For the canonical problem $W(t)$ is a white noise so that Eq. (9.36) can be rewritten as

$$R_N(\tau) = \int_{-\infty}^{\infty} h_R(\lambda_1) \int_{-\infty}^{\infty} h_R(\lambda_2) \frac{N_0}{2} \delta(\lambda_1 - \lambda_2 - \tau) d\lambda_1 d\lambda_2 \tag{9.37}$$

Using the sifting property of the delta function gives

$$R_N(\tau) = \frac{N_0}{2} \int_{-\infty}^{\infty} h_R(\lambda_1) h_R(\lambda_1 - \tau) d\lambda_1 = \frac{N_0}{2} V_{h_R}(\tau) \tag{9.38}$$

where $V_{h_R}(\tau)$ is the correlation function of the deterministic impulse response $h_R(t)$ as defined in Chapter 2. This demonstrates that the output correlation function from a linear filter with a white noise input is only a function of the filter impulse response and the white noise spectral density.

The frequency domain description of the random process $N(t)$ is equally simple. The power spectral density is given as

$$S_N(f) = \mathcal{F}\{R_N(\tau)\} = \frac{N_0}{2} \mathcal{F}\{V_{h_R}(\tau)\} = \frac{N_0}{2} G_{h_R}(f) = \frac{N_0}{2} |H_R(f)|^2 \tag{9.39}$$

The average output noise power is given using either Eq. (9.38) or Eq. (9.39) as

$$\sigma_N^2 = E[N^2(t)] = R_N(0) = \frac{N_0}{2} \int_{-\infty}^{\infty} |H_R(f)|^2 df = \frac{N_0}{2} \int_{-\infty}^{\infty} |h_R(t)|^2 dt \tag{9.40}$$

It should be noted that Eq. (9.40) is another example of Parseval's theorem in signal analysis.

The result in Eq. (9.39) is a special case of a more general result. The general result is

$$S_N(f) = S_W(f)|H_R(f)|^2 = S_W(f) G_{h_R}(f) \tag{9.41}$$

This can be obtained by taking the Fourier transform of Eq. (9.36) as this double integral can be reformulated as a double convolution. It is clear that by substituting $S_W(f) = N_0/2$ into Eq. (9.41) gives the result in Eq. (9.39). Consequently, the output spectral density of a linear time-invariant filter is equal to the product of the input power spectral density and the energy spectrum of the impulse response (the magnitude squared of the transfer function). This characteristic enables an easy characterization of the impact of a filter in the frequency domain.

EXAMPLE 9.15

This example considers the sampled wideband Gaussian noise shown in Figure 9.7(b) being put into a lowpass filter of bandwith 1600 Hz. This lowpass filter will significantly

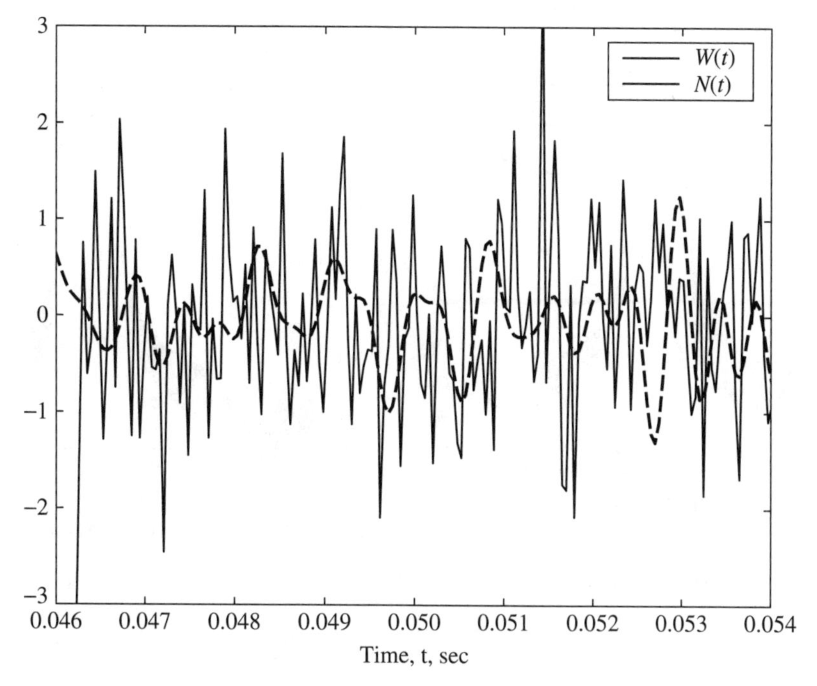

Figure 9.16 Sample paths of wideband and filtered noise. Filter bandwidth = 1600 Hz.

smooth out the sampled noise and reduce the variance of the output process. Two sample functions; one for the input wideband noise and one for the filtered noise, are shown in Figure 9.16. The lowering of the output power of the noise and the smoothing in the time variations that is expected to be produced at the lowpass filter output are readily apparent in this figure. The output measured PSD of the filtered noise is shown in Figure 9.17(b). This measured PSD validates Eq. (9.39), as this PSD is clearly the result of multiplying the measured input noise spectral density (modeled in the text as a white noise) with the lowpass filter energy spectrum as seen in Figure 9.17(a). This is another great example in communication engineering where the theory exactly predicts the measurement and why the theory has found such utility in practice. As a final point, the histogram of the output samples, shown in Figure 9.18, again demonstrates that the output noise is well modeled as a zero mean Gaussian random process.

EXAMPLE 9.16

Consider an AWGN with a one-sided spectral density of N_0 being input into an ideal lowpass filter of bandwidth W, i.e.,

$$H_R(f) = \begin{cases} 1 & |f| \leq W \\ 0 & \text{elsewhere} \end{cases} \tag{9.42}$$

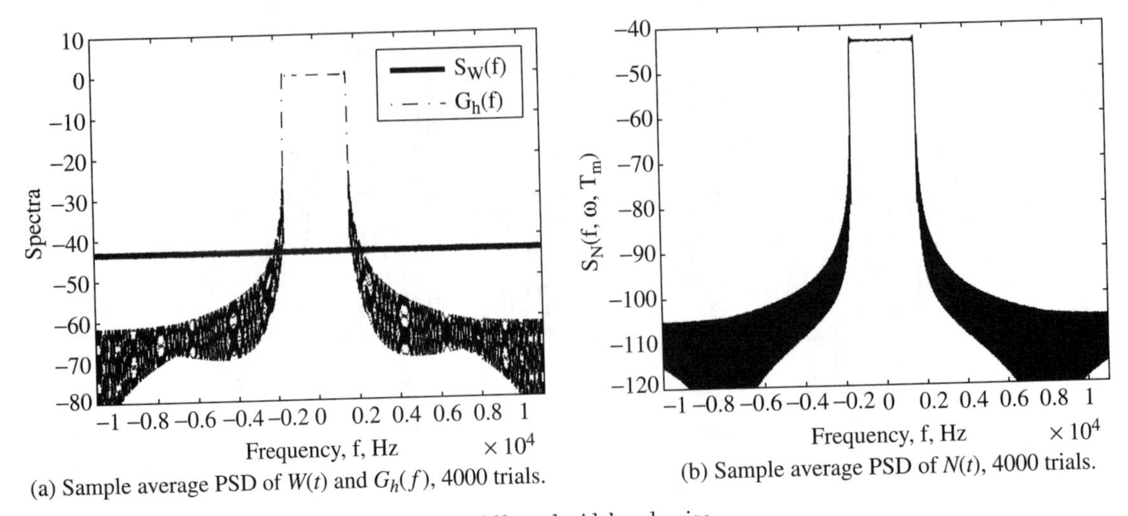

(a) Sample average PSD of $W(t)$ and $G_h(f)$, 4000 trials.

(b) Sample average PSD of $N(t)$, 4000 trials.

Figure 9.17 Measured spectral characteristics of filtered wideband noise.

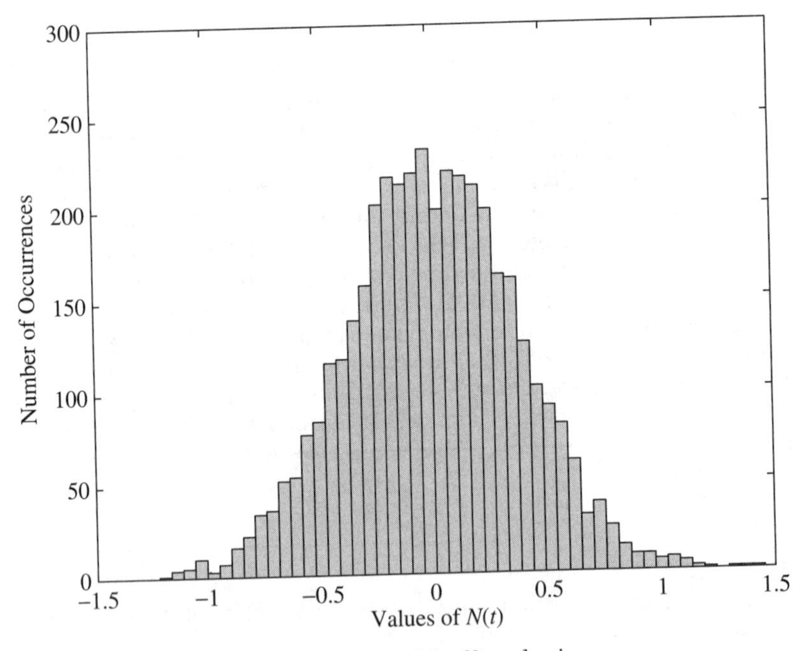

Figure 9.18 A histogram of the samples of the filtered noise.

The output spectral density is given as

$$S_N(f) = \begin{cases} \dfrac{N_0}{2} & |f| \leq W \\ 0 & \text{elsewhere} \end{cases} \tag{9.43}$$

and the average output power is given as $\sigma_N^2 = N_0 W$.

Engineers noted the simple form for the average output noise power expression in Example 9.16 and decided to parameterize the output noise power of all filters by a single number, the noise equivalent bandwidth.

Definition 9.6 The noise equivalent bandwidth of a filter is

$$B_N = \frac{1}{2|H_R(0)|^2} \int_{-\infty}^{\infty} |H_R(f)|^2 df$$

For an arbitrary filter the average output noise power will be $\sigma_N^2 = N_0 B_N |H_R(0)|^2$. Note, for a constant $|H_R(0)|$ the smaller the noise equivalent bandwidth the smaller the noise power. This implies that noise power can be minimized by making B_N as small as possible. In the homework problems we will show that the output signal to noise ratio is not a strong function of $|H_R(0)|$ so that performance in the presence of noise is well parameterized by B_N. In general, $H_R(f)$ is chosen as a compromise between complexity of implementation, signal distortion, and noise mitigation.

> **Point 7:** A stationary, white, and Gaussian noise of two-sided spectral density $N_0/2$ when passed through a filter will produce a stationary Gaussian noise whose power spectral density is given as
>
> $$S_N(f) = \frac{N_0}{2}|H_R(f)|^2$$

9.6 The Solution of the Canonical Problem

Putting together the results of the previous five sections leads to a solution of the canonical problem posed in this chapter.

First, the characterization of one sample, $N(t_s)$, of the random process $N(t)$ is considered. This case requires a four step process summarized as

1. Identify N_0 and $H_R(f)$.
2. Compute $S_N(f) = \frac{N_0}{2}|H_R(f)|^2$.
3. $\sigma_N^2 = E[N^2(t)] = R_N(0) = \frac{N_0}{2}\int_{-\infty}^{\infty} |H_R(f)|^2 df = N_0 B_N |H_R(0)|^2$
4. $f_{N(t_s)}(n_1) = \frac{1}{\sqrt{2\pi\sigma_N^2}} \exp\left(-\frac{n_1^2}{2\sigma_N^2}\right)$

EXAMPLE 9.17

Consider two ideal lowpass filters with

$$H_{R(i)}(f) = \begin{cases} 1 & |f| \leq W_i \\ 0 & \text{elsewhere} \end{cases} \tag{9.44}$$

with $W_1 = 400$ Hz and $W_2 = 6400$ Hz and a noise spectral density of

$$N_0 = \frac{1}{1600}$$

Step 2 gives

$$S_{N_i}(f) = \begin{cases} \frac{1}{3200} & |f| \leq W_i \\ 0 & \text{elsewhere} \end{cases} \tag{9.45}$$

Step 3 gives $\sigma_{N_1}^2 = 0.25$ and $\sigma_{N_2}^2 = 4$. The average noise power between the two filter outputs is different by a factor of 16. The PDFs in step 4 are plotted in Figure 9.19.

Second, the characterization of two samples, $N(t_1)$ and $N(t_2)$, of the random process $N(t)$ is considered. This case requires a five step process summarized as

1. Identify N_0 and $H_R(f)$.
2. Compute $S_N(f) = \frac{N_0}{2}|H_R(f)|^2$.

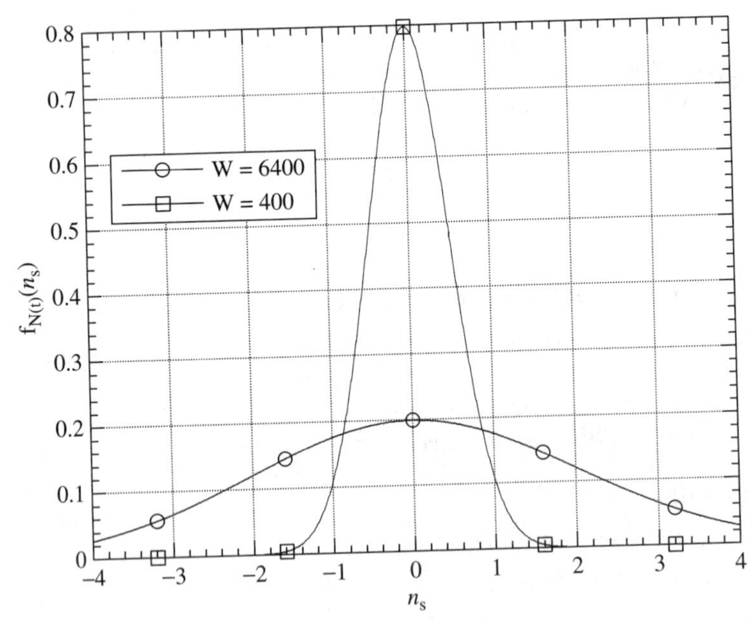

Figure 9.19 The PDF of a filter output sample for $W = 400$ and $W = 6400$ Hz.

3. $R_N(\tau) = \mathcal{F}^{-1}\{S_N(f)\}$
4. $\sigma_N^2 = R_N(0)$ and $\rho_N(\tau) = R_N(\tau)/\sigma_N^2$
5. $f_{N(t_1)N(t_2)}(n_1, n_2) = \frac{1}{2\pi\sigma_N^2\sqrt{(1-\rho_N(\tau)^2)}} \exp\left[\frac{-1}{2\sigma_N^2(1-\rho_N(\tau)^2)}\left(n_1^2 - 2\rho_N(\tau)n_1n_2 + n_2^2\right)\right]$

EXAMPLE 9.18

Consider two ideal lowpass filters with

$$H_{R(i)}(f) = \begin{cases} 1 & |f| \leq W_i \\ 0 & \text{elsewhere} \end{cases} \tag{9.46}$$

with $W_1 = 400$ Hz and $W_2 = 6400$ Hz and a noise spectral density of

$$N_0 = \frac{1}{1600}$$

Step 2 gives

$$S_{N_i}(f) = \begin{cases} \frac{1}{3200} & |f| \leq W_i \\ 0 & \text{elsewhere} \end{cases} \tag{9.47}$$

Step 3 gives $R_{N_1}(\tau) = 0.25\text{sinc}(800\tau)$ and $R_{N_2}(\tau) = 4\text{sinc}(12{,}800\tau)$. Step 4 has the average noise power between the two filter outputs being different by a factor of 16 since $\sigma_{N_1}^2 = 0.25$ and $\sigma_{N_2}^2 = 4$. For an offset of τ seconds the correlation coefficients are $\rho_{N_1}(\tau) = \text{sinc}(800\tau)$ and $\rho_{N_2}(\tau) = \text{sinc}(12{,}800\tau)$. The PDFs in step 5 are plotted in Figure 9.20(a) and Figure 9.20(b) for $\tau = 250$ μs. With the filter having a bandwidth of 400 Hz the variance of the output samples is less and the correlation between samples is greater. This higher correlation implies that the probability that the two random variables take significantly different values is very small. This correlation is

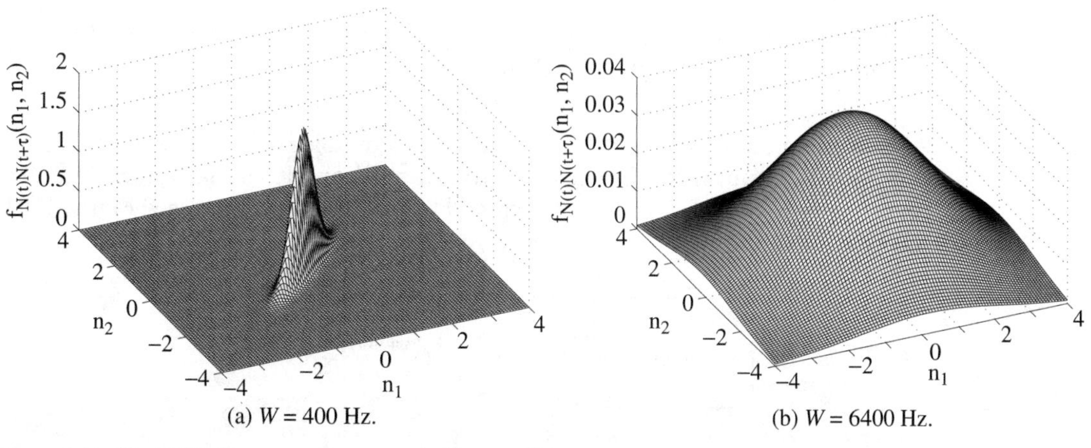

(a) $W = 400$ Hz.

(b) $W = 6400$ Hz.

Figure 9.20 The PDF of two samples separated by $\tau = 250$ μs.

evident due to the knife edge like joint PDF. These two joint PDFs show the characteristics in an analytical form that were evident in the noise sample paths shown in Figure 9.10.

Similarly, three or more samples of the filter output could be characterized in a very similar fashion. The tools developed in this chapter give a student the ability to analyze many of the important properties of noise that are of interest in communication systems design.

9.7 Homework Problems

Problem 9.1. For this problem consider the canonical block diagram shown in Figure 9.21 with $s_i(t) = \cos(2\pi(200)t)$ and with $W(t)$ being an additive white Gaussian noise with two-sided spectral density of $N_0/2$. Assume $H_R(f)$ is an ideal lowpass filter with bandwidth of 400 Hz. Two samples are taken from the filter output at time $t = 0$ and $t = \tau$, i.e., $Y_o(0)$ and $Y_o(\tau)$.

(a) Choose N_0 such that the output noise power, $E[N^2(t)] = 1$.

(b) $Y_o(0)$ is a random variable. Compute $E[Y_o(0)]$ and $E[Y_o^2(0)]$, and $\text{var}(Y_o(0))$.

(c) $Y_o(\tau)$ is a random variable. Compute $E[Y_o(\tau)]$ and $E[Y_o^2(\tau)]$, and $\text{var}(Y_o(\tau))$.

(d) Find the correlation coefficient between $Y_o(0)$ and $Y_o(\tau)$ i.e.,

$$\rho_{Y_o}(\tau) = \frac{E[(Y_o(0) - E[Y_o(0)])(Y_o(\tau) - E[Y_o(\tau)])]}{\sqrt{\text{var}(Y_o(0))\text{var}(Y_o(\tau))}} \tag{9.48}$$

(e) Plot the joint density function, $f_{Y_o(0)Y_o(\tau)}(y_1, y_2)$, of these two samples for $\tau = 2.5$ ms.

(f) Plot the joint density function, $f_{Y_o(0)Y_o(\tau)}(y_1, y_2)$, of these two samples for $\tau = 25$ μs.

(g) Compute the time average signal to noise ratio where the instantaneous signal power is defined as $E[Y_o(t)]^2$.

Problem 9.2. Real valued additive white Gaussian noise (two-sided spectral density $N_0/2$) is input into a linear-time-invariant filter with a real valued impulse response $h_R(t)$ where $h_R(t)$ is constrained to have a unit energy (see Figure 9.21 where $s_i(t) = 0$).

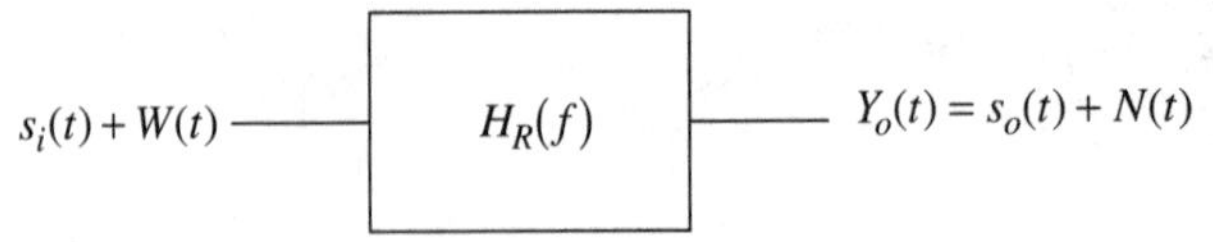

$s_i(t) + W(t)$ —— [$H_R(f)$] —— $Y_o(t) = s_o(t) + N(t)$

Figure 9.21 Block diagram for signals, noise, and linear systems.

(a) Calculate the output correlation function, $R_N(\tau)$.

(b) The output signal is sampled at a rate of f_s. Give conditions on the impulse response, $h_R(t)$, that will make these samples independent. Give one example of a filter where the output samples will be independent.

(c) Give the PDF of one sample, e.g., $f_{N(t_1)}(n_1)$.

(d) Give the PDF of any two sample, e.g., $f_{N(t_1)N(t_1+k/f_s)}(n_1, n_2)$ where k is a nonzero integer and the filter satisfies the conditions in (b).

(e) Give an expression for $P(N(t_1) > 2)$.

Problem 9.3. Real valued additive white Gaussian noise, $W(t)$ with a two-sided spectral density $N_0/2 = 0.1$ is input into a linear-time-invariant filter with a transfer function of

$$H(f) = \begin{cases} 2 & |f| \leq 10 \\ 0 & \text{elsewhere} \end{cases} \tag{9.49}$$

The output is denoted $N(t)$.

(a) What is the correlation function, $R_W(\tau)$, that is used to model $W(t)$?

(b) What is $E[N(t)]$?

(c) Calculate the output power spectral density, $S_N(f)$.

(d) What is $E[N^2(t)]$?

(e) Give the PDF of one sample, e.g., $f_{N(t_1)}(n_1)$.

(f) Give an expression for $P(N(t_1) > 3)$.

Problem 9.4. For this problem consider the canonical block diagram shown in Figure 9.21 with $s_i(t) = A$ and with $W(t)$ being an additive white Gaussian noise with two-sided spectral density of $N_0/2$. Assume a 1-Ω system and break the filter response into the DC gain term, $H_R(0)$ and the normalized filter response, $H_N(f)$, i.e., $H_R(f) = H_R(0)H_N(f)$.

(a) What is the input signal power?

(b) What is the output signal power, P_s?

(c) What is the average output noise power, σ_N^2?

(d) Show that the output SNR $= P_s/\sigma_N^2$ is not a function of $H_R(0)$.

(e) Give the output SNR as a function of A, N_0, and B_N.

Problem 9.5. For the canonical problem given in Figure 9.1, the output noise spectral density is given as

$$S_N(f) = \begin{cases} N_0 f^2 & |f| \leq W \\ 0 & \text{elsewhere} \end{cases} \tag{9.50}$$

(a) What is the average ouput noise power?

(b) Give a possible $H_R(f)$ that would result in this output noise?

(c) Think of a way to implement this $H_R(f)$ using a lowpass filter cascaded with another linear system that is used in the demodulation of an analog modulation.

Problem 9.6. Show that $S_N(f) = S_W(f)|H_R(f)|^2$ by taking the Fourier transform of Eq. (9.36).

Problem 9.7. Consider the following linear system with a signal, $s_i(t)$, plus a white Gaussian noise, $W(t)$, as inputs.

$$Y(t) = s_o(t) + N(t) = \int_{t-T_i}^{t} x(\lambda)d\lambda = \int_{t-T_i}^{t} (s_i(\lambda) + W(\lambda))d\lambda \tag{9.51}$$

where T_i is defined as the integration time.

(a) What is the impulse response of this linear system, $h_r(t)$.

(b) Show that

$$\sigma_N^2 = \frac{N_0}{2}\int_{-\infty}^{\infty} |h_r(t)|^2 dt \tag{9.52}$$

(c) If the input to the filter is a DC term plus the AWGN, i.e.,

$$Y(t) = \int_{t-T_i}^{t} (A + W(\lambda))d\lambda \tag{9.53}$$

find the output SNR$= P_s/\sigma_N^2$ as a function of T_i.

(d) For the case considered in (c) show that SNR grows monotonically with T_i.

(e) If the input to the filter is a sinusoid signal plus the AWGN, i.e.,

$$Y(t) = \int_{t-T_i}^{t} (\sqrt{2}A\cos(2\pi f_m t) + W(\lambda))d\lambda \tag{9.54}$$

find the output SNR as a function of T_i.

(f) For the case considered in (e) show that SNR does not grows monotonically with T_i and find the optimum T_i for each f_m.

Problem 9.8. Real valued additive white Gaussian noise, $W(t)$ with a two-sided spectral density $N_0/2 = 0.1$ is input into a linear-time-invariant filter with a transfer function of

$$H_R(f) = \begin{cases} 2 & |f| \leq W \\ 0 & \text{elsewhere} \end{cases} \tag{9.55}$$

The output is denoted $N(t)$.

(a) What is the correlation function, $R_W(\tau)$, that is used to model $W(t)$?

(b) What is $E[N(t)]$?

(c) Calculate the output power spectral density, $S_N(f)$.

(d) Select W to give $E[N^2(t)] = 10$.

(e) Give the PDF of one sample, e.g., $f_{N(t_1)}(n_1)$ when $E[N^2(t)] = 10$.

(f) Given an expression for $P(N(t_1) > 3)$ when $E[N^2(t)] = 10$.

Problem 9.9. As simple model of a radar receiver consider the following problem. A radar sends out a pulse and listens for the return at the receiver whose filtering is represented with an impulse response $h_R(t)$. The situation when no target is present is represented with the input being modeled as a real valued additive white Gaussian noise, $Y(t) = W(t)$, ($W(t)$ has a two-sided spectral density $N_0/2$) being input into a linear-time-invariant filter with a real valued impulse response $h_R(t)$. Denote the output as $V(t)$. The situation where a target is present can be represented as the same filter with a input $Y(t) = A + W(t)$. A target detection decision is given by a simple threshold test at time t

$$V(t) \underset{\text{Target Absent}}{\overset{\text{Target Present}}{\gtrless}} \gamma \tag{9.56}$$

(a) For the case when no target is present, completely characterize a sample of the process $V(t)$.

(b) For a particular time t, find a value of the threshold, γ, where the probability that a target will be declared present when one is not present is 10^{-5}.

(c) For the case when a target is present completely characterize a sample of the process $V(t)$.

(d) For the value of the threshold selected in (b) find how large A should be to correctly detect a target present with a probability of detection greater than 0.9 for any value of t.

Problem 9.10. A real valued additive white Gaussian noise, $W(t)$, (two-sided spectral density $N_0/2$) is input into a linear-time-invariant filter with a real valued impulse response $h_R(t)$. Assume $h_R(t)$ represents an ideal lowpass filter with bandwidth B_T and $N(t)$ is the output of the filter.

(a) Find $R_N(\tau)$.

(b) Choose B_T such that $E[N^2(t)] = 1$.

(c) For the random process $Z(t) = N(t) - N(t - t_1)$ find $R_Z(\tau)$.

(d) Select a value of t_1 such that $E[Z^2(t)] = 0.5$ for the B_T computed in (b).

Problem 9.11. **(Adapted from PD)** Prof. Fitz likes to gamble but he is so busy thinking up problems to torment students that he can never find time to get to Las Vegas. So instead he decides to gamble with the students in his communication classes. The game has real valued additive white Gaussian noise (two-sided spectral density $N_0/2 = 1$), $W(t)$, input into a linear-time-invariant filter with a real valued impulse response

$$h_R(t) = \begin{cases} 2 & 0 \leq t \leq 0.5 \\ -2 & 0.5 < t \leq 1 \\ 0 & \text{elsewhere} \end{cases} \tag{9.57}$$

Prof. Fitz samples the noise, $N(t)$, at a time of his choosing, t_0, and notes the sign of the noise sample. The student playing the game gets to choose a later time, $t_0 + \tau$, $\tau > 0$ to sample the noise and if the sign of the noise sample is the opposite of Prof. Fitz's sample then the student gets out of the final exam.

(a) Calculate the output correlation function, $R_N(\tau)$.

(b) Prof. Fitz samples at $t_0 = 1$, find $E[N^2(1)]$.

(c) Give the probability density function of Prof. Fitz's sample, e.g., $f_{N(t_0)}(n_1)$.

(d) If you are playing the game and Prof. Fitz samples the noise at $t_0 = 1$, select a sample time to optimize your odds of getting out of the test. Justify your answer. Could you do better if you were allowed to sample the process at a time before Prof. Fitz samples the process?

(e) If Prof. Fitz gets a realization $N(1) = 2$ in his sample, compute the probability of your getting out of the test for the optimized sample time you gave in (d).

Problem 9.12. Consider the model in Figure 9.1 where the noise is zero mean white Gaussian with a one-sided noise spectral density of N_0 W/Hz and filter has an impulse response of

$$h_R(t) = \begin{cases} \exp[-at] & t \geq 0 \\ 0 & \text{elsewhere} \end{cases} \tag{9.58}$$

where $a > 0$.

(a) Find $E[N^2(t_s)]$.

(b) Find $P(N(t_s) \leq 1)$.

(c) Find the noise equivalent bandwidth of this filter, B_N.

(d) Find $P(N(t_s + \tau) \leq 0 | N(t_s) = 1)$. *Hint:* Eq. (3.23) might be useful.

Problem 9.13. Noise in communication systems are often modeled with the system in Figure 9.22. The noise source consists of an ideal voltage noise source and a source resistance. If the noise source produces a voltage of $N(t)$ where $N(t)$ is a stationary noise with $E[N^2(t)] = P_N$.

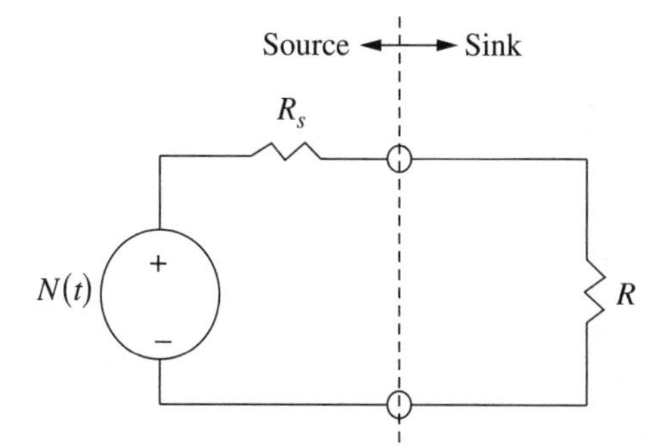

Figure 9.22 A noise source and sink.

(a) Find the average power delivered to the sink as a function of R_s and R.

(b) What is the value of R that would maximize this power and what would be the corresponding maximum average power delivered.

Problem 9.14. Given Figure 9.23 and $W(t)$ being an additive white Gaussian noise ($R_W(\tau) = N_0/2\delta(\tau)$).

(a) Express the probability density function (PDF) of the random variable $N_1(t_1)$ for a fixed t_1 in terms of $h_1(t)$ and N_0.

(b) Find the $E[N_1(t_1)N_2(t_1)]$ for a fixed t_1 in terms of $h_1(t)$, $h_2(t)$, and N_0. Specify the joint PDF of $N_1(t_1)$ and $N_2(t_1)$.

(c) If $h_1(t)$ is given in Figure 9.24 then find an impulse response for $h_2(t)$ such that $N_1(t_1)$ and $N_2(t_1)$ for a fixed t_1 are independent random variables with equal variances *Hint:* The answer is not unique.

Problem 9.15. For this problem consider the canonical block diagram shown in Figure 9.21 with $s_i(t) = A\cos(200\pi t)$ and with $W(t)$ being an additive white Gaussian noise with two-sided spectral density of $N_0/2 = 0.01$. $H_R(f)$ is a filter with a transfer function given in Figure 9.25.

(a) What is $s_o(t)$ and what is P_{s_o}?

(b) If two samples are taken from $W(t)$, e.g., $W(t_1)$ and $W(t_2)$, what spacing in the time samples, $t_2 - t_1$, must be maintained to have two samples that are statistically independent random variables?

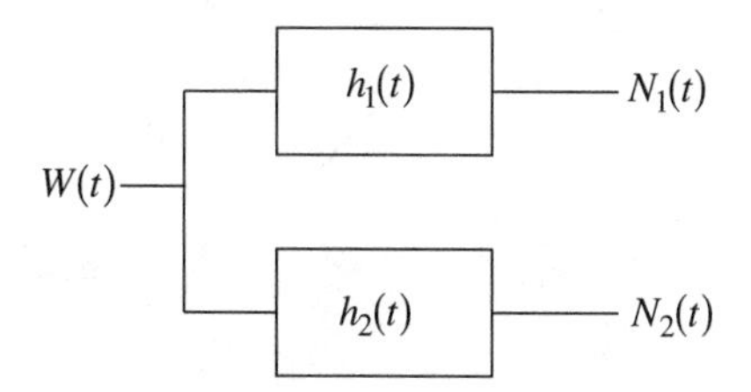

Figure 9.23 White noise into a linear filter, 4000 trials.

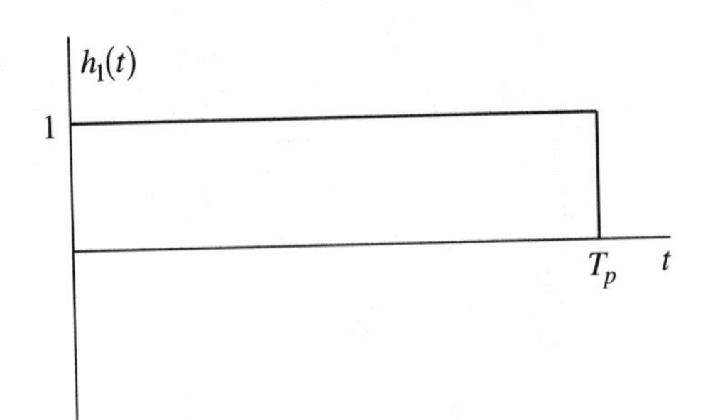

Figure 9.24 An example impulse response.

(c) What is the spectral density of $N(t)$, $S_N(f)$?

(d) What is the noise power, $E[N^2(t)]$?

(e) Choose A such that the output SNR $= 10$.

Problem 9.16. Consider the model in Figure 9.1 where the noise is zero mean white Gaussian with a one-sided noise spectral density of N_0 W/Hz and filter is an ideal bandpass filter of bandwidth B_T that has a transfer function of

$$H_R(f) = \begin{cases} 3 & f_c - \frac{B_T}{2} \leq |f| \leq f_c + \frac{B_T}{2} \\ 0 & \text{elsewhere} \end{cases} \tag{9.59}$$

Assume $f_c \geq B_T$.

(a) Find $S_N(f)$.

(b) Find $E[N^2(t_s)]$.

(c) Find $R_N(\tau)$.

(d) Find $P(N(t_s) \leq 1)$.

(e) Find $P(N(t_s + \tau) \leq 0 | N(t_s) = 1)$. *Hint:* Eq. (3.23) might be useful.

(f) Find the value of τ that maximizes $P(N(t_s + \tau) \leq 0 | N(t_s) = 1)$.

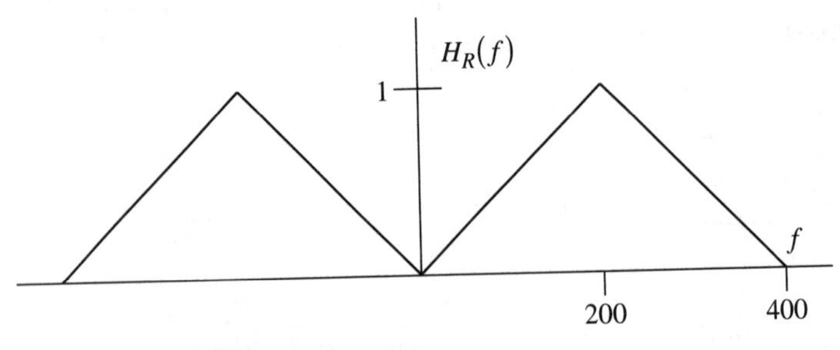

Figure 9.25 Transfer function of $H_R(f)$.

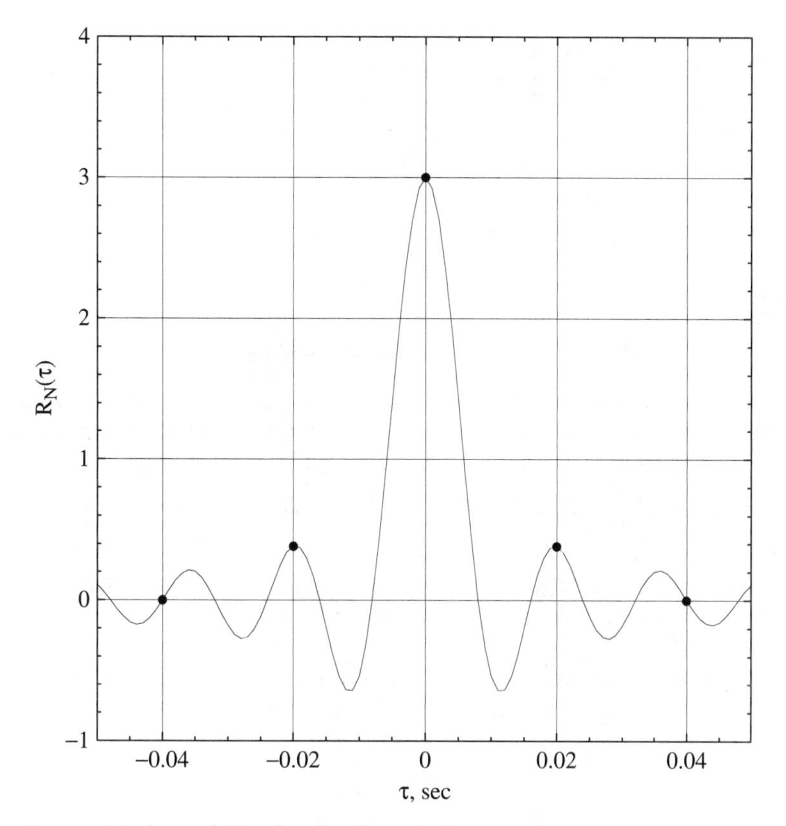

Figure 9.26 A correlation function for a stationary noise.

Problem 9.17. A zero mean, stationary, Gaussian noise, $N(t)$, is characterized by the correlation function given in Figure 9.26.

(a) Give the average power of this noise.

(b) Give the joint probability density function of two samples of this noise, $N(1)$ and $N(1.02)$.

(c) Find $P(N(1) \geq 4)$.

(d) If this noise is the output of an ideal lowpass filter with $H_R(0) = 1$ driven by white noise, estimate the filter bandwidth and two-sided noise spectral density of the noise, $N_0/2$.

Problem 9.18. Two noise processes, $N_1(t)$ and $N_2(t)$, are zero mean, stationary, Gaussian proceses, and have the following correlation functions:

$$R_{N_1}(\tau) = 100\text{sinc}\left(\frac{\tau}{4}\right) \qquad R_{N_2}(\tau) = 50\text{sinc}(25\tau) \tag{9.60}$$

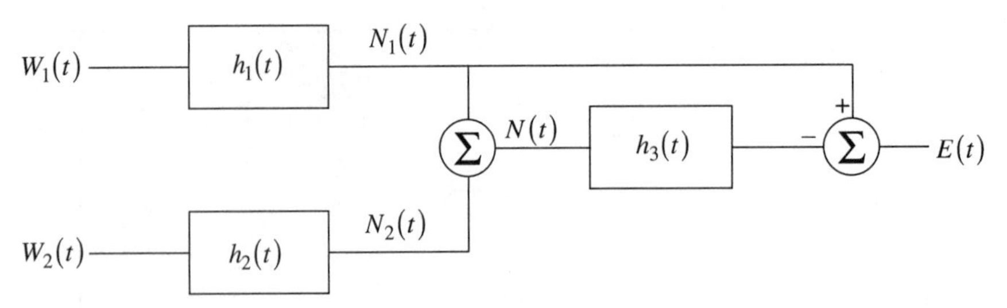

Figure 9.27 A system block diagram.

(a) Which random process has a greater power? Why?

(b) Two samples are taken from $N_2(t)$, $N_2(t_1)$, and $N_2(t_1+\tau)$. What is the smallest value of τ such that the two samples are orthogonal random variables?

(c) For $N_1(t)$, what is the correlation coefficient corresponding to $\tau = 1$? Give the joint PDF of two samples of $N_1(t)$ taken 1 second apart.

(d) Find $P(N_1(t) < -10)$.

(e) Using only linear devices (amplifiers and filters), generate a random process statistically identical to $N_1(t)$ from $N_2(t)$.

Problem 9.19. Consider the block diagram in Figure 9.27 where $W_i(t)$ are independent white Gaussian noises with $R_{W_i}(\tau) = N_0/2\delta(\tau), \quad i = 1, 2$.

(a) Find $f_{N(t_0)}(n)$.

(b) Find $P(N(t_0) > -1)$.

(c) Find $f_{N(t_0)N(t_0-\tau)}(n_1, n_2)$.

(d) Find $f_{N(t_0)}(n|N(t_0 - \tau) = n_1)$.

(e) Find the simplest expression for $S_E(f)$.

Problem 9.20. Two random processes are defined as

$$N_1(t) = X\cos(2\pi f_o t) + Y \tag{9.61}$$

and

$$N_2(t) = X\cos(2\pi f_o t) + Y\sin(2\pi f_o t) \tag{9.62}$$

where X and Y are independent zero mean jointly Gaussian random variables with unity variance and f_o is a deterministic constant.

(a) Are these processes Gaussian processes?

(b) For each of the processes that are Gaussian given a joint density function of the samples of the random process taken at $t = 0$ and $t = 1/(4f_o)$.

(c) Are these processes stationary? *Hint:* Trigonometry formulas will be useful here.

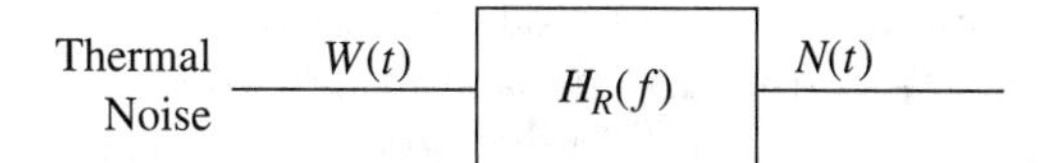

Figure 9.28 The noise model.

Problem 9.21. Consider an additive white Gaussian noise input into a linear-time-invariant system as given in Figure 9.28. Suppose that

$$H_R(f) = \begin{cases} A & |f| \leq W \\ 0 & \text{elsewhere} \end{cases} \tag{9.63}$$

and that $S_W(f) = \frac{N_0}{2}$.

(a) Choose A such that $\sigma^2_{N(t)} = 1$.

(b) If $W = 100$ Hz and $N(t_0) = 2$ find the conditional mean $E[N(t_0 - \tau)| N(t_0) = 2]$. *Hint:* See Section 3.3.3.

(c) If $W = 100$ Hz, $N(t_0) = 2$, and $\tau > 0.01$ find the value of τ that will minimize

$$\text{var}(N(t_0 - \tau)|N(t_0) = 2) \tag{9.64}$$

Problem 9.22. Often in engineering practice message signals are modeled as random processes. Assume $M(t)$ is a Gaussian random process with the power spectrum given in Figure 9.29.

(a) What is the power of this signal, P_m?

(b) Find the PDF of one sample of this message signal.

(c) What is the correlation function of the message signal, $R_M(\tau)$?

(d) Find and plot the PDF of two samples of this message signal taken $\tau = 1/200$ seconds apart.

(e) Find and plot the PDF of two samples of this message signal taken $\tau = 0.25/200$ seconds apart.

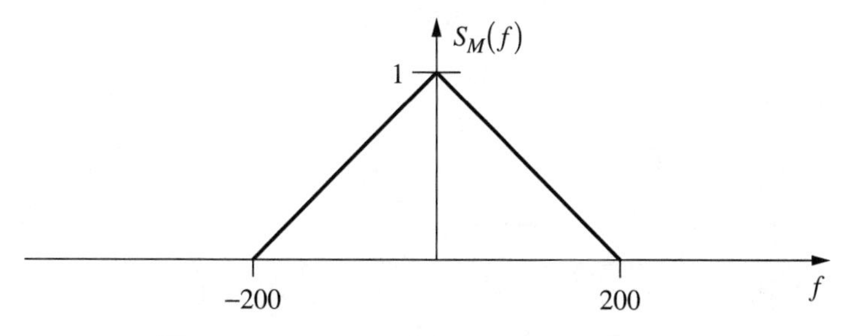

Figure 9.29 The power spectrum of the message signal.

Problem 9.23. This chapter came to the conclusion that noise in a communication system is well modeled as Gaussian and stationary. In a similar modeling exercise provide a detailed justification of whether it is appropriate to model the following random processes as Gaussian and/or stationary

(a) The number of victories in a year for the Ohio State University men's American football team

(b) The number of skiers per day at the Vail Colorado ski resort

(c) The number of oxygen molecules in one liter of air in your classroom

(d) The number of women engineers graduating from Purdue University in a year

(e) The number of cars per hour passing a point on a Los Angeles freeway

Problem 9.24. Consider a model with a zero mean Gaussian and stationary noise, $N_1(t)$, input into a linear filter with an impulse response $h(t)$ and a transfer function of $H(f)$. Denote the output noise $N_2(t)$.

(a) Show that the cross–correlation between the input and the output of the filter is given as

$$R_{N_1N_2}(\tau) = E[N_1(t)N_2(t-\tau)] = \int_{-\infty}^{\infty} R_{N_1}(\tau+\lambda)h(\lambda)d\lambda \tag{9.65}$$

(b) Show the cross spectrum between the input and output is

$$S_{N_1N_2}(f) = \mathcal{F}\{R_{N_1N_2}(\tau)\} = S_{N_1}(f)H^*(f) \tag{9.66}$$

Problem 9.25. Active electronic devices produce more noise than passive devices (e.g., a resistor) and this can decrease the output SNR from the device. An ideal model of an amplifier in a communication system is given in Figure 9.30.

(a) In the ideal model of an amplifier with $P_s = 10$, $N_0 = 0.001$ and with $H_R(f)$ being an ideal lowpass filter of bandwidth, $B = 1000$ Hz, find the output SNR, i.e.,

$$SNR_{ID} = \frac{P_{s_{in}}}{P_N} \tag{9.67}$$

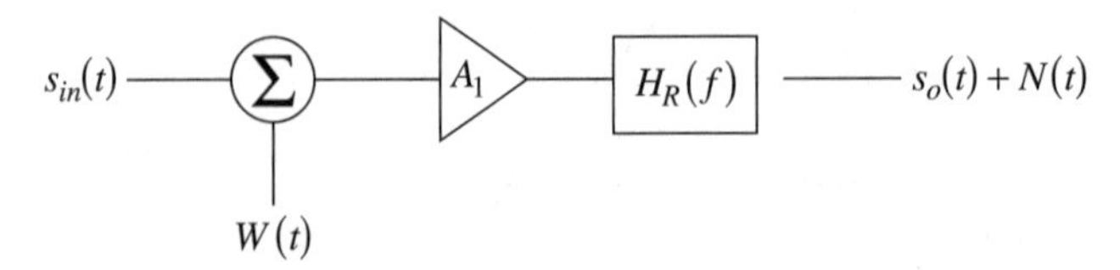

Figure 9.30 An idealized model of an active device.

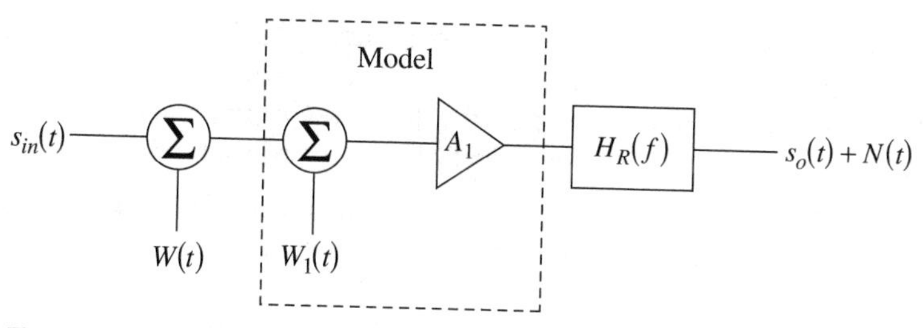

Figure 9.31 A model of an active device accounting for the extra noise generated in the device.

(b) In fact, for an active device the output SNR is given as

$$SNR_o = \frac{SNR_{ID}}{F} \tag{9.68}$$

where $F > 1$ is defined as the noise figure of the device. Show that this nonideal active device can be modeled as shown in Figure 9.31 where $W_1(t)$ is an additive white Gaussian noise independent of $W(t)$. Find the spectral density of $W_1(t)$ as a function of F.

(c) Figure 9.32 shows a model of two active devices cascaded in a communication system. Show that the cascaded noise figure is $F_T = F_1 + \frac{F_2-1}{A_1^2}$, i.e.,

$$SNR_o = \frac{SNR_{ID}}{F_T} = \frac{SNR_{ID}}{F_1 + \frac{F_2-1}{A_1^2}} \tag{9.69}$$

Problem 9.26. Gaussian random variables are convenient in communication systems analysis because only a finite number of parameters are needed to fully characterize the statistical nature of the random variables. Samples taken from a stationary Gaussian random process require even fewer parameters to characterize.

(a) If N_1 and N_2 are two jointly Gaussian zero mean random variables how many parameters and what are the parameters needed to specify the joint distribution.

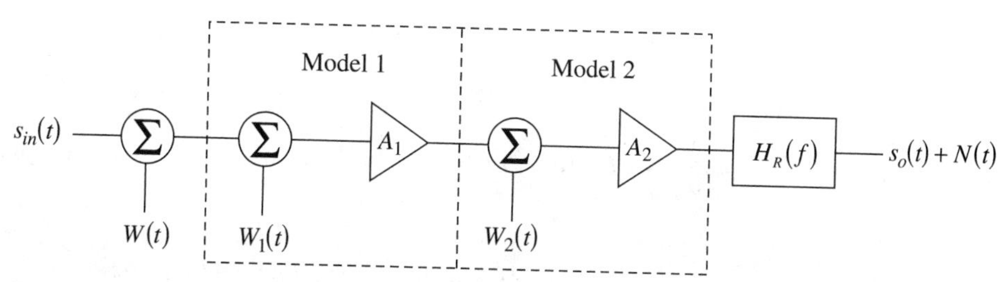

Figure 9.32 A model for a cascade of two active devices.

(b) If $N_1 = N(t)$ and $N_2 = N(t+\tau)$ are two jointly Gaussian zero mean random variables obtained by sampling a stationary zero mean bandpass Gaussian process at time t and $t+\tau$, how many parameters and what are the parameters needed to specify the joint distribution.

(c) If N_1, N_2, and N_3 are three jointly Gaussian zero mean random variables how many parameters and what are the parameters needed to specify the joint distribution.

(d) If $N_1 = N(t)$, $N_2 = N(t+\tau)$, and $N_3 = N(t-\tau)$ are three jointly Gaussian zero mean random variables obtained by sampling a stationary zero mean bandpass Gaussian process at times $t, t+\tau$, and $t-\tau$, how many parameters and what are the parameters needed to specify the joint distribution.

9.8 Example Solutions

Problem 9.4.

(a) $P_{s_i} = A^2$

(b) Since the input signal is a DC signal ($f = 0$) and has a Fourier transform $S_i(f) = A\delta(f)$ the output is $s_o(t) = AH_R(0)$. Consequently, $P_s = A^2|H_R(0)|^2$.

(c)

$$\sigma_N^2 = \frac{N_0}{2}\int_{-\infty}^{\infty}|H_R(f)|^2 df = \frac{N_0}{2}|H_R(0)|^2\int_{-\infty}^{\infty}|H_N(f)|^2 df \tag{9.70}$$

(d)

$$SNR = \frac{P_s}{\sigma_N^2} = \frac{2A^2}{N_0}\left[\int_{-\infty}^{\infty}|H_N(f)|^2\, df\right]^{-1} \tag{9.71}$$

(e) Using the definition of B_N gives

$$B_N = \frac{1}{2|H_R(0)|^2}\int_{-\infty}^{\infty}|H_R(f)|^2 df = \frac{1}{2|H_R(0)|^2}\int_{-\infty}^{\infty}|H_R(0)|^2|H_N(f)|^2 df \tag{9.72}$$

Therefore, $B_N = \frac{1}{2}\int_{-\infty}^{\infty}|H_N(f)|^2\, df$ and $\text{SNR} = \frac{A^2}{N_0 B_N}$.

Problem 9.20.

(a) Note that $N_1(t)$ and $N_2(t)$ are both zero mean Gaussian processes with correlation functions

$$R_{N_i}(\tau) = \frac{N_0}{2}\int_{-\infty}^{\infty} h_i(t)h_i(t-\tau)dt \tag{9.73}$$

Since $N(t) = N_1(t) + N_2(t)$ and $N_1(t)$ is independent of $N_2(t)$ we have

$$\sigma_N^2 = R_N(0) = \text{var}(N(t)) = \text{var}(N_1(t)) + \text{var}(N_2(t))$$
$$= \frac{N_0}{2}\int_{-\infty}^{\infty} (|h_1(t)|^2 + |h_2(t)|^2)dt \tag{9.74}$$

and

$$f_{N(t)}(n) = \frac{1}{\sqrt{2\pi\sigma_N^2}} \exp\left[-\frac{n^2}{2\sigma_N^2}\right] \tag{9.75}$$

(b)

$$P(N(t) > -1) = \int_{-1}^{\infty} f_{N(t)}(n)dn = \frac{1}{2} + \frac{1}{2}\text{erf}\left(\frac{1}{\sqrt{2\text{var}(N(t))}}\right) \tag{9.76}$$

(c) Since $N(t)$ is a zero mean process the PDF is characterized with

$$R_N(0) = \sigma_N^2 \qquad \rho_N(\tau) = \frac{R_N(\tau)}{\sigma_N^2} = \frac{N_0}{2\sigma_N^2}\int_{-\infty}^{\infty} (h_1(t)h_1(t-\tau) + h_2(t)h_2(t-\tau))dt \tag{9.77}$$

Using these results in

$$f_{N(t)N(t-\tau)}(x,y) = \frac{1}{2\pi\sigma_N^2\sqrt{1-\rho_N^2(\tau)}}$$
$$\times \exp\left(-\frac{1}{2\sigma_N^2(1-\rho_N^2(\tau))}(x^2 - 2\rho_N(\tau)xy + y^2)\right) \tag{9.78}$$

(d) Recall that

$$f_{N(t)}(x|N(t-\tau) = y) = \frac{1}{\sigma_N\sqrt{2\pi(1-\rho_N^2(\tau))}}$$
$$\times \exp\left(-\frac{1}{2\sigma_N^2(1-\rho_N^2(\tau))}(x - \rho_N(\tau)y)^2\right) \tag{9.79}$$

(e) The effective transfer function from input 1 to the output is

$$H_{E1}(f) = H_1(f)(1 - H_3(f)) \tag{9.80}$$

The effective transfer function from input 2 to the output is

$$H_{E2}(f) = -H_2(f)H_3(f) \tag{9.81}$$

Since both of the input noises, $W_i(t)$, are independent the output power spectrum is given as

$$S_E(f) = \frac{N_0}{2}|H_1(f)(1 - H_3(f))|^2 + \frac{N_0}{2}| - H_2(f)H_3(f)|^2 \tag{9.82}$$

9.9 Miniprojects

Goal: To give exposure

- to a small scope engineering design problem in communications.
- to the dynamics of working with a team.
- to the importance of engineering communication skills (in this case oral presentations).

Presentation: The forum will be similar to a design review at a company (only much shorter). The presentation will be of 5 minutes in length with an overview of the given problem and solution. The presentation will be followed by questions from the audience (your classmates and the professor). Each team member should be prepared to give the presentation.

9.9.1 Project 1

Project Goals: To use your engineering knowledge of random processes to solve a practical engineering problem.

You have a voice signal with average power of −10 dBm and bandwidth of 4 kHz that you want to transmit over a cable from one building on campus to another building on campus. The system block diagram is shown in Figure 9.33. The noise in the receiver electronics is accurately modeled as an AWGN with a one-sided noise spectral density of $N_0 = -174$ dBm/Hz. The cable is accurately modeled as a filter with the following impulse response

$$h_c(t) = L_p \delta(t - \tau_p) \tag{9.83}$$

where L_p is the cable loss. You are using cable with a loss of 2 dB/1000 ft. How long of a cable can be laid and still achieve at least 10 dB SNR? If the signal was changed to a video signal with −10 dBm average power and bandwidth of 6 MHz, how long of a cable can be laid and still achieve at least 10 dB SNR?

9.9.2 Project 2

Project Goals: To use the knowledge of random process theory to identify the characteristics of an electronic signal.

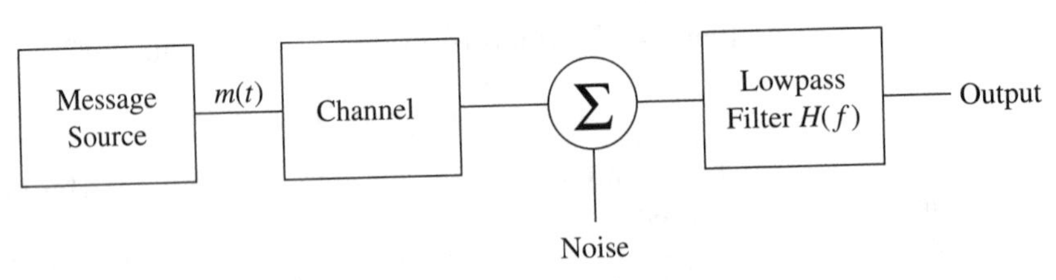

Figure 9.33 Block diagram of a baseband (cable) communication system.

You have been given a radio receiver that was recovered from a crash site of an alien spacecraft and are part of a project team that has been asked to understand this radio. The technician on your project has sampled a waveform, $N(t)$, from the radio output at a sample rate of $f_s = 10$ MHz and put it in the file *aliennoise.mat*. Examine the file and determine whether you think the signal can be characterized as Gaussian and/or stationary. If the process is well modeled as Gaussian and stationary then estimate the correlation function, $R_N(\tau)$, and the power spectral denisty, $S_N(f)$, of this process. Determine what bandwidth contains 99% of the signal power.

Chapter

10

Noise in Bandpass Communication Systems

Noise in communication systems is produced from a filtering operation on the wideband noise that appears at the receiver input. Most communication occurs in a fairly limited bandwidth around some carrier frequency. Since Chapter 9 showed that the input noise in a communication system is well modeled as an additive white Gaussian noise, it makes sense to limit the effects of this noise by filtering around this carrier frequency. Consequently, the canonical model for noise in a bandpass communication system is given in Figure 10.1. The white noise, $W(t)$, models a wideband noise which is present at the receiver input. This noise can come from a combination of the receiver electronics, man-made noise sources, or galactic noise sources. This noise is often well modeled as a zero mean, stationary, and Gaussian random process with a power spectral density of $N_0/2$ W/Hz. The filter, $H_R(f)$, represents the frequency response of the receiver system. This frequency response is often designed to have a bandpass characteristic to match the transmission band and is bandpass around the carrier frequency, f_c. Chapter 9 showed that $N_c(t)$ is a stationary Gaussian process with a power spectral density (PSD) of

$$S_{N_c}(f) = |H_R(f)|^2 \frac{N_0}{2} \tag{10.1}$$

The noise, $N_c(t)$, will consequently have all it's power concentrated around the carrier frequency, f_c. Noise with this characteristic is denoted **bandpass noise**. The average power of the noise is given as

$$E\left[N_c^2(t)\right] = \sigma_N^2 = \int_{-\infty}^{\infty} S_{N_c}(f)df = \frac{N_0}{2}\int_{-\infty}^{\infty} |H_R(f)|^2 df \tag{10.2}$$

This noise in a bandpass communication system will then be passed through an I/Q downconverter to produce an in-phase noise, $N_I(t)$, and a quadrature

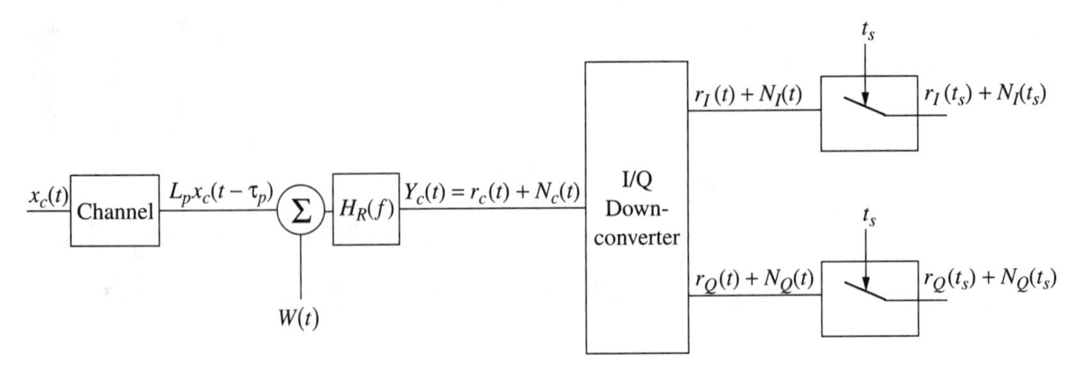

Figure 10.1 The canonical model for communication system analysis.

noise, $N_Q(t)$. This chapter provides the necessary tools to characterize the low-pass noise processes, $N_I(t)$ and $N_Q(t)$, that result in this situation.

EXAMPLE 10.1

Consider a receiver system with an ideal bandpass filter response of bandwidth B_T centered at f_c, i.e.,

$$H_R(f) = \begin{cases} A & ||f| - f_c| \leq \frac{B_T}{2} \\ 0 & \text{elsewhere} \end{cases} \tag{10.3}$$

where A is a real, positive constant. The bandpass noise will have a spectral density of

$$S_{N_c}(f) = \begin{cases} \dfrac{A^2 N_0}{2} & ||f| - f_c| \leq \frac{B_T}{2} \\ 0 & \text{elsewhere} \end{cases} \tag{10.4}$$

Consequently

$$E\left[N_c^2(t)\right] = \sigma_N^2 = \int_{-\infty}^{\infty} S_{N_c}(f)df = A^2 N_0 B_T \tag{10.5}$$

These tools will enable an analysis of the performance of bandpass communication systems in the presence of noise. This ability to analyze the effects of noise is what distinguishes the competent communication systems engineer from the hardware designers and the technicians they work with. The simplest analysis problem is examining a particular point in time, t_s, and characterizing the resulting noise samples, $N_I(t_s)$ and $N_Q(t_s)$, to extract a parameter of interest, e.g., average signal-to-noise ratio (SNR). Bandpass communication system performance can be characterized completely if probability density functions (PDF) of the noise samples of interest can be identified, e.g.,

$$f_{N_I}(n_1), \qquad f_{N_I(t)N_I(t+\tau)}(n_1, n_2), \quad \text{and/or} \quad f_{N_Q(t)N_I(t+\tau)}(n_1, n_2) \tag{10.6}$$

To accomplish this analysis task this chapter first introduces notation for discussing the bandpass noise and the lowpass noise. The resulting lowpass noise processes are shown to be zero mean, jointly stationary, and jointly Gaussian. Consequently, the PDFs like those detailed in Eq. (10.6) will entirely be a function of the second order statistics of the lowpass noise processes. These second order statistics can be produced from the PSD given in Eq. (10.1).

EXAMPLE 10.2

If a sampled white Gaussian noise like that considered in Figure 9.7(b) with a measured spectrum like that given in Figure 9.14(a) is put through a bandpass filter a bandpass noise will result. For the case of a bandpass filter with a center frequency of 6500 Hz and a bandwidth of 2000 Hz one sample path of the measured output PSD, $S_{N_c}(f, \omega_1, T_m)$, is shown in Figure 10.2(a). An estimate of the PSD using an average of 4000 sample paths is shown in Figure 10.2(b). The validity of Eq. (10.1) is clearly evident in this output PSD as this PSD can be viewed as the multiplication of the input white PSD and a bandpass transfer function. One resulting sample function of this output bandpass noise, $N_c(t, \omega_1)$ is given in Figure 10.3(a). A histogram of 4000 output samples taken at $t = 0.0045$ seconds of this noise process, $N_c(0.0045)$, is shown in Figure 10.3(b). This histogram demonstrates that a bandpass noise as is typically seen in bandpass communication systems is well modeled as a zero mean Gaussian random process.

> **Point 1:** In bandpass communications the input noise, $N_c(t)$, is a stationary Gaussian random process with power spectral density given in Eq. (10.1). The effect of noise on the performance of a bandpass communication system can be analyzed if the PDFs of the noise samples of interest can be characterized as in Eq. (10.6).

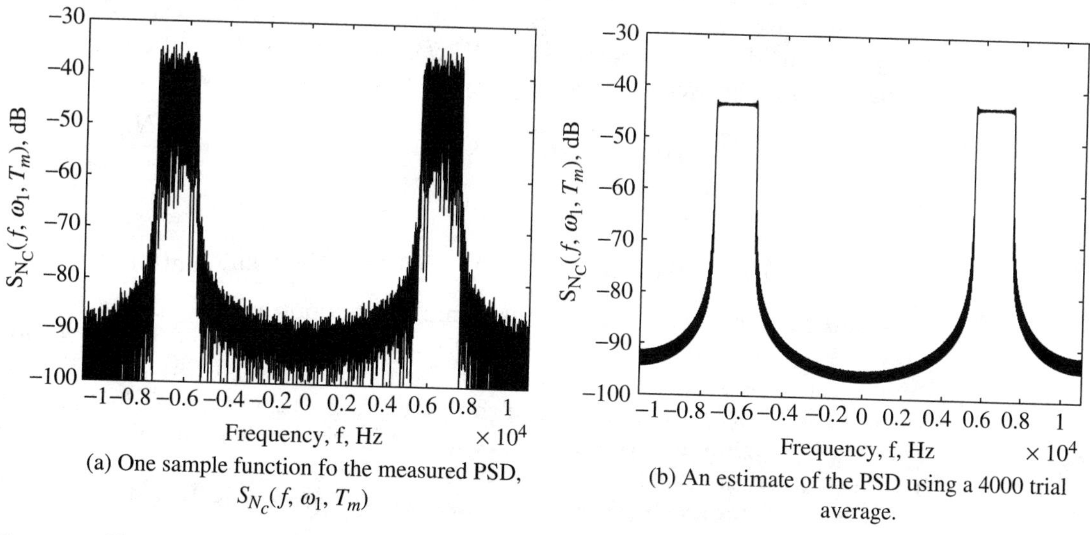

(a) One sample function fo the measured PSD, $S_{N_c}(f, \omega_1, T_m)$

(b) An estimate of the PSD using a 4000 trial average.

Figure 10.2 The measured spectral characteristics of bandpass filtered white noise.

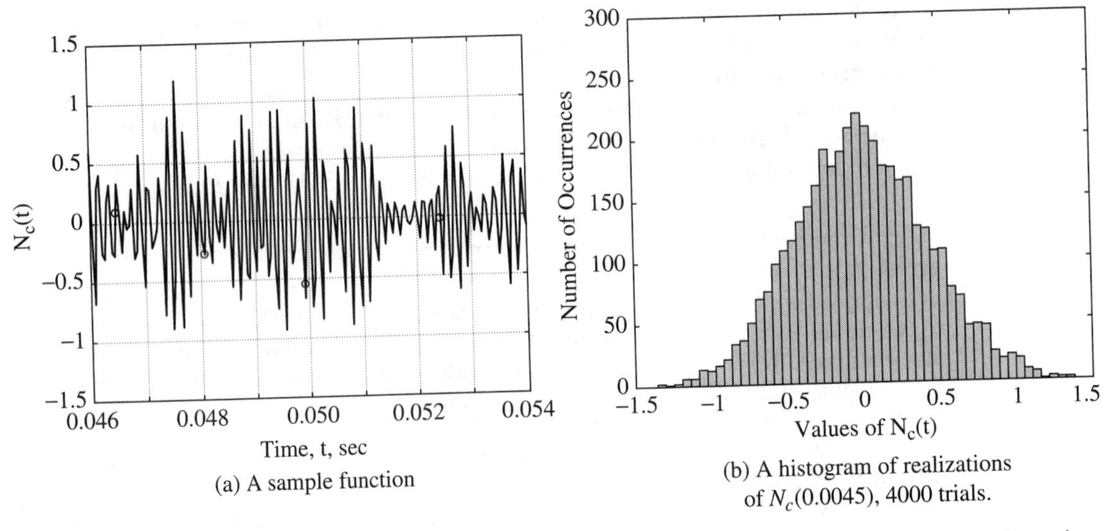

(a) A sample function

(b) A histogram of realizations of $N_c(0.0045)$, 4000 trials.

Figure 10.3 Further characterization of bandpass noise resulting from bandpass filtering a white Gaussian noise.

10.1 Bandpass Random Processes

A bandpass random process will have the same form as a bandpass signal. Consequently, it can be written as

$$N_c(t) = N_I(t)\sqrt{2}\cos(2\pi f_c t) - N_Q(t)\sqrt{2}\sin(2\pi f_c t) \tag{10.7}$$

$$= N_A(t)\sqrt{2}\cos(2\pi f_c t + N_P(t)) \tag{10.8}$$

$N_I(t)$ in Eq. (10.7) is normally referred to as the **in-phase (I)** component of the noise and $N_Q(t)$ is normally referred to as the **quadrature (Q)** component of the bandpass noise. The **amplitude** of the bandpass noise is $N_A(t)$ and the **phase** of the bandpass noise is $N_P(t)$. As in the deterministic case the transformation between the two representations are given by

$$N_A(t) = \sqrt{N_I(t)^2 + N_Q(t)^2} \qquad N_P(t) = \tan^{-1}\left[\frac{N_Q(t)}{N_I(t)}\right]$$

and

$$N_I(t) = N_A(t)\cos(N_P(t)) \qquad N_Q(t) = N_A(t)\sin(N_P(t))$$

A bandpass random process has sample functions which appear to be a sinewave of frequency f_c with a slowly varying amplitude and phase. An example of a bandpass random process is shown in Figure 10.3(a). As in the deterministic signal case, a method of characterizing a bandpass random process which is independent of the carrier frequency is desired. The complex envelope representation provides such a vehicle.

The **complex envelope** of a bandpass random process is defined as

$$N_z(t) = N_I(t) + jN_Q(t) = N_A(t)\exp[jN_P(t)] \tag{10.9}$$

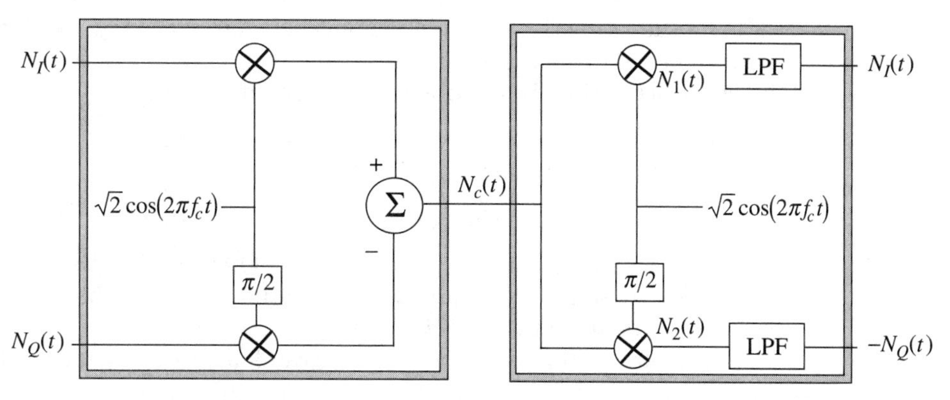

Figure 10.4 The transformations between bandpass noise and the baseband components.

The original bandpass random process can be obtained from the complex envelope by

$$N_c(t) = \sqrt{2}\Re[N_z(t)\exp[j2\pi f_c t]]$$

Since the complex exponential is a deterministic function, the complex random process $N_z(t)$ contains all the randomness in $N_c(t)$. In a similar fashion as a bandpass signal, a bandpass random process can be generated from its I and Q components and a complex baseband random process can be generated from the bandpass random process. Figure 10.4 shows these transformations. Figure 10.5 shows a bandpass noise and the resulting in-phase and quadrature components that are output from a downconverter structure shown in Figure 10.4.

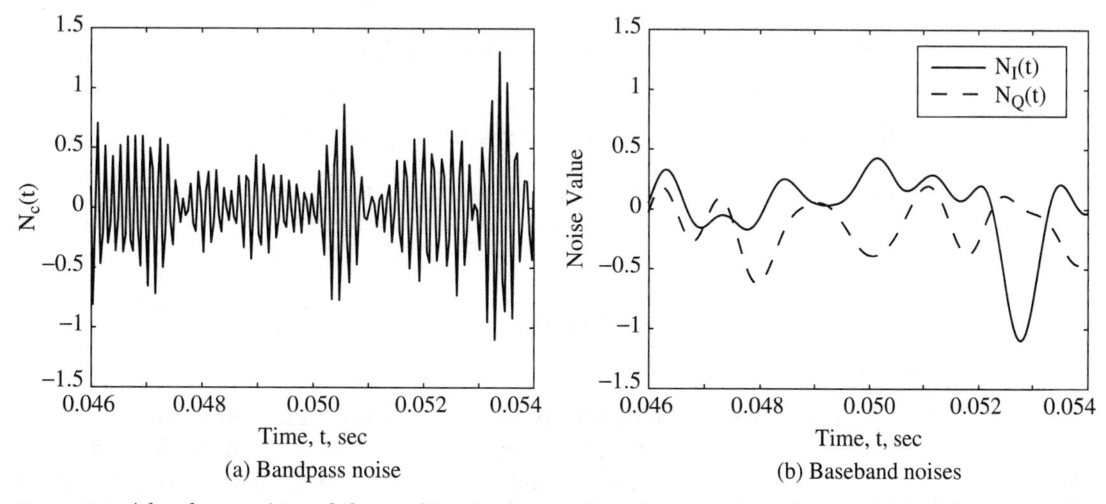

Figure 10.5 A bandpass noise and the resulting in-phase and quadrature noises. $B_T = 2000$ Hz and $f_c = 6500$ Hz.

Correlation functions are important in characterizing random processes. Autocorrelation functions were a focus of Chapter 9. For bandpass processes the crosscorrelation function will be important as well.

Definition 10.1 Given two real random processes, $N_I(t)$ and $N_Q(t)$, the crosscorrelation function between these two random processes is given as

$$R_{N_I N_Q}(t_1, t_2) = E[N_I(t_1) N_Q(t_2)] \tag{10.10}$$

The crosscorrelation function is a measure of how similar the two random processes behave. In an analogous manner to the discussion in Chapter 9 a crosscorrelation coefficient can be defined.

The correlation function of a bandpass random process, which is derived using Eq. (10.7), is given by

$$\begin{aligned} R_{N_c}(t_1, t_2) &= E[N_c(t_1) N_c(t_2)] \\ &= 2R_{N_I}(t_1, t_2) \cos(2\pi f_c t_1) \cos(2\pi f_c t_2) - 2R_{N_I N_Q}(t_1, t_2) \cos(2\pi f_c t_1) \\ &\quad \times \sin(2\pi f_c t_2) - 2R_{N_Q N_I}(t_1, t_2) \sin(2\pi f_c t_1) \cos(2\pi f_c t_2) \\ &\quad + 2R_{N_Q}(t_1, t_2) \sin(2\pi f_c t_1) \sin(2\pi f_c t_2) \end{aligned} \tag{10.11}$$

Consequently, the correlation function of the bandpass noise is a function of both the correlation function of the two lowpass noise processes and the crosscorrelation between the two lowpass noise processes.

Definition 10.2 The correlation function of the complex envelope of a bandpass random process is

$$R_{N_z}(t_1, t_2) = E[N_z(t_1) N_z^*(t_2)] \tag{10.12}$$

Using the definition of the complex envelope given in Eq. (10.9) produces

$$R_{N_z}(t_1, t_2) = R_{N_I}(t_1, t_2) + R_{N_Q}(t_1, t_2) + j[-R_{N_I N_Q}(t_1, t_2) + R_{N_Q N_I}(t_1, t_2)] \tag{10.13}$$

The correlation function of the bandpass signal, $R_{N_c}(t_1, t_2)$, is derived from the complex envelope, via

$$R_{N_c}(t_1, t_2) = 2E(\Re[N_z(t_1) \exp[j2\pi f_c t_1]] \Re[N_z^*(t_2) \exp[-j2\pi f_c t_2]]) \tag{10.14}$$

This complicated function can be simplified in some practical cases. The case when the bandpass random process, $N_c(t)$, is a stationary random process is one of them.

10.2 Characteristics of the Complex Envelope

10.2.1 Three Important Results

This section shows that the lowpass noise, $N_I(t)$ and $N_Q(t)$ derived from a bandpass noise, $N_c(t)$, are zero mean, jointly Gaussian, and jointly stationary when $N_c(t)$ is zero mean, Gaussian, and stationary. These characteristics simplify the description of $N_I(t)$ and $N_Q(t)$ considerably. Important quantities in this analysis will be

$$N_1(t) = N_c(t)\sqrt{2}\cos(2\pi f_c t)$$

$$N_2(t) = -N_c(t)\sqrt{2}\sin(2\pi f_c t) \qquad (10.15)$$

which are the outputs of the multipliers in the down converter in Figure 10.4.

Property 10.1 If the bandpass noise, $N_c(t)$, has a zero mean, then $N_I(t)$ and $N_Q(t)$ both have a zero mean.

Proof: This property's validity can be proved by considering how $N_I(t)$ (or $N_Q(t)$) is generated from $N_c(t)$ as shown in Figure 10.4. The I component of the noise is expressed as

$$N_I(t) = N_1(t) * h_L(t) = \sqrt{2}\int_{-\infty}^{\infty} h_L(t-\tau)N_c(\tau)\cos(2\pi f_c \tau)d\tau$$

where $h_L(t)$ is the impulse response of the lowpass filter. Neither $h_L(t)$ nor the cosine term are random, so the linearity property of the expectation operator can be used to obtain

$$E(N_I(t)) = \sqrt{2}\int_{-\infty}^{\infty} h_L(t-\tau)E(N_c(\tau))\cos(2\pi f_c \tau)d\tau = 0$$

The same ideas holds for $N_Q(t)$. □

This property is important since the input thermal noise to a communication system is typically zero mean; consequently, the I and Q components of the resulting bandpass noise will also be zero mean.

Definition 10.3 Two random processes $N_I(t)$ and $N_Q(t)$ are jointly Gaussian random processes if any set of samples taken from the two processes are a set of joint Gaussian random variables.

Property 10.2 If the bandpass noise, $N_c(t)$, is a Gaussian random process then $N_I(t)$ and $N_Q(t)$ are jointly Gaussian random processes.

Proof: The detailed proof techniques are beyond the scope of this course but are contained in most advanced texts concerning random processes (e.g., [DR87]). A sketch of the ideas needed in the proof is given here. A random process which is a product of a deterministic waveform and a Gaussian random process is still a Gaussian random process. Hence $N_1(t)$ and $N_2(t)$ are jointly Gaussian random processes. $N_I(t)$ and $N_Q(t)$ are also jointly Gaussian processes since they are linear functionals of $N_1(t)$ and $N_2(t)$ (i.e., $N_I(t) = N_1(t) * h_L(t)$). □

Again this property implies that the I and Q components of the bandpass noise in most communication systems will be well modeled as Gaussian random processes. It is easy to deduce that many of the samples of $N_I(t)$ (and similarly $N_Q(t)$) are Gaussian random variables. Sampling $N_c(t)$ at $t = k/(2f_c)$ where k is an integer gives $N_c(k/(2f_c)) = (-1)^k N_I(k/(2f_c))$. Since $N_c(t)$ is a

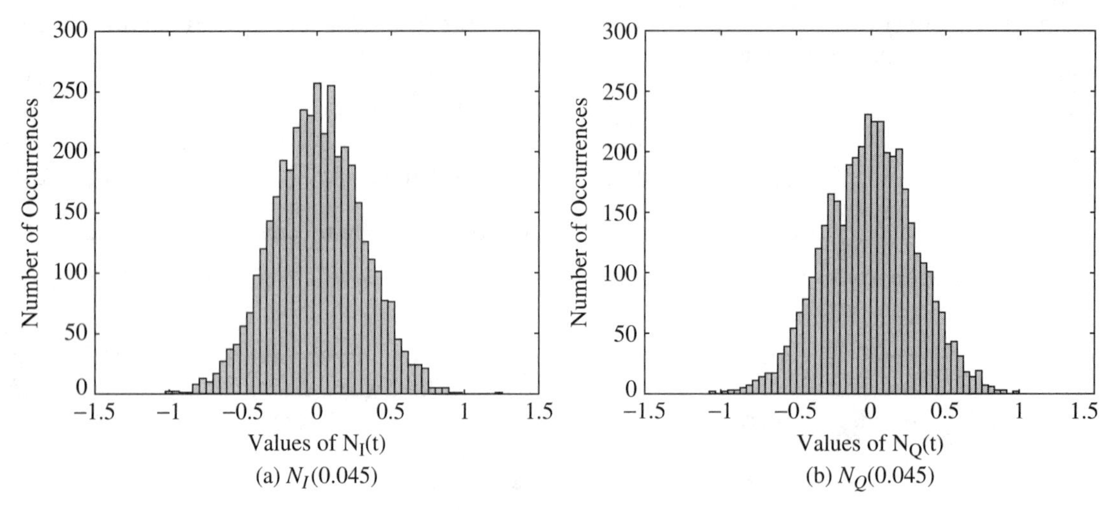

Figure 10.6 Histogram of samples of the random variables $N_I(0.045)$ and $N_Q(0.045)$ after downconverting a bandpass noise process.

Gaussian random process these samples are Gaussian random variables and consequently many samples of the lowpass processes are also Gaussian random variables. Property 10.2 simply implies that all jointly considered samples of $N_I(t)$ and $N_Q(t)$ are jointly Gaussian random variables.

EXAMPLE 10.3

Consider the previous bandpass filtered noise example where $f_c = 6500$ Hz and the bandwidth is 2000 Hz. Figure 10.6 shows histogram of samples of the random variable $N_I(0.045)$ and $N_Q(0.045)$ (i.e., the lowpass processes sampled at $t = 45$ ms) after downconversion of a bandpass noise process. Again it is apparent from this figure that the lowpass noise is well modeled as a zero mean Gaussian random process. It is interesting to note that at least in this example samples from $N_I(t)$ and $N_Q(t)$ appear to have a common distribution.

Property 10.3 If a bandpass signal, $N_c(t)$, is a stationary Gaussian random process, then $N_I(t)$ and $N_Q(t)$ are also jointly stationary, jointly Gaussian random processes.

Proof: Define the random process $N_{z1}(t) = N_1(t) + jN_2(t) = N_c(t)\sqrt{2} \times \exp[-j2\pi f_c t]$. Since $N_c(t)$ is a stationary Gaussian random process then $R_{N_{z1}}(t_1, t_2) = 2R_{N_c}(\tau)\exp[-j2\pi f_c \tau] = R_{N_{z1}}(\tau)$ where $\tau = t_1 - t_2$. Since $N_z(t) = N_{z1}(t) * h_L(t)$, if $N_{z1}(t)$ is stationary then the stationarity of the output complex envelope, $N_z(t)$, is due to Property 9.11. Using the stationarity of $N_z(t)$ with the result in Eq. (10.13) gives the following relationship

$$R_{N_z}(\tau) = R_{N_I}(t_1, t_2) + R_{N_Q}(t_1, t_2) + j\left[-R_{N_I N_Q}(t_1, t_2) + R_{N_Q N_I}(t_1, t_2)\right] \quad (10.16)$$

which implies that both the real and imaginary part of $R_{N_z}(\tau)$ must only be functions of τ so that

$$R_{N_I}(t_1,t_2)+R_{N_Q}(t_1,t_2)=g_1(\tau) \qquad -R_{N_I N_Q}(t_1,t_2)+R_{N_Q N_I}(t_1,t_2)=g_2(\tau) \tag{10.17}$$

Alternately, since $N_c(t)$ has the form given in Eq. (10.11) a rearrangement (by using trigonometric identities) with $t_2 = t_1 - \tau$ gives

$$\begin{aligned} R_{N_c}(\tau) = & \left[R_{N_I}(t_1,t_2)+R_{N_Q}(t_1,t_2)\right]\cos(2\pi f_c\tau)+\left[R_{N_I}(t_1,t_2)-R_{N_Q}(t_1,t_2)\right] \\ & \times\cos(2\pi f_c(2t_2+\tau))+\left[R_{N_I N_Q}(t_1,t_2)-R_{N_Q N_I}(t_1,t_2)\right]\sin(2\pi f_c\tau) \\ & -\left[R_{N_I N_Q}(t_1,t_2)+R_{N_Q N_I}(t_1,t_2)\right]\sin(2\pi f_c(2t_2+\tau)) \end{aligned} \tag{10.18}$$

Since $N_c(t)$ is Gaussian and stationary this implies that the right-hand side of Eq. (10.18) is a function only of the time difference, τ and not the absolute time t_2. Consequently, the factors multiplying the sinusoidal terms having arguments containing t_2 must be zero. Consequently, a second set of constraints is

$$R_{N_I}(t_1,t_2)=R_{N_Q}(t_1,t_2) \qquad R_{N_I N_Q}(t_1,t_2)=-R_{N_Q N_I}(t_1,t_2) \tag{10.19}$$

The only way for Eq. (10.17) and Eq. (10.19) to be satisfied is if

$$R_{N_I}(t_1,t_2)=R_{N_Q}(t_1,t_2)=R_{N_I}(\tau)=R_{N_Q}(\tau) \tag{10.20}$$

$$R_{N_I N_Q}(t_1,t_2)=-R_{N_Q N_I}(t_1,t_2)=R_{N_I N_Q}(\tau)=-R_{N_Q N_I}(\tau) \tag{10.21}$$

Since all correlation functions and crosscorrelation functions are functions of τ then $N_I(t)$ and $N_Q(t)$ are jointly stationary. □

Point 2: If the input noise, $N_c(t)$, is a zero mean, stationary, Gaussian random process then $N_I(t)$ and $N_Q(t)$ are zero mean, jointly Gaussian, and jointly stationary.

10.2.2 Important Corollaries

This section discusses several important corollaries to the important results derived in the last section. The important result from the last section is summarized as follows: if $N_c(t)$ is zero mean, Gaussian and stationary then $N_I(t)$ and $N_Q(t)$ are zero mean, jointly Gaussian, and jointly stationary.

Property 10.4

$$R_{N_I}(\tau)=R_{N_Q}(\tau) \tag{10.22}$$

Surprisingly, this property, given in Eq. (10.20), implies that both $N_I(t)$ and $N_Q(t)$ behave in a statistically identical manner. Consequently, the power of the noise in each component is identical, i.e., $E[N_I^2(t)] = E[N_Q^2(t)] = \sigma_{N_I}^2$. This is empirically confirmed by examining the histograms in Figure 10.6.

Property 10.5

$$R_{N_I N_Q}(\tau) = -R_{N_I N_Q}(-\tau) \tag{10.23}$$

This property, due to Eq. (10.21), implies $R_{N_I N_Q}(\tau)$ is an odd function with respect to τ.

Property 10.6 $R_{N_z}(\tau) = 2R_{N_I}(\tau) - j2R_{N_I N_Q}(\tau)$

Proof: This is shown by using Eq. (10.22) and Eq. (10.23) in Eq. (10.13). □

This implies that the real part of $R_{N_z}(\tau)$ is an even function of τ and the imaginary part of $R_{N_z}(\tau)$ is an odd function of τ.

Property 10.7

$$R_{N_c}(\tau) = 2R_{N_I}(\tau)\cos(2\pi f_c\tau) + 2R_{N_I N_Q}(\tau)\sin(2\pi f_c\tau) = \Re\left[R_{N_z}(\tau)\exp(j2\pi f_c\tau)\right] \tag{10.24}$$

Proof: This is shown by using Eq. (10.22) and Eq. (10.23) in Eq. (10.18). □

This property implies there is a simple relationship between the correlation function of the stationary Gaussian bandpass noise and the correlation function of the complex envelope of the bandpass noise. This relationship has significant parallels to the relationship between bandpass and the baseband signals given in Chapter 4.

Property 10.8 $\text{var}(N_c(t)) = \sigma_N^2 = \text{var}(N_I(t)) + \text{var}(N_Q(t)) = 2\text{var}(N_I(t)) = \text{var}(N_z(t)) = 2\sigma_{N_I}^2$

Proof: This is shown by using Eq. (10.22) in Eq. (10.24) for $\tau = 0$. □

This property states that the power in the bandpass noise is the same as the power in the complex envelope. This power in the bandpass noise is also the sum of the powers in the two lowpass noises which comprise the complex envelope.

Property 10.9 For the canonical problem considered in this chapter, $N_I(t)$ and $N_Q(t)$ are completely characterized by the functions $R_{N_I}(\tau)$ and $R_{N_I N_Q}(\tau)$.

Proof: Any joint PDF of samples taken from jointly stationary and jointly Gaussian processes are completely characterized by the first- and second-order moments. Note, first that $N_I(t)$ and $N_Q(t)$ are zero mean processes. Consequently, the characterization of the process only requires the identification of the variance, $\sigma_{N_I}^2 = R_{N_I}(0)$, and the correlation coefficient between samples. The correlation coefficient between samples from the same process is given as $\rho_{N_I}(\tau) = R_{N_I}(\tau)/R_{N_I}(0)$ while the correlation coefficient between samples taken from $N_I(t)$ and $N_Q(t)$ is $\rho_{N_I N_Q}(\tau) = R_{N_I N_Q}(\tau)/R_{N_I}(0)$. □

EXAMPLE 10.4
If one sample from the in-phase noise is considered then

$$f_{N_I}(n_i) = \frac{1}{\sqrt{2\pi\sigma_{N_I}^2}} \exp\left(-\frac{n_i^2}{2\sigma_{N_I}^2}\right) \tag{10.25}$$

where $\sigma_{N_I}^2 = R_{N_I}(0)$.

EXAMPLE 10.5
If two samples from the in-phase noise are considered then

$$f_{N_I(t)N_I(t-\tau)}(n_1, n_2) = \frac{1}{2\pi\sigma_{N_I}^2\sqrt{\left(1-\rho_{N_I}^2(\tau)\right)}} \exp\left[\frac{-\left(n_1^2 - 2\rho_{N_I}(\tau)n_1n_2 + n_2^2\right)}{2\sigma_{N_I}^2\left(1-\rho_{N_I}^2(\tau)\right)}\right]$$

where $\rho_{N_I}(\tau) = R_{N_I}(\tau)/R_{N_I}(0)$.

EXAMPLE 10.6
If one sample each from the in-phase noise and the quadrature noise are considered then

$$f_{N_I(t)N_Q(t-\tau)}(n_1, n_2) = \frac{1}{2\pi\sigma_{N_I}^2\sqrt{\left(1-\rho_{N_IN_Q}^2(\tau)\right)}} \exp\left[\frac{-\left(n_1^2 - 2\rho_{N_IN_Q}(\tau)n_1n_2 + n_2^2\right)}{2\sigma_{N_I}^2\left(1-\rho_{N_IN_Q}^2(\tau)\right)}\right]$$

where $\rho_{N_IN_Q}(\tau) = R_{N_IN_Q}(\tau)/\sqrt{R_{N_I}(0)R_{N_Q}(0)} = R_{N_IN_Q}(\tau)/R_{N_I}(0)$.

Property 10.10 The random variables $N_I(t_s)$ and $N_Q(t_s)$ are independent random variables for any value of t_s.

Proof: From Eq. (10.23) we know $R_{N_IN_Q}(\tau)$ is an odd function. Consequently, $R_{N_IN_Q}(0) = 0$. Since $N_I(t)$ and $N_Q(t)$ are jointly Gaussian and orthogonal random variables then they are also independent. □

Any joint PDF of samples of $N_I(t_s)$ and $N_Q(t_s)$ taken at the same time have the simple PDF

$$f_{N_I(t)N_Q(t)}(n_1, n_2) = \frac{1}{2\pi\sigma_{N_I}^2} \exp\left[\frac{-\left(n_1^2 + n_2^2\right)}{2\sigma_{N_I}^2}\right] \tag{10.26}$$

This simple PDF will prove useful in our performance analysis of bandpass communication systems.

Point 3: Since $N_I(t)$ and $N_Q(t)$ are zero mean, jointly Gaussian, and jointly stationary then a complete statistical description of $N_I(t)$ and $N_Q(t)$ is available from $R_{N_I}(\tau)$ and $R_{N_I N_Q}(\tau)$.

10.3 Spectral Characteristics

At this point we need a methodology to translate the known PSD of the bandpass noise Eq. (10.1) into the required correlation function, $R_{N_I}(\tau)$, and the required crosscorrelation function, $R_{N_I N_Q}(\tau)$. To accomplish this goal we need a definition.

Definition 10.4 For two random processes $N_I(t)$ and $N_Q(t)$ whose crosscorrelation function is given as $R_{N_I N_Q}(\tau)$ the cross spectral density is

$$S_{N_I N_Q}(f) = \mathcal{F}\{R_{N_I N_Q}(\tau)\} \tag{10.27}$$

Property 10.11 The PSD of $N_z(t)$ is given by

$$S_{N_z}(f) = \mathcal{F}\{R_{N_z}(\tau)\} = 2S_{N_I}(f) - j2S_{N_I N_Q}(f) \tag{10.28}$$

where $S_{N_I}(f)$ and $S_{N_I N_Q}(f)$ are the power spectrum of $N_I(t)$ and the crosspower spectrum of $N_I(t)$ and $N_Q(t)$, respectively.

Proof: This is seen taking the Fourier transform of $R_{N_z}(\tau)$ as given in Property 10.6. □

Property 10.12 $S_{N_I N_Q}(f)$ is purely an imaginary function and an odd function of f.

Proof: This is true since $R_{N_I N_Q}(\tau)$ is an odd real valued function (see Property 2.6). A power spectral density must be real and positive so by examining Eq. (10.28) it is clear $S_{N_I N_Q}(f)$ must be imaginary for $S_{N_z}(f)$ to be a valid power spectral density. □

Property 10.13 The even part of $S_{N_z}(f)$ is due to $S_{N_I}(f)$ and the odd part is due to $S_{N_I N_Q}(f)$.

Proof: A spectral density of a real random process is always even. $S_{N_I}(f)$ is a spectral density of a real random process. By Property 10.12, $S_{N_I N_Q}(f)$ is a purely imaginary and odd function of frequency. □

Property 10.14

$$S_{N_I}(f) = \frac{S_{N_z}(f) + S_{N_z}(-f)}{4} \qquad S_{N_I N_Q}(f) = \frac{S_{N_z}(-f) - S_{N_z}(f)}{j4}$$

Proof: This is a trivial result of Property 10.13 □

Consequently, Property 10.14 provides a simple method to compute $S_{N_I}(f)$ and $S_{N_I N_Q}(f)$ once $S_{N_z}(f)$ is known. $S_{N_z}(f)$ can be computed from $S_{N_c}(f)$ given in Eq. (10.1) in a simple way as well.

Property 10.15

$$S_{N_c}(f) = \frac{1}{2}S_{N_z}(f - f_c) + \frac{1}{2}S_{N_z}(-f - f_c)$$

Proof: Examining Eq. (10.24) and the frequency translation theorem of the Fourier transform, the spectral density of the bandpass noise, $N_z(t)$, is expressed as

$$S_{N_c}(f) = 2S_{N_I}(f) * \left[\frac{1}{2}\delta(f - f_c) + \frac{1}{2}\delta(f + f_c)\right]$$
$$+ 2S_{N_I N_Q}(f) * \left[\frac{1}{2j}\delta(f - f_c) - \frac{1}{2j}\delta(f + f_c)\right]$$

where $*$ again denotes convolution. This equation can be rearranged to give

$$S_{N_c}(f) = \left[S_{N_I}(f - f_c) - jS_{N_I N_Q}(f - f_c)\right] + \left[S_{N_I}(f + f_c) + jS_{N_I N_Q}(f + f_c)\right] \tag{10.29}$$

Noting that due to Property 10.12

$$S_{N_z}(-f) = 2S_{N_I}(f) + j2S_{N_I N_Q}(f) \tag{10.30}$$

Equation (10.29) reduces to the result in Property 10.15. □

This is a very fundamental result. Property 10.15 states that the power spectrum of a bandpass random process is simply derived from the power spectrum of the complex envelope and vice versa. For positive values of f, $S_{N_c}(f)$ is obtained by translating $S_{N_z}(f)$ to f_c and scaling the amplitude by 0.5 and for negative values of f, $S_{N_c}(f)$ is obtained by flipping $S_{N_z}(f)$ around the origin, translating the result to $-f_c$, and scaling the amplitude by 0.5. Likewise $S_{N_z}(f)$ is obtained from $S_{N_c}(f)$ by taking the positive frequency PSD which is centered at f_c and translating it to baseband ($f = 0$) and multiplying it by 2. Property 10.15 also demonstrates in another manner that the average power of the bandpass and baseband noises are identical since the area under the PSD is the same (this was previously shown in Property 10.8).

EXAMPLE 10.7

Example 10.1 showed a receiver system with an ideal bandpass filter that had a bandpass noise PSD of

$$S_{N_c}(f) = \begin{cases} \frac{A^2 N_0}{2} & \left||f| - f_c\right| \leq \frac{B_T}{2} \\ 0 & \text{elsewhere} \end{cases} \tag{10.31}$$

Moving the positive frequency portion of the spectrum at f_c to baseband and doubling the height gives

$$S_{N_z}(f) = \begin{cases} A^2 N_0 & |f| \leq \frac{B_T}{2} \\ 0 & \text{elsewhere} \end{cases} \tag{10.32}$$

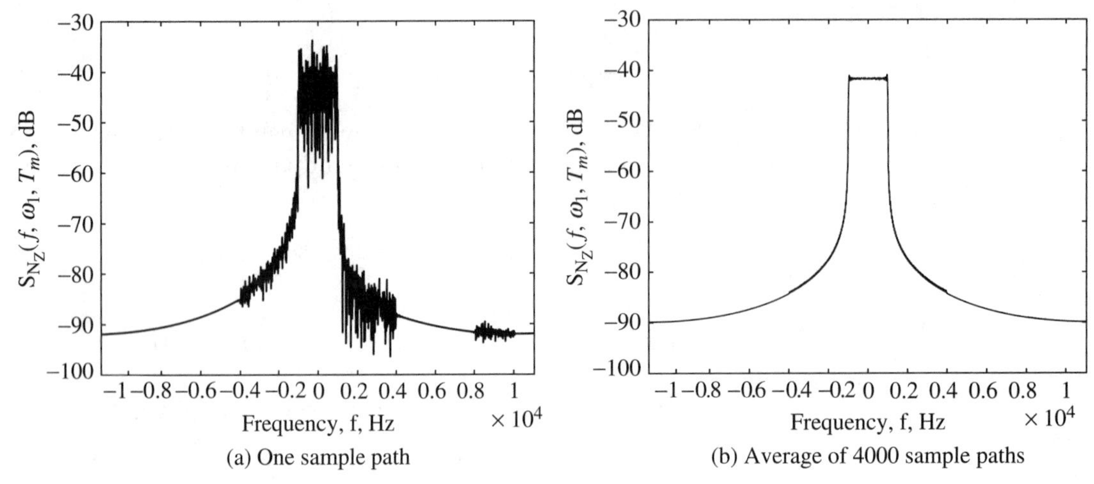

Figure 10.7 The measured PSD of the complex envelope of a bandpass process resulting from bandpass filtering a white noise.

Since there is no odd part to this PSD

$$S_{N_I}(f) = \begin{cases} \frac{A^2 N_0}{2} & |f| \leq \frac{B_T}{2} \\ 0 & \text{elsewhere} \end{cases} \qquad S_{N_I N_Q}(f) = 0 \quad \square \tag{10.33}$$

Again considering the previous example of bandpass filtered white noise with $f_c = 6500$ Hz and a bandwidth of 2000 Hz a resulting measured power spectral density of the complex envelope is given in Figure 10.7. This measured PSD demonstrates the validity of the analytical results given in Eq. (10.36).

EXAMPLE 10.8

Consider an example PSD that might appear in a receiver for SSB-AM transmission given as

$$S_{N_c}(f) = \begin{cases} \frac{A^2 N_0}{2} & 0 \leq |f| - f_c \leq B_T \\ 0 & \text{elsewhere} \end{cases} \tag{10.34}$$

Moving the positive frequency portion of the spectrum at f_c to baseband and doubling the height gives

$$S_{N_z}(f) = \begin{cases} A^2 N_0 & 0 \leq f \leq B_T \\ 0 & \text{elsewhere} \end{cases} \tag{10.35}$$

Finding the even and odd parts of this PSD using Property 10.14 gives

$$S_{N_I}(f) = \begin{cases} \frac{A^2N_0}{4} & |f| \leq B_T \\ 0 & \text{elsewhere} \end{cases} \qquad S_{N_IN_Q}(f) = \begin{cases} \frac{A^2N_0}{j4} & -B_T \leq f \leq 0 \\ \frac{-A^2N_0}{j4} & 0 \leq f \leq B_T \\ 0 & \text{elsewhere} \end{cases} \tag{10.36}$$

It is clear from this example that the random processes $N_I(t)$ and $N_Q(t)$ are in some way correlated. The exact correlation can be computed by an inverse Fourier transform. The other interesting characteristic resulting from this example is that the same bandpass noise spectrum resulted in significantly different baseband noise characteristics dependent on the selected value of f_c in the downconverter. As one example of the different characteristics, the baseband noise in this example has twice the bandwidth of the noise in the previous example while the bandpass noise in each case had the same bandwidth.

Point 4: $S_{N_I}(f)$ and $S_{N_IN_Q}(f)$ can be computed in a straightforward fashion from $S_{N_c}(f)$ given in Eq. (10.1). An inverse Fourier transform will produce the functions $R_{N_I}(\tau)$ and $R_{N_IN_Q}(\tau)$. From $R_{N_I}(\tau)$ and $R_{N_IN_Q}(\tau)$ a complete statistical description of the complex envelope noise process can be obtained.

10.4 The Solution of the Canonical Bandpass Problem

The tools and results are now in place to completely characterize the complex envelope of the bandpass noise typically encountered in a bandpass communication system. First, the characterization of one sample, $N_I(t_s)$, of the random process $N_I(t)$ is considered (or equivalently $N_Q(t_s)$). This case requires a six step process summarized as

1. Identify N_0 and $H_R(f)$.
2. Compute $S_{N_c}(f) = \frac{N_0}{2}|H_R(f)|^2$.
3. Compute $S_{N_z}(f)$.
4. Compute

$$S_{N_I}(f) = \frac{S_{N_z}(f) + S_{N_z}(-f)}{4}$$

5. $\sigma_{N_I}^2 = R_{N_I}(0) = \int_{-\infty}^{\infty} S_{N_I}(f)df$
6. $f_{N_I(t_s)}(n_1) = \frac{1}{\sqrt{2\pi\sigma_{N_I}^2}} \exp\left(-\frac{n_1^2}{2\sigma_{N_I}^2}\right) = f_{N_Q(t_s)}(n_1)$

The only difference between this process and the process used for lowpass processes as highlighted in Section 9.6 is steps 3–4. These two steps simply are the transformation of the bandpass PSD into the PSD for one channel of the lowpass complex envelope noise.

EXAMPLE 10.9

The previous examples showed a receiver system with an ideal bandpass filter had after the completion of steps 1–4

$$S_{N_I}(f) = \begin{cases} \frac{A^2 N_0}{2} & |f| \leq \frac{B_T}{2} \\ 0 & \text{elsewhere} \end{cases} \tag{10.37}$$

Consequently, $\sigma_{N_I}^2 = R_{N_I}(0) = \frac{A^2 N_0 B_T}{2}$.

Second, the characterization of two samples from one of the channels of the complex envelope, $N_I(t_1)$ and $N_I(t_2)$, is considered. This case requires a seven step process summarized as

1. Identify N_0 and $H_R(f)$.
2. Compute $S_{N_c}(f) = \frac{N_0}{2}|H_R(f)|^2$.
3. Compute $S_{N_z}(f)$.
4. Compute

$$S_{N_I}(f) = \frac{S_{N_z}(f) + S_{N_z}(-f)}{4}$$

5. $R_{N_I}(\tau) = \mathcal{F}^{-1}\{S_{N_I}(f)\}$
6. $\sigma_{N_I}^2 = R_{N_I}(0)$ and $\rho_{N_I}(\tau) = R_{N_I}(\tau)/\sigma_{N_I}^2$
7. $f_{N_I(t_1)N_I(t_2)}(n_1, n_2) = \frac{1}{2\pi\sigma_{N_I}^2\sqrt{(1-\rho_{N_I}^2(\tau))}} \exp\left[\frac{-1}{2\sigma_{N_I}^2(1-\rho_{N_I}^2(\tau))}\left(n_1^2 - 2\rho_{N_I}(\tau)n_1 n_2 + n_2^2\right)\right]$

The only difference between this process and the process used for lowpass processes as highlighted in Section 9.6 is step 3–4. These two steps again are the transformation of the bandpass PSD into the PSD for one channel of the lowpass complex envelope noise.

EXAMPLE 10.10

The previous examples showed a receiver system with an ideal bandpass filter had after the completion of steps 1–4

$$S_{N_I}(f) = \begin{cases} \frac{A^2 N_0}{2} & |f| \leq \frac{B_T}{2} \\ 0 & \text{elsewhere} \end{cases} \tag{10.38}$$

Consequently, step 5 gives $\sigma_{N_I}^2 = R_{N_I}(0) = \frac{A^2 N_0 B_T}{2}$ and $R_{N_I}(\tau) = \frac{A^2 N_0 B_T}{2}\text{sinc}(B_T \tau)$. This implies that $\rho_{N_I}(\tau) = \text{sinc}(B_T \tau)$.

Finally, the characterization of two samples one from each of the channels of the complex envelope, $N_I(t_1)$ and $N_Q(t_2)$, is considered. This case requires a seven step process summarized as

1. Identify N_0 and $H_R(f)$.
2. Compute $S_{N_c}(f) = \frac{N_0}{2}|H_R(f)|^2$.
3. Compute $S_{N_z}(f)$.
4. Compute

$$S_{N_I}(f) = \frac{S_{N_z}(f) + S_{N_z}(-f)}{4} \qquad S_{N_I N_Q}(f) = \frac{S_{N_z}(-f) - S_{N_z}(f)}{j4}$$

5. $R_{N_I N_Q}(\tau) = \mathcal{F}^{-1}\{S_{N_I N_Q}(f)\}$
6. $\sigma_{N_I}^2 = R_{N_I}(0)$ and $\rho_{N_I N_Q}(\tau) = R_{N_I N_Q}(\tau)/\sigma_{N_I}^2$
7.

$$f_{N_I(t_1)N_Q(t_2)}(n_1, n_2) = \frac{1}{2\pi\sigma_{N_I}^2\sqrt{(1-\rho_{N_I N_Q}^2(\tau))}} \times \exp\left[\frac{-1}{2\sigma_{N_I}^2(1-\rho_{N_I N_Q}^2(\tau))}(n_1^2 - 2\rho_{N_I N_Q}(\tau)n_1 n_2 + n_2^2)\right]$$

The only difference between this process and the process for two samples from the same channel is the computation of the crosscorrelation function, $R_{N_I N_Q}(\tau)$.

EXAMPLE 10.11

The previous examples showed a receiver system with an ideal bandpass filter had after the completion of steps 1–4

$$S_{N_I}(f) = \begin{cases} \frac{A^2 N_0}{2} & |f| \leq \frac{B_T}{2} \\ 0 & \text{elsewhere} \end{cases} \qquad S_{N_I N_Q}(f) = 0 \tag{10.39}$$

Consequently, step 5 gives $\sigma_{N_I}^2 = R_{N_I}(0) = \frac{A^2 N_0 B_T}{2}$ and $R_{N_I N_Q}(\tau) = 0$. This implies that $N_I(t_1)$ and $N_Q(t_2)$ are independent random variables regardless of the values of t_1 and t_2.

Similarly, three or more samples from either channel of the complex envelope could be characterized in a very similar fashion. The tools developed in this chapter give a student the ability to analyze many of the important properties of noise that are of interest in a bandpass communication system design.

10.5 Complex Additive White Gaussian Noise

A complex additive white Gaussian noise can be used to accurately model bandpass noise in communication systems. Consider a bandpass noise generated by an ideal bandpass filter, e.g.,

$$H_R(f) = \begin{cases} 1 & f_c - B_i \leq |f| \leq f_c + B_i \\ 0 & \text{elsewhere} \end{cases} \tag{10.40}$$

The baseband (complex envelope) power spectrum of the bandpass noise, $S_{N_z}(f)$, and signal will look like that in Figure 10.8(a). Since it is, in general, difficult to build a filter at a carrier frequency that is spectrally compact, the noise spectrum will typically be wider than the signal spectrum. The baseband processing will normally cut the noise bandwidth down to something close to W. A "white" noise that has the PSD as given in Figure 10.8(b) will have the exact same output noise characteristics after baseband processing because

1. the baseband processing will only pass the noise in the message bandwidth, W, and
2. the bandpass noise has a complex envelope that has a PSD that is constant, like the white noise, over this band.

Consequently, an accurate model for noise in a bandpass communication system is given as

$$Y_z(t) = r_z(t) + W_z(t) \tag{10.41}$$

where $r_z(t)$ is the received signal and $W_z(t)$ is the model for the corrupting noise where

$$S_{W_z}(f) = N_0 \qquad R_{W_z}(\tau) = N_0\delta(\tau) \tag{10.42}$$

This model will be denoted complex additive white Gaussian noise in the sequel.

A complex AWGN is intended to model a bandpass noise whose bandwidth is larger than the received signal bandwidth. The advantage of the complex

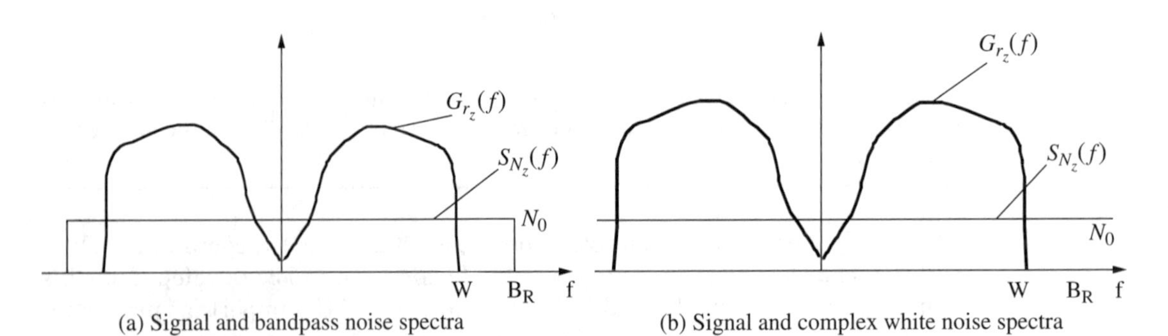

Figure 10.8 Exact and model power spectral densities.

AWGN model is that the analysis and algorithm development can proceed in a much cleaner fashion than with a more realistic noise model while not introdcing significant modeling errors. This model will be used exclusively in the sequel and the accuracy of the approximation will be explored in the homework.

10.6 Conclusion

Bandpass communication systems will be corrupted by bandpass noise. This chapter shows that the bandpass noise can be represented with a baseband complex noise. This baseband noise is in many cases of interest a zero mean, Gaussian, and stationary random process. A method to completely characterize a bandpass noise was presented. A simplified model for bandpass noise that will be frequently used in the sequel was also presented. With these tools in place it is now possible to evaluate the resultant message signal reconstruction fidelity of the various demodulation structures that have been proposed.

10.7 Homework Problems

Problem 10.1. What conditions on the bandpass filter characteristic of the receiver, $H_R(f)$, must be satisfied such that $N_I(t_1)$ and $N_Q(t_2)$ are independent random variables for all t_1 and t_2 when the input is AWGN?

Problem 10.2. This problem considers noise that might be seen in a vestigial sideband demodulator. Consider a bandpass stationary Gaussian noise at the input to a demodulator with a spectrum given as

$$S_{N_c}(f) = \begin{cases} 2 & f_c - 1000 \leq |f| \leq f_c + 3000 \\ 0 & \text{elsewhere} \end{cases} \tag{10.43}$$

Assume operation is in a 1-Ω system.

(a) What bandpass filter, $H_R(f)$, would produce this spectrum from a white noise input with a two-sided noise spectral density of $N_0/2 = 0.5$?

(b) What is the spectral density of $N_I(t)$?

(c) What is $E[N_I^2(t)]$?

(d) Give the joint PDF of $N_I(t_0)$ and $N_Q(t_0)$ in a 1-Ω system for a fixed t_0.

(e) Compute $S_{N_I N_Q}(f)$.

Problem 10.3. This problem considers noise which might be seen in a single sideband demodulator. Consider a bandpass stationary Gaussian noise at the input to a demodulator with a spectrum given as

$$S_{N_c}(f) = \begin{cases} 3.92 \times 10^{-3} & f_c \leq |f| \leq f_c + 3000 \\ 0 & \text{elsewhere} \end{cases} \tag{10.44}$$

Assume operation is in a 1-Ω system.

(a) What is the spectral density of $N_I(t)$?

(b) What is $E[N_I^2(t)]$?

(c) Give the joint PDF of $N_I(t_0)$ and $N_Q(t_0)$.

(d) Give the joint PDF of $N_I(t_1)$ and $N_Q(t_1 - \tau)$.

(e) Plot the PDF in (d) for $\tau = 0.0001$ and $\tau = 0.001$.

Problem 10.4. Consider a bandpass stationary Gaussian noise at the input to a demodulator with a spectrum given as

$$S_{N_c}(f) = \begin{cases} 0.001 & f_c - 2000 \leq |f| \leq f_c + 2000 \\ 0 & \text{elsewhere} \end{cases} \tag{10.45}$$

Assume operation is in a 1-Ω system.

(a) Find and plot the joint density function of $N_I(t_0)$ and $N_Q(t_0)$, $f_{N_I(t_0)N_Q(t_0)}(n_i, n_q)$.

(b) Consider the complex random variable $\tilde{N}_z(t_0) = \tilde{N}_I(t_0) + j\tilde{N}_Q(t_0) = N_z(t_0) \times \exp(-j\phi_p)$ and find a one–to–one mapping such that

$$\tilde{N}_I(t_0) = g_1(N_I(t_0), N_Q(t_0)) \qquad \tilde{N}_Q(t_0) = g_2(N_I(t_0), N_Q(t_0)) \tag{10.46}$$

(c) Detail the inverse mapping

$$N_I(t_0) = h_1(\tilde{N}_I(t_0), \tilde{N}_Q(t_0)) \qquad N_Q(t_0) = h_2(\tilde{N}_I(t_0), \tilde{N}_Q(t_0)) \tag{10.47}$$

(d) Using the results of Section 3.3.4 show that $f_{\tilde{N}_I(t_0)\tilde{N}_Q(t_0)}(n_i, n_q) = f_{N_I(t_0)N_Q(t_0)} \times (n_i, n_q)$. In other words, the noise distribution is unchanged by a phase rotation. This result is very important for the study of coherent receiver performance in the presence of noise detailed in the sequel.

Problem 10.5. Show for a stationary Gaussian bandpass noise, $N_c(t)$, that $|S_{N_I N_Q}(f)| \leq S_{N_I}(f)$. If $|S_{N_I N_Q}(f_0)| = S_{N_I}(f_0)$ for some f_0 then this places a constraint on either $H_R(f_c + f_0)$ or $H_R(f_c - f_0)$. What is this constraint?

Problem 10.6. For a stationary bandpass noise with complex envelope $N_z(t) = N_I(t) + jN_Q(t)$ that has a baseband autocorrelation function of $R_{N_I}(\tau)$ and a baseband crosscorrelation function of $R_{N_I N_Q}(\tau)$

(a) Find $E[|N_z(t)|^2]$.

(b) Find $E[N_z^2(t)]$.

(c) Find $E[N_z(t)N_z^*(t - \tau)]$.

(d) Find $E[N_z(t)N_z(t - \tau)]$.

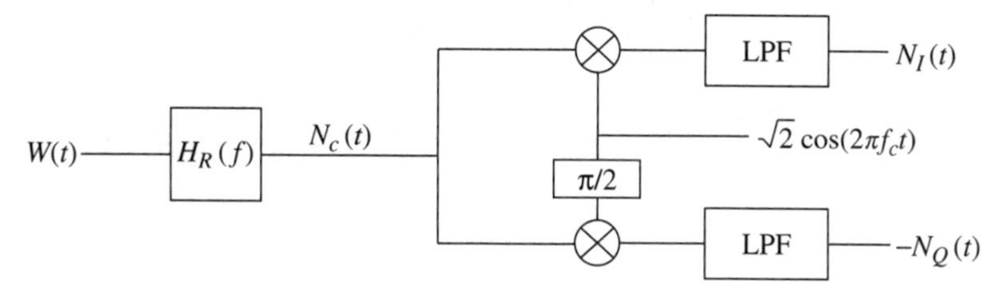

Figure 10.9 Noise and the receiver frontend processor.

Problem 10.7. Given the block diagram in Figure 10.9 where $W(t)$ is an AWGN with two-sided spectral density $N_0/2$ and $H_R(f)$ is a bandpass filter with a passband including f_c, find the conditions on the filter $H_R(f)$ such that

(a) $N_I(t)$ and $N_Q(t)$ have the same correlation function.

(b) $R_{N_I}(\tau) = R_{N_Q}(\tau) = \frac{\sin(\pi B\tau)}{\pi B\tau}$ and $R_{N_I N_Q}(\tau) = 0$.

Problem 10.8. Real valued additive white Gaussian noise, $W(t)$, with a one-sided spectral density $N_0 = 0.01$ is input into a bandpass receiver which has a block diagram shown in Figure 10.9. The noise power spectrum output from the down converter is measured as

$$S_{N_z}(f) = \begin{cases} 0.25 & |f| \le 100 \\ 0 & \text{elsewhere} \end{cases} \tag{10.48}$$

(a) Find $E\left[|N_z(t)|^2\right]$.

(b) Find $S_{N_I}(f)$ and $S_{N_I N_Q}(f)$.

(c) What is $|H_R(f)|$?

(d) Give the probability density function of one sample of $N_I(t)$, i.e., $f_{N_I(t)}(n)$.

(e) Compute $P(N_I(t) > 4)$.

Problem 10.9. Real valued additive white Gaussian noise, $W(t)$, with a two-sided spectral density $N_0/2$ is input into a demodulator for signal sideband amplitude modulation which has a block diagram shown in Figure 10.11. The time-invariant filter has a transfer function of

$$H_R(f) = \begin{cases} 2 & f_c \le |f| \le f_c + W \\ 0 & \text{elsewhere} \end{cases} \tag{10.49}$$

(a) Find the power spectral density of $N_c(t)$, $S_{N_c}(f)$?

(b) Find $S_{N_I}(f)$?

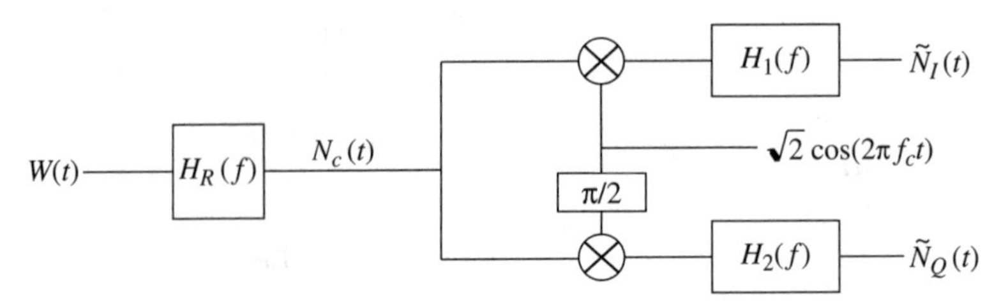

Figure 10.10 A bandpass noise model.

(c) Calculate $R_{N_I}(\tau)$.

(d) Detail out the form for the probability density function (PDF) of one sample from $N_I(t)$, $f_{N_I(t_s)}(n)$.

(e) Find $S_{N_I N_Q}(f)$.

(f) Find the joint PDF of samples taken from $N_I(t)$ and $N_Q(t)$ taken at the same time, $f_{N_I(t_s)N_Q(t_s)}(n_i, n_q)$.

Problem 10.10. Consider the model given in Figure 10.10 where $W(t)$ is a real AWGN with one–sided power spectral density of N_0. $H_1(f)$ and $H_2(f)$ are two potentially different filters. This problem investigates how the characteristics of the bandpass noise change if the filter in the I channel is different from the filter in the Q channel.

(a) Find the power spectral density of $\tilde{N}_I(t)$, $S_{\tilde{N}_I}(f)$, in terms of $S_{N_I}(f)$.

(b) Show that

$$R_{\tilde{N}_I \tilde{N}_Q}(\tau) = \int_{-\infty}^{\infty} h(\lambda_1) \int_{-\infty}^{\infty} R_{N_I N_Q}(\tau + \lambda_2 - \lambda_1) h_2(\lambda_2) d\lambda_2 d\lambda_1 \tag{10.50}$$

(c) Show that

$$S_{\tilde{N}_I \tilde{N}_Q}(f) = S_{N_I N_Q}(f) H_2^*(f) H_1(f) \tag{10.51}$$

(d) Will $\tilde{N}_I(t)$ always be independent of $\tilde{N}_Q(t)$ as when the filters are the same? If not then give an example?

(e) Give conditions on $H_1(f)$ and $H_2(f)$ such that $S_{\tilde{N}_I \tilde{N}_Q}(f)$ would be imaginary and odd as is the case for $S_{N_I N_Q}(f)$. If these conditions were satisfied would $\tilde{N}_I(t)$ always be independent of $\tilde{N}_Q(t)$?

Problem 10.11. Show that if $N_z(t)$ is a stationary complex envelope of a bandpass noise and $\tilde{N}_z(t) = N_z(t)\exp[j\phi_p]$ then

(a) $R_{\tilde{N}_I}(\tau) = R_{\tilde{N}_Q}(\tau) = R_{N_I}(\tau)$

(b) $R_{\tilde{N}_I \tilde{N}_Q}(\tau) = R_{N_I N_Q}(\tau)$

Problem 10.12. Real valued additive white Gaussian noise, $W(t)$, with a two-sided spectral density $N_0/2$ is input into a demodulator which has a block diagram

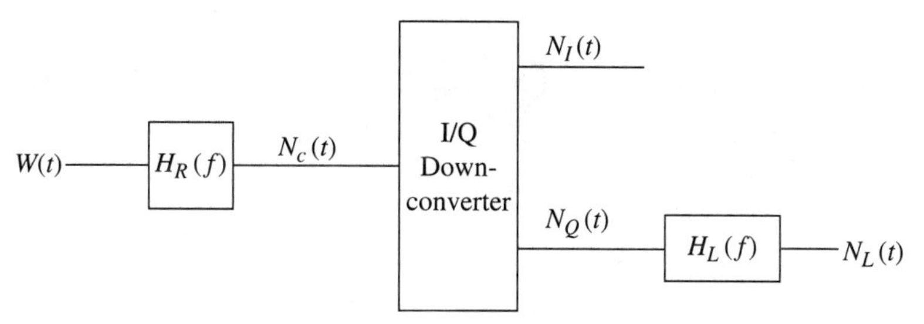

Figure 10.11 Demodulator block diagram.

shown in Figure 10.11. The time-invariant filter has a transfer function of

$$H_R(f) = \begin{cases} 2 & f_c - 100 \le |f| \le f_c + 100 \\ 0 & \text{elsewhere} \end{cases} \tag{10.52}$$

(a) Find the power spectral density of $N_c(t)$, $S_{N_c}(f)$.

(b) $E[N_c^2(t)] = 2$, what is N_0?

(c) Find the power spectral density of $N_I(t)$, $S_{N_I}(f)$ when $E[N_c^2(t)] = 2$.

(d) Find $E[N_I^2(t)]$ for the value of N_0 obtained in (b).

(e) Assume that $H_L(f)$ is an ideal lowpass filter with $H_L(0) = 1$. Choose the bandwidth of the filter such that $E[N_L^2(t)] = \frac{1}{2}E[N_I^2(t)]$.

Problem 10.13. An interesting relationship exists between the power spectral density of a phase noise and the distribution of the instantaneous frequency deviation. This problem explores this relationship, which has proven useful in practice, using the techniques introduced in this chapter and in the development of angle modulation. Assume a random process is characterized as

$$N_z(t) = A_c \exp[jN_p(t)] \tag{10.53}$$

where $N_p(t)$ is a phase noise generated in, for example, a frequency synthesizer.

(a) Show that

$$R_{N_z}(t,\tau) = E[N_z(t)N_z^*(t-\tau)] \approx A_c^2 E[\exp(j2\pi F_d(t)\tau)] \tag{10.54}$$

where $F_d(t) = \frac{1}{2\pi}\frac{d}{dt}N_p(t)$.

(b) If $F_d(t)$ is a stationary random process, show that

$$R_{N_z}(t,\tau) \approx R_{N_z}(\tau) \approx A_c^2 \Phi_{F_d}(\tau) \tag{10.55}$$

where $\Phi_{F_d}(t)$ is the characteristic function[1] of a sample of the random process $F_d(t)$.

[1]The characteristic function of a random variable was introduced in Problem 3.24 in this text but is a common tool in the analysis of noise [LG89, Hel91].

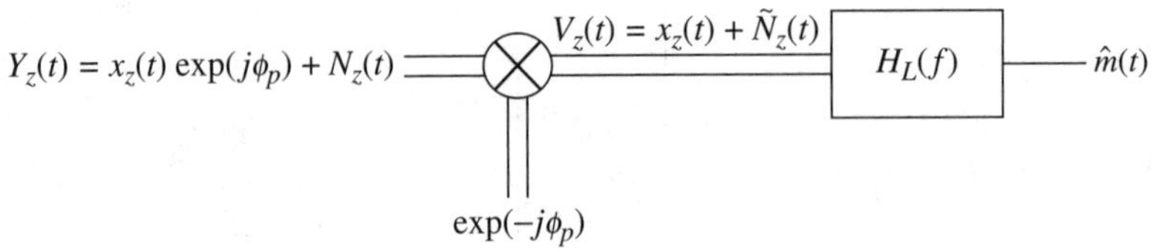

$Y_z(t) = x_z(t)\exp(j\phi_p) + N_z(t)$

$V_z(t) = x_z(t) + \tilde{N}_z(t)$

$H_L(f)$

$\hat{m}(t)$

$\exp(-j\phi_p)$

Figure 10.12 A demodulator block diagram.

(c) Finally show that

$$S_{N_z}(f) \approx A_c^2 f_{F_d}(f) \tag{10.56}$$

where $f_{F_d}(f)$ is the PDF of the random variable $F_d(t)$.

(d) Make an engineering assessment of the validity and the shortcomings of the approximations in this problem.

Problem 10.14. It is clear from Example 10.7 and Example 10.8 that a baseband noise of varying bandwidth can be produced depending on the f_c that is chosen in the downconverter. For reasonable choices of f_c what are the maximum and minimum bandwidths that a lowpass noise process would possess for a bandpass bandwidth of B_T.

Problem 10.15. In a DSB-AM receiver modeled with Figure 10.12, the following complex envelope is received

$$Y_z(t) = \cos(200\pi t)\exp[j\phi_p] + N_z(t) \tag{10.57}$$

Assume $\hat{m}(t) = m_e(t) + \tilde{N}_I(t)$ and for simplicity that the bandpass noise, $N_z(t)$ is characterized as a white noise

$$S_{N_z}(f) = N_0 \tag{10.58}$$

The lowpass filter has a transfer function of

$$H_L(f) = \frac{1}{1 + j\frac{f}{f_l}} \tag{10.59}$$

(a) Prove $\tilde{N}_I(t)$ is a stationary random process.

(b) Compute $m_e(t)$ (the signal at the filter output).

(c) Compute $E[\tilde{N}_I^2(t)]$ as a function of f_l. *Hint:* Power can be computed in the time and frequency domain.

(d) Compute the output SNR as a function of f_l.

(e) What value of f_l optimizes the output SNR?

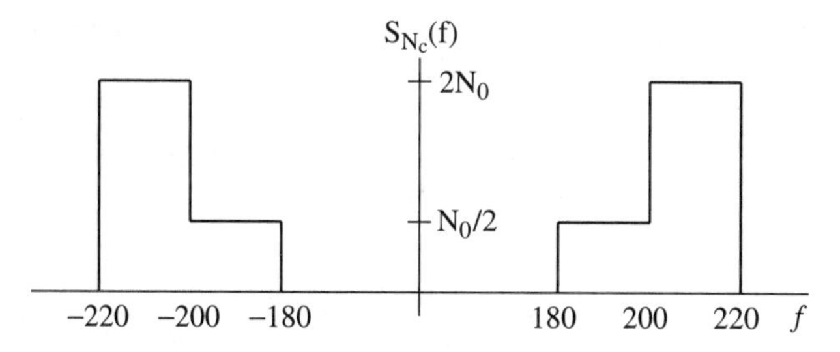

Figure 10.13 An example bandpass noise power spectrum.

Problem 10.16. A computer can do a lot of tedious computations that you do not want to do by hand. To this end repeat the Problem 10.3 except with the bandpass noise characterized with

$$S_{N_z}(f) = \begin{cases} N_0 & |f| \leq 4000 \\ 0 & \text{elsewhere} \end{cases} \tag{10.60}$$

Hint: The simplest approach may not be the same as in Problem 10.15.

Problem 10.17. **(JG)** A WSS bandpass random process, $N_c(t)$, has power spectral density, $S_{N_c}(f)$, shown in Figure 10.13.

(a) Sketch $S_{N_z}(f)$, the spectral density of the complex envelope.

(b) Sketch $S_{N_I}(f)$, the spectral density of the in-phase processes.

(c) Sketch $S_{N_I N_Q}(f)$, the cross spectral density between the in-phase and quadrature processes.

Problem 10.18. Real valued additive white Gaussian noise, $W(t)$ with a one-sided spectral density $N_0 = 0.002$ is input into a bandpass receiver which has a block diagram shown in Figure 10.14. $H_R(f)$ is an ideal bandpass filter centered at carrier frequency f_c with bandwidth 6000 Hz, i.e.,

$$H_R(f) = \begin{cases} 1 & f_c - 3000 \leq |f| \leq f_c + 3000 \\ 0 & \text{elsewhere} \end{cases} \tag{10.61}$$

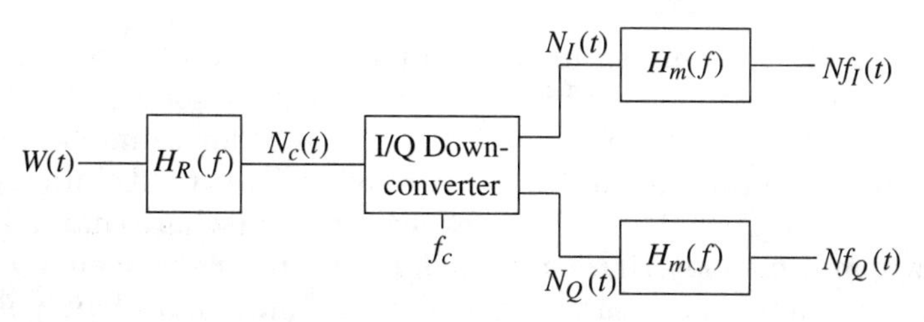

Figure 10.14 Demodulator block diagram.

The baseband filter, $H_m(f)$, has a cosine pulse frequency response, i.e.,

$$H_m(f) = \begin{cases} 1 & |f| \leq 750 \\ \cos\left(\frac{\pi}{1000}(|f| - 750)\right) & 750 \leq |f| \leq 1250 \\ 0 & \text{elsewhere} \end{cases} \qquad (10.62)$$

(a) Find $S_{N_I}(f)$.

(b) Find $S_{N_I N_Q}(f)$.

(c) Find $S_{\tilde{N}_Q}(f)$.

(d) Find the PDF of $\tilde{N}_Q(t_0)$.

(e) Compute $P(\tilde{N}_Q(t) > 4)$.

Problem 10.19. Gaussian random variables are convenient in communication systems analysis because only a finite number of parameters are needed to fully characterize the statistical nature of the random variables. Samples taken from a stationary Gaussian bandpass random process require even fewer parameters to characterize.

(a) If N_1 and N_2 are two jointly Gaussian zero mean random variables how many parameters and what are the parameters needed to specify the joint distribution.

(b) If $N_1 = N_I(t)$ and $N_2 = N_Q(t)$ are two jointly Gaussian zero mean random variables obtained by sampling a stationary zero mean bandpass Gaussian process at time t, how many and what are the parameters needed to specify the joint distribution.

(c) If N_1, N_2, N_3, and N_4 are four jointly Gaussian zero mean random variables how many parameters and what are the parameters needed to specify the joint distribution.

(d) If $N_1 = N_I(t)$, $N_2 = N_Q(t)$, $N_3 = N_I(t + \tau)$, and $N_4 = N_Q(t + \tau)$ are four jointly Gaussian zero mean random variables obtained by sampling a stationary zero mean bandpass Gaussian process at times t and $t + \tau$, how many parameters and what are the parameters needed to specify the joint distribution.

Problem 10.20. In communication systems where multiple users are to be supported within a common architecture, a prevalent source of interference is adjacent channel interference (ACI). ACI occurs, for example, in a channelized system when the transmitted spectra of a user extends into the channel of an adjacent user. Consider an FM broadcast system where channels are spaced 200 kHz apart and a interference transmission spectrum given in Figure 10.15. Assume this interference can be modeled as a stationary bandpass Gaussian noise, $N_c(t)$, for this problem.

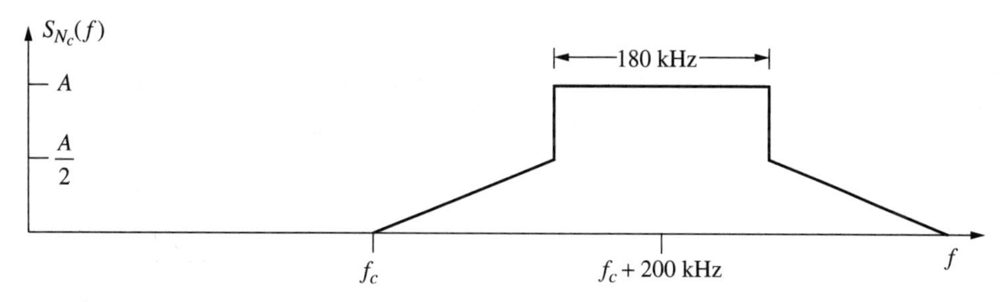

Figure 10.15 An interference spectrum in a channelized system.

(a) Choose the value of A such that the total interference power is 10 W in a 1-Ω system.

(b) If this interference is processed by a radio with an ideal unity gain bandpass filter of 180 kHz bandwidth centered at f_c and then downconverted to baseband, find the resulting $S_{N_z}(f)$ at the output of the radio receiver.

(c) What is the total power of this interference that appears at the output of the radio receiver?

(d) Find $S_{N_I}(f)$ and $S_{N_I N_Q}(f)$.

(e) Find $E[N_I^2(t)]$.

10.8 Example Solutions

Problem 10.2.

(a) We know $N_0 = 1$ and

$$S_{N_c}(f) = \frac{N_0}{2}|H_c(f)|^2 \tag{10.63}$$

therefore,

$$|H_R(f)|^2 = \begin{cases} 4 & f_c - 1000 \le |f| \le f_c + 3000 \\ 0 & \text{elsewhere} \end{cases} \tag{10.64}$$

Consequently,

$$H_R(f) = \begin{cases} 2e^{j\theta(f)} & f_c - 1000 \le |f| \le f_c + 3000 \\ 0 & \text{elsewhere} \end{cases} \tag{10.65}$$

where $\theta(f)$ is an arbitrary phase for the filter transfer function.

(b) We know

$$S_{N_c}(f) = \frac{1}{2}S_{N_z}(f - f_c) + \frac{1}{2}S_{N_z}(-f - f_c) \tag{10.66}$$

Therefore,

$$S_{N_z}(f) = \begin{cases} 4 & -1000 \le f \le 3000 \\ 0 & \text{elsewhere} \end{cases} \tag{10.67}$$

Solving for $S_{N_I}(f)$ gives

$$S_{N_I}(f) = \frac{S_{N_z}(f) + S_{N_z}(-f)}{4} \tag{10.68}$$

Therefore,

$$S_{N_I}(f) = \begin{cases} 1 & -3000 \le f \le -1000 \\ 2 & -1000 \le f \le 1000 \\ 1 & 1000 \le f \le 3000 \\ 0 & \text{elsewhere} \end{cases} \tag{10.69}$$

(c) $E[N_I^2(t)] = \int_{-\infty}^{\infty} S_{N_I}(f)df = 8000$

(d) Using Problems 10.10 and 10.24

$$f_{N_I(t)N_Q(t)}(n_1, n_2) = \frac{1}{16000\,\pi} \cdot \exp\left[-\frac{n_1^2 + n_2^2}{16000}\right] \tag{10.70}$$

(e) Solving for $S_{N_I N_Q}(f)$ gives

$$S_{N_I N_Q}(f) = \frac{S_{N_z}(-f) - S_{N_z}(f)}{j4} \tag{10.71}$$

Therefore,

$$S_{N_I N_Q}(f) = \begin{cases} -j & -3000 \le f \le -1000 \\ j & 1000 \le f \le 3000 \\ 0 & \text{elsewhere} \end{cases} \tag{10.72}$$

Problem 10.9.

(a) The results of Chapter 9 give the bandpass power spectral density as

$$S_{N_c}(f) = \frac{N_0}{2}|H_R(f)|^2 = \begin{cases} 2N_0 & f_c \le |f| \le f_c + W \\ 0 & \text{elsewhere} \end{cases} \tag{10.73}$$

(b) The baseband noises $N_I(t)$ and $N_Q(t)$ are zero mean, jointly stationary and jointly Gaussian random processes. First convert the bandpass spectrum to baseband

$$S_{N_z}(f) = \begin{cases} 4N_0 & 0 \le f \le W \\ 0 & \text{elsewhere} \end{cases} \tag{10.74}$$

The formula for $S_{N_I}(f)$ is

$$S_{N_I}(f) = \frac{S_{N_z}(f) + S_{N_z}(-f)}{4} \tag{10.75}$$

which gives

$$S_{N_I}(f) = \begin{cases} N_0 & -W \leq |f| \leq W \\ 0 & \text{elsewhere} \end{cases} \tag{10.76}$$

(c) Recall that $R_{N_I}(\tau) = \mathcal{F}^{-1}(S_{N_I}(f))$ so that we have

$$R_{N_I}(\tau) = 2N_0 W \operatorname{sinc}(2W\tau) \tag{10.77}$$

(d) Recall that $N_I(t_s)$ is a zero mean Gaussian random variable with $\sigma_N^2 = R_{N_I}(0) = 2N_0W$. The PDF is then given as

$$f_{N_I(t_s)}(n) = \frac{1}{\sqrt{4\pi N_0 W}} \exp\left[-\frac{n^2}{4N_0W}\right] \tag{10.78}$$

(e) The cross spectrum is given as

$$S_{N_I N_Q} = \frac{S_{N_z}(-f) - S_{N_z}(f)}{j4} = \begin{cases} -jN_0 & -W \leq f \mid \leq 0 \\ jN_0 & 0 \leq f \mid \leq W \\ 0 & \text{elsewhere} \end{cases} \tag{10.79}$$

(f) The joint PDF of $N_I(t_s)$ and $N_Q(t_s)$ will always be that of zero mean, uncorrelated, jointly Gaussian random variables as $R_{N_I N_Q}(0) = 0$. Since the variance is again $\sigma_N^2 = R_{N_I}(0) = 2N_0W$ the PDF is given as

$$f_{N_I(t_s)N_Q(t_s)}(n_1, n_2) = \frac{1}{4\pi N_0 W} \exp\left[-\frac{n_1^2 + n_2^2}{4N_0W}\right] \tag{10.80}$$

10.9 Miniprojects

Goal: To give exposure

- to a small scope engineering design problem in communications.
- to the dynamics of working with a team.
- to the importance of engineering communication skills (in this case oral presentations).

Presentation: The forum will be similar to a design review at a company (only much shorter). The presentation will be of 5 minutes in length with an overview of the given problem and solution. The presentation will be followed by questions from the audience (your classmates and the professor). All team members should be prepared to give the presentation.

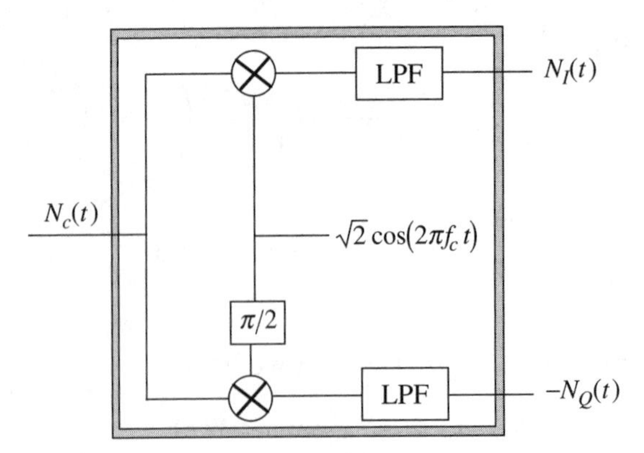

Figure 10.16 Block diagram of an I/Q downconverter with noise.

10.9.1 Project 1

Project Goals: To use the theory of bandpass noise to troubleshoot an anomalous behaviour in a communication system.

A colleague at your company, *Wireless.com*, is working on characterizing the noise in the front end of an intermediate frequency (IF) receiver that the company is trying to make its first billion on. The design carrier frequency, f_c, was not documented well by the designer (he left for another startup!). The frequency is known to lie somewhere between $f_c = 2.5$ kHz and $f_c = 8$ kHz. Your colleague is getting some very anomalous results in his testing and has come to you since he knows you were taught communications theory by a world renowned professor. The bandpass noise output from the receiver is processed during testing in a programmable I/Q downconverter with a carrier frequency $\tilde{f}_c$, as shown in Figure 10.16. The lowpass filter in the I/Q downconverter is programmed to have a cutoff frequency of $\tilde{f}_c$. The anomalous results your colleague sees are

- if he chooses $\tilde{f}_c = 5000$ Hz then the output noise, $N_I(t)$ and $N_Q(t)$, has a bandwidth of 5000 Hz,
- if $\tilde{f}_c = 4000$ Hz then the output noise, $N_I(t)$ and $N_Q(t)$, has a bandwidth of 4000 Hz.

Your colleague captured and stored a sample function of the noise and it is available for downloading in the file `noizin.mat`. Examining this file will be useful to complete this project.

(a) Explain why the output noise bandwidth changes as a function $\tilde{f}_c$.

(b) Assuming that DSB-AM is the design modulation for the receiver, try and identify the probable design carrier frequency.

Chapter

11

Fidelity in Analog Demodulation

This chapter examines the performance of all the different analog modulations that were considered in this text in the presence of noise. This analysis allows a comparison of the resulting fidelity of message reconstruction in terms of the resulting signal to noise ratio (SNR). This chapter will first examine the transmission of information without modulation and use this as a baseline for comparison. Each modulation will then be compared to this baseline and a figure of merit for each scheme can be computed.

11.1 Unmodulated Signals

A first evaluation that is useful is the resulting fidelity of an unmodulated message signal reconstruction in the presence of an AWGN, $W(t)$. This situation is shown in Figure 11.1. This situation represents a direct connection of the message signal over a wire/cable from the message source to the message sink. The filter, $H_R(f)$, represents the signal processing at the receiver to limit the noise power. We consider this situation to give a baseline for the fidelity comparisons in the sequel.

For analog communications the fidelity metric used in this textbook will be SNR. Recall the output SNR in the system in Figure 11.1 is defined as

$$\text{SNR} = \frac{P_{x_z}}{P_N} = \frac{A^2 P_m}{P_N} \tag{11.1}$$

The noise power, P_N, can be calculated with methods described in Chapter 9. Specifically

$$P_N = \sigma_N^2 = N_0 B_N |H_R(0)|^2 = \frac{N_0}{2}\int_{-\infty}^{\infty} |H_R(f)|^2\, df \tag{11.2}$$

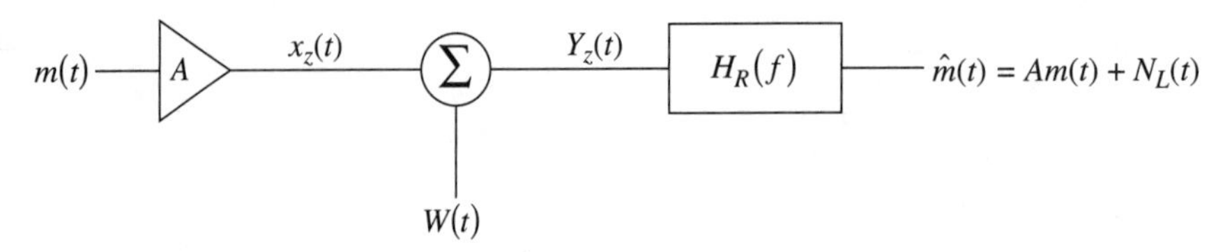

Figure 11.1 A model for baseband transmission.

For ease of discussion and comparison this chapter will consider only ideal lowpass filters, e.g.,

$$H_R(f) = \begin{cases} 1 & |f| \leq B_T \\ 0 & \text{elsewhere} \end{cases} \tag{11.3}$$

Clearly the smaller B_T, the smaller the noise power, but B_T cannot be made too small or else signal distortion will occur and contribute to $N_L(t)$. Again for this chapter to simplify the presentation we will assume $B_T = W$ and consequently signal distortion is not an issue. The output SNR is defined as

$$\text{SNR}_o = \frac{P_{x_z}}{N_0 W} = \frac{A^2 P_m}{N_0 W} = \text{SNR}_b \tag{11.4}$$

Equation (11.4) is surprisingly simple but it is still instructive to interpret it. In general, the noise spectral density is determined by the environment that you are communicating in and is not something the communication system designer has control over. For a fixed noise spectral density the SNR can be made larger by a communication system designer by increasing the message power, P_m, decreasing the loss in the transmission, here A, or decreasing the filter bandwidth, W. Some interesting insights include

- Transmitting farther distances (smaller A) will produce lower output SNR.
- Transmitting wider bandwidth signal (video vs. audio) will produce lower output SNR.

In general we will find the same trade-offs exist with modulated systems as well. Note, this text denotes the output SNR achieved with baseband (unmodulated) transmission as SNR_b. This will be a common reference point to compare analog communication system's fidelity in message reconstruction.

EXAMPLE 11.1

Consider the example of a message signal given as

$$m(t) = \cos(2\pi f_m t) \tag{11.5}$$

and a channel gain of $A = 1$ corrupted by a wideband noise, $W(t)$, with a one-sided spectral density $N_0 = 1/8000$. For $f_m = 500$ Hz the plot of one sample path of the input

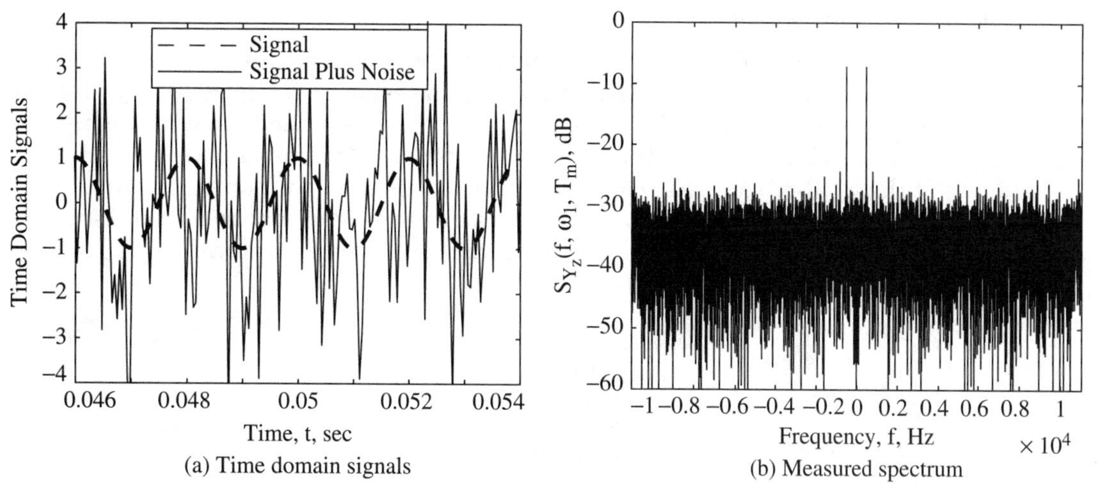

(a) Time domain signals (b) Measured spectrum

Figure 11.2 The input signal plus wideband noise for the example baseband transmission system.

signal plus noise, $Y_z(t, \omega)$, and the input measured power spectrum, $S_{Y_z}(f, \omega_1, T_m)$, are given in Figure 11.2. If this signal is input into a LPF filter with a bandwidth of 1000 Hz the output signal and spectrum are shown in Figure 11.3. It is clear from these figures that the variance of the noise is reduced and the time characteristics of the noise are also modified. In this example the output SNR = 6 dB.

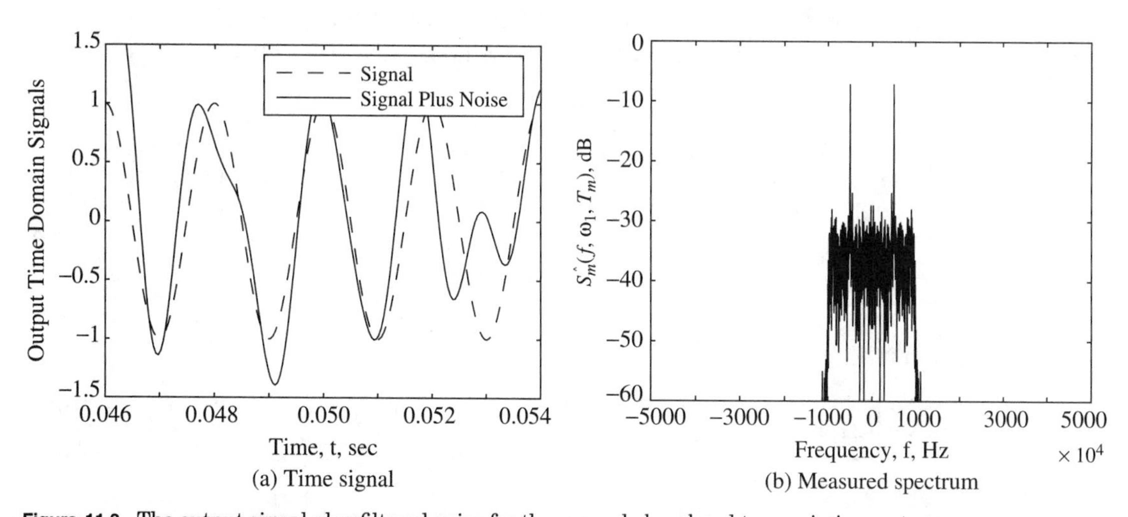

(a) Time signal (b) Measured spectrum

Figure 11.3 The output signal plus filtered noise for the example baseband transmission system.

11.2 Bandpass Demodulation

Now let us consider the SNR evaluation of bandpass modulated message signal in the presence of an AWGN. This situation is shown in Figure 11.4. The signal components and the noise components were characterized in the prior chapters so now an integration of this information is needed to provide a characterization of the SNR of bandpass communication systems. Again for ease of discussion and comparison this chapter will consider only ideal bandpass filters with a bandwidth exactly matched to the transmitted signal bandwidth, e.g.,

$$H_R(f) = \begin{cases} 1 & f_c - \frac{B_T}{2} \leq |f| \leq f_c + \frac{B_T}{2} \\ 0 & \text{elsewhere} \end{cases} \tag{11.6}$$

The output complex envelope has the form

$$Y_z(t) = x_z(t)\exp(j\phi_p) + N_z(t) \tag{11.7}$$

where $N_z(t) = N_I(t) + jN_Q(t)$. The power of the transmitted signal, $x_z(t)$, will again be denoted P_{x_z}. The resulting lowpass noise at the output of the downconverter is well characterized via the tools developed in Chapter 10. For example, for the filter given in Eq. (11.6) it can be shown that

$$S_{N_I}(f) = \begin{cases} \dfrac{N_0}{2} & |f| \leq \frac{B_T}{2} \\ 0 & \text{elsewhere} \end{cases} \tag{11.8}$$

and $S_{N_I N_Q}(f) = 0$. An inverse Fourier transform gives $R_{N_I}(\tau) = \frac{N_0 B_T}{2} \times \text{sinc}(B_T\tau)$. The variance of the lowpass noise is $\text{var}(N_I(t)) = \sigma_{N_I}^2 = \frac{N_0 B_T}{2}$ and the joint PDF is

$$f_{N_I(t)N_Q(t)}(n_1, n_2) = \frac{1}{2\pi\sigma_N^2}\exp\left[\frac{-\left(n_1^2+n_2^2\right)}{2\sigma_N^2}\right] = \frac{1}{\pi N_0 B_T}\exp\left[\frac{-\left(n_1^2+n_2^2\right)}{N_0 B_T}\right] \tag{11.9}$$

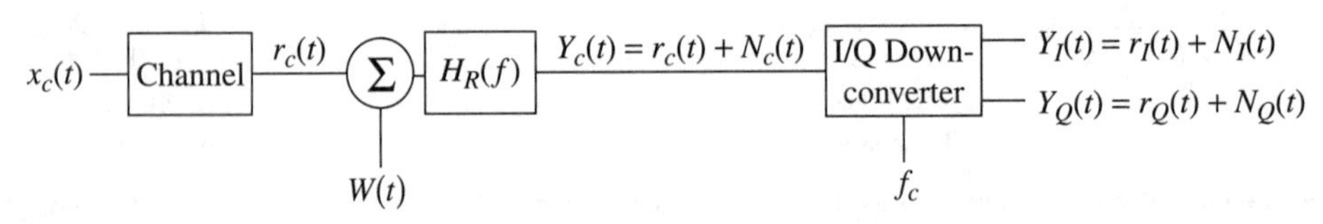

Figure 11.4 A model for bandpass modulated transmission.

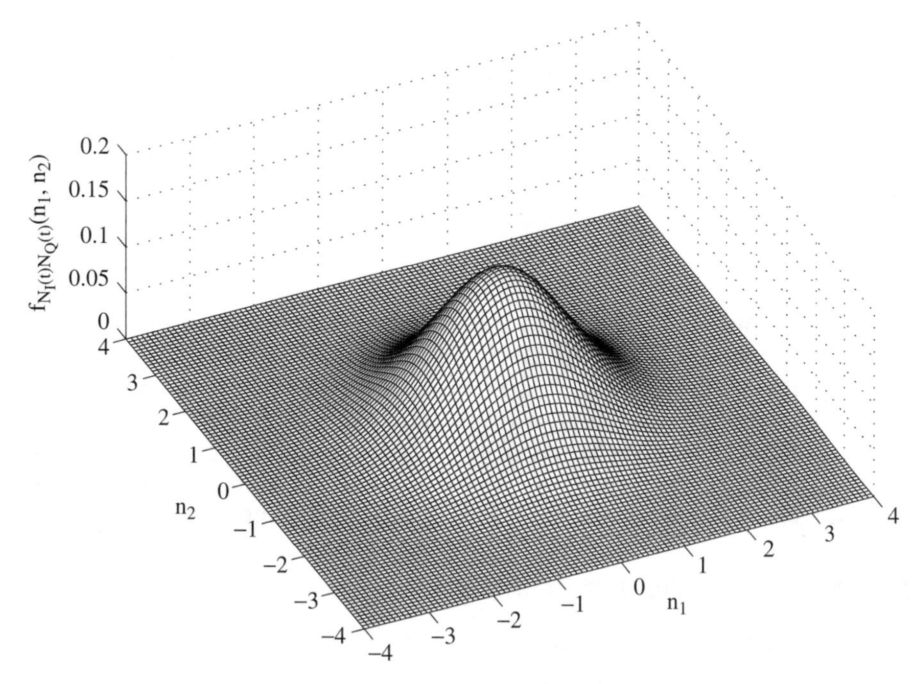

Figure 11.5 Plots of the joint PDF of $N_I(t)$ and $N_Q(t)$ for $N_0 B_T = 2$.

A plot of the density function for $N_0 B_T = 2$ is shown in Figure 11.5. The input SNR to the demodulator is defined as

$$\mathrm{SNR}_i = \frac{P_{x_z}}{P_{N_z}} = \frac{P_{x_z}}{N_0 B_T} = \mathrm{SNR}_b \frac{W}{B_T} = \mathrm{SNR}_b \nu_B \tag{11.10}$$

Consequently, the input SNR to the demodulator is both a function of the baseband SNR and the spectral efficiency of the transmission. This is important to note that broadcast FM typically has a lower input SNR than does broadcast AM since FM's spectral efficiency is worse than AM's spectral efficiency.

11.2.1 Coherent Demodulation

Many of the analog modulations considered in this text are demodulated with coherent demodulation structures and it is worth examining the noise characteristics of coherent demodulation structures. The coherent demodulator has a block diagram as given in Figure 11.6. The derotated noise, $\tilde{N}_z(t) = N_z(t)\exp(-j\phi_p)$ in the coherent demodulator has the same statistical characteristics as $N_z(t)$. This is easy to show mathematically by

$$\begin{aligned} R_{\tilde{N}_z}(\tau) &= E[\tilde{N}_z(t)\tilde{N}_z^*(t-\tau)] = E[N_z(t)\exp(-j\phi_p)N_z^*(t-\tau)\exp(j\phi_p)] \\ &= E[N_z(t)N_z^*(t-\tau)] = R_{N_z}(\tau) \end{aligned} \tag{11.11}$$

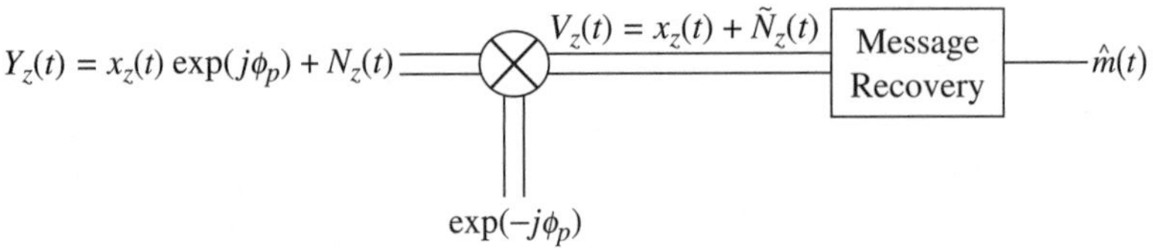

$Y_z(t) = x_z(t)\exp(j\phi_p) + N_z(t)$ — $V_z(t) = x_z(t) + \tilde{N}_z(t)$ — Message Recovery — $\hat{m}(t)$; $\exp(-j\phi_p)$

Figure 11.6 The coherent demodulation structure.

Equation (11.11) shows that rotating the noise does not change the statistical characteristics of the noise. Intuitively this makes sense because in examining the PDF plotted in Figure 11.5, the joint PDF of the lowpass noise is clearly circularly symmetric. Any rotation of this PDF would not change the form of the PDF. This is mathematically proven in Problem 10.4. This circularly symmetric noise characteristic and the results of Chapter 10 will allow us to rigorously characterize the performance of coherent demodulation schemes.

11.3 Coherent Amplitude Demodulation

11.3.1 Coherent Demodulation

Given the tools developed so far the fidelity analysis of coherent AM demodulation systems is straightforward. Using the results of Section 11.2.1 we know that the baseband derotated noise is going to be zero mean, stationary, and Gaussian with a PSD given as

$$S_{\tilde{N}_I}(f) = \begin{cases} \dfrac{N_0}{2} & |f| \leq \frac{B_T}{2} \\ 0 & \text{elsewhere} \end{cases} \tag{11.12}$$

Coherent DSB-AM

DSB-AM uses coherent demodulation. Recall the complex envelope form of DSB-AM is $x_z(t) = A_c m(t)$ so consequently the representative block diagram for a DSB-AM demodulator in the presence of noise is given in Figure 11.7. The output of the *real* operator has the form $V_I(t) = A_c m(t) + \tilde{N}_I(t)$ and the output spectrum of the signal and the noise is shown in Figure 11.8(a). This signal

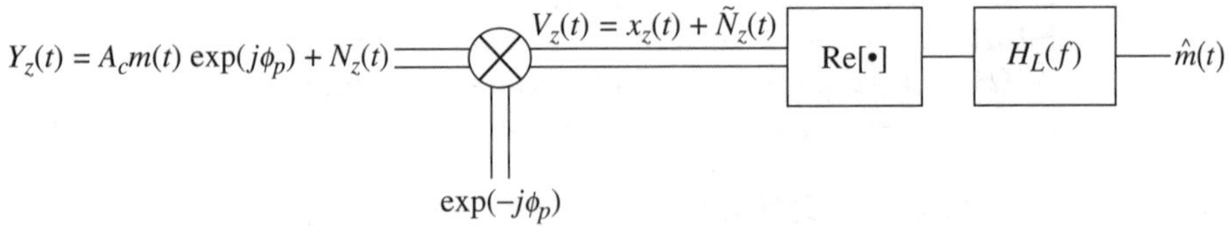

$Y_z(t) = A_c m(t)\exp(j\phi_p) + N_z(t)$ — $V_z(t) = x_z(t) + \tilde{N}_z(t)$ — Re[•] — $H_L(f)$ — $\hat{m}(t)$; $\exp(-j\phi_p)$

Figure 11.7 The coherent demodulation structure for DSB-AM.

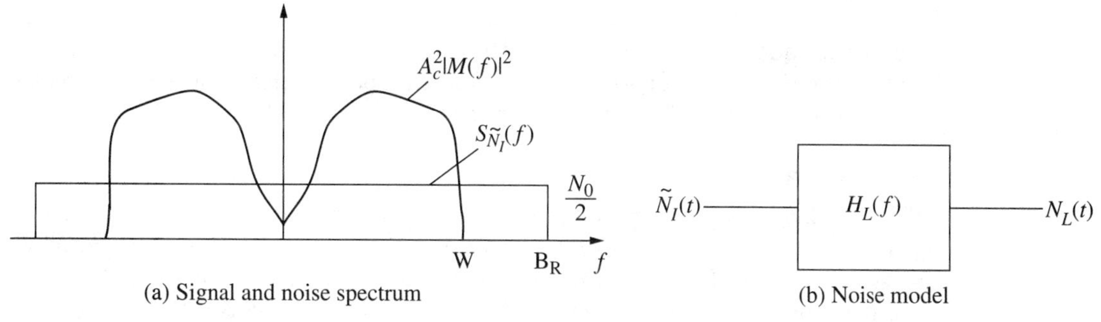

Figure 11.8 The DSB-AM demodulator noise processing.

and noise model is exactly the same form as seen in Section 11.1, consequently the filter, $H_L(f)$, is subject to the same trade-offs between minimizing the noise power and minimizing the signal distortion.

The filter, $H_L(f)$, maximizes the SNR without signal distortion by being a lowpass filter of bandwidth W. For simplicity in discussion the lowpass filter will be assumed to be an ideal lowpass filter and to have the transfer function

$$H_L(f) = \begin{cases} 1 & |f| \leq W \\ 0 & \text{elsewhere} \end{cases} \tag{11.13}$$

The resulting message estimate for DSB-AM has the form $\hat{m}(t) = A_c m(t) + N_L(t)$. The distorting noise, $N_L(t)$, is produced by a lowpass filtering of $\tilde{N}_I(t)$ as shown in Figure 11.7(b). This output noise has a PSD given as

$$S_{N_L}(f) = S_{\tilde{N}_I}(f)\,|H_L(f)|^2 = S_{N_I}(f)\,|H_L(f)|^2 = \begin{cases} \frac{N_0}{2} & |f| \leq W \\ 0 & \text{elsewhere} \end{cases} \tag{11.14}$$

The output noise power is $\text{var}(N_L(t)) = N_0 W$ and the output SNR is

$$\text{SNR}_o = \frac{A_c^2 P_m}{N_0 W} = \frac{P_{x_z}}{N_0 W} = \text{SNR}_b \tag{11.15}$$

Recall the E_T is the transmission efficiency defined as $\text{SNR}_o = E_T\,\text{SNR}_b$, so that it is clear here that $E_T = 1$ so that coherent demodulation of DSB-AM has the same performance as an unmodulated signal. Since DSB-AM has $E_T = 1$, two interesting insights are

1. The output SNR is exactly the same as the case where the signal was transmitted unmodulated. This implies that a coherent bandpass modulation and demodulation process can result in no loss of performance compared to baseband transmission while greatly expanding the spectrum that can be used for transmission.

2. Again the two things that affect the output SNR are the transmitted power[1] and the message signal bandwidth. Consequently, transmitting video will cost more in terms of resources (transmitted power) than transmitting voice for the same level of output performance.

Coherent SSB-AM

SSB-AM also uses coherent demodulation. Recall the complex envelope form of SSB-AM is $x_z(t) = A_c(m(t) + jm(t) * h_Q(t))$, so, consequently, the representative block diagram of SSB-AM in noise is very similar to that in Figure 11.7. The output of the *real* operator has the form $V_I(t) = A_c m(t) + \tilde{N}_I(t)$. This form is exactly the same form as DSB-AM. The resulting message estimate for DSB-AM has the form $\hat{m}(t) = A_c m(t) + N_L(t)$ where for the bandpass filter given in Eq. (11.6)

$$S_{N_L}(f) = S_{\tilde{N}_I}(f)\,|H_L(f)|^2 = \begin{cases} \frac{N_0}{2} & |f| \le W \\ 0 & \text{elsewhere} \end{cases} \tag{11.16}$$

The output noise power is $\text{var}(N_L(t)) = N_0 W$ and the output signal to noise ratio is

$$\text{SNR}_o = \frac{A_c^2 P_m}{N_0 W} \tag{11.17}$$

Recall that the transmitted power of SSB-AM is $P_{x_z} = 2A_c^2 P_m$ so this implies that

$$\text{SNR}_o = \frac{P_{x_z}}{2N_0 W} = \frac{\text{SNR}_b}{2} \tag{11.18}$$

With this method of signal processing for demodulation of SSB-AM it is clear that $E_T = 0.5$ and it appears, at first glance, that SSB-AM is less efficient in using the transmitted signal power than is an unmodulated system.

This loss in efficiency and fidelity of the message reconstruction can be recovered by better signal processing in the receiver. The methodology to improve this efficiency is apparent when examining the signal and noise spectrum of an upper sideband transmission shown in Figure 11.9. The bandpass filter in Eq. (11.6) is passing all signal and noise within $B_T/2$ Hz of the carrier frequency. This bandwidth is required in DSB-AM but clearly the receiver filter can be reduced in bandwidth and have a form

$$H_R(f) = \begin{cases} 1 & f_c \le |f| \le f_c + \frac{B_T}{2} \\ 0 & \text{elsewhere} \end{cases} \tag{11.19}$$

[1]Also the channel attenuation but we have chosen to normalize this to unity to simplify the discussion

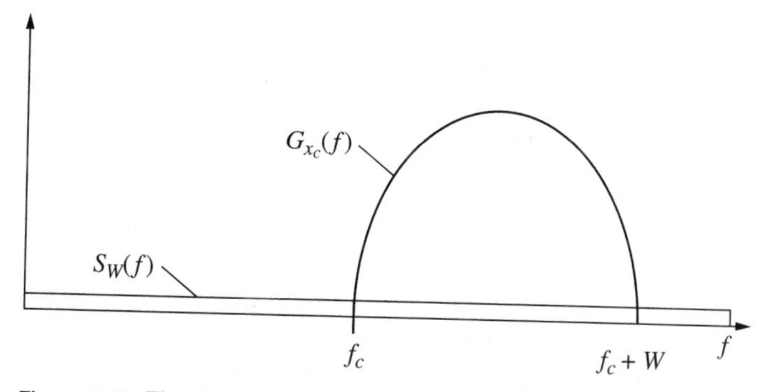

Figure 11.9 The signal and noise spectrum in SSB-AM reception.

The bandpass noise PSD then becomes

$$S_{N_c}(f) = \begin{cases} \frac{N_0}{2} & f_c \leq |f| \leq f_c + \frac{B_T}{2} \\ 0 & \text{elsewhere} \end{cases} \tag{11.20}$$

Using the results of Chapter 10 gives

$$S_{N_I}(f) = \frac{S_{N_z}(f) + S_{N_z}(-f)}{4} = \begin{cases} \frac{N_0}{4} & |f| \leq \frac{B_T}{2} \\ 0 & \text{elsewhere} \end{cases} \tag{11.21}$$

The effect of the choosing a filter that only passes the frequencies where the modulated signal is present instead of a filter like that in Eq. (11.6) is a reduction by a factor of two in the output noise spectral density. By choosing the LPF to have bandwidth W the output SNR is then given as

$$\text{SNR}_o = \frac{2A_c^2 P_m}{N_0 W} = \frac{2P_{x_z} E_T}{N_0 W} = \frac{P_{x_z}}{N_0 W} = \text{SNR}_b \tag{11.22}$$

The output SNR is exactly the same as the case where the signal was transmitted unmodulated so that $E_T = 1$. This implies SSB-AM can also result in no loss of performance compared to a baseband or unmodulated transmission using only the W Hz of bandwidth.

11.4 Noncoherent Amplitude Demodulation

LC-AM uses noncoherent demodulation. Recall the complex envelope form of LC-AM is $x_z(t) = A_c(1 + am(t))$ with a chosen to allow distortion-free envelope detection. Consequently, the representative block diagram of LC-AM demodulation in noise is given in Figure 11.10. Noting

$$\begin{aligned} Y_z(t) &= \exp(j\phi_p)(A_c[1 + am(t)] + N_z(t)\exp(-j\phi_p)) \\ &= \exp(j\phi_p)(A_c[1 + am(t)] + \tilde{N}_z(t)) \\ &= \exp(j\phi_p)(A_s(t) + \tilde{N}_z(t)) \end{aligned} \tag{11.23}$$

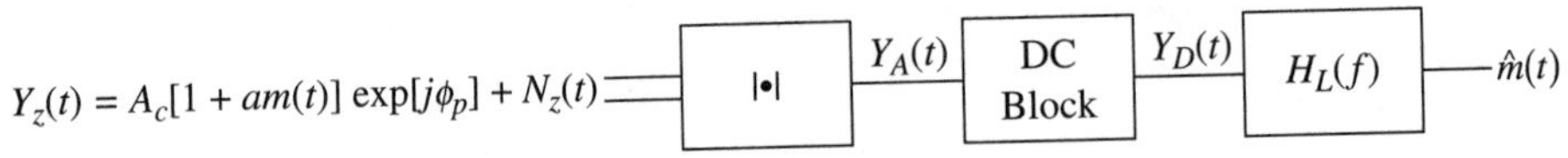

$Y_z(t) = A_c[1 + am(t)]\exp[j\phi_p] + N_z(t)$ → $|\bullet|$ → $Y_A(t)$ → DC Block → $Y_D(t)$ → $H_L(f)$ → $\hat{m}(t)$

Figure 11.10 The noncoherent demodulation structure for LC-AM.

where $A_s(t)$ is the amplitude modulated signal. This representation is interesting because the noise $\tilde{N}_z(t)$ can be viewed as having the real component in-phase with the received signal and the imaginary component in quadrature with the received signal. A vector diagram of the output complex envelope from the downconverter at one point in time, $t = t_s$, is given in Figure 11.11. Since $A_s(t) = A_c[1+am(t)]$ is a real valued function, the phase of the received signal, $r_z(t)$, is entirely due to ϕ_p. The value of $Y_A(t_s)$ is represented in Figure 11.11 by the length of the vector $Y_z(t_s)$. Considering this model, the output of the envelope detector has the form

$$Y_A(t) = |Y_z(t)| = \sqrt{(A_c[1+am(t)] + \tilde{N}_I(t))^2 + \tilde{N}_Q^2(t)}$$

$$= \sqrt{(A_s(t) + \tilde{N}_I(t))^2 + \tilde{N}_Q^2(t)} \tag{11.24}$$

Using the results of Section 3.3.4, a transformation of random variables could be accomplished and the output could be completely characterized as a Ricean

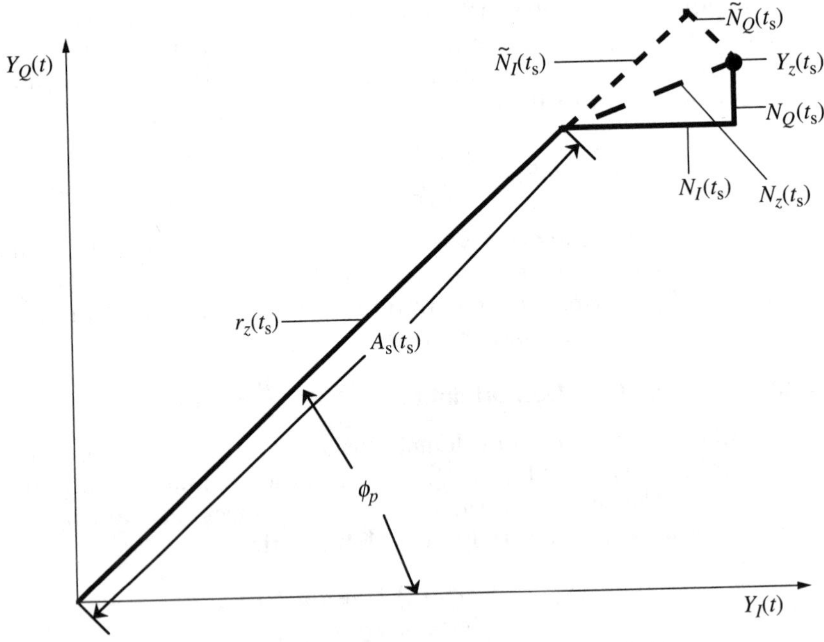

Figure 11.11 A vector diagram representing one time sample of $Y_z(t)$.

random variable [DR87]. While rigorous the procedure lacks intuition into the effects of noise on noncoherent AM demodulation and will not be pursued in this text.

To gain some insight into the operation of a LC-AM system a large SNR approximation is invoked. At high SNR, $A_c \gg |\tilde{N}_I(t)|$ and $A_c \gg |\tilde{N}_Q(t)|$ which implies (see Problem 2.13) that

$$Y_A(t) \approx \sqrt{(A_c[1 + am(t)] + \tilde{N}_I(t))^2} = A_c(1 + am(t)) + \tilde{N}_I(t) \tag{11.25}$$

Consequently, at high SNR only the noise in-phase with the carrier will make a difference in the output envelope and the output looks much like that given in for DSB-AM with the exception that the signal format is different. The output of the DC block will be $y_D(t) = A_c am(t) + \tilde{N}_I(t) = m_e(t) + \tilde{N}_I(t)$ and with a lowpass filter bandwidth of W and using the same tools as above the output SNR is

$$\text{SNR}_o = \frac{P_{m_e}}{P_{\tilde{N}_I}} = \frac{A_c^2 a^2 P_m}{N_0 W} \tag{11.26}$$

Recall the definition of the message to carrier power ratio (MCPR) introduced in Chapter 6 so that

$$\text{SNR}_o = \frac{A_c^2 \text{MCPR}}{N_0 W} \tag{11.27}$$

Since

$$P_{x_z} = A_c^2(1 + \text{MCPR}) = A_c^2\left(1 + a^2 P_m\right) \tag{11.28}$$

this implies that

$$\text{SNR}_o = \text{SNR}_b \frac{\text{MCPR}}{1 + \text{MCPR}} = \frac{P_{m_e}}{P_{x_z}} \text{SNR}_b \tag{11.29}$$

This implies that $E_T = \frac{\text{MCPR}}{1+\text{MCPR}} = \frac{P_{m_e}}{P_{x_z}}$ or, in words, the transmission efficiency of LC-AM is the ratio of the power delivered at the demodulator output to the transmitted power. The important thing to remember is that MCPR $\ll 1$ for LC-AM to enable envelope detection so that the transmission efficiency of LC-AM is also small. Consequently, at high SNR the only loss in performance in using a noncoherent demodulator for LC-AM is due to the wasted power allocated to the carrier signal. The average MCPR of commercial LC-AM is usually less than 10% so the loss in SNR that occurs in using LC-AM with envelope detection versus DSB-AM with coherent detection is greater than 10 dB! This 10 dB lower fidelity is only compensated by the much lower complexity of the envelope detector. This trade-off of fidelity versus complexity was deemed acceptable in the days before transistors. Today no competent communication engineer would give up 10 dB for so little gain in ease of implementation.

As SNR_i gets smaller a threshold effect will be experienced in envelope detection. The threshold effect occurs at SNRs below 10 dB so the analysis present in this text is valid above a 10 dB input SNR and the details of the threshold effect should be considered carefully when operating at lower SNR. This impact of large noise and the resulting performance characteristics are not considered in this introductory text but many excellent references cover this topic. One example is [SBS66].

11.5 Angle Demodulations

The analysis of the performance of angle modulation demodulation algorithms is quite complicated and a rich literature exists (e.g., [SBS66, Ric48]). The tack that will be taken in this section is to simplify the analysis to provide the student a good overview of the effects of noise on demodulation.

11.5.1 Phase Modulation

Phase modulation (PM) uses a noncoherent direct phase demodulator. Recall the complex envelope of the received PM signal is

$$r_z(t) = A_c \exp[j(k_p m(t) + \phi_p)] = A_c \exp(j\varphi_s(t)) \tag{11.30}$$

where $\varphi_s(t)$ represents the overall received phase of the signal. Consequently, the representative block diagram of PM demodulation in noise is given in Figure 11.12. The output complex envelope from the downconverter at one point in time, $t = t_s$, is represented as in Figure 11.13 for an angle modulated signal. The output of the phase detector has the form

$$Y_P(t) = \tan^{-1}(Y_Q(t), Y_I(t)) = k_p m(t) + \phi_p + \varphi_e(t) = \varphi_s(t) + \varphi_e(t) \tag{11.31}$$

where $\varphi_e(t)$ is the phase noise produced by the corrupting noise. Defining $\breve{N}_z(t) = N_z(t)\exp(-j\varphi_s(t))$ in a similar way as was done for coherent demodulation it can be shown that

$$\varphi_e(t) = \tan^{-1}(\breve{N}_Q(t), A_c + \breve{N}_I(t)). \tag{11.32}$$

Again using the results of Section 3.3.4, a transformation of random variables could be accomplished and the output could be completely characterized [DR87]. While rigorous the procedure lacks intuition into the effects of noise on noncoherent PM demodulation.

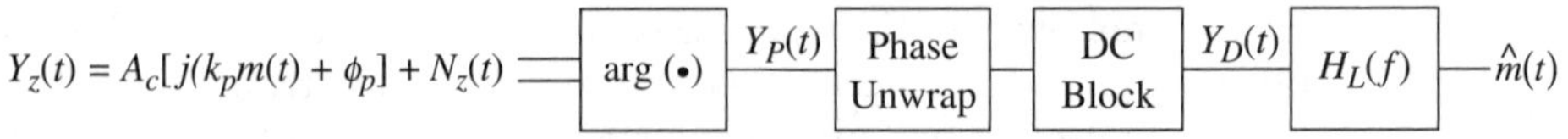

$Y_z(t) = A_c[j(k_p m(t) + \phi_p] + N_z(t)$ → arg (•) → $Y_P(t)$ → Phase Unwrap → DC Block → $Y_D(t)$ → $H_L(f)$ → $\hat{m}(t)$

Figure 11.12 The noncoherent demodulation structure for PM.

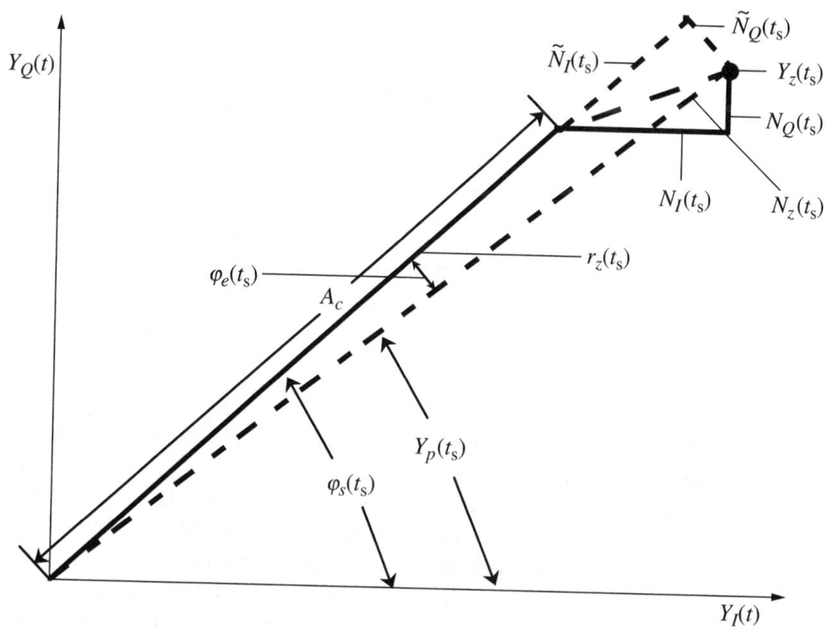

Figure 11.13 A vector diagram representing one time sample of $Y_z(t)$ for angle modulation.

Again to gain some insight into the operation of a PM system a large SNR approximation is invoked. At high SNR $A_c \gg |\breve{N}_I(t)|$ and $A_c \gg |\breve{N}_Q(t)|$, which implies (see Problem 2.13) that

$$\varphi_e(t) \approx \tan^{-1}\left(\frac{\breve{N}_Q(t)}{A_c}\right) \approx \frac{\breve{N}_Q(t)}{A_c} \tag{11.33}$$

It is important to note that the inverse tangent in Eq. (11.33) is the more common one argument inverse tangent function. The important insight gained from this high SNR approximation is that the phase noise produced is a function of the noise in quadrature to the modulated signal. At high SNR the output phase noise is approximately Gaussian, zero mean (i.e., $E[\breve{N}_Q(t)] = 0$) and the output noise variance is a function of both the input noise and the input signal (i.e., $E[\varphi_e^2(t)] = E[\breve{N}_Q^2(t)]/A_c^2$). The following approximation is also valid at high SNR [SBS66]

$$R_{\breve{N}_Q}(t, \tau) = R_{N_Q}(\tau) = R_{N_I}(\tau) \tag{11.34}$$

Consequently, the output phase noise is Gaussian, zero mean and has an approximate PSD of $S_{\varphi_e}(f) = S_{N_I}(f)/A_c^2$. The signal and noise spectrum are as shown in Figure 11.14(a).

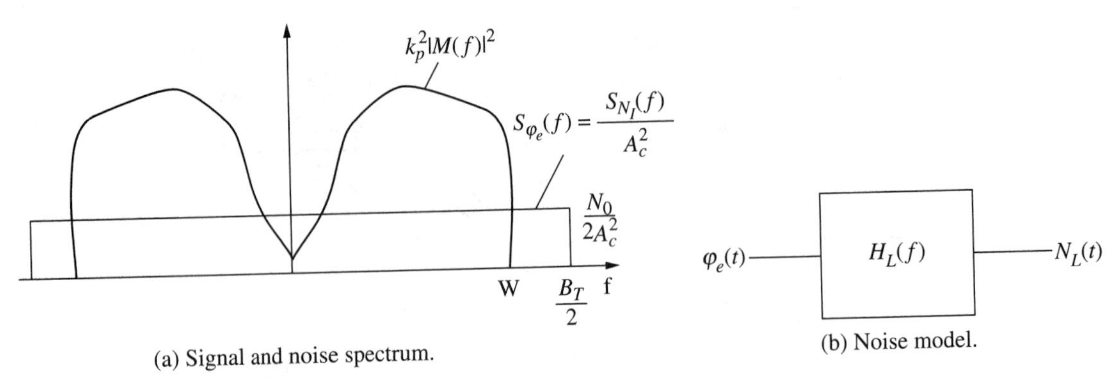

(a) Signal and noise spectrum.

(b) Noise model.

Figure 11.14 The PM demodulator noise processing.

The effect of the noise due to the remainder of the processing needed to recover the message can also be characterized. The DC block removes the term due to ϕ_p and this section will assume the DC block is very small notch in frequency and does not impact the noise power significantly. Since $B_T > 2W$ for PM the noise power can be reduced by filtering the output of the DC block. The model for the noise processing in PM is modeled with Figure 11.14(b). Assuming the lowpass filter is ideal and has a bandwidth of W then output message estimate has the form $\hat{m}(t) = k_p m(t) + N_L(t)$ where for the bandpass filter given in Eq. (11.6)

$$S_{N_L}(f) = S_{\varphi_e}(f)\,|H_L(f)|^2 = \frac{S_{N_I}(f)}{A_c^2}\,|H_L(f)|^2 = \begin{cases} \frac{N_0}{2A_c^2} & |f| \leq W \\ 0 & \text{elsewhere} \end{cases} \tag{11.35}$$

The output noise power is $\text{var}(N_L(t)) = N_0 W / A_c^2$ and the output signal to noise ratio is (recall $P_{x_z} = A_c^2$ for PM)

$$\text{SNR}_o = \frac{A_c^2 k_p^2 P_m}{N_0 W} = k_p^2 P_m \frac{P_{x_z}}{N_0 W} = k_p^2 P_m \text{SNR}_b \tag{11.36}$$

The transmission efficiency is $E_T = k_p^2 P_m$. The resulting SNR performance for PM has four interesting insights

1. Again two parameters that affect SNR_o are the transmitted power and the message signal bandwidth (much like AM systems). In addition, there is a third parameter $E_T = k_p^2 P_m$ that can be chosen by the designer.
2. At high SNR the effective performance can be improved an arbitrary amount by increasing k_p. If $k_p^2 P_m > 1$, then using a nonlinear modulation like PM allows the designer to actually get a performance better than that achieved at baseband.
3. Recall k_p also increases the output bandwidth of a PM signal. Hence PM allows an engineer to trade spectral efficiency for SNR performance. Such a trade-off is not possible with AM.

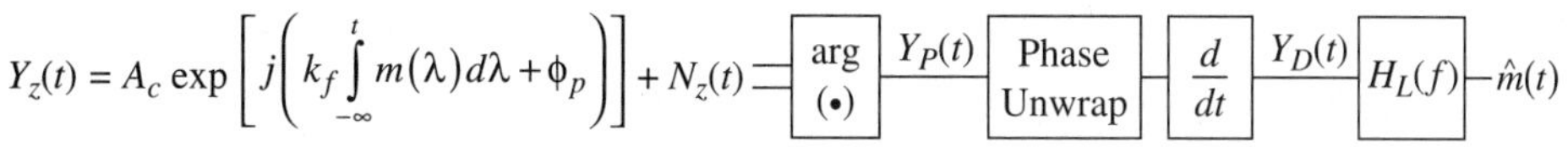

$$Y_z(t) = A_c \exp\left[j\left(k_f \int_{-\infty}^{t} m(\lambda)d\lambda + \phi_p \right)\right] + N_z(t) \rightarrow \boxed{\arg(\bullet)} \xrightarrow{Y_P(t)} \boxed{\text{Phase Unwrap}} \rightarrow \boxed{\frac{d}{dt}} \xrightarrow{Y_D(t)} \boxed{H_L(f)} \rightarrow \hat{m}(t)$$

Figure 11.15 The noncoherent demodulation structure for FM.

4. For large values of k_p phase unwrapping becomes increasingly difficult. Consequently, there is a practical limit to the size of k_p and hence a limit to the gain in SNR one can achieve compared to an unmodulated system.

As SNR_i gets smaller a threshold effect will be experienced in PM demodulation and that will not be discussed in this introductory text. An interested reader should, for example, see [SBS66].

11.5.2 Frequency Modulation

FM has a noncoherent demodulation structure much like that of PM. Recall the complex envelope of the received FM signal is

$$r_z(t) = A_c \exp\left[j\left(k_f \int_{-\infty}^{t} m(\lambda)d\lambda + \phi_p \right)\right] = A_c \exp(j\varphi_s(t)) \tag{11.37}$$

where $\varphi_s(t)$ represents the overall received phase of the signal. The demodulation structure for FM is shown in Figure 11.15. The output of the phase detector has the form

$$Y_P(t) = \tan^{-1}(Y_Q(t), Y_I(t)) = k_f \int_{-\infty}^{t} m(\lambda)d\lambda + \phi_p + \varphi_e(t) = \varphi_s(t) + \varphi_e(t) \tag{11.38}$$

Again defining $\breve{N}_z(t) = N_z(t)\exp(-j\varphi_s(t))$ it can be shown that

$$\varphi_e(t) = \tan^{-1}(\breve{N}_Q(t), A_c + \breve{N}_I(t)) \tag{11.39}$$

Again a complete statistical characterization is possible but lacks the intuition that we want to achieve in this course.

The effect of the noise due to the remainder of the processing needed to recover the message can also be characterized using similar approximations as was done for PM. We can use the same high SNR approximations as with PM to gain insight into the performance of FM. Recall at high SNR

$$\varphi_e(t) \approx \tan^{-1}\left(\frac{\breve{N}_Q(t)}{A_c}\right) \approx \frac{\breve{N}_Q(t)}{A_c} \tag{11.40}$$

so that the noise at the output of the direct phase detector is well modeled as Gaussian, zero mean and has a PSD of $S_{\varphi_e}(f) = S_{N_I}(f)/A_c^2$. Since the derivative operation is a linear system the processing after the phase conversion nonlinearity can be represented as the cascade of two linear systems as shown in Figure 11.16(a). The derivative operation removes the phase offset, ϕ_p so

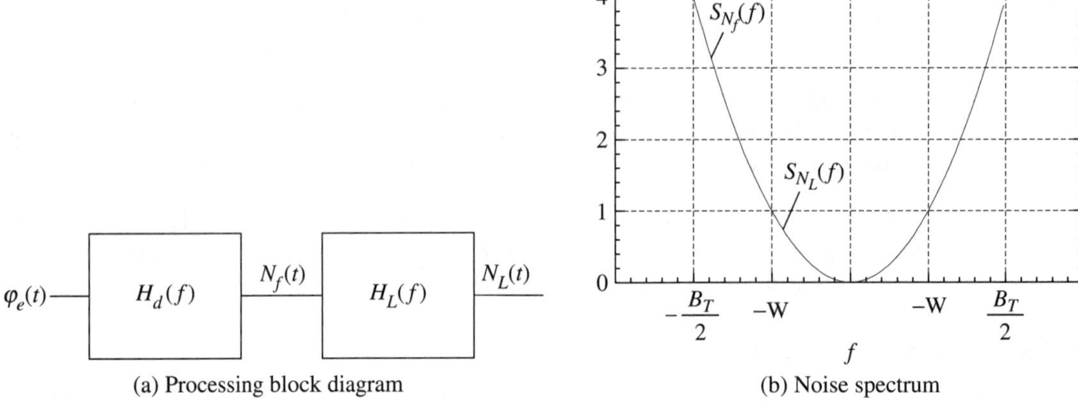

Figure 11.16 Noise processing in an FM demodulator.

that $Y_D(t) = k_f m(t) + d\varphi_e(t)/dt = k_f m(t) + N_f(t)$. The derivative is a linear system with a transfer function of

$$H_d(f) = j2\pi f \tag{11.41}$$

so at high SNR the output power spectrum of $N_f(t)$ is approximated as

$$S_{N_f}(f) = \frac{S_{N_I}(f)}{A_c^2}|H_d(f)|^2 = \begin{cases} \dfrac{2N_0\pi^2 f^2}{A_c^2} & |f| \le \frac{B_T}{2} \\ 0 & \text{elsewhere} \end{cases} \tag{11.42}$$

This output spectrum is plotted in Figure 11.16(b). Interestingly the derivative operation has colored the noise (PSD is not flat with frequency) as the derivative operation tends to accentuate high frequencies and attenuate low frequencies. The next section will demonstrate how engineers used this colored noise to their advantage in high fidelity FM based communications. The final output after the lowpass filter can also be characterized using the tools developed in this course. The final output after an ideal filter with a bandwidth of W has the form $\hat{m}(t) = k_f m(t) + N_L(t)$ where

$$S_{N_L}(f) = \begin{cases} \dfrac{2N_0\pi^2 f^2}{A_c^2} & |f| \le W \\ 0 & \text{elsewhere} \end{cases} \tag{11.43}$$

The resulting signal power is $P_s = k_f^2 P_m$ and the noise power is given as

$$P_N = \int_{-\infty}^{\infty} S_{N_L}(f)df = \frac{2N_0\pi^2}{A_c^2}\int_{-W}^{W} f^2 df = \frac{4N_0\pi^2 W^3}{3A_c^2} \tag{11.44}$$

so that the resulting output SNR is

$$\mathrm{SNR}_o = \frac{3A_c^2 k_f^2 P_m}{4N_0\pi^2 W^3} = \frac{3k_f^2 P_m}{4\pi^2 W^2}\frac{P_{x_z}}{N_0 W} = \frac{3k_f^2 P_m}{4\pi^2 W^2}\mathrm{SNR}_b \tag{11.45}$$

Here it is clear that $E_T = \frac{3k_f^2 P_m}{4\pi^2 W^2}$. Since k_f is a parameter chosen by the designer the transmission efficiency of FM is again selectable. Recalling that $k_f = 2\pi f_k$, where f_k is the frequency deviation constant (measured in Hz/V), gives a more succinct form for the output SNR as

$$\mathrm{SNR}_o = \frac{3k_f^2 P_m}{4\pi^2 W^2}\mathrm{SNR}_b = \frac{3f_k^2 P_m}{W^2}\mathrm{SNR}_b \tag{11.46}$$

The resulting SNR performance for FM has two interesting insights

1. At high SNR the effective performance can improve an arbitrary amount by increasing f_k. Recall f_k also increases the output bandwidth of a FM signal. Hence FM also allows an engineer to trade spectral efficiency for SNR performance. Edwin Armstrong was the first person to realize that this trade-off was possible.
2. The output noise from an FM demodulator is not white (flat with frequency). The sequel will demonstrate how this colored noise characteristic can be exploited.

As SNR_i gets smaller a threshold effect will be experienced in FM demodulation but that will not be discussed in this introductory text. An interested reader should, for example, see [SBS66].

11.6 Improving Fidelity with Pre-Emphasis

The insights gained in analyzing the noise in the previous section have helped engineers understand techniques to improve the output SNR in angle demodulations. The three insights that led to this improved technique are

1. The noise at the output of the differentiator is not white so that noise power is larger at higher frequencies.
2. The power of the noise tends to be distributed differently over frequency than the power of the signal. A fairly typical example of this different distribution is shown in Figure 11.17.
3. The received signal power, $P_{x_z} = P_{r_z}$, is not a function of the message signal.

Consequently, if one can process the signal at the transmitter and receiver in such a way that the signal is unchanged and that the high frequency components of the noise are de-emphasized then a significant SNR advantage can be gained. The way to achieve this gain is shown in Figure 11.18 and has been denoted pre-emphasis and de-emphasis in the communications literature.

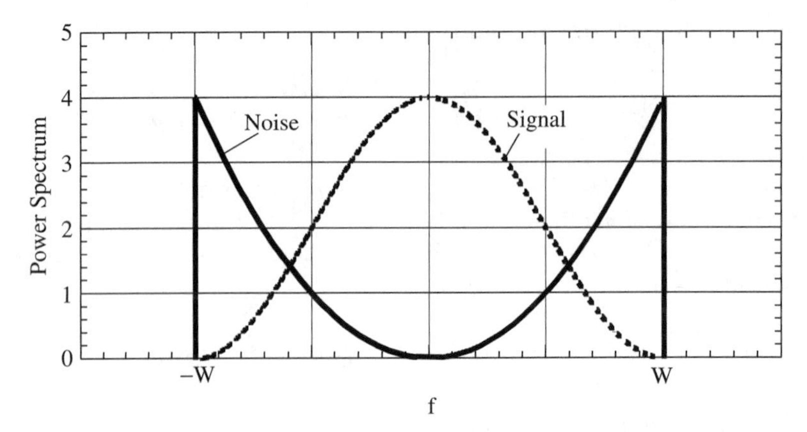

Figure 11.17 Examples of the signal and noise spectrum at the output of an FM demodulator.

The idea is to pre-emphasize the high frequencies of the signal before FM modulation with a linear filter, $H_E(f)$. This pre-emphasis will not change the transmitted signal power which will still be $P_{x_z} = A_c^2$. This pre-emphasis will impact the transmitted spectrum some as this adding gain to the high frequency components of the message spectrum could increase the peak frequency deviation. At the receiver the de-emphasis filter, $H_D(f)$, is chosen to give an overall unity gain to the message signal, e.g.,

$$|H_E(f)H_D(f)| = 1 \qquad |f| \leq W \tag{11.47}$$

Denoting the input and output of the de-emphasis filter as

$$\hat{m}_E(t) = m_E(t) + N_L(t) \qquad \hat{m}(t) = m(t) + N_D(t) \tag{11.48}$$

it is clear this de-emphasis filter will further decrease the output noise power since the PSD of $N_D(t)$ will be given as

$$S_{N_D}(f) = S_{N_L}(f)\,|H_D(f)|^2 \tag{11.49}$$

The actual amount of SNR gain will be a function of the specific filter transfer functions and the spectral characteristics of the message signal. In practical systems this gain is typically over 10 dB. Some specific examples will be explored in the homework.

It is interesting to note that commercial analog "FM" broadcast in the United States uses pre-emphasis and de-emphasis. Since a pre-emphasis circuit effectively is behaving like a differentiator, the true transmitted signal in broadcast

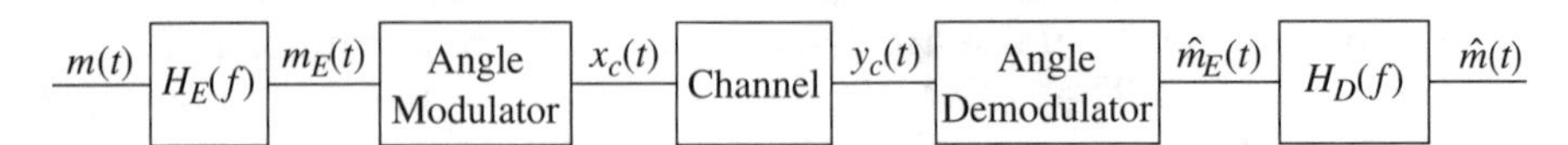

Figure 11.18 Block diagram of pre-emphasis and de-emphasis in FM systems.

radio is actually a hybrid of phase modulation and frequency modulation. It is interesting that this modulation is known in popular culture as FM while, in fact, it is not a true frequency modulation of the message signal.

11.7 Final Comparisons

A summary of the important characteristics of analog modulations is given in Table 11.1. At the beginning of Chapter 5 the performance metrics for analog communication systems were given as: complexity, fidelity, and spectral efficiency.

In complexity the angle modulations offer the best characteristics. Transmitters for all of the analog modulations are reasonable simple to implement. The most complicated transmitters are needed for VSB-AM and SSB-AM. The complexity is needed to achieve the bandwidth efficiency. The angle modulations have a transmitted signal with a constant envelope. This characteristic makes designing the output power amplifiers much simpler since the linearity of the amplifier is not a significant issue. Receivers for LC-AM and angle modulation can be implemented in simple noncoherent structures. Receivers for DSB-AM and VSB-AM require the use of a more complex coherent structure. SSB-AM additionally will require a transmitted reference for automatic coherent demodulation and operation.

In spectral efficiency SSB-AM offers the best characteristics. SSB-AM is the only system that achieves 100% efficiency. DSB-AM and LC-AM are 50% efficient. VSB-AM offers performance somewhere between SSB-AM and DSB-AM. Angle modulations are the least bandwidth efficient. At best they achieve $E_B = 50\%$, and often much less.

In terms of fidelity there is a variety of options. Coherent demodulation offers performance at the level that could be achieved with baseband transmission. Noncoherent demodulation schemes can have a threshold effect at low SNR but above threshold provide very good performance. The only difference between coherent demodulation and noncoherent demodulation of LC-AM at high SNR is the transmission inefficiency needed to make envelope detection possible.

TABLE 11.1 A comparison of the trade-offs in analog modulation

Modulation	Spectral Efficiency	TX Complexity	RX Complexity	Fidelity
DSB-AM	50%	moderate	moderate	$E_T = 1$
LC-AM	50%	moderate	small	$E_T \approx$ MCPR Experiences a threshold effect
VSB-AM	$> 50\%$	large	large	$0.5 \leq E_T \leq 1$
SSB-AM	100%	large	large	$0.5 \leq E_T = 1$
PM	$< 50\%$	small	moderate	Can trade-off BW for E_T by choice of k_p Experiences a threshold effect
FM	$< 50\%$	small	moderate	Can trade-off BW for E_T by choice of k_f Experiences a threshold effect Colored noise in demodulation can be exploited

Angle modulation when operated, above threshold offers a trade-off between output SNR at the demodulator and the transmitted signal bandwidth. In general, angle modulations offer the ability to trade bandwidth efficiency for output SNR. This ability is not available in AM modulation and is the reason FM is used for high fidelity broadcast in the United States. Finally, the noise spectrum at the output of an FM demodulator is nonwhite. This nonwhite spectral characteristic can be taken advantage of by appropriate signal processing to improve the output SNR even further. This idea of pre-emphasis and de-emphasis is an excellent example of communication engineers getting significant performance advantages for little cost in complexity by simply understanding a little communication theory.

11.8 Homework Problems

Problem 11.1. For DSB-AM with a coherent demodulation, if the receiver phase reference has an error of ϕ_e compute the output SNR. How big can ϕ_e be and yet limit the degradation in SNR to less than 3 dB from the optimum performance?

Problem 11.2. Consider a bandpass Gaussian noise at the input to a demodulator with a spectrum given as

$$\begin{aligned} S_{N_c}(f) &= 0.01 \qquad f_c - 200 \leq |f| \leq f_c + 300 \\ &= 0 \qquad \text{elsewhere} \end{aligned} \tag{11.50}$$

in the system given in Figure 11.4. Assume that the modulation is DSB-AM with

$$r_c(t) = A_c m(t)\sqrt{2}\cos(2\pi f_c t + \phi_p) \tag{11.51}$$

where $A_c = 1$ and the message signal has a $W = 60$ Hz bandwidth and power $P_m = 10$ W. Using the signals $Y_I(t)$ and $Y_Q(t)$ as inputs and assuming you know ϕ_p

(a) Define $\tilde{N}_z(t) = N_z(t)\exp(-j\phi_p)$. Give the joint PDF of $\tilde{N}_I(t_0)$ and $\tilde{N}_Q(t_0)$, $f_{\tilde{N}_I(t_0)\tilde{N}_Q(t_0)}(n_1, n_2)$ in a 1-Ω system for a fixed t_0.

(b) Find the best demodulator structure that leaves the signal undistorted and maximizes the SNR.

(c) Give the maximum SNR.

Problem 11.3. The following complex envelope is received

$$Y_z(t) = (\cos(200\pi t) + \sin(20\pi t))\exp[j\pi/7] + N_z(t) \tag{11.52}$$

The bandpass noise is characterized with

$$S_{N_z}(f) = \begin{cases} 0.005 & |f| \leq 4000 \\ 0 & \text{elsewhere} \end{cases} \tag{11.53}$$

(a) What kind of modulation is being used and what is the message signal (up to a constant).

(b) Sketch the block diagram of the demodulation processing assuming that the demodulator knows $\phi_p = \pi/7$.

(c) If the lowpass filtering in the message recovery portion of the demodulator is ideal and unit gain with 4000 Hz bandwidth compute the output SNR.

(d) What ideal lowpass filter bandwidth would optimize the output SNR?

(e) Could an ideal bandpass filter at the output of the demodulator improve performance? If so give the optimum filter specifications and the resulting SNR.

Problem 11.4. In discussing noise in angle modulation we had defined a derotated noise

$$\breve{N}_z(t) = N_z(t) \exp(-j\varphi_s(t)) \tag{11.54}$$

This problem will examine the approximations made in the text and see how they can be justified. Assume the bandpass filter at the receiver front end is ideal and has a bandwidth of $B_T = 2(f_p + W)$ and is symmetric around f_c so that $R_{N_I N_Q}(\tau) = 0$.

(a) Find $R_{N_I}(\tau)$.

(b) Show that $\breve{N}_Q(t) = N_Q(t)\cos(\varphi_s(t)) - N_I(t)\sin(\varphi_s(t))$.

(c) Compute $R_{\breve{N}_Q}(t_1, \tau) = E[\breve{N}_Q(t_1)\breve{N}_Q(t_1 - \tau)]$.

(d) Since the result in (c) is a function of $\varphi_s(t_1) - \varphi_s(t_1 + \tau)$ show that $\varphi_s(t_1) - \varphi_s(t_1 + \tau) \approx 2\pi f_d(t_1)\tau$.

(e) Show that only when the instantaneous frequency deviation is large, will $R_{\breve{N}_Q}(t_1, \tau)$ be much different than $R_{N_Q}(\tau)$.

(f) Plot $R_{\breve{N}_Q}(t_1, \tau)$ and $R_{N_Q}(\tau)$ for $f_d(t_1) = f_p$ and $f_d(t_1) = f_p/2$.

Problem 11.5. A student from another university claims that the optimal receiver for SSB-AM is given in Figure 11.19. The student argues that the quadrature

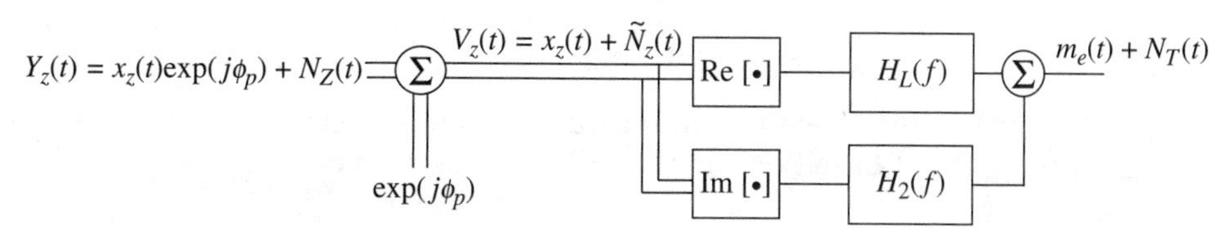

Figure 11.19 An alternative demodulator for SSB-AM.

component contains information related to the message signal and you should use this information in the optimum demodulator. You begin to doubt the education you received and set about to verify whether Prof. Fitz is right in what he presented in this chapter or if the other student got a better education at the other university. For this problem assume the filter at the front end of the radio is given as Eq. (11.19) and that the noise output from $H_2(f)$ is denoted $\breve{N}_Q(t)$.

(a) Show that for the case of upper sideband transmission via SSB-AM if $H_1(f)$ is a lowpass filter of bandwidth W and that

$$H_2(f) = \begin{cases} j\,\mathrm{sgn}(f) & |f| \leq W \\ 0 & \text{elsewhere} \end{cases} \tag{11.55}$$

then $m_e(t) = 2A_c m(t)$. (It doesn't look good for Prof. Fitz as the signal in the alternative demodulator is twice as big what Prof. Fitz presented!)

(b) Denote $\breve{N}_I(t)$ as the output of $H_1(f)$ and $\breve{N}_Q(t)$ as the output of $H_2(f)$ and show that

$$\mathrm{var}(N_T(t)) = \mathrm{var}(\breve{N}_I(t)) + \mathrm{var}(\breve{N}_Q(t)) + 2R_{\breve{N}_I\breve{N}_Q}(0) \tag{11.56}$$

(c) Compute $E[\breve{N}_I^2(t)]$.

(d) Compute $E[\breve{N}_Q^2(t)]$.

(e) What is $S_{\breve{N}_I\breve{N}_Q}(f)$? (Feel free to use the results in Problem 10.10.)

(f) What is $R_{\breve{N}_I\breve{N}_Q}(0)$?

(g) What is the output SNR? Which structure would you choose for implementing a SSB-AM demodulator Figure 11.19 or Figure 11.7 and why?

Problem 11.6. This problem considers the demodulation of stereo broadcast considered in Problem 7.24.

(a) Postulate a demodulation structure for stereo broadcast that has as outputs $\hat{m}_S(t)$ and $\hat{m}_D(t)$. Assume a coherent demodulation for the DSB-AM signal is possible[2].

(b) Using the assumption that the power of the sum and difference signals are both P_m calculate the output SNR for the estimation of $m_S(t)$.

(c) Using the assumption that the power of the sum and difference signals are both P_m calculate the output SNR for the estimation of $m_D(t)$. The output SNRs are different, why?

(d) Using these results postulate a mechanism to trigger the stereo indicator light based on the SNR differences seen on the two channels.

---

[2]This coherent demodulation is achieved by processing the pilot tone.

Problem 11.7. Consider an FM system with $W = 15$ kHz, $P_m = 1$, $\max|m(t)| = 3$, and an input noise characterized as

$$S_W(f) = 10^{-6} \tag{11.57}$$

Assume the receiver filter is an ideal bandpass filter with a bandwidth equal to Carson's bandwidth.

(a) Define the bandpass SNR for an angle modulated signal to be

$$\text{SNR}_{bp} = \frac{A_c^2}{E\left[N_c^2(t)\right]} \tag{11.58}$$

and find the value of A_c (as a function of f_k) needed to provide $\text{SNR}_{bp} =$ 10 dB (so that the demodulator will be operating above threshold and the small noise approximation is valid).

(b) With the A_c as specified in (a) choose an f_k such that $\text{SNR}_o = 30$ dB.

(c) Broadcast radio in the United States was originally designed with only one audio channel of 15 kHz such that $D = 5$, for the system parameters given in this problem with A_c as specified in (a) what would the SNR_o for this original broadcast standard?

Problem 11.8. Assume a communication system uses DSB-AM and the received signal is corrupted by a white Gaussian noise with a one-sided noise spectral density of $N_0 = 10^{-6}$. Specify the optimum demodulator and find the received signal power, P_{x_z}, that would be needed to achieve a SNR = 10 dB for

(a) $W = 4$ kHz

(b) $W = 15$ kHz

(c) $W = 4.5$ MHz

Problem 11.9. A simple pre-emphasis filter which is commonly used in practice is given by

$$H_{pe}(f) = 1 + \frac{jf}{f_0} \tag{11.59}$$

where $f_0 = 2.1$ kHz. The corresponding de-emphasis filter is a simple lowpass filter having the form

$$H_{de}(f) = \frac{1}{1 + \frac{jf}{f_0}} \tag{11.60}$$

(a) Defining I_{pde} as the noise power improvement ratio when using pre-emphasis/de-emphasis, calculate I_{pde} for this filter combination. Sorry to say this will require a little calculus!

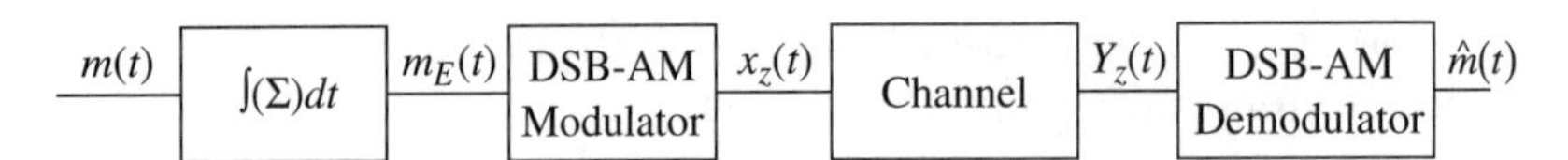

Figure 11.20 A pre-emphasis and de-emphasis system for DSB-AM.

(b) For the case of mono FM broadcast ($f_p = 75$ kHz, $W = 15$ kHz) and an bandpass SNR of

$$\frac{P_{x_z}}{B_T N_0} \tag{11.61}$$

calculate the output signal to noise ratio with and without pre-emphasis/de-emphasis.

(c) For the same parameters as in (b), calculate the value of

$$\text{SNR}_b = \frac{P_{x_z}}{W N_0} \tag{11.62}$$

Recall that this is the optimum output SNR for an AM system.

(d) What is the performance improvement of FM modulation (above threshold) with and without pre-emphasis/de-emphasis over DSB-AM modulation?

Problem 11.10. You have just started your job at Asprocket Systems and you and a senior Asprocket engineer have been given the job of improving the performance of a DSB-AM radio station that will be deployed by the Outer Boehtavian government for news spinning (propaganda). The senior engineer said that one way to improve noise performance is to use pre-emphasis and de-emphasis in the radios and proposed the system shown in Figure 11.20. You have a bad feeling about this as in your high powered education you were never taught about pre-emphasis and de-emphasis being used in AM systems. Analyze the system design by the senior Asprocket engineer and decide whether it is useful or not. Justisty your answers by computing the output SNR$=\frac{A_c^2 P_m}{P_N}$ and comparing that to $\text{SNR}_b = \frac{P_{x_z}}{P_N}$. For simplicity of computation assume $m(t)$ is a random process with a power spectrum

$$S_M(f) = \begin{cases} \frac{P_m}{2W} & |f| \le W \\ 0 & \text{elsewhere} \end{cases} \tag{11.63}$$

Problem 11.11. **(JG)** Consider the baseband communication system, shown in Figure 11.21, where the signal and noise are both modeled as zero mean,

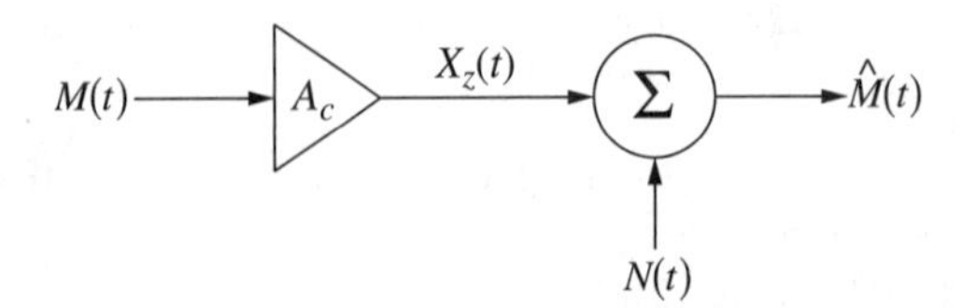

Figure 11.21 A baseband noise system.

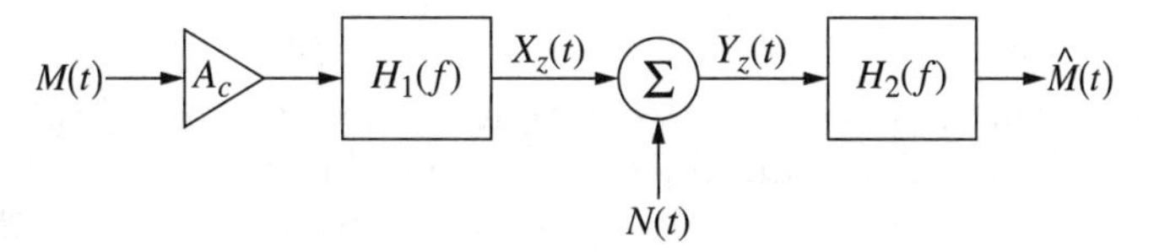

Figure 11.22 A baseband noise system with transmitter and receiver filtering.

WSS, random processes with power spectral densities $S_M(f)$ and $S_N(f)$, respectively.

$$S_M(f) = \begin{cases} \frac{1}{2W} & |f| < W \\ 0 & |f| > W \end{cases} \qquad S_N(f) = \begin{cases} aN_0 & |f| < \frac{W}{2} \\ bN_0 & \frac{W}{2} < |f| < W \\ 0 & |f| > W \end{cases} \tag{11.64}$$

(a) What is the SNR in the received signal $\hat{M}(t)$?

(b) The system is modified by including transmit and receive filters as shown in Figure 11.22. Design $H_2(f)$ to whiten the noise process for $|f| < W$.

(c) Given your choice of $H_2(f)$, design $H_1(f)$ to achieve distortionless transmission.

(d) What is the SNR improvement in the new system versus the old system?

Problem 11.12. Consider a situation where a message signal is modeled as a stationary Gaussian random process with a PSD given in Figure 11.23. Assume this signal is corrupted by an additive white Gaussian noise with a one–sided spectral density of N_0. This combined signal plus noise is put into an ideal lowpass filter that has a bandwidth of W. Choose W to optimize the output SNR and give the value of this optimum SNR.

Problem 11.13. Consider the situation presented in Example 11.1. Replace the LPF with another filter that would achieve an output with SNR = 12 dB.

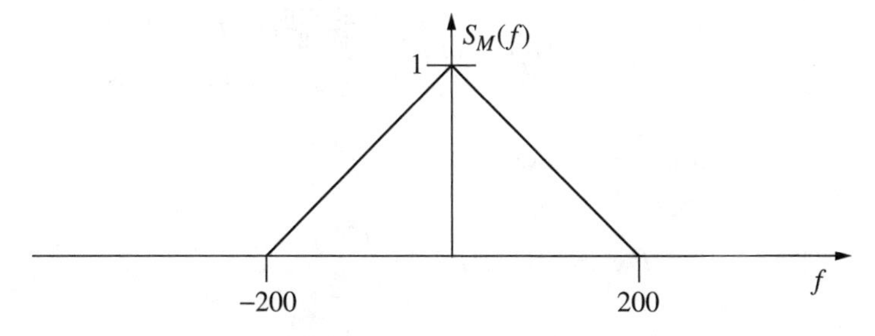

Figure 11.23 The power spectrum of the message signal.

11.9 Example Solutions

(a) The modulation is DSB-AM. Recall that $Y_z(t) = x_z(t)\exp[j\phi_p] + N_z(t)$ and the message signal is real. Since there is no time varying angle the signal is clearly not an angle modulated. There is no DC offset in the complex envelope so it is not LC-AM. The received signal spectrum is two sided so no quadrature modulation was implemented. If DSB-AM has been implemented then the message signal is $m(t) = A(\cos(200\pi t) + \sin(20\pi t))$ where $A = 1/A_c$.

(b) See Figure 11.7 with $\phi_p = \pi/7$.

(c) Using the results of Chapter 9 it is known that

$$S_{N_L}(f) = S_{N_I}(f)|H_L(f)|^2 \tag{11.65}$$

The results of Chapter 10 we have

$$S_{N_I}(f) = \frac{S_{N_z}(f) + S_{N_z}(-f)}{4} \tag{11.66}$$

and finally

$$S_{N_L}(f) = S_{N_I}(f)|H_L(f)|^2 = \begin{cases} \frac{.005}{2}, & |f| \leq 4000 \\ 0, & \text{elsewhere} \end{cases}$$

Consequently, since the noise power is given as

$$P_N = E\left[N_L^2(t)\right] = \int_{-\infty}^{\infty} S_{N_L}(f)df = 20. \tag{11.67}$$

The output signal power is easily computed using Parseval's theorem to be $P_s = 0.5 + 0.5$. The resulting SNR $= 0.05$.

(d) Since the message bandwidth is 100 Hz, so the ideal lowpass filter has 100 Hz bandwidth. The new output noise spectral density is

$$S_{N_L} = S_{N_I}|H_L(f)|^2 = \begin{cases} \frac{.005}{2}, & |f| \leq 100 \\ 0, & \text{elsewhere} \end{cases}$$

and the resulting noise power is

$$P_N = E\left[N_L^2(t)\right] = \int_{-\infty}^{\infty} S_{N_L}(f)df = 0.5. \tag{11.68}$$

The resulting SNR in this case is SNR $= 2$.

(e) The message bandwidth is 100 Hz, but we don't have any signal from 0 to 10 Hz, if we use a bandpass filter from 10 to 100 Hz we can improve SNR.

$$S_{N_L} = S_{N_I}|H_L(f)|^2 = \begin{cases} \frac{.005}{2}, & 10 \leq |f| \leq 100 \\ 0, & \text{elsewhere} \end{cases}$$

$$SNR = \frac{0.5 + 0.5}{0.35} = 2.22$$

Problem 11.5. This architecture was actually proposed in another textbook and supposedly used in practice. The analysis of this problem is a classic example of how understanding the theory of random processes can ensure that you are not fooled by incorrect hueristic arguements in engineering. The answer is "yes", there is information in the quadrature channel of a SSB-AM demodulated signal about the message but in terms of possible achieved SNR it is redundant information.

(a) Denote the noiseless signal out of $H_1(f)$ as $m_1(t)$ then $m_1(t) = A_c m(t)$. Also denote the noiseless signal out of $H_2(f)$ as $m_2(t)$ then the output Fourier transform is given as

$$M_2(f) = A_c H_2(f) M_h(f) = A_c\,(j\,\text{sgn}(f))\,(-j\,\text{sgn}(f))\,M(f) = A_c M(f) \tag{11.69}$$

Consequently, $m_e(t) = m_1(t) + m_2(t) = 2A_c m(t)$.

(b) Since $N_T(t) = \breve{N}_I(t) + \breve{N}_Q(t)$ it is straightforward to show that

$$\begin{aligned} E\left[N_T^2(t)\right] &= E\left[(\breve{N}_I(t) + \breve{N}_Q(t))^2\right] \\ &= E\left[\breve{N}_I^2(t)\right] + E\left[\breve{N}_Q^2(t)\right] + 2E\left[\breve{N}_I(t)\breve{N}_Q(t)\right] \\ &= \text{var}(\breve{N}_I(t)) + \text{var}(\breve{N}_Q(t)) + 2R_{\breve{N}_I\breve{N}_Q}(0) \end{aligned} \tag{11.70}$$

(c) From the text we have that

$$S_{\breve{N}_z}(f) = \begin{cases} N_0 & 0 \leq f \leq W \\ 0 & \text{elsewhere} \end{cases} \tag{11.71}$$

and

$$S_{\breve{N}_I}(f) = S_{\breve{N}_Q}(f) = \begin{cases} \frac{N_0}{4} & -W \leq f \leq W \\ 0 & \text{elsewhere} \end{cases} \tag{11.72}$$

Since from Chapter 9 it is known that

$$S_{\breve{N}_I}(f) = S_{\breve{N}_I}(f)\,|H_1(f)|^2 \tag{11.73}$$

this implies that

$$\operatorname{var}(\breve{N}_I(t)) = \int_{-\infty}^{\infty} S_{\breve{N}_I}(f)df = \frac{N_0 W}{2} \tag{11.74}$$

(d) Similarly,

$$S_{\breve{N}_Q}(f) = S_{\breve{N}_I}(f)\,|H_2(f)|^2 \tag{11.75}$$

so that

$$\operatorname{var}(\breve{N}_Q(t)) = \int_{-\infty}^{\infty} S_{\breve{N}_Q}(f)df = \frac{N_0 W}{2} \tag{11.76}$$

(e) From Chapter 10 it is known that

$$S_{\breve{N}_I \breve{N}_Q}(f) = \frac{S_{N_z}(-f) - S_{N_z}(f)}{j4} = \begin{cases} \dfrac{jN_0 \operatorname{sgn}(f)}{4} & -W \le f \le W \\ 0 & \text{elsewhere} \end{cases} \tag{11.77}$$

Problem 10.10 gives the result

$$S_{\breve{N}_I \breve{N}_Q}(f) = S_{\breve{N}_I \breve{N}_Q}(f) H_2^*(f) H_1(f) = \begin{cases} \dfrac{jN_0 \operatorname{sgn}(f)}{4} - j\operatorname{sgn}(f) & -W \le f \le W \\ 0 & \text{elsewhere} \end{cases} \tag{11.78}$$

$$= \begin{cases} \dfrac{N_0}{4} & -W \le f \le W \\ 0 & \text{elsewhere} \end{cases} \tag{11.79}$$

(f) Recall that

$$R_{\breve{N}_I \breve{N}_Q}(0) = \int_{-\infty}^{\infty} S_{\breve{N}_I \breve{N}_Q}(f)df = \frac{N_0 W}{2} \tag{11.80}$$

(g) The output SNR is

$$\mathrm{SNR}_o = \frac{P_{m_e}}{\operatorname{var}(N_T(t))} = \frac{4A_c^2 P_m}{2N_0 W} = \frac{2A_c^2 P_m}{N_0 W} \tag{11.81}$$

This output SNR is exactly the same as for the demodulation architecture given in the text and yet it is more complicated than the architecture given in class. Consequently, a good engineer would prefer the architecture given in class.

11.10 Miniprojects

Goal: To give exposure

- to a small scope engineering design problem in communications.
- to the dynamics of working with a team.
- to the importance of engineering communication skills (in this case oral presentations).

Presentation: The forum will be similar to a design review at a company (only much shorter) The presentation will be of 5 minutes in length with an overview of the given problem and solution. The presentation will be followed by questions from the audience (your classmates and the professor). All team members should be prepared to give the presentation.

11.10.1 Project 1

Project Goals: To examine the validity of the large SNR approximation that was made in the analysis of the demodulation fidelity of LC-AM.

Consider a received signal of the form

$$Y_z(t) = A_s + N_z(t) \tag{11.82}$$

where $SNR = \frac{A_s^2}{2\sigma_{N_I}^2}$. As a minimum plot a sample histogram with at least 10,000 points of the resulting values of the random variable $Y_A(t)$ for $SNR =$ 10 dB and $SNR = 0$ dB. Make an assessment of the SNR where the approximation given in Eq. (11.25) is valid.

11.10.2 Project 2

Project Goals: To examine the validity of the large SNR approximation that was made in the analysis of the demodulation fidelity of angle modulations.

Consider a received signal of the form

$$Y_z(t) = A_s + N_z(t) \tag{11.83}$$

where $SNR = \frac{A_s^2}{2\sigma_{N_I}^2}$. As a minimum plot a sample histogram with at least 10,000 points of the resulting values of the random variable $Y_P(t)$ for $SNR = 10$ dB and $SNR = 0$ dB. Make an assessment of the SNR where the approximation given in Eq. (11.33) is valid.

Part 4

Fundamentals of Digital Communication

Chapter

12

Digital Communication Basics

The problem that is of interest in this text is point-to-point binary data communications. The system model for such a communication system is given in Figure 12.1. A source of binary encoded data with a total of K_b bits is present and it is desired to transmit this data to a binary data sink across a physical channel. The data output by the source is represented by a $K_b \times 1$ dimensional vector, $\vec{I}$, whose components take values 0,1. The 2^{K_b} different words will be enumerated with an index $i = 0, \ldots, 2^{K_b} - 1$. It should be noted that in general each of these K_b bit words is not equally likely to be sent. A probability π_i will be associated with the sending of each word. This information vector to be sent is then mapped into one of 2^{K_b} analog waveforms represented by the waveform $x_i(t)$. Denote T_p as the largest support of all the $x_i(t)$, then the transmission rate or bit rate of the system is $W_b = K_b/T_p$ bits per second. The transmitted waveform is put through a channel of some sort and corrupted by noise or interference. The composite received signal, $Y_c(t)$, is then used to estimate which one of the possible vectors led to the transmitted waveform. This estimate is denoted $\hat{\vec{I}}$ and this estimate is passed to the data sink.

It should be noted that point-to-point data communications is but the simplest abstraction of the data communications reality. This point-to-point communication is often referred to as physical layer communications. There are many higher layers of abstraction in data communication. These higher layers deal with concepts like how do multiple users access a shared communication media, how is a communication path established in a multiple hop network, how do applications communicate across the network. These issues are typically covered in the curriculum in a course on communication networks. Typical textbooks that explore the details of communication networks are [KR04, LGW00, Tan02]. The student interested in the details of networked communication should take these courses or read these textbooks.

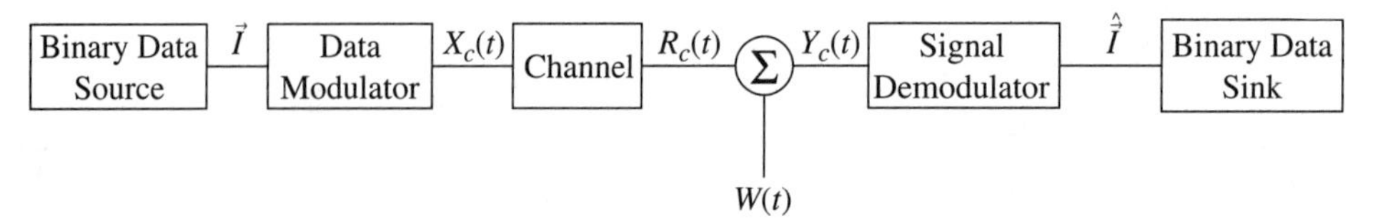

Figure 12.1 A model for point-to-point data communication.

12.1 Digital Transmission

In exactly the same way as in analog communication, digital communication has a modulator and demodulator. The modulator produces an analog signal that depends on the digital data to be communicated. This analog signal is transmitted over a channel (a cable or radio propagation). The demodulator takes the received signal and constructs an estimate of the transmitted digital data. There are two types of digital communications that are implemented in practice: Baseband data communications as exemplified by compact disc recordings or magnetic recording (e.g., computer hard drives) and carrier modulated data communications as exemplified by computer voiceband modems or radio modems used in wireless local area computer networks. In this text we want to have a unifying framework that enables the material that is learned to apply to either baseband or bandpass data communications. As with analog communications, the complex envelope notation is used to achieve this goal. The only caveat that needs to be stated is that baseband data communication will always have a zero imaginary component, while for bandpass communications the imaginary component of the complex envelope might be nonzero.

12.1.1 Digital Modulation

Definition 12.1 Digital modulation is a transformation of $\vec{I}$ into a complex envelope, $X_z(t)$.

This transformation is equivalent to transforming $m(t)$ into a bandpass signal, $X_c(t)$. In a slight deviation for digital communications, as compared to analog communications, the transmitted waveform will be considered a random process. The randomness in this transmitted waveform will be the random information bits, $\vec{I}$, that are being transmitted. The digital modulation process, $X_z(t) = \Gamma_m(\vec{I})$ is represented in Figure 12.2. It should be emphasized

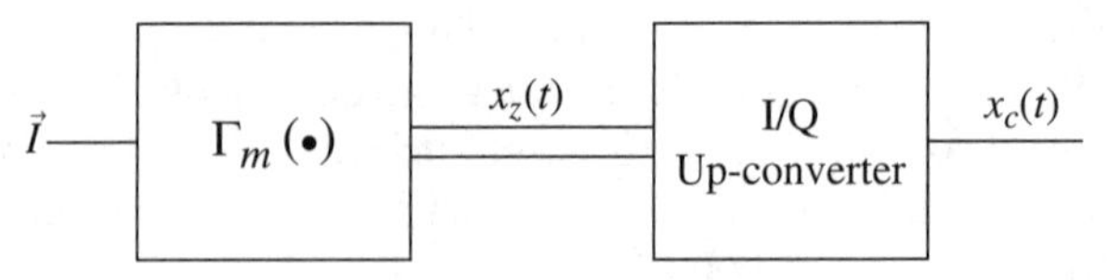

Figure 12.2 The digital modulation process. Note the IQ upconverter is given in Figure 4.4.

that digital communication is achieved by producing and transmitting analog waveforms. There is no situation where communication takes place that this transformation from digital to analog does not occur. Similarly, it should be emphasized that the modulator must be capapble of generating 2^{K_b} continuous time waveforms to represent each of the possible binary data vectors produced by the binary data source. When necessary, the possible data vector will be enumerated as $\vec{I} = i.\ i \in \{0, \ldots, 2^{K_b} - 1\}$ and the possible transmitted waveforms will be enumerated with $X_z(t) = x_i(t)$.

EXAMPLE 12.1

Transmission of two bits requires $2^2 = 4$ waveforms. An example of a signal set that can transmit 2 bits of information is given as

$$x_0(t) = \begin{cases} \sin\left(\dfrac{4\pi t}{T_p}\right) & 0 \le t \le T_p \\ 0 & \text{elsewhere} \end{cases} \qquad x_1(t) = \begin{cases} 1 & 0 \le t \le T_p \\ 0 & \text{elsewhere} \end{cases} \tag{12.1}$$

$$x_2(t) = \begin{cases} -\sin\left(\dfrac{\pi t}{T_p}\right) & 0 \le t \le T_p \\ 0 & \text{elsewhere} \end{cases} \qquad x_3(t) = \begin{cases} \dfrac{2t}{T_p} & 0 \le t \le T_p/2 \\ 2 - \dfrac{2t}{T_p} & T_p/2 \le t \le T_p \\ 0 & \text{elsewhere} \end{cases} \tag{12.2}$$

The transmission rate of this signal set is $W_b = 2/T_p$.

12.1.2 Digital Demodulation

Definition 12.2 Digital demodulation is a transformation of $Y_c(t)$ into estimates of the transmitted bits, $\hat{\vec{I}}$.

Demodulation takes the received signal, $Y_c(t)$, and downconverts this signal to the baseband signal, $Y_z(t)$. The baseband signal is then processed to produce an estimate of the transmitted data vector, $\vec{I}$. This estimate will be denoted $\hat{\vec{I}}$. Again, for this part of the text the channel output is always assumed to be

$$R_c(t) = L_p X_c(t - \tau_p)$$

where L_p is the propagation loss and τ_p is the propagation time delay. Define $\phi_p = -2\pi f_c \tau_p$ so that the channel output is given as

$$\begin{aligned} R_c(t) &= \sqrt{2} L_p X_A(t - \tau_p) \cos\left(2\pi f_c(t - \tau_p) + X_P(t - \tau_p)\right) \\ &= \Re[\sqrt{2} L_p X_z(t - \tau_p) \exp[j\phi_p] \exp[j2\pi f_c t]] \end{aligned} \tag{12.3}$$

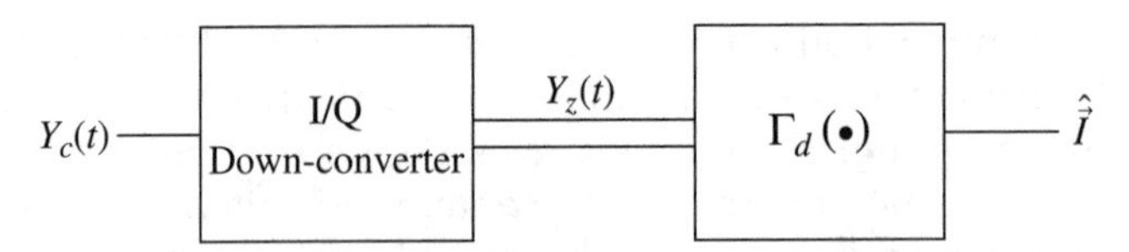

Figure 12.3 The digital demodulation process. Note the IQ down-converter is given in Figure 4.4.

It is obvious from Eq. (12.3) that the received complex envelope is $R_z(t) = L_p X_z(t - \tau_p) \exp[j\phi_p]$. It is important to note that a time delay in a carrier modulated signal will produce a phase offset. The demodulation process conceptually is a down conversion to baseband and a reconstruction of the transmitted signal from $Y_z(t)$. The block diagram for the demodulation process is seen in Figure 12.3.

Demodulation is the process of producing an $\hat{\vec{I}}$ from $Y_z(t)$ via a function $\Gamma_d(Y_z(t))$. The remainder of the discussion on digital communications in this text will focus on identifying $\Gamma_m(\vec{I})$ (modulators) and $\Gamma_d(Y_z(t))$ (demodulators) and assessing the fidelity of message reconstruction of these modulators and demodulators in the presence of noise. It is worth noting at this point that the word modem is actually an engineering acronym for a device that was both a modulator and a demodulator. The term modem has become part of the English language and is now synonymous with any device that is used to transmit digital data (computer modem, cable modem, wireless modem, etc.).

12.2 Performance Metrics for Digital Communication

In evaluating the efficacy of various designs in this text the performance metrics commonly applied in engineering design must be examined. The most commonly used metrics for digital communications are

- Fidelity – This metric typically measures how often data transmission errors are made given the amount of transmitted power.
- Complexity – This metric almost always translates directly into cost.
- Bandwidth Efficiency – This metric measures how much bandwidth a modulation uses to implement the communication.

12.2.1 Fidelity

Fidelity in digital communication is reflected in how often transmission errors occur as a function of the SNR. Transmission errors can be either bit errors (1 bit in error) or frame errors (any error in a message or packet). The application often determines the appropriate error metric. With data transmission at a fixed transmit power, P_{x_c}, the reliability of any data communication can be increased by lowering the speed of the data communication. A lower speed transmission normally implies the signal bandwidth will be smaller so that the receiver bandwidth will be smaller and consequently the SNR can be made higher. Data communication engineers strive for a SNR measure that is not an explicit function of the transmission bandwidth. The accepted measure in engineering practice is the ratio of the average received energy per bit over the noise spectral density, E_b/N_0.

Definition 12.3 The average energy per bit for a received signal, $R_z(t)$, where K_b bits are transmitted is

$$E_b = \frac{E\left[E_{R_z}\right]}{K_b} = \frac{1}{K_b}E\left[\int_{-\infty}^{\infty} |R_z(t)|^2\,dt\right] = \frac{L_p^2}{K_b}E\left[\int_{-\infty}^{\infty} |X_z(t)|^2\,dt\right] \tag{12.4}$$

Hence throughout this text we shall parameterize performance by E_b/N_0 as that is the industry standard. It should be noted that the expectation or average in Eq. (12.4) is over the random transmitted data bits. Since the random transmitted bits are discrete random variables the expectation will be a summation versus the probability of each of the possible transmitted words. Consequently, the general form for the average received energy per bit is

$$E_b = \frac{L_p^2}{K_b}\sum_{i=0}^{2^{K_b}-1}\pi_i\left[\int_{-\infty}^{\infty}|x_i(t)|^2\,dt\right] = \frac{L_p^2}{K_b}\sum_{i=0}^{2^{K_b}-1}\pi_i E_i \tag{12.5}$$

where E_i is the energy of waveform $x_i(t)$ and π_i is the probability that the i th word was transmitted.

EXAMPLE 12.2

For the waveforms considered in Example 12.1 with equal likely a priori bits, the average energy per bit is

$$E_b = \frac{L_p^2\left(E_0+E_1+E_2+E_3\right)}{8} = \frac{T_p L_p^2\left(0.5+1+0.5+0.5\right)}{8} = \frac{5T_pL_p^2}{16} \tag{12.6}$$

12.2.2 Complexity

Complexity is a quantity that requires engineering judgment to estimate. Often the cost of a certain level of complexity changes over time. A good example of this trade-off changing over a short period of time was seen in the land mobile telephony market in the 1990s when early in the decade many people resisted a move to a standard based on code division multiple access technologies based on the cost and complexity of the handheld phones. By the end of the decade the proposed telecommunication standards had become much more complex but the advances in circuit technology allowed low cost implementations.

12.2.3 Bandwidth Efficiency

The bandwidth efficiency of a communication system is typically a measure of how well the system is using the bandwidth resource. In this text the bit rate of communication is denoted W_b bits/second and the transmission bandwidth is denoted B_T. Bandwidth costs money to acquire and the owners of this bandwidth want to communicate at as high a data rate as possible. Examples are licenses to broadcast radio signals or the installation of copper wires to connect two points. Hence bandwidth efficiency is very important for people who try to make money selling communication resources. For instance, if one communication system has a bandwidth efficiency that is twice the bandwidth efficiency of a second system then the first system can support twice the users on the same bandwidth. Twice the number of users implies twice the revenue. The measure of bandwidth efficiency for digital communications that will be used in this text is denoted spectral efficiency and is defined as

$$\eta_B = \frac{W_b}{B_T} \text{ bits/s/Hz}$$

The goal of this section is to associate a spectral characteristic or a signal bandwidth with a digital modulation. This spectral characteristic determines the bandwidth that a radio needs to have to support the transmission, as well as the spectral efficiency of a digital transmission scheme. The way this will be done in this text is to note that if the data being transmitted is known then the transmitted signal, $x_z(t)$, is a deterministic energy signal. The spectral characterization of deterministic energy signals is given by the energy spectrum

$$G_{x_z}(f) = |X_z(f)|^2 = \mathcal{F}\left\{V_{x_z}(\tau)\right\} = \mathcal{F}\left\{\int_{-\infty}^{\infty} x_z(t)x_z^*(t-\tau)dt\right\} \tag{12.7}$$

Building upon this spectral characterization in the same way as was done for Gaussian random processes (see Chapter 9), the function that will be used throughout this text to describe the spectral characteristics of a transmitted digitally modulated signal is the average energy spectrum per bit.

Definition 12.4 The average energy spectrum per bit for a digitally modulated signal, $X_z(t)$, where K_b bits are transmitted is

$$D_{X_z}(f) = \frac{E\left[G_{X_z}(f)\right]}{K_b} \tag{12.8}$$

This definition of average energy spectrum per bit (measured in Joules per bit per Hertz) is consistent with the definition of power spectral density for random processes as given in Definition 9.5 (measured in Joules per second per Hertz). It should be noted that the expectation or average in Eq. (12.8) is over the random transmitted data bits. Since the random transmitted bits are discrete random variables the expectation will be a summation versus the probability of each of the possible transmitted words. Consequently, the general form for the average energy spectrum per bit is

$$D_{X_z}(f) = \frac{1}{K_b} \sum_{i=0}^{2^{K_b}-1} \pi_i G_{x_i}(f) \tag{12.9}$$

It should be noted that the use of the Rayleigh energy theorem gives

$$E_b = L_p^2 \int_{-\infty}^{\infty} D_{x_z}(f) df \tag{12.10}$$

so that $D_{x_z}(f)$ can also be interpreted as the spectral density of the transmitted energy per bit. The transmission bandwidth of a digitial communication signal, denoted B_T, can be obtained from the average energy spectrum per bit, $D_{X_z}(f)$.

EXAMPLE 12.3

For the waveforms considered in Example 12.1 the individual energy spectrum of each waveform is given as

$$G_{x_0}(f) = \frac{T_p^{\,2}}{4} |\mathrm{sinc}(fT_p - 2)\exp[-j\pi f T_p] - \mathrm{sinc}(fT_p + 2)\exp[-j\pi f T\,]|^2 \tag{12.11}$$

$$G_{x_1}(f) = T_p^{\,2}\mathrm{sinc}(fT_p)^2$$

$$G_{x_2}(f) = \frac{T_p^{\,2}}{4} |\mathrm{sinc}(fT_p - 0.5)\exp[-j\pi f T_p + j\pi/2]$$

$$-\mathrm{sinc}(fT_p + 0.5)\exp[-j\pi f T\, - j\pi/2]|^2$$

$$G_{x_3}(f) = \frac{T_p^{\,2}}{2}\left(\mathrm{sinc}\left(\frac{fT_p}{2}\right)\right)^4$$

With equal likely a priori bits, the average energy per bit is

$$D_{X_z}(f) = \frac{1}{8}\sum_{i=0}^{3} G_{x_i}(f). \tag{12.12}$$

This energy spectrum is plotted in Figure 12.4. This text will often use the 3 dB energy bandwidth, B_3, as a way to quantify the spectral occupancy of a digital modulation. The 3 dB bandwidth of this 4-ary modulation is $B_3 = 0.86/T_p$ and consequently the spectral efficiency is $\eta_B = 2.32$ bits/s/Hz.

12.2.4 Other Important Characteristics

Many times in communication applications other issues besides bandwidth efficiency, complexity, and performance are important. For example, for a handheld mobile device the size, weight, and battery usage are important for the user.

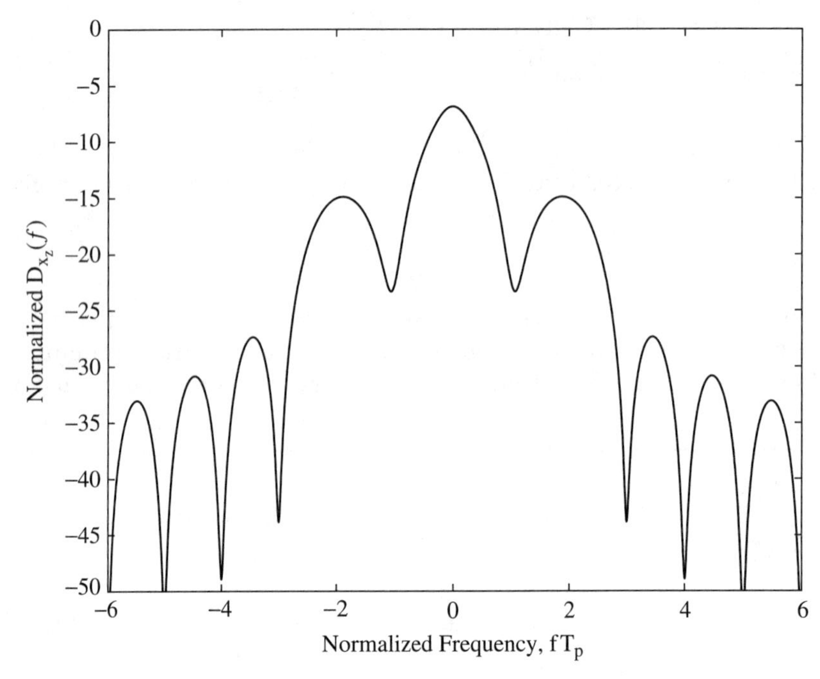

Figure 12.4 The average energy spectrum per bit for the signal set in Example 12.1.

Often in wireless communications energy efficiency of the algorithms are of paramount importance. Often to increase the talk time for a mobile phone an algorithm will give up performance for using less energy. One important issue in mobile devices is the linearity of final amplifier before the antenna. A high power linear amplifier is both expensive and consumes larger amounts of current, so this is not a desirable characteristic for a mobile device. On the other hand, high power is often needed to communicate with a remote base station or a satellite. Likewise nonlinear amplifiers, which are more energy efficient, often produce unacceptable distortion or spectral regrowth. In many handheld devices modulations are chosen to minimize the requirement on the linearity of the power amplifier. These issues will not be a major focus of this text like the bandwidth efficiency, complexity, and fidelity, but are mentioned as they are often involved in performance trade-offs for system design in practice.

12.3 Some Limits on Performance of Digital Communication Systems

Digital communications is a relatively unique field in engineering in that there is a theory that gives some performance limits for data transmission. The body of work that provides us with these fundamental limits is information theory. The founder of information theory was Claude Shannon [Sha48] and Shannon had built his theory on other seminal work that included [Har28]. While this text cannot derive all the important results from information theory it will attempt to highlight the important results in information theory that relate to the material in this text: physical layer transmission of digital information. A course in information theory is highly recommended [CT92] to students who want a deep level of understanding of data communications.

An important contribution of Claude Shannon was to identify that every channel had an associated capacity, C and reliable (in fact error-free) transmission is possible on the channel when $W_b < C$. A channel of significant interest for a majority of this text is the channel which experiences an additive white Gaussian noise (AWGN) distortion. For this AWGN channel when the signal uses a transmission bandwidth B_T, Shannon identified the capacity as [Sha48]

$$C = B_T \log_2(1 + SNR). \tag{12.13}$$

This immediately leads to a constraint on the spectral efficiency that can be reliably achieved

$$\eta_B < \log_2(1 + SNR) \tag{12.14}$$

Equation (12.14) unfortunately states that to achieve a linear increase in spectral efficiency a communication engineer must provide exponentially greater received SNR. Hence in most communication system applications the spectral efficiencies achieved are usually less than 15 bits/s/Hz (often much less).

Further insight into the problem is gained by reformulating Eq. (12.14). First, the received noise power is directly a function of the transmission bandwidth. If we assume an ideal bandpass filter of bandwidth B_T, the results of Chapter 9 give the noise power and SNR as

$$P_N = N_0 B_T \qquad SNR = \frac{P_s}{N_0 B_T} \tag{12.15}$$

Recall that most communication system engineers like to quantify performance with E_b/N_0 and that $P_s = E_b W_b$ so that Eq. (12.14) becomes

$$\eta_B < \log_2\left(1 + \frac{E_b}{N_0}\frac{W_b}{B_T}\right) = \log_2\left(1 + \frac{E_b}{N_0}\eta_B\right) \tag{12.16}$$

The achievable spectral efficiency versus E_b/N_0 is represented in Figure 12.5. The line in Figure 12.5 represents the solutions to the equation

$$\eta_B = \log_2\left(1 + \frac{E_b}{N_0}\eta_B\right) \tag{12.17}$$

For a given E_b/N_0, information theory indicates that reliable communication at spectral efficiencies below the line in Figure 12.5 are achievable while spectral efficiencies above the line are not achievable. Throughout the remainder of the text the goal will be to give an exposition on how to design communication

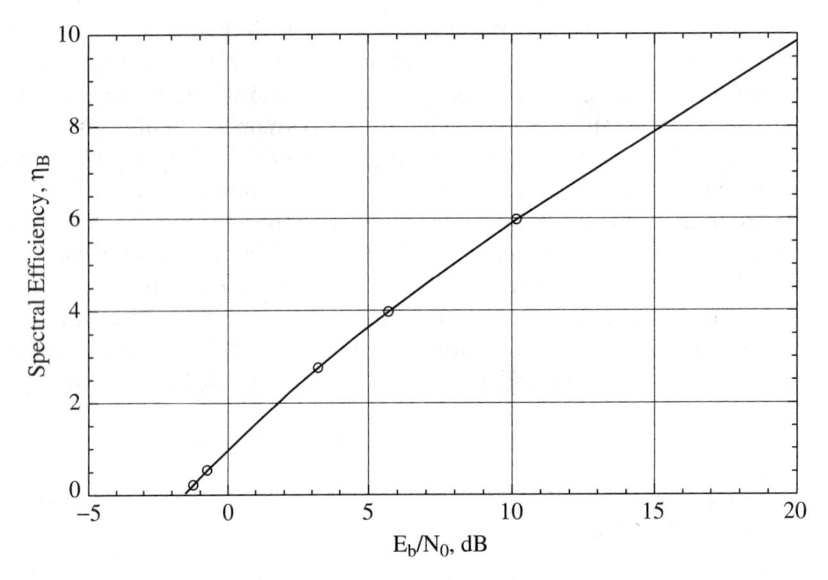

Figure 12.5 Maximum achievable spectral efficiency.

systems that have operating points which can approach this ultimate performance given in Figure 12.5.

The results in Figure 12.5 provide some interesting insights for how communication systems should be designed. In situations where bandwidth is the most restricted resource the goal then is to drive the received E_b/N_0 to as large a value as possible. For example, many telecommunication systems have designed operating points where $E_b/N_0 > 10$ dB. In situations where E_b is the most restricted resource it is possible to still achieve reliable communication by reducing the spectral efficiency. For example, communications with deep space probes is limited by the amount of power that can be received. Communication systems for deep space communication are designed most often to have relatively low bit rates and by setting $\eta_B < 1$. Also you can see from Figure 12.5 that there is a limit on how small E_b/N_0 can be made and still maintain reliable communications. This minimum is $E_b/N_0 = \ln 2 = -1.59$ dB. These results from information theory provide benchmarks by which we can calibrate performance as we progress in our understanding of digital communication theory.

12.4 Conclusion

In the roughly 50 years since Shannon proposed his mathematical theory of communication, engineers have been hard at work trying to achieve the implied limits given by Shannon's work. An example of a progression of the technology for voiceband computer modems in shown in Table 12.1. It should be noted that a voiceband modem can support a transmission bandwidth of $B_T = 3.2 -$ 4 kHz. Data communications started with modems of modest performance and

TABLE 12.1 The evolution of computer voiceband modems

Modem Standard	Year	Data Rate	η_B
V.21	1962	300 bps	0.09375
V.22	1970	1200 bps	0.375
V.22bis	1976	2400 bps	0.75
V.32	1984	9600 bps	3.0
V.32bis	1991	14400 bps	4.5
V.34	1995	28800 bps	9.0
V.90	1998	56000 bps	17.5

complexity. The performance of modems evolved in an ever increasing trajectory of performance and complexity. As the capability to process signals in electronic circuits dropped in price, engineers quickly used that added ability to make cost-effective modems that had increasing bandwidth efficiency and performance. The remainder of this text will try to give some flavor of the journey that has happened in the broader field of data communications in pursuit of Shannon's bound as the cost of producing high performance circuitry has dropped.

12.5 Homework Problems

Problem 12.1. You have been asked to build a digital communication system that transmits $K_b = 3$ bits of information. The specification given to you by your supervisor is to achieve at least a $W_b = 1$ Mbits/s transmission rate.

(a) How many different waveforms are needed to transmit $K_b = 3$ bits of information?

(b) Specify a set of waveforms that will transmit $K_b = 3$ bits with a transmission rate of $W_b = 1$ Mbits/s.

(c) Assume each of the 3-bit words are equally likely and compute the 3 dB bandwidth of the transmission specified in your answer to (b).

(d) What is the spectral efficiency of your design in (b)?

Problem 12.2. Show that the minimum E_b/N_0 needed to support reliable communication is

$$\left(\frac{E_b}{N_0}\right)_{\min} = \ln 2 \tag{12.18}$$

Problem 12.3. The touch-tone dialing in a telephone is a form of data communication. When a key is pressed on a telephone two tones are generated, i.e.,

$$x_i(t) = A\cos(2\pi f_i(1)t) + A\cos(2\pi f_i(2)t) \tag{12.19}$$

This type of modulation is referred to as dual tone multiple frequency (DTMF) modulation. The DTMF tones can send $K_b = 4$ bits even though there are only 12 keys on the phone. The modulation mappings are shown in Table 12.2.

TABLE 12.2 The dual tone multiple frequency (DTMF) modulation mappings

$f_i(1)$	$f_i(2)$ 1209 Hz	1336 Hz	1477 Hz	1633 Hz
697 Hz	i = 0 1	i = 1 ABC 2	i = 2 DEF 3	i = 3 A
770 Hz	i = 4 GHI 4	i = 5 JKL 5	i = 6 MNO 6	i = 7 B
852 Hz	i = 8 PRS 7	i = 9 TUV 8	i = 10 WXY 9	i = 11 C
941 Hz	i = 12 *	i = 13 oper 0	i = 14 #	i = 15 D

To simplify the problem assume $T_p = 1$ second and each word of the 16 total words that can be transmitted is equally likely. Note also that the modulation is given as a real signal.

(a) Plot the transmitted signal when person tries to call the operator ($i = 13$) for $A = 1$ over the interval [0, 0.1].

(b) Transmission on telephone lines are often thought of as having $f_c = 1200$ Hz. Give the simplest form of the complex envelope when a person tries to call the operator ($i = 13$).

(c) What is E_b as a function of A.

(d) Find and plot $D_{X_z}(f)$.

Problem 12.4. Frequency shift keying is a modulation that sends the word of information by transmitting a carrier pulse of one of $M = 2^{K_b}$ frequencies. This is an obvious simple signalling scheme and one used in many early modems. The signal set is given as

$$x_i(t) = \begin{cases} A\exp[j2\pi f_d(2i - M + 1)t] & 0 \leq t \leq T_p \\ 0 & \text{otherwise} \end{cases} \tag{12.20}$$

where f_d is known as the frequency deviation. The frequency difference between adjacent frequency pulses in the signal set is $2f_d$. For the problem assume $K_b = 3$, $f_d = 1$, $T_p = 1$, and that each of the eight words is equally likely to be transmitted.

(a) Plot the transmitted signal when $\vec{I} = 4$ with $f_c = 10$.

(b) What is E_b as a function of A.

(c) Find and plot $D_{X_z}(f)$.

Problem 12.5. Phase shift keyed modulation sends the word of information by transmitting a carrier pulse of one of $M = 2^{K_b}$ phases. This is an obvious simple signalling scheme and one used in many modems. The first form of the signal set one might consider is given as

$$x_i(t) = \begin{cases} A\exp\left[j\dfrac{\pi(2i+1)}{M}\right] & 0 \le t \le T_p \\ 0 & \text{elsewhere} \end{cases} \tag{12.21}$$

The phases have been chosen uniformly spaced around the unit circle. This is known as natural mapping PSK. For the problem assume $K_b = 3$, $T_p = 1$, and that each of the eight words is equally likely to be transmitted.

(a) Plot the transmitted signal when $\vec{I} = 4$ with $f_c = 10$.

(b) What is E_b as a function of A.

(c) Find and plot $D_{X_z}(f)$.

Problem 12.6. The U.S. National Aeronautics and Space Administration (NASA) sent the Pathfinder probe to the planet Mars. This probe had a 10-W transmitter and this resulted in a received signal power on Earth of $P_s = -145$ dBm. Assume the receiver noise spectral density is $N_0 = -170$ dBm/Hz.

(a) What is the highest transmission rate that could possibly be achieved between the probe on Mars and the receiver on Earth?

(b) If NASA did not want to interfere with other transmissions in space and wanted to be bandwidth efficient and achieve $\eta_B = 1$ bit/s/Hz, what is the highest transmission rate that can be achieved?

Problem 12.7. In a communication system the received average power is $P_s = -80$ dBm and the noise spectral density is $N_0 = -170$ dBm/Hz. An English text message has 128 possible characters.

(a) What is the lower bound to shortest time that a 4-character text message could be sent in this communication system?

(b) If this text message system was to be sent over a commercial system with a requirement of achieving at least $\eta_B = 2$ bits/s/Hz what is the lower bound to the shortest transmission time that can be achieved.

Problem 12.8. A satellite communication system can support two types of users: (1) voice and (2) video. Voice users need to support transmission rates of $W_b =$ 5 kbps and video users need to support transmission rates of $W_b = 4$ Mbps. The demodulation protocols are such that each user needs $E_b/N_0 = 4$ dB to achieved the desired fidelity.

(a) The voice and video user will need a different average signal power to support their application. Give the ratio of the required received average signal power of the two users.

(b) The different received signal powers discussed in (a) are achieved by giving each user a different size antenna. If the power gain of an antenna is proportional to the area of the antenna and the voice user has an antenna 0.5 m in diameter, how large, in diameter, will the antenna for the video user have to be?

Problem 12.9. If the telephone network was well modeled by an AWGN channel with $B_T = 3.2$ kHz and an $E_b/N_0 = 30$ dB what information transmission rate could reliably be supported? Given the information in Table 12.1 could the telephone network be accurately modeled as an AWGN channel with $B_T = 3.2$ kHz and an $E_b/N_0 = 30$ dB?

12.6 Example Solutions

Problem 12.10. Rewriting Eq. (12.17) gives

$$2^{\eta_B} = \left(1 + \frac{E_b}{N_0}\eta_B\right) \tag{12.22}$$

or equivalently

$$\frac{E_b}{N_0} = \frac{2^{\eta_B} - 1}{\eta_B} \tag{12.23}$$

The lowest received energy per bit that allows reliable communication is

$$\left(\frac{E_b}{N_0}\right)_{\min} = \lim_{\eta_B \to 0} = \frac{2^{\eta_B} - 1}{\eta_B} \tag{12.24}$$

Using L'Hopital's rule this reduces to

$$\left(\frac{E_b}{N_0}\right)_{\min} = \ln 2 \tag{12.25}$$

12.7 Miniprojects

Goal: To give exposure

- to a small scope engineering design problem in communications.
- to the dynamics of working with a team.
- to the importance of engineering communication skills (in this case oral presentations).

Presentation: The forum will be similar to a design review at a company (only much shorter). The presentation will be of 5 minutes in length with an overview of the given problem and solution. The presentation will be followed by questions from the audience (your classmates and the professor). Each team member should be prepared to make the presentation on the due date.

12.7.1 Project 1

Project Goals: Design a signal set that achieves a specified data rate and spectral efficiency.

The standard computer modem in 1962 was capable of a $W_b = 300$ bps transmission rate on an analog telephone line. The analog telephone line for this project is assumed to pass frequencies $600 \leq f \leq 3000$. Design a digital modulation for $K_b = 4$ capable of transmitting at greater or equal to $W_b = 300$ bps with 98% of the average energy per bit contained in the range $600 \leq f \leq 3000$. The following items need to be submitted with your solution to verify your design

(a) A time plot of the bandpass waveform, $x_{12}(t)$, that will be sent when $\vec{I} = 12$.

(b) A plot of the bandpass average energy spectrum.

(c) Computer code that computes the average energy per bit in $600 \leq f \leq 3000$.

Chapter

13

Optimal Single Bit Demodulation Structures

13.1 Introduction

In this chapter we consider the transmission and demodulation of one bit of information transmitted on an all-pass channel. This chapter will demonstrate how statistical decision theory is useful in the design of digital communications. The treatment here is less rigorous than one might see in an engineering book on detection (e.g., [Poo88]). The treatment given here is my synthesis of ideas already presented by many authors before. A particular favorite of mine is [Web87].

Here a set of notation is formulated. The bit of information will be denoted I with $I = 0, 1$. The discussion will assume that the bit to be sent is a random variable with $P(I = 0) = \pi_0$ and $P(I = 1) = \pi_1$. The transmitted waveform is $X_z(t)$. If $I = 0$ then $X_z(t) = x_0(t)$ and if $I = 1$ then $X_z(t) = x_1(t)$. The support[1] of $x_0(t)$ and $x_1(t)$ is in the interval $[0, T_p]$. In other words the signals $x_0(t)$ and $x_1(t)$ are both energy waveforms no longer than T_p in length. Consequently the bit rate is defined to be $W_b = 1/T_p$.

EXAMPLE 13.1

Figure 13.1 shows an example of two waveforms for one bit transmission. These two waveforms are explicitely given as

$$x_0(t) = \begin{cases} \sin\left(\dfrac{4\pi t}{T_p}\right) & 0 \leq t \leq T_p \\ 0 & \text{elsewhere} \end{cases} \qquad x_1(t) = \begin{cases} 1 & 0 \leq t \leq T_p \\ 0 & \text{elsewhere} \end{cases} \tag{13.1}$$

---

[1]The support of a function is the domain of the function where the range is nonzero valued.

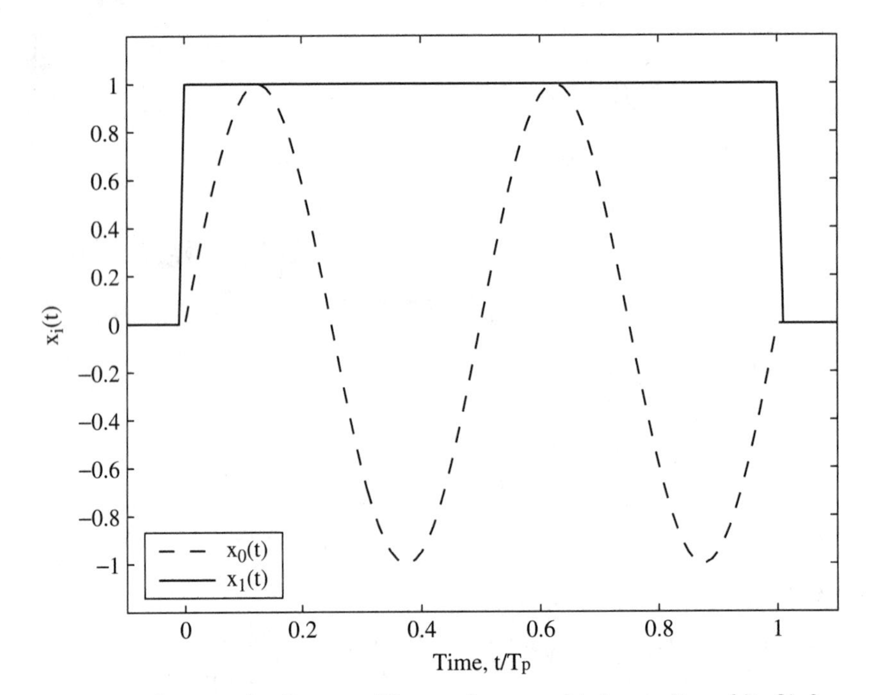

Figure 13.1 An example of two possible waveforms used to transmit one bit of information.

For the channel that was considered in the analog portion of the text the output received signal can be accurately modeled with

$$Y_z(t) = L_p X_z(t - \tau_p) \exp[j\phi_p] + W_z(t) \tag{13.2}$$

where $W_z(t)$ is a **complex** additive white Gaussian noise (AWGN) with $S_{W_z}(f) = N_0$ that models a thermal noise at the receiver front end[2], L_p is the propagation loss, τ_p is the propagation delay, and ϕ_p is a propagation induced phase shift. This book will be focused on **coherent** digital communications. Coherent communications implies that the receiver knows perfectly the distortion that can occur in transmission (with the model in Eq. (13.2) the exact value of L_p, τ_p, and ϕ_p). For coherent communication, the model in Eq. (13.2) can complicate the ideas of digital modulation and demodulation design so we will adopt a simpler one, i.e.,

$$Y_z(t) = X_z(t) + W_z(t) \tag{13.3}$$

The model in Eq. (13.3) assumes the transmitted signal is received undistorted. There is little loss in generality in considering this model for coherent communications since each of the parameters L_p, τ_p, and ϕ_p in Eq. (13.2) has a simple

[2]See Section 10.5 for a motivation of this noise model.

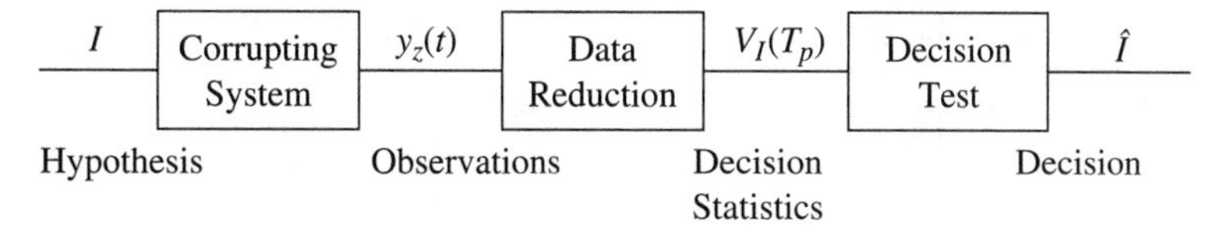

Figure 13.2 The problem formulation for hypothesis testing.

separable change for the optimal demodulator. A homework problem explores this difference. The energy per bit in this simple binary transmission example is

$$E_b = \pi_0 E_0 + \pi_1 E_1 \tag{13.4}$$

13.1.1 Statistical Hypothesis Testing

Digital communications borrows a vast majority of its theory from the well-developed theory of statistical hypothesis testing. A hypothesis test comes about when a person is faced with the problem of making a definite decision with respect to an uncertain hypothesis which is known only through its observable consequences. A statistical hypothesis test is an algorithm based on a set of observations to decide on the alternative (for or against the hypothesis) which minimizes certain risks [Leh86].

The general statistical hypothesis testing problem formulation is given in Figure 13.2. In general there is some system that disrupts the ability to make decisions. The output of that system is the raw observations. These raw observations are processed in some way to make the decision. The decision making process is usually formulated by getting either an analytical or an empirical understanding of how likely each hypothesis is and how the observations relate to the hypothesis.

EXAMPLE 13.2

Imagine you have just been hired at Westwood University as the men's basketball coach. Half the team, upset about the firing of the previous coach, quit the team and you have a big game against your archrival, Spoiled Children Institute of Technology (SCIT), in three days. You are desperate to find Division I basketball players and have started searching around campus (hypothesis: basketball player or no basketball player). Your assistant coach had the brilliant insight that basketball players tend to be taller than average and you decided to base your entire decision on the height of individuals (decision statistic: height). The assistant went out and measured 100 basketball players and 100 nonbasketball players and gave you the histograms in Figure 13.3. Given the problem formulation three questions can be asked

1. If a person is 5′4″ tall (64 inches) could he be a basketball player? The answer is yes but it is highly unlikely so you would probably decide against asking a 5′4″-tall person to join the team.

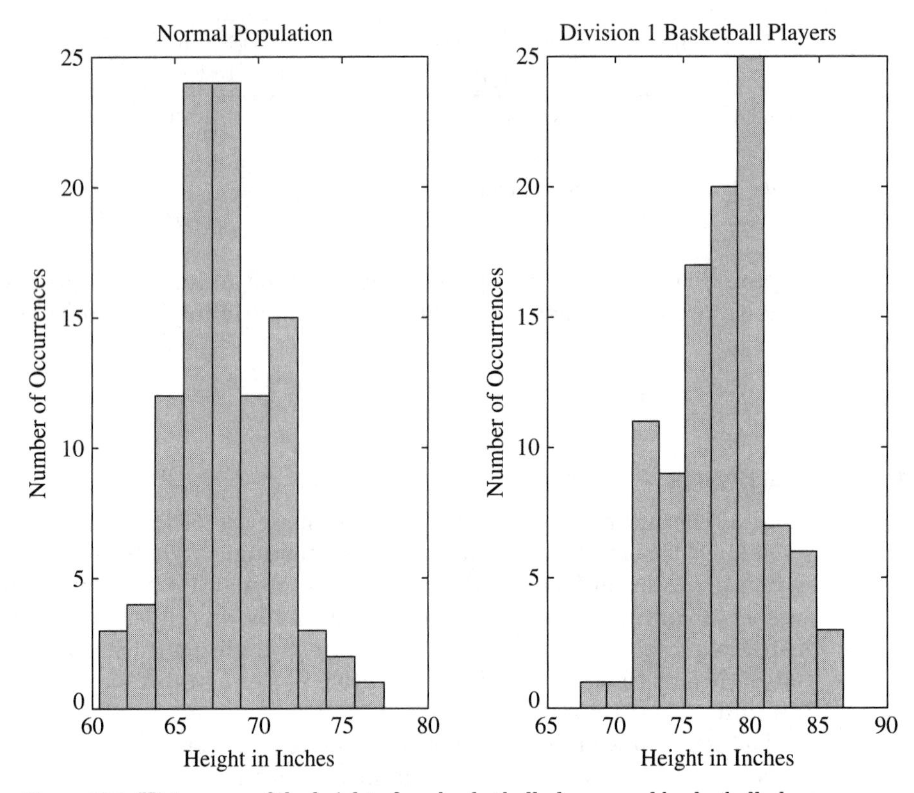

Figure 13.3 Histograms of the height of nonbasketball players and basketball players.

2. If a person is 7′0″ tall (84 inches) could he not be a basketball player? Again the answer is "yes" but it is highly unlikely so you would probably invite all 7′0″ people to join the team.
3. If a person is 6′0″ tall (72 inches) would you invite him to be on the team? There seems to be about an equal probability, judging from Figure 13.3, that a nonbasketball playing person would be 6′0″ tall as there is that a basketball player would be 6′0″ tall. The best decision would likely be that 6′0″ tall person not be invited since there are many more nonbasketball playing people than basketball players and hence it is more probable that a 6′0″-tall person is not a basketball player.

This example shows intuitively how decisions are made, how statistics are formed and how the probability of each hypothesis should impact the decision. These elements are all in the digital communications problem.

It is clear from the previous example of trying to decide on basketball players that both what is observed in a statistical test and what are the prior distributions on the possible outcomes can both significantly impact the decision that is made. To capture these characteristics statisticians have defined two important quantities.

Definition 13.1 The a priori probability is the probability associated with a possible hypothesis before any experiments are completed.

EXAMPLE 13.3

For the binary signaling in Example 13.1 it will be assumed that each bit value is equally likely a priori, i.e., $\pi_0 = \pi_1 = 0.5$.

EXAMPLE 13.4

Continuing with Example 13.2, a rough estimate would be that the probability a randomly chosen person on the Westwood University campus is a Division I basketball player is $\pi_0 = 0.001$.

Definition 13.2 The a posteriori probability (APP) is the probability associated with a possible hypothesis that takes into account any observed experimental outcomes and the a priori probability.

A priori probabilities and a posteriori probabilities have a prominent role in the theory of digital communications and in fact form the basis of most modern data modem technology.

13.1.2 Statistical Hypothesis Testing in Digital Communications

The optimum structure for demodulation of known transmitted signals in the presence of the white noise (i.e., the model in Eq. (13.3)) is shown in Figure 13.4. The structure consists of a linear filter where the real output of this linear filter is sampled and subject to a threshold test. The justification for why this structure is optimum will be left to a course on detection theory (e.g., [Poo88]). Linear filters, samplers, and threshold tests are some of the most common electrical components available. The fact that the combination of a linear-time-invariant filter, a sampler, and a threshold test forms the best single bit digital communications demodulator is quite striking. The sample time can be arbitrary but

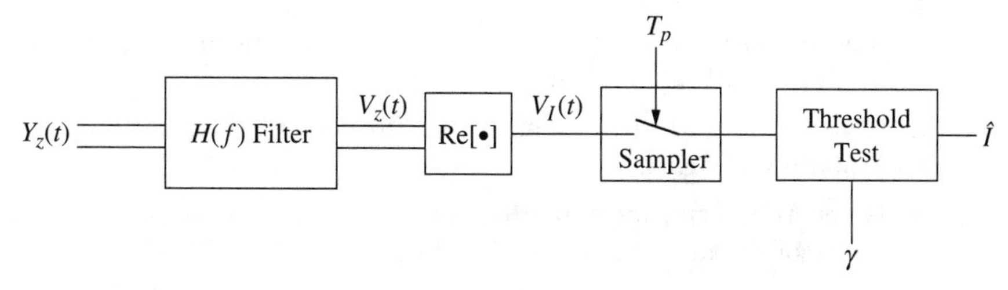

Figure 13.4 The optimum demodulator for a single bit transmission in AWGN.

here is chosen to be at T_p, the end of the transmitted pulse. Hopefully after the entire optimum demodulation structure is examined in detail the reasons for this selection will become clear. For clarity, a threshold test is a component that provides the following logical operation: if

$$V_I(T_p) > \gamma \qquad \text{then} \qquad \hat{I} = 1 \tag{13.5}$$

else

$$V_I(T_p) \le \gamma \qquad \text{then} \qquad \hat{I} = 0 \tag{13.6}$$

Threshold tests or comparators are easily implemented in electronic circuits.

Definition 13.3 A statistic is any processing of data to produce a number that represents the data but is of lower dimensionality than the original data.

This definition matches with how statistics are used in a wide variety of applications (e.g., mean and variance of grades). In the case considered here $V_I(T_p)$ is one number that represents the entire continuous time received waveform, $Y_z(t)$ and is a statistic for deciding about I.

Definition 13.4 A sufficient statistic is a statistic where making a decision or estimate based only on the statistic results in no loss in information/performance in estimation or detection compared to using the full data record.

In the binary detection problem, an optimal decision about I can be made by only considering $V_I(T_p)$, consequently $V_I(T_p)$ is a sufficient statistic for the entire observed waveform, $Y_z(t)$.

13.1.3 Digital Communications Design Problem

The digital communications design problem consists of identifying the components of the optimal demodulator such that a good trade-off can be achieved between fidelity, complexity, and spectral efficiency. The approach taken in this text to understanding this trade-off for single bit demodulation is to work through five simple design tasks

1. Given $x_0(t), x_1(t)$ and $H(f)$, design the optimum threshold test (find γ).
2. Given $x_0(t)$ and $x_1(t)$ and $H(f)$ with the optimum threshold test, compute fidelity of the bit estimation. The fidelity metric of interest will be the probability of making a bit decision error.
3. Given $x_0(t)$ and $x_1(t)$, design the filter, $H(f)$ that minimizes the bit error probability (BEP).
4. Given the optimum demodulator, design $x_0(t)$ and $x_1(t)$ to optimize the fidelity of the message reconstruction.
5. Design $x_0(t)$ and $x_1(t)$ to have desired spectral characteristics.

13.2 Minimum Probability of Error Bit Demodulation

The goal in this section is to address Design Task 1. We are looking to fix $x_0(t)$, $x_1(t)$, and $H(f)$ and find a threshold, γ, that gives the optimum error probability. Our criterion for optimal demodulation will be based on which bit is more probable given an observed output from the receiver, $V_I(T_p) = v_I$. Recall this text maintains the convention that capital letters denote random variables and lower case letters will denote observed realizations of those random variables. This situation of trying to map a random variable into one of two possible decisions on a bit is a situation where this convention is necessary to clarify the actual underlying mathematics. $V_I(T_p)$ and $\hat{I}$ are random variables but the optimum mapping from the observations $V_I(T_p) = v_I$ to the decision $\hat{I}(v_I)$ is a deterministic function.

EXAMPLE 13.5

Consider the signals shown in Figure 13.1 with a truncated ideal lowpass filter with design bandwidth of $B_T = 4/T_p$, i.e.,

$$h(t) = \begin{cases} 2B_T \operatorname{sinc}(2B_T(t - \tau_h)) & 0 \le t \le 2\tau_h \\ 0 & \text{elsewhere} \end{cases} \tag{13.7}$$

where $\tau_h = 3/T_p$. Note for simplicity of displaying the signals at various points of the demodulator we have chosen both the transmitted signals and the filter to be real valued. The signals and the noiseless output of the filter are plotted in the Figure 13.5. The vertical line represents a sampling time when the waveforms are relatively easy to distinguish and this sample time will be used in this example as we examine Design Task 1.

The decision will be made based on which value of the transmitted bit is more probable (likely) given the observed value v_I. This type of demodulation is known as maximum *a posteriori* bit demodulation (MAPBD). A MAPBD computes the *a posteriori* probability (APP) for each possible hypothesis and then chooses the hypothesis whose APP is largest. The MAPBD chooses the most likely hypothesis after both the a priori information and the observations are analyzed. MAPBD can be shown to be the minimum probability of error demodulation scheme by using the theory of Bayes detection [Leh86, Web87]. Mathematically this can be stated as

$$P(I = 1|v_I) \underset{\hat{I}=0}{\overset{\hat{I}=1}{\gtrless}} P(I = 0|v_I) \tag{13.8}$$

This decoding rule first computes the APP of each possible value of the transmitted bit, $P(I = i|v_I)$, $i = 0, 1$, and then makes a decision based on which APP is the largest. This procedure of computing an APP based on the observed channel outputs is a common theme in modern digital communications.

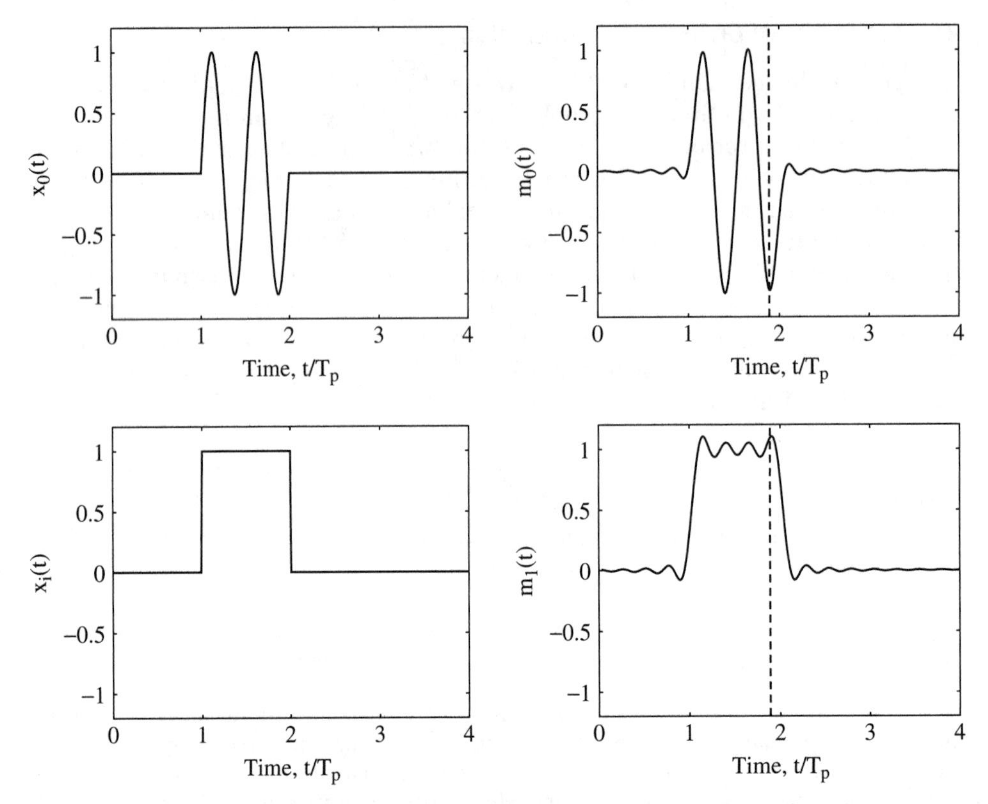

Figure 13.5 The transmitted signals and the noiseless output of $H(f)$ for Example 13.5.

These probabilities can be computed with Bayes rule (mixed form)

$$P(I=1|v_I) = \frac{f_{V_I|I}(v_I|I=1)P(I=1)}{f_{V_I}(v_I)}$$
$$P(I=0|v_I) = \frac{f_{V_I|I}(v_I|I=0)P(I=0)}{f_{V_I}(v_I)} \tag{13.9}$$

where $f_{V_I|I}(v_I|I=i)$ is the conditional PDF of $V_I(T_p)$ when $I=i$ is transmitted and $f_{V_I}(v_I)$ is the unconditional PDF of $V_I(T_p)$. When discussing detection problems the notation $\pi_0 = P(I=0)$ and $\pi_1 = P(I=1)$ is often used [Poo88] and this text will adopt this notation as well. When the common term present in Eq. (13.9) is cancelled from both sides of Eq. (13.8) the decoding rule becomes

$$f_{V_I|I}(v_I|I=1)\pi_1 \underset{\hat{I}=0}{\overset{\hat{I}=1}{\gtrless}} f_{V_I|I}(v_I|I=0)\pi_0 \tag{13.10}$$

Figure 13.6 shows the block diagram for this MAPBD threshold test. The forming of the threshold test is essentially the finding of which of two decision

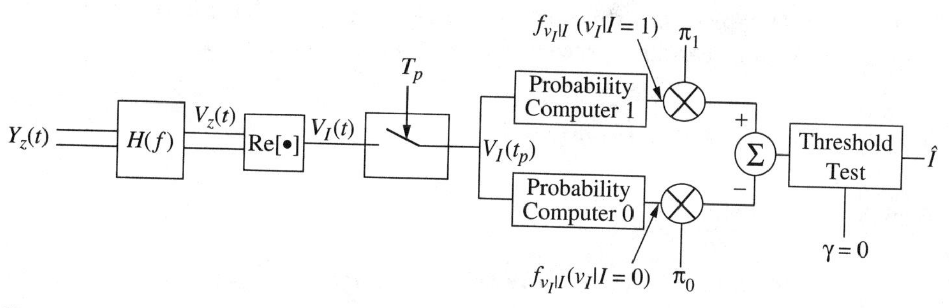

Figure 13.6 The maximum a posteriori bit demodulator.

statistics are larger. This is a common theme that will weave itself through the remainder of the text. To prepare the reader for this thread, the decision test in Eq. (13.10) can be restated as

$$\hat{I} = \arg\max_{i=0,1} f_{V_I|I}(v_I|I = i)\pi_i \tag{13.11}$$

The optimum test can often be simplified from the form given in Eq. (13.10) but the discussion throughout the course will often return to a demodulator format based on the form in Eq. (13.10). Another form of the test that is often used is known as the likelihood ratio test [Poo88] given as

$$\frac{f_{V_I|I}(v_I|I = 1)}{f_{V_I|I}(v_I|I = 0)} \underset{\hat{I}=0}{\overset{\hat{I}=1}{\gtrless}} \frac{\pi_0}{\pi_1} \tag{13.12}$$

From a practical point of view this maximum a posteriori bit demodulator is not interesting if the computation of $f_{V_I|I}(v_I|I = i)$, $i = 0, 1$ is not simple. Fortunately, the form of the receiver for the single bit demodulator can be significantly simplified.

13.2.1 Characterizing the Filter Output

The filter output is straightforward to characterize since conditioned on the transmitted signal, the input is a known signal plus white Gaussian noise. If $I = 0$ and denoting $h(t)$ as the filter impulse response, then

$$\begin{aligned} V_{z,0}(t) &= \int_{-\infty}^{\infty} (x_0(\tau) + W_z(\tau))h(t-\tau)d\tau \\ &= \int_{-\infty}^{\infty} x_0(\tau)h(t-\tau)d\tau + \int_{-\infty}^{\infty} W_z(\tau)h(t-\tau)d\tau \\ &= m_0(t) + N_z(t). \end{aligned} \tag{13.13}$$

The term $m_0(t)$ represents the output of the filter when $x_0(t)$ is input. Recall $N_z(t)$ is a Gaussian random process that is characterized with

$$R_{N_z}(\tau) = N_0 V_h(\tau) = N_0 \int_{-\infty}^{\infty} h(t)h^*(t-\tau)dt \qquad \text{and} \qquad S_{N_z}(f) = N_0|H(f)|^2 \tag{13.14}$$

Likewise if $I = 1$ then

$$\begin{aligned} V_{z,1}(t) &= \int_{-\infty}^{\infty} (x_1(\tau) + W_z(\tau))h(t-\tau)d\tau \\ &= \int_{-\infty}^{\infty} x_1(\tau)h(t-\tau)d\tau + \int_{-\infty}^{\infty} W_z(\tau)h(t-\tau)d\tau \\ &= m_1(t) + N_z(t) \end{aligned} \tag{13.15}$$

where $m_1(t)$ represents the output of the filter when $x_1(t)$ is input.

Output time samples of this filter are easily characterized. If $I = 0$, the sample taken at time T_p is given as

$$V_{0,z}(T_p) = m_0(T_p) + N_z(T_p)$$

and the real part of this sample is given as

$$V_{0,I}(T_p) = m_{0,I} + N_I(T_p)$$

where $m_{0,I} = \Re[m_0(T_p)]$. Consequently, when $I = 0$, $V_I(T_p)$ is a Gaussian random variable (RV) with a PDF given as

$$f_{V_I|I}(v_I|I=0) = \frac{1}{\sqrt{2\pi\sigma_{N_I}^2}} \exp\left[-\frac{(v_I - m_{0,I})^2}{2\sigma_{N_I}^2}\right] \tag{13.16}$$

where $\sigma_{N_I}^2 = \text{var}(N_I(T_p))$. Since $N_I(T_p)$ is the output of a linear filter it is easy to show using the results from Chapters 9 and 10 that $2\sigma_{N_I}^2 = \int_{-\infty}^{\infty} N_0|H(f)|^2 df$. In a similiar fashion

$$f_{V_I|I}(v_I|I=1) = \frac{1}{\sqrt{2\pi\sigma_{N_I}^2}} \exp\left[-\frac{(v_I - m_{1,I})^2}{2\sigma_{N_I}^2}\right] \tag{13.17}$$

where $m_{1,I} = \Re[m_1(T_p)]$. Consequently, the MAPBD given in Eq. (13.10) requires the computation of two Gaussian PDFs having the same variance but different means. The means of these two PDFs are a function of the demodulation filter and the two possible transmitted signals representing the two possible values of the bit.

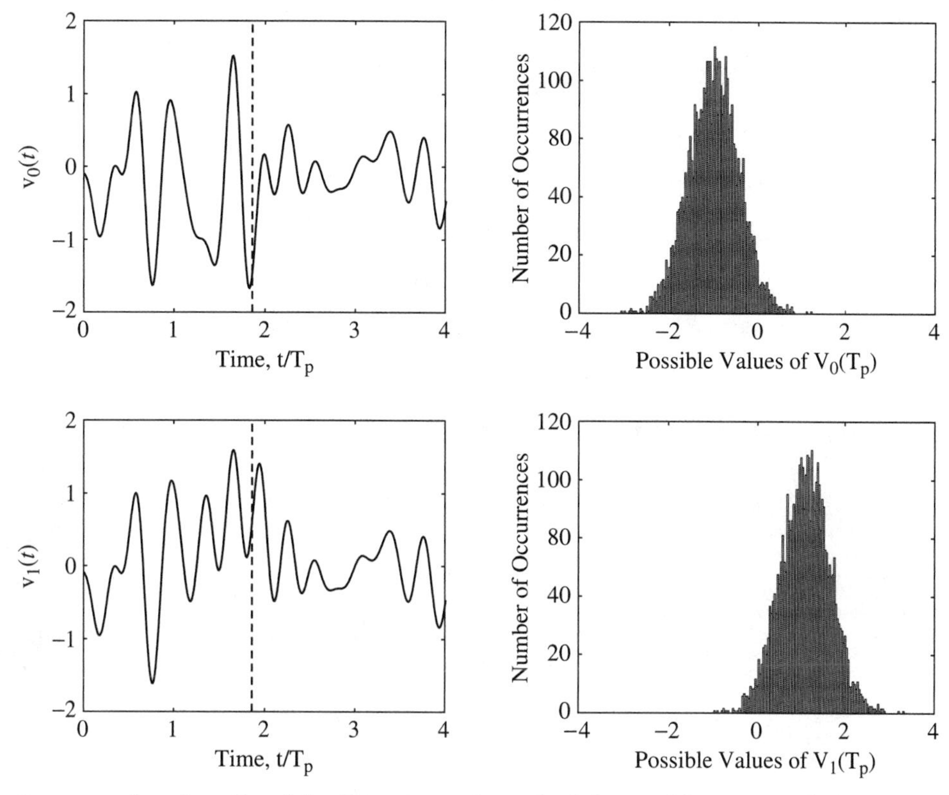

Figure 13.7 Sample paths of the filter outputs for each of the possible transmitted signals and histograms of the filter output samples.

EXAMPLE 13.6

Continuing with Example 13.5, Figure 13.7 shows a sample path of $V_0(t)$ (top left) and a sample path of $V_1(t)$ (bottom left) with $E_b/N_0 = 10$ dB. These sample paths clearly behave like a signal, $m_i(t)$, plus a noise. The same sample point as considered in Example 13.5 is again shown with the vertical line. Histograms of the sampled filter output for 4000 random trials of this experiment when each of the two hypotheses are true ($I = 0$ at the top right and $I = 1$ at the bottom right) are also shown Figure 13.7. This histogram shows empirically that the Gaussian PDF is correct model for $f_{V_I|I}(v_I|I = i)$.

13.2.2 Uniform A Priori Probability

There are many ways to simplify the receiver structure from this point but the one that provides the most intuition and is most practical is to assume a uniform *a priori* probability (i.e., $\pi_0 = \pi_1 = 0.5$). Uniform *a priori* probabilities often arise in practice since this is the goal of source coding [CT92]. Uniform *a priori*

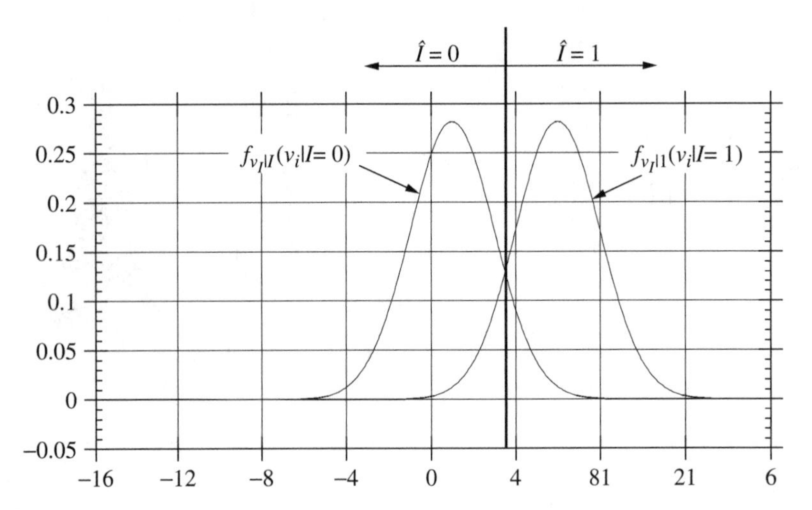

Figure 13.8 An example of the resulting test given by Eq. (13.18). $m_{0,I} = 1$, $m_{1,I} = 6$, and $\sigma^2_{N_I} = 2$.

probabilities also produce a threshold test which is not a function of the *a priori* probabilities. The form of the optimal test when the *a priori* probability is not uniform is explored in the homework. When $\pi_0 = 0.5$, the optimum decoder, often denoted the maximum likelihood bit demodulator (MLBD), is

$$\frac{1}{\sqrt{2\pi\sigma^2_{N_I}}}\exp\left[-\frac{(v_I - m_{1,I})^2}{2\sigma^2_{N_I}}\right] \underset{\hat{I}=0}{\overset{\hat{I}=1}{\gtrless}} \frac{1}{\sqrt{2\pi\sigma^2_{N_I}}}\exp\left[-\frac{(v_I - m_{0,I})^2}{2\sigma^2_{N_I}}\right] \tag{13.18}$$

Consequently, this MLBD computes two Gaussian PDFs (having the same variance and different means) and makes a decision based on which of the two PDFs takes a higher value. This is the maximum likelihood principle [Poo88]. The monotonic nature of the Gaussian PDF implies that this results in a simple threshold test on $v_I(T_p)$. Figure 13.8 shows pictorially how this decoder operates.

The MLBD has a very interesting geometric interpretation. Cancelling common terms and taking log[3] gives

$$(v_I - m_{1,I})^2 \underset{\hat{I}=1}{\overset{\hat{I}=0}{\gtrless}} (v_I - m_{0,I})^2 \tag{13.19}$$

Recall the definition of Euclidean distance in an N dimensional vector space.

[3]Taking the log does not change the decision rule as $\log(x)$ is a monotonic function.

Definition 13.5 Euclidean distance in an N dimensional vector space between two points $\vec{x}$ and $\vec{y}$ is

$$\delta_E(\vec{x}, \vec{y}) = \sqrt{\sum_{i=1}^{N} (x_i - y_i)^2} \tag{13.20}$$

For simplicity of notation the squared Euclidean distance is also defined.

Definition 13.6 The squared Euclidean distance in an N dimensional vector space between two points $\vec{x}$ and $\vec{y}$ is

$$\Delta_E(\vec{x}, \vec{y}) = \sum_{i=1}^{N} (x_i - y_i)^2 \tag{13.21}$$

Consequently, the MLBD can be rewritten as

$$\Delta_E(v_I, m_{1,I}) \underset{\hat{I}=1}{\overset{\hat{I}=0}{\gtrless}} \Delta_E(v_I, m_{0,I}) \tag{13.22}$$

This implies that the maximum likelihood bit decision is based on whether v_I is geometrically closer to $m_{1,I}$ or $m_{0,I}$. This geometric interpretation of decision rules will arise several more times in the context of demodulation for digital communications.

The decoder can further be simplified to a very simple threshold test. Assuming $m_{1,I} > m_{0,I}$, completing square, and doing some algebra reduces the optimum decision rule to

$$v_I \underset{\hat{I}=0}{\overset{\hat{I}=1}{\gtrless}} \gamma = \frac{m_{1,I} + m_{0,I}}{2} \tag{13.23}$$

The MLBD compares the observed filter output, v_I with a threshold which is the arithmetic average of the two filter outputs corresponding to the possible transmitted signals in the absence of noise. For the example considered in Figure 13.8 where $m_{0,I} = 1$, $m_{1,I} = 6$, and $\sigma^2_{N_I} = 2$ the optimum threshold is $\gamma = 3.5$. It is interesting to note that the noise variance does not enter into the MLBD structure. It will be left as a student exercise to compute the threshold test for the case $m_{1,I} < m_{0,I}$. It is equally simple. The block diagram for the simplest form for the MLBD is shown in Figure 13.9.

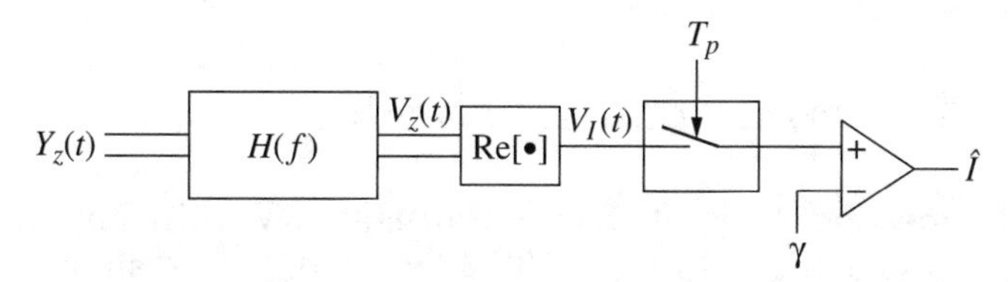

Figure 13.9 The simple form of the MLBD.

EXAMPLE 13.7

For Example 13.1 it is apparent by examining Figure 13.7 that $m_0(T_p) = -0.98$ and $m_1(T_p) = 1.09$. These values result in a threshold test given as

$$v_I \underset{\hat{I}=0}{\overset{\hat{I}=1}{\gtrless}} \gamma = 0.055 \tag{13.24}$$

Design Task 1 is complete; we have found the optimum threshold test given $x_0(t)$, $x_1(t)$, and $H(f)$ for the uniform prior distribution case. Results for a nonuniform prior distribution are similiar.

13.3 Analysis of Demodulation Fidelity

Recall Design Task 2 is given $x_0(t)$ and $x_1(t)$ and $H(f)$ with the MLBD threshold test, γ, compute the fidelity of the message reconstruction for the optimum demodulator. The fidelity metric of interest will be the probability of making a bit decision error and this is often referred to as the bit error probability (BEP).

Definition 13.7 Bit error probability (BEP) is

$$P_B(E) = P(\hat{I} \neq I)$$

The BEP can be computed using total probability as

$$P_B(E) = P(\hat{I} = 1|I = 0)\pi_0 + P(\hat{I} = 0|I = 1)\pi_1 \tag{13.25}$$

Recall that when $m_{1,I} > m_{0,I}$, the MAPBD has the form

$$v_I \underset{\hat{I}=0}{\overset{\hat{I}=1}{\gtrless}} \gamma \tag{13.26}$$

so that

$$P(\hat{I} = 1|I = 0) = P(V_I(T_p) > \gamma|I = 0) \tag{13.27}$$

and

$$P(\hat{I} = 0|I = 1) = P(V_I(T_p) < \gamma|I = 1) \tag{13.28}$$

Since conditioned on I, $V_I(T_p)$ is a Gaussian RV with known mean and variance, the probabilities in Eqs. (13.27) and (13.28) are simple to compute.

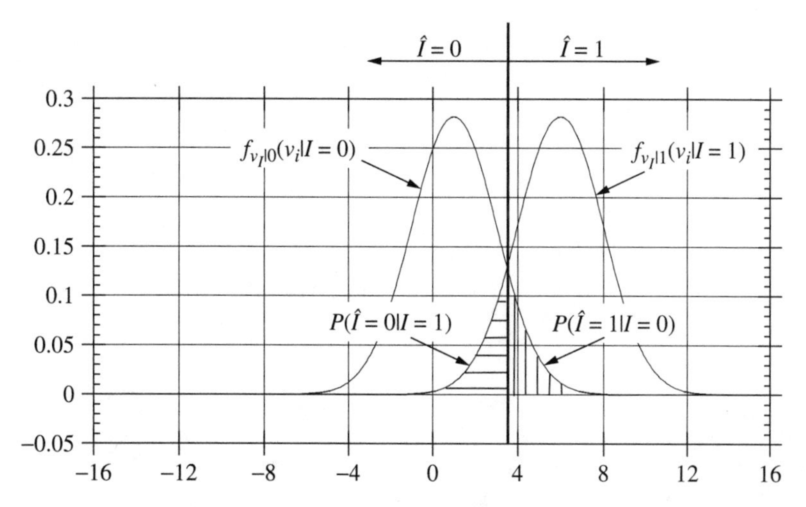

Figure 13.10 The tails of the Gaussian distribution that are the conditional bit error probability. This is for a test given by Eq. (13.18) with $m_{0,I} = 1$, $m_{1,I} = 6$, and $\sigma^2_{N_I} = 2$.

These probabilities in Eqs. (13.27) and (13.28) are simply the area under the tails of two Gaussian PDFs. Examples of the tails are illustrated in Figure 13.10.

13.3.1 Erf Function

This text will compute the probability that a Gaussian RV lies in an interval using the erf() and erfc() functions [Ae72]. While probabilities of this form are expressed using different functions by various authors, this text uses the erf() and erfc() because it is commonly available in math software packages (e.g., Matlab).

Definition 13.8 The erf function is

$$\text{erf}(z) = \frac{2}{\sqrt{\pi}} \int_0^z e^{-t^2} dt \tag{13.29}$$

The cumulative distribution function of a Gaussian RV, X, with mean m_X and variance σ^2_X is then given as

$$F_X(x) = \frac{1}{2} + \frac{1}{2}\text{erf}\left(\frac{x - m_X}{\sqrt{2}\sigma_X}\right)$$

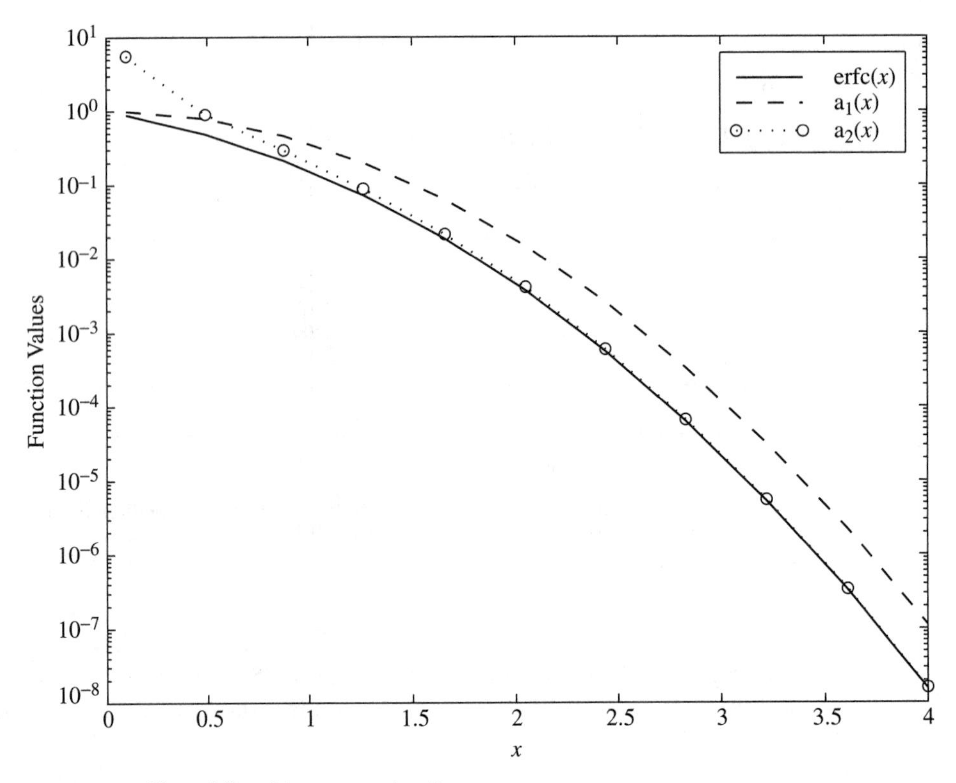

Figure 13.11 The erfc() and two approximations.

Three properties of the erf function important for finding probabilities associated with Gaussian RVs are given as

$$\mathrm{erf}(\infty) = 1$$

$$\mathrm{erfc}(z) = 1 - \mathrm{erf}(z)$$
$$\mathrm{erf}(-z) = -\mathrm{erf}(z)$$

Some approximations to the erfc(x) are available and they provide insight into the probability of error in digital communications. Two upperbounds are

$$\mathrm{erfc}(x) \le \exp[-x^2] = a_1(x) \tag{13.30}$$

$$\mathrm{erfc}(x) \le \frac{1}{\sqrt{\pi}x}\exp[-x^2] = a_2(x) \tag{13.31}$$

These approximations and erfc(x) are plotted in Figure 13.11.

13.3.2 Uniform A Priori Probability

Considering the case of the uniform a priori probabilities ($\pi_0 = \pi_1 = 0.5$) with the MLBD demonstrates the simplicity of this probability of error calculation.

Examining Eq. (13.27) and using Eq. (13.16) gives

$$P(\hat{I} = 1|I = 0) = \int_{\gamma}^{\infty} f_{V_I|I}(v_I|I = 0)dv_I$$
$$= \int_{\gamma}^{\infty} \frac{1}{\sqrt{2\pi\sigma_{N_I}^2}} \exp\left[-\frac{(v_I - m_{0,I})^2}{2\sigma_{N_I}^2}\right] dv_I \qquad (13.32)$$

This probability can be simplified using the following change of variables

$$t = \frac{v_I - m_{0,I}}{\sqrt{2}\sigma_{N_I}} \qquad dt = \frac{dv_I}{\sqrt{2}\sigma_{N_I}} \qquad (13.33)$$

to give

$$P(\hat{I} = 1|I = 0) = \frac{1}{2}\text{erfc}\left(\frac{\gamma - m_{0,I}}{\sqrt{2}\sigma_{N_I}}\right) = \frac{1}{2}\text{erfc}\left(\frac{m_{1,I} - m_{0,I}}{2\sqrt{2}\sigma_{N_I}}\right) \qquad (13.34)$$

where the value for γ given in Eq. (13.23) is used to get the last expression. Similiar manipulations give

$$P(\hat{I} = 0|I = 1) = \frac{1}{2}\text{erfc}\left(\frac{m_{1,I} - m_{0,I}}{2\sqrt{2}\sigma_{N_I}}\right) \qquad (13.35)$$

and consequently the BEP is also

$$P_B(E) = \frac{1}{2}\text{erfc}\left(\frac{m_{1,I} - m_{0,I}}{2\sqrt{2}\sigma_{N_I}}\right) \qquad (13.36)$$

Design Task 2 is complete; we have found the BEP of the MLBD given $x_0(t)$, $x_1(t)$ and $H(f)$ for the uniform prior distribution case. Results for a nonuniform prior distribution are similiar.

EXAMPLE 13.8

Continuing Example 13.1 it is apparent that

$$P_B(E) = \frac{1}{2}\text{erfc}\left(\frac{2.07}{2\sqrt{2}\sigma_{N_I}}\right) \qquad (13.37)$$

Recall that

$$\sigma_{N_I}^2 = \frac{N_0}{2}\int_{-\infty}^{\infty} |h(t)|^2 dt \qquad (13.38)$$

Some observations about the MLBD performance are in order here.

Definition 13.9 The effective signal–to–noise (SNR) ratio for MLBD is

$$\eta = \left(\frac{m_{1,I} - m_{0,I}}{2\sqrt{2}\sigma_{N_I}}\right)^2 = \frac{(m_{1,I} - m_{0,I})^2}{8\sigma_{N_I}^2}$$

Note the following

- $P_B(E) = \frac{1}{2}\text{erfc}(\sqrt{\eta})$.
- $P_B(E)$ is monotone decreasing in η, in fact Eq. (13.30) indicates $P_B(E)$ behaves like $\frac{1}{2}\exp[-\eta]$ for moderately large η.
- Maximizing η will minimize $P_B(E)$.
- For a fixed $x_0(t)$ and $x_1(t)$ the only quantity that affects the value of η, which is under control of the communication system designer, is the filter response ($h(t)$ or $H(f)$).

13.4 Filter Design

Considering the BEP characteristic highlighted in the previous section leads to Design Task 3. This task is stated; given $x_0(t)$ and $x_1(t)$ design the filter, $H(f)$, that minimizes the BEP or equivalently maximizes η.

13.4.1 Maximizing Effective SNR

To solve this problem η must be expressed in terms of $H(f)$. To this end note

$$2\sigma_{N_I}^2 = N_0 \int_{-\infty}^{\infty} |H(f)|^2 df \qquad (13.39)$$

and

$$\begin{aligned} m_1(T_p) - m_0(T_p) &= \int_{-\infty}^{\infty} (x_1(\lambda) - x_0(\lambda))h(T_p - \lambda)d\lambda \\ &= \mathcal{F}^{-1}\{(X_1(f) - X_0(f))H(f)\}|_{t=T_p} \\ &= \int_{-\infty}^{\infty} (X_1(f) - X_0(f))H(f)\exp[j2\pi f T_p]df \qquad (13.40) \end{aligned}$$

To get a more compact notation we define

$$b_{10}(t) = x_1(t) - x_0(t) \qquad B_{10}(f) = X_1(f) - X_0(f)$$

Consequently,

$$\eta = \left(\frac{m_{1,I} - m_{0,I}}{2\sqrt{2}\sigma_{N_I}}\right)^2 = \frac{\left(\Re\left\{\int_{-\infty}^{\infty} B_{10}(f)H(f)\exp[j2\pi f T_p]df\right\}\right)^2}{4N_0\int_{-\infty}^{\infty} |H(f)|^2 df} \qquad (13.41)$$

It should be noted that both the signal power (the numerator of Eq. (13.41)) and the noise power (the denominator of Eq. (13.41)) are a function of $H(f)$.

The form of $H(f)$ that maximizes the effective SNR is not readily apparent in examining Eq. (13.41) but desired results can be obtain by using Schwarz's inequality.

Theorem 13.1 **(Schwarz's Inequality)** For two functions $X(f)$ and $Y(f)$ where $\int_{-\infty}^{\infty} |X(f)|^2 df < \infty$ and $\int_{-\infty}^{\infty} |Y(f)|^2 df < \infty$.

$$\left| \int_{-\infty}^{\infty} X(f)Y^*(f) df \right|^2 \leq \int_{-\infty}^{\infty} |X(f)|^2 df \int_{-\infty}^{\infty} |Y(f)|^2 df$$

where equality holds only if $X(f) = AY(f)$ where A is a complex constant.

Making the assignment

$$X(f) = H(f) \qquad Y(f) = B_{10}^*(f) \exp[-j2\pi f T_p] \tag{13.42}$$

and noting

$$(\Re(X))^2 \leq |X|^2 = (\Re(X))^2 + (\Im(X))^2 \tag{13.43}$$

Schwarz's inequality results in the following inequality for the effective SNR

$$\eta = \frac{\left(\Re\left\{\int_{-\infty}^{\infty} B_{10}(f)H(f) \exp[j2\pi f T_p] df\right\}\right)^2}{4N_0 \int_{-\infty}^{\infty} |H(f)|^2 df} \leq \frac{1}{4N_0} \int_{-\infty}^{\infty} |B_{10}(f)|^2 df \tag{13.44}$$

η can be maximized by selecting (the equality condition in Schwarz's Inequality)

$$H(f) = CB_{10}^*(f) \exp[-j2\pi f T_p] \tag{13.45}$$

where C is a real constant. It should be noted that the constant now needs to be real so that

$$\Im\left\{\int_{-\infty}^{\infty} B_{10}(f)H(f) \exp[j2\pi f T_p] df\right\} = 0 \tag{13.46}$$

such that

$$\left(\Re\left\{\int_{-\infty}^{\infty} B_{10}(f)H(f) \exp[j2\pi f T_p] df\right\}\right)^2 = \left|\left\{\int_{-\infty}^{\infty} B_{10}(f)H(f) \times \exp[j2\pi f T_p] df\right\}\right|^2 \tag{13.47}$$

Schwarz's inequality has provided two powerful results: the optimum filter (Eq. (13.45)) and the maximum effective signal to noise ratio (Eq. (13.44)) for binary digital communications. From the perspective of BEP the constant C makes no difference so without loss of generality the remainder of the discussion will assume $C = 1$. (C cancels out of the SNR given in Eq. (13.41)).

13.4.2 The Matched Filter

Since the optimum filter has the form in the frequency domain given in Eq. (13.45) the impulse response is given as

$$h(t) = \mathcal{F}^{-1}\{B_{10}^*(f)\exp[-j2\pi f T_p]\} = b_{10}^*(T_p - t) = x_1^*(T_p - t) - x_0^*(T_p - t) \tag{13.48}$$

For MLBD the effective signal is $b_{10}(t) = x_1(t) - x_0(t)$ and the optimum filter impulse response is effectively a time reversed, time shifted, conjugate of this effective signal. This filter is known as the **matched filter** since it is matched to the effective signal and the form for this SNR maximizing filter was first identified by North [Nor43]. Design Task 3 is complete; given $x_0(t)$ and $x_1(t)$, the filter, $H(f)$ that minimizes the BEP has been found.

EXAMPLE 13.9

Consider the matched filter for the signals shown in Figure 13.1, the matched filter is shown in Figure 13.12. It should be noted that having the sample time at T_p makes the matched filter a causal filter. The two possible transmitted signals and the noiseless outputs from the matched filter are plotted in the Figure 13.13. The vertical line represents $t = T_p$.

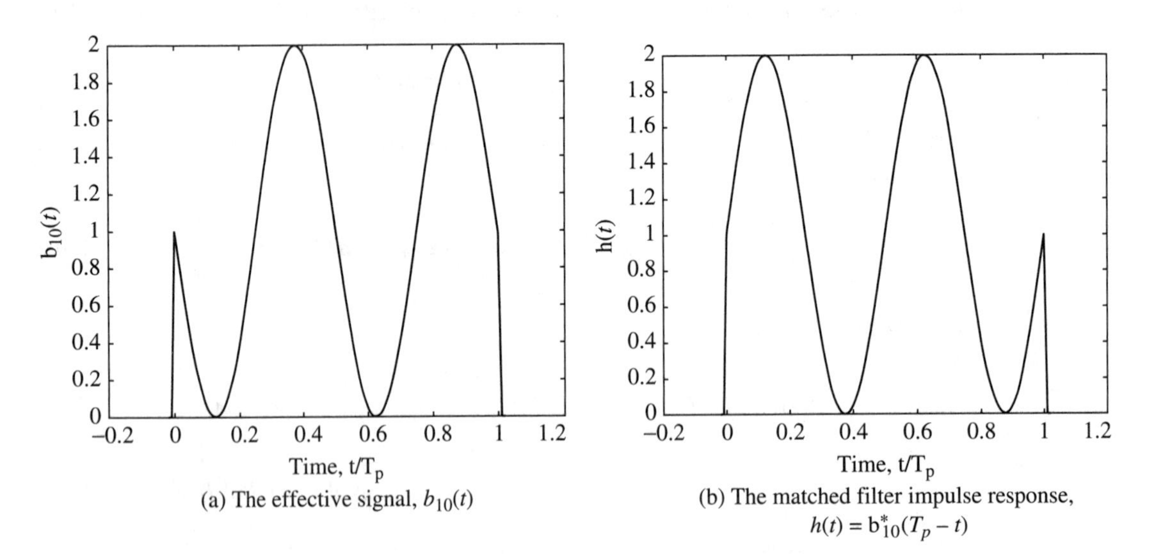

Figure 13.12 The effective signal, $b_{10}(t)$ and the matched filter impulse response, $b_{10}^*(T_p - t)$ for the two example waveforms considered in Figure 13.1.

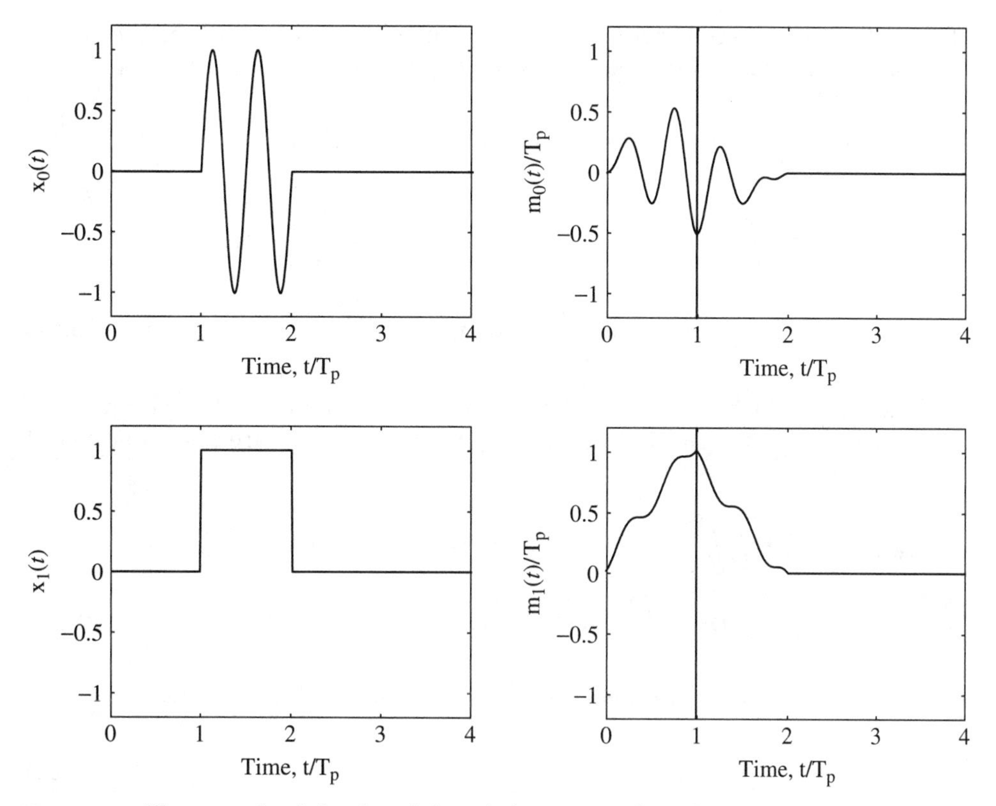

Figure 13.13 The transmitted signals and the noiseless output of matched filter for Example 13.9.

13.4.3 MLBD with the Matched Filter

More insight into digital communication receiver design can be achieved by looking in detail at the resulting demodulation structure when using a matched filter demodulator. First examine the matched filter output,

$$V_z(T_p) = \int_{-\infty}^{\infty} Y_z(\tau)h(T_p - \tau)d\tau = \int_{-\infty}^{\infty} Y_z(\tau)(x_1^*(\tau) - x_0^*(\tau))d\tau \tag{13.49}$$

Examining Eq. (13.49) shows that the optimum filter correlates the input signal plus noise, $Y_z(t)$, with the conjugate of the difference between the possible transmitted signals. Consequently, the optimum filter is often denoted and implemented as a correlation operation.

The output noise, $N_z(t)$, is the same regardless of which bit is transmitted so it will be characterized first. Recall the output noise is given as

$$N_z(T_p) = \int_{-\infty}^{\infty} W_z(\tau)h(T_p - \tau)d\tau = \int_{-\infty}^{\infty} W_z(\tau)(x_1^*(\tau) - x_0^*(\tau))d\tau \tag{13.50}$$

Since $W_z(t)$ is a complex AWGN, the output noise, $N_z(T_p)$ will be a complex Gaussian RV. Using standard theory the ouput variance of this RV is given as

$$
\begin{aligned}
2\sigma_{N_I}^2 &= N_0 \int_{-\infty}^{\infty} |h(\tau)|^2 d\tau = N_0 \int_{-\infty}^{\infty} |x_1^*(T_p - \tau) - x_0^*(T_p - \tau)|^2 d\tau \\
&= N_0 \left(\int_{-\infty}^{\infty} |x_1(\tau)|^2 d\tau + \int_{-\infty}^{\infty} |x_0(\tau)|^2 d\tau - 2\Re \left\{ \int_{-\infty}^{\infty} x_1(\tau) x_0^*(\tau) d\tau \right\} \right) \\
&= N_0 \left(E_1 + E_0 - 2\Re \left\{ \int_{-\infty}^{\infty} x_1(\tau) x_0^*(\tau) d\tau \right\} \right) \qquad (13.51)
\end{aligned}
$$

In a similiar fashion the conditional means of the matched filter outputs for the two possible transmitted signals, $m_1(T_p)$ and $m_0(T_p)$, are easily derived. If $I = 1$ then

$$
\begin{aligned}
m_1(T_p) &= \int_{-\infty}^{\infty} x_1(\tau) h(T_p - \tau) d\tau = \int_{-\infty}^{\infty} x_1(\tau)(x_1^*(\tau) - x_0^*(\tau)) d\tau \\
&= E_1 - \int_{-\infty}^{\infty} x_1(\tau) x_0^*(\tau) d\tau \qquad (13.52)
\end{aligned}
$$

The conditional mean of the decision statistic when $I = 1$ is

$$
m_{1,I} = E_1 - \Re \left\{ \int_{-\infty}^{\infty} x_1(\tau) x_0^*(\tau) d\tau \right\} \qquad (13.53)
$$

Likewise, if $I = 0$ then

$$
\begin{aligned}
m_0(T_p) &= \int_{-\infty}^{\infty} x_0(\tau) h(T_p - \tau) d\tau = \int_{-\infty}^{\infty} x_0(\tau)[x_1^*(\tau) - x_0^*(\tau)] d\tau \\
&= \int_{-\infty}^{\infty} x_0(\tau) x_1^*(\tau) d\tau - E_0 = \left(\int_{-\infty}^{\infty} x_1(\tau) x_0^*(\tau) d\tau \right)^* - E_0 \qquad (13.54)
\end{aligned}
$$

The conditional mean of the decision statistic when $I = 0$ is

$$
m_{0,I} = \Re \left\{ \int_{-\infty}^{\infty} x_1(\tau) x_0^*(\tau) d\tau \right\} - E_0 \qquad (13.55)
$$

The optimum decision threshold with matched filter processing is

$$
\gamma = \frac{m_{1,I} + m_{0,I}}{2} = \frac{E_1 - E_0}{2} \qquad (13.56)
$$

Consequently, when matched filter processing is utilized all important quantities in the receiver are a function of the energy of the two signals used to represent the bit values, E_1 and E_0, and $\int_{-\infty}^{\infty} x_1(\tau) x_0^*(\tau) d\tau$.

Since the quantity $\int_{-\infty}^{\infty} x_1(\tau)x_0^*(\tau)d\tau$ appears many times throughout discussions of digital communications systems a definition will be introduced to simplify the notation.

Definition 13.10 The signal correlation coefficient between two deterministic signals, $x_i(t)$ and $x_j(t)$, is

$$\rho_{ij} = \frac{\int_{-\infty}^{\infty} x_i(\tau)x_j^*(\tau)d\tau}{\sqrt{E_i E_j}}$$

Using this definition gives

$$2\sigma_{N_I}^2 = N_0(E_1 + E_0 - 2\sqrt{E_1 E_0}\Re\{\rho_{10}\}) \quad (13.57)$$

$$m_{1,I} = E_1 - \sqrt{E_1 E_0}\Re\{\rho_{10}\} \quad (13.58)$$

$$m_{0,I} = \sqrt{E_1 E_0}\Re\{\rho_{10}\} - E_0 \quad (13.59)$$

Using these simplified forms for the important parameters in the MLBD with matched filter processing allows us to gain some insight into the signal design problem.

EXAMPLE 13.10

Continuing with Example 13.9, Figure 13.14 shows a sample path of $V_0(t)$ (top left) and a sample path of $V_1(t)$ (bottom left) for the matched filter with $E_b/N_0 = 10$ dB. These sample paths clearly behave like a signal, $m_i(t)$, plus a noise. A histogram of 4000 samples of the filter output are also shown Figure 13.14. This histogram shows imperically that the Gaussian PDF is correct model for $f_{V_I|i}(v_I|I = i)$. Since the decision threshold is halfway between the two means of the Gaussian PDF this figure also shows that the matched filter produces a better bit error performance than the nonoptimal filter considered in Example 13.8. It should also be noted that $E_0 = 0.5T_p$, $E_1 = T_p$, and $\rho_{10} = 0$ and consequently $m_{1,I} = E_1$ and $m_{0,I} = -E_0$ and these results are reflected in the histograms of the matched filter output signals shown in Figure 13.14.

13.4.4 More Insights on the Matched Filter

Two questions often arise with students after they read the previous sections. The two questions are

1. Why is the $\Re[\bullet]$ operator needed in the demodulator?
2. Could you do better by looking at more than one sample?

Understanding the answer to these questions and the optimal demodulator is obtained by looking at the characteristics of the signals at the matched filter outputs. First let's break the matched filter into two filters as shown in Figure 13.15 and ignore the noise to gain some insight. The two filters are now

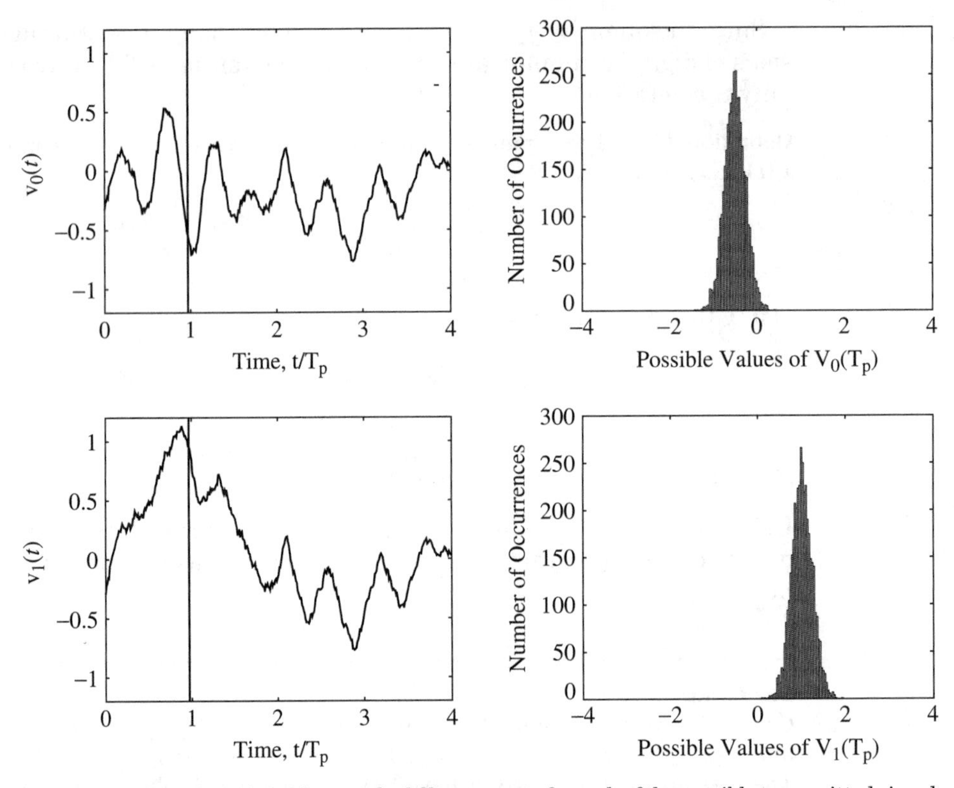

Figure 13.14 Sample paths of the matched filter outputs for each of the possible transmitted signals and histograms of the matched filter output samples.

matched to each of the two possible transmitted waveforms and the output signals when $x_j(t)$ is transmitted are

$$v_{0,j}(t) = \int_0^{T_p} x_j(\lambda)x_0^*(T_p - t + \lambda)d\lambda \tag{13.60}$$

$$v_{1,j}(t) = \int_0^{T_p} x_j(\lambda)x_1^*(T_p - t + \lambda)d\lambda \tag{13.61}$$

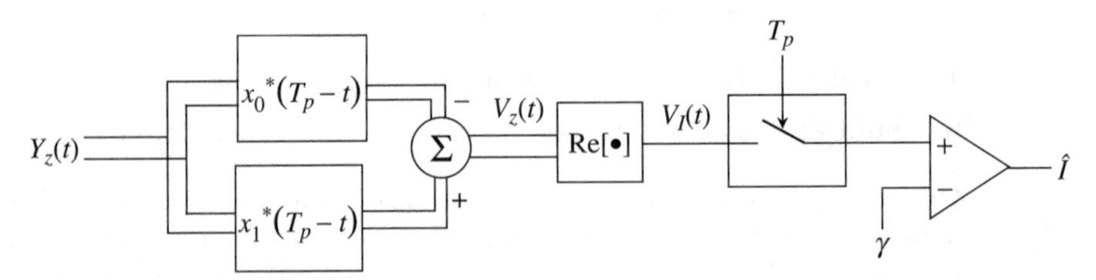

Figure 13.15 The two filter representation of the matched filter to the effective signal.

Note that at $t = T_p$

$$v_{j,j}(T_p) = \int_0^{T_p} |x_j(\lambda)|^2 d\lambda \tag{13.62}$$

so that the signal received at each time is derotated into a positive real signal before the integration is performed. So in the filter matched to the transmitted signal, it is clear the output sample will be $v_{jj}(T_p) = E_j + j0$, while in the filter not matched to the signal sent the output sample will be $v_{\bar{j}j}(T_p) = \rho_{j\bar{j}}\sqrt{E_j E_{\bar{j}}}$. Recall that $|\rho_{j\bar{j}}| \leq 1$ so $v_{jj}(T_p)$ can be viewed as the signal of interest out of the matched filter and $v_{\bar{j}j}(T_p)$ can be viewed as an interference. The signal of interest is a real valued signal since the matched filter perfectly derotates the signal to which it is matched. This perfect derotation characteristic of the matched filter is why only the real part of the output of the matched filter is sufficient for optimum demodulation.

It is also clear in looking at Eq. (13.60) that the matched filter uses the entire received signal to produce the decision statistic. Since the received signal is correlated with a version of the transmitted signal, the matched filter output is essentially a combination of the received signal over the entire support of the transmitted signal. The optimal filter is taking a large number of samples of the input signal from the channel over the support of the transmitted signal and combining them in an optimal manner. This view of the matched filter processing makes it apparent why only one sample out of the matched filter is needed. Consequently, using the matched filter demodulator makes the real part of the matched filter output sampled at $t = T_p$ a sufficient statistic for demodulation. The simplicity and optimality of the demodulator are striking.

13.5 Signal Design

Design Task 4, the signal design task, can now be addressed. This signal design task is stated as: given the optimum demodulator, design $x_0(t)$ and $x_1(t)$ to optimize BEP. The results of the previous section indicate that BEP of the binary detector is a function of E_1, E_0, and ρ_{10}. Fidelity of digital communication systems is most often quantified by the average energy per transmitted bit given as

$$E_b = \frac{E_1 + E_0}{2}$$

Throughout the remainder of the text E_b will be used in characterizing fidelity of message reconstruction.

To bring further insight to the discussion it is useful to introduce a definition.

Definition 13.11 The Euclidean square distance between two continuous time signals, $x_i(t)$ and $x_j(t)$ is

$$\Delta_E(i,j) = \int_{-\infty}^{\infty} |x_i(t) - x_j(t)|^2 dt$$

It should be noted that

$$\Delta_E(i,j) = \int_{-\infty}^{\infty} |x_i(t)|^2 dt + \int_{-\infty}^{\infty} |x_j(t)|^2 dt - \int_{-\infty}^{\infty} x_i(t)x_j^*(t)dt$$
$$- \int_{-\infty}^{\infty} x_j(t)x_i^*(t)dt \tag{13.63}$$
$$= E_i + E_j - 2\sqrt{E_i E_j}\Re\{\rho_{ij}\} \tag{13.64}$$

The BEP optimization and signal design task can now be formulated using the Euclidean square distance. The effective SNR with matched filter processing can be expressed with the Rayleigh energy theorem as

$$\eta = \frac{1}{4N_0}\int_{-\infty}^{\infty} |B_{10}(f)|^2 df = \frac{1}{4N_0}\int_{-\infty}^{\infty} |b_{10}(t)|^2 dt \tag{13.65}$$

Using the concept of the Euclidean square distance this effective SNR reduces to

$$\eta = \frac{1}{4N_0}\int_{-\infty}^{\infty} |x_1^*(T_p - t) - x_0^*(T_p - t)|^2 dt = \frac{\Delta_E(1,0)}{4N_0} \tag{13.66}$$

Consequently, the BEP for matched filtering processing is given as

$$P_B(E) = \frac{1}{2}\text{erfc}\left(\sqrt{\frac{\Delta_E(1,0)}{4N_0}}\right) \tag{13.67}$$

The preceding results demonstrate that the only parameter necessary to consider to optimize a binary signal set is the square Euclidean distance between the two possible transmitted signals. Equivalently, the square Euclidean distance is determined by the signal energies and the signal correlation coefficient.

Some examples will help illustrate the ideas of signal design. The "uneducated" masses typically think that digital communications occurs with the following waveforms

$$x_{1A}(t) = \begin{cases} \sqrt{\dfrac{2E_b}{T_p}} & 0 \le t \le T_p \\ 0 & \text{elsewhere} \end{cases} \tag{13.68}$$

$$x_{0A}(t) = 0 \tag{13.69}$$

This particular signal set has $E_1 = 2E_b$, $E_0 = 0$, and $\rho_{10} = 0$. The resulting squared Euclidean distance is $\Delta_E(1,0) = 2E_b$ and the BEP is given as

$$P_B(E) = \frac{1}{2}\text{erfc}\left(\sqrt{\frac{E_b}{2N_0}}\right) \tag{13.70}$$

BEP can be improved by increasing the energy in $x_0(t)$ and simultaneously decreasing the ρ_{10}. For example the binary set of waveforms

$$x_{0B}(t) = \begin{cases} \sqrt{\dfrac{E_b}{T_p}} & 0 \leq t \leq T_p \\ 0 & \text{elsewhere} \end{cases} \tag{13.71}$$

$$x_{1B}(t) = \begin{cases} -\sqrt{\dfrac{E_b}{T_p}} & 0 \leq t \leq T_p \\ 0 & \text{elsewhere} \end{cases} \tag{13.72}$$

has $E_1 = E_b$, $E_0 = E_b$, and $\rho_{10} = -1$ and $\Delta_E(1,0) = 4E_b$. For the same average energy, signal set B provides the same BEP as signal set A in an environment where the noise spectral density is twice as high. As communication engineers we are interested in the performance of the systems we design for varying noise levels hence signal set B is denoted as having a 3-dB performance gain over signal set A. Figure 13.16 shows the resulting square Euclidean distance versus signal energy for a fixed ρ_{10} and with $E_b = 1$. It is clear the signal set considered in Eq. (13.71) and Eq. (13.72), $E_1 = E_0 = E_b$ and $\rho_{10} = -1$ and $\Delta_E(1,0) = 4E_b$, provides the best BEP performance. The situation where

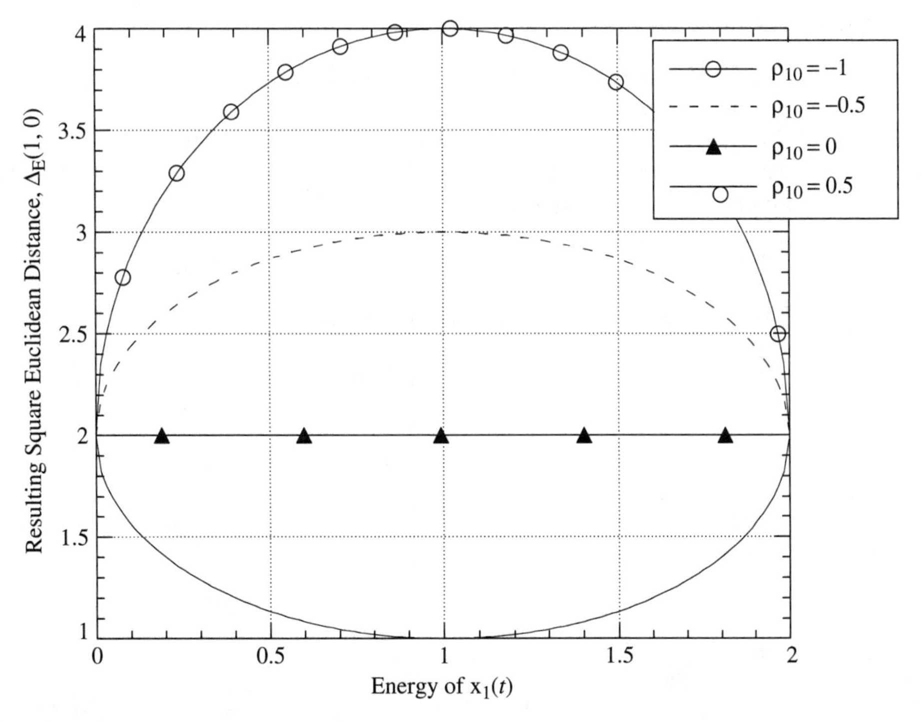

Figure 13.16 Square Euclidean distance versus E_1. $E_b = 1$.

$x_0(t) = -x_1(t)$ is common in digital communications and it often referred to as **antipodal** signaling. Also, in general, higher energy and lower signal correlation will produce better BEP.

EXAMPLE 13.11

Returning to the signal set introduced in Example 13.1 it is apparent that

$$E_0 = \frac{T_p}{2} \qquad E_1 = T_p \qquad \rho_{10} = 0 \tag{13.73}$$

It is apparent that $\Delta_E(1,0) = 1.5T_p$ and the BEP for the optimum demodulator for this signal set is

$$P_B(E) = \frac{1}{2}\operatorname{erfc}\left(\sqrt{\frac{1.5T_p}{4N_0}}\right) \tag{13.74}$$

Design Task 4 is complete; we have found the best signal designs for matched filter MLBD. The interesting characteristics of the signal design problem is that BEP is only a function of the squared Euclidean distance and that the resulting optimum signal design is not unique. There are an infinite number of signal sets that achieve the same BEP. Further optimization will have to be accomplished considering other parameters besides BEP.

13.6 Spectral Characteristics

The final design task is to understand the relationship between the digital communication waveforms and the resulting bandwidth. For single bit transmission the average energy spectral density is straightforward to compute. Given the framework used here and the definition of the average energy spectrum density given in Eq. (12.8) it is apparent that

$$D_{X_z}(f) = \pi_0 G_{x_0}(f) + \pi_1 G_{x_1}(f) \tag{13.75}$$

For example, for the signal set considered in the last section given as

$$x_0(t) = \begin{cases} \sqrt{\dfrac{E_b}{T_p}} & 0 \le t \le T_p \\ 0 & \text{elsewhere} \end{cases} \tag{13.76}$$

$$x_1(t) = \begin{cases} -\sqrt{\dfrac{E_b}{T_p}} & 0 \le t \le T_p \\ 0 & \text{elsewhere} \end{cases} \tag{13.77}$$

has

$$G_{x_1}(f) = E_b T_p \left(\frac{\sin(\pi f T_p)}{\pi f T_p}\right)^2 \qquad G_{x_0}(f) = E_b T_p \left(\frac{\sin(\pi f T_p)}{\pi f T_p}\right)^2 \tag{13.78}$$

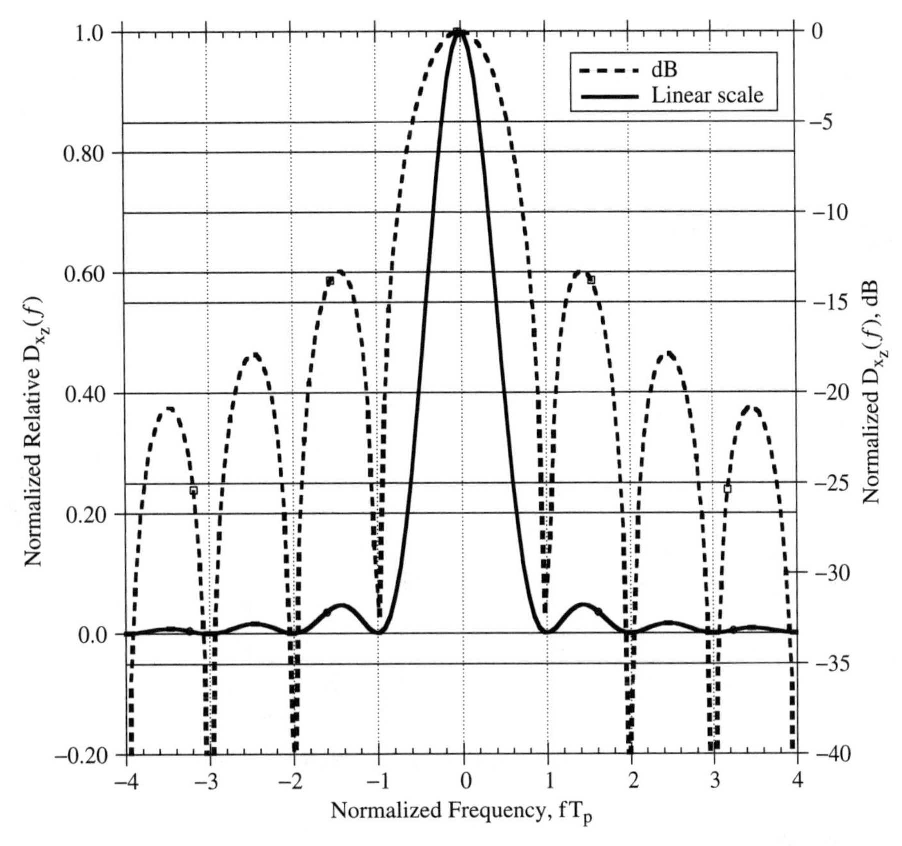

Figure 13.17 The average energy spectrum. $\pi_0 = \pi_1 = 0.5$.

The resulting energy spectrum for this case is plotted in Figure 13.17 for both linear scale and dB scale for $E_bT_p = 1$. It should be noted that the bandwidth in any definition is inversely proportional to the length of the pulse. Engineering intuition says that faster transmission rates (smaller T_p) requires more bandwidth. The average energy spectrum gives a way to quantify engineering intuition.

EXAMPLE 13.12

Returning to Example 13.1 the individual energy spectrum are given as

$$G_{x_0}(f) = \frac{T_p^2}{4}|\text{sinc}(fT_p - 2)\exp[-j\pi fT_p] - \text{sinc}(fT_p + 2)\exp[-j\pi f T]|^2 \quad (13.79)$$

$$G_{x_1}(f) = T_p^2\text{sinc}(fT_p)^2$$

The average energy spectrum for $\pi_0 = 0.5$ is plotted in Figure 13.18. The 3dB bandwidth of this signal set is $B_3 = 0.917/T_p$.

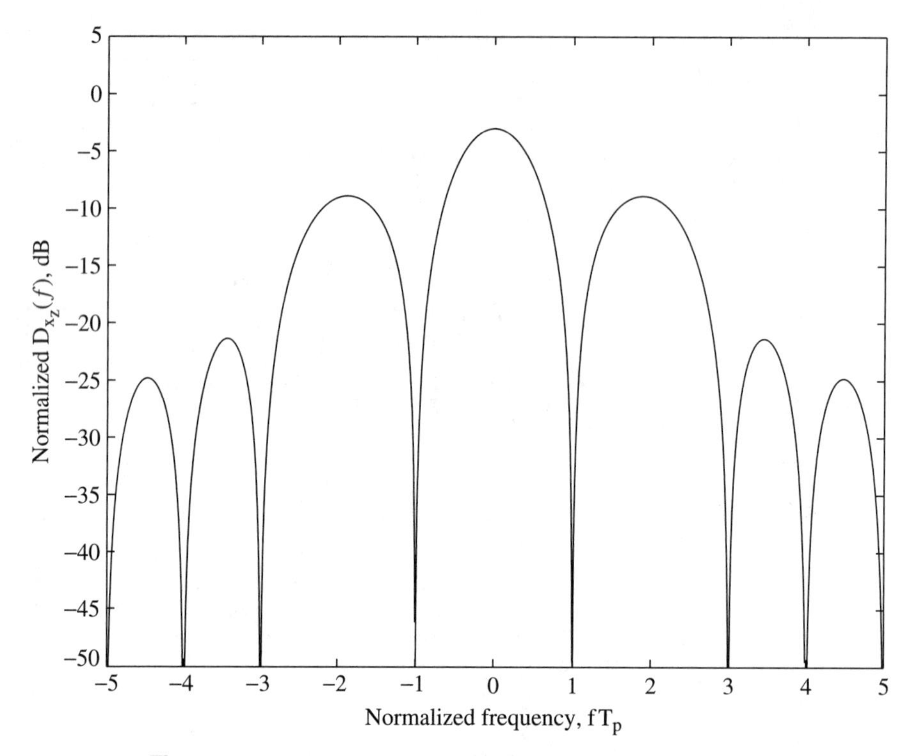

Figure 13.18 The average energy spectrum per bit for the signal set in Example 13.1.

13.7 Examples

The chapter is concluded by considering two obvious and classical examples of carrier modulated digital communication: binary frequency shift keying (BFSK) and binary phase shift keying (BPSK).

13.7.1 Frequency Shift Keying

BFSK modulation sends the bit of information by transmitting a carrier pulse of one of two frequencies. This is an obvious simple signalling scheme and one used in many early modems. The signal set is given as

$$x_0(t) = \begin{cases} \sqrt{\frac{E_b}{T_p}} \exp[j2\pi f_d t] & 0 \leq t \leq T_p \\ 0 & \text{elsewhere} \end{cases} \tag{13.80}$$

$$x_1(t) = \begin{cases} \sqrt{\frac{E_b}{T_p}} \exp[-j2\pi f_d t] & 0 \leq t \leq T_p \\ 0 & \text{elsewhere} \end{cases} \tag{13.81}$$

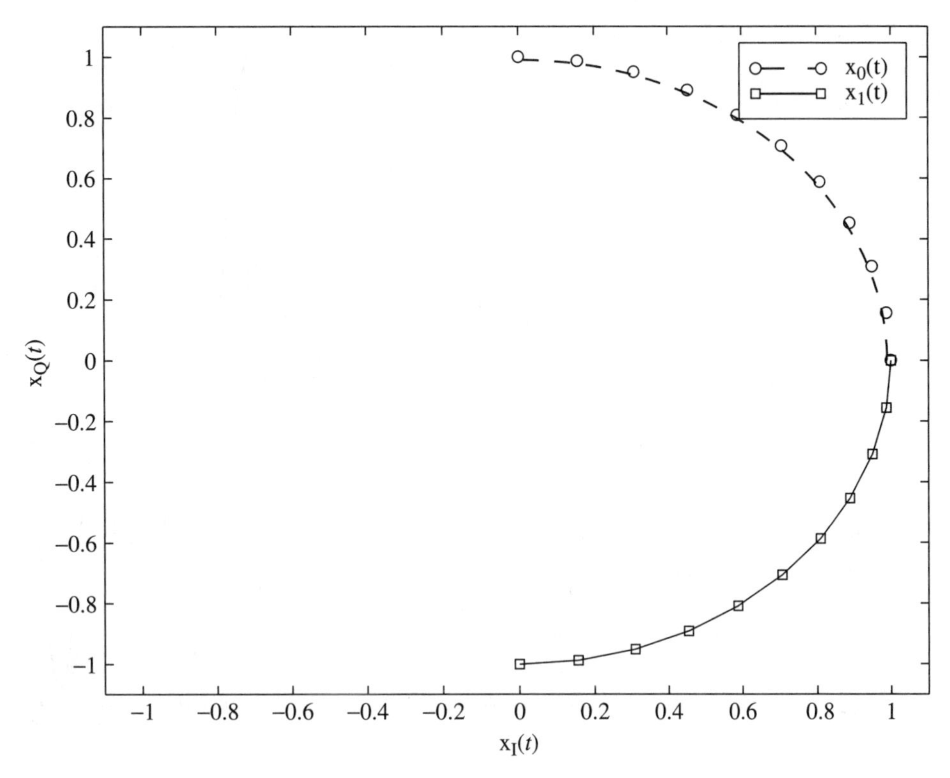

Figure 13.19 The vector diagram for the two BFSK waveforms for $0 \leq t \leq T_p$. $f_d T_p = 0.25$ and $E_b = T_p$.

where f_d is known as the frequency deviation. The frequency difference between the two carrier pulses is $2f_d$. It is apparent that each waveform in a BFSK signal set has equal energy and in this example the energy has been set to $E_1 = E_0 = E_b$. The vector diagram for the two waveforms for BFSK with $f_d = 0.25/T_p$ and $E_b = T_p$ is shown in Figure 13.19. The plots of the bandpass waveforms for $f_d = 1/T_p$ and $f_c = 3/T_p$ and $E_b = T_p$ are shown in Figure 13.20. This subsection will consider the optimum demodulator and the design of optimum signal sets for BFSK.

BFSK is interesting to investigate as the resulting characteristics are counter to many engineer's intuition. The matched filter impulse response is given as

$$\begin{aligned} h(t) &= x_1^*(T_p - t) - x_0^*(T_p - t) \\ &= j2\sqrt{\frac{E_b}{T_p}} \sin(2\pi f_d(T_p - t)) \qquad 0 \leq t \leq T_p \\ &= 0 \qquad \text{elsewhere} \end{aligned} \tag{13.82}$$

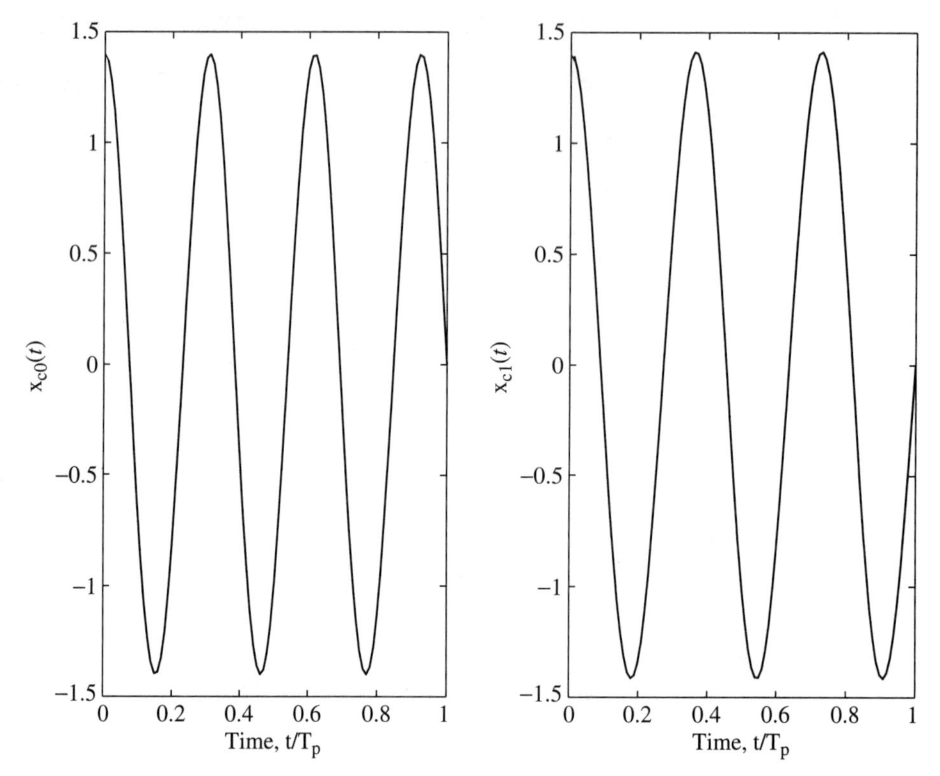

Figure 13.20 The bandpass BFSK waveforms. $f_d T_p = 1$, $f_c T_p = 3$, and $E_b = T_p$.

Note since $\Re\{x_1(t)\} = \Re\{x_0(t)\}$ the optimum demodulator only uses the imaginary part of the received signal to differentiate between the two bits, i.e.,

$$V_I(T_p) = -2\sqrt{\frac{E_b}{T_p}} \int_0^{T_p} Y_Q(t)\sin(2\pi f_d t)dt \tag{13.83}$$

For $\pi_0 = \pi_1 = 0.5$ (MLBD), the optimum threshold is given as

$$\gamma = \frac{E_1 - E_0}{2} = 0 \tag{13.84}$$

Consequently, the threshold test is very simple for the MLBD of a BFSK modulation, i.e.,

$$\int_0^{T_p} Y_Q(t)\sin(2\pi f_d t)dt \underset{\hat{I}=1}{\overset{\hat{I}=0}{\gtrless}} 0 \tag{13.85}$$

The important parameter for determining the BEP of the optimum demodulator is the square Euclidean distance. The only parameter that is not specfied in the BFSK model is the frequency difference between the two tones in the modulation. The parameter f_d determines this difference and also will determine the Euclidean distance. In general,

$$\Delta_E(1,0) = E_1 + E_0 - 2\Re\{\sqrt{E_1 E_0}\rho_{10}\}$$

For BFSK $E_0 = E_1 = E_b$ and

$$\rho_{10} = \frac{\int_{-\infty}^{\infty} x_1(t)x_0^*(t)dt}{E_b} = \frac{1}{T_p}\int_0^{T_p} \exp[-j4\pi f_d t]dt$$

$$= \frac{\sin(4\pi f_d T_p)}{4\pi f_d T_p} - j\frac{\cos(4\pi f_d T_p) - 1}{4\pi f_d T_p} \tag{13.86}$$

Consequently, the Euclidean distance is given as

$$\Delta_E(1,0) = 2E_b\left(1 - \frac{\sin(4\pi f_d T_p)}{4\pi f_d T_p}\right) \tag{13.87}$$

and the BEP performance is

$$P_B(E) = \frac{1}{2}\text{erfc}\left(\sqrt{\frac{\Delta_E(1,0)}{4N_0}}\right) = \frac{1}{2}\text{erfc}\left(\sqrt{\frac{E_b}{2N_0}\left(1 - \frac{\sin(4\pi f_d T_p)}{4\pi f_d T_p}\right)}\right) \tag{13.88}$$

Figure 13.21 has a plot of the resulting correlation and the Euclidean distance as a function of f_d.

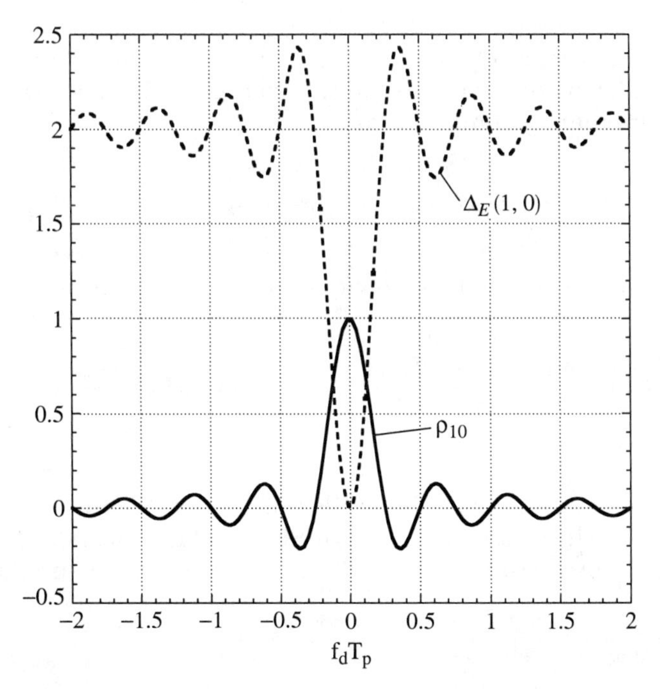

Figure 13.21 Correlation coefficient and Euclidean squared distance for BFSK.

The frequency deviation, f_d, significantly affects the BEP. The best BEP is obtained for $f_d \approx \frac{3}{8T_p}$ which produces $\Re\{\rho_{10}\} \approx -0.21$. In other words, the performance of coherent BFSK is optimized when the two frequencies used in the modulation are different by approximately $0.75/T_p$ Hz. This performance with $\Delta_E(1,0) = 2.42E_b$ is 2.2 dB from the optimal modulation, i.e., $\Delta_E(1,0) = 4E_b$. Hence BFSK is not a modulation of choice when the highest fidelity of message reconstruction is the driving constraint. Often engineering intuition leads one to believe that BEP is optimized when the frequency difference is large. At large frequency offsets the signal set converges to being orthogonal ($\Re\{\rho_{10}\} = 0$) and this produces 0.8 dB degradation in the BEP compared to the optimal BFSK. Orthogonal BFSK is often used for a variety of practical reasons. Coherent demodulation for $\rho_{10} = 0$ produces a BEP of

$$P_B(E) = \frac{1}{2}\text{erfc}\left(\sqrt{\frac{E_b}{2N_0}}\right) \tag{13.89}$$

The BEP is plotted versus E_b/N_0 in Figure 13.22 for the case of minimum BEP unconstrained signaling (denoted BPSK[4]), minimum BEP BFSK and orthogonal BFSK. The minimum f_d that still achieves an orthogonal signal set is $f_d = 0.25/T_p$. This form of BFSK modulation has some practical advantages when transmitting more than one bit of information and hence has been dubbed minimum shift keying (MSK) modulation. This modulation will be examined in more detail in the sequel.

The spectral efficiency of BFSK can be evaluated by looking at the average energy spectral density per bit. Recall the average energy spectral density per bit is given for binary modulations as

$$D_{X_z}(f) = \pi_0 G_{x_0}(f) + \pi_1 G_{x_1}(f) \tag{13.90}$$

The energy spectrum of the two individual waveforms is given as

$$G_{x_0}(f) = E_b T_p\left(\frac{\sin(\pi(f - f_d)T_p)}{\pi(f - f_d)T_p}\right)^2 \quad G_{x_1}(f) = E_b T_p\left(\frac{\sin(\pi(f + f_d)T_p)}{\pi(f + f_d)T_p}\right)^2 \tag{13.91}$$

The resulting energy spectrum for BFSK is plotted in Figure 13.23 for various f_d for $E_b T_p = 1$. The bandwidth of the BFSK signal is a function of both f_d and T_p. A smaller T_p produces a wider bandwidth while a larger f_d produces a wider

[4]Why, will be apparent next section.

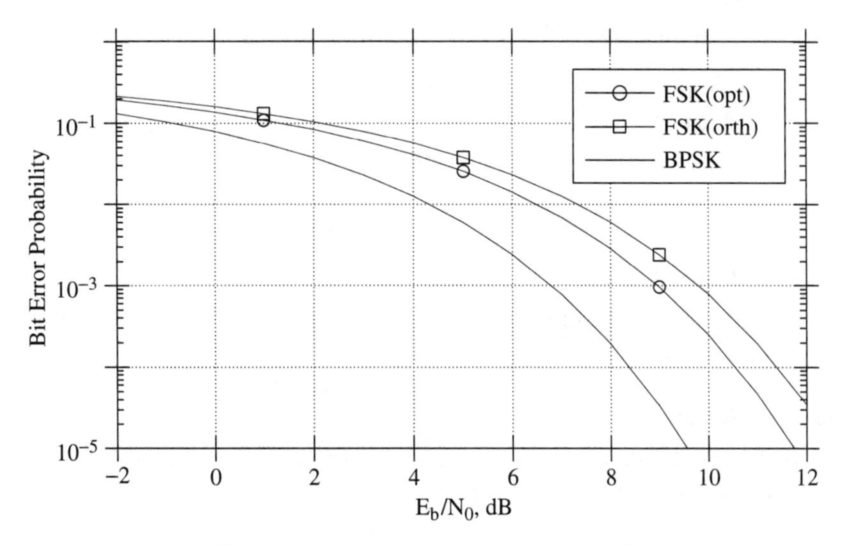

Figure 13.22 Probablity of error for some example binary modulations.

bandwidth. It should be noted that with this form of BFSK the transmission rate is $W_b = 1/T_p$. In examining Figure 13.23 the bandwidth of BFSK for the optimal BEP is about $B_T = 1.5/T_p$ (i.e., 3 dB bandwidth) so that the spectral efficiency of the optimal BEP form of BFSK is about $\eta_B = 0.67$ bit/s/Hz. MSK has $B_T = 1.0/T_p$ (i.e., 3 dB bandwidth) and $\eta_B = 1.0$ bit/s/Hz.

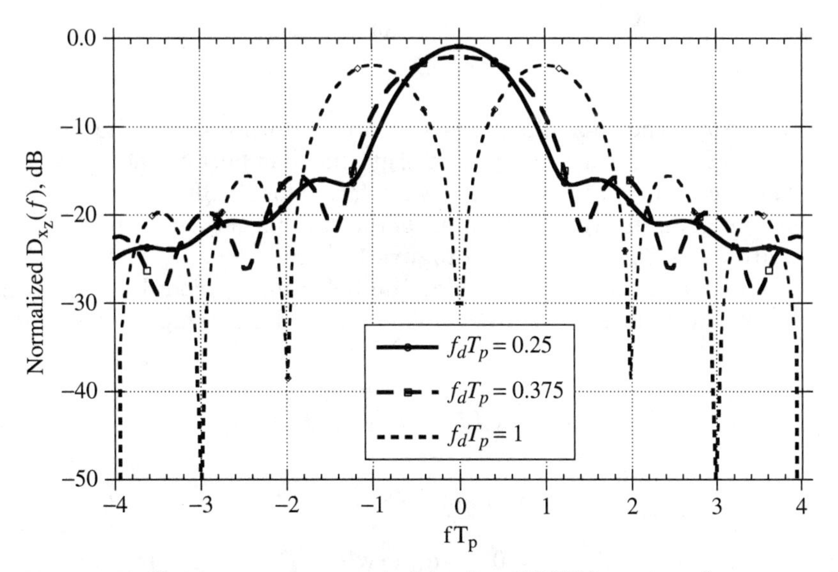

Figure 13.23 The average energy spectral density of BFSK for various f_d. $\pi_0 = \pi_1 = 0.5$.

The advantages of BFSK are summarized as

- Very simple to generate. Simply gate one of two oscillators on depending on the bit to be sent.
- Simple demodulation structure.

The disadvantages of BFSK are summarized as

- BEP is not optimum.
- Best BEP performance requires a spectral efficiency of less than 1 bit/s/Hz.

13.7.2 Phase Shift Keying

BPSK modulation sends the bit of information by transmitting a carrier pulse of one of two phases. The signal set is given as

$$\begin{aligned} x_0(t) &= \sqrt{\frac{E_b}{T_p}} \qquad 0 \leq t \leq T_p \\ &= 0 \qquad \text{otherwise} \end{aligned} \tag{13.92}$$

$$\begin{aligned} x_1(t) &= \sqrt{\frac{E_b}{T_p}} \exp[j\theta] \qquad 0 \leq t \leq T_p \\ &= 0 \qquad \text{otherwise} \end{aligned} \tag{13.93}$$

where θ is known as the phase deviation. It is apparent that each waveform in a BPSK signal set has equal energy that has here been set to $E_1 = E_0 = E_b$. The vector diagram for the two waveforms for BPSK with $\theta = \pi$ and $E_b = T_p$ is shown in Figure 13.24. The plots of the bandpass waveforms for $\theta = \pi$, $E_b = T_p$, and $f_c = 3/T_p$ are shown in Figure 13.25.

This subsection will consider the optimum demodulator and the design of optimum signal sets for BPSK. The matched filter impulse response is given as

$$\begin{aligned} h(t) &= x_1^*(T_p - t) - x_0^*(T_p - t) \\ &= \sqrt{\frac{E_b}{T_p}}(\exp(-j\theta) - 1) \qquad 0 \leq t \leq T_p \\ &= 0 \qquad \text{otherwise} \end{aligned} \tag{13.94}$$

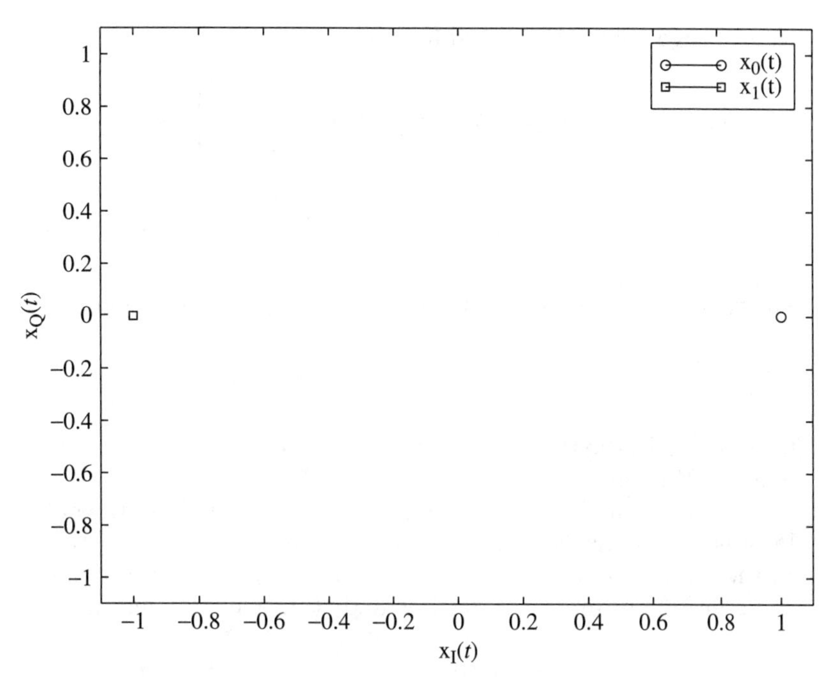

Figure 13.24 The vector diagram for the two BPSK waveforms for $0 \leq t \leq T_p$. $\theta = \pi$ and $E_b = T_p$.

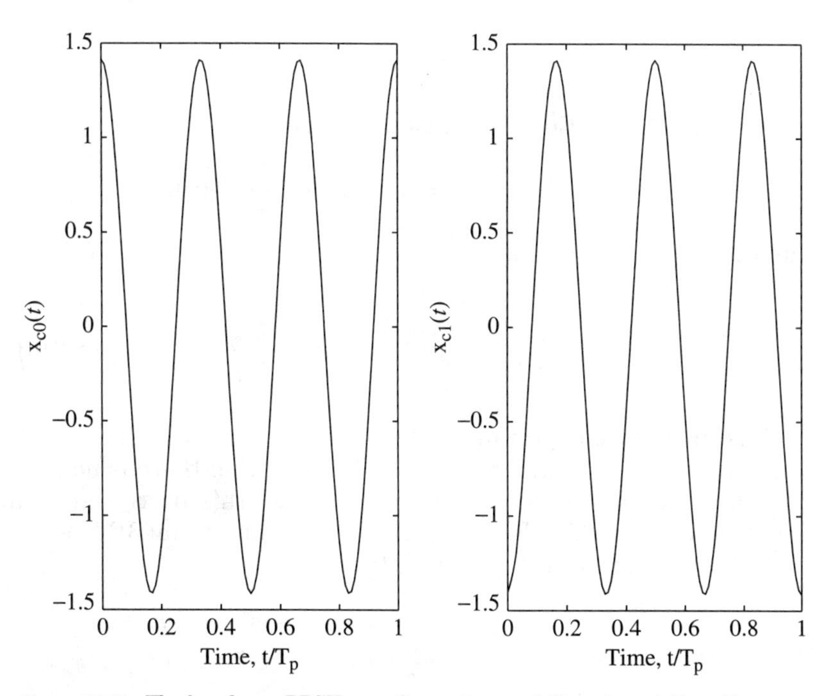

Figure 13.25 The bandpass BPSK waveforms. $\theta = \pi$, $f_c T_p = 3$, and $E_b = T_p$.

The voltage for the threshold test is

$$V_I(T_p) = \Re\left\{(\exp(-j\theta) - 1)\sqrt{\frac{E_b}{T_p}}\int_0^{T_p} Y_z(t)dt\right\}$$

$$= (\cos(\theta) - 1)\sqrt{\frac{E_b}{T_p}}\int_0^{T_p} Y_I(t)dt + \sin(\theta)\sqrt{\frac{E_b}{T_p}}\int_0^{T_p} Y_Q(t)dt \quad (13.95)$$

For $\pi_0 = \pi_1 = 0.5$ (MLBD), the optimum threshold is given as

$$\gamma = \frac{E_1 - E_0}{2} = 0 \quad (13.96)$$

Conseqeuntly, the threshold test is very simple (compare to zero) for the MLBD of a BPSK modulation.

The important parameter for determining the BEP of the optimum demodulator is the square Euclidean distance. The only parameter that is not specfied in the BPSK model is the phase difference between the two signals. The parameter θ determines the Euclidean distance. Recall again,

$$\Delta_E(1, 0) = E_1 + E_0 - 2\Re\{\sqrt{E_1 E_0}\rho_{10}\}$$

For BPSK $E_0 = E_1 = E_b$ and

$$\rho_{10} = \exp(j\theta) \quad (13.97)$$

Consequently, the Euclidean distance is given as

$$\Delta_E(1, 0) = 2E_b(1 - \cos(\theta))$$

and the BEP performance is

$$P_B(E) = \frac{1}{2}\mathrm{erfc}\left(\sqrt{\frac{\Delta_E(1, 0)}{4N_0}}\right) = \frac{1}{2}\mathrm{erfc}\left(\sqrt{\frac{E_b}{2N_0}(1 - \cos(\theta))}\right) \quad (13.98)$$

It is obvious that optimum BEP is obtained for $\theta = \pi$. The optimum signal set has $x_1(t) = -x_0(t)$. Frankly, in BPSK modulation there is no reason to choose anything other than $\theta = \pi$ so in the sequel this particular modulation will be denoted as BPSK. This signal set gives optimum BEP performance. The optimum threshold test simplifies to

$$\int_0^{T_p} Y_I(t)dt \underset{\hat{I}=1}{\overset{\hat{I}=0}{\gtrless}} 0 \quad (13.99)$$

The BEP is

$$P_B(E) = \frac{1}{2}\text{erfc}\left(\sqrt{\frac{E_b}{N_0}}\right) \tag{13.100}$$

The bit error probability is plotted versus E_b/N_0 in Figure 13.22. The spectral efficiency of BPSK can be evaluated by looking at the average energy spectral density and this is plotted in Figure 13.17. It should be noted that with this form of BPSK the transmission rate is $W_b = 1/T_p$. In examining Figure 13.17 the bandwidth of BPSK is about $B_T = 1/T_p$ (i.e., 3 dB bandwidth) so that the spectral efficiency of BPSK is about $\eta_B = 1$ bit/s/Hz.

The advantages of BPSK are summarized as

- Very simple to generate
- Simple demodulation structure
- Optimum BEP performance

The one disadvantage of the BPSK signal set discussed in this section is that the spectral characteristics might not be all that is desired. Methods to improve the spectral characteristics will be discussed in the homework and in Chapter 16.

13.7.3 Discussion

This section introduced two example modulations to transmit 1 bit of information: BFSK and BPSK. These two modulations are the most common used binary modulations in engineering practice. BPSK has the advantages in BEP performance and spectral efficiency. BFSK has some advantages in complexity that become more striking when synchronization issues are considered. As a final point it is worth comparing the performance of these two modulations with the upperbounds provided by information theory (see Section 12.3). For the discussion in this text we will denote reliable communication as being an error rate of 10^{-5}. BPSK requires approximately $E_b/N_0 = 9.5$ dB to achieve reliable communication while the optimum error probability form of BFSK requires approximately $E_b/N_0 = 11.7$ dB. MSK (the orthogonal form of FSK) achieves reliable communications at $E_b/N_0 = 12.5$ dB. These three operating points and the upperbound on the possible performance are plotted in Figure 13.26. It is clear from this figure that BPSK and BFSK result in performance far from the upperbound but that should not concern the reader as there is still a lot of communication theory to explore.

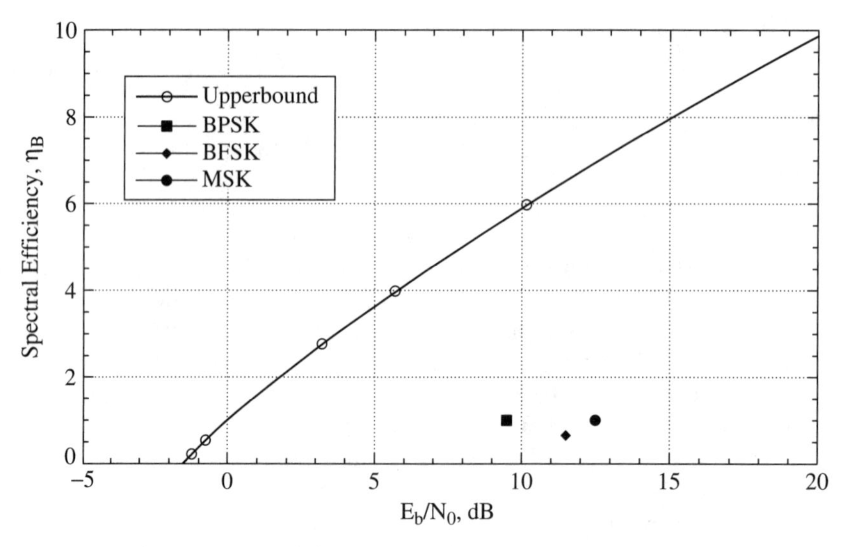

Figure 13.26 A comparison of the spectral efficiency of BPSK and BFSK with the upperbound.

13.8 Homework Problems

Problem 13.1. Consider a binary digital communication system with equally likely bits that is corrupted by an additive white Gaussian noise of two-sided spectral density of $N_0/2$. The two possible transmitted signals are of the form

$$\begin{aligned} x_0(t) &= \sqrt{\frac{E_b}{T_p}} \exp[j(2\pi f_0 t + \theta_0)] \qquad 0 \le t \le T_p \\ &= 0 \qquad \text{elsewhere} \\ x_1(t) &= \sqrt{\frac{E_b}{T_p}} \exp[j(2\pi f_1 t + \theta_1)] \qquad 0 \le t \le T_p \\ &= 0 \qquad \text{elsewhere} \end{aligned} \tag{13.101}$$

(a) For $f_0 = 10$ Hz, $\theta_0 = 0$ and $f_c = 50$ Hz plot $x_c(t)$ when $I = 0$.

(b) Give the average energy spectrum per bit $D_{x_z}(f)$ (as a function of f_0 and f_1).

(c) Detail out the optimum demodulator. Give impulse responses for any filters and specify any threshold tests.

(d) Compute ρ_{10}.

(e) If $f_0 = f_1 = 0$, choose values for θ_0 and θ_1 to optimize the BEP.

(f) If $\theta_0 = 0$ and $\theta_1 = 90^\circ$ choose values for f_0 and f_1 to optimize the BEP.

Problem 13.2. For MAPBD with $m_{0,I} = 1, m_{1,I} = 6$, and $\sigma_{N_I}^2 = 2$ (see Figure 13.8)

(a) Find the simplest form for MAPBD as a function of π_0. *Hint:* It is also a threshold test.

(b) Compute and plot the $P_B(E)$ as a function of π_0.

(c) Note that $\pi_0 = 0.5$ corresponds to MLBD. Plot the BEP performance if the MLBD was used for $\pi_0 \neq 0.5$ and compare to the results in (b).

(d) Let $\sigma_{N_I}^2$ be arbitrary and interpret the resulting MAPBD derived in a) when $\sigma_{N_I}^2$ gets large (large noise in observation) and when $\sigma_{N_I}^2$ gets small (little noise in observation). Does the detector characteristics follow intuition? Why?

Problem 13.3. Consider BFSK considered in Section 13.7.1, and find the exact frequency separation which optimizes BEP. Note, it is not exactly $f_d = \frac{3}{8T_p}$.

Problem 13.4. Consider BFSK considered in Section 13.7.1 on an AWGN channel, and specify for $\pi_0 \neq 0.5$.

(a) The optimum demodulator

(b) The optimum demodulator BEP

(c) $D_{x_z}(f)$

Consider the special case of $\pi_0 = 0.25$ and $E_b/N_0 = 10$ dB in detail and produce numeric results for parts (b) and (c).

Problem 13.5. In Problem 2.12 three pulse shapes were presented. Assume 1 bit is to be transmitted on an AWGN channel and $\pi_0 = 0.5$.

(a) Pick two out of the three pulse shapes to represent the possible value of the bit being transmitted in a single shot binary system such that the probability of bit error is minimized. Give the resulting correlation coefficient and probability of error when the one-sided noise spectral density is N_0.

(b) Antipodal signalling is the optimal binary communications design. Choose one of the three pulse shapes to use in an antipodal binary communication system that would make the most efficient use of bandwidth? Justify your answer.

Problem 13.6. Ethernet uses a biphase modulation to transmit data bits. Biphase modulation is a baseband linear modulation where each bit, I, is transmitted with $x_0(t) = u(t)$ or $x_1(t) = -u(t)$ where the pulse shape, $u(t)$ given in Figure13.27. Assume an AWGN channel and equally likely data bits.

(a) Compute the resulting BEP.

(b) If $1/T_p = 10$ MHz and the Ethernet signal at the transmitter has a peak value of 1 volt (assume a 1-Ω system) and the value of $N_0 = -160$ dBm/Hz,

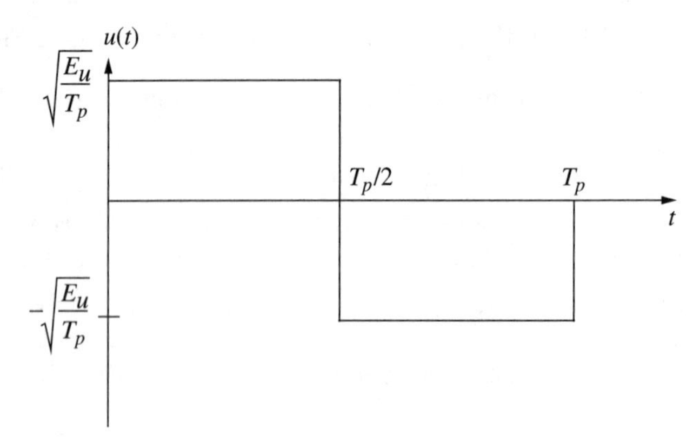

Figure 13.27 The pulse used in Ethernet.

find the amount of cable loss, L_c, that can be tolerated and still achieve a 10^{-9} error rate. Note these conditions imply that in Figure 13.27 $\sqrt{E_u/T_p} = L_c$.

(c) Compute the average energy spectrum per bit for the Ethernet modulation.

(d) Postulate why this pulse shape was chosen for transmission over cables.

Problem 13.7. Recall that VSB modulation has the form $x_z(t) = x_I(t) + j(h_v(t) * x_I(t))$. Assume that $x_I(t)$ is chosen from a BPSK signal set where the duration of the carrier pulse is given as T and that the support for the quadrature modulation $x_Q(t) = h_v(t) * x_I(t)$ is given as $T_p > T$. Find the simplest form of the optimum demodulator and compute the resulting BEP.

Problem 13.8. A bandpass binary digital communication system with equally likely bits was designed to have $x_z(t) = d_i u(t)$ where $d_0 = 1$ and $d_1 = -1$. The hardware implementation resulted in the actual transmitted modulation symbols being $d_0 = 1$ and $d_1 = \exp[j\frac{7\pi}{8}]$.

(a) If the demodulator was built to be optimum for the design waveforms, what would be the resulting probability of error.

(b) Give the optimum demodulation structure for the actual transmitted waveforms.

(c) Give the BEP for the optimum demodulator for the actual transmitted waveforms.

Problem 13.9. It is possible to construct a binary signaling set with $\rho_{10} \neq -1$ which has better BEP than a signal set with $\rho_{10} = -1$. Construct two signal sets (one with $\rho_{10} \neq -1$ and one with $\rho_{10} = -1$) with the same E_b such that the $\rho_{10} \neq -1$ signal set has a better BEP when the corrupting noise in an AWGN than the $\rho_{10} = -1$ signal set.

Problem 13.10. In this problem we consider a form of binary combined phase and frequency modulation. Assume equally likely data bits and an AWGN at the receiver that corrupts the transmission. The two possible baseband transmitted signals are

$$x_0(t) = \sqrt{\frac{E_b}{T_p}} \exp\left[\frac{j\pi t}{4T_p}\right] \qquad 0 \le t \le T_p$$

$$= 0 \qquad \text{elsewhere} \tag{13.102}$$

$$x_1(t) = \sqrt{\frac{E_b}{T_p}} \exp\left[-\frac{j\pi t}{4T_p} + j\theta\right] \qquad 0 \le t \le T_p$$

$$= 0 \qquad \text{elsewhere} \tag{13.103}$$

(a) For $f_c = 4/T_p$ plot the bandpass signal corresponding to $x_0(t)$.

(b) Find the demodulator that minimizes the probability of bit error for a single bit transmission given

$$Y_z(t) = X_z(t) + W_z(t)$$

where $W_z(t)$ is a complex additive white Gaussian noise of one-sided spectral density N_0.

(c) Give an exact expression for the BEP in terms of E_b, θ, and N_0 for this signal set and optimum demodulator. What is the value of θ that minimizes the probability of error?

(d) Compute the average energy spectrum per bit. What is the spectral efficiency using the 3 dB bandwidth?

Problem 13.11. You are called in by Nokey to be a consultant to the design of a binary digital communications system for use in an AWGN channel. With the assumpion of equally likely bits, one waveform chosen was

$$x_0(t) = \sqrt{2}\sin\left(\frac{2\pi t}{T_p}\right) \qquad 0 \le t \le T_p$$

$$= 0 \qquad \text{elsewhere}$$

The engineers at Nokey are hotly debating between the following two waveforms as the second waveform.

$$x_{1a}(t) = \sqrt{2}\cos\left(\frac{2\pi t}{T_p}\right) \qquad 0 \le t \le T_p$$

$$= 0 \qquad \text{elsewhere}$$

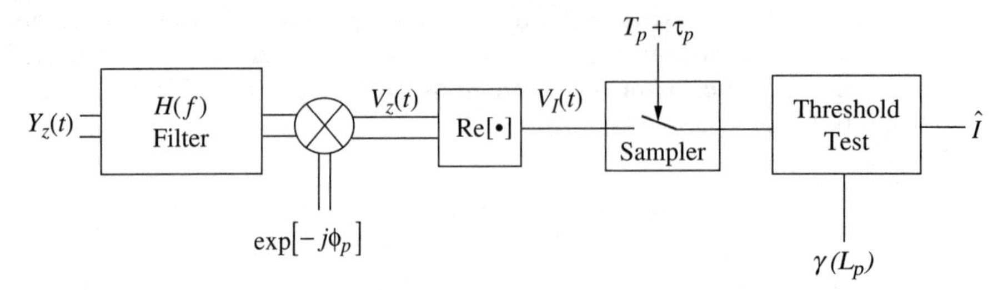

Figure 13.28 The optimal demodulator for the signal model in Eq. (13.2).

and

$$
\begin{aligned}
x_{1b}(t) &= -1 \qquad 0 \le t \le \frac{T_p}{2} \\
&= 0 \qquad \text{elsewhere}
\end{aligned}
$$

(a) Select the better of the two waveforms for bit error performance and give the reasons for your selection.

(b) Give the optimum demodulation structure for both pairs of waveforms. Detail out the matched filter response(s) and threshold tests for each possible pair of waveforms.

(c) Which set of waveforms would have better spectral characteristics? Explain why.

Problem 13.12. Show that optimum demodulation for the signal model given in Eq. (13.2) is given by the system shown in Figure 13.28. Compute the value of $\gamma(L_p)$.

Problem 13.13. In certain situations it is useful to hop to many different frequencies during transmission of a bit (e.g., to avoid intentional or unintentional jamming). Consider such a communication system where two hops are made during one bit transmission so that the transmitted signal has the form

$$
X_z(t) = \begin{cases} D_z \sqrt{\dfrac{E_b}{T_p}} \exp[j2\pi f_1 t] & 0 \le t \le \frac{T_p}{2} \\ D_z \sqrt{\dfrac{E_b}{T_p}} \exp[j2\pi f_2 t] & \frac{T_p}{2} < t \le T_p \\ 0 & \text{elsewhere} \end{cases} \tag{13.104}
$$

where $D_z = d_i$ for $I = i$. Assume each bit value is equally likely and that the received signal is of the form

$$
Y_z(t) = X_z(t) + W_z(t) \tag{13.105}
$$

where $W_z(t)$ is a complex additive white Gaussian noise of one-sided spectral density N_0.

(a) Detail out the optimum demodulator and simplify as much as possible.

(b) Give the BEP as a function of d_0, d_1, f_1, and f_2.

(c) Optimize the BEP with a selection of d_0, d_1, f_1, and f_2 under the constraint that $E_b = (E_0 + E_1)/2$.

(d) Is it possible to select frequencies f_1 and f_2, $f_1 \neq f_2$ such that $X_P(t)$ is continuous. If yes, give an example.

(e) Find and plot $D_{x_z}(f)$ for $f_1 = -2/T_p$ and $f_2 = 1.5/T_p$.

Problem 13.14. In this problem we consider a form of baseband binary modulation where the signals are restricted to be real and positive valued. Assume equally likely data bits and an additive white Gaussian noise at the receiver that corrupts the transmission. Two possible sets of baseband transmitted signals are signal set A

$$\begin{aligned} x_{0A}(t) &= C_A \qquad 0 \leq t \leq T_p \\ &= 0 \qquad \text{elsewhere} \end{aligned} \tag{13.106}$$

$$x_{1A}(t) = 0 \tag{13.107}$$

and signal set B

$$\begin{aligned} x_{0B}(t) &= C_B \cos\left(\frac{\pi t}{2T_p}\right) \qquad 0 \leq t \leq T_p \\ &= 0 \qquad \text{elsewhere} \end{aligned} \tag{13.108}$$

$$\begin{aligned} x_{1B}(t) &= C_B \sin\left(\frac{\pi t}{2T_p}\right) \qquad 0 \leq t \leq T_p \\ &= 0 \qquad \text{elsewhere} \end{aligned} \tag{13.109}$$

(a) What is the average energy per bit for these two signal sets (e.g., $E_b(A)$ and $E_b(B)$) as a function of C_A and C_B?

(b) Find the demodulator that minimizes the probability of bit error for each signal set for $E_b = 1$ in a single bit transmission given

$$Y_z(t) = X_z(t) + W_z(t)$$

where $W_z(t)$ is an additive white Gaussian noise of one-sided spectral density N_0.

(c) Which signaling scheme produces better BEP for $E_b(A) = E_b(B) = 1$. Quantify the increase in SNR which would be necessary in the worse performing scheme to give the two schemes equal BEP.

(d) How much loss in BEP performance is incurred for the best of these two schemes compared to the case of optimum signaling with $E_b = 1$ where the signals are allowed to take positive and negative values. Give an example set of two waveforms that achieves this optimum BEP.

(e) Find the average energy spectrum per bit for signal set A. Give a signal set defined over the same interval $[0, T_p]$ with the same BEP as signal set A but with better spectral characteristics. Be specific about exactly which spectral characteristics are better.

Problem 13.15. This problem is concerned with the transmission of 1 bit of information, I, by the waveforms $x_0(t)$ and $x_1(t)$ that have energy $\tilde{E}_{x_0}$ and $\tilde{E}_{x_1}$, respectively, during the time interval $[0, T_p]$. Assume demodulation is to be done in the presence of an additive white Gaussian noise, $W_z(t)$, with one-sided spectral density of N_0. A common impairment in radio circuits is a DC offset added to the signal (most often in either analog up or down conversion). Consequently, the received signal has the form

$$Y_z(t) = C + x_i(t) + W_z(t) \tag{13.110}$$

where C is a known complex constant.

(a) If $C = 0$ what are the optimum waveforms when the received average energy per bit is constrained to be E_b? Sketch the maximum likelihood bit demodulator. Give the probability of bit error as a function of E_b and N_0.

(b) For $C \neq 0$ find the received E_1 and the received E_0 and the average received energy per bit, E_b, as a function of C.

(c) Consider a maximum likelihood bit demodulator. For $C \neq 0$ what are the optimum waveforms when the received energy per bit is constrained to be E_b? Sketch the simplest maximum likelihood bit demodulator. Detail out the filters and the threshold test.

(d) Show a judicious choice of transmitted waveforms leads to a demodulator that is not a function of C.

(e) For $C \neq 0$ and the optimum waveforms from (c) give the probability of bit error as a function of E_b, C, and N_0.

(f) For $C \neq 0$ formulate the maximum a posteriori bit demodulator.

(g) For $C \neq 0$ specify a set of waveforms $x_0(t)$ and $x_1(t)$ such that the spectral efficiency can be argued to be near optimum.

Problem 13.16. A binary digital communication system uses the two waveforms in Figure 13.29 to communicate a bit of information. Assume these signals are corrupted by an additive white Gaussian noise with a two-sided spectral density of $N_0/2$ and that the bits are equally likely.

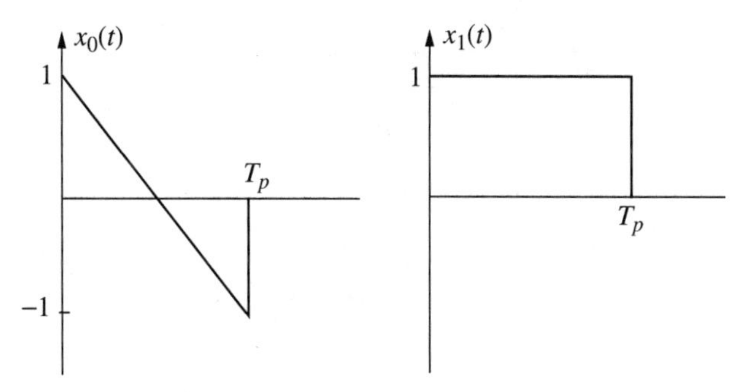

Figure 13.29 A binary signal set.

(a) Sketch the simplest form for optimum bit demodulator.

(b) Calculate E_0, E_1, and ρ_{10}.

(c) Calculate the Euclidean square distance between $x_0(t)$ and $x_1(t)$, $\Delta_E(1, 0)$.

(d) What is the resulting bit error probability.

(e) Keep one of either $x_0(t)$ or $x_1(t)$ and select a third signal $x_2(t)$ such that $|x_2(t)| \leq 1$ and $x_2(t)$ is of length T_p to be used to transmit 1 bit of information such that the resulting BEP for the optimum demodulation is as low as possible.

Problem 13.17. Digital communication waveforms can often be viewed as analog communication waveforms with specific message waveforms.

(a) To this end show that BFSK is equivalent to FM modulating one of two waveforms, i.e.,

$$x_i(t) = A_c \exp\left[j2\pi f_k \int_{-\infty}^{t} m_i(\lambda) d\lambda\right] \tag{13.111}$$

where

$$m_0(t) = \begin{cases} 1 & 0 \leq t \leq T_p \\ 0 & \text{elsewhere} \end{cases} \qquad m_1(t) = \begin{cases} -1 & 0 \leq t \leq T_p \\ 0 & \text{elsewhere} \end{cases} \tag{13.112}$$

Identify f_k and A_c.

(b) To this end show that BPSK is equivalent to DSB-AM modulating one of the two waveforms given in Eq. (13.112). Identify A_c.

Problem 13.18. In the wireless local network protocol denoted IEEE 802.11b the lowest rate modulation is a binary modulation using the two waveforms given in Figure 13.30. Assume $\pi_0 = 0.5$ and that $\tau_1 = 1/11\ \mu$s, $\tau_2 = 2/11\ \mu$s, and $\tau_3 = 3/11\ \mu$s.

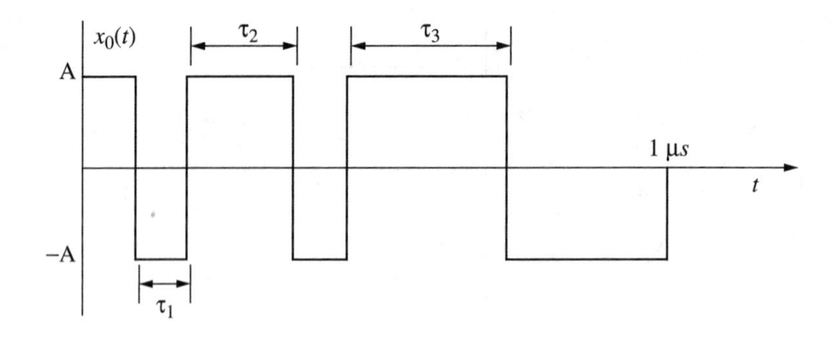

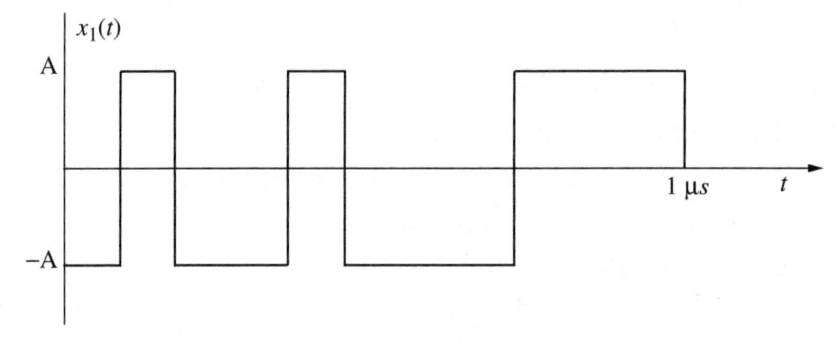

Figure 13.30 The two signals used in a binary modulation for IEEE 802.11b.

(a) Choose A such that the energy per bit is E_b.

(b) What is the bit rate?

(c) Keeping $x_0(t)$, can you choose a new waveform different than $x_1(t)$ that will have the same energy and will improve the BEP? If you can then show an example.

(d) Assume equally likely bits and a complex additive white Gaussian noise with a one-sided power spectral density N_0 corrupts the received signal. Find the simplest form of the optimum demodulator.

(e) Find $D_{x_z}(f)$. Using the 3 dB bandwidth what is the spectral efficiency of this modulation?

Problem 13.19. In Example 12.1 four example waveforms were introduced to transmit 2 bits of information. Pick two of these four waveforms to transmit 1 bit of information such that the probability of bit error for the optimum demodulator is minimized. Assume that $\pi_0 = 0.5$. For the two signals selected,

(a) What is the energy per bit, E_b.

(b) What is the probability of error when $N_0 = 0.1$.

(c) Plot $D_{x_z}(f)$.

(d) Plot the matched filter impulse response.

Problem 13.20. You have decided to try to predict the outcome of the next presidential race by using your electrical engineering knowledge learned your communication theory class. You decide to use one outcome of an approval rating poll for the current president in the summer before the election as the observation. This observation is a random variable V. The minimum error rate decision rule was given as

$$f_{V|1}(v|I=1)\pi_1 \underset{\hat{I}=0}{\overset{\hat{I}=1}{\gtrless}} f_{V|0}(v|I=0)\pi_0 \tag{13.113}$$

where π_0 is the a priori probability that the challenging party wins and π_1 is the a priori probability the incumbent party wins. After careful analysis of the polling data you realize that

$$\begin{aligned} V &= 53 + N \qquad I = 1 \text{ (when the incumbent party wins)} \\ V &= 49 + N \qquad I = 0 \text{ (when the challenging party wins)} \end{aligned} \tag{13.114}$$

where N is a zero mean Gaussian random variable with $E[N^2] = 4$.

(a) If $\pi_0 = 0.4$ find π_1.

(b) If $\pi_0 = 0.4$ and $v = 51$ is observed, make the minimum probability of error decision.

(c) If $\pi_0 = 0.5$ and $v = 50$ is observed, make the minimum probability of error decision.

(d) If $\pi_0 = 0.5$ find the probability of error in the optimum test when the incumbent wins.

(e) If $v = 50$ find the region for the values of π_0 where the decision will be opposite of that found in (c).

Problem 13.21. The two waveforms in Figure 13.31 are to be used to send 1 bit. Assume that $\pi_0 = 0.5$.

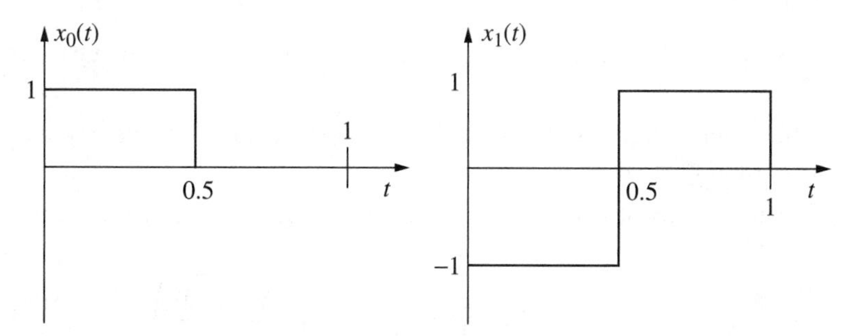

Figure 13.31 Two waveforms to send 1 bit.

(a) Compute E_0 and E_1.

(b) Specify the optimum demodulator structure. Clearly specify any threshold test, filter impulse responses, and/or sample times.

(c) Compute the Euclidean distance between $x_0(t)$ and $x_1(t)$.

Problem 13.22. Using the Rayleigh energy theorem show that

$$\Re[\rho_{10}] = \frac{\Re\left[\int_{-\infty}^{\infty} X_1(f)X_0^*(f)df\right]}{\sqrt{E_1 E_0}} \tag{13.115}$$

Problem 13.23. In a manufacturing assembly line the error in placing an object in a test fixture is modeled as a random variable, V_I. If the robot is properly calibrated then the error is well modeled as

$$f_{V_I}(v|C) = \begin{cases} \frac{1}{2} & -1 \leq v \leq 1 \\ 0 & \text{elsewhere} \end{cases} \tag{13.116}$$

If the robot is not calibrated then the error is well modeled as

$$f_{V_I}(v|NC) = \frac{1}{\sqrt{2\pi}} \exp\left[-\frac{v^2}{2}\right] \tag{13.117}$$

Assume the robot is calibrated roughly 75% of the time ($\pi_C = 0.75$).

Hint: Since Eq. (13.116) is not a Gaussian PDF, a little thought will be needed to generalize the results of this chapter to answer the questions.

(a) If you observe a $V_i = v$ give a decision algorithm based on the value of v that would give the minimum probability of error in determining if the system was uncalibrated.

(b) If the robot is calibrated only 25% of the time the decision algorithm will change. Precisely define the new decision algorithm for $\pi_C = 0.25$. Give the intuition of why the decision algorithm has changed so dramatically.

13.9 Example Solutions

Problem 13.10.

(a) The plot of the bandpass signal corresponding to $x_0(t)$ is shown in Figure 13.32.

(b) The block diagram of the optimum demodulator is exactly as given in the text (see Figure 13.33).

(c) Recall for an optimal demodulator of a binary signal set that the BEP is given as

$$P_B(E) = \frac{1}{2}\text{erfc}\left(\sqrt{\frac{\Delta_E(1,0)}{4N_0}}\right) \tag{13.118}$$

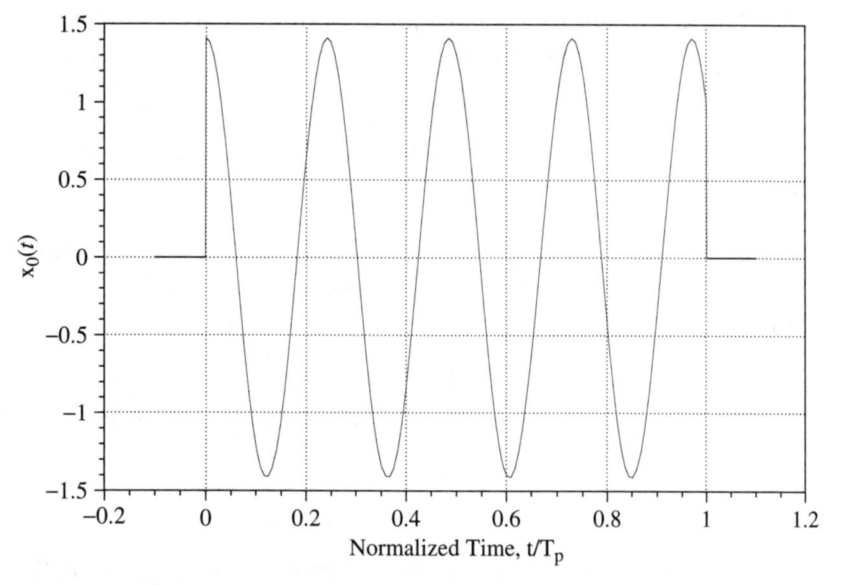

Figure 13.32 The bandpass signal corresponding to $x_0(t)$.

The Euclidean square distance for this signal set is

$$\Delta_E(1,0) = \int_0^{T_p} |x_1(t) - x_0(t)|^2 dt \tag{13.119}$$

$$= 2E_b - \frac{2E_b}{T_p}\int_0^{T_p} \Re\left\{\exp\left(j\frac{\pi t}{2T_p} - j\theta\right)\right\} dt \tag{13.120}$$

$$= 2E_b - \frac{2E_b}{T_p}\Re\left\{\exp(-j\theta)\int_0^{T_p} \exp\left(j\frac{\pi t}{2T_p}\right) dt\right\} \tag{13.121}$$

$$= 2E_b - 2E_b\Re\left\{\exp(-j\theta)\left(\frac{\sin(\pi/2)}{\pi/2} - \frac{\cos(\pi/2) - 1}{\pi/2}\right)\right\} \tag{13.122}$$

$$= 2E_b - \frac{4E_b}{\pi}\Re\{\exp(-j\theta)(1+j)\} \tag{13.123}$$

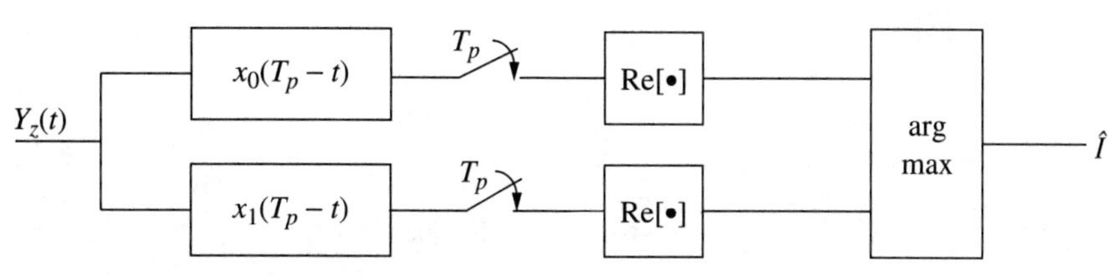

Figure 13.33 Demodulator block diagram.

The maximum Euclidean square distance is achieved for $\theta = -3\pi/4$. Note this best performance is

$$\Delta(1,0) = 2E_b\left(1 + \frac{2\sqrt{2}}{\pi}\right) = 3.8E_b < 4E_b \tag{13.124}$$

so that the performance is worse than BPSK but better than the best BFSK signaling scheme as introduced in class.

(d)

$$G_{x_0}(f) = E_b T_p(\text{sinc}(fT_p - \pi/8))^2 \tag{13.125}$$

$$G_{x_1}(f) = E_b T_p(\text{sinc}(fT_p + \pi/8))^2 \tag{13.126}$$

$$D_{X_z}(f) = \frac{1}{2}(G_{x_0}(f) + G_{x_1}(f)) \tag{13.127}$$

This energy spectrum is given in Figure 13.34. Examining this figure gives $B_3 = 0.88$.

Problem 13.13. This signaling scheme is known as fast frequency hopping.

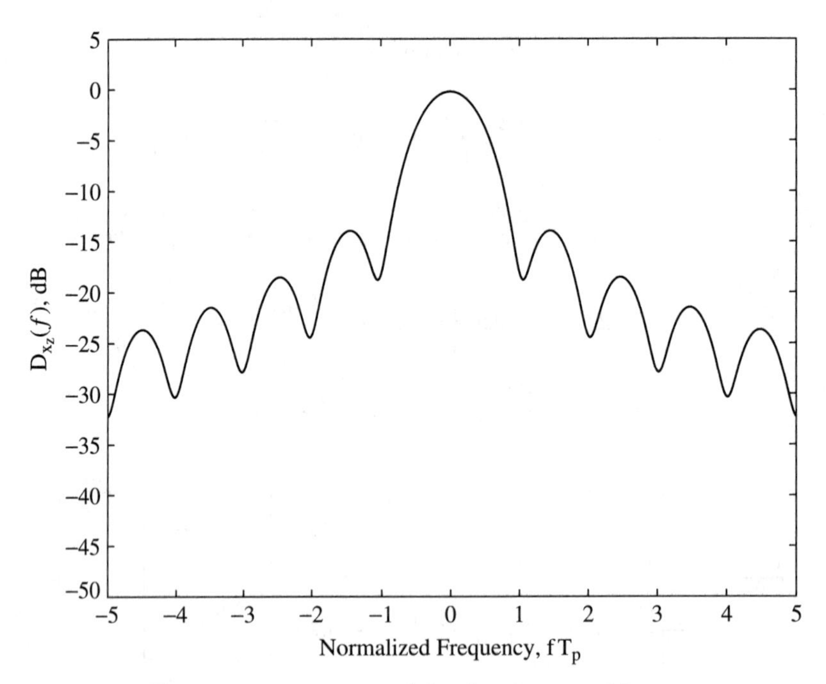

Figure 13.34 The average energy spectral density. $\pi_0 = \pi_1 = 0.5$.

(a) This modulation has the form $X_i(t) = d_i u(t)$. Hence only one filter is needed and this is given as

$$Q = \int Y_z(t) u^*(t) dt \tag{13.128}$$

$$= \int_0^{\frac{T_p}{2}} Y_z(t) \sqrt{\frac{E_b}{T_p}} \exp[-j2\pi f_1 t] dt + \int_{\frac{T_p}{2}}^{T_p} Y_z(t) \sqrt{\frac{E_b}{T_p}} \exp[-j2\pi f_2 t] dt \tag{13.129}$$

The ML decision metrics can be formulated as

$$E_i = |d_i|^2 E_u = |d_i|^2 E_b \tag{13.130}$$

$$T_0 = \Re[d_0^* Q] - \frac{|d_0|^2 E_b}{2} \tag{13.131}$$

$$T_1 = \Re[d_1^* Q] - \frac{|d_1|^2 E_b}{2} \tag{13.132}$$

(b)

$$P_B(E) = \frac{1}{2} \operatorname{erfc}\left[\sqrt{\frac{\Delta_E(1,0)}{4N_0}}\right] \tag{13.133}$$

$$\text{where} \quad \Delta_E(1,0) = |d_0 - d_1|^2 E_b \tag{13.134}$$

(c) Since $d_0 = 1$ and $d_0 = -1$ this produces $\Delta_E(1,0) = 4E_b$.

(d) Only point of possible discontinuity is at $t = T_p/2$.

$$2\pi f_1 \frac{T_p}{2} = 2\pi f_2 \frac{T_p}{2} \tag{13.135}$$

$$2\pi f_1 \frac{T_p}{2} = 2\pi f_2 \frac{T_p}{2} + 2\pi n \qquad \text{where } n \text{ is an integer} \tag{13.136}$$

$$f_2 = f_1 + \frac{2n}{T_p} \tag{13.137}$$

(e) The average energy spectrum is given as

$$D_{x_z}(f) = \frac{1}{2}(G_{X_0}(f) + G_{X_1}(f)) \tag{13.138}$$

Since $X_i(t) = d_i u(t)$, the individual energy spectrums are given as

$$G_{X_0}(f) = |X_0(f)|^2 = |d_0|^2 |U(f)|^2 = |U(f)|^2 = G_{X_1}(f) \tag{13.139}$$

Note that $u(t) = u_1(t) + u_2(t)$ where

$$u_1(t) = \begin{cases} \sqrt{\frac{E_b}{T_p}} \exp[j2\pi f_1 t] & 0 \leq t \leq T_p/2 \\ 0 & \text{elsewhere} \end{cases} \tag{13.140}$$

and

$$u_2(t) = \begin{cases} \sqrt{\frac{E_b}{T_p}} \exp[j2\pi f_1 t] & T_p/2 \leq t \leq T_p \\ 0 & \text{elsewhere} \end{cases} \tag{13.141}$$

Taking the Fourier transform gives

$$U_1(f) = \sqrt{E_b T_p} \exp\left[-j\frac{\pi f T_p}{2}\right] \operatorname{sinc}\left(\frac{(f - f_1)T_p}{2}\right) \tag{13.142}$$

and

$$U_2(f) = \sqrt{E_b T_p} \exp\left[-j\frac{3\pi f T_p}{2}\right] \operatorname{sinc}\left(\frac{(f - f_2)T_p}{2}\right) \tag{13.143}$$

Finally

$$|U(f)|^2 = |U_1(f) + U_2(f)|^2 = D_{x_z}(f) \tag{13.144}$$

See Matlab code and plot in Figure 13.35.

```
%
% Generating the spectrum of FH
% Author: M. Fitz
% Last modified: 1/28/04
%
close all
clear all
numpts=2048;
f_1=2;
f_2=-2;
freq=linspace(-4,4, numpts+1);
g_pr=zeros(1,numpts+1);
dum1=exp(-j*pi*freq/2).*sinc(freq-f_1*ones(1,numpts+1));
dum2=exp(-j*3*pi*freq/2).*sinc(freq-f_2*ones(1,numpts+1));
g_pr=(dum1+dum2).*conj(dum1+dum2);
g_prdb=10*log10(g_pr);
figure(1)
plot(freq,g_prdb)
axis([-4 4 -50 5])
xlabel('Normalized frequency, fT_p')
ylabel('Normalized energy spectrum per bit')
hold off
```

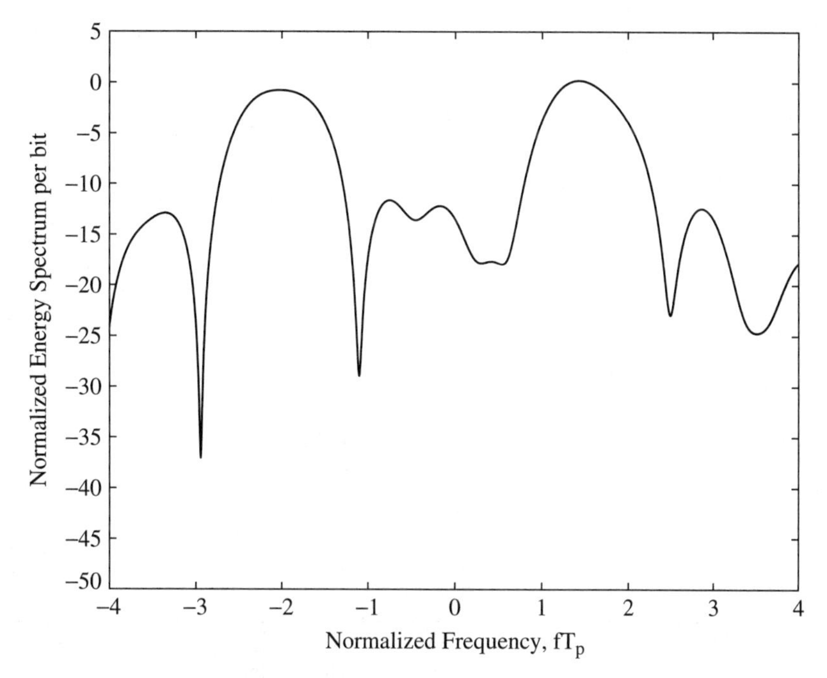

Figure 13.35 Spectrum of a frequency hopped signal.

13.10 Miniprojects

Goal: To give exposure

- to a small scope engineering design problem in communications.
- to the dynamics of working with a team.
- to the importance of engineering communication skills (in this case oral presentations).

Presentation: The forum will be similar to a design review at a company (only much shorter). The presentation will be of 5 minutes in length with an overview of the given problem and solution. The presentation will be followed by questions from the audience (your classmates and the professor). Each team member should be prepared to give the presentation.

13.10.1 Project 1

Project Goals: This project will design a signal set to meet a typical spectral emissions mask and compute the probability of error for both bandpass and baseband matched filtering in the presence of sample time error to show the advantages of processing signals at baseband.

Design a real valued pulse shape, $u(t)$, that can be used with binary pulse amplitude modulation having the bandpass form

$$x_{c0}(t) = u(t)\sqrt{2}\cos(2\pi f_c t) \qquad x_{c1}(t) = -u(t)\sqrt{2}\cos(2\pi f_c t)$$

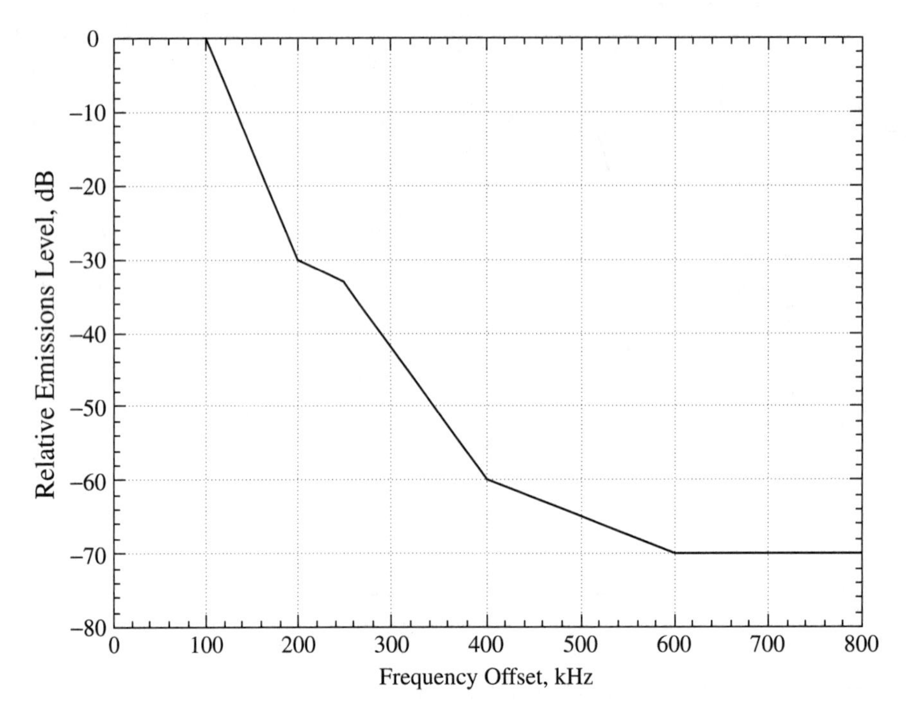

Figure 13.36 Spectral emissions mask.

that will meet the spectral emissions mask given in Figure 13.36 but the pulse must not extend for longer than 40 μs in time. The carrier frequency is 1 MHz. This spectral emissions mask is the one used for Global System for Mobile communications (GSM) handsets. GSM is a second generation cellular telephony standard. So this problem is one of great practical interest.

We have discussed one type of demodulator structures for bandpass signals; baseband matchd filters. This structure match filters to the bandpass signal after creating the complex envelope of the received signal (see Figure 13.37). It is also possible to build a matched filter to the bandpass signal itself ($h(t) = x_{1c}(T_p - t) - x_{0c}(T_p - t)$)

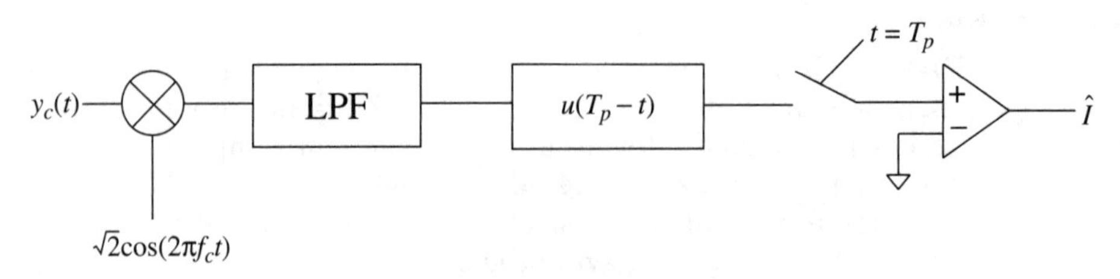

Figure 13.37 The baseband demodulator for binary pulse amplitude modulation.

(a) Derive the form for and plot the output of the bandpass matched filter for all time when $x_c(t) = u(t)\sqrt{2}\cos(2\pi f_c t)$.

(b) Derive the form for and plot the output of the baseband matched filter when $x_c(t) = u(t)\sqrt{2}\cos(2\pi f_c t)$.

(c) Compute the E_b/N_0 required to produce a 10^{-5} error probability for the optimum demodulator.

(d) If the bandpass matched filter is sampled at $T_p + 2 \times 10^{-7}$ instead of T_p compute the E_b/N_0 required to produce a 10^{-5} error probability.

(e) If the baseband matched filter is sampled at $T_p + 2 \times 10^{-7}$ instead of T_p compute the E_b/N_0 required to produce a 10^{-5} error probability.

(f) How big would the timing error have to be for the baseband matched filter to achieve the same answer as in (d)?

13.10.2 Project 2

Project Goals: Engage in an implementation of an optimum demodulator of a binary signal set.

Consider baseband binary signal set (with sample frequency $f_s = 22{,}050$ Hz) and the received bandpass signal ($f_c = 3500$ Hz) given in `sbproj2data.mat`.

(a) Plot the two baseband signals and determine the bit rate, W_b.

(b) Sketch a block diagram of the optimum demodulator for these two signals.

(c) Make the optimum bit decision and compute all the sufficient statistics of the demodulation. *Hint:* If implementing the demodulator at baseband make sure to account for the impact of the lowpass filters in the I/Q down-converter on the possible baseband signals.

Chapter 14

Transmitting More Than One Bit

People who buy communication systems rarely want to only transmit 1 bit. Consequently, there is a practical need to generalize the results from the previous chapter to the transmission of multiple bits. The problem considered here is that we transmit K_b bits by selecting one of $M = 2^{K_b}$ waveforms. In the literature this is often denoted M-ary digital communications. The demodulator has the task of deciding which of these waveforms was sent in the presence of an additive noise. This chapter will consider a variety of problems associated with transmitting more than 1 bit.

14.1 A Reformulation for 1 Bit Transmission

The optimum demodulator can be reformulated in several ways. Many of these reformulations give insight for receiver structures we will consider in the sequel. Recall from Chapter 13 that the MLBD, when $Y_z(t) = y_z(t)$ is observed, has the form

$$v_I(T_p) = \Re\left[\int_{-\infty}^{\infty} y_z(\tau)(x_1^*(\tau) - x_0^*(\tau))d\tau\right] \underset{\hat{I}=0}{\overset{\hat{I}=1}{\gtrless}} \frac{E_1 - E_0}{2} \tag{14.1}$$

Rearranging Eq. (14.1) gives

$$\Re\left[\int_{-\infty}^{\infty} y_z(\tau)x_1^*(\tau)d\tau\right] - \frac{E_1}{2} \underset{\hat{I}=0}{\overset{\hat{I}=1}{\gtrless}} \Re\left[\int_{-\infty}^{\infty} y_z(\tau)x_0^*(\tau)d\tau\right] - \frac{E_0}{2} \tag{14.2}$$

The form of the MLBD in Eq. (14.2) which forms a decision statistic for each possible signal by computing the output of a matched filter to that transmitted signal and adding an energy correction term will frequently appear in the sequel. In fact, to simplify the discussion the following definition is useful.

Definition 14.1 The maximum likelihood metric for $I = i$ is

$$T_i = \Re\left[\int_{-\infty}^{\infty} y_z(\tau)x_i^*(\tau)d\tau\right] - \frac{E_i}{2} \tag{14.3}$$

Using this definition allows a reformulation of the MLBD as

$$T_1 \underset{\hat{I}=0}{\overset{\hat{I}=1}{\gtrless}} T_0 \tag{14.4}$$

Equation (14.4) gives a particularly insightful form of the decision process: form the maximum likelihood metric for each bit value and choose to decode the bit value corresponding to the largest metric.

The MAPBD can similarly take other useful forms. Multiplying both sides of Eq. (14.2) by $2/N_0$ and taking the exponential gives another form for the MLBD, i.e.,

$$\exp\left[\frac{2}{N_0}\Re\left[\int_{-\infty}^{\infty} y_z(\tau)x_1^*(\tau)d\tau\right] - \frac{E_1}{N_0}\right] \underset{\hat{I}=0}{\overset{\hat{I}=1}{\gtrless}} \exp\left[\frac{2}{N_0}\Re\left[\int_{-\infty}^{\infty} y_z(\tau)x_0^*(\tau)d\tau\right] - \frac{E_0}{N_0}\right] \tag{14.5}$$

The justification of why the constant term is $2/N_0$ is provided in a book on detection theory [Poo88]. Using the maximum likelihood metric the MLBD simplifies to

$$\exp\left[\frac{2T_1}{N_0}\right]\pi_1 \underset{\hat{I}=0}{\overset{\hat{I}=1}{\gtrless}} \exp\left[\frac{2T_0}{N_0}\right]\pi_0 \tag{14.6}$$

The generalization to unequal priors has the MAPBD taking the form

$$\exp\left[\frac{2}{N_0}\Re\left[\int_{-\infty}^{\infty} y_z(\tau)x_1^*(\tau)d\tau\right] - \frac{E_1}{N_0}\right]\pi_1 \underset{\hat{I}=0}{\overset{\hat{I}=1}{\gtrless}}$$

$$\exp\left[\frac{2T_1}{N_0}\right] \underset{\hat{I}=0}{\overset{\hat{I}=1}{\gtrless}} \exp\left[\frac{2T_0}{N_0}\right] \tag{14.7}$$

In fact, this demodulator structure can be viewed as being equivalent to computing the APP for each of the two possible bits given the receiver input and then selecting the bit value that corresponds to this maximum, i.e.,

$$\frac{\exp\left[\frac{2T_1}{N_0}\right]\pi_1}{K} = P(I=1|y_z(t)) \underset{\hat{I}=0}{\overset{\hat{I}=1}{\gtrless}} P(I=0|y_z(t)) = \frac{\exp\left[\frac{2T_0}{N_0}\right]\pi_0}{K} \tag{14.8}$$

where $y_z(t)$ is the observed receiver input and K is a constant that is not a function of I. Forms for the optimal detector given in Eq. (14.5) and Eq. (14.7) will be used often in the sequel to form optimum detectors in some special cases.

14.2 Optimum Demodulation

Now we want to consider the digital communication system design problem of transmitting more than one bit of information. In the sequel, K_b shall denote the number of bits to be transmitted and the bit sequence is denoted

$$\vec{I} = [I(1)\, I(2)\, \ldots\, I(K_b)]^T \tag{14.9}$$

where $I(k)$ take values 0,1. For simplicity of notation a particular sequence of bits can be designated by the numeric value the bits represent, i.e., $\vec{I} = i$, $i \in \{0, \ldots, M-1\}$ where $M = 2^{K_b}$. To represent the M values the bit sequence can take, M different analog waveforms should be available for transmission. Denote by $x_i(t)$, $i \in \{0, \ldots, M-1\}$ as the waveform transmitted when $\vec{I} = i$ is to be transmitted. Here again we will assume the analog waveforms have support on $t \in [0, T_p]$. Given this problem formulation we want to extend the results obtained in the previous chapter. It is worth noting that this problem formulation gives a bit rate of $W_b = K_b/T_p$ bits per second. The average energy per bit is given as

$$E_b = \frac{1}{K_b} \sum_{i=0}^{M-1} \pi_i E_i \tag{14.10}$$

where again E_i is the energy of waveform $x_i(t)$.

EXAMPLE 14.1

Consider the case of $K_b = 3$, the bits and the words are enumerated as

$I(1)$	$I(2)$	$I(3)$	i
0	0	0	0
0	0	1	1
0	1	0	2
0	1	1	3
1	0	0	4
1	0	1	5
1	1	0	6
1	1	1	7

EXAMPLE 14.2

Here we consider a modulation known as pulse width modulation with $K_b = 2$. This modulation embeds information in the width of the pulse that is transmitted. This is a popular baseband modulation in certain applications where positive logic level signals are desired in the transmitter. The transmitted signals are given as

$$x_0(t) = \begin{cases} \sqrt{\frac{32E_b}{10T_p}} & 0 \le t \le T_p/4 \\ 0 & \text{elsewhere} \end{cases} \qquad x_1(t) = \begin{cases} \sqrt{\frac{32E_b}{10T_p}} & 0 \le t \le T_p/2 \\ 0 & \text{elsewhere} \end{cases} \tag{14.11}$$

$$x_2(t) = \begin{cases} \sqrt{\frac{32E_b}{10T_p}} & 0 \le t \le 3T_p/4 \\ 0 & \text{elsewhere} \end{cases} \qquad x_3(t) = \begin{cases} \sqrt{\frac{32E_b}{10T_p}} & 0 \le t \le T_p \\ 0 & \text{elsewhere} \end{cases} \tag{14.12}$$

A student exercise worth completing is to show that energy per bit for this signal set is E_b.

14.2.1 Optimum Word Demodulation Receivers

A first receiver to be considered is the maximum a posteriori word demodulator (MAPWD). This MAPWD can again be shown to be the minimum word error probability receiver using Bayes detection theory [Leh86, Web87]. A straightforward generalization of Eq. (14.8) leads to

$$\hat{\vec{I}} = \arg \max_{i \in \{0,\ldots,M-1\}} P(\vec{I} = i | y_z(t)) \tag{14.13}$$

where the arg max notation refers to the particular binary word that has the maximum APP. This is the obvious generalization of binary MAP detection to the M-ary decision problem. Using the results of the previous section this MAP decoding rule becomes

$$\hat{\vec{I}} = \arg \max_{i \in \{0,\ldots,M-1\}} \exp\left[\frac{2}{N_0}\Re\left[\int_{-\infty}^{\infty} y_z(\tau)x_i^*(\tau)d\tau\right] - \frac{E_i}{N_0}\right]\pi_i$$

$$= \arg \max_{i \in \{0,\ldots,M-1\}} \exp\left[\frac{2}{N_0}\Re[v_i(T_p)] - \frac{E_i}{N_0}\right]\pi_i \tag{14.14}$$

$$= \arg \max_{i \in \{0,\ldots,M-1\}} \exp\left[\frac{2T_i}{N_0}\right]\pi_i \tag{14.15}$$

where

$$v_i(t) = \int_{-\infty}^{\infty} y_z(\tau)x_i^*(T_p - t + \tau)d\tau \tag{14.16}$$

is denoted the ith matched filter output and

$$E_i = \int_{-\infty}^{\infty} |x_i(t)|^2 dt \tag{14.17}$$

is the energy of the ith analog waveform. The ith matched filter output when sampled at $t = T_p$ again will give a correlation of the received signal with the ith possible transmitted signal. Here T_i represents the maximum likelihood

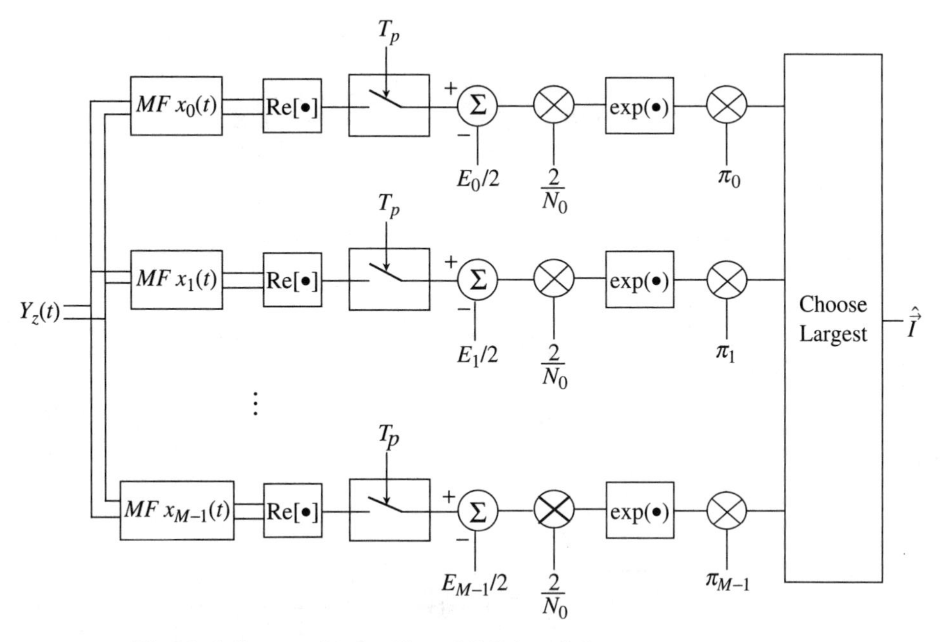

Figure 14.1 The block diagram for the M-ary MAP demodulator.

metric for the word $\vec{I} = i$. For future use it is important to reemphasize that

$$\exp\left[\frac{2T_i}{N_0}\right]\pi_i = \exp\left[\frac{2}{N_0}\Re[v_i(T_p)] - \frac{E_i}{N_0}\right]\pi_i = KP(\vec{I} = i|y_z(t)) \qquad (14.18)$$

where K is a constant that is not a function of $\vec{I}$. The block diagram for the optimum MAPWD is shown in Figure 14.1.

The demodulator in the case of equal priors, i.e., $\pi_i = 1/M, \ \forall i$, can be greatly simplified. We will denote this demodulator as the maximum likelihood word demodulator (MLWD). Since all terms have an equal π_i this common term can be canceled from each term in the decision rule and since the ln(•) function is monotonic the MLWD is given as

$$\begin{aligned}\hat{\vec{I}} &= \arg \max_{i\in\{0,\ldots,M-1\}} \Re[v_i(T_p)] - \frac{E_i}{2} \\ &= \arg \max_{i\in\{0,\ldots,M-1\}} T_i \end{aligned} \qquad (14.19)$$

The maximum likelihood metric is computed for each of the possible transmitted signals. Decoding is accomplished by selecting the binary word associated with the largest maximum likelihood metric. The block diagram for the optimum MLWD is shown in Figure 14.2.

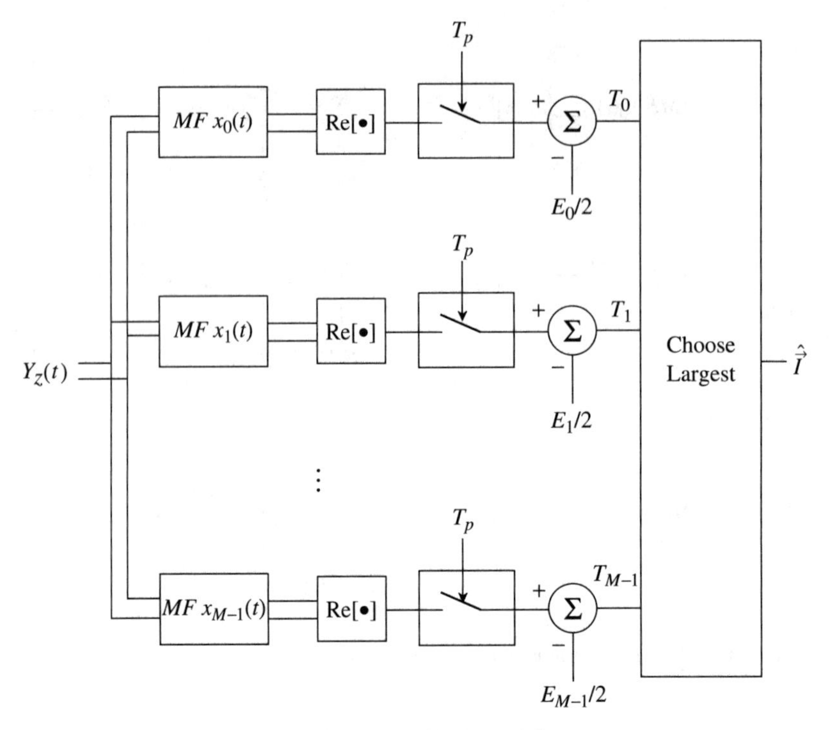

Figure 14.2 The block diagram for the MLWD demodulator.

EXAMPLE 14.3

This example returns to the 4-ary example first introduced in Example 12.1. Since the possible transmitted signals are real, the four matched filter outputs are given as

$$v_0(T_p) = \int_0^{T_p} \sin\left(\frac{4\pi\tau}{T_p}\right) y_z(\tau)d\tau \qquad v_1(T_p) = \int_0^{T_p} y_z(\tau)d\tau$$

$$v_2(T_p) = -\int_0^{T_p} \sin\left(\frac{\pi\tau}{T_p}\right) y_z(\tau)d\tau$$

$$v_3(T_p) = \int_0^{T_p/2} \frac{2\tau}{T_p} y_z(\tau)d\tau + \int_{T_p/2}^{T_p} \left(2 - \frac{2\tau}{T_p}\right) y_z(\tau)d\tau \tag{14.20}$$

The energy of each of the transmitted signals is given as

$$E_0 = T_p/2 \qquad E_1 = T_p \qquad E_2 = T_p/2 \qquad E_3 = T_p/3 \tag{14.21}$$

The maximum likelihood metrics for this modulation are given as

$$T_0 = \Re[v_0(T_p)] - T_p/2 \qquad T_1 = \Re[v_1(T_p)] - T_p$$

$$T_2 = \Re[v_2(T_p)] - T_p/2 \qquad T_3 = \Re[v_3(T_p)] - T_p/3 \tag{14.22}$$

EXAMPLE 14.4

Returning to the pulse width modulation first introduced in Example 14.2, the four matched filter outputs are given as

$$v_0(T_p) = \sqrt{\frac{32E_b}{10T_p}} \int_0^{T_p/4} y_z(\tau)d\tau \qquad v_1(T_p) = \sqrt{\frac{32E_b}{10T_p}} \int_0^{T_p/2} y_z(\tau)d\tau$$

$$v_2(T_p) = \sqrt{\frac{32E_b}{10T_p}} \int_0^{3T_p/4} y_z(\tau)d\tau \qquad v_3(T_p) = \sqrt{\frac{32E_b}{10T_p}} \int_0^{T_p} y_z(\tau)d\tau \tag{14.23}$$

The energy of each of the transmitted signals is given as

$$E_0 = \frac{8E_b}{10} \qquad E_1 = \frac{16E_b}{10} \qquad E_2 = \frac{24E_b}{10} \qquad E_3 = \frac{32E_b}{10} \tag{14.24}$$

The maximum likelihood metrics for this modulation are given as

$$T_0 = \sqrt{\frac{32E_b}{10T_p}} \int_0^{T_p/4} y_I(\tau)d\tau - \frac{4E_b}{10} \qquad T_1 = \sqrt{\frac{32E_b}{10T_p}} \int_0^{T_p/2} y_I(\tau)d\tau - \frac{8E_b}{10}$$

$$T_2 = \sqrt{\frac{32E_b}{10T_p}} \int_0^{3T_p/4} y_I(\tau)d\tau - \frac{12E_b}{10} \qquad T_3 = \sqrt{\frac{32E_b}{10T_p}} \int_0^{T_p} y_I(\tau)d\tau - \frac{16E_b}{10} \tag{14.25}$$

The important thing to notice for both MAPWD and MLWD is that the optimal demodulator complexity increases exponentially with the number of bits transmitted. The number of matched filters required in each demodulator is $M = 2^{K_b}$, consequently the complexity of these demodulation schemes is $O(2^{K_b})$. The notation $O(x)$ implies the complexity of the algorithm is proportional to x, i.e., a constant times x. This complexity is obviously unacceptable if large files of data are to be transmitted. To make data communications practical, ways will have to be developed that make the complexity linear in the number of bits sent, i.e., $O(K_b)$. This chapter will continue to look at M-ary signaling schemes as they often are used in practice (in conjunction with other complexity reduction techniques) to achieve either lower error rates or higher bandwidth efficiency.

14.2.2 Analysis of Demodulation Fidelity

The analysis of the fidelity of the message reconstruction in an M-ary demodulator is much more complicated than that of the binary detector but many of the ideas are similar. Two important quantities should be related at this point: the energy per symbol and the energy per bit. The average symbol energy is denoted E_s and this is given as

$$E_s = \sum_{i=0}^{M-1} E_i \pi_i \tag{14.26}$$

E_s represents the average energy transmitted to communicate the K_b bits. The case of equal priors reduces to

$$E_s = \frac{1}{M} \sum_{i=0}^{M-1} E_i \tag{14.27}$$

The average energy per bit is then given as

$$E_b = \frac{E_s}{K_b} \tag{14.28}$$

Communication engineers frequently parametrize fidelity of demodulation with both E_s and E_b.

EXAMPLE 14.5

Returning again to the pulse width modulation introduced in Example 14.2 we have

$$E_b = \frac{E_0 + E_1 + E_2 + E_3}{8} = \left(\frac{8E_b}{80} + \frac{16E_b}{80} + \frac{24E_b}{80} + \frac{32E_b}{80} \right) \tag{14.29}$$

We again start the process of understanding the fidelity of demodulation by characterizing the likelihood metrics.

Likelihood Metrics

The likelihood metrics, T_i, $i \in \{0, \ldots, M-1\}$ are matched filter outputs with an energy correction term. Conditioned on the transmitted signal the likelihood metrics are Gaussian random variables. For instance, if $x_j(t)$ is assumed transmitted then denote the ith conditional likelihood metric as

$$\begin{aligned} T_{i|j} &= \Re\left[\int_{-\infty}^{\infty} Y_z(t) x_i^*(t) dt \right] - \frac{E_i}{2} \\ &= \Re\left[\int_{-\infty}^{\infty} x_j(t) x_i^*(t) dt \right] + N_I^{(i)} - \frac{E_i}{2} \end{aligned} \tag{14.30}$$

where $N_I^{(i)}$ is a real zero mean Gaussian random variable given as

$$N_I^{(i)} = \Re\left[\int_{-\infty}^{\infty} W_z(t) x_i^*(t) dt \right] \tag{14.31}$$

with

$$\text{var}(N_I^{(i)}) = \frac{E_i N_0}{2}. \tag{14.32}$$

It should be noted that

$$\begin{aligned} T_{i|j} &= \Re[\rho_{ji} \sqrt{E_i E_j}] - \frac{E_i}{2} + N_I^{(i)} \\ T_{j|j} &= \frac{E_j}{2} + N_I^{(j)} \end{aligned} \tag{14.33}$$

Finally, if two conditional likelihood metrics, $T_{i|j}$ and $T_{k|j}$ are considered jointly, they will be jointly Gaussian random variables with a correlation coefficient of (see Problem 14.14)

$$\rho = \frac{E\left[N_I^{(i)} N_I^{(k)}\right]}{\frac{N_0}{2}\sqrt{E_i E_k}} = \Re[\rho_{ik}] \tag{14.34}$$

This correlation between the matched filter outputs will be explored in more depth in the homework.

Probability of Word Error

The probability of making a demodulation error is denoted the word error probability, $P_W(E)$. This probability can be written via total probability as

$$P_W(E) = \sum_{j=0}^{M-1} P(\hat{\vec{I}} \neq j | \vec{I} = j)\pi_j \tag{14.35}$$

When the priors are unequal the nonlinear nature of MAPWD in Eq. (14.14) makes the computation of $P(\hat{\vec{I}} \neq j | \vec{I} = j)$ very difficult. Not many results exist in the literature discussing $P_W(E)$ calculations for MAPWD.

The case of equal priors has a demodulator with a more exploitable structure. For MLWD the word error probability is

$$P_W(E) = \frac{1}{M}\sum_{j=0}^{M-1} P(\hat{\vec{I}} \neq j | \vec{I} = j) \tag{14.36}$$

The conditional probability of word error is given as

$$\begin{aligned} P(\hat{\vec{I}} \neq j | \vec{I} = j) &= P\left(\max_{j \neq i} T_i > T_j \,\middle|\, \vec{I} = j\right) \\ &= P\left(\max_{j \neq i} T_{i|j} > T_{j|j}\right) \\ &= P\left(\bigcup_{\substack{i=0 \\ j \neq i}}^{M-1} \{T_{i|j} > T_{j|j}\}\right) \end{aligned} \tag{14.37}$$

In general, the probability in Eq. (14.37) is quite tough to compute. While each of the $T_{i|j}$, $i = 1, M$ is a Gaussian random variable, computing the distribution of the maximum of correlated Gaussian random variables is nontrivial. For a significant number of modulations with small M the probability in Eq. (14.37) is computable with a reasonable amount of work on a computer. Some of these examples will be explored in the sequel and in homework problems. For a general modulation and large M the problem is much tougher. One special case where a general result is available is the case of an orthogonal modulation, i.e., $\rho_{ij} = 0, \ \forall i \neq j$. Error rate analysis for an orthogonal signal set is explored in the homework.

14.2.3 Union Bound

A common upperbound on the probability in Eq. (14.37) has found great utility in the analysis of communication systems. This so-called union bound is simply stated in general form as

$$P\left(\bigcup_{i=1}^{N} A_i\right) \leq \sum_{i=1}^{N} P(A_i) \tag{14.38}$$

where A_i are arbitrary events. The idea of the union bound is simply understood if one considers the Venn diagram of Figure 14.3. Consider the probability of each event, A_i to be proportional to the area in the Venn diagram. The union bound given in Eq. (14.38) counts the area in $A_1 \bigcap A_2$, $A_1 \bigcap A_3$, and $A_2 \bigcap A_3$ twice and the area of $A_1 \bigcap A_2 \bigcap A_3$ three times. The union bound is satisfied with equality if the events are disjoint.

The union bound can now be used to upperbound the word error probability in an M-ary optimum word demodulator. Using the union bound in Eq. (14.37) gives

$$\begin{aligned} P(\hat{\vec{I}} \neq j | \vec{I} = j) &= P\left(\bigcup_{\substack{i=0 \\ j \neq i}}^{M-1} \{T_{i|j} > T_{j|j}\}\right) \\ &\leq \sum_{\substack{i=0 \\ j \neq i}}^{M-1} P(T_{i|j} > T_{j|j}) \end{aligned} \tag{14.39}$$

$P(T_{i|j} > T_{j|j})$ is the probability that the ith maximum likelihood metric for the transmitted signal is greater than the maximum likelihood metric corresponding to the actual transmitted signal. This probability is often denoted the pairwise error probability (PWEP).

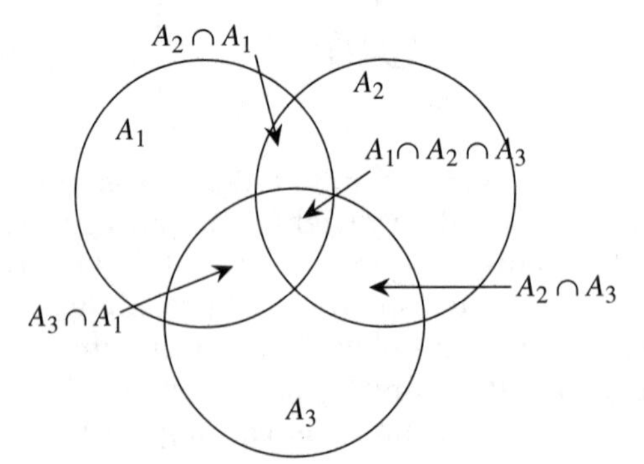

Figure 14.3 A Venn diagram to explain the union bound.

The PWEP has a form readily analyzed. It should first be noted that the PWEP is exactly the same error rate computation that was done in Chapter 13 for binary modulations. This derivation is repeated here in its more general form to give another perspective on the problem. Recall the form for the decision metrics are

$$T_{i|j} = \Re[\rho_{ji}\sqrt{E_i E_j}] - \frac{E_i}{2} + N_I^{(i)}$$

$$T_{j|j} = \frac{E_j}{2} + N_I^{(j)} \tag{14.40}$$

Consequently, the PWEP is given as

$$P(T_{i|j} > T_{j|j}) = P\left(\Re[\rho_{ji}\sqrt{E_i E_j}] - \frac{E_i}{2} + N_I^{(i)} > \frac{E_j}{2} + N_I^{(j)}\right)$$

$$= P\left(N_{ij} > \frac{E_i + E_j}{2} - \Re[\rho_{ji}\sqrt{E_i E_j}]\right) \tag{14.41}$$

$$= P\left(N_{ij} > \frac{\Delta_E(i,j)}{2}\right) \tag{14.42}$$

where $N_{ij} = N_I^{(i)} - N_I^{(j)}$. Since N_{ij} is a zero mean Gaussian random variable the PWEP is simply expressed as the tail probability of a Gaussian random variable. Since

$$N_{ij} = \Re\left[\int_{-\infty}^{\infty} W_z(t)[x_i^*(t) - x_j^*(t)]dt\right] \tag{14.43}$$

the variance is given as

$$\text{var}(N_{ij}) = \frac{N_0}{2}[E_i + E_j - 2\Re[\rho_{ji}\sqrt{E_i E_j}]] = \frac{N_0 \Delta_E(i,j)}{2} \tag{14.44}$$

Using this variance in Eq. (14.41) gives

$$P(T_{i|j} > T_{j|j}) = \frac{1}{2}\text{erfc}\left(\sqrt{\frac{E_i + E_j - 2\Re[\rho_{ji}\sqrt{E_i E_j}]}{4N_0}}\right)$$

$$= \frac{1}{2}\text{erfc}\left(\sqrt{\frac{\Delta_E(i,j)}{4N_0}}\right) \tag{14.45}$$

where again

$$\Delta_E(i,j) = \int_{-\infty}^{\infty} |x_j(t) - x_i(t)|^2 dt \tag{14.46}$$

is the squared Euclidean distance between $x_i(t)$ and $x_j(t)$. It should be noted that the PWEP is exactly the same as the binary error probability given in the previous chapter.

The PWEP can be combined with Eq. (14.39) to give a union bound to the conditional word error probability. The conditional probability of error given that $x_j(t)$ was transmitted is now upperbounded by

$$P(\hat{\vec{I}} \neq j | \vec{I} = j) \leq \sum_{i \neq j} \frac{1}{2} \text{erfc}\left(\sqrt{\frac{\Delta_E(i,j)}{4N_0}}\right) \tag{14.47}$$

Definition 14.2 The conditional union bound is the upperbound on the conditional word error probability and is denoted

$$P_{WUB}(E|\vec{I} = j) = \sum_{i \neq j} \frac{1}{2} \text{erfc}\left(\sqrt{\frac{\Delta_E(i,j)}{4N_0}}\right) \tag{14.48}$$

This upperbound on the conditional word error probability is formed by computing and summing the pairwise error probability for every other codeword besides the postulated transmitted symbol.

The conditional union bound can be combined with Eq. (14.36) to provide an overall bound on the word error rate. An upperbound to the word error probability is now given as

$$\begin{aligned} P_W(E) \leq P_{WUB}(E) &= \sum_{j=0}^{M-1} \frac{1}{M} P_{WUB}(E|\vec{I} = j) \\ &= \sum_{j=0}^{M-1} \sum_{i \neq j} \frac{1}{2M} \text{erfc}\left(\sqrt{\frac{\Delta_E(i,j)}{4N_0}}\right) \end{aligned} \tag{14.49}$$

and this upperbound is often referred to as the union bound in the literature. It is important to realize that this sum is over all possible transmitted signals and all possible decoding errors for each of the possible transmitted signals.

EXAMPLE 14.6

We can enumerate all the Euclidean distances for the 4-ary modulation example introduced in Example 12.1 by examining the possible cross correlations, i.e.,

$$0 = \int_0^{T_p} x_0(t)x_1(t)dt = \int_0^{T_p} x_0(t)x_2(t)dt = \int_0^{T_p} x_0(t)x_3(t)dt \tag{14.50}$$

$$\int_0^{T_p} x_1(t)x_2(t)dt = -\frac{2T_p}{\pi} \tag{14.51}$$

$$\int_0^{T_p} x_1(t)x_3(t)dt = \frac{T_p}{2} \tag{14.52}$$

$$\int_0^{T_p} x_2(t)x_3(t)dt = -\frac{4T_p}{\pi^2} \tag{14.53}$$

This results in an enumeration of the possible Euclidean square distances that is detailed in the following table

$\Delta_E(i, j)$

	$\vec{I} = i$			
$\vec{I} = j$	0	1	2	3
0	0	$\frac{3T_p}{2}$	T_p	$\frac{5T_p}{6}$
1	$\frac{3T_p}{2}$	0	$\left(\frac{3}{2} + \frac{4}{\pi}\right)T_p$	$\frac{T_p}{3}$
2	T_p	$\left(\frac{3}{2} + \frac{4}{\pi}\right)T_p$	0	$\left(\frac{5}{6} + \frac{8}{\pi^2}\right)T_p$
3	$\frac{5T_p}{6}$	$\frac{T_p}{3}$	$\left(\frac{5}{6} + \frac{8}{\pi^2}\right)T_p$	0

This enumeration of the squared Euclidean distances between all pairs of words enable the union bound to be computed for this signal set.

EXAMPLE 14.7

We can enumerate all the Euclidean distances for the pulse width modulation first introduced in Example 14.2 in a table

$\Delta_E(i, j)$

	$\vec{I} = i$			
$\vec{I} = j$	0	1	2	3
0	0	$\frac{8E_b}{10}$	$\frac{16E_b}{10}$	$\frac{24E_b}{10}$
1	$\frac{8E_b}{10}$	0	$\frac{8E_b}{10}$	$\frac{16E_b}{10}$
2	$\frac{16E_b}{10}$	$\frac{8E_b}{10}$	0	$\frac{8E_b}{10}$
3	$\frac{24E_b}{10}$	$\frac{16E_b}{10}$	$\frac{8E_b}{10}$	0

This enumeration of the squared Euclidean distances between all pairs of words enable the union bound to be computed for this signal set.

The important points resulting from the union bound are

1. Squared Euclidean distance is again an important quantity in understanding the fidelity of message reconstruction.
2. The conditional union bound and the union bound typically have the form

$$P_{WUB}(E|\vec{I}=j) = \sum_{k=1}^{N_j} \frac{A_{dj}(k)}{2} \text{erfc}\left(\sqrt{\frac{\Delta_{Ej}(k)}{4N_0}}\right) \tag{14.54}$$

$$P_W(E) \leq P_{WUB}(E) = \sum_{k=1}^{N} \frac{A_d(k)}{2M} \text{erfc}\left(\sqrt{\frac{\Delta_E(k)}{4N_0}}\right) \tag{14.55}$$

where the sum enumerates the $N_j \leq M-1$ and $N \leq M(M-1)/2$ possible different squared Euclidean distances, $\Delta_{Ej}(k)$, $k = 1, N_j$ and $\Delta_E(k)$, $k = 1, N$, respectively. Note that $A_{dj}(k)$ and $A_d(k)$ are the number of signal pairs having a squared Euclidean distance between them of $\Delta_{Ej}(k)$ and $\Delta_E(k)$, respectively. It should also be noted that $\sum_{k=1}^{N_j} A_{dj}(k) = M-1$ and $\sum_{k=1}^{N} A_d(k) = M(M-1)$. The combination of the enumeration of the squared Euclidean distance, $\Delta_E(k)$, and the number of signal pairs $A_d(k)$ for each distance is often denoted the squared Euclidean distance spectrum weight. All sets of pairs $\{\Delta_E(k), A_d(k)\}$, $k = 1, N$ is often denoted the squared Euclidean distance spectrum of a signal set. Similiarly, a conditional Euclidean distance spectrum can be defined.

EXAMPLE 14.8

For pulse width modulation there are two unique conditional Euclidean distance spectra. For $\vec{I} = 0$ there are three terms in the conditional distance spectrum, i.e.,

$$N_0 = 3 \qquad \{\Delta_{E0}(k) A_{d0}(k)\} \in \left[\left\{\frac{8E_b}{10}, 1\right\}, \left\{\frac{16E_b}{10}, 1\right\}, \left\{\frac{24E_b}{10}, 1\right\}\right] \tag{14.56}$$

and for $\vec{I} = 1$ there are two terms, i.e.,

$$N_1 = 2 \qquad \{\Delta_{E0}(k) A_{d0}(k)\} \in \left[\left\{\frac{8E_b}{10}, 2\right\}, \left\{\frac{16E_b}{10}, 1\right\}\right] \tag{14.57}$$

which implies, for example, that

$$P_{WUB}(E|\vec{I}=0) = \frac{1}{2}\text{erfc}\left(\sqrt{\frac{8E_b}{40N_0}}\right) + \frac{1}{2}\text{erfc}\left(\sqrt{\frac{16E_b}{40N_0}}\right) + \frac{1}{2}\text{erfc}\left(\sqrt{\frac{24E_b}{40N_0}}\right) \tag{14.58}$$

Pulling all the conditional distance spectra together we have $N = 3$ with $\Delta_E(1) = \frac{8E_b}{10}$, $\Delta_E(2) = \frac{16E_b}{10}$, and $\Delta_E(3) = \frac{24E_b}{10}$ and $A_d(1) = 6$, $A_d(2) = 4$, and $A_d(3) = 2$.

Consequently, the union bound is expressed as

$$P_W(E) \leq P_{WUB}(E) = \frac{3}{4}\text{erfc}\left(\sqrt{\frac{8E_b}{40N_0}}\right) + \frac{1}{2}\text{erfc}\left(\sqrt{\frac{16E_b}{40N_0}}\right) + \frac{1}{4}\text{erfc}\left(\sqrt{\frac{24E_b}{40N_0}}\right) \tag{14.59}$$

---

3. The bound is often dominated by the minimum squared Euclidean distance of the signal set, $\Delta_E(\text{min})$, where

$$\Delta_E(\text{min}) = \min_{\substack{j \in \{0,\dots,M-1\} \\ i \in \{0,\dots,M-1\} \\ i \neq j}} \Delta_E(i,j) \tag{14.60}$$

   Since $\text{erfc}(x) \sim \exp[-x^2]$ it is usually the case that

$$\text{erfc}\left(\sqrt{\frac{\Delta_E(\text{min})}{4N_0}}\right) \gg \text{erfc}\left(\sqrt{\frac{\Delta_E(i,j)}{4N_0}}\right) \quad \forall \Delta_E(i,j) \neq \Delta_E(\text{min}) \tag{14.61}$$

4. The union bound is usually tight at high SNR. The bound over counts the cases when more than one maximum likelihood metric for a nontransmitted word is greater than the maximum likelihood metric for the transmitted word. At high SNR the probability of two or more metrics being greater than the true metric is very small. Consequently, the overbounding probability is very small.
5. Using items 3 and 4 allows one to deduce that

$$P_W(E) \approx \frac{A_d(\text{min})}{2M}\text{erfc}\left(\sqrt{\frac{\Delta_E(\text{min})}{4N_0}}\right) \tag{14.62}$$

   is a good approximation to the true error rate at high SNR.

EXAMPLE 14.9

For pulse width modulation a reasonable estimate of performance at high SNR is expressed as

$$P_W(E) = \frac{3}{4}\text{erfc}\left(\sqrt{\frac{8E_b}{40N_0}}\right) \tag{14.63}$$

Similarly, for the 4-ary modulation introduced in Example 12.1 a reasonable estimate of performance at high SNR is expressed as

$$P_W(E) = \frac{1}{4}\text{erfc}\left(\sqrt{\frac{T_p}{12N_0}}\right) \tag{14.64}$$

A concept that arises often in error rate analysis is the concept of geometric uniformity [For91].

Definition 14.3 A geometrically uniform signal set is one in which the conditional distance spectrum of each of the possible transmitted signals is the same.

The important characteristic of a geometrically uniform signal set is

$$P_{WUB}(E|\vec{I}=0) = P_{WUB}(E|\vec{I}=j) = P_{WUB}(E) \qquad (14.65)$$

This characteristic implies that the number of terms in the union bound that needs to be considered reduces from $M(M-1)$ to $M-1$ as only one transmitted code word needs to be considered. Note that pulse width modulation considered in Example 14.4 is not a geometrically uniform signal set, but $x_0(t)$ and $x_3(t)$ have the same conditional distance spectrum as do $x_1(t)$ and $x_2(t)$.

14.2.4 Signal Design

Signal design for optimizing the fidelity of message reconstruction of an M-ary modulation is typically implemented with a max-min approach. Since the minimum squared Euclidean distance will dominate the word error rate of an optimum demodulator, the best signal design will happen when the minimum distance between all pairs of the M signals is maximized. This is known as the max-min design criteria for digital communications. In general, signal design in digital communications is a complex endeavor with many open problems.

The spectral characteristics of an M-ary signal set can be characterized with a straightforward extension of the techniques used for binary signals. Recall from Eq. (12.8) that the average energy spectrum per bit for a transmitted signal, $X_z(t)$, where K_b bits are transmitted is

$$D_{x_z}(f) = \frac{E[G_{x_z}(f)]}{K_b} \qquad (14.66)$$

The average here again is over random data bits that are being transmitted.

EXAMPLE 14.10

The average energy spectrum per bit for pulse width modulation can be obtained by identifying the energy spectrum for each of the possible transmitted waveforms. For pulse width modulation this is given as

$$G_{x_0}(f) = \frac{T_p^2}{16}\text{sinc}(fT_p/4)^2 \qquad G_{x_1}(f) = \frac{T_p^2}{4}\text{sinc}(fT_p/2)^2 \qquad (14.67)$$

$$G_{x_2}(f) = \frac{9T_p^2}{16}\text{sinc}(3fT_p/4)^2 \qquad G_{x_3}(f) = T_p^2\text{sinc}(fT_p)^2$$

If the signals are equally likely then the average energy spectrum for pulse width modulation is given in Figure 14.4. Since $W_b = 2/T_p$ and using the 3 dB bandwidth ($B_T \approx 1/T_p$) the spectral efficiency is $\eta_B = 2$ bits/s/Hz.

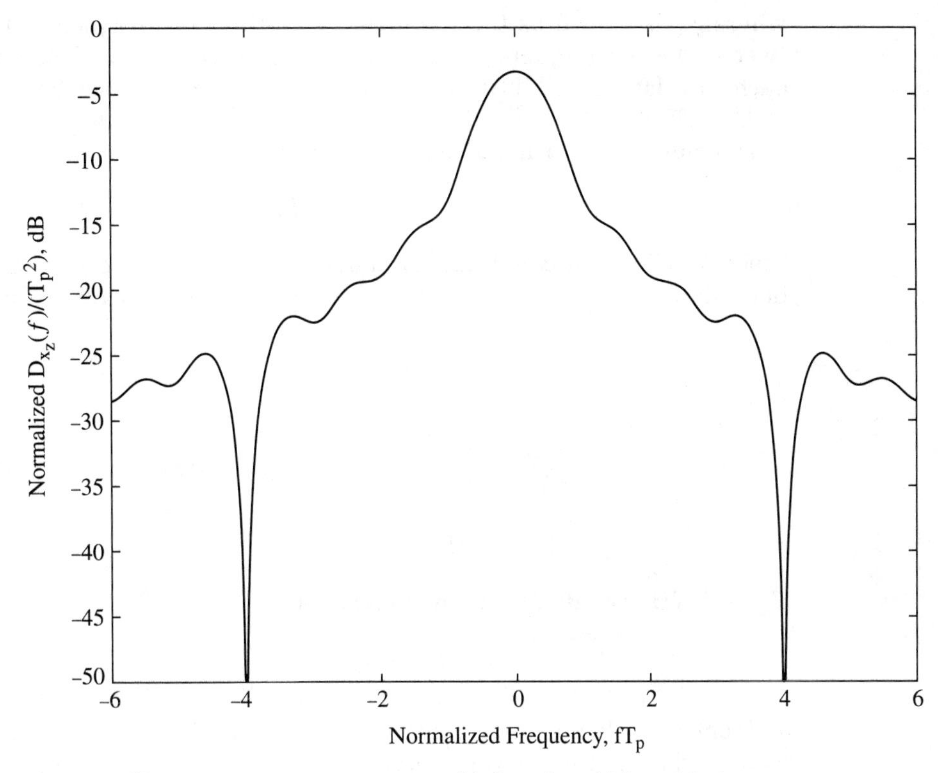

Figure 14.4 The average energy spectrum per bit for pulse width modulation.

14.3 Examples

This chapter is concluded by considering two obvious and important examples of M-ary carrier modulated digital communication: M-ary frequency shift keying (MFSK) and M-ary phase shift keying (MPSK). In these examples it is assumed that $\pi_i = 1/M,\ i \in \{0, \ldots, M-1\}$.

14.3.1 *M*-ary FSK

MFSK modulation sends the word of information by transmitting a carrier pulse of one of M frequencies. This is an obvious simple signaling scheme and one used in many early modems. The signal set is given as

$$x_i(t) = \begin{cases} \sqrt{\dfrac{K_b E_b}{T_p}} \exp[j2\pi f_d(2i - M + 1)t] & 0 \le t \le T_p \\ 0 & \text{otherwise} \end{cases} \tag{14.68}$$

where f_d is known as the frequency deviation. The frequency difference between adjacent frequency pulses in the signal set is $2f_d$. It is apparent that each waveform in a MFSK signal set has equal energy that has here been set to $E_s = K_b E_b$.

The matched filter impulse response is given as

$$h_i(t) = x_i^*(T_p - t) \tag{14.69}$$

Since MFSK is an equal energy signal set the energy correction term is not needed in the demodulator and the ML decision metrics are given as

$$\begin{aligned} T_i &= \Re\left[\int_0^{T_p} Y_z(t)x_i^*(t)dt\right] \\ &= \sqrt{\frac{K_b E_b}{T_p}}\Re\left[\int_0^{T_p} Y_z(t)\exp[-j2\pi f_d(2i - M + 1)t]dt\right] \\ i &\in \{0,\ldots,M-1\} \end{aligned} \tag{14.70}$$

The MLWD demodulator is then given as

$$\hat{I} = \arg \max_{i\in\{0,\ldots,M-1\}} T_i \tag{14.71}$$

This optimum demodulator computes M matched filter outputs and then selects as a word estimate the word value that corresponds to the largest real part of all the matched filter outputs.

The word error rate performance of MFSK is a function of the frequency spacing, $2f_d$. Using the results from BFSK, i.e., (13.87), the pairwise squared Euclidean distance is given as $\Delta_E(i,j) = 2K_bE_b(1 - \frac{\sin(4\pi f_d(i-j)T_p)}{4\pi f_d(i-j)T_p})$. Consequently, for MFSK the placement of the tones is a bit more delicate problem. The union bound is given as

$$P_{WUB}(E) = \sum_{n=1}^{M-1} \frac{M-n}{M}\operatorname{erfc}\left(\sqrt{\frac{K_bE_b}{2N_0}\left(1 - \frac{\sin(4\pi f_d n T_p)}{4\pi f_d n T_p}\right)}\right) \tag{14.72}$$

The selection of the frequency spacing, f_d, becomes a tricky problem for $M > 2$ as one needs to balance all the terms in the union bound to optimize the error rate. The MFSK signal set is not in general a geometrically uniform signal set.

For the special case of $\Re[\rho_{ij}] = 0$ both the $P_W(E)$ can be easily computed and a simple form for the union bound results. This is known as orthogonal MFSK and is achieved if $f_d = \frac{n}{4T_p}$ where n is a positive integer. The closed form probability of error expression for orthogonal MFSK is explored in the homework and plotted in Figure 14.5 for $M = 2, 4, 8, 16$. The important thing to realize for MFSK is that for large enough E_b/N_0, the $P_W(E)$ is monotonically decreasing with M. In fact, one can show that the error rate can be made arbitrarily small when $E_b/N_0 > \ln(2)$ [Sha48]. Consequently, the performance

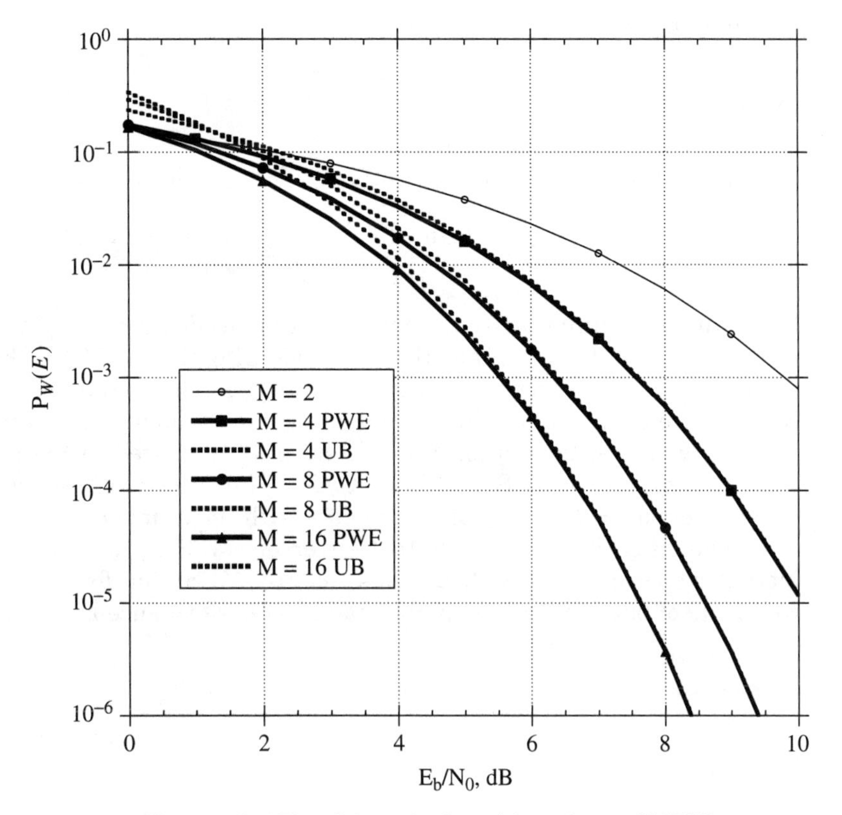

Figure 14.5 The true $P_W(E)$ and the union bound for orthogonal MFSK.

of a digital communication system can be improved by transmitting more bits or having more waveforms to choose from. This result is also counter normal engineering intuition. Shortly, we will show that this performance improvement is achieved only at the cost of an increase in the bandwidth of the signal. The union bound for orthogonal MFSK is

$$P_W(E) \leq \frac{M-1}{2} \operatorname{erfc}\left(\sqrt{\frac{K_b E_b}{2N_0}}\right) \tag{14.73}$$

For $\Re[\rho_{ij}] = 0$ MFSK is a geometrically uniform signal set. Figure 14.5 also plots the union bound for MFSK. It is obvious that the union bound converges to the true error probability at high SNR. For the particular case of MFSK the bound is tight enough to be indiscernibly different on the graph for $E_b/N_0 > 6$ dB. This characteristic for the union bound holds for most signal sets optimally demodulated in the presence of an AWGN.

The average energy spectrum per bit is again used to characterize the spectral efficiency. Recall the average energy spectral density per bit is given for M-ary modulations as

$$D_{x_z}(f) = \frac{1}{K_b} \sum_{i=0}^{M-1} \pi_i G_{x_i}(f) \tag{14.74}$$

Recall that the energy spectrum of the individual waveforms is given as

$$G_{x_i}(f) = K_b E_b T_p \left(\frac{\sin(\pi(f - f_d(2i - M + 1))T_p)}{\pi(f - f_d(2i - M + 1))T_p} \right)^2 \tag{14.75}$$

Recall that the minimum frequency separation needed to achieve an orthogonal modulation is $f_d T_p = 0.25$ and that by considering Eq. (14.74) and Eq. (14.75) it is obvious that the spectral content is growing proportional to $B_T = f_d(2^{K_b+1})$. An example of each of the individual energy spectrums (dotted lines) and the average energy spectrum for 8FSK (solid line) is plotted in Figure 14.6. The transmission rate of MFSK is $W_b = K_b/T_p$. The spectral efficiency then is approximately $\eta_B = K_b/2^{K_b-1}$ and decreases with the number of bits transmitted or equivalently decreases with M. Conversely MFSK provides monotonically increasing performance with M. Consequently, MFSK has found use in practice when lots of bandwidth is available and good performance is required.

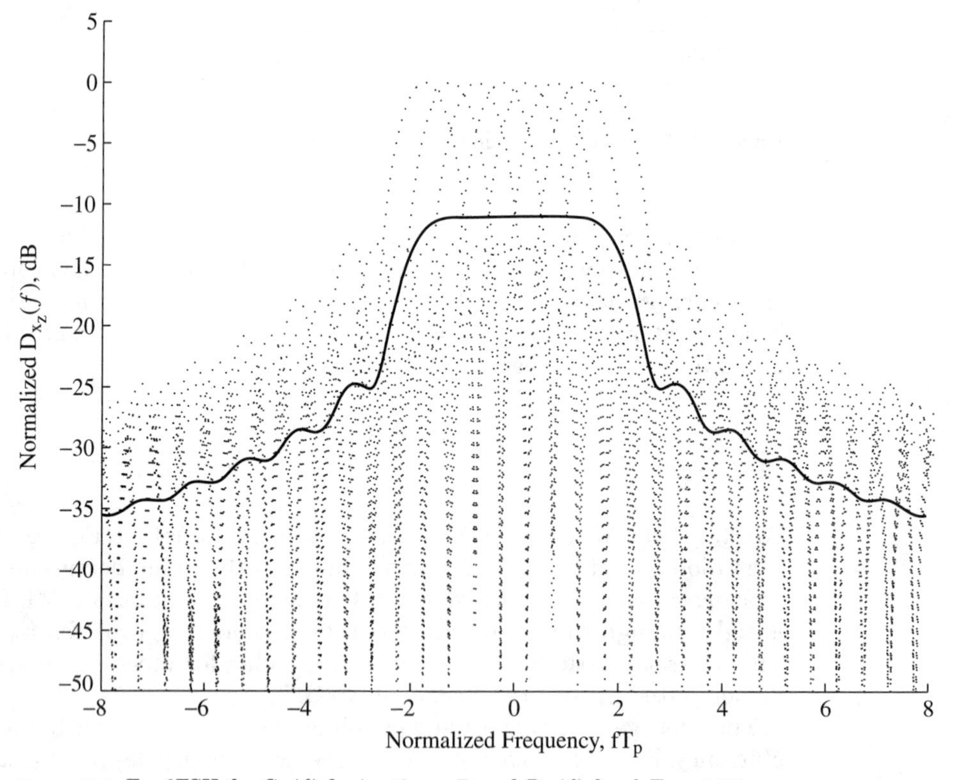

Figure 14.6 For 8FSK the $G_{x_i}(f)$ for $i \in \{0, \ldots, 7\}$ and $D_{x_z}(f)$ for $f_d T_p = 0.25$.

The advantages of MFSK are summarized as

- Very simple to generate. Simply gate one of M oscillators on depending on the word to be sent.
- Performance (error rate) improves monotonically with K_b. This is counter intuition as normal intuition would expect performance degrading with increasing K_b.

The disadvantages of MFSK are summarized as

- The bandwidth increases exponentially with K_b, hence the spectral efficiency of MFSK decreases with K_b.
- Complexity increases exponentially with K_b.

An additional important point demonstrated by this example

- Union bound is simple to compute and is a tight bound at moderate to high SNR.

14.3.2 *M*-ary PSK

MPSK modulation sends the word of information by transmitting a carrier pulse of one of M phases. This is an obvious simple signaling scheme and one used in many modems. The form of the signal set is given as

$$x_i(t) = \begin{cases} \sqrt{\dfrac{K_b E_b}{T_p}} \exp[j\theta(i)] & 0 \leq t \leq T_p \\ 0 & \text{elsewhere} \end{cases} \tag{14.76}$$

where $\theta(i)$ is a mapping from the information words into a phase. The phases in MPSK are normally uniformly spaced around the unit circle in engineering practice. The mapping of information word values into phases is not unique. Two common mappings exist in communications practice: natural mapping, $\theta_N(i)$, and Gray mapping, $\theta_G(i)$ [Gra53]. Natural mapping assigns the phase sequentially with the binary numeric representation of the information word. Unfortunately, if natural mapping is used then words which map to adjacent phases and hence close Euclidean distances can have many bit differences. Gray coding is a technique mitigating the effects of word errors on the $P_B(E)$ by mapping of binary words into phases in a way to where adjacent phases are only different in 1 bit. An example of the two mappings for $M = 4$ is shown in Table 14.1. Since natural mapping is straightforward to enumerate functionally, i.e.,

$$\theta_N(i) = \frac{\pi(2i + 1)}{M} \tag{14.77}$$

TABLE 14.1 The two most common $M = 4$ PSK mappings

$\vec{I}$	$I(1)\,I(2)$	$\theta_N(i)$	$\theta_G(i)$
0	00	$\frac{\pi}{4}$	$\frac{\pi}{4}$
1	01	$\frac{3\pi}{4}$	$\frac{3\pi}{4}$
2	10	$\frac{5\pi}{4}$	$\frac{7\pi}{4}$
3	11	$\frac{7\pi}{4}$	$\frac{5\pi}{4}$

the remainder of this section will consider natural mapping exclusively. The sequel will return to consider Gray mapping in more detail for QPSK.

Since MPSK is an equal energy signal set the energy correction term is not needed in the demodulator and the ML decision metrics are given as

$$T_i = \Re\left[\int_0^{T_p} Y_z(t)x_i^*(t)dt\right] = \sqrt{\frac{K_b E_b}{T_p}}\Re\left[\int_0^{T_p} Y_z(t)\exp\left[-j\frac{\pi(2i+1)}{M}\right]dt\right]$$

$$= \Re\left[\exp\left[-j\frac{\pi(2i+1)}{M}\right]\sqrt{\frac{K_b E_b}{T_p}}\int_0^{T_p} Y_z(t)dt\right]$$

$$= \Re\left[\exp\left[-j\frac{\pi(2i+1)}{M}\right]Q\right] \qquad i \in \{0, \ldots, M-1\} \tag{14.78}$$

where

$$Q = \sqrt{\frac{K_b E_b}{T_p}}\int_0^{T_p} Y_z(t)dt \tag{14.79}$$

is denoted the pulse shape matched filter output. The MLWD computes the output of one filter, Q, derotates this value by each of the possible transmitted phasors and picks the signal that gives the largest real value. This decision rule is equivalent to picking the transmitted phase that is closest to the phase of the pulse shape matched filter so that

$$\hat{\vec{I}} = \arg\max_{i\in\{0,\ldots,M-1\}} T_i = \arg\min_{i\in\{0,\ldots,M-1\}}\left|Q_p - \frac{\pi(2i+1)}{M}\right| \tag{14.80}$$

where $Q_p = \arg\{Q\}$.

Surprisingly this optimum demodulator has only one filter output to compute. This filter output is then processed to produce the decision metric. The word decision is then made on the basis of which of these metric is the largest. The decision rule reduces down to the establishing of decision regions in the complex plane for the pulse matched filter output Q. The decision regions for 4PSK are shown in Figure 14.7a. The fact that the optimum M-ary demodulator only has one filter is due to the transmitted signal set having the form

$$x_i(t) = d_i u(t) \tag{14.81}$$

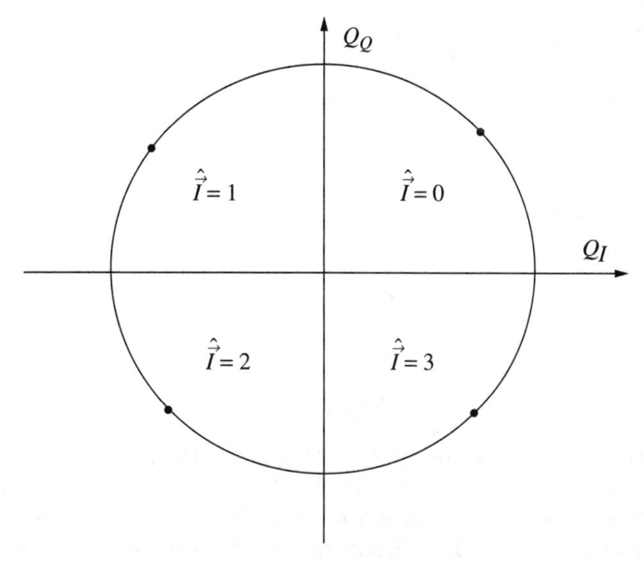

Figure 14.7a The decision regions for natural mapping 4PSK.

where

$$u(t) = \begin{cases} \sqrt{\dfrac{K_b E_b}{T_p}} & 0 \le t \le T_p \\ 0 & \text{elsewhere} \end{cases} \tag{14.82}$$

is the pulse shape. Modulations which have the form of Eq. (14.81) are often denoted linear modulations by communications engineers. Since linear modulations greatly reduce the demodulator complexity they will be explored in detail in the sequel.

The resulting error rate of MPSK is a function of the phase spacing, $\frac{2\pi}{M}$. The squared Euclidean distance is given as $\Delta_E(i, j) = 2K_b E_b(1 - \cos(\frac{2\pi(i-j)}{M}))$. Note since cosine is a periodic function, MPSK is a geometrically uniform signal set. The union bound for MPSK is given as

$$P_W(E) \le \frac{1}{2}\text{erfc}\left(\sqrt{\frac{K_b E_b}{N_0}}\right) + \sum_{i=1}^{M/2-1} \text{erfc}\left(\sqrt{\frac{K_b E_b}{2N_0}\left[1 - \cos\left(\frac{2\pi i}{M}\right)\right]}\right) \tag{14.83}$$

The MFSK signal set was not in general a geometrically uniform signal set but here we see that the MPSK signal set is geometrically uniform when the phases of the modulation symbols are uniformly distributed around the unit circle.

It is interesting to note that for MPSK the union bound can actually be tightened compared to the general result given in Eq. (14.83). This tightening of the union bound is due to the property on the next page.

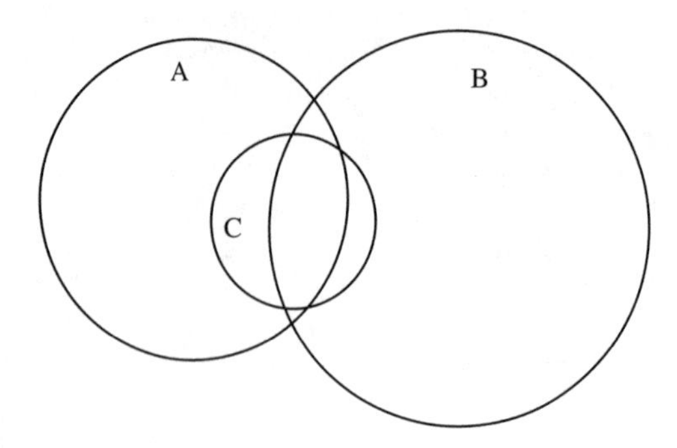

Figure 14.7b An event that does not need to be considered in the union bound computation.

Property 14.1 $P(A \cup B \cup C) = P(A \cup B)$ if $C \subset A \cup B$.

This property implies that if the set C is a subset of the union of the sets A and B then C can be ignored in computing a union bound. For digital communications this implies that if any pairwise error event is a subset of the union of two (or more) pairwise error probabilities then that pairwise error event does not need to be considered in the union bound computation. This concept is shown in Figure 14.7b with a Venn diagram.

Since the MPSK signal set is geometrically uniform we can consider $x_0(t)$ to be the transmitted signal without loss of generality. The pairwise error probability is then given as

$$\{T_{i|0} > T_{0|0}\} = \left\{\Re\left[\exp\left[-j\frac{\pi(2i+1)}{M}\right]Q\right] > \Re\left[\exp\left[-j\frac{\pi}{M}\right]Q\right]\right\}$$

$$= \left\{\left|Q_p - \frac{\pi(2i+1)}{M}\right| < \left|Q_p - \frac{\pi}{M}\right|\right\} \tag{14.84}$$

This reduces to

$$\{T_{i|0} > T_{0|0}\} = \left\{\frac{\pi(i+1)}{M} \le Q_p \le \frac{\pi(M+i+1)}{M}\right\} \tag{14.85}$$

Given this form of the pairwise error probability it can be seen that

$$\{T_{1|0} > T_{0|0}\} \cup \{T_{M-1|0} > T_{0|0}\} = \left\{\frac{2\pi}{M} \le Q_p \le 2\pi\right\} \tag{14.86}$$

Consequently, for any $i \in \{2, \ldots, M-2\}$

$$\{T_{i|0} > T_{0|0}\} \subset \{T_{1|0} > T_{0|0}\} \cup \{T_{M-1|0} > T_{0|0}\} \tag{14.87}$$

This implies that a tighter union bound for MPSK than Eq. (14.83) is given as

$$P_W(E) \le \text{erfc}\left(\sqrt{\frac{K_b E_b}{2N_0}\left[1 - \cos\left(\frac{2\pi}{M}\right)\right]}\right) \tag{14.88}$$

This tightened union bound only includes the minimum squared Euclidean distance terms. For MPSK these two pairwise error probability terms include all the possible ways an error can occur.

MPSK word error probability performance can be computed exactly. If we assume that $x_k(t)$ was transmitted, the conditional matched filter output is given as

$$Q = \exp\left[j\frac{\pi(2k+1)}{M}\right]K_bE_b + N_z \tag{14.89}$$

where N_z is a complex Gaussian noise with a variance of

$$\mathrm{var}(N_z) = N_0K_bE_b \tag{14.90}$$

Considering the decision regions and the form of the conditional matched filter output, the probability of word error can be computed. This idea is explored in the homework problems. The resulting probability of word error is plotted in Figure 14.8 for $M = 2, 4, 8, 16$. In contrast to MFSK the probability of error

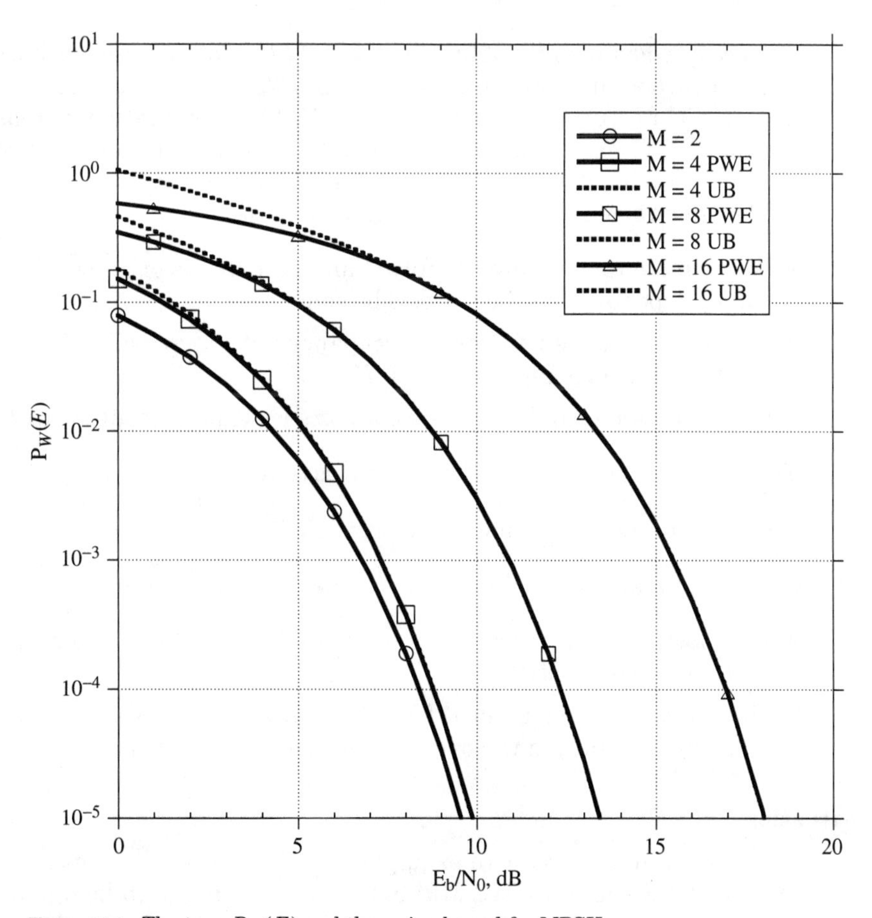

Figure 14.8 The true $P_W(E)$ and the union bound for MPSK.

increases with M for MPSK. The error rate degrades about 4 dB for each doubling of M. The union bound is also plotted in Figure 14.8 for $M = 2, 4, 8, 16$. The union bound is again asymptotically tight with moderate SNR. This again demonstrates the utility of a union bound error rate analysis.

The average energy spectrum per bit is again used to characterize the spectral efficiency of MPSK. Recall the average energy spectral density per bit is given for M-ary modulations as

$$D_{x_z}(f) = \frac{1}{K_b} \sum_{i=0}^{M-1} \pi_i G_{x_i}(f) \tag{14.91}$$

Recall that the energy spectrum of the individual waveforms is given as

$$\begin{aligned} G_{x_i}(f) &= K_b E_b T_p \left| \exp\left[j\frac{\pi(2i+1)}{M} \right] \right|^2 \left(\frac{\sin(\pi f T_p)}{\pi f T_p} \right)^2 \\ &= K_b E_b T_p \left(\frac{\sin(\pi f T_p)}{\pi f T_p} \right)^2 \end{aligned} \tag{14.92}$$

The occupied bandwidth of MPSK, $B_T = 1/T_p$, does not increase with M while the bit rate does increase with M, $W_b = K_b/T_p$ to provide a spectral efficiency of $\eta_B = K_b$. Unfortunately, the error rate of MPSK modulation increases with increasing M. Consequently, MPSK modulations are of interest in practice when the available bandwidth is small and the SNR is large.

The advantages of MPSK are summarized as

- Very simple to generate. Simply change the phase of an oscillator to one M values depending on the word to be sent.
- No increase in the bandwidth occupancy with increasing K_b. Consequently, spectral efficiency increases with K_b.
- Demodulation complexity does not increase exponentially with K_b.

The disadvantages of MPSK are summarized as

- Error rate increases monotonically with K_b.

Additional important points demonstrated by this example

- The most general union bound is simple to compute and is a tight bound at moderate to high SNR.
- The demodulator structure of MPSK allows a tighter union bound to be computed by considering the overlap in the decision regions.

14.3.3 Discussion

This section introduced two example modulations to transmit K_b bits of information: MFSK and MPSK. MPSK has the advantage in being able to supply an increasing spectral efficiency with K_b at the cost of requiring more E_b/N_0

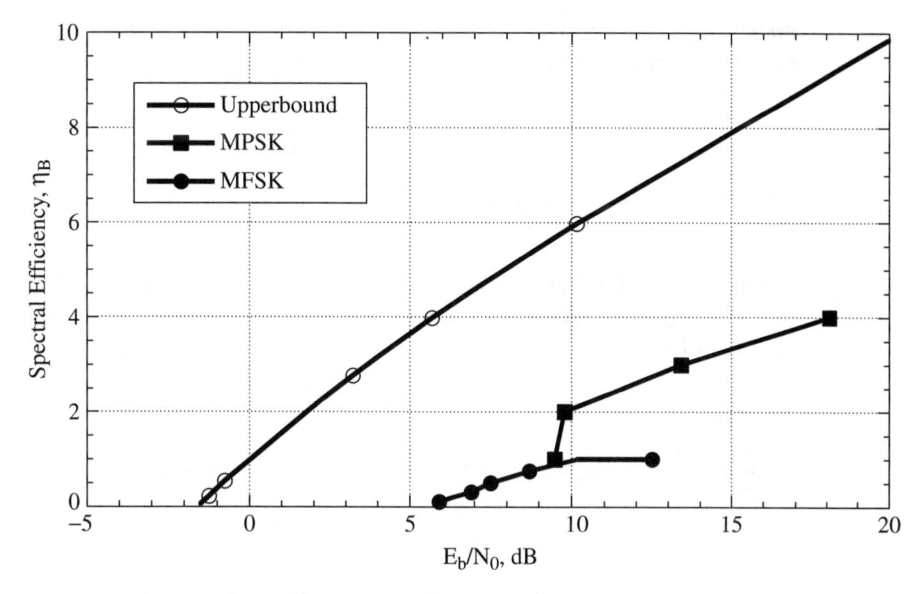

Figure 14.9 A comparison of the spectral efficiency of MPSK and MFSK with the upperbound.

to achieve the same error rate. MFSK can provide improved error rate performance with K_b but at a cost of a loss of spectral efficiency. Also the decoding complexity of MPSK is significantly less than the decoding complexity of MFSK. As a final point it is worth comparing the spectral efficiency performance of these two modulations with the upperbounds provided by information theory (see Section 12.3). As before we will denote reliable communication as being an error rate of 10^{-5}. The operating points of MFSK and MPSK and the upperbound on the possible performance are plotted in Figure 14.9. It is clear from this graph that different modulations give us a different set of points in a performance versus spectral efficiency trade-off. Also the two examples we considered in this chapter have a performance much lower than the upperbound provided by information theory. This is still not too disturbing as we have significantly more digital communication theory to explore.

14.4 Homework Problems

Problem 14.1. Find the probability of word error in MLWD for an arbitrary M for the case of equal energy orthogonal signaling, $\Re[\rho_{ij}] = 0,\ \forall i \neq j$. *Hint*: Condition on a value of the correct likelihood metric and find the conditional probability of word error (a distribution function of a max(•) random variable). Next average over the density function of correct likelihood metric. Compute numeric values and plot for $M = 4, 8, 16$ for values of $E_b/N_0 = 0 - 10$ dB. The first derivation of this result is available in [Kot60] but it is worth it to try to obtain the solution yourself before checking the reference. What value of f_d achieves orthogonality for the MFSK example considered in this chapter for a given T_p.

Problem 14.2. Show that if you design an equicorrelated ($\Re[\rho_{ij}] = \rho \leq 0,\ \forall i \neq j$), equal energy M-ary signaling scheme then

$$\rho \geq -\frac{1}{M-1}$$

Interpret what this means as M gets large. *Hint*: Consider a composite signal that is a sum of all M signals.

Problem 14.3. The PDF of the phase error between the signal phase and the measured phase, $E_P = Y_P - x_P$, of the complex envelope of a carrier modulated signal in bandpass Gaussian noise, $Y_z = x_z + N_z$, is given as

$$f_{E_P}(e_p)$$
$$= \begin{cases} \frac{\exp[-P]}{2\pi} + \sqrt{\frac{P}{4\pi}}\cos(e_p)\exp[-P\sin^2(e_p)](1+\operatorname{erf}(\sqrt{P}\cos(e_p))) & -\pi \leq e_p \leq \pi \\ 0 & \text{elsewhere} \end{cases} \tag{14.93}$$

where $P = \frac{x_A^2}{\mathrm{var}(N_z)}$.

(a) Plot this PDF for $P = 10$ dB.

(b) Use this result to compute the word error probability for M-ary PSK. Compute numeric values and plot for $M = 4, 8, 16$ for values of $E_b/N_0 =$ 0–20 dB.

Problem 14.4. A simple 4-ary modulation is given as

$$x_i(t) = d_i(1)u(t) + d_i(2)u(t-T)$$

where $u(t)$ is a pulse shape with energy E_u having support on $[0, T_u]$ and $T_u > T$, $i = 2 \times I(2) + I(1)$, and $d_i(l) = (-1)^{I(l)}$. This signal is to be detected in the presence of an AWGN with one-sided spectral density of N_0 and the words are a priori equally likely, $\pi_i = 0.25\ i \in \{0, \ldots, 3\}$.

(a) What is the length of the transmission, T_p?

(b) Demodulation for this modulation is a function of the correlation function of the pulse shape, $u(t)$, evaluated at T, $V_u(T)$. Find and detail out the optimum word error demodulator as a function of $V_u(T)$.

(c) Find the union bound to the optimum word error demodulator probability of word decision error.

(d) Define the following matched filter outputs

$$Q(1) = \int_0^{T_p} y_z(t)u^*(t)dt \qquad Q(2) = \int_0^{T_p} y_z(t+T)u^*(t)dt$$

Find a simple form for the optimum demodulators given above as a function of $Q(1)$ and $Q(2)$. $Q(1)$ and $Q(2)$ are sufficient statistics for optimal detection.

(e) If $Q(1) = 0.5$, $Q(2) = -0.1$, $N_0 = 0.3$, $V_u(T) = 0.1$, and $E_u = 1$ what is the optimum word?

Problem 14.5. Show via the union bound that M-ary orthogonal FSK signaling can have a $P_W(E)$ arbitrarily small by a proper selection of K_b as long as $E_b/N_0 > 2\ln(2)$. Actually this statement is true for $E_b/N_0 > \ln(2)$ [Sha48] but the result is harder to prove (but worth an attempt by the interested student!)

Problem 14.6. 8-ary phase shift keying (8PSK) has a received signal of the form

$$Y_z(t) = x_i(t) + W_z(t) = d_i u(t) + W_z(t) \tag{14.94}$$

where $W_z(t)$ is a complex AWGN and if $\vec{I} = i$ then $d_i = \exp(\frac{j\pi i}{4})$ $\quad i \in \{0, \ldots, 7\}$. Assume $\pi_i = \frac{1}{8}$ $\quad i \in \{0, \ldots, 7\}$.

(a) Find the minimum probability of word error demodulator.

(b) Find and plot the tightest union bound to the probability of word error for the demodulator found in (a) for $E_b/N_0 = 0{-}13$ dB.

(c) Gray coding assigns bit patterns to M-ary signal points in such a way that adjacent signals (in Euclidean space) only have 1 bit different. Find a Gray code mapping for this 8PSK signal set.

(d) Show that with the demodulator in (a) and the bit to symbol mapping derived in (c) that the resulting BEP is

$$P_B(E) = \frac{1}{3}\left[\operatorname{erfc}\left(\sqrt{\frac{3E_b}{N_0}}\sin\left(\frac{\pi}{8}\right)\right) + \operatorname{erfc}\left(\sqrt{\frac{3E_b}{N_0}}\cos\left(\frac{\pi}{8}\right)\right)\right.$$
$$\left. \times \left[\frac{1}{2} + \frac{1}{2}\operatorname{erf}\left(\sqrt{\frac{3E_b}{N_0}}\sin\left(\frac{\pi}{8}\right)\right)\right]\right] \tag{14.95}$$

Problem 14.7. Consider a 4-ary communication waveform that achieves $W_b = 2/T_p$ having the form shown in Figure 14.10. Assume all words have equal priors and the corrupting noise is an AWGN with a one-sided spectral density of N_0.

(a) Identify the MLWD structure.

(b) Give the union bound for the performance.

(c) Identify the 3 dB bandwidth of the transmitted waveform and the spectral efficiency.

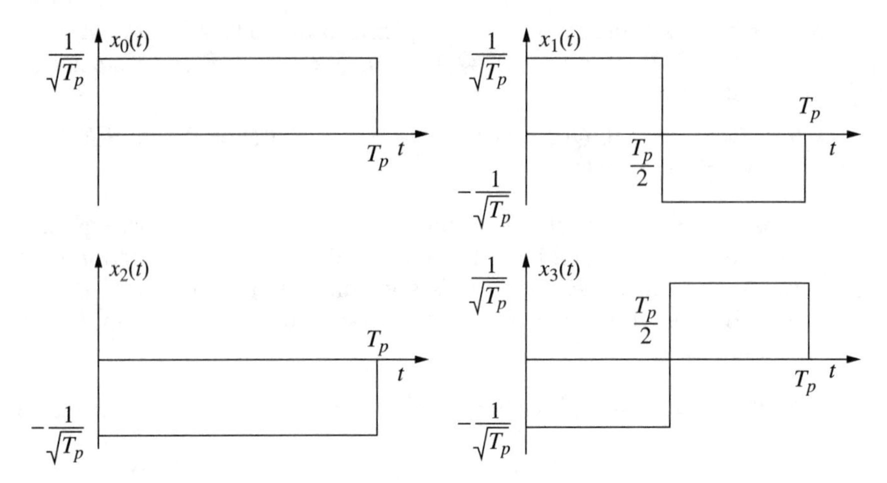

Figure 14.10 A 4-ary modulation.

A proposed approach to increase the throughput of digital communications waveforms like that shown in Figure 14.10 is variable phase shift keying (VPSK) [Wal97]. In variable phase shift keying the transition time between different voltage levels can be modulated to provide more possible waveforms. For example, Figure 14.11 shows a waveform with three possible transition times.

(d) Design a VPSK signal set based on Figure 14.10 that has $W_b = 3/T_p$. *Hint*: You need eight waveforms and only two of four waveforms in Figure 14.10 have transitions.

(e) Optimize performance of the signal set proposed in (d) as a function of the shift parameter, τ_s and compare it to the original waveform.

(f) Compute the spectral efficiency (using 3 dB bandwidth again) for this case of optimal performance.

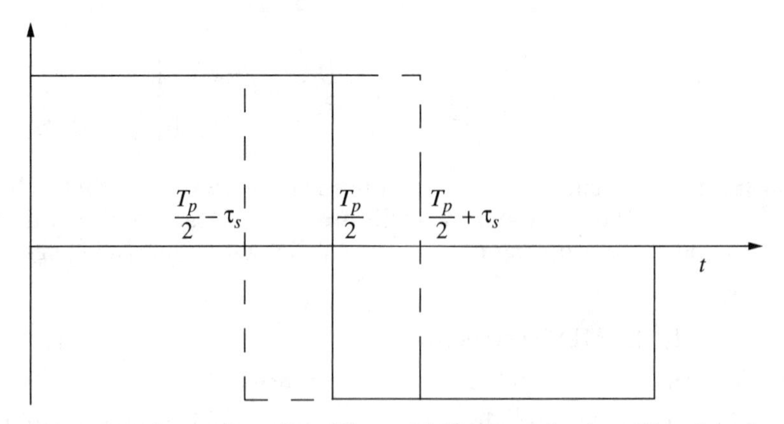

Figure 14.11 A waveform from Figure 14.10 where the transition time is modulated.

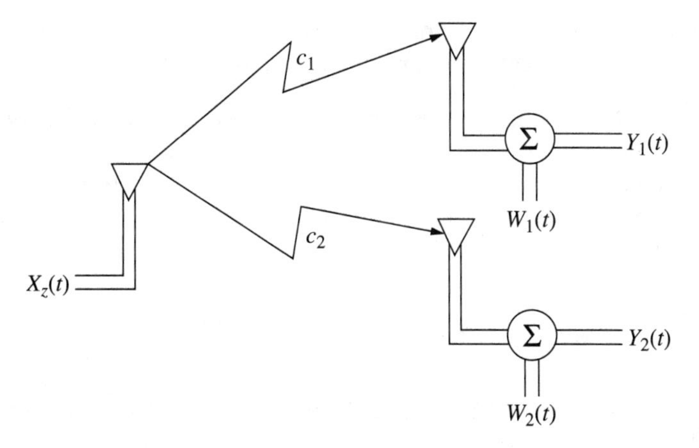

Figure 14.12 Observing a signal on multiple antennas.

Problem 14.8. In wireless communications it is often useful to observe a received waveform on more than one receiver antenna to improve the reliability of the decision. A situation like this is depicted in Figure 14.12 for $L_r = 2$ antennas where the received signal at the jth antenna is given as

$$Y_j(t) = c_j X_z(t) + W_j(t) \tag{14.96}$$

where C_j is a complex constant that represents jth channel distortion and $W_j(t)$ is a white Gaussian noise with $R_{W_j}(\tau) = N_0\delta(\tau)$. Assume that $W_1(t)$ is independent of $W_2(t)$. Assume that c_1 and c_2 are known at the demodulator and that $\pi_0 = 0.5$.

(a) Optimal decisions are based on APPs, i.e., $P(I = m|y(t))$. Bayes rule tells us that

$$P(I = m|y) = \frac{f_Y(y|I = m)\pi_m}{f_Y(y)} \tag{14.97}$$

The results from detection theory tell us that

$$\frac{f_Y(y|I = m)}{f_Y(y)} = C \exp[2T_m/N_0] \tag{14.98}$$

How would the computation of the APP change if two observations with independent noises are obtained.

(b) Using the results from (a) for a given $x_0(t)$ and $x_1(t)$ find the minimum probability of error receiver obtained by observing $Y_j(t) = y_j(t) \quad j = 1, 2$. Simplify as much as possible. The resulting receiver structure is known as maximal ratio combining [Bre59] and essentially gives a matched filter in space and time.

(c) Compute the error rate performance of the optimum receiver as a function of c_1, c_2, and $\Delta_E(0, 1)$.

(d) Knowing the values of c_1 and c_2 what is the signaling scheme that would minimize the bit error rate for a fixed transmitted energy per bit.

Problem 14.9. In this problem we examine a binary communication system where three decisions are possible, $\hat{I} = 1, \hat{I} = 0, \hat{I} = E$ where E represents a no decision (erasure). The two possible baseband transmitted signals are $x_0(t)$ and $x_1(t) = -x_0(t)$ with $\pi_0 = \pi_1 = 0.5$. Recall the optimum binary demodulator is of the form

$$P(I = 1|v_I) \underset{\hat{I}=0}{\overset{\hat{I}=1}{\gtrless}} P(I = 0|v_I) \tag{14.99}$$

where v_I is the output of the matched filter. The generalization considered in this problem is

$$\begin{aligned} \hat{I} &= 1 \qquad P(I = 1|v_I) > 0.75 \\ \hat{I} &= E \qquad 0.25 < P(I = 1|v_I) \leq 0.75 \\ \hat{I} &= 0 \qquad P(I = 1|v_I) \leq 0.25 \end{aligned} \tag{14.100}$$

(a) Assume $y_z(t) = x_z(t) + W_z(t)$ where $W_z(t)$ is a complex additive white Gaussian with $R_{W_z}(\tau) = N_0\delta(\tau)$, then the demodulator in Eq. (14.100) simplifies to a threshold test as in the binary case. Identify the decision statistic and the two decision thresholds.

(b) Give an expression for the bit error probability $P(\hat{I} \neq I, \hat{I} \neq E)$ in terms of E_b, and N_0 for this signal set and demodulator. Plot this performance in comparison to the traditional binary demodulator.

Problem 14.10. Consider the MFSK example given in the text. Use the union bound to find the optimum frequency spacing for

(a) $M = 4$ at $E_b/N_0 = 8$ dB

(b) $M = 8$ at $E_b/N_0 = 9$ dB

For each case plot the union bound and compare it to the union bound for $\Re[\rho_{ij}] = 0$. A computer might be a friend in this problem.

Problem 14.11. The touch-tone dialing in a telephone is a form of M-ary communication. When a key is pressed on a telephone two tones are generated, i.e.,

$$x_i(t) = A\cos(2\pi f_i(1)t) + A\cos(2\pi f_i(2)t) \tag{14.101}$$

This type of modulation is referred to as dual tone multiple frequency (DTMF) modulation. The DTMF tones can send $K_b = 4$ bits even though there are only 12 keys on the phone. The modulation mappings are shown in Table 14.2.

TABLE 14.2 The dual tone multiple frequency (DTMF) modulation mappings

	$f_i(2)$			
$f_i(1)$	1209 Hz	1336 Hz	1477 Hz	1633 Hz
697 Hz	i = 0 1	i = 1 ABC 2	i = 2 DEF 3	i = 3 A
770 Hz	i = 4 GHI 4	i = 5 JKL 5	i = 6 MNO 6	i = 7 B
852 Hz	i = 8 PRS 7	i = 9 TUV 8	i = 10 WXY 9	i = 11 C
941 Hz	i = 12 *	i = 13 oper 0	i = 14 #	i = 15 D

To simplify the problem assume $T_p = 1$ second. Note also that the modulation is given as a real signal.

(a) Plot the transmitted signal when person tries to call the operator ($i = 13$) for $A = 1$ over the interval $[0, 0.1]$.

(b) Transmission on telephone lines are often thought of as having $f_c = 1200$ Hz. Give the simplest form of the complex envelope when a person tries to call the operator ($i = 13$).

(c) What is E_i $i \in \{0, \ldots, 15\}$.

(d) Detail out the MLWD. Can you get away with only 8 filters instead of 16? If so show the structure.

(e) Compute and plot the union bound to the probability of word error for $E_b/N_0 = 0$–10 dB. Note, with a 16-ary modulation a computer will be your friend in this problem.

Problem 14.12. Consider a 4-ary communication waveform set that achieves $W_b = 2/T_p$ having the form shown in Figure 14.13. Assume all words are equally likely and that the corrupting noise is an AWGN with a one-side spectral density of N_0.

(a) Compute the value of A and B such that $E_i = 2E_b, i = 0, 3$.

(b) Identify the MLWD structure.

(c) Show that only two filters are needed to implement this structure.

(d) Compute the union bound to the probability of word error.

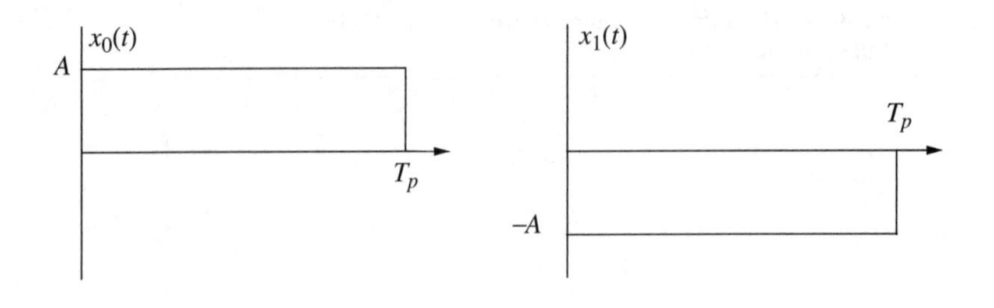

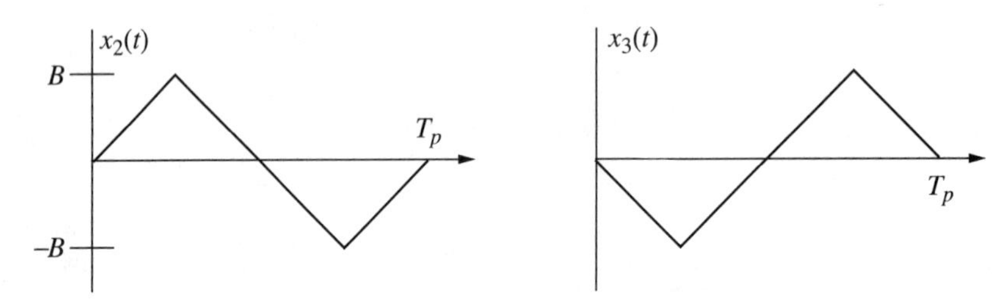

Figure 14.13 The waveforms for a 4-ary communication system.

Problem 14.13. Three bits of information are transmitted with an 8-ary modulation that has been called pulse position modulation (PPM) and is characterized with

$$\begin{aligned} x_0(t) &= u(t) \qquad x_1(t) = u(t - T_p/8) \\ x_2(t) &= u(t - T_p/4) \qquad x_3(t) = u(t - 3T_p/8) \\ x_4(t) &= u(t - T_p/2) \qquad x_5(t) = u(t - 5T_p/8) \\ x_6(t) &= u(t - 3T_p/4) \qquad x_7(t) = u(t - 7T_p/8) \end{aligned} \tag{14.102}$$

where

$$u(t) = \begin{cases} \sqrt{\dfrac{24E_b}{T_p}} & 0 \le t \le T_p/8 \\ 0 & \text{elsewhere} \end{cases} \tag{14.103}$$

Recall for comparison that 8-ary orthogonal frequency shift keying (8FSK) has a minimum $f_d = \frac{1}{4T_p}$ and a transmitted signal given as

$$x_i(t) = \begin{cases} \sqrt{\dfrac{3E_b}{T_p}} \exp\left[\dfrac{j\pi t(i - 3.5)}{T_p}\right] & 0 \le t \le T_p \\ 0 & \text{elsewhere} \end{cases} \tag{14.104}$$

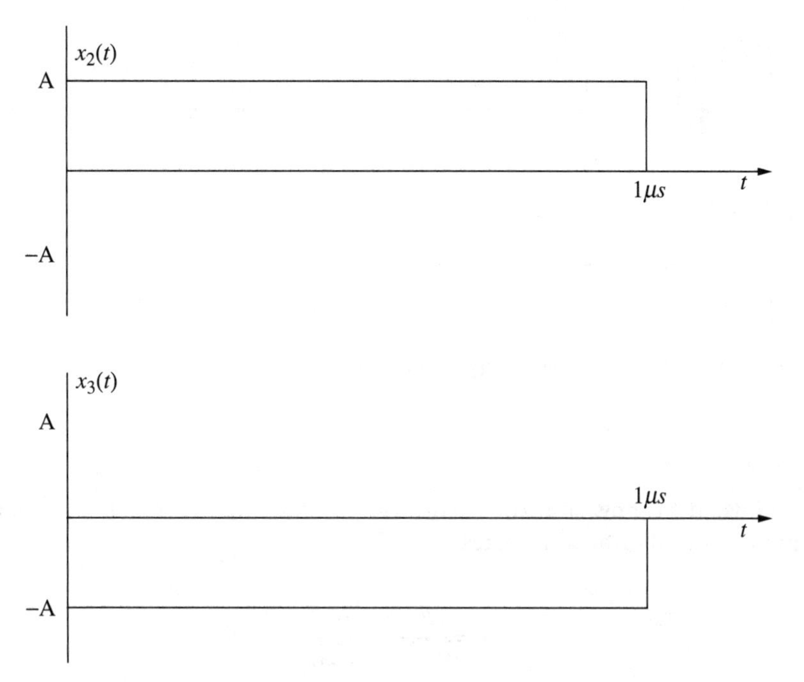

Figure 14.14 Two more waveforms for a 4-ary modulation.

(a) Detail out the maximum likelihood word demodulator for 8PPM. Are there any simplifications that are possible due to using PPM?

(b) Find the union bound to the word error performance for 8PPM. How will this performance compare to 8FSK?

(c) Find the average energy spectrum per bit for 8PPM. How does this spectrum compare to 8FSK?

(d) As a communication engineer assess the advantages and disadvantages of 8PPM versus 8FSK.

Problem 14.14. In the wireless local network protocol denoted IEEE 802.11b the lowest rate modulation is a binary modulation using the two waveforms given in Figure 13.30. A modified system chooses to send $K_b = 2$ equally likely bits by adding two more waveforms as shown in Figure 14.14

(a) Select A such that the energy per bit is E_b.

(b) Show how to compute T_2.

(c) Compute the union bound for this 4-ary modulation.

Problem 14.15. Recall that

$$\Re[W_z(t)x_i^*(t)] = W_I(t)x_{i,I}(t) + W_Q(t)x_{i,Q}(t) \tag{14.105}$$

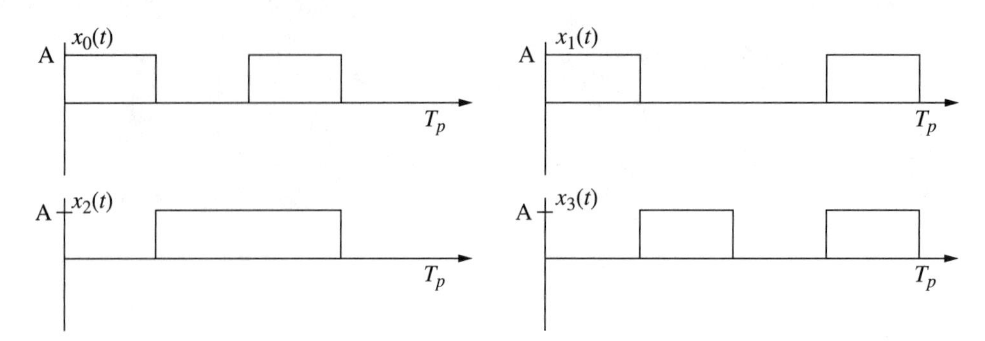

Figure 14.15 Waveforms for a 4-ary modulation.

and use it to show that the correlation coefficient between the noise in different maximum likelihood metrics is

$$\rho = \frac{E\left[N_I^{(i)} N_I^{(k)}\right]}{\frac{N_0}{2}\sqrt{E_i E_k}} = \Re[\rho_{ik}] \tag{14.106}$$

Problem 14.16. The four waveforms in Figure 14.15 are to be used to send $K_b = 2$ bits. Assume the received signal is distorted by an AWGN with a one–sided spectral density of N_0 and that all the signals are a priori equally likely.

(a) Select A such that the energy per bit is E_b.

(b) What is the average energy spectrum per bit?

(c) Detail out the MLWD. It is possible to have the MLWD be implemented with one filter that is sampled at four different time instances. Give the form for this filter.

(d) Compute the union bound for this 4-ary modulation.

(e) Would this modulation perform better or worse than 4-ary orthogonal FSK in terms of word error probability?

(f) Would this modulation have better or worse spectral efficiency than 4-ary orthogonal FSK?

Problem 14.17. Consider a military communication system where bandwidth efficiency is not a primary driver. A decision was made to limit the choice of modulations to MFSK. Large packets of data would be sent by streaming MFSK symbols in time. A team of engineers decided on $M = 16$ for a system design. Formulate a valid set of reasons why this team selected $M = 16$ when trading off the available performance versus complexity. Would the "optimality" of the selection be a function of time?

14.5 Example Solutions

Problem 14.12.

(a) Given $E_i = 2E_b$, where $i = 0, 1, 2, 3$ and we note that,

$$E_i = \int_{-\infty}^{\infty} |x_i|^2 dt = \int_{0}^{T_p} |x_i|^2 dt \tag{14.107}$$

Thus, $E_0 = E_1 = A^2 T_p = 2E_b$ and $E_2 = E_3 = B^2 \frac{T_p}{2} = 2E_b$. Then,

$$A = \sqrt{\frac{2E_b}{T_p}} \tag{14.108}$$

$$B = \sqrt{\frac{4E_b}{T_p}} \tag{14.109}$$

(b) The MLWD is given in Figure 14.2.

(c) Notice that $E_0 = E_1 = E_2 = E_3$ and that,

$$x_0(t) = -x_1(t) \tag{14.110}$$

$$x_2(t) = -x_3(t) \tag{14.111}$$

Therefore, only two matched filters are needed and MLWD structure could be simplified as shown in Figure 14.16.

(d) Recall that union bound to the probability of word error is,

$$P_W(E) \leq P_{WUB}(E) = \sum_{k=1}^{N} \frac{A_d(k)}{2M} \operatorname{erfc}\left(\sqrt{\frac{\Delta_E(k)}{4N_0}}\right) \tag{14.112}$$

where N is the number of all different square Euclidean distances, $A_d(k)$ is the number of signal pairs having a square Euclidean distance of $\Delta_E(k)$ and M is 2^{K_b} which $K_b = 2$ in this case.
We can enumerate all the Euclidean distances for this 4-ary communication system by calculating $\Delta_E(i, j)$, i.e.,

$$\Delta_E(i, j) = \int_{-\infty}^{\infty} |x_j(t) - x_i(t)|^2 dt \tag{14.113}$$

and results are shown in Table 14.3.

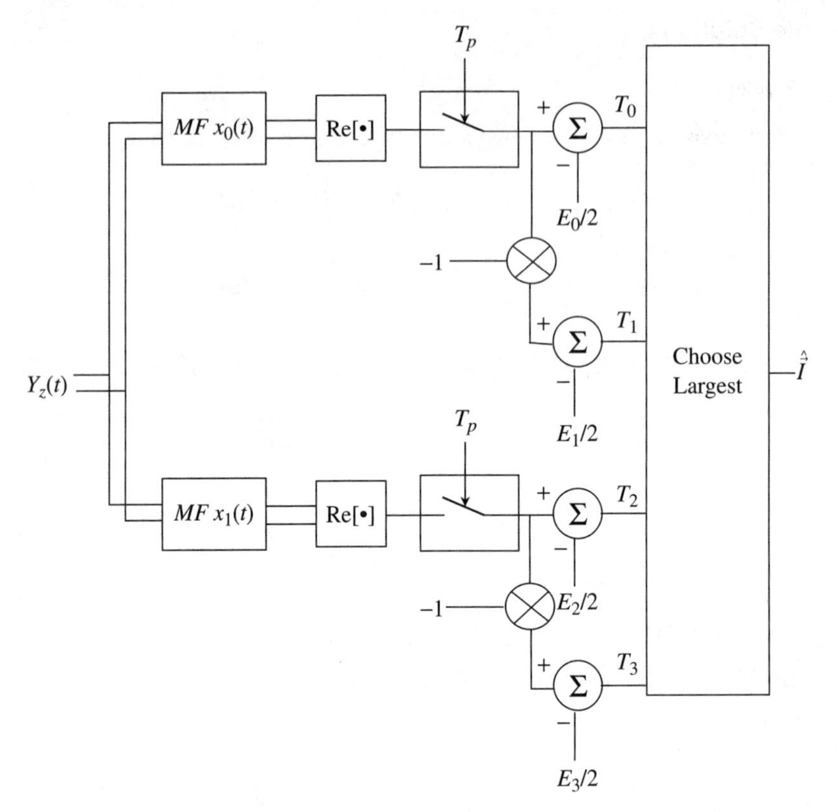

Figure 14.16 The MLWD using only two filters.

Thus, we have $N = 2$ with $\Delta_E(1) = 8E_b$ and $\Delta_E(2) = 4E_b$, also $A_d(1) = 4$ and $A_d(2) = 8$. Consequently, the union bound is expressed as,

$$\begin{aligned} P_W(E) &\le P_{WUB}(E) \\ &= \frac{1}{2}\mathrm{erfc}\left(\sqrt{\frac{2E_b}{N_0}}\right) + \mathrm{erfc}\left(\sqrt{\frac{E_b}{N_0}}\right) \end{aligned} \tag{14.114}$$

TABLE 14.3 $\Delta_E(i, j)$

	$\vec{I} = i$			
$\vec{I} = j$	0	1	2	3
0	0	$8E_b$	$4E_b$	$4E_b$
1	$8E_b$	0	$4E_b$	$4E_b$
2	$4E_b$	$4E_b$	0	$8E_b$
3	$4E_b$	$4E_b$	$8E_b$	0

Problem 14.16.

(a) $2E_b = \frac{A^2 T_p}{2}$ so that $A = 2\sqrt{\frac{E_b}{T_p}}$.

(b) Defining

$$u_4(t) = \begin{cases} 1 & 0 \leq t \leq T_p/4 \\ 0 & \text{elsewhere} \end{cases} \tag{14.115}$$

then it is apparent that

$$x_i(t) = u_4(t - \tau_{1,i}) + u_4(t - \tau_{2,i}) \tag{14.116}$$

Consequently, the Fourier transform is given as

$$\begin{aligned} X_i(f) &= \frac{AT_p}{4} \text{sinc}\left(\frac{f t_p}{4}\right) \exp[-j2\pi f \tau_{1,i}] \\ &\quad + \frac{AT_p}{4} \text{sinc}\left(\frac{f t_p}{4}\right) \exp[-j2\pi f \tau_{2,i}] \end{aligned} \tag{14.117}$$

$$= \frac{AT_p}{4} \text{sinc}\left(\frac{f t_p}{4}\right) (\exp[-j2\pi f \tau_{1,i}] + \exp[-j2\pi f \tau_{2,i}]) \tag{14.118}$$

The energy spectrum of the individual waveforms is then given as

$$G_{x_i}(f) = \frac{A^2 T_p^2}{16} \left(\text{sinc}\left(\frac{f t_p}{4}\right)\right)^2 [2 + 2\cos(2\pi f (\tau_{1,i} - \tau_{2,i}))] \tag{14.119}$$

The average energy spectrum is then given as

$$\begin{aligned} D_{x_z}(f) &= \frac{A^2 T_p^2}{32} \left(\text{sinc}\left(\frac{f t_p}{4}\right)\right)^2 \\ &\quad \times \left[4 + 2\cos(\pi f T_p) + \cos\left(\frac{\pi f T_p}{2}\right) + \cos\left(\frac{3\pi f T_p}{2}\right)\right] \end{aligned} \tag{14.120}$$

(c) The ML decision metric is

$$T_i = \Re\left[\int_0^{T_p} Y_z(t) x_i(t) dt\right] \qquad i = 0, \cdots, 3 \tag{14.121}$$

and the decoded bit is

$$\hat{I} = \arg \max_{i=0,\cdots,3} T_i \tag{14.122}$$

It should be noted that this modulation is an equal energy modulation scheme and no energy correction scheme is required. The one filter has an impulse response of $u_4(t)$ so that

$$Q(t) = \int Y_z(\lambda)u_4(t-\lambda)d\lambda \tag{14.123}$$

The sufficient statistics for demodulation are the four samples from the filter given as

$$Q_1 = Q(T_p/4) = \int_0^{T_p/4} Y_z(t)dt \quad Q_2 = Q(T_p/2) = \int_{T_p/4}^{T_p/2} Y_z(t)dt \tag{14.124}$$

$$Q_3 = Q(3T_p/4) = \int_{T_p/2}^{3T_p/4} Y_z(t)dt \quad Q_4 = Q(T_p) = \int_{3T_p/4}^{T_p} Y_z(t)dt \tag{14.125}$$

The ML metrics are given as

$$T_0 = \Re[Q_1 + Q_3] \quad T_1 = \Re[Q_1 + Q_4] \quad T_2 = \Re[Q_2 + Q_3] \quad T_3 = \Re[Q_2 + Q_4] \tag{14.126}$$

(d) A quick examination of the waveforms shows the signal set is geometrically uniform so that

$$P_{WUB}(E) = \text{erfc}\left(\sqrt{\frac{E_b}{2N_0}}\right) + \frac{1}{2}\text{erfc}\left(\sqrt{\frac{E_b}{N_0}}\right) \tag{14.127}$$

(e) Recall the union bound for orthogonal 4FSK is

$$P_{WUB}(E) = \frac{3}{2}\text{erfc}\left(\sqrt{\frac{E_b}{N_0}}\right) \tag{14.128}$$

minimum squared Euclidean distance between signals for 4FSK is 3 dB larger than the minimum squared Euclidean distance for the signal set considered in this problem.

(f) The spectral efficiency of 4FSK is approximately $\eta_B = 1$ while it is apparent from Eq. (14.120) the spectral efficiency of the considered modulation is approximately $\eta_B = 0.5$. Given the results in (e) and (f) there appears to be no relative advantage to this modulation compared to 4FSK.

14.6 Miniprojects

Goal: To give exposure

- to a small scope engineering design problem in communications.
- to the dynamics of working with a team.
- to the importance of engineering communication skills (in this case oral presentations).

Presentation: The forum will be similar to a design review at a company (only much shorter). The presentation will be of 5 minutes in length with an overview of the given problem and solution. The presentation will be followed by questions from the audience (your classmates and the professor). Each team member should be prepared to give the presentation.

14.6.1 Project 1

Project Goals: Engage in an implementation of an optimum demodulator of a 4-ary signal set.

Consider baseband binary signal set (with sample frequency $f_s = 22{,}050$ Hz) and the received bandpass signal ($f_c = 3500$ Hz) given in `mbproj1data.mat`.

(a) Plot the four baseband signals. Find E_b. Can you identify the modulation that is being used to send $K_b = 2$ bits?

(b) Sketch a block diagram of the optimum demodulator for this signal set for $\pi_i = 0.25,\ i = 0, 1, 2, 3$.

(c) Make the optimum word decision and compute all the sufficient statistics of the demodulation for $\pi_i = 0.25,\ i = 0, 1, 2, 3$.

(d) Make the optimum word decision and compute all the sufficient statistics of the demodulation for

$$\pi_0 = 0.25 \qquad \pi_1 = 0.35 \qquad \pi_2 = 0.05 \qquad \pi = 0.35. \tag{14.129}$$

and $N_0 = 1/4410$.

Chapter

15

Managing the Complexity of Optimum Demodulation

The results of the previous chapter have shown that the idea of transmitting and demodulating many bits of information is a direct extension of the ideas for transmitting and demodulating 1 bit of information. The biggest impediment to using these concepts in practical communication systems is the fact that the optimal demodulator complexity grows exponentially with the number of bits transmitted ($O(2^{K_b})$). The goal for a practical system has to be a demodulator with complexity that grows linearly with the number of bits ($O(K_b)$).

Fortunately, the example of M-ary phase shift keyed (MPSK) modulation considered in Chapter 14 gave one insight into how complexity of optimal demodulation can be made practical. MPSK has a form $x_i(t) = d_i u(t)$ where a modulation symbol, d_i, is linearly modulated on a pulse shape, $u(t)$. When a modulation takes this form only one matched filter needs to be computed for the optimum demodulator. The complexity of the demodulator is then greatly reduced.

Additionally if bits can be modulated in such a way that each bit can be decoded independently from the other bits then this would allow the demodulators introduced in the last two chapters to be used in parallel for each bit. If this decoupling of each bit decision can be achieved it would allow the complexity of the optimal demodulator to have a complexity that is linear in the number of bits transmitted. This chapter will show that if each bit is modulated onto an orthogonal waveform then this decoupling of the bit decisions is possible.

The goal of this chapter is to show that both linear modulations and orthogonal modulations are well-used techniques in digital communications to manage the complexity of packet data transmission.

15.1 Linear Modulations

Linear modulation has the form

$$x_i(t) = d_i \sqrt{E_b} u(t) \quad i \in \{0, \ldots, M-1\} \tag{15.1}$$

The function of time $u(t)$ is known as the pulse shape. In this text it will be assumed that $u(t)$ has unit energy, $E_u = 1$. The modulation symbol, d_i, is, in general, complex valued function of the transmitted information word, $D_z = a(\vec{I})$, and $d_i = a(i) \in \Omega_d$. The function $a(i)$ is known as the constellation mapping. A commonly used graphic for interpreting linear modulations is the constellation plot. A constellation plot for 4PSK modulation, $\Omega_d = \{\sqrt{2}e^{\frac{j\pi}{4}}, \sqrt{2}e^{\frac{j3\pi}{4}}, \sqrt{2}e^{\frac{-j\pi}{4}}, \sqrt{2}e^{\frac{-j3\pi}{4}}\}$, is shown in Figure 15.1. For a consistent energy normalization the constellation will always be chosen such that

$$\sum_{i=0}^{M-1} |d_i|^2 \pi_i = K_b \tag{15.2}$$

EXAMPLE 15.1

For $K_b = 3$ an example linear modulation is given with $\Omega_d = \{A(1+j), jA, A(-1+j), -A, 0, A, -jA, A(-1-j)\}$. To achieve the desired normalization

$$A = \sqrt{\frac{12}{5}} \tag{15.3}$$

The constellation plot is given in Figure 15.2.

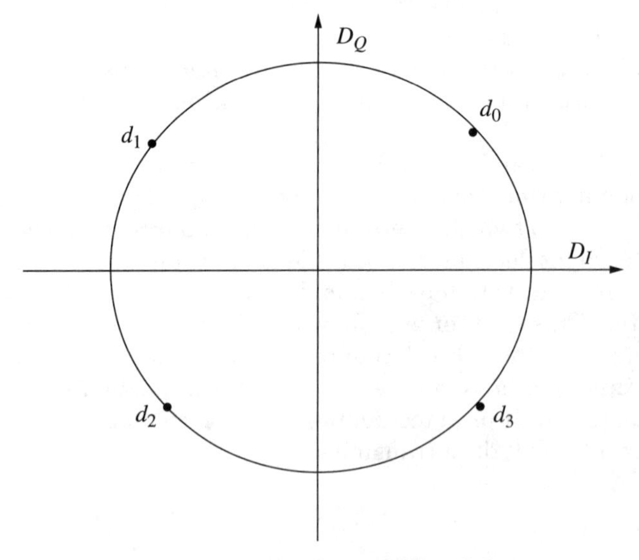

Figure 15.1 The constellation plot of 4PSK modulation.

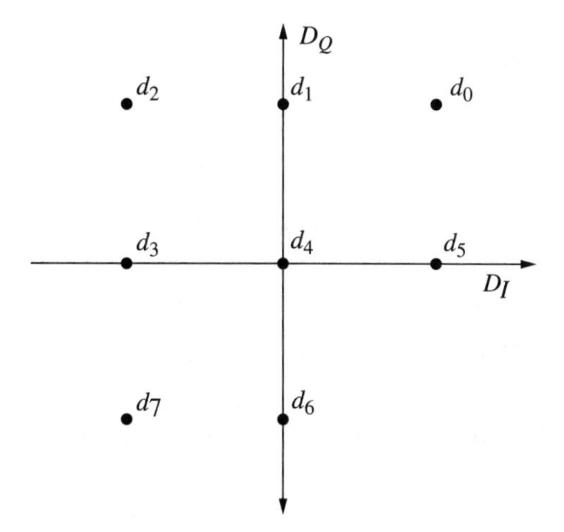

Figure 15.2 The constellation plot for an example 8-ary modulation.

15.1.1 MLWD for Linear Modulation

The MLWD for a linear modulation has a greatly simplified form. Recall the optimum demodulator has the form

$$\hat{\hat{I}} = \arg \max_{i \in \{0,\ldots,M-1\}} \Re[V_i(T_p)] - \frac{E_i}{2} \tag{15.4}$$

where the ith matched filter output has the form

$$\begin{aligned} V_i(T_p) &= \int_0^{T_p} Y_z(t) x_i^*(t) dt \\ &= d_i^* \sqrt{E_b} \int_0^{T_p} Y_z(t) u^*(t) dt \\ &= d_i^* \sqrt{E_b} Q \end{aligned} \tag{15.5}$$

where Q is denoted the pulse shape matched filter output and is given as

$$Q = \int_0^{T_p} Y_z(t) u^*(t) dt \tag{15.6}$$

Note the matched filter can be viewed as a fixed filter with an impulse response $u^*(T_p - t)$ that is sampled at time T_p. A block diagram of the demodulator for linear modulations is shown in Figure 15.3. Several observations can be made about this structure.

- The number of filtering operations in the demodulation has been reduced from $M = 2^{K_b}$ to one when compared to the demodulator for a general M-ary modulation.

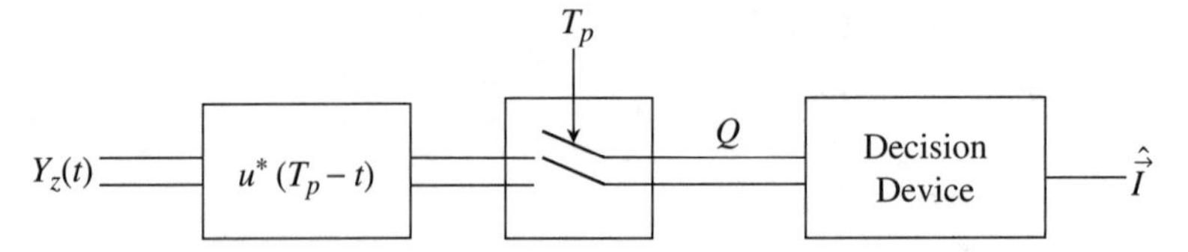

Figure 15.3 The block diagram of a demodulator for linear modulation.

- The decision device can still potentially have a complexity that is $O(2^{K_b})$.
- The energy of the ith transmitted signal is $E_i = |d_i|^2 E_b$.

The optimum demodulator for linear modulation has an intuitively pleasing geometric interpretation. The optimum demodulator can be expressed as

$$\hat{I} = \arg \max_{i \in \{0,\ldots,M-1\}} \sqrt{E_b} \Re[d_i^* Q] - \frac{|d_i|^2 E_b}{2} \tag{15.7}$$

Multiplying by 2 and adding a constant with respect to the possible transmitted symbols, $-|Q|^2$, gives

$$\begin{aligned} \hat{I} &= \arg \max_{i \in \{0,\ldots,M-1\}} -|Q|^2 + 2\Re[d_i^* Q \sqrt{E_b}] - |d_i|^2 E_b \\ &= \arg \max_{i \in \{0,\ldots,M-1\}} -\left|Q - d_i \sqrt{E_b}\right|^2 \\ &= \arg \min_{i \in \{0,\ldots,M-1\}} \left|Q - d_i \sqrt{E_b}\right|^2 \end{aligned} \tag{15.8}$$

Consequently, the optimum demodulator computes Q and finds the possible transmitted constellation point $d_i \sqrt{E_b}$ which has the minimum squared Euclidean distance to Q. This is known by communications engineers as a minimum distance decoder. This minimum distance decoder induces decision regions in the complex plane. These decision regions represent the matched filter output values that correspond to a most likely transmitted symbol. The decision regions for 4PSK MLWD are shown in Figure 15.4. In the sequel the decision regions will be denoted A_i where if $Q \in A_i$ then $\hat{I} = i$ represents the MLWD output. It should be noted that the decision regions are bounded by the lines that are defined as the equal distance points from each pair of constellation points. For instance, with QPSK there are four such lines and each decision region is bounded by two of these lines.

EXAMPLE 15.2

Considering Example 15.1 the decision regions are defined as in Figure 15.5. The decision regions here tend to have rectangular forms as the constellation is defined on a grid. The one notable exception is in the bottom right corner where there is a point missing from the grid.

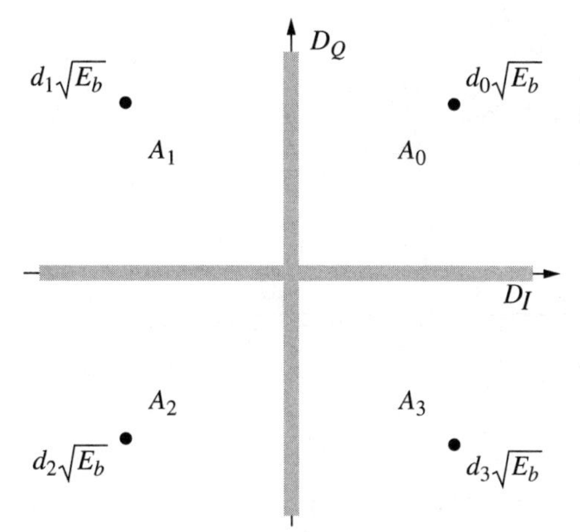

Figure 15.4 The minimum distance decision regions for 4PSK.

Recall the matched filter output when the ith word was transmitted has the form

$$Q = \int_0^{T_p} y_z(t)u^*(t)dt = \int_0^{T_p} x_i(t)u^*(t)dt + \int_0^{T_p} W_z(t)u^*(t)dt$$
$$= d_i\sqrt{E_b} + N_z(T_p) \tag{15.9}$$

With this view the signal portion of the matched filter output corresponds to the transmitted symbol scaled by the bit energy, $\sqrt{E_b}$, and the noise causes a random translation of the signal in the complex plane. The idea of a minimum

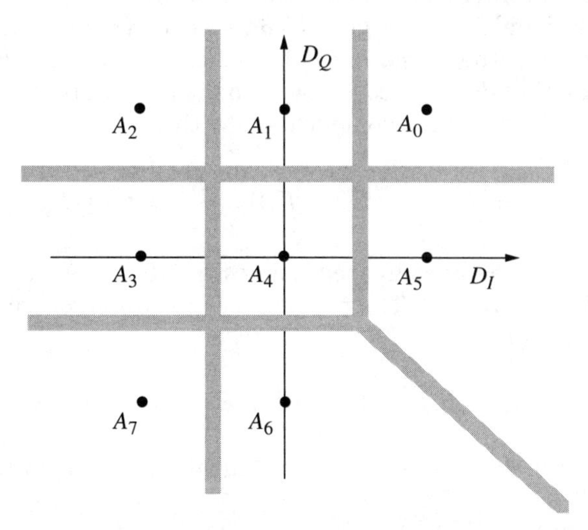

Figure 15.5 The decision regions for the linear modulation given in Example 15.1.

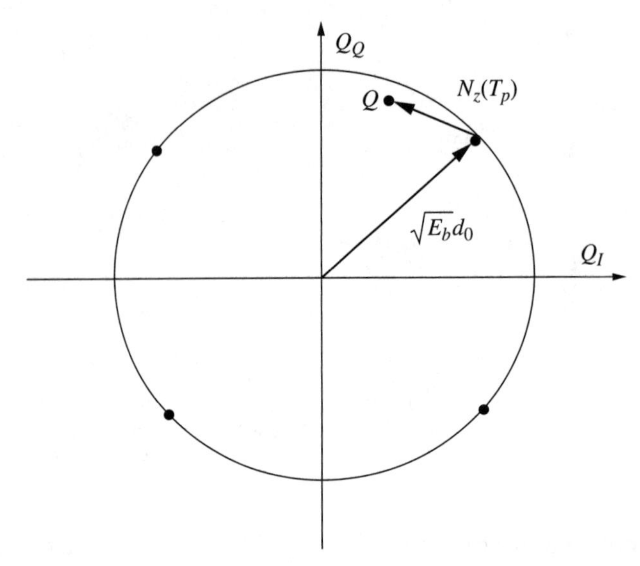

Figure 15.6 An example geometric interpretation of the signal and noise in linear modulation. $D_z = d_0 = \sqrt{2}e^{j\pi/4} = 1 + j$.

distance decoder is simply trying to find which signal set would correspond to the smallest noise magnitude. For example, if d_0 from the 4PSK constellation in Figure 15.1 is the transmitted symbol then the matched filter output would have a representation as given in Figure 15.6.

15.1.2 Error Rate Evaluation for Linear Modulation

The symbol error probability calculation also has a nice geometric interpretation. Comparing the signal form in Figure 15.6 to decision regions in Figure 15.4 an error will occur when the noise, $N_z(T_p)$, pushes the matched filter output into a decision region that does not correspond to the transmitted symbol. Since $N_z(T_p)$ is a complex Gaussian random variable with

$$\text{var}(N_z(T_p)) = E_u N_0 = N_0 \qquad \text{var}(N_I(T_p)) = \frac{N_0}{2} = \text{var}(N_Q(T_p)) \tag{15.10}$$

the PDF of the noise is easily computed using Eq. (10.26). An example noise PDF is plotted in Figure 15.7(a). The probability of word error conditioned on the ith symbol being sent is

$$P_W(E|\vec{I} = i) = P(Q \notin A_i|\vec{I} = i) = P(d_i\sqrt{E_b} + N_z(T_p) \notin A_i) \tag{15.11}$$

Consequently, this probability is simply an integral of the noise PDF, i.e.,

$$P_W(E|i) = \int_{R_i} f_{N_z}(n_z)dn_z = \int 1_{R_i}(n_z)f_{N_z}(n_z)dn_z \tag{15.12}$$

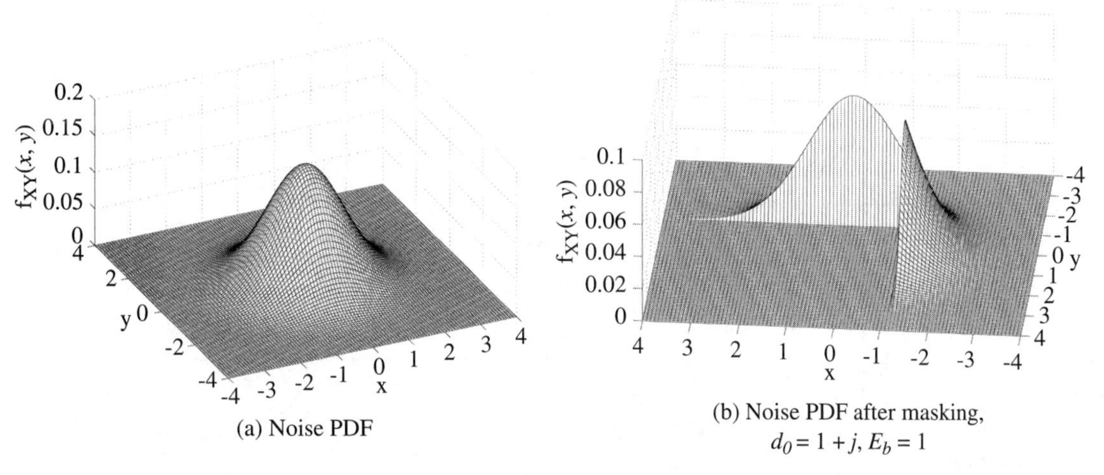

Figure 15.7 A noise PDF corresponding to $\text{var}(N_z(T_p)) = 1$ and the masking needed to compute the word error rate for QPSK modulation.

where the indicator function is used to mask the PDF only in the regions where noise will cause an error. Figure 15.7(b) shows the masked PDF for a 4PSK decision regions given in Figure 15.4 for $d_0 = \sqrt{2}\exp(j\frac{\pi}{4})$.

EXAMPLE 15.3

For the linear modulation presented in Example 15.1 several of the integration regions for conditional word error rates can be computed in close form. For example, the conditional word error rate for $\vec{I} = 4$ is given as

$$P_W(E|\vec{I} = 4) = 1 - \int_{-b}^{b}\int_{-b}^{b} f_{N_z}(n_z)dn_z \tag{15.13}$$

where $b = \frac{A\sqrt{E_b}}{2}$. Likewise, the conditional error rate for $\vec{I} = 1$ is

$$P_W(E|\vec{I} = 1) = 1 - \int_{-b}^{b}\int_{-b}^{\infty} f_{N_z}(n_z)dn_z \tag{15.14}$$

The conditional decision region for $\vec{I} = 5$ is a bit more complex but it is still easily specified and the integration can be completed.

The $P_W(E)$ for a given K_b can be a significant function of how the constellation points are placed in the complex plane. For example, consider two common linear modulations, pulse amplitude modulation (PAM) and phase shift keyed (PSK) modulation that have been implemented often in engineering practice. PSK uses only the phase to transmit information while PAM uses only the amplitude to transmit information. Examples of the constellation plots for these two modulations are shown in Figure 15.8(a) for $K_b = 2$.

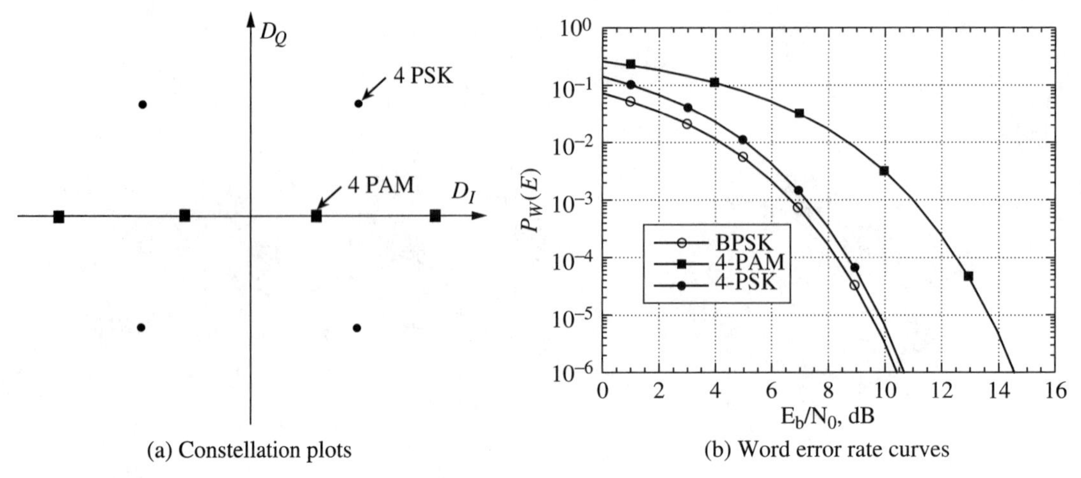

Figure 15.8 Constellation diagrams and word error rate curves comparing 4PAM and 4PSK.

Note 4-ary pulse amplitude modulation (4PAM) is characterized with $\Omega_d = \{\pm\sqrt{2}/\sqrt{5}, \pm 3\sqrt{2}/\sqrt{5}\}$. It is clear from the constellation plot that the points of QPSK are farther apart than the points for 4PAM and hence QPSK will be able to send 2 bits of information with more fidelity than 4PAM. The $P_W(E)$ for each of these linear modulations are given for $K_b = 1, 2$ as (see problems)

$$P_W(E) = \frac{1}{2}\text{erfc}\left(\sqrt{\frac{E_b}{N_0}}\right) \qquad \text{BPSK/BPAM} \tag{15.15}$$

$$P_W(E) = \text{erfc}\left(\sqrt{\frac{E_b}{N_0}}\right) - \left(\frac{1}{2}\text{erfc}\left(\sqrt{\frac{E_b}{N_0}}\right)\right)^2 \qquad \text{4PSK} \tag{15.16}$$

$$P_W(E) = \frac{3}{4}\text{erfc}\left(\sqrt{\frac{2E_b}{5N_0}}\right) \qquad \text{4PAM} \tag{15.17}$$

These curves are plotted in Figure 15.8(b). It is clear from Figure 15.8(b) that the manner in which constellation points are chosen in sending K_b bits with a linear modulation can have a significant impact on the error rate.

The union bound for the word error probability of a linear modulation is straightforward to compute. The union bound can be formed by noting that the pairwise Euclidean squared distance is given as

$$\Delta_E(i, j) = E_b |d_i - d_j|^2 \tag{15.18}$$

Consequently, the union bound can be formed by looking only at the constellations. The union bound is not a function of the pulse shape used in the linear modulation.

EXAMPLE 15.4

Considering Example 15.1 the complete union bound would have to consider seven different distances for each postulated transmitted signal. Just like MPSK in Chapter 14 this union bound can be tightened. For example, for $\vec{I} = 0$ the conditional union bound is

$$P_{WUB}(E|\vec{I} = 0) = \frac{1}{2}\text{erfc}\left(\sqrt{\frac{\Delta_E(1,0)}{4N_0}}\right) + \frac{1}{2}\text{erfc}\left(\sqrt{\frac{\Delta_E(5,0)}{4N_0}}\right) \tag{15.19}$$

for $\vec{I} = 1$ the conditional union bound is

$$P_{WUB}(E|\vec{I} = 1) = \frac{1}{2}\text{erfc}\left(\sqrt{\frac{\Delta_E(0,1)}{4N_0}}\right) + \frac{1}{2}\text{erfc}\left(\sqrt{\frac{\Delta_E(2,1)}{4N_0}}\right) + \frac{1}{2}\text{erfc}\left(\sqrt{\frac{\Delta_E(4,1)}{4N_0}}\right) \tag{15.20}$$

for $\vec{I} = 4$ the conditional union bound is

$$P_{WUB}(E|\vec{I} = 4) = \frac{1}{2}\text{erfc}\left(\sqrt{\frac{\Delta_E(1,4)}{4N_0}}\right) + \frac{1}{2}\text{erfc}\left(\sqrt{\frac{\Delta_E(3,4)}{4N_0}}\right) + \frac{1}{2}\text{erfc}\left(\sqrt{\frac{\Delta_E(5,4)}{4N_0}}\right) + \frac{1}{2}\text{erfc}\left(\sqrt{\frac{\Delta_E(6,4)}{4N_0}}\right) \tag{15.21}$$

and for $\vec{I} = 5$ the conditional union bound is

$$P_{WUB}(E|\vec{I} = 5) = \frac{1}{2}\text{erfc}\left(\sqrt{\frac{\Delta_E(0,5)}{4N_0}}\right) + \frac{1}{2}\text{erfc}\left(\sqrt{\frac{\Delta_E(4,5)}{4N_0}}\right) + \frac{1}{2}\text{erfc}\left(\sqrt{\frac{\Delta_E(6,5)}{4N_0}}\right) \tag{15.22}$$

Combining all the conditional union bounds and noting that adjacent points have a Euclidean distance of $\Delta_E(1,0) = \sqrt{\frac{12E_b}{5}}$ and points separated on a diagonal have a distance of $\Delta_E(6,5) = \sqrt{\frac{24E_b}{5}}$ the union bound is given as

$$P_{WUB}(E) = \frac{5}{4}\text{erfc}\left(\sqrt{\frac{3E_b}{5N_0}}\right) + \frac{1}{8}\text{erfc}\left(\sqrt{\frac{6E_b}{5N_0}}\right) \tag{15.23}$$

So an additional benefit of linear modulation beside the simple MLWD is that the $P_W(E)$ can always be easily computed. The actual word error probability calculation might require a little computer work (two dimensional numerical integration) but this effort is not unreasonable for a modern communication engineer. The full union bound is easily computed. Tighter union bounds are easily identified by looking at the decision regions. While often a general M-ary modulation must resort to a full union bound computation to characterize the $P_W(E)$ this is not necessary for linear modulation.

15.1.3 Spectral Characteristics of Linear Modulation

For linear modulations the average energy spectrum per bit is straightforward to compute. Using the definition of the energy spectrum per bit gives

$$D_{x_z}(f) = \frac{E\left[G_{x_z}(f)\right]}{K_b} = G_u(f)\frac{E_b E\left[|D_z|^2\right]}{K_b} = G_u(f)\frac{E_b \sum_{i=0}^{M-1} |d_i|^2 \pi_i}{K_b} = E_b G_u(f) \tag{15.24}$$

The important point to note about Eq. (15.24) is that the average energy spectrum per bit is entirely (up to a constant multiple) a function of the pulse shape spectrum, $G_u(f)$. Transmitting more bits (increasing K_b) has no effect on the occupied spectrum. Consequently, with a bandwidth efficient choice of the pulse shape, $u(t)$, linear modulation is a very effective modulation for bandlimited channels. A design of a pulse shape that will meet a specified spectral mask is explored in the projects at the end of Chapter 13.

15.1.4 Example: Square Quadrature Amplitude Modulation

This section will be concluded with a discussion of square quadrature amplitude modulation (QAM) which is a commonly used linear modulation in engineering practice. A square QAM constellation consists of a square of M_1 points by M_1 points with the M_1 points in each dimension placed on a equally spaced grid. Normally the square is chosen such that $M_1 = 2^{K_b/2}$ so that an integer number of bits is used to specify the constellation in each dimension. This constellation is considered in detail in the homework problems and the important results to highlight here are that the spectral efficiency increases logarithmically with M_1 and that the word error probability of this modulation is upperbounded as

$$P_W(E) \leq 1 - \left(\text{erf}\left(\sqrt{\frac{6\log_2(M_1)}{M_1^2 - 1}\frac{E_b}{N_0}}\right)\right)^2 \tag{15.25}$$

A plot of this bound to the word error probability is given in Figure 15.9 for various M_1. Much like was seen earlier in MPSK the word error performance significantly degrades for a fixed SNR as the spectral efficiency is increased. To keep the same level of performance the signal to noise ratio must be increased by 4 dB as the spectral efficiency increases by 2 bits/s/Hz.

15.1.5 Summary of Linear Modulation

Important points about linear modulation are

- Demodulation complexity of linear modulation does not increase exponentially with the number of bits transmitted since only one matched filter output needs to be computed and simple decision regions can be identified. Linear modulation is often used because it has a low demodulation complexity.

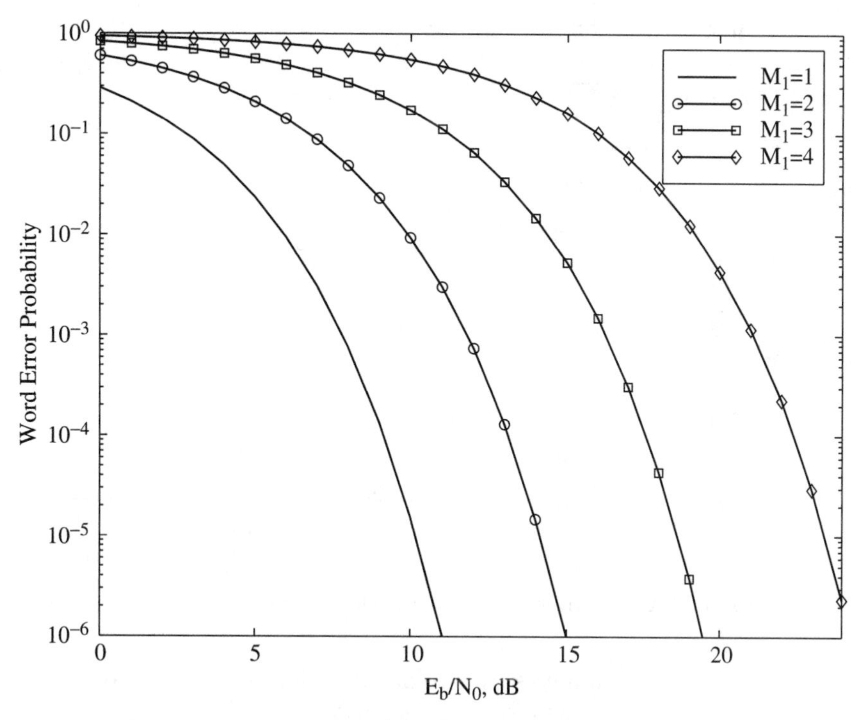

Figure 15.9 The word error probability bound for square QAM.

- The energy spectrum per bit of a linear modulation is entirely a function of the pulse shape that is used. Bandwidth efficient pulse shapes are often used in conjunction with M-ary linear modulation when bandwidth efficiency is of paramount importance. This will be examined in more detail in Chapter 16.
- Fidelity in message reconstruction gets significantly worse as M gets larger. This can be seen by examining the error rate of linear modulations in both Figure 15.8(b), Figure 15.9, and Figure 14.8. Consequently, linear modulation provides a classic trade-off in bandwidth efficiency (bandwidth efficiency is proportional to K_b) versus fidelity (fidelity significantly degrades with increasing K_b).
- The selection of the constellation points in an M-ary linear modulation can significantly affect the resulting error rate. This can be seen by examining the performance of 4PAM and 4PSK (both $K_b = 2$) in Figure 15.8(b). 4PAM has a 3-dB worse performance than 4PSK since the 4PAM constellation points have not been placed as effectively in the complex plane as the 4PSK constellation points. A more general demonstration of this point is shown in Figure 15.10 where the performance of the square QAM is compared to Shannon's bound and the other modulations introduced up until this point. Clearly a QAM modulation by placing constellation points more judiciously in the complex plane can provide better performance at a common spectral efficiency than an MPSK modulation. The interesting characteristic of QAM is

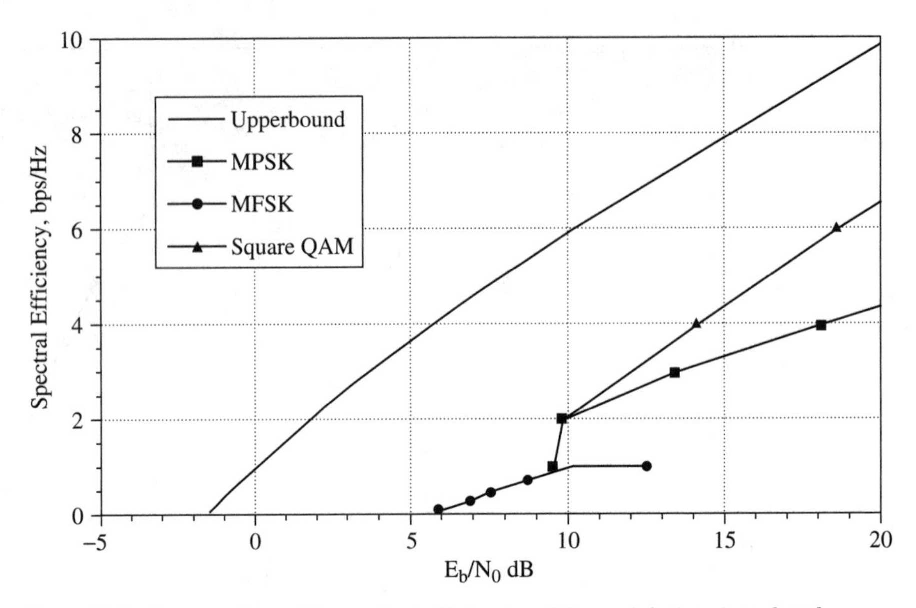

Figure 15.10 A comparison of the spectral efficiencies of the modulations introduced.

that it performance seems to parallel Shannon's upperbound as spectral efficiency increases while MPSK's performance fall away from Shannon's bound as spectral efficiency increases. An interesting paper that investigates optimal and suboptimal constellations is [FGL+84].

It is interesting to note that computer modems introduced in Chapter 12 have migrated to linear modulation over time. The increase in transmission data rate is achieved by an increase in Baud rate[1] and an increase in the size of the constellation of the linear modulation used in the modem. Examples in this evolution are documented in Table 15.1. It should be noted that in early modems the data rate had the following characteristic

$$W_b = \log_2(M) \times \text{ Baud rate} \tag{15.26}$$

TABLE 15.1 Evolution of Baud rates and constellation sizes in voiceband computer modems

Modem Standard	Data Rate	Baud Rate	M
V.22	1,200 bps	600 Hz	4
V.22bis	2,400 bps	600 Hz	16
V.32	9,600 bps	2,400 Hz	32
V.32bis	14,400 bps	2,400 Hz	128
V.34	28,800 bps	3,200 Hz	896

[1]Baud rate will be discussed later in this chapter when we introduce the idea of stream modulations.

As modem standards became more sophisticated (V.32 and subsequent generations) redundancy was added to the signaling to improve performance. An overview of how this redundancy helps improve the fidelity of message reconstruction is the focus of Chapter 17.

15.2 Orthogonal Modulations

The goal at this point is to identify a constraint that can be put on signal designs such that optimum demodulation can be formulated with a complexity of $O(K_b)$. This linear complexity can be achieved if each bit can be demodulated in an independent fashion. There is a simple constraint on the ML metric that will enable optimal bit-by-bit decisions. This section will focus on the case of equal priors ($\pi_i = 1/M$) but extension to the case of unequal priors is also possible. When there are K_b bits sent then if the ML metric can be expressed in an additive form as

$$T_i = \sum_{k=1}^{K_b} T_{m_k}^{(k)} \tag{15.27}$$

where $\{\vec{I}\} = \{I(1) = m_1, \cdots, I(K_b) = m_{K_b}\}$ then each bit can be demodulated in an independent fashion. In other words, if each T_i can be broken up into a sum of terms which are individually only a function of a single bit position, k, and bit value for that position, $I(k) = m_k$ then independent bit demodulation can be implemented. This sum form means that finding the maximum likelihood is found by a series of binary maximizations on each term of the summation, i.e.,

$$\hat{\vec{I}} = [\hat{I}(1)\ldots\hat{I}(K_b)] = \arg\max_{i\in\{0,\ldots,M-1\}} T_i = \left[\arg\max_{m_1} T_{(m_1)}^{(1)} \ldots \arg\max_{m_{K_b}} T_{(m_{K_b})}^{(K_b)}\right] \tag{15.28}$$

This bit-by-bit maximization has a complexity that is $O(K_b)$ as opposed to $O(2^{K_b})$ that is the general case.

Many of the modulation schemes that are used in practice for their reduced complexity optimal demodulation are due to the following property.

Property 15.1 (Orthogonal Modulations) If K_b bits are transmitted by summing up K_b waveforms each used to transmit a single bit, i.e.,

$$x_i(t) = \sum_{l=1}^{K_b} x(m_l, l, t) \tag{15.29}$$

where again $\vec{I} = i = [m_1\ m_2\ \ldots\ m_{K_b}]$ and $x(m_l, l, t)$ is the waveform used to transmit $I(l) = m_l$ then the ML metric can be put in the additive form required for simplified bit detection if the waveforms used to transmit each of the bits are orthogonal, i.e.,

$$\Re\left[\int_{-\infty}^{\infty} x(m_l, l, t)x^*(m_k, k, t)dt\right] = 0 \qquad \forall k \neq l, m_l = 0, 1, m_k = 0, 1 \tag{15.30}$$

Proof: The ML metric consists of two components: the matched filter output and the energy correction term. The matched filter term for the transmitted signal that has the form in Eq. (15.29) is given as

$$\Re\left[\int_{-\infty}^{\infty} y_z(t)x_i^*(t)dt\right] = \Re\left[\int_{-\infty}^{\infty} y_z(t)\left(\sum_{l=1}^{K_b} x(m_l,l,t)\right)^* dt\right]$$

$$= \sum_{l=1}^{K_b} \Re\left[\int_{-\infty}^{\infty} y_z(t)x^*(m_l,l,t)dt\right] \tag{15.31}$$

Consequently, just due to the form in Eq. (15.29) the matched filter output has the additive form required for simplified detection. The remaining term is the energy correction term given for Eq. (15.29) as

$$\frac{E_i}{2} = \frac{1}{2}\int_{-\infty}^{\infty} |x_i(t)|^2 dt = \frac{1}{2}\int_{-\infty}^{\infty} \sum_{l=1}^{K_b} x(m_l,l,t)\left(\sum_{k=1}^{K_b} x(m_k,k,t)\right)^* dt$$

$$= \frac{1}{2}\sum_{l=1}^{K_b}\int_{-\infty}^{\infty} |x(m_l,l,t)|^2 dt + \frac{1}{2}\sum_{l=1}^{K_b}\sum_{\substack{k=1\\k\neq l}}^{K_b}\int_{-\infty}^{\infty} x(m_l,l,t)(x(m_k,k,t))^* dt \tag{15.32}$$

$$= \frac{1}{2}\sum_{l=1}^{K_b} E_{x(m_l,l)} + \sum_{l=2}^{K_b}\sum_{k=1}^{l-1}\Re\left[\int_{-\infty}^{\infty} x(m_l,l,t)(x(m_k,k,t))^* dt\right] \tag{15.33}$$

where $E_x(m_l,l)$ is the energy of $x(m_l,l,t)$, i.e.,

$$E_x(m_l,l) = \int_0^{T_p} |x(m_l,l,t)|^2 dt \tag{15.34}$$

Clearly the second term in Eq. (15.33) will go to zero if the orthogonality condition holds and then the energy correction term will also have an additive form. Using these results it is apparent that

$$T_{m_k}^{(k)} = \Re\left[\int_{-\infty}^{\infty} y_z(t)x^*(m_k,k,t)dt\right] - \frac{E_x(m_k,k)}{2} \quad \square \tag{15.35}$$

An interesting characteristic of orthogonal modulations is that the decision metric has exactly the same form as the single bit detector presented in Chapter 13. Using the above notation for orthogonal modulation we see that the optimal decision rule for each bit is now given as

$$\exp\left[\frac{2T_1^{(k)}}{N_0}\right]\pi_1 \underset{\hat{I}(k)=0}{\overset{\hat{I}(k)=1}{\gtrless}} \exp\left[\frac{2T_0^{(k)}}{N_0}\right]\pi_0 \tag{15.36}$$

The MLBD has the form

$$T_1^{(k)} \underset{\hat{I}(k)=0}{\overset{\hat{I}(k)=1}{\gtrless}} T_0^{(k)}$$

$$\Re[V_1(k)] - \frac{1}{2}E_{x(1,k)} \underset{\hat{I}(k)=0}{\overset{\hat{I}(k)=1}{\gtrless}} \Re[V_0(k)] - \frac{1}{2}E_{x(0,k)} \tag{15.37}$$

where

$$V_{m_k}(k) = \int_{-\infty}^{\infty} Y_z(t)x^*(m_k, k, t)\,dt \tag{15.38}$$

is the matched filter to the waveform used to transmit $I(k) = m_k$. This structural characteristic of orthogonal modulations implies that K_b bits can optimally be demodulated by K_b binary demodulators operated in parallel. These binary demodulators have the same form as discussed in Chapter 13 and is the reason why so much effort was spent characterizing the single bit demodulator. Well over 95% of the communications systems in the world utilize orthogonality in some form so Chapters 13 and 14 are fundamental material in the understanding of digital communication. With orthogonal modulation optimal demodulation now has a complexity of $O(K_b)$ as desired in practical communication systems.

While orthogonality seems to offer a solution to the complexity issue, the orthogonality exacts a price in the fidelity of demodulation. Since the orthogonality reduces the demodulation to exactly that of a single bit developed in Chapter 13 we can deduce the performance is exactly that of the single bit demodulator. Consequently, the bit error probability performance is lower bounded by

$$P_B(E, k) = \frac{1}{2}\mathrm{erfc}\left(\sqrt{\frac{\Delta_E(1, 0, k)}{4N_0}}\right) \geq \frac{1}{2}\mathrm{erfc}\left(\sqrt{\frac{E_b}{N_0}}\right) \tag{15.39}$$

where $\Delta_E(1, 0, k)$ is the Euclidean square distance between the two waveforms used to send the kth bit, i.e.,

$$\Delta_E(1, 0, k) = \int_{-\infty}^{\infty} |x(1, k, t) - x(0, k, t)|^2\,dt \tag{15.40}$$

The orthogonality implies each decision is independent. This independence implies that the word error probability for an orthogonal modulation when each bit is sent with waveforms that have the same Euclidean square distance becomes

(see Problem 15.12)

$$P_W(E) = 1 - \prod_{k=1}^{K_b}(1 - P_B(E,k))$$

$$= 1 - (1 - P_B(E))^{K_b} \geq 1 - \left(1 - \frac{1}{2}\text{erfc}\left(\sqrt{\frac{E_b}{N_0}}\right)\right)^{K_b} \quad (15.41)$$

The unfortunate consequence of orthogonality is that the performance of single bit demodulation is far from Shannon's bound as detailed in Chapter 13.

Orthogonal modulations offer an interesting trade-off for digital communication system designers. Orthogonality allows the complexity of optimal demodulation to be $O(K_b)$ as would be desired when K_b is large. For this reason almost all communication systems in practice use orthogonality in some form. Orthogonality with independent bit modulation waveforms limits the achievable performance to be far from Shannon's bound and is not desirable in modern communication systems. In practice orthogonality is still used because there are techniques that can be used in conjunction with orthogonal modulations to move performance close to Shannon's bound. These techniques will be briefly overviewed in Chapter 17

15.3 Orthogonal Modulation Examples

15.3.1 Orthogonal Frequency Division Multiplexing

A commonly used modulation that admits a simple optimal bit demodulation is orthogonal frequency division multiplexing (OFDM) [Cha66]. OFDM has found utility in telephone, cable, and wireless modems. With OFDM each of the K_b bits is independently modulated on a separate subcarrier frequency and the subcarrier frequencies are chosen to ensure the orthogonality. The format for an OFDM signal is

$$X_z(t) = \begin{cases} \sum_{l=1}^{K_b} D_z(l)\sqrt{\frac{E_b}{T_p}}\exp[j2\pi f_d(2l - K_b - 1)t] & 0 \leq t \leq T_p \\ 0 & \text{elsewhere} \end{cases} \quad (15.42)$$

where $D_z(l) = a(I(l))$ and $2f_d$ is the separation between adjacent frequencies that are used to transmit the information. The transmission rate of this form of OFDM is $W_b = K_b/T_p$ bits/s. In examining Eq. (15.42) it is clear that in this form of OFDM a binary linear modulation is used on each of the K_b different subcarrier frequencies. For clarity of discussion the remainder of the section will assume the linear modulation is BPSK (i.e., $a(0) = 1$ and $a(1) = -1$). A more general form of OFDM could use any type of linear modulation and any number of bits per subcarrier. For example, 4 bits could be transmitted per frequency using a 16QAM modulation (see Section 15.1.4 and Problem 15.1). Some of the generalizations of OFDM will be explored in the homework.

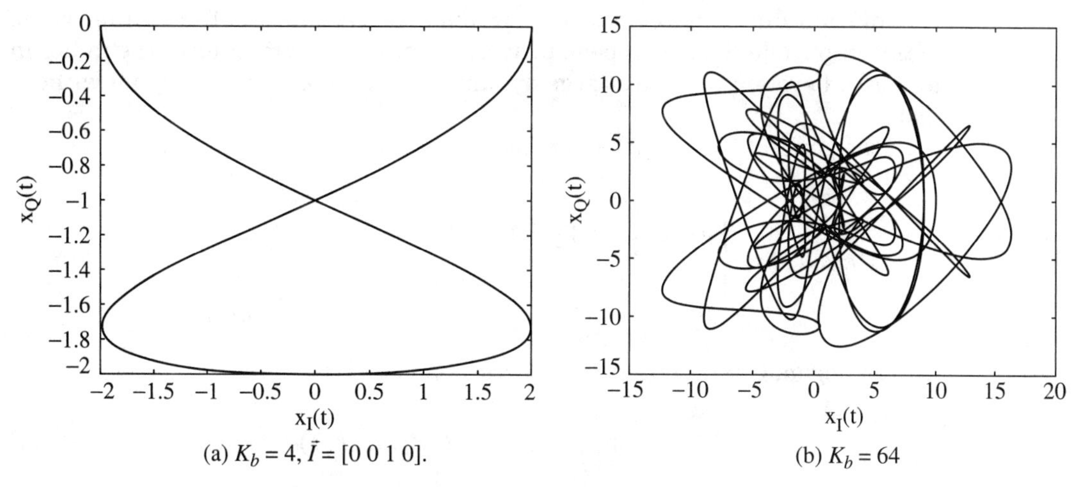

(a) $K_b = 4, \vec{I} = [0\ 0\ 1\ 0]$. (b) $K_b = 64$

Figure 15.11 Example vector diagrams of OFDM transmissions.

The OFDM transmitted waveform is a sum of K_b complex sinusoids. This transmitted waveform will have a complex envelope that changes significantly over the transmission time as the K_b complex sinusoids change in phase relative to each other. The larger the value of K_b the larger this variation over the transmission time will be. Figure 15.11 shows the vector diagrams of some example OFDM transmitted waveforms. The vector diagram clearly becomes more complex as more bits are transmitted. In addition, the peak value of the amplitude of the complex envelope increases with K_b. For example, Figure 15.12 shows $x_A(t)$ of some example OFDM transmitted waveforms. An OFDM waveform has

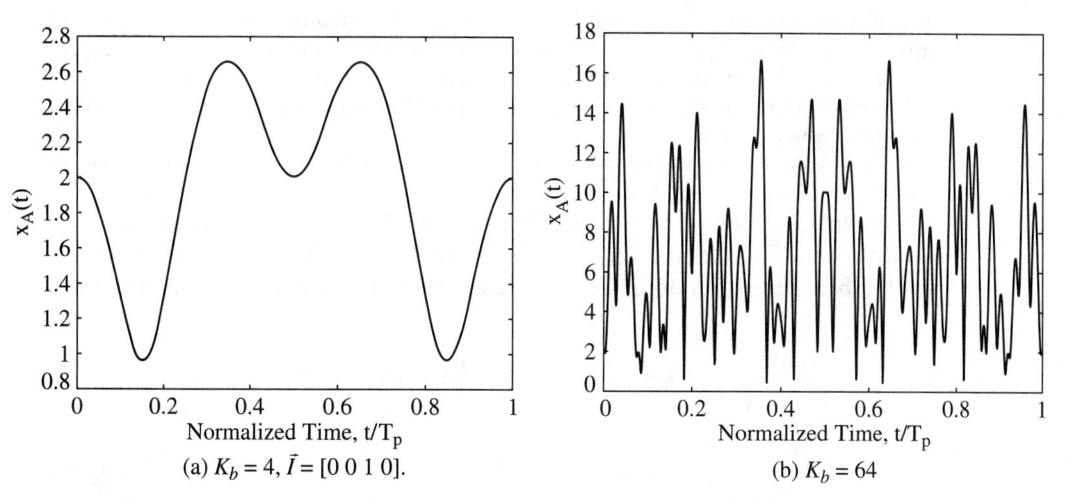

(a) $K_b = 4, \vec{I} = [0\ 0\ 1\ 0]$. (b) $K_b = 64$

Figure 15.12 Example $x_A(t)$ for OFDM transmissions.

a significant difference between the peaks in amplitude and the average value of the amplitude. This high peak to average ratio (PAPR)[2] requires the radios in an OFDM system to have a large dynamic range to process the signal without distortion.

In the general framework of orthogonal modulations OFDM has set

$$x(m_l, l, t) = \begin{cases} d_{m_l}\sqrt{\dfrac{E_b}{T_p}}\exp[j2\pi f_d(2l - K_b - 1)t] & 0 \leq t \leq T_p \\ 0 & \text{elsewhere} \end{cases} \tag{15.43}$$

The orthogonality condition then corresponds to

$$\Re\left[\int_{-\infty}^{\infty} x(m_l, l, t)x^*(m_k, k, t)dt\right] = 0 \tag{15.44}$$

$$\Re\left[d_{m_l}d^*_{m_k}\frac{E_b}{T_p}\int_0^{T_p}\exp[j4\pi f_d(l-k)t]dt\right] = 0 \tag{15.45}$$

Using the result of Eq. (13.86) shows that the minimum spacing with an arbitrary complex constellation for orthogonality is achieved with $f_d = 1/(2T_p)$. If the constellation is restricted to have real values (e.g., BPSK) then the minimum frequency spacing is achieved with $f_d = 1/(4T_p)$.

EXAMPLE 15.5

The wireless local area network standard denoted IEEE 802.11a uses OFDM. The 802.11a standard uses OFDM that can be represented with $K_b = 53$ with $T_p = 3.2\ \mu$s. The frequency spacing of the subcarrier tones is set with $f_d = 156.25$ kHz because complex constellations are used in some modes of transmission. The potential transmission rate of the 802.11a waveform with BPSK modulation would be 16.25 MHz. The top transmission speed of 54 Mbps is achieved by using a QAM modulation on each subcarrier. QAM constellations are investigated Section 15.1.4 and in the problems at the end of the chapter.

The optimal demodulator has the form of K_b parallel single bit optimal demodulators. This demodulator is shown in Figure 15.13. Restricting ourselves to BPSK modulation on each subcarrier frequency the MLBD has the form

$$\Re\left[\int_0^{T_p} Y_z(t)\sqrt{\frac{1}{T_p}}\exp[-j2\pi f_d(2k - K_b - 1)t]dt\right] = \Re[Q(k)] \underset{\hat{I}(k)=1}{\overset{\hat{I}(k)=0}{\gtrless}} 0 \tag{15.46}$$

[2]See the definition of PAPR in Chapter 5.

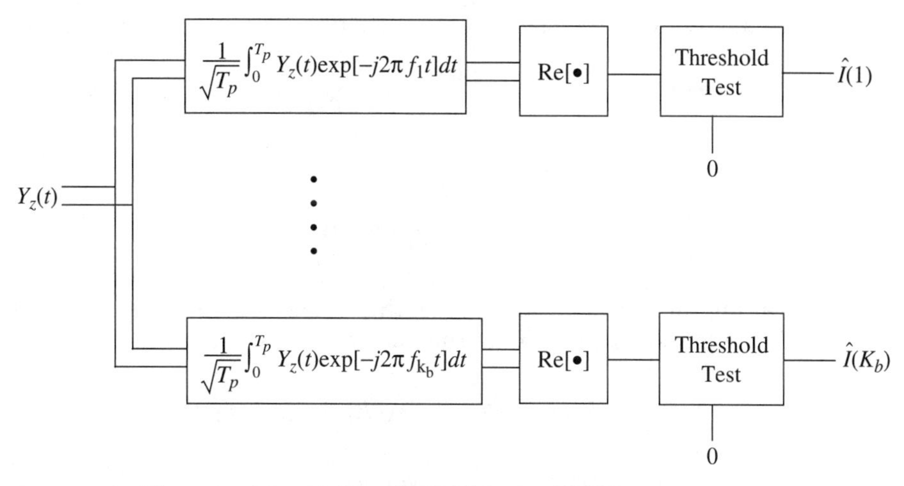

Figure 15.13 The optimal demodulator for OFDM using BPSK.

Since OFDM is using a linear modulation on each subcarrier we will use $Q(k)$ to denote the matched filter output for the kth bit or subcarrier. The demodulator computes a filter output for each bit, $k \in \{1, \ldots, K_b\}$, and hence the complexity of the OFDM optimum demodulator is $O(K_b)$ as opposed to the $O(2^{K_b})$ for an arbitrary modulation that transmits K_b bits of information. For future reference here we note that $Q(k)$ can be viewed at the Fourier transform of $Y_z(t)$ evaluated at $f = f_k = f_d((2k - K_b - 1)$. The optimal OFDM demodulator evaluates the real part of the Fourier transform at K_b points symmetrically spaced around $f = 0$ and uses these Fourier transform values in a threshold test. Since $\Re[Q(k)] = D_z(k)\sqrt{E_b} + N_I(k)$ where $\text{var}(N_I(k)) = \frac{N_0}{2}$, the bit error probability performance of OFDM using BPSK on each subcarrier is clearly

$$P_B(E) = \frac{1}{2}\text{erfc}\left(\sqrt{\frac{E_b}{N_0}}\right) \tag{15.47}$$

The optimal demodulator for OFDM is essentially K_b single bit demodulators implemented in parallel (one for each subcarrier). Because of the orthogonality of the subcarriers the fidelity achieved by an OFDM demodulator is the same as the demodulator for a single bit transmitted in isolation (see Chapter 13).

Spectral Characteristics of OFDM

Recall that the average energy spectrum per bit is

$$D_{x_z}(f) = \frac{E[G_{x_z}(f)]}{K_b} \tag{15.48}$$

To simplify the notation needed in this discussion we will make the following definition.

Definition 15.1 The Fourier transform of the unit energy rectangular pulse function

$$u_r(t) = \begin{cases} \dfrac{1}{\sqrt{T_p}} & 0 \le t \le T_p \\ 0 & \text{elsewhere} \end{cases} \tag{15.49}$$

is

$$U_r(f) = \sqrt{T_p}\frac{\sin(\pi f T_p)}{\pi f T_p}\exp[-j\pi f T_p] \tag{15.50}$$

Taking the Fourier transform of the OFDM signal in Eq. (15.42) and using the frequency shift property of the Fourier transform gives

$$X_z(f) = \sum_{l=1}^{K_b} D_z(l)\sqrt{E_b}U_r(f - f_d(2l - K_b - 1)) \tag{15.51}$$

The energy spectrum is then given as

$$G_{X_z}(f) = E_b\sum_{l_1=1}^{K_b}\sum_{l_2=1}^{K_b} D_z(l_1)D_z^*(l_2)U_r(f - f_d(2l_1 - K_b - 1))U_r^*(f - f_d(2l_2 - K_b - 1)) \tag{15.52}$$

The average energy spectrum per bit is then

$$D_{X_z}(f) = \frac{E_b}{K_b}E\left[\sum_{l_1=1}^{K_b}\sum_{l_2=1}^{K_b} D_z(l_1)D_z^*(l_2) \right.$$
$$\left. \times U_r(f - f_d(2l_1 - K_b - 1))U_r^*(f - f_d(2l_2 - K_b - 1))\right] \tag{15.53}$$

The average energy spectrum per bit can be greatly simplified when BPSK modulation is used on each frequency in OFDM. Recall each bit is assumed to be random and independently distributed. Here it will be further assumed that the bits are identically distributed with $P(I(k) = 0) = \pi_0$ and $P(I(k) = 1) = \pi_1$ for $k = 1, \ldots, K_b$. Examining the term corresponding to a fixed l_1 and l_2, $l_1 \neq l_2$, in Eq. (15.53) (one of K_b^2 terms in the summation) gives

$$\begin{aligned} &E[D_z(l_1)D_z^*(l_2)]U_r(f - f_d(2l_1 - K_b - 1))U_r^*(f - f_d(2l_2 - K_b - 1)) \\ &\quad = (\pi_0\pi_0 + \pi_1\pi_1 - \pi_0\pi_1 - \pi_1\pi_0)U_r(f - f_d(2l_1 - K_b - 1)) \\ &\quad\quad \times U_r^*(f - f_d(2l_2 - K_b - 1)) = (\pi_0 - \pi_1)^2 \\ &\quad\quad \times U_r(f - f_d(2l_1 - K_b - 1))U_r^*(f - f_d(2l_2 - K_b - 1)) \end{aligned} \tag{15.54}$$

Examining the term corresponding to $l_1 = l_2$, in Eq. (15.53) gives

$$\begin{aligned} &E\left[|D_z(l_1)|^2\right]|U_r(f - f_d(2l_1 - K_b - 1))|^2 \\ &\quad = (\pi_0 + \pi_1)|U_r(f - f_d(2l_1 - K_b - 1))|^2 \\ &\quad = |U_r(f - f_d(2l_1 - K_b - 1))|^2 \end{aligned} \tag{15.55}$$

The average energy spectrum per bit is then given as

$$D_{X_z}(f) = \frac{E_b}{K_b}\sum_{l_1=1}^{K_b}|U_r(f - f_d(2l_1 - K_b - 1))|^2 + \frac{E_b}{K_b}(\pi_1 - \pi_0)^2 \times \sum_{l_2=1}^{K_b}\sum_{\substack{l_1=1\\ l_1\neq l_2}}^{K_b} U_r(f - f_d(2l_1 - K_b - 1))U_r^*(f - f_d(2l_2 - K_b - 1)) \tag{15.56}$$

Further simplifications occur if each bit is equally likely. For equally likely bits the second term in Eq. (15.56) becomes zero. The spectrum in this case becomes

$$D_{X_z}(f) = \frac{E_b}{K_b}\sum_{l=1}^{K_b}|U_r(f - f_d(2l - K_b - 1))|^2 \tag{15.57}$$

This spectrum is plotted in Figure 15.14 (the solid line) for $K_b = 4$ and $f_d = 0.5/T_p$. The dashed lines in Figure 15.14 represent the energy spectrum of each modulated subcarrier of the OFDM waveform. The conclusions we can

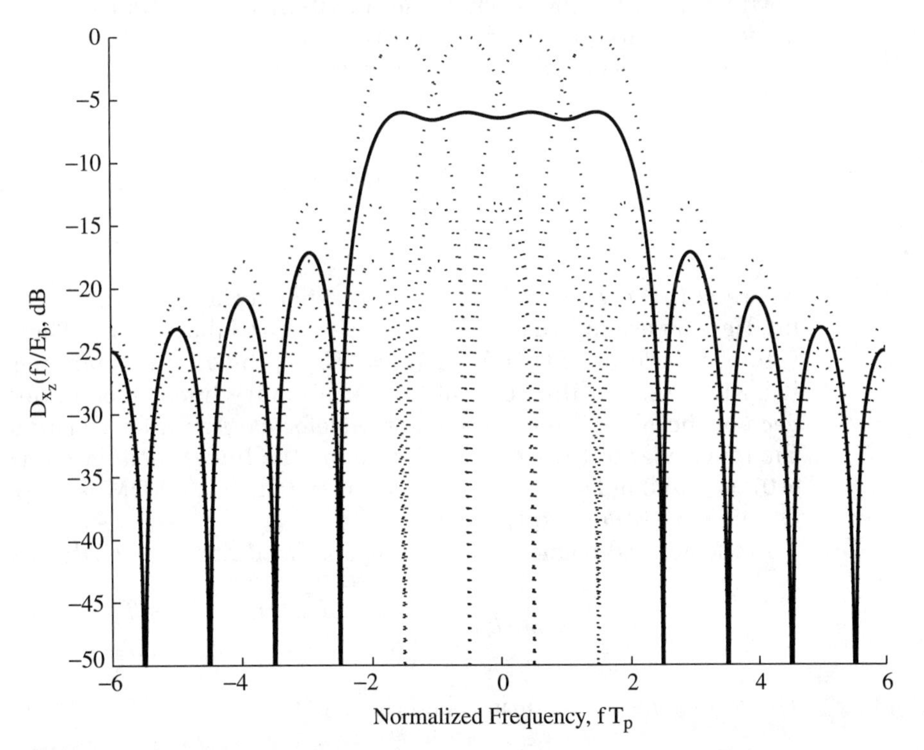

Figure 15.14 Average energy spectrum per bit for OFDM. $K_b = 4$, equally likely bits and BPSK modulation, with $f_d = 1/(2T_p)$.

draw about the average energy spectrum of OFDM is that the bandwidth occupancy for this type of an OFDM is proportional to $1/T_p$ and to K_b. Recall that the transmission rate is $W_b = K_b/T_p$ bits/s. Consequently, the transmission spectral efficiency is in the neighborhood of 1 bit/s/Hz, though the exact number will be a function of how the engineering bandwidth is defined.

In conclusion OFDM provides a method to implement multiple bit transmission with good performance and reasonable complexity and spectral efficiency. The bit error probability performance of the OFDM scheme highlighted in this section gives the same bit error probability performance as BPSK used in isolation. The spectral efficiency is roughly 1 bit/s/Hz. The complexity of the optimum receiver is $O(K_b)$. Consequently, OFDM has found significant utility in engineering practice. One issue with OFDM is that the signal has a large PAPR and this characteristic can require a high performance and costly radio system to get high fidelity in message reconstruction.

15.3.2 Orthogonal Code Division Multiplexing

A second commonly used modulation that admits a simple optimal bit demodulation is orthogonal code division multiplexing (OCDM). OCDM is often used in cellular radio communication and in satellite communication on the downlink. For OCDM, each of the K_b bits is independently modulated on the same carrier frequency with an orthogonal waveform. This orthogonal waveform is often termed the spreading waveform. The typical format for an OCDM signal is

$$X_z(t) = \begin{cases} \sum_{l=1}^{K_b} D_z(l)\sqrt{E_b}s_l(t) & 0 \le t \le T_p \\ 0 & \text{elsewhere} \end{cases} \tag{15.58}$$

where $D(l) = a(I(l))$ and $s_l(t)$ is often denoted the spreading signal for the lth bit. Here we assume that both $E[|D(l)|^2] = 1$ and that $E_{s_l} = 1$. The transmission rate of this form of OCDM is $W_b = K_b/T_p$ bits/s. In examining Eq. (15.58) it is clear that in this form of OCDM a binary linear modulation is used on each of the K_b different spreading waveforms. Again for clarity of discussion the remainder of the section will assume the linear modulation is BPSK (i.e., $a(0) = 1$ and $a(1) = -1$). It should be noted that OFDM is a special case of OCDM with $s_l(t) = \exp[j2\pi f_l t]/\sqrt{T_p}$.

In the general framework of orthogonal modulations OCDM has set

$$x(m_l, l, t) = \begin{cases} d_{m_l}\sqrt{E_b}s_l(t) & 0 \le t \le T_p \\ 0 & \text{elsewhere} \end{cases} \tag{15.59}$$

The orthogonality condition then corresponds to

$$\Re\left[\int_{-\infty}^{\infty} x(m_l, l, t)x^*(m_k, k, t)dt\right] = E_b\Re\left[d_{m_l}d^*_{m_k}\int_0^{T_p} s_l(t)s_k^*(t)dt\right] = 0 \tag{15.60}$$

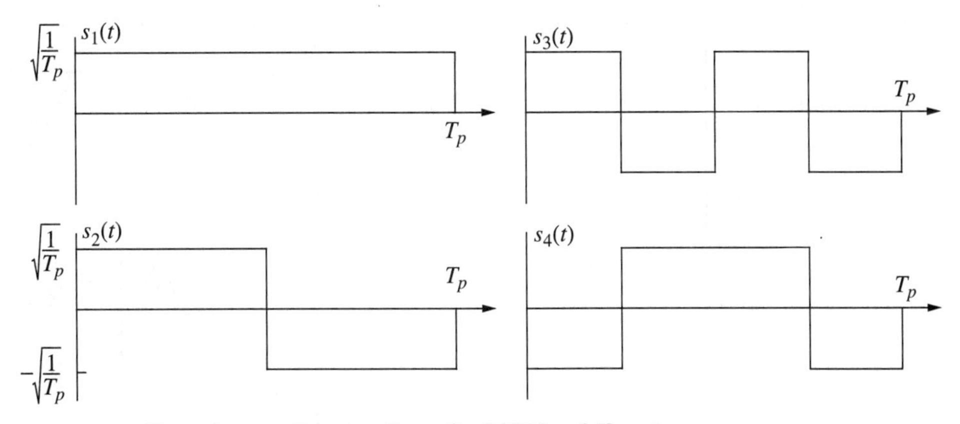

Figure 15.15 Example spreading waveforms for OCDM and $K_b = 4$.

The remaining question is how to find a set of K_b orthogonal waveforms. There is a wide variety of ways to construct these waveforms and one example is given in Figure 15.15 for $K_b = 4$. Certainly this construction is not unique. The time waveform of OCDM can have significant variations across the transmission time. For instance Figure 15.16 shows the time plot of the output envelope[3] of two possible transmitted waveforms for the spreading waveforms introduced in Figure 15.15 for $K_b = 4$. In both cases the transmitted energy is equal but how that energy is spread across the transmit time is very much a function of the transmitted information. An OCDM waveform has a large PAPR. The PAPR will increase as K_b increases. This high PAPR requires the radios in an OCDM system to have a large dynamic range to process the signal without distortion.

EXAMPLE 15.6

One of the standards for mobile telephones is denoted IS-95. IS-95 is primarily a voice communication system and voice communications can be achieved with transmission rates of less than 10 kHz with modern voiceband source encoders. IS-95 uses OCDM as the forward link modulation with $K_b = 64$. The spreading waveforms used to send each bit are Walsh functions which are discussed in Problem 15.22 (This problem shows how Walsh functions are derived from Hadamard matrices). In IS-95 $T_p = 52.08\bar{3}$ μs so if BPSK modulation was used a bit rate of 1.2288 Mbps could be achieved. In voice mobile telephony data rates only need to support voiceband speech transmission. Hence IS-95 uses each of the spreading waveforms to transmit a bit to potentially 64 different users. This use of one physical layer radio link to support many information transactions is known as multiple access communications. Each spreading waveform is simultaneously capable of supporting a 19.2 kbps transmission rate for each user which is a data rate greater than necessary for modern speech encoders.

---

[3]Because the spreading waveforms are all real $x_A(t) = |x_I(t)|$.

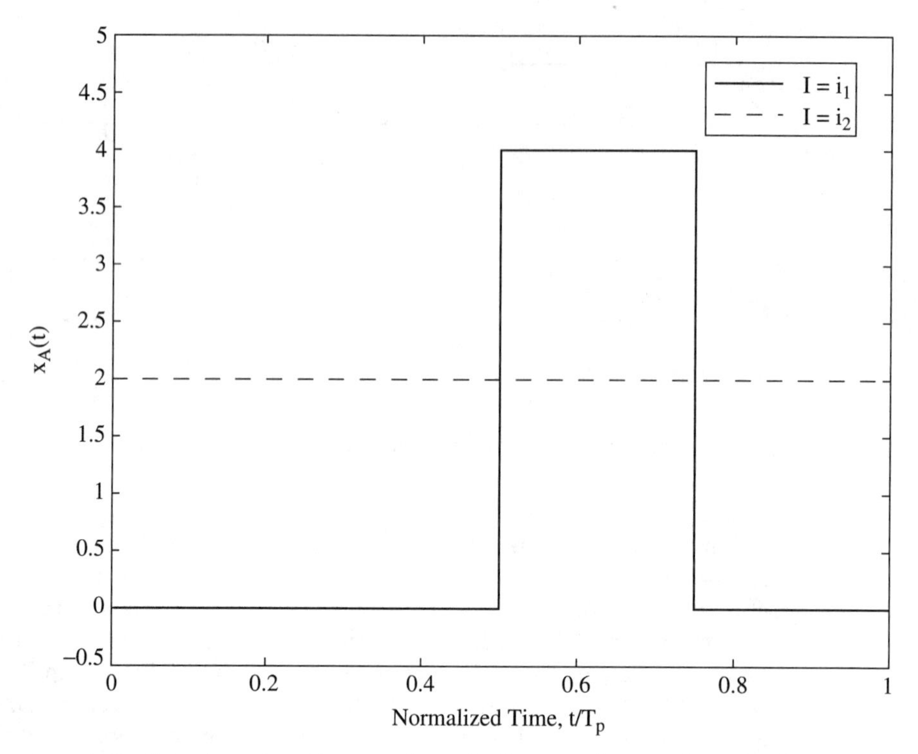

Figure 15.16 Output envelope time waveforms for the spreading waveforms in Figure 15.15. $i_1 = [0\ 1\ 0\ 0]$ and $i_2 = [0\ 0\ 0\ 0]$.

The optimal demodulator again has the form of K_b parallel single bit optimal demodulators. This demodulator is shown in Figure 15.17. Restricting ourselves to BPSK modulation on each spreading waveform the MLBD has the form

$$\Re\left[\int_0^{T_p} Y_z(t)s_k^*(t)\,dt\right] = \Re[Q(k)] \underset{\hat{I}(k)=1}{\overset{\hat{I}(k)=0}{\gtrless}} 0 \tag{15.61}$$

The demodulator computes a filter output for each bit, $k \in \{1, \ldots, K_b\}$, and hence the complexity of optimal OCDM demodulation is again $O(K_b)$ as opposed to $O(2^{K_b})$ for an arbitrary modulation that transmits K_b bits of information. Since $\Re[Q(k)] = D_z(k)\sqrt{E_b} + N_I(k)$ where $\text{var}(N_I(k)) = \frac{N_0}{2}$, the bit error probability performance of OCDM using BPSK on each spreading waveform is again

$$P_B(E) = \frac{1}{2}\text{erfc}\left(\sqrt{\frac{E_b}{N_0}}\right) \tag{15.62}$$

The optimal demodulator for OCDM is again K_b single bit demodulators implemented in parallel (one for each orthogonal spreading waveform). Because of the orthogonality of the spreading waveforms the fidelity achieved by an OCDM

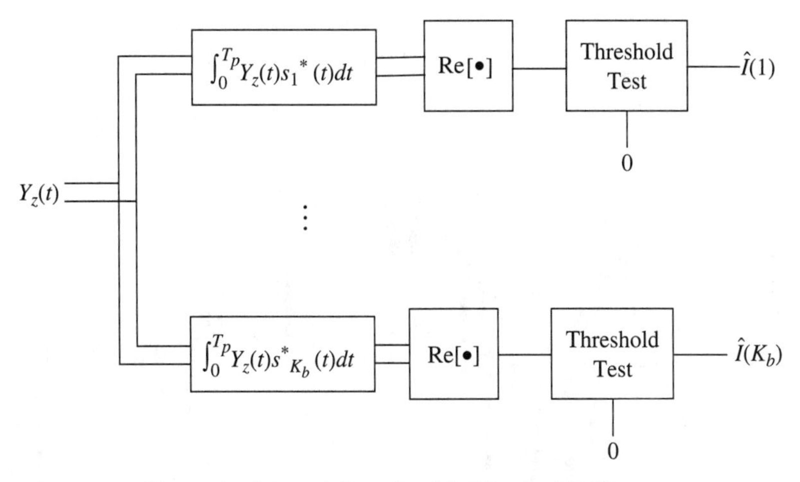

Figure 15.17 The optimal demodulator for OCDM using BPSK.

demodulator is the same as the demodulator for a single bit transmitted in isolation (see Chapter 13).

Spectral Characteristics of OCDM

Again the average energy spectrum per bit is

$$D_{x_z}(f) = \frac{E\left[G_{x_z}(f)\right]}{K_b} = \frac{E_b}{K_b} E\left[\sum_{l=1}^{K_b}\sum_{k=1}^{K_b} D_z(l)D_z^*(k)S_l(f)S_k^*(f)\right] \tag{15.63}$$

Using the results of Section 15.3.1 and assuming for simplicity that each bit is equally likely and BPSK modulation is used, the spectrum in this case becomes

$$D_{X_z}(f) = \frac{E_b}{K_b}\sum_{l=1}^{K_b} |S_l(f)|^2 \tag{15.64}$$

where $S_l(f) = \mathcal{F}\{s_l(t)\}$. The important thing to note here is that the spectrum of OCDM waveforms is directly proportional to the spectrum of the chosen spreading waveforms. The spectrum of the set of spreading signals chosen in Figure 15.15 (the solid line) is shown in Figure 15.18. The dashed lines in Figure 15.18 represent the energy spectrum of each modulated spreading signal of the OFDM waveform. The conclusions we can draw about the average energy spectrum of OCDM is that the bandwidth occupancy for this type of an OCDM is proportional to $1/T_p$ and to K_b. The bandwidth is greater than if a single bit was sent in isolation since K_b orthogonal waveforms need to be constructed for each of the K_b transmitted bits.

In conclusion OCDM provides a method to implement multiple bit transmission with good performance and reasonable complexity and spectral efficiency. The bit error probability performance of the OCDM scheme highlighted in this section gives the same bit error probability performance as BPSK used in isolation. The spectral efficiency is roughly 1 bit/s/Hz. The complexity of the optimum

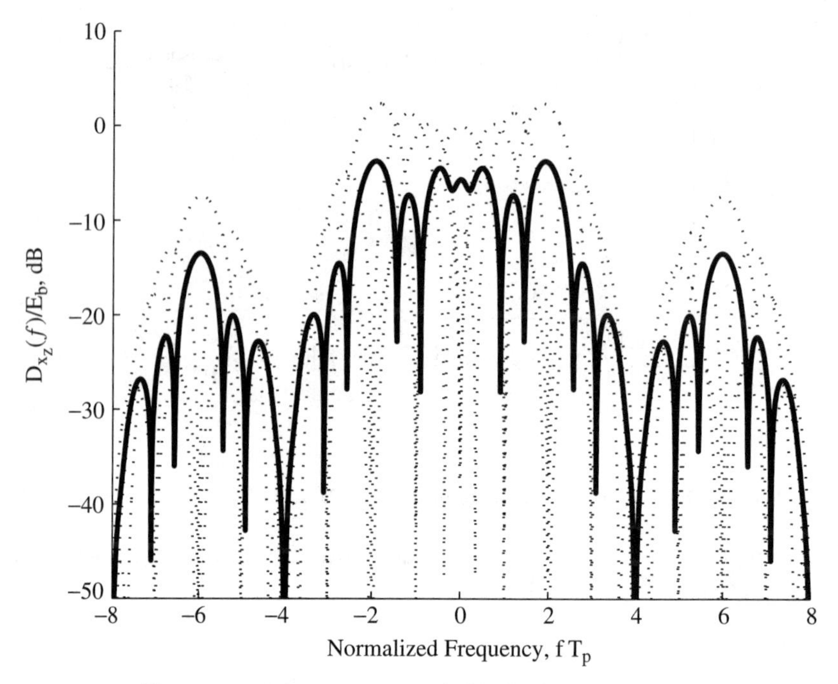

Figure 15.18 The average energy spectrum per bit for the $K_b = 4$ example OCDM in Figure 15.15.

receiver is $O(K_b)$. OCDM is a generalization of OFDM. Consequently, OCDM has found significant utility in engineering practice. One issue with OCDM is that the signal has a large PAPR and this characteristic can require a high performance and costly radio system to get high fidelity in message reconstruction.

15.3.3 Binary Stream Modulation

A third commonly used modulation that admits a simple optimal bit demodulation is orthogonal time division multiplexing. This is perhaps the most intuitive form of orthogonal modulation. Orthogonal time division multiplexing is used in a vast majority of digital communication systems in some form. The idea is simple, data is streamed in time (1 bit following another) and for the remainder of this text this orthogonal time division multiplexing will be referred to as **stream** modulation. With a stream modulation each of the K_b bits is independently modulated on the same carrier frequency with a time shifted waveform. The typical format for a stream modulation using linear modulation is

$$X_z(t) = \sum_{l=1}^{K_b} D_z(l)\sqrt{E_b}u(t-(l-1)T) \tag{15.65}$$

where $D(l) = a(I(l))$, T is known as the symbol[4] or bit time in stream modulations, and $u(t)$ is the unit energy pulse of length T_u. The transmission rate of stream modulation is $W_b = K_b/T_p$ bits/s. It should be noted that $T_p = (K_b - 1)T + T_u$ and for large K_b that $W_b \approx 1/T$.

Definition 15.2 The Baud rate of a stream modulation is $W_{Bau} = 1/T$.

The time shifted pulses are repeated every T seconds. Hence the rate at which bits (or more generally multiple bit symbols) are put into the channel is the Baud rate. In examining Eq. (15.65) it is clear that in this stream modulation a binary linear modulation is used in each of the K_b different time intervals. Again for clarity of discussion the remainder of the section will assume the linear modulation is BPSK (i.e., $a(0) = 1$ and $a(1) = -1$).

In the general framework of orthogonal modulations stream modulation has set

$$x(m_l, l, t) = \begin{cases} d_{m_l}\sqrt{E_b}u(t - (l-1)T) & 0 \le t \le T_p \\ 0 & \text{elsewhere} \end{cases} \tag{15.66}$$

where again $u(t)$ is denoted the pulse shape. The orthogonality condition then corresponds to

$$\begin{aligned} 0 &= \Re\left[\int_{-\infty}^{\infty} x(m_l, l, t)x^*(m_k, k, t)dt\right] \\ &= E_b\Re\left[d_{m_l}d^*_{m_k}\int_{-\infty}^{\infty} u(t-(l-1)T)u^*(t-(k-1)T)dt\right] \\ &= E_b\Re[d_{m_l}d^*_{m_k}V_u((k-l)T)] \end{aligned} \tag{15.67}$$

where $V_u(\tau)$ is the pulse shape correlation function defined in Chapter 2 as

$$V_u(\tau) = \int_{-\infty}^{\infty} u(t)u^*(t-\tau)dt. \tag{15.68}$$

This orthogonal time shift condition is often known as Nyquist's criterion for zero intersymbol interference (ISI) [Nyq28a]. The remaining question is how to design orthogonal time shifted waveforms. There is a wide variety of ways to construct these waveforms but the simplest way is to limit $u(t)$ to only have support on $[0, T]$. For example, if $K_b = 4$ one can choose $T = T_p/4$ and have the set of time shifted waveforms as given in Figure 15.19. Certainly this construction is not unique. It is interesting to note with the waveforms chosen in Figure 15.19 the amplitude of the transmitted signal will be constant. The ability to more carefully control peak to average power ratio is one advantage of stream modulation.

[4]A generalization not considered in this chapter but frequently used in practice is the idea of a sending multiple bit symbol (instead of a 1 bit symbol considered here) every T seconds in a stream modulation. Essentially this is a combination of linear modulation and stream modulation.

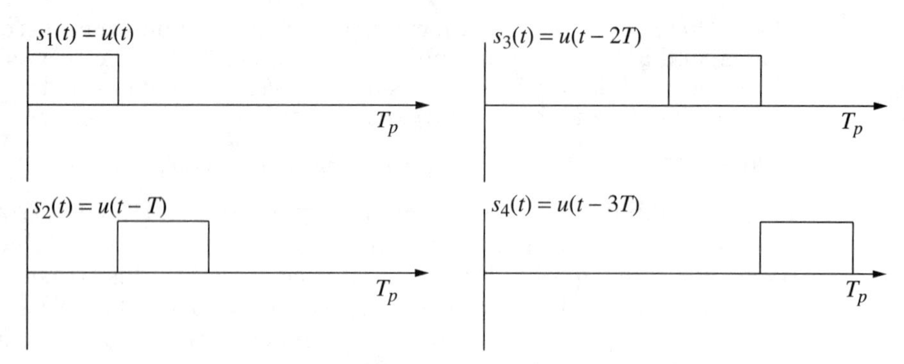

Figure 15.19 An example of time shifted pulses for a linear stream modulation and $K_b = 4$.

The optimal demodulator again has the form of K_b parallel single bit optimal demodulators. Since the sample modulation format is repeated in time the required filtering operation can be implemented with one filter sampled serially in time. This demodulator is shown in Figure 15.20. Restricting ourselves to BPSK modulation on each time shifted pulse the optimum demodulator has the form

$$\Re\left[\int_0^{T_p} Y_z(t)u^*(t-(k-1)T)dt\right] = \Re[Q(k)] \underset{\hat{I}(k)=1}{\overset{\hat{I}(k)=0}{\gtrless}} 0 \tag{15.69}$$

The demodulator has one filter output whose output is sampled for each bit, $k \in \{1, \ldots, K_b\}$, and hence the complexity of the stream modulation optimum demodulator is again $O(K_b)$ as opposed to the $O(2^{K_b})$. The pulse shape matched filter output will be referenced frequently in the sequel hence we formally define it.

Definition 15.3 Recall the pulse shape matched filter has an impulse response $h_o(t) = u^*(T_u - t)$, so that the pulse shape matched filter output is

$$Q_u(t) = \int_0^{T_p} Y_z(\lambda)u^*(\lambda + T_u - t)d\lambda \tag{15.70}$$

where T_u is the pulse shape time support.

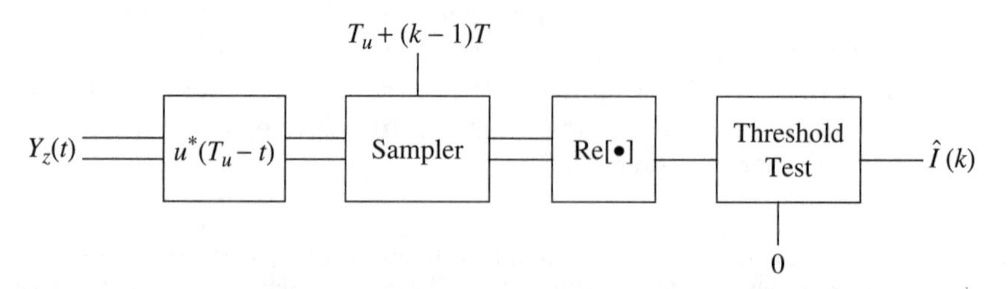

Figure 15.20 The optimal demodulator for stream modulations using BPSK.

Using this definition it is easy to see that

$$Q(k) = Q_u(T_u + (k-1)T) \tag{15.71}$$

A characteristic of linear stream modulation that makes for an efficient modem implementation is that only one filter is needed to implement the demodulator. This advantage of stream modulation over other orthogonal modulations is much more significant when the filter is implemented with analog circuits. This pulse shape matched filter only needs to be sampled at different times to obtain the sufficient statistics for demodulation. When filters are implemented with digital cirucits the computational complexity required for demodulation of all orthogonal modulations is about the same. This is the reason that stream modulations dominated digital modem implementations up untill the 1980's as most processing was done with analog circuitry. As Moore's Law kicked in and implementations became predominantly digital other modulations like OFDM and OCDM became practical.

Finally, again the performance of stream modulation is upperbounded by the performance of binary modulations transmitted in isolation. This is true since $\Re[Q(k)] = D_z(k)\sqrt{E_b} + N_I(k)$ where $\text{var}(N_I(k)) = \frac{N_0}{2}$, the bit error probability performance of stream modulation using BPSK is

$$P_B(E) = \frac{1}{2}\text{erfc}\left(\sqrt{\frac{E_b}{N_0}}\right) \tag{15.72}$$

Spectral Characteristics of Stream Modulations

Again the average energy spectrum per bit is

$$D_{x_z}(f) = \frac{E[G_{x_z}(f)]}{K_b} = \frac{E_b}{K_b} E\left[\sum_{l=1}^{K_b}\sum_{k=1}^{K_b} D_z(l) D_z^*(k) U(f) U^*(f) \exp[-j2\pi f(l-k)T]\right] \tag{15.73}$$

Using the results of Section 15.3.1 and assuming for simplicity that each bit is equally likely and BPSK modulation is used, the spectrum in this case becomes

$$D_{X_z}(f) = \frac{E_b}{K_b}\sum_{l=1}^{K_b} |U(f)|^2 = E_b G_u(f) \tag{15.74}$$

where $U(f) = \mathcal{F}\{u(t)\}$ and $G_u(f) = |U(f)|^2$. The important thing to note here is that the spectrum of linear stream modulation is directly proportional to the spectrum of the chosen pulse shape. The spectrum of the linear stream modulation defined in Figure 15.19 is shown in Figure 15.21. The conclusions we can draw about the average energy spectrum of stream modulation is that the bandwidth occupancy is also proportional to $1/T_p$ and to K_b.

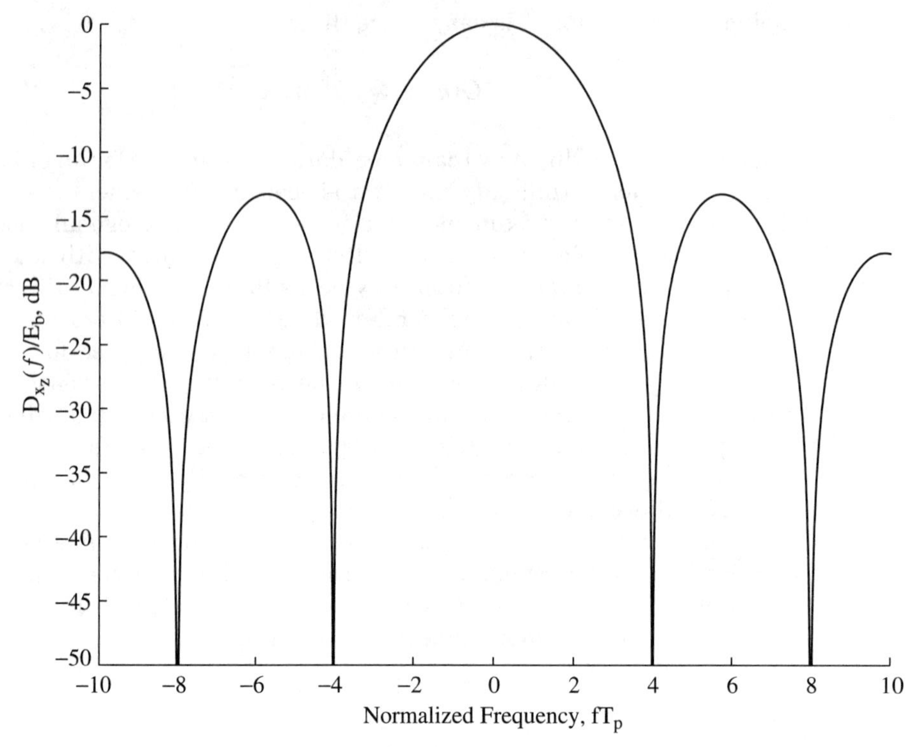

Figure 15.21 The average energy spectrum per bit for stream modulation for the example pulse shape given in Figure 15.19.

15.4 Conclusions

This chapter looked at two signal design techniques that simplify the optimal demodulator structures: linear modulation and orthogonal modulation. Linear modulation simplifies the demodulator by only requiring one matched filter. Linear modulation provides reduced performance but no bandwidth expansion as K_b grows larger. Consequently, linear modulations often find utility when spectral efficiency is at a premium and the expected SNR is high. Orthogonal modulation is a technique where K_b bits are transmitted by using orthogonal waveforms for each bit. The orthogonality of the individual bit waveforms enables each bit to be detected optimally with a complexity identical to the situation where the bit was transmitted in isolation. Examples of orthogonal modulation included in this chapter were modulations where orthogonality was obtained by frequency spacing (OFDM), obtained by complex waveforms (OCDM), and obtained by time spacing (stream modulation). Both linear modulation and orthogonal modulation provide optimal demodulation complexity that scales better with K_b than does general signaling. Using signals designed as either a linear or an orthogonal modulation produces a fidelity upperbounded by the error rate of BPSK. Linear modulation is advantageous as it allows for spectral efficiencies greater than 1 bit/s/Hz and is typically why it is used in practice. As a final note all deployed communication systems use stream modulation in some form even when using OFDM or OCDM. It is also estimated that greater than 90% of the digital communication systems use linear modulations in some form.

15.5 Homework Problems

Problem 15.1. 16 quadrature amplitude modulation (QAM) is a linear modulation with $D_i = D_{Ii} + jD_{Qi}$ where $D_{Ii}, D_{Qi} \in \{-3A, -A, A, 3A\}$ that can be used to transmit 4 bits per symbol ($K_b = 4$).

(a) Find the value of A such that E_b is the average energy per bit.

(b) Find the decision regions corresponding to the values of Q for MLWD of 16QAM.

(c) Is 16QAM a geometrically uniform signal set?

(d) Compute the probability of word error for MLWD, $P_W(E)$, for 16QAM and plot for $E_b/N_0 = 0$, 15 dB.

(e) Compute the full union bound to the probability of word error for MLWD for 16QAM and plot for $E_b/N_0 = 0$, 15 dB. *Hint:* The union bound requires the computation of 15 distances for each of the 16 possible transmitted signals so it might be useful to use the computer.

(f) It is possible to eliminate some terms in the union bound for 16QAM as was done for MPSK. Identify the union bound having the smallest number of terms for 16QAM.

(g) Using the union bound result and assuming a pulse shape is chosen such that $B_T = 1/T_p$ plot the 16QAM spectral efficiency in Figure 14.9.

Problem 15.2. Consider the two 8-ary ($K_b = 3$) linear modulations whose constellations are given in Figure 15.22. One of the 8-ary constellations has four points each placed on two concentric circles. The other 8-ary constellation has eight points selected from the 16QAM (see Problem 15.1) constellation in a checkerboard fashion.

a) Determine the value of each constellation point such that E_b is the energy per bit and $\Delta_1 = \Delta_2$.

b) Find the decision regions corresponding to the values of Q for MLWD of these two 8-ary modulations.

c) Which constellation has a better performance? Why? How does the performance of the best constellation compare to the 8-ary linear modulation introduced in Example 15.1?

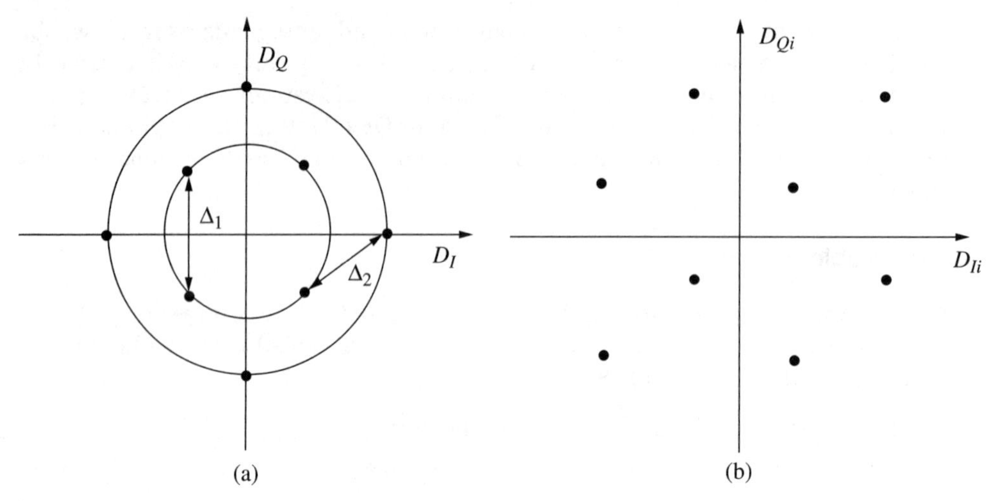

Figure 15.22 Two 8-ary constellation plots for linear modulations.

Problem 15.3. This problem examines the probability of word error for MLWD of some common linear modulations.

(a) Show that the probability of word error for 4PSK is given as

$$P_W(E) = \operatorname{erfc}\left(\sqrt{\frac{E_b}{N_0}}\right) - \left(\frac{1}{2}\operatorname{erfc}\left(\sqrt{\frac{E_b}{N_0}}\right)\right)^2 \tag{15.75}$$

(b) 4-ary equally spaced pulse amplitude modulation (PAM) has a constellation that is characterized with $\Omega_d = \{\pm\sqrt{2/5}, \pm 3\sqrt{2/5}\}$. Show that the probability of word error for 4PAM is given as

$$P_W(E) = \frac{3}{4}\operatorname{erfc}\left(\sqrt{\frac{2E_b}{5N_0}}\right) \tag{15.76}$$

(c) Plot the spectral efficiency versus E_b/N_0 of 4PAM on Figure 14.9 using a word error rate of 10^{-5}.

Problem 15.4. This problem examines the union bound to the probability of word error for MLWD of some common linear modulations.

(a) Show that the full union bound for the word error probability of 4PSK is given as

$$P_W(E) = \operatorname{erfc}\left(\sqrt{\frac{E_b}{N_0}}\right) + \frac{1}{2}\operatorname{erfc}\left(\sqrt{\frac{2E_b}{N_0}}\right) \tag{15.77}$$

(b) There is a way to tighten the union bound for 4PSK, give the tightest union bound to the word error probability.

(c) Show that the full union bound for the word error probability of 4PAM ($\Omega_d = \{\pm\sqrt{2/5}, \pm 3\sqrt{2/5}\}$) is given as

$$P_W(E) = \frac{3}{4}\text{erfc}\left(\sqrt{\frac{2E_b}{5N_0}}\right) + \frac{1}{2}\text{erfc}\left(\sqrt{\frac{8E_b}{5N_0}}\right) + \frac{1}{4}\text{erfc}\left(\sqrt{\frac{18E_b}{5N_0}}\right) \quad (15.78)$$

(d) There is a way to tighten the union bound for 4PAM, give the tightest union bound to the word error probability

Show that these results are tight to the true values given in Problem 15.3 at high SNR.

Problem 15.5. Imagine that digital computers were first developed with five logic levels to match the number of fingers on the typical human hand. Digital communications would have evolved to be the transmission of quinits (having five possible values) as opposed to the transmission of bits. In this problem you will consider what this course might have been like if this alternate evolution had occurred.

Consider a 5-ary modulation of the form $X_z(t) = D_z u(t)$ corrupted by an additive white Gaussian noise with one-sided spectral density of N_0. Consider two possible signal sets:

i	$d_i^{(1)}$	$d_i^{(2)}$
0	1	A
1	$\exp\left(\frac{j2\pi}{5}\right)$	jA
2	$\exp\left(\frac{j4\pi}{5}\right)$	$-A$
3	$\exp\left(\frac{j6\pi}{5}\right)$	$-jA$
4	$\exp\left(\frac{j8\pi}{5}\right)$	0

where A is a positive real constant.

(a) Choose the value of A such that both signal sets have the same average E_s.

(b) Give the demodulator that minimizes the word error probability and define the decision regions for each of the two possible signal sets.

(c) Is either signal set geometrically uniform?

(d) Compute the full union bound to $P_W(E)$ for each of the two signal sets.

(e) Simplify the union bound as much as possible for each signal set by eliminating terms from the full union bound if possible.

(f) Use numerical integration to compute the actual $P_W(E)$.

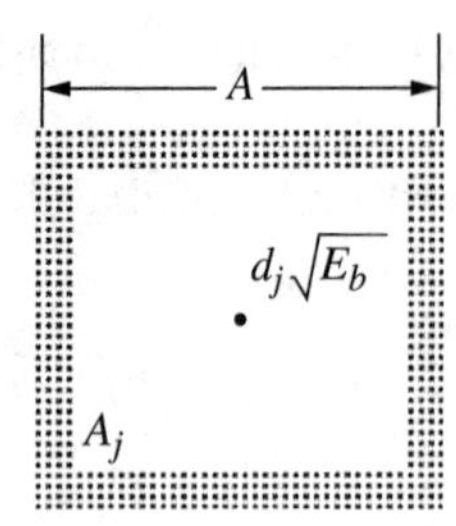

Figure 15.23 A square decision region.

Problem 15.6. Consider one of the 5-ary modulations from the previous problem of the form $X_z(t) = D_z u(t)$ with equally likely symbols and corrupted by an additive white Gaussian noise with one-sided spectral density of N_0.

i	d_i
0	1
1	$\exp\left(\frac{j2\pi}{5}\right)$
2	$\exp\left(\frac{j4\pi}{5}\right)$
3	$\exp\left(\frac{j6\pi}{5}\right)$
4	$\exp\left(\frac{j8\pi}{5}\right)$

(a) What is the average energy spectral density per quinit, $D_{x_z}(f)$?

(b) If $u(t) = \sqrt{\frac{E_u}{W}} \frac{\sin(\pi W t)}{\pi W t}$ give$D_{x_z}(f)$.

(c) If this particular communication system's electronics was particularly susceptible to 60 Hz interference from the power supply, could you come up with a pulse shape that produced a communication signal that did not have significant spectral content at 60 Hz. Assume that $T_p < 0.05$.

Problem 15.7. Assume signals are sent with a linear modulation and the resulting decision region for $\vec{I} = j$ is characterized as being square as in Figure 15.23. Show that

$$P_W(E|\vec{I} = j) = 1 - \left(\operatorname{erf}\left(\frac{A}{2\sqrt{N_0}}\right)\right)^2 \tag{15.79}$$

Problem 15.8. A problem examining Alamouti space-time signaling [Ala98]. Consider a 4-ary digital communication system with equally likely words where the received signal is corrupted by an additive white Gaussian noise of two-sided spectral density of $N_0/2$. The system uses two antennas as shown in Figure 15.24 and the four possible transmitted signals are shown in Figure 15.25. The received signal is of the form

$$y_z(t) = c_1 x_{i1}(t) + c_2 x_{i2}(t) + W_z(t) = x_i(t) + W_z(t)$$

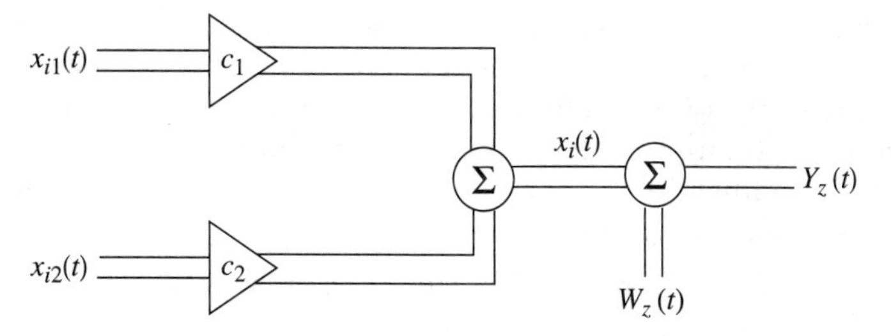

Figure 15.24 The two antenna communication system.

where c_1 and c_2 are known constants at the demodulator. The signal set is defined with

$$\begin{aligned} x_{01}(t) &= u_1(t) \quad x_{11}(t) = u_2(t) \quad x_{21}(t) = u_4(t) \quad x_{31}(t) = u_3(t) \\ x_{02}(t) &= u_2(t) \quad x_{12}(t) = u_3(t) \quad x_{22}(t) = u_1(t) \quad x_{32}(t) = u_4(t) \end{aligned}$$

(a) Show that this system results in an equal received energy signal set for any values of c_1 and c_2. Compute E_s.

(b) Detail out the optimum word demodulator (MLWD). Give impulse responses for any filters and simplify the structure as much as possible.

(c) Compute the square Euclidean distance between each of the signal pairs.

(d) Defining

$$u_r(t) = \begin{cases} \dfrac{1}{\sqrt{T_p}} & 0 \le t \le \frac{T_p}{2} \\ 0 & \text{elsewhere} \end{cases} \tag{15.80}$$

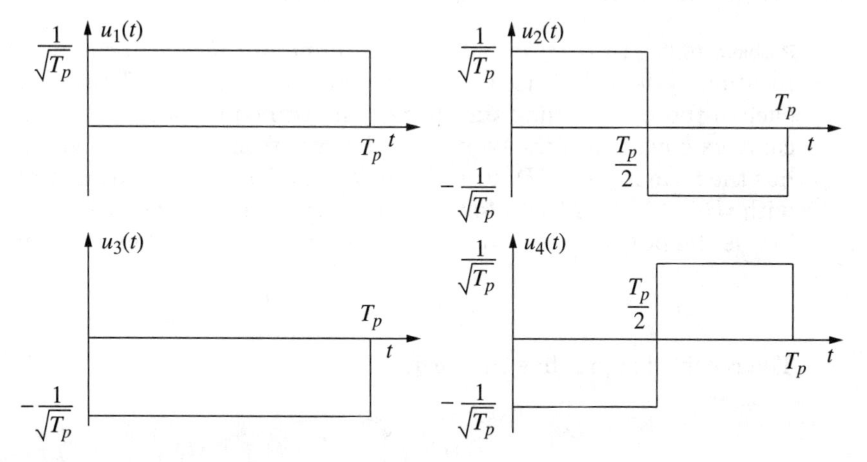

Figure 15.25 The transmitted waveforms.

show that $x_i(t) = d_i(1)s_1(t)+d_i(2)s_2(t)$, where $d_i(l)\, l = 1, 2$ represents BPSK modulation symbols, $s_1(t) = (c_1 u_r(t)+c_2 u_r(t-T_p/2)$, and $s_2(t)$ is orthogonal to $s_1(t)$. Identify $s_2(t)$ as a function of c_1, c_2, and $u_r(t)$. This implies each bit can be demodulated in an independent fashion and the demodulator for each bit simplifies to a binary threshold test. Derive the demodulator for $I(2)$ when this orthogonality condition exists.

(e) Compute the $P_B(E)$ for the MLBD for $I(1)$ and a given c_1 and c_2 conditioned on $\vec{I} = 0$.

Problem 15.9. Consider a demodulation scheme for a binary linear stream modulation where Nyquist's criterion for zero ISI holds. Prove that the real part of the noise sample taken from the matched filter for bit k_1, $k_1 \in \{1, \dots, K_b\}$, $N_I(k_1)$, is independent of the real part of the noise sample taken from the matched filter for bit k_2, $k_2 \in \{1, \dots, K_b\}$, $N_I(k_2)$ where $k_1 \neq k_2$.

Problem 15.10. In Problem 2.11 three pulse shapes are presented and this problem considers their use in a binary stream modulation of the form

$$X_z(t) = \sum_{l=1}^{K_b} x_{I(l)}(t - (l-1)T) \tag{15.81}$$

(a) If $x_0(t) = u_1(t)$ and $x_1(t) = u_2(t)$ find the smallest value of T where Nyquist's criterion is still satisfied.

(b) If $x_0(t) = u_1(t)$ and $x_1(t) = u_2(t)$ plot the resulting transmitted waveform for $K_b = 10$ and $\vec{I} = [0\ 1\ 1\ 1\ 1\ 0\ 0\ 1\ 0\ 0]$ for the value of T selected in (a).

(c) If $x_0(t) = u_1(t)$ and $x_1(t) = u_2(t)$ plot the resulting matched filter output waveform both signals for $K_b = 10$ and $\vec{I} = [0\ 1\ 1\ 1\ 1\ 0\ 0\ 1\ 0\ 0]$ for the value of T selected in (a).

(d) Repeat (a), (b), and (c) for the case when $x_0(t) = u_1(t)$ and $x_1(t) = u_3(t)$.

Problem 15.11. In this problem we examine an M-ary linear wireless communication system that uses multiple transmit antennas. This problem assumes each of the L_t antennas simultaneously transmits one bit of information. The complex envelope of the transmitted signal from the ith antenna, $i = 1, \dots, L_t$, has the form $X_{zi}(t) = D_i u(t)$ where $D_i = a_i(I_i)$ is a complex modulation symbol with $|D_i| \leq 1$, I_i is the information bit transmitted from the ith antenna, and $u(t)$ is the pulse shape. The constellation mapping is denoted with

$$a_i(0) = d_{0i} \qquad a_i(1) = d_{1i} \tag{15.82}$$

The received signal has the form

$$Y_z(t) = \sum_{i=1}^{L_t} c_i X_{zi}(t) + W(t) \tag{15.83}$$

where c_i are the complex channel gains for the ith antenna and $W(t)$ is a complex AWGN of one-sided spectral density of N_0, i.e. $R_W(\tau) = N_0\delta(\tau)$.

(a) What is the bandpass signal corresponding to $X_{z1}(t)$ when a carrier frequency of f_c is used?

(b) Show

$$Y_z(t) = D_z u(t) + W(t) \tag{15.84}$$

and find D_z. How many values can D_z take and what are they?

(c) Assume $L_t = 2$ and that c_1 and c_2 are known at the receiver. Formulate the optimum demodulator and simplify as much as possible.

(d) Assume BPSK modulation is used on each antenna $d_{0i} = 1$ and $d_{1i} = -1$ and compute the resulting union bound on the word error probability as a function of c_1 and c_2.

(e) If it is known before transmission that $c_1 = 1$ and $c_2 = \frac{1}{\sqrt{2}}(1 + j)$ postulate a binary modulation $D_i = a_i(I_i)$ for each antenna that would result in the best word error probability performance.

Problem 15.12. You have been asked to design a data communication system for the Zwertians. The Zwertians are people from a distant solar system who have the unique characteristic of having 3 fingers. Consequently, their number system is entirely base 3 and units of information are available in trinits (1 of 3 values). The Zwertians want a linear modulation, consequently a trinit is transmitted with the complex envelope

$$X_z(t) = D_z u(t) \tag{15.85}$$

where the pulse, $u(t)$, has energy $E_u = 1$. To complete your design you must specify an optimum demodulator and a 3-ary modulation scheme. Assume each value of the transmitted trinit is equally likely. The received signal is corrupted by an additive white Gaussian noise of one-sided spectral density N_0.

(a) Assume the three possible transmitted symbols are denoted $d_i \quad i = 1, 2, 3$. Find the demodulator that minimizes the probability of symbol error.

(b) Find the union bound to the symbol error in terms of N_0, and $d_i \quad i = 1, 2, 3$ for the optimum demodulator decision when $D_z = d_1$.

(c) Assume that the average transmitted energy per symbol is constrained to be $E_s = 1$, find the optimum values of $d_i \quad i = 2, 3$ when $d_1 = 1$.

Problem 15.13. Show that if K_b bits are transmitted on orthogonal waveforms with BPSK modulation then the word error probability is given as

$$P_W(E) = 1 - \left(1 - \frac{1}{2}\text{erfc}\left(\sqrt{\frac{E_b}{N_0}}\right)\right)^{K_b} \tag{15.86}$$

Using the characteristics of the binomial coefficients to arrange in Eq. (15.86) a sum like the union bound. From this sum can you identify the most probable error event for an orthogonal modulation and how many bit errors occur in the word. For example, if $K_b = 4$ and $\vec{I} = [0\ 0\ 1\ 0]$ find all most probable error code words.

Problem 15.14. Consider $K_b = 6$ bits being transmitted with OFDM. Compute the output spectrum of the transmitted signal with the closest possible carrier spacing, sketch the optimum demodulator and estimate the resulting optimum word error probability performance for

(a) 6 carrier frequencies each using BPSK modulation

(b) 3 carrier frequencies each using Gray coded 4PSK modulation

(c) 2 carrier frequencies each using 8PSK modulation

Make a comparison between each of these transmission strategies in terms of performance and spectral efficiency.

Problem 15.15. The OFDM transmitted signal is a sum of complex sinusoids and in general will be complex valued. In certain special cases the transmitted signal will be real valued only. When does this occur? Identify all the situations when this happens for $K_b = 4$ and BPSK modulation on each carrier.

Problem 15.16. Identify two other spreading waveforms of equal energy that are mutually orthogonal for an OCDM system with $K_b = 3$ besides

$$s_1(t) = \begin{cases} 1 & 0 \le t < \frac{T_p}{3} \\ \exp\left[\frac{j2\pi}{3}\right] & \frac{T_p}{3} \le t < \frac{2T_p}{3} \\ \exp\left[\frac{j4\pi}{3}\right] & \frac{2T_p}{3} \le t \le T_p \\ 0 & \text{elsewhere} \end{cases} \tag{15.87}$$

A goal in finding these additional two waveforms is to make $D_{x_z}(f)$ as compact as possible.

Problem 15.17. Identify four spreading waveforms of length T_p that are equal energy and are mutually orthogonal for an OCDM system with $K_b = 4$, and that are more spectrally efficient than the example spreading waveforms shown in Figure 15.15.

Problem 15.18. There is an alternate form for PWEP first introduced by Craig [Cra91] that proves to be useful in many situations and this problem will lead the student through the derivation. Consider a matched filter output Q for BPSK modulation when $D_z = a(1) = -1$ is transmitted. The probability of error in this case is given as (see Section 15.1.2)

$$P_B(E|I=1) = \int_{R_1} f_{N_z}(n_z)dn_z = \int 1_{R_1}(n_z) f_{N_z}(n_z)dn_z \tag{15.88}$$

where $R_1 = \{N_z : N_I \geq \sqrt{E_b}\}$. Use the form $N_z(t) = N_A(t)\exp[jN_P(t)]$ to express the above integral in polar coordinates and derive

$$P_B(E|I=1) = \frac{1}{\pi}\int_0^{\frac{\pi}{2}} \exp\left[\frac{-E_b}{N_0\cos^2(n_p)}\right] dn_p \tag{15.89}$$

Problem 15.19. The following is a pulse shape that is used in communication systems

$$u(t) = \begin{cases} \sqrt{\dfrac{2}{T_u}} \sin\left(\dfrac{\pi t}{T_u}\right) & 0 \leq t \leq T_u \\ 0 & \text{elsewhere} \end{cases} \tag{15.90}$$

Consider that this pulse is used in a binary linear stream modulation of the form

$$X_z(t) = \sum_{l=1}^{K_b} D_z(l)\sqrt{E_b}u(t-(l-1)T) \tag{15.91}$$

where each bit value is equally likely and the received signal has the form $Y_z(t) = X_z(t) + W(t)$ with $W(t)$ is an AWGN with one-sided spectral density of N_0.

(a) Compute $V_u(\tau)$.

(b) How fast can the symbol or bit rate, $W_b = 1/T$, be and still have the MLBD simplify to a binary threshold test? Provide the form of the MLBD when BPSK modulation is used (i.e., $a(0) = d_0 = 1$ and $a(1) = d_1 = -1$).

(c) What is the resulting performance of the optimal bit detector found in (b).

(d) Compute and plot $D_{x_z}(f)$.

At the maximum W_b for the waveform Eq. (15.91), a colleague of yours claims that you can double the transmission rate of the system in (b) and still have optimal bit demodulators that are simple threshold tests. The claim is that this is achieved by sending a separate binary stream modulation in the Q-channel that is offset in time from the original one by $T_u/2$. The new transmitted signal (called minimum shift keying (MSK) by your colleague) has the form

$$\tilde{X}_z(t) = \sum_{l=1}^{K_b} D_{z1}(l)\sqrt{E_b}u(t-(l-1)T) + jD_{z2}(l)\sqrt{E_b}u\left(t-(l-1)T - \frac{T}{2}\right) \tag{15.92}$$

(e) Show that $\tilde{X}_z(t)$ can be put in a form as given in Eq. (15.29). Check the orthogonality conditions to verify your colleague's assertion.

(f) For the time interval $\frac{T}{2} \leq t \leq T$ find $\tilde{X}_A(t)$ and $\tilde{X}_P(t)$ as a function of $D_{z1}(1)$ and $D_{z2}(1)$.

(g) Given the results in (e) can the waveform in Eq. (15.92) be interpreted as an FSK waveform?

h) Compute $D_{x_z}(f)$ for MSK.

Problem 15.20. Consider a 4-ary modulation of the form $X_z(t) = D_z\sqrt{E_b}u(t)$ corrupted by an additive white Gaussian noise with one-sided spectral density of N_0. Consider two possible signal sets:

i	$d_i^{(1)}$	$d_i^{(2)}$
0	$\sqrt{2}$	$A\sqrt{3}$
1	$j\sqrt{2}$	jA
2	$-\sqrt{2}$	$-A\sqrt{3}$
3	$-j\sqrt{2}$	$-jA$

where A is a positive real constant.

(a) Choose the value of A such that both signal sets have the same average E_s.

(b) Give the demodulator that minimizes the word error probability and define the decision regions for each of the two possible signal sets.

(c) Compute the union bound to $P_W(E)$ for each of the two signal sets.

(d) Decide which signal set for a common E_s will give better performance with the optimum demodulator.

Problem 15.21. Show in OFDM that if $f_d = 1/(2T_p)$ and K_b is odd that $X_z(0) = X_z(T_p)$. Show that this is not necessarily true for K_b even.

Problem 15.22. Define a 2×2 matrix

$$\mathbf{H}_2 = \begin{bmatrix} 1 & 1 \\ 1 & -1 \end{bmatrix} \tag{15.93}$$

and a recursion defined as

$$\mathbf{H}_{2^n} = \begin{bmatrix} \mathbf{H}_{2^{n-1}} & \mathbf{H}_{2^{n-1}} \\ \mathbf{H}_{2^{n-1}} & -\mathbf{H}_{2^{n-1}} \end{bmatrix} \tag{15.94}$$

These sets of matrices are known as Hadamard matrices. Show that if $K_b = 2^n$ then the K_b spreading waveforms generated as

$$s_i(t) = \sum_{j=1}^{2^n} \left[\mathbf{H}_{2^n}\right]_{ij} u_{2^n}\left(t - (j-1)\frac{T_p}{2^n}\right) \qquad i = 1, 2^n \tag{15.95}$$

where

$$u_{2^n}(t) = \begin{cases} \frac{1}{\sqrt{T_p}} & 0 \leq t \leq \frac{T_p}{2^n} \\ 0 & \text{elsewhere} \end{cases} \tag{15.96}$$

constituted a set of spreading waveforms for an OCDM system. Specifically sketch the waveforms and show orthogonality for $K_b = 2$ and $K_b = 8$. For $K_b = 4$ this set of spreading waveforms is given in Figure 15.15. A proof by induction for the general case would be most complete.

Problem 15.23. In the wireless local network protocol denoted IEEE 802.11b the lowest rate modulation is a binary modulation using the two waveforms given in Figure 15.30.

(a) If the two waveforms in Figure 15.30 are to be used in a binary stream modulation, how fast can the symbol or bit rate, $W_b = 1/T$, be and still have the maximum likelihood bit demodulator simplify to a binary threshold test for each bit? *Hint:* The answer is $W_b > 1,000,000$ Hz.

(b) Assume the two waveforms in Figure 15.30 were to be used in an orthogonal code division multiplexed system to send 1 bit. Find another spreading waveform, $s_2(t)$, for the case $K_b = 2$ defined on $[0, 1\mu s]$ with $E_{s_2} = E_b$ such that when BPSK modulation is used with $s_2(t)$ the maximum likelihood bit demodulator simplifies to a binary threshold test for each bit.

Problem 15.24. 8PAM is a linear modulation used to transmit $K_b = 3$ bits of information that is characterized with $\Omega_d \in \{\pm A, \pm 3A, \pm 5A, \pm 7A\}$. Assume each word is equally likely and the received signal is corrupted by an additive white Gaussian noise with one-sided spectral density of N_0.

(a) For the case where $E_u = 1$, find the value of A such that $E_s = 3E_b$.

(b) Define the decision regions for 8PAM that operate on the output of the pulse shape matched filter, Q.

(c) Find the probability of word error conditioned on $D_z = A$.

(d) A closed form expression exists for the probability of word error. Find it.

Problem 15.25. Plot the average energy spectrum of the example OFDM system presented in this chapter with $\pi_0(l) = 0.25\ l = 1, \ldots, K_b$ with $K_b = 6$.

Problem 15.26. Plot the average energy spectrum of the example stream modulation system presented in this chapter with $\pi_0(l) = 0.25\ l = 1, \ldots, K_b$ with $K_b = 6$.

Problem 15.27. $K_b = 3$ bits are to be transmitted via a linear modulation using a unit energy real valued pulse shape, $u(t)$, and a constellation given in Figure 15.2.

(a) Appropriately normalize the constellation so that the transmitted energy per bit is E_b.

(b) For a carrier frequency of f_c give the form of the bandpass waveform transmitted when $\vec{I} = 0$.

(c) Detail out the MLWD and simplified to as great a degree as possible.

(d) The spectral efficiency of a digital modulation is an important characteristic in engineering design. If you were given an explicite form for $u(t)$ give a methodology to find the spectral efficiency.

(e) Give an expression (does not need to be evaluated) for the conditional word error probability for the MLWD when $\vec{I} = 5$.

(f) Identify the smallest number of pairwise error probabilities that need be computed to provide a union bound to the conditional word error probability for the MLWD when $\vec{I} = 5$.

Problem 15.28. In this problem we consider a derivative of the signaling scheme used in Ethernet local area networks. Assume equally likely data bits. Ethernet signaling uses a linear stream modulation, i.e.,

$$X_c(t) = \sum_{l=1}^{K_b} D(l)\sqrt{E_b}u(t-(l-1)T) \tag{15.97}$$

with the pulse shape in Figure 15.26 where $E_u = 1$ is assumed. Note that Ethernet is a baseband modulation scheme so that we use the notation $X_c(t)$ versus $X_z(t)$ and that $D(l)$ must be real valued. Assume that $Y_c(t) = X_c(t) + W(t)$ where $W(t)$ is an additive white Gaussian noise of one-sided spectral density N_0 and that $T = T_u$.

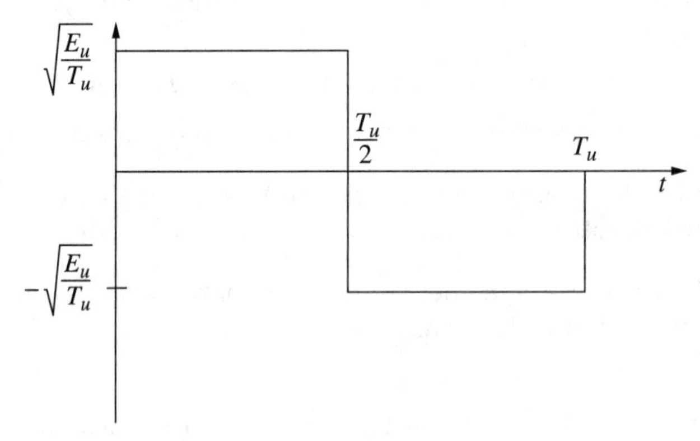

Figure 15.26 The pulse for Ethernet signaling.

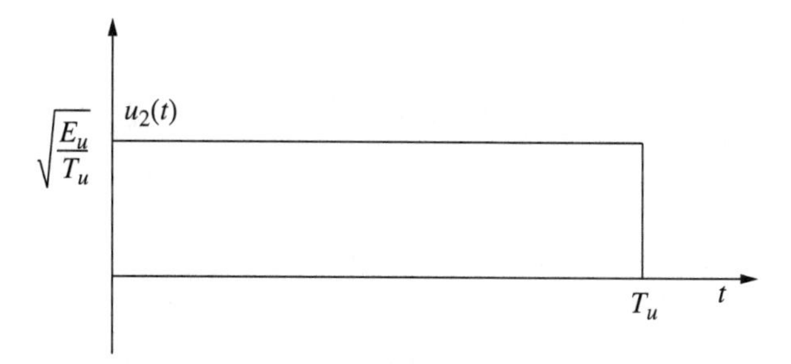

Figure 15.27 The alternative pulse for doubling the rate of Ethernet.

(a) Find the optimum demodulator.

(b) Find modulation symbol mappings from $I(l)$ to $D(l)$ that optimizes performance.

(c) Give an exact expression for the bit error probability in terms of E_b and N_0 for this optimum signal set and demodulator.

(d) Compute the average energy spectrum per bit, $D_{x_z}(f)$ and the spectral efficiency.

You are asked to double the transmission rate of Ethernet by transmitting two bits per symbol, $I_1(l)$ and $I_2(l)$, and the following options are given

(i) A linear modulation using $u(t)$ with a 4-ary modulation characterized with $D(l) = \pm 3A, \pm A$

(ii) A nonlinear modulation of the form

$$X_c(t) = \sum_{l=1}^{K_b} D_1(l)\sqrt{E_b}u(t-(l-1)T) + \sum_{l=1}^{K_b} D_2(l)\sqrt{E_b}u_2(t-(l-1)T) \quad (15.98)$$

where $D_i(l) = \pm 1$ and $u_2(t)$ is a second pulse shape given in Figure 15.27.

(e) Choose A such that each of the two signal sets have the same average energy per bit, E_b.

(f) Give the MLWD for each option.

(g) Find the word error probility or a union bound to the word error probability in terms of E_b, N_0 for both options.

(h) Compare the spectral efficiency of each scheme by computing the average energy spectrum per bit, $D_{x_z}(f)$ of both options.

Problem 15.29. In a system that transmits $K_b = 1$ bit of information the transmitted waveform has the form

$$X_z(t) = D_1\sqrt{E_b}u_s(t) \quad (15.99)$$

where

$$D_1 = (-1)^{I(1)} \qquad u_s(t) = \begin{cases} A\sin\left(\frac{\pi t}{T_p}\right) & 0 \le t \le T_p \\ 0 & \text{elsewhere} \end{cases} \tag{15.100}$$

The received signal is well modeled as $Y_z(t) = X_z(t) + W_z(t)$ where $W_z(t)$ is a complex white Gaussian noise with a one-sided power spectral density of N_0.

(a) Give a value of A such that the received energy per bit is E_b.

(b) Compute the spectral efficiency of this transmission scheme using the 3 dB transmission bandwidth.

(c) Give the MLBD and compute the performance of the MLBD.
It is possible to transmit 2 bits of information ($K_b = 2$) with waveforms of the form

$$X_z(t) = D_1\sqrt{E_b}u_s(t) + D_2\sqrt{E_b}u_c(t) \tag{15.101}$$

where

$$D_i = (-1)^{I(i)} \qquad u_c(t) = \begin{cases} A\cos\left(\frac{\pi t}{T_p}\right) & 0 \le t \le T_p \\ 0 & \text{elsewhere} \end{cases} \tag{15.102}$$

This modulation became known as Q^2PSK [SB89] and sparked significant interest in the 1990s.

(d) Is this an orthogonal modulation?

(e) Compute the spectral efficiency of this $K_b = 2$ transmission using the 3 dB transmission bandwidth.

(f) Detail out the MLWD architecture and compute the performance of this MLWD.

Problem 15.30. A commonly used linear modulation for satellite communications is known as 12/4QAM [MO98]. The constellation diagram for 12/4QAM is shown in Figure 15.28. Assume each of the symbols is equally likely for this problem (i.e., $\pi_i = 1/16$).

(a) Give constraints on the two circle radii, δ_1 and δ_2, such that $E\left[|D_z|^2\right] = 4$.

(b) Find the MLWD and define decsion regions for this modulation.

(c) Is the modulation geometrically uniform?

(d) Find the simplest form for the union bound to the probability of word error.

(e) Optimize the fidelity of demodulation by a selection of δ_1 and δ_2, such that $E\left[|D_z|^2\right] = 4$ for $P_{WUB}(E) < 10^{-4}$.

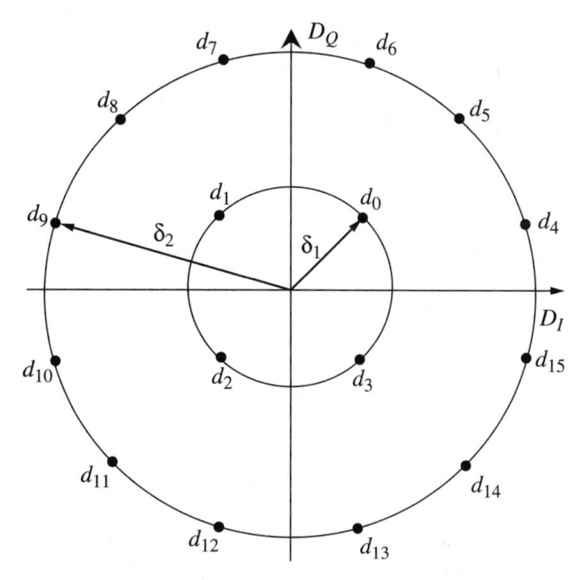

Figure 15.28 The constellation diagram for 12/4QAM.

Problem 15.31. A commonly used linear modulation for spectrally efficient data communication is known as 32Cross. 32Cross is constructed by first having a 36 point constellation where points are on a rectangular grid with six points in each dimension, i.e., $d_{Ii} \in \{-5A, -3A, -1A, 1A, 3A, 5A\}$ and $d_{Qi} \in \{-5A, -3A, -1A, 1A, 3A, 5A\}$ and removing the (highest power) four corner points.

(a) Plot the constellation and give a guess as to why the modulation is known as 32Cross.

(b) Give the value of A such that $E[|D_z|^2] = 5$.

(c) Find the MLWD and define decsion regions for this modulation.

(d) Is the modulation geometrically uniform?

(e) Find the simplest form for the union bound to the probability of word error.

Problem 15.32. Show that if two spreading waveforms are orthogonal in the time domain, i.e.,

$$\int_{-\infty}^{\infty} s_l(t)s_k^*(t)dt = 0 \tag{15.103}$$

then they are also orthogonal in the frequency domain

$$\int_{-\infty}^{\infty} S_l(f)S_k^*(f)df = 0 \tag{15.104}$$

Hint: Consider Eq. (2.67).

Problem 15.33. Consider two competing FSK based systems with rectangular pulse shapes that transmit $K_b = 3$ bits of information with a transmission rate of $W_b = 10$ kHz. The first is a binary FSK waveform where the 3 bits are streamed orthogonally in time and the second is an 8FSK system that transmits the bits with an M-ary modulation.

(a) What is the value of T_p that will achieve the desired $W_b = 10$ kHz.

(b) For the two systems compute the 3 dB bandwidth.

(c) For the two systems compute the word error probability.

(d) Which system would you recommend using? Why?

Problem 15.34. Consider the set of spreading waveforms in Figure 15.15. Show that the spectral efficiency can be doubled with no loss in performance by setting

$$s_{l+4}(t) = j s_l(t). \tag{15.105}$$

Problem 15.35. While going to school at Ohio State University you lent your digital communication book to your friend at University of Michigan and he, after a brief read, got very excited and immediately founded a start-up company, *WolverineCom.com*, and built a modem that sent a signal

$$X_z(t) = \sqrt{E_b} \sum_{l=1}^{3} D_z(l) s_l(t) \tag{15.106}$$

where $s_l(t)$ are unit energy spreading waveforms and $D_z(k) = (-1)^{I(k)}$ is from a BPSK signal set and each bit is independent and equally likely. Unfortunately, he did not read the notes carefully and did not ensure that the $s_l(t)$ are orthogonal to each other. The correlation of the waveforms is captured by the matrix

$$\mathbf{G} = \begin{bmatrix} 1 & 0 & 0 \\ 0 & 1 & 0.9 \\ 0 & 0.9 & 1 \end{bmatrix} \tag{15.107}$$

where $[\mathbf{G}]_{ij} = \int_0^{T_p} s_i(t) s_j^*(t)\, dt$. After struggling to get the demodulation part of the modem to work on a frequency flat channel corrupted by an AWGN, he has come to you with the offer of stock options if you can help him.

a) Form the MLWD and show that $Q(k) = \int_0^{T_p} Y_z(t) s_k^*(t) dt \quad k = 1, 2, 3$ are sufficient statistics for optimum word demodulation.

(b) Find the minimum Euclidean squared distance for this modulation and identify two transmitted words that achieve this distance. Would you invest in WolverineCom.com?

(c) Find the simplest form for the maximum likelihood demodulator for $I(1)$ based on observations $q(1)$, $q(2)$, and $q(3)$.

Problem 15.36. Square quadrature amplitude modulation (QAM) is a commonly used linear modulation in engineering practice. A square QAM constellation consists of a square of M_1 points by M_1 points where the M_1 points in each dimension are placed on an equal spaced grid.

(a) Plot square QAM constellations for $M_1 = 2, 4, 8$.

(b) If constellation points for a constellation of size M_1^2 are located at

$$\begin{aligned} d_{In} &= B(2m - M_1 + 1) \quad m = 0, \ldots, M_1 - 1 \\ d_{Qm} &= B(2n - M_1 + 1) \quad n = 0, \ldots, M_1 - 1 \end{aligned} \tag{15.108}$$

show that

$$B = \sqrt{\frac{6\log_2(M_1)}{M_1^2 - 1}} \tag{15.109}$$

will normalize the constellation such that $E\left[|D_z|^2\right] = 2\log_2(M_1)$. *Hint:*

$$\sum_{i=1}^{N} i = \frac{(N+1)N}{2} \qquad \sum_{i=1}^{N} i^2 = \frac{(N+1)(2N+1)N}{6} \tag{15.110}$$

(c) With square QAM decisions can be made independently on the bits modulating the I channel and the Q channel. What conditions must hold for this to be true.

(d) Show the word error probability of square QAM is upperbounded by

$$P_W(E) \leq 1 - \left(\operatorname{erf}\left(B\sqrt{\frac{E_b}{N_0}}\right)\right)^2 \tag{15.111}$$

Hint: Use Problem 15.7.

Problem 15.37. A modulation often used in practice is known as staggered QPSK. Staggered QPSK with a bit rate of $W_b = 1/T$ where K_b is even has the form

$$X_z(t) = \sum_{l=1}^{K_b/2} D_I(l)u_r(t-(l-1)T) + j\sum_{l=1}^{K_b/2} D_Q(l)u_r\left(t-(l-1)T - \frac{T}{2}\right) \tag{15.112}$$

where $D_I(k) = (-1)^{I(k)}$ and $D_Q(k) = (-1)^{I(k+K_b/2)}$ where $k = 1, \ldots, K_b/2$ and $u_r(t)$ is a unit energy rectangular pulse.

(a) This modulation is an orthogonal modulation. Prove this by showing the waveforms associated with each bit are orthogonal to the waveforms associated with the other bits.

(b) Give a block diagram for a simple optimum detector.

(c) Compute the bit error probability for the optimum detector.

(d) Compute the word error probability for the optimum detector.

15.6 Example Solutions

Problem 15.20.

(a) Note that the first signal set is 4-ary PSK and has and average constellation energy of $E_{D_1} = 2$. The second signal set has the energy of

$$E_{D_2} = \frac{A^2}{4}[3+1+1+3] = 2A^2 \tag{15.113}$$

Consequently, the value of A needed to achieve equal energy is $A = 1$.

(b) Since the modulations are linear, the sufficient statistic for demodulation is the matched filter output

$$Q = \int_0^{T_p} y_z(t)u^*(t)dt \tag{15.114}$$

The decision rule is given as

$$\hat{\hat{I}} = \arg \min_{i=1,\ldots,3} |Q - \sqrt{E_b}d_i|^2 \tag{15.115}$$

The decision regions for these minimum distance decoders are shown in Figure 15.29.

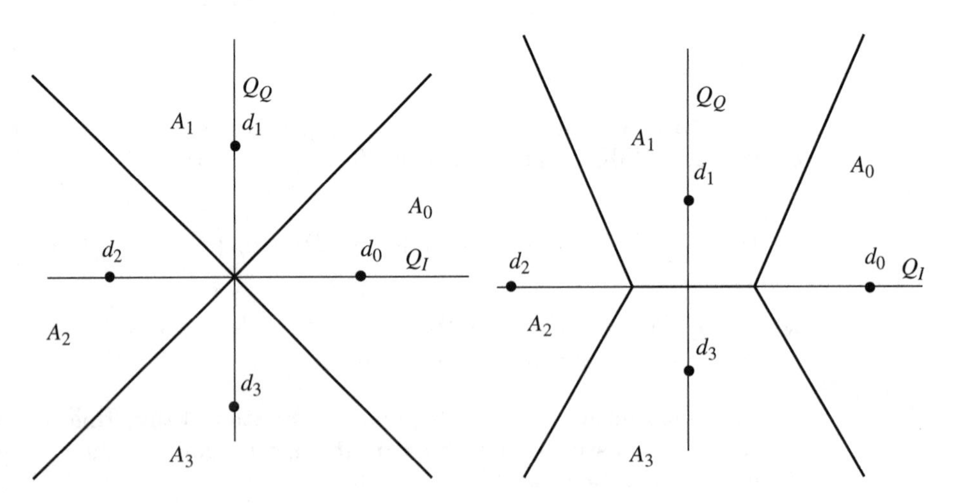

Figure 15.29 The two minimum distance decoders.

(c) The resulting distance spectrum for the first signal set is

$\Delta_E(i, j)$

	$\vec{I} = i$			
$\vec{I} = j$	0	1	2	3
0	0	$4E_b$	$8E_b$	$4E_b$
1	$4E_b$	0	$4E_b$	$8E_b$
2	$8E_b$	$4E_b$	0	$4E_b$
3	$4E_b$	$8E_b$	$4E_b$	0

Hence the full union bound for the first signal set is given as

$$P_{WUB}(E) = \text{erfc}\left(\sqrt{\frac{E_b}{N_0}}\right) + \frac{1}{2}\text{erfc}\left(\sqrt{\frac{2E_b}{N_0}}\right) \tag{15.116}$$

A tighter union bound is given as

$$P_{WUB}(E) = \text{erfc}\left(\sqrt{\frac{E_b}{N_0}}\right) \tag{15.117}$$

The distance spectrum for the second signal set is

$\Delta_E(i, j)$

	$\vec{I} = i$			
$\vec{I} = j$	0	1	2	3
0	0	$4E_b$	$12E_b$	$4E_b$
1	$4E_b$	0	$4E_b$	$4E_b$
2	$12E_b$	$4E_b$	0	$4E_b$
3	$4E_b$	$4E_b$	$4E_b$	0

Hence the full union bound for the second signal set is given as

$$P_{WUB}(E) = \frac{5}{4}\text{erfc}\left(\sqrt{\frac{E_b}{N_0}}\right) + \frac{1}{4}\text{erfc}\left(\sqrt{\frac{3E_b}{N_0}}\right) \tag{15.118}$$

(d) The first signal set has a minimum distance of $\Delta_E(\text{min}) = 4E_b$ with a multiplicity of 8. The second signal set has a minimum distance of $\Delta_E(\text{min}) = 4E_b$ but a multiplicity of 10. The greater multiplicity of the minimum distance error event means that the second signal set will have worse performance.

Problem 15.28.

(a) Ethernet is clearly using a stream modulation. When $T = T_u$ the pulse shape used in Ethernet satisfies Nyquist's criterion for zero ISI. Consequently, the matched filter to the pulse shape sampled at times $t_s = kT$ $k = 1, \ldots, K_b$ will form the sufficient statistics for demodulation. For an arbitrary real constellation the optimum demodulator for each bit has the form

$$\hat{I}(k) = \arg\max_{m_k=0,1} d_{m_k} \int_0^{T_p} y_c(t)u(t-(k-1)T)dt - \frac{E_b(d_{m_k})^2}{2}$$

$$= \arg\max_{m_k=0,1} d_{m_k} Q(k) - \frac{E_b(d_{m_k})^2}{2} \qquad (15.119)$$

(b) Clearly $d_0 = 1$ and $d_1 = -1$ would optimize performance.

(c) Since the output of the matched filter is

$$Q(k) = D(k)\sqrt{E_b} + N_z(k) \qquad (15.120)$$

and using (15.15) gives

$$P_B(E) = \frac{1}{2}\text{erfc}\left(\sqrt{\frac{E_b(d_0 - d_1)^2}{4N_0}}\right) \qquad (15.121)$$

(d) The pulse shape spectrum is given as

$$U(f) = \sqrt{T_u}\,\text{sinc}\left(\frac{fT_u}{2}\right)\exp\left[\frac{-j\pi f T_u}{2}\right]$$

$$-\sqrt{T_u}\,\text{sinc}\left(\frac{fT_u}{2}\right)\exp\left[\frac{-j3\pi f T_u}{2}\right] \qquad (15.122)$$

The corresponding energy spectrum is plotted in Figure 15.30 and the spectral efficiency of Ethernet (using 3 dB bandwidth) is $\eta_B = 1/2.32 = 0.43$ bit/s/Hz. Matlab code is appended.

(e) Recall we want

$$\frac{1}{M}\sum_{i=0}^{3} d_i^2 = K_b \qquad (15.123)$$

so that $A = \sqrt{\frac{2}{5}}$.

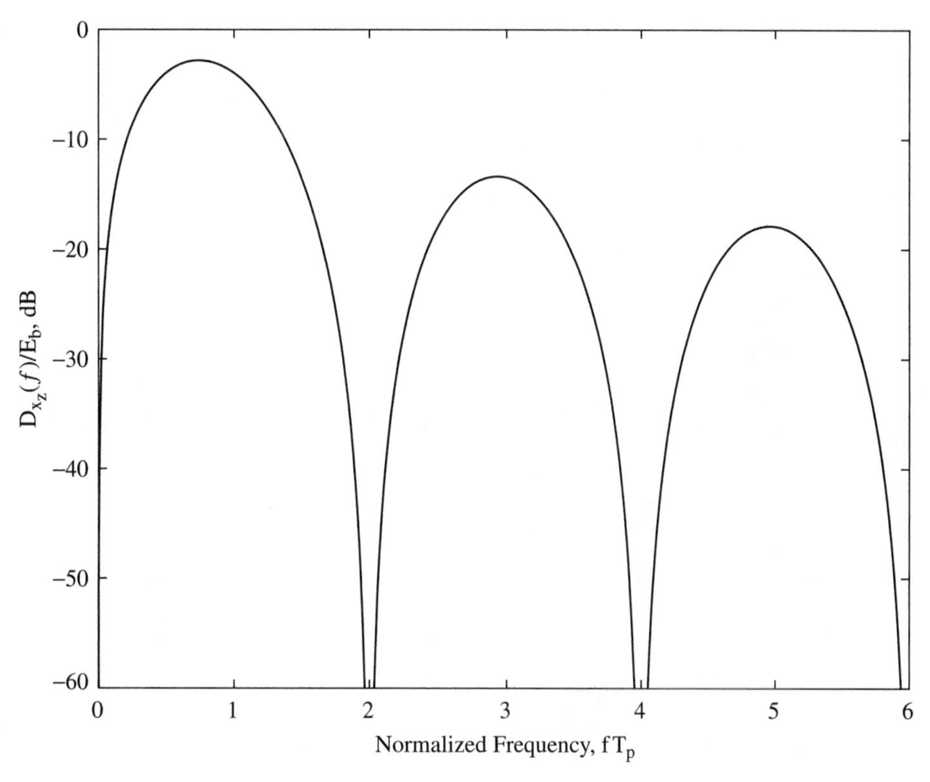

Figure 15.30 $D_{x_z}(f)$ for $W_b = 1/T$.

(f) It is apparent that for the second alternative that the orthogonality condition still holds so that we have

$$\hat{I}_i(k) = \arg\max_{m_k=0,1} d_{m_k} \int_0^{T_p} y_c(t)u_i(t-(k-1)T)dt - \frac{E_b(d_{m_k})^2}{2} \quad i = 1, 2 \tag{15.124}$$

For the first alternative the symbols are still orthogonal so each symbol can be decoded independently and combined via

$$\hat{\hat{I}}(k) = \arg\max_{i=0,3} d_i \int_0^{T_p} y_c(t)u(t-(k-1)T)dt - \frac{E_b(d_i)^2}{2} \tag{15.125}$$

(g) Noting the orthogonality of the modulation and using (15.40) for the second option we have

$$P_W(E) = 1 - \left(1 - \frac{1}{2}\operatorname{erfc}\left(\sqrt{\frac{E_b(d_0 - d_1)^2}{4N_0}}\right)\right)^{2K_b} \tag{15.126}$$

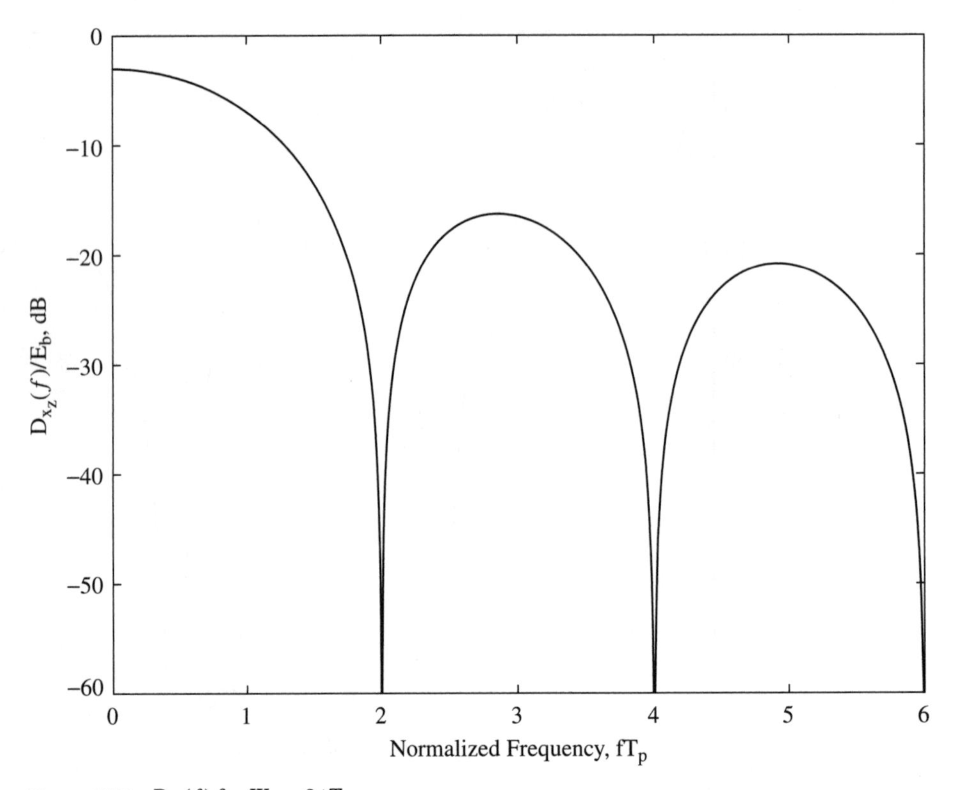

Figure 15.31 $D_{x_z}(f)$ for $W_b = 2/T$.

Again noting the orthogonality of the symbol modulation and using the result in Eq. (15.17) we have for the first option

$$P_W(E) = 1 - \left(1 - \frac{3}{4}\text{erfc}\left(\sqrt{\frac{2E_b}{5N_0}}\right)\right)^{K_b} \tag{15.127}$$

(h) With linear modulation the spectrum will not be changed so the spectrum for the first option will have the same form as in Figure 15.30. The resulting spectral efficiency is $\eta_B = 0.86$ bit/s/Hz. The spectrum for the second option, using similar techniques as in the text and assuming BPSK, is given as

$$D_{x_z}(f) = \frac{1}{2}(G_{u_1}(f) + G_{u_2}(f)) \tag{15.128}$$

The resulting average energy spectrum is given as Figure 15.31. The resulting spectral efficiency is $\eta_B = 0.86$ bit/s/Hz. Consequently, the resulting spectral efficiency is roughly the same.

```
%
% Solution for Problem mc.28
% Author: M. Fitz
% Last modified: 10/14/06
%
close all
clear all
%
% Using with numpts points over 0<f<xaxislim
%
numpts=10000;
xaxislim=6;
freq=linspace(0.00001,xaxislim,numpts);
%
% part d)
%
uf1=0.5*sinc(freq/2).*exp(-j*pi*(freq)/2);
uf2=-0.5*sinc(freq/2).*exp(-j*3*pi*(freq)/2);
gu1f=(abs(uf1+uf2)).^2;
figure(1)
plot(freq,10*log10(gu1f))
axis([0 xaxislim -60 0])
hs=subplot(1,1,1)
set(hs,'Fontsize',14)
xlabel('Normalized Frequency, f  T_{p}', 'Fontsize',16)
ylabel('D_{x_z}(f) / E_b, dB','Fontsize',16)
%
% part h)
%
uf3=sinc((freq)).*exp(-j*pi*freq);
gu2f=(abs(uf3)).^2;
dxzf=0.5*(gu1f+gu2f);
figure(2)
%plot(freq,10*log10(gu1f), freq, 10*log10(gu2f),freq,10*log10(dxzf))
plot(freq,10*log10(dxzf))
axis([0 xaxislim -60 0])
hs=subplot(1,1,1)
set(hs,'Fontsize',14)
xlabel('Normalized Frequency, f  T_{p}', 'Fontsize',16)
ylabel('D_{x_z}(f) / E_b, dB','Fontsize',16)
```

15.7 Miniprojects

Goal: To give exposure

- to a small scope engineering design problem in communications.
- to the dynamics of working with a team.
- to the importance of engineering communication skills (in this case oral presentations).

Presentation: The forum will be similar to a design review at a company (only much shorter) The presentation will be of 5 minutes in length with an overview of the given problem and solution. The presentation will be followed by questions from the audience (your classmates and the professor). Each team member should be prepared to give the presentation.

15.7.1 Project 1

Project Goals: Engage in an implementation of an optimum demodulator of an 8-ary signal set that uses linear modulation.

Consider linear modulation and pulse shape (with sample frequency $f_s = 22{,}050$ Hz) and the received bandpass signal ($f_c = 3500$ Hz) given in `mcproj1data.mat`. **Any computer code (e.g., Matlab) should be turned in with the project write-up.**

(a) Plot the constellation. Find E_b.

(b) Sketch a block diagram of the optimum demodulator for this signal set for $\pi_i = 0.125,\ i = 0, \ldots, 7$.

(c) Make the optimum word decision and compute all the sufficient statistics of the demodulation for $\pi_i = 0.125,\ i = 0, \ldots, 7$.

15.7.2 Project 2

Project Goals: Build an OFDM transmission system that meets an engineering goal.

Build a digital representation of an OFDM transmitted signal where the sampling frequency is given as $f_s = 22{,}050$ Hz. The system should transmit $K_b = 16$ bits with 1 bit on each subcarrier, have a carrier frequency of $f_c = 1800$ Hz and achieve a transmission rate of $W_b = 2000$ Hz. **Any computer code (e.g., Matlab) should be turned in with the project write-up.**

Specifically a completed design will require the student to

(a) Plot one sample path of the transmitted signal, i.e., $x_i(t)$ for some $i \in \{0, \ldots, 65535\}$. The actual 16 bits will be supplied by the instructor.

(b) Compute and plot $D_{X_z}(f)$ and compare it to a measured energy spectrum of the sample path generated in (a).

(c) Run a significant number of trials where the value of $\vec{I}$ is randomly chosen and compute the PAPR of this OFDM waveform.

15.7.3 Project 3

Project Goals: Build a demodulator for an OCDM transmission system.

Consider a $K_b = 8$ OCDM system (with sample frequency $f_s = 22{,}050$ Hz) with the set of spreading waveforms and the received baseband signal given in `mcproj3data.mat`. These spreading waveforms are as given in Problem 15.22. **Any computer code (e.g., Matlab) should be turned in with the project write-up.**

(a) What is the transmission rate of this system, W_b?

(b) Sketch a block diagram of the optimum demodulator for this signal set for equally likely and independent bits.

(c) Make the optimum bit decision and compute all the sufficient statistics of the demodulation for equally likely and independent bits.

Chapter

16

Spectrally Efficient Data Transmission

16.1 Spectral Containment

In digital communications it is often desirable to get many users in a band of frequencies. Examples include channels in broadcast or cable television and phone calls in mobile telephony. In this case it is often undesirable for emissions from one user to interfere with the transmission from another user. In the modulations introduced in Chapter 15 the transmitted spectra have significant out of band power due to the sinc(fT_p) characteristic in the spectrum. This sinc(fT_P) characteristic is due to the using of a rectangular pulse in each of these modulations. This out of band power can produce significant interference among users.

To prevent significant adjacent channel interference a spectral mask is often imposed on the transmission. Figure 16.1(a) shows a typical spectral mask that must be met by transmitter electronics. Within the transmission bandwidth, B_T, the spectrum is not regulated. Outside the transmission bandwidth the output power is slowly tapered off until a large attenuation is achieved at the next adajacent channel. This mask is the spectral emissions mask for the narrowband radio services band as defined by the Federal Communications Commission [Com04]. Forcing a transmitter to obey this type of emissions mask allows more channels to operate with a better performance in a given frequency band. Meeting this emission mask will require a more sophisticated signal design than has been investigated up to this point in the text. For example, Figure 16.1(b) shows the mask in comparison to the spectrum of a digital modulation that uses a rectangular pulse shape like the examples introduced in Chapter 15. Clearly a rectangular pulse shape is not well motivated in this application due to the high sidelobes in frequency. These high sidelobes will cause significant interference to adjacent channel users. This is especially true when a near-far situation exists. An example of a near-far situation would be a

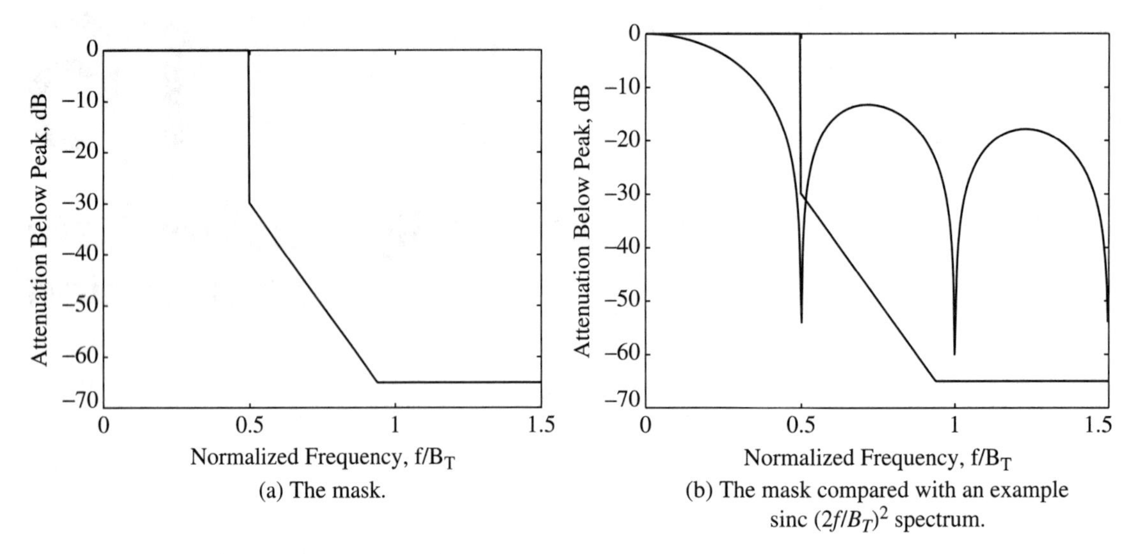

Figure 16.1 A typical spectral emission mask.

communication system trying to demodulate a digital television signal from a remote station while simultaneously being close to an antenna broadcasting on a different channel (carrier frequency). Keeping the interference from the close by station low will be important in maintaining a high fidelity of demodulation for the remote station. Designing pulse shapes that meet a spectral mask (one for a mobile telephone) was investigated in Project 13.1. Meeting this emission mask and maintaining the orthogonality that enables low complexity demodulation as was introduced in Chapter 15 is the focus of this chapter.

Another situation that often arises is where the physical channel has significant spectral shaping characteristics. The best example of this is digital communications on analog telephone lines. The original analog telephone lines were designed to work with 4 kHz baseband signals and to minimize the effects of aliasing there is a sharp cutoff filter in each phone line. Consequently, it is important to find communication techniques that can accomodate this frequency selective characteristic of the channel. For example, Figure 16.2 shows what might be a typical communication system model. If the channel is known at the transmitter then this knowledge can be included as part of the signal design, i.e., the design problem then focuses on specifying the form of $X_z(t)$ or $X_c(t)$ to achieve a desired form for $R_z(t)$ or $R_c(t)$. Being able to design a communication system that can operate in that known environment is another goal of this chapter.

$X_z(t)$ → $h_z(t)$ → $R_z(t)$ → Σ → $Y_z(t)$; $W_z(t)$ → Σ

Figure 16.2 A frequency selective channel model.

16.2 Squared Cosine Pulse

The following Fourier transform pair is frequently used in bandwidth efficient communications.

Definition 16.1 The squared cosine pulse family is characterized as

$$p_c(x, T_z) = \begin{cases} 1 & |x| \le \dfrac{(1-\alpha)T_z}{2} \\ \cos^2\left(\dfrac{\pi}{2\alpha T_z}\left(|x| - \dfrac{(1-\alpha)T_z}{2}\right)\right) & \dfrac{(1-\alpha)T_z}{2} \le |x| \le \dfrac{(1+\alpha)T_z}{2} \\ 0 & \text{elsewhere} \end{cases} \tag{16.1}$$

where $0 \le \alpha \le 1$. α is the parameter that indexes this family of pulse shapes. The Fourier transform of the squared cosine pulse family is

$$P_c(y, T_z) = \mathcal{F}\{p_c(x, T_z)\} = \frac{\cos(\pi\alpha y T_z)}{1-(2\alpha y T_z)^2}\, T_z \text{sinc}(yT_z) \tag{16.2}$$

It should be noted dummy variables of x and y are used instead of the traditional t and f in this development. The reason is that this pulse family will be used as both a time pulse and a frequency pulse. Figure 16.3 shows a plot of the time function and the Fourier transform. When $\alpha = 0$ then $p_c(x, T_z)$ becomes the rectangular pulse of width T_z. When $\alpha > 0$ then the Fourier transform dies off much quicker than the Fourier transform of the rectangular pulse due to the tapering at the edge of the pulse. This characteristic is useful for making the transmissions more bandwidth efficient and meeting a spectral emissions mask.

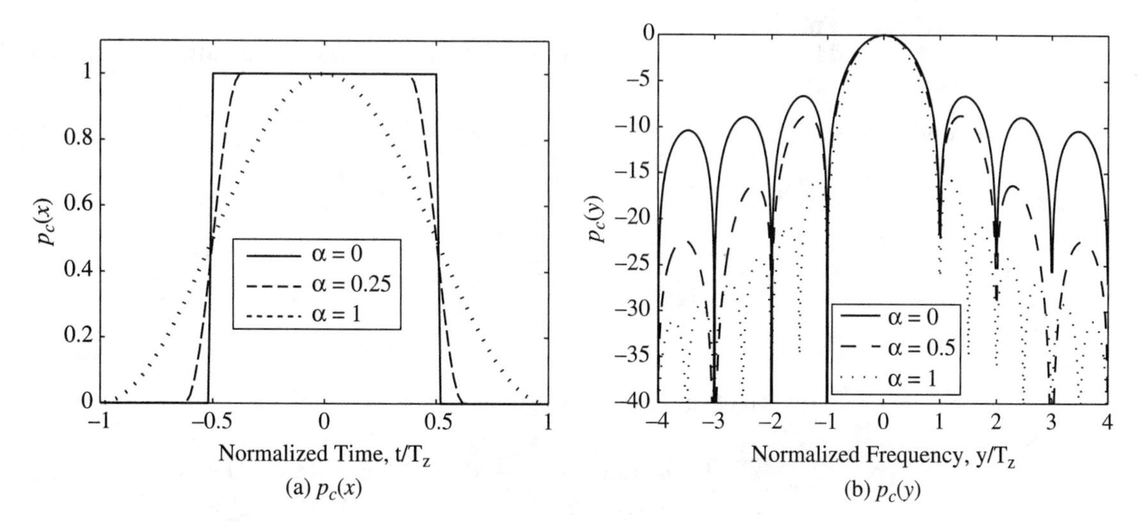

(a) $p_c(x)$ (b) $p_c(y)$

Figure 16.3 The "time" and "frequency" characteristics of the squared cosine pulse.

The squared cosine pulse shape has two additional important characteristics that make it useful for bandwidth efficient digital communications.

- For $|x| \leq T_z$ the squared cosine pulse satisfies

$$p_c(x, T_z) + p_c(x - T_z, T_z) + p_c(x + T_z, T_z) = 1 \tag{16.3}$$

- $P_c(\frac{n}{T_z}, T_z) = 0$ for n an integer and $n \neq 0$.

The first characteristic indicates that the area under $p_c(x, T_z)$ is constant for any value of α even though the length of the pulse, $(1+\alpha)T_z$, is changing with α. This characteristic is due to the fact that $\cos(x - \pi/2) = \sin(x)$ and that $\cos^2(x) + \sin^2(x) = 1$. The second characteristic indicates the Fourier transform goes to zero at the same values of frequency as the Fourier transform of the rectangular pulse. The zeros of the Fourier transform are due to the $T_z\text{sinc}(yT_z)$ term.

The squared cosine pulse family is often referred to as the raised cosine pulse in the literature. The reason for this can be seen in the case of $\alpha = 1$ where the pulse is given using a trigonometric identity as

$$p_c(x, T_z) = \begin{cases} \cos^2\left(\frac{\pi x}{2T_z}\right) & |x| \leq T_z \\ 0 & \text{elsewhere} \end{cases} = \begin{cases} \frac{1}{2} + \frac{1}{2}\cos\left(\frac{\pi x}{T_z}\right) & |x| \leq T_z \\ 0 & \text{elsewhere} \end{cases} \tag{16.4}$$

The "raised cosine" terminology evolved in the communications literature from the fact that with $\alpha = 1$ the 1/2 term raises the cosine term. Why this text breaks with tradition and uses the notation "squared cosine" family should be apparent by how this pulse is used in achieving spectrally efficient pulses for orthogonal modulations.

A second Fourier transform pair is also frequently used in bandwidth efficient communications.

Definition 16.2 The cosine pulse family is characterized as

$$u_c(x, T_z) = \begin{cases} 1 & |x| \leq \frac{(1-\alpha)T_z}{2} \\ \cos\left(\frac{\pi}{2\alpha T_z}\left(|x| - \frac{(1-\alpha)T_z}{2}\right)\right) & \frac{(1-\alpha)T_z}{2} \leq |x| \leq \frac{(1+\alpha)T_z}{2} \\ 0 & \text{elsewhere} \end{cases} \tag{16.5}$$

where $0 \leq \alpha \leq 1$. The Fourier transform of the cosine pulse family is

$$U_c(y, T_z) = \frac{T_z(1-\alpha)\text{sinc}(yT_z(1-\alpha))}{1 - (4\alpha yT_z)^2} + \frac{4T_z\alpha\cos(\pi yT_z(1+\alpha))}{\pi\left(1 - (4\alpha yT_z)^2\right)} \tag{16.6}$$

It should be noted for clarity that $u_c(x, T_z) = \sqrt{p_c(x, T_z)}$. This pulse is also often described in the literature as a square root raised cosine pulse. Finally, neither of the two important characteristics of the squared cosine pulse family hold true for the cosine pulse family.

16.3 Spectral Shaping in OFDM

OFDM is a modulation that independently modulates data bits on different subcarrier frequencies. For OFDM a subcarrier refers to a frequency offset compared to the true carrier frequency, f_c. A simple demodulator results if the modulation on each of the subcarrier frequencies are orthogonal. With the rectangular pulse shape, $u_r(t)$, assumed in Chapter 15, the orthogonality condition between the signal for subcarrier l and subcarrier k was shown to be

$$\Re\left[D_z(l)D_z^*(k)\frac{E_b}{T_p}\int_{-\infty}^{\infty}|u_r(t)|^2\exp[j4\pi f_d(l-k)t]dt\right]=0 \qquad \forall k\neq l \tag{16.7}$$

The spectrum is also a direct function of the shaping pulse. The spectrum in the case of equally likely bits independently modulated on each carrier becomes

$$D_{X_z}(f)=\frac{E_b}{K_b}\sum_{l=1}^{K_b}|U_r(f\ -\ f_d(2l-K_b-1))|^2 \tag{16.8}$$

The desired goal is to identify a shaping pulse, $u_s(t)$, that both maintains the orthogonality condition of Eq. (16.7) and gives better out of band spectral emission characteristics than the rectangular pulse shape.

The results of Section 16.2 can be used to identify spectral efficient transmission strategies for OFDM. To generalize OFDM we can use an arbitrary pulse shape, $u_s(t)$. This implies that the transmitted signal has the form

$$X_z(t)=\sum_{l=1}^{K_b}D_z(l)\sqrt{\frac{E_b}{T_p}}u_s(t)\exp[j2\pi f_d(2l-K_b-1)t] \tag{16.9}$$

and the orthogonality condition becomes

$$\Re\left[D_z(l)D_z^*(k)\frac{E_b}{T_p}\int_{-\infty}^{\infty}|u_s(t)|^2\exp[j4\pi f_d(l-k)t]dt\right]=0 \tag{16.10}$$

Defining $p(t)=|u_s(t)|^2$ and restating the orthogonality condition as

$$\Re[D_z(l)D_z^*(k)P(2f_d(k-l))]=0 \tag{16.11}$$

it is immediately obvious that orthogonality holds for an arbitrary pulse shape and an arbitrary constellation only if

$$P(2f_d(k-l))=0 \qquad \forall\, l\neq k \tag{16.12}$$

Stated in words, the Fourier transform of $p(t) = |u_s(t)|^2$ must go to zero at integer multiples of $2f_d$. Considering the characteristics of the squared cosine pulse it is apparent that a selection of $p(t) = p_c(t, 1/(2f_d))$ will give the desired orthogonality. This implies that a bandwidth efficient implementation of OFDM has the form

$$X_{zs}(t) = \sum_{l=1}^{K_b} D_z(l)\sqrt{\frac{E_b}{T_p}}u_c(t)\exp[j2\pi f_d(2l - K_b - 1)t] = u_c(t)X_{zu}(t) \quad (16.13)$$

where $X_{zu}(t)$ is the original OFDM modulation without shaping. Using this form of spectral shaping is convenient since it only has to be done on the transmitted signal after modulation.

This pulse shaping in OFDM has two negative aspects. When keeping the subcarrier spacings fixed in an OFDM system the shaping will lower the data rate and spectral efficiency. For a subcarrier spacing of $1/T_p$ the transmission length goes from T_p with no shaping to $(1+\alpha)/T_p$ with shaping. This lengthening of the pulse leads to a proportional reduction in data rate and spectral efficiency. Second, the matched filter for each subcarrier must use the shaped pulse in computing $Q(k)$. This shaping, while not difficult, adds more complexity. The matched filter uses the pulse shape which is not equivalent to taking a Fourier transform of the received signal as is needed with the unshaped pulse.

Good spectral efficiency and spectral compactness can be achieved with a small α in the case when K_b is large. The smaller the value of α, the smaller the region in time over which the the OFDM signal needs to be shaped and the higher the spectral efficiency. Figure 16.4(a) shows a plot of the output

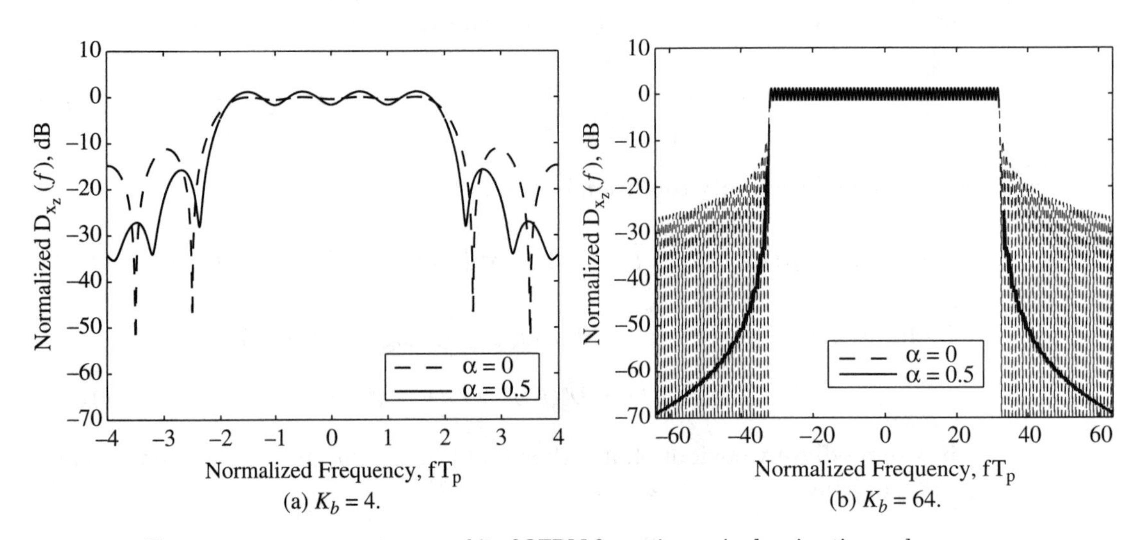

Figure 16.4 The average energy spectrum per bit of OFDM for various raised cosine time pules.

spectrum for OFDM for $K_b = 4$ for various values of α. It should be noted that the cosine pulse shape does not allow a stringent spectral mask to be met with a small K_b, but with larger K_b the spectrum becomes quite compact. This can be observed in Figure 16.4(b) that shows a plot of the output spectrum for OFDM for $K_b = 64$ for various values of α. This is advantageous as most practical OFDM systems use at least 64 subcarriers so that a small amount of spectral shaping is sufficient to meet a typical frequency mask.

It should be noted that this pulse shaping is not precisely used in practice. In general, OFDM modems are used in frequency selective channels and these frequency selective channels make orthogonality hard to achieve. The spectral shaping techniques that are used in practice are similar to that developed here in the sense that shaped pulses are used in lieu of rectangular pulses where a majority of the shaping is done at the pulse boundaries (equivalent to a small α).

16.4 Spectral Shaping in Linear Stream Modulations

Linear stream modulation is the most prevalent form of digital communications so spectral shaping of linear modulations is an important topic. Recall linear stream modulations have the form

$$X_z(t) = \sum_{l=1}^{K_b} D_z(l)\sqrt{E_b}u(t - (l-1)T) \tag{16.14}$$

The orthogonality condition that needs to be satisfied such that simple demodulation is possible is given as

$$\Re[D_z(l)D_z^*(k)V_u((k-l)T)] = 0 \tag{16.15}$$

where $V_u(\tau)$ is the pulse shape correlation function defined in Chapter 2, i.e.,

$$V_u(\tau) = \int_{-\infty}^{\infty} u(t)u^*(t-\tau)dt \tag{16.16}$$

The average energy spectrum per bit of linear stream modulation for the case of equally likely symmetric constellations was shown in Chapter 15 to be

$$D_{X_z}(f) = E_b|U(f)|^2 = E_b G_u(f) \tag{16.17}$$

Putting together Eqs. (16.15) and (16.17) shows that it is desired that a good pulse shape for linear stream modulations both must have a compact spectrum and an autocorrelation function that goes to zeros at integer multiples of the symbol time.

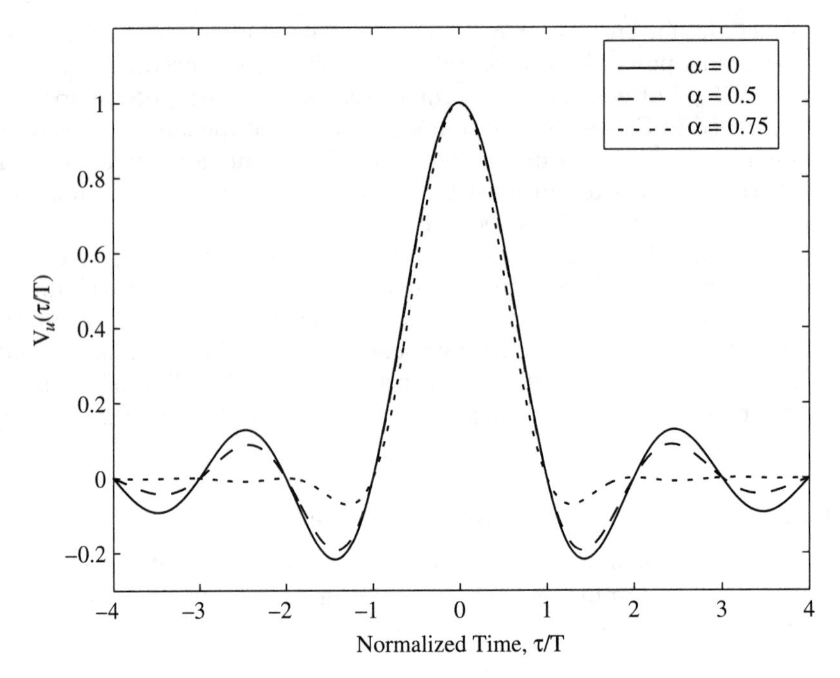

Figure 16.5 The autocorrelation of a spectral cosine pulse.

The square cosine function again has the desired characteristics due to the periodic zeros property of $P_c(y, T_z)$. Using Eq. (16.2) we can set

$$V_u(\tau) = P_c(\tau, 1/T) = \frac{\cos\left(\frac{\pi\alpha\tau}{T}\right)}{1-\left(\frac{2\alpha\tau}{T}\right)^2}\operatorname{sinc}\left(\frac{\tau}{T}\right) \tag{16.18}$$

where $0 \leq \alpha \leq 1$. The $\operatorname{sinc}(\frac{\tau}{T})$ term in Eq. (16.18) is what ensures that Nyquist's criterion for zero ISI is satisfied in the matched filter output. A plot of this pulse autocorrelation is shown in Figure 16.5. The reason this function satisfies Nyquist's criterion for zero ISI is clearly due to the periodic zero crossings. If the autocorrelation function of the pulse in a linear modulation is chosen as in Eq. (16.18) then the average energy spectrum bit of the linear stream modulation is then given as

$$D_{x_z}(f) = E_b G_u(f) = E_b p_c(f, 1/T)$$

$$= \begin{cases} E_b T & |f| \leq \dfrac{1-\alpha}{2T} \\ E_b T \cos^2\left(\dfrac{\pi T}{2\alpha}\left(|f| - \dfrac{(1-\alpha)}{2T}\right)\right) & \dfrac{1-\alpha}{2T} \leq |f| \leq \dfrac{1+\alpha}{2T} \\ 0 & \text{elsewhere} \end{cases} \tag{16.19}$$

This spectrum is clearly bandlimited (i.e., $B_T \approx \frac{1+\alpha}{T}$) and this text denotes the pulse/transform pair in Eqs. (16.18) and (16.19) as a spectral squared cosine family of pulse shapes. In direct analog to the discussion about the squared cosine pulse family, this pulse pair is often denoted in the literature as a spectral raised cosine pulse family. α is usually denoted the excess bandwidth factor as it determines how much the ideal bandwidth expands from the minimum, $B_{Tmin} = \frac{1}{T}$.

The actual pulse used in the linear modulation is a spectral cosine pulse family. Since

$$G_u(f) = \mathcal{F}\{P_c(\tau, 1/T)\} = |U(f)|^2, \tag{16.20}$$

it is clear that the pulse spectrum is given as

$$U(f) = u_c(f, 1/T) = \sqrt{p_c(f, 1/T)}$$
$$= \begin{cases} \sqrt{T} & |f| \leq \dfrac{(1-\alpha)}{2T} \\ \sqrt{T}\cos\left(\dfrac{\pi T}{2\alpha}\left(|f| - \dfrac{(1-\alpha)}{2T}\right)\right) & \dfrac{(1-\alpha)}{2T} \leq |f| \leq \dfrac{(1+\alpha)}{2T} \\ 0 & \text{elsewhere} \end{cases} \tag{16.21}$$

The pulse itself is given as

$$u(t) = U_c(t, 1/T) = \frac{1-\alpha}{\sqrt{T}} \frac{\operatorname{sinc}\left(\frac{(1-\alpha)t}{T}\right)}{1 - \left(\frac{4\alpha t}{T}\right)^2} + \frac{4\alpha \cos\left(\frac{\pi(1+\alpha)t}{T}\right)}{\pi\sqrt{T}\left(1 - \left(\frac{4\alpha t}{T}\right)^2\right)} \tag{16.22}$$

Figure 16.6(a) shows plots of the spectral cosine pulse shape for various values of α. It should be noted that these pulses have infinite time support but the pulses die off to zero at large arguments. The rate of decay of the pulses away from the peak of the pulse in this family increases with α. A typical time waveform for linear stream modulations is shown in Figure 16.6(b).

For practical applications approximations are used in generating the pulse shape for a linear stream modulation. The infinite time support of the spectral cosine pulse implies that approximations to these pulses must be used in practice. The simplest approach to implementation a practical pulse is to window or truncate the spectral cosine pulse in time to produce a finite time support. Again we will denote the truncated pulse length with T_u. The truncation implies that the orthogonality condition will be violated for some or all symbol time offsets. With an appropriate choice of window function, the amount of ISI produced can be made acceptably small. The design projects at the end of the chapter will explore windowing of a spectral cosine pulse for use in stream modulations. Also as detailed in Eq. (16.22) the spectral cosine pulse is an anticausal pulse.

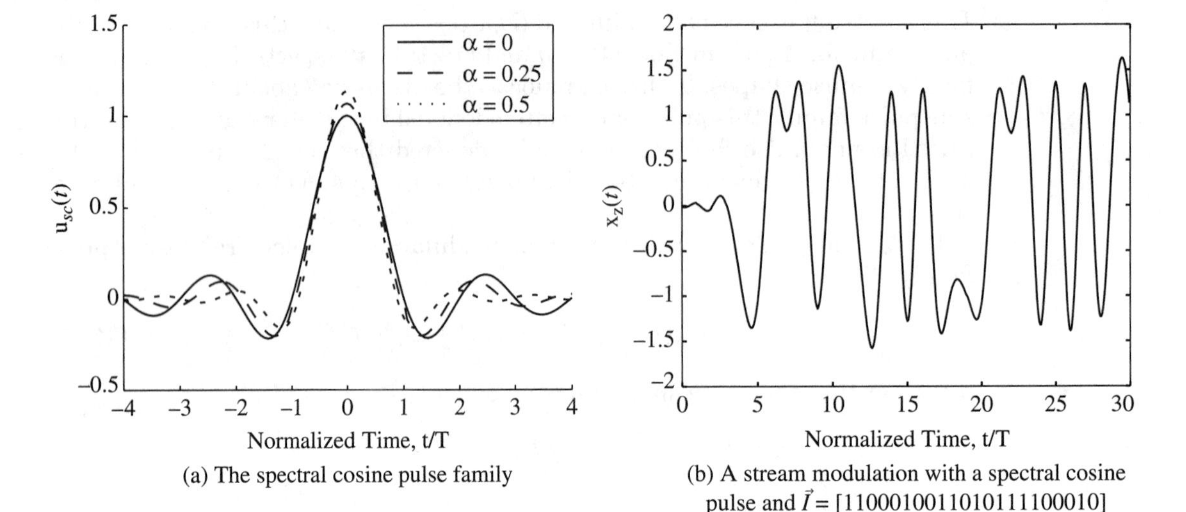

(a) The spectral cosine pulse family

(b) A stream modulation with a spectral cosine pulse and $\vec{I}$ = [11000100110101111000010]

Figure 16.6 Spectrally efficient stream modulation using BPSK, $\alpha = 0.3$, and $T_u = 8T$.

Once the pulse is truncated there is no impact on performance in making the pulse causal as both the autocorrelation function of the pulse and the energy spectrum of the pulse are invariant to a delay in the pulse. A wide range of practical communication systems use the spectral cosine pulse and in fact all vector signal analyzers have this pulse programmed as a standard mode.

Recall the demodulator for a linear stream modulation samples the output of a filter matched to the pulse shape for each transmitted symbol. The demodulator structure is shown in Figure 16.7. Recall that the matched filter output has the form

$$Q(t) = \int_{-\infty}^{\infty} Y_z(\lambda) u^*(\lambda - t + T_u) d\lambda \tag{16.23}$$

The matched filter output when sampled at $t = T_u + (k-1)T$ will have the form

$$Q(k) = D_z(k)\sqrt{E_b} + \sqrt{E_b} \sum_{l \neq k} D_z(l) V_u(k-l) + N_z(k) \tag{16.24}$$

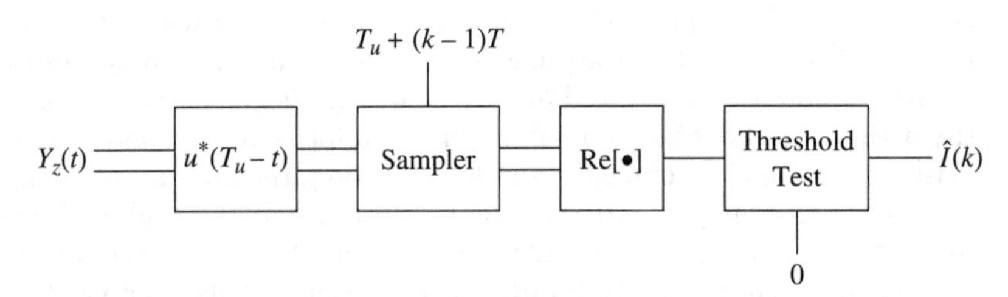

Figure 16.7 The optimal demodulator for stream modulations using BPSK.

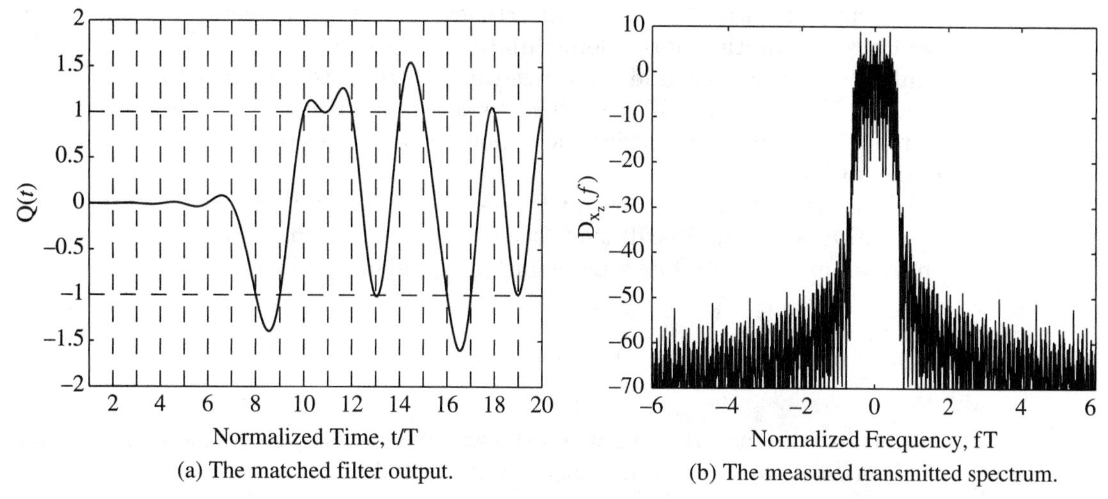

(a) The matched filter output.

(b) The measured transmitted spectrum.

Figure 16.8 Spectrally efficient stream modulation using BPSK, $\alpha = 0.3$, and $T_u = 8T$.

It is clear that if Nyquist's criterion holds for the pulse shape, $u(t)$, then the kth data bit can be detected without interference from other data bits ($n \neq k$). Figure 16.8(a) shows the noiseless output of the matched filter for a spectrally efficient pulse where $T_u = 8T$ and $E_b = 1$. The vertical lines in Figure 16.8(a) represent the sample times and the horizontal lines represent the modulation symbol values. It is apparent from this figure that even though 8 symbols are overlapped at any time, designing the pulses to achieve the orthogonality condition allows each BPSK modulation symbol to be detected without interference from the other pulses. Using this shaping on the pulse also achieves the desired spectral efficiency as shown in Figure 16.8(b). This figure is a measurement of the spectrum of the transmitted signal and demonstrates the desired spectral containment.

16.5 Testing Orthogonal Modulations

While orthogonal modulations are used in a wide variety of applications, it is not always possible to keep the waveforms orthogonal at the receiver. The nonorthogonality of the received waveforms can be produced by a wide variety of distortions that arise in practice. A partial list of the causes of the loss in orthogonality is

- A channel that is frequency selective
- Phase noise in the up and downconverters in the radio system of a modem
- Nonlinear distortion in the radio system of the modem
- Complexity constraints on the modem implementation
- Noise in the sample timing location for the matched filter output in the modem

In many situations this loss in orthogonality is significant enough to warrant a reformulation of the demodulation algorithms. In many other cases the goal is to characterize and minimize the distortion that causes the nonorthogonality. This section will introduce some common tools used by communication engineers to characterize the nonorthogonality of received waveforms in a demodulator.

When orthogonality is lost at the demodulator this results in interference from other symbols. Recall that OCDM, which is the most general case of orthogonal modulation, has a transmitted signal of the form

$$X_z(t) = \sqrt{E_b} \sum_{l=1}^{K_b} D_z(l) s_l(t) \tag{16.25}$$

For orthogonal modulations when the waveforms for each bit have maintained their orthogonality then the output of the matched filter for the kth bit is

$$Q(k) = \int_0^{T_p} Y_z(t) s_l^*(t) dt = D_z(k)\sqrt{E_b} + N_z(k) \tag{16.26}$$

If distortion has happened in the communication system then the matched filter outputs will have the form

$$Q(k) = \int_0^{T_p} Y_z(t) s_l^*(t) dt = \sqrt{E_b} \sum_{l=1}^{K_b} D_z(l) g(k,l) + N_z(k) \tag{16.27}$$

Hence the distortion in the received waveform causes interference from other symbols in the matched filter output of the desired symbol. Measuring and characterizing the amount of intersymbol interference (ISI) is the goal of this section.

Recall that communication waveforms have three dimensions (I, Q, and time) and most of the tools in use in practice are methods to represent the three dimensions of a communication waveform in two dimensions. Examples that we will explore here are

1. The scatter plot (I vs. Q in the matched filter output)
2. The vector diagram for stream modulations
3. The eye diagram (I vs. time or Q vs. time in the matched filter output) for stream modulation.

These three tools are used frequently in engineering practice.

16.5.1 The Scatter Plot

The scatter plot simply plots the I and Q points of the matched filter output. In the absence of noise and ISI the scatter plot should just be a scaled constellation plot as in this case $Q(k) = \sqrt{E_b} D_z(k)$. For example, the scatter plot of

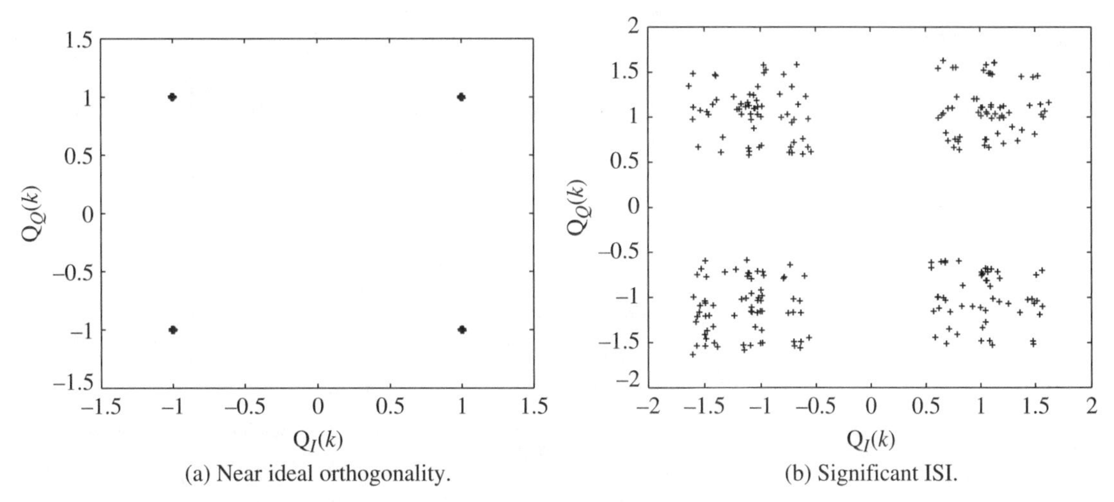

(a) Near ideal orthogonality.

(b) Significant ISI.

Figure 16.9 Scatter plots for QPSK modulation.

a QPSK modulation when the amount of nonorthogonality is small is shown in Figure 16.9(a). A scatter plot for a case where some significant distortion is present in the received signal is shown in Figure 16.9(b). The ISI present in the matched filter output is manifested as a spreading of each of the constellation points. While the distortion shown in Figure 16.9(b) appears to not be significant enough to cause errors in the absence of noise it should be apparent that the addition of noise with this distortion will significantly degrade the error rate performance. The scatter plot is a quick visual way for a communication engineer to get a handle on the amount of nonorthogonality in a modulation.

An often used measure of distortion for communication engineers is the error vector magnitude (EVM). The error vector for the matched filter output is defined to be $E_q(k) = Q(k) - \sqrt{E_b}D_z(k)$ and a vector diagram demonstrating this concept is shown in Figure 16.10(a). Communication engineers often quote the statistic of EVM. EVM is the average error vector magnitude compared to the largest normalized constellation point. There are many definitions of EVM but the most often used definition is RMS-EVM. Precisely RMS-EVM is defined as

$$\text{RMS-EVM} = \frac{\sqrt{\sum_{l=1}^{K_b} |E_q(l)|^2}}{\sqrt{K_b}\max_{i=0,\ldots,M-1} |\sqrt{E_b}d_i|} \tag{16.28}$$

and is most often reported in percentage. Most communication systems that use orthogonal modulations strive to achieve an EVM of less than 5%. Systems which use large constellations (64QAM or 256QAM) like digital television often have EVM requirements less than 3%. Most high performance vector analyzer type test equipment have automated EVM computations for common modulations used in accepted telecommunications standards. For example, Figure 16.10(b) shows an example test suite for a radio using one of the 64QAM

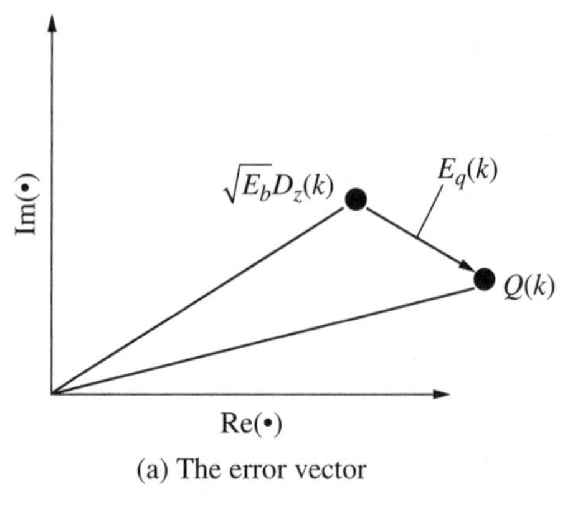

(a) The error vector

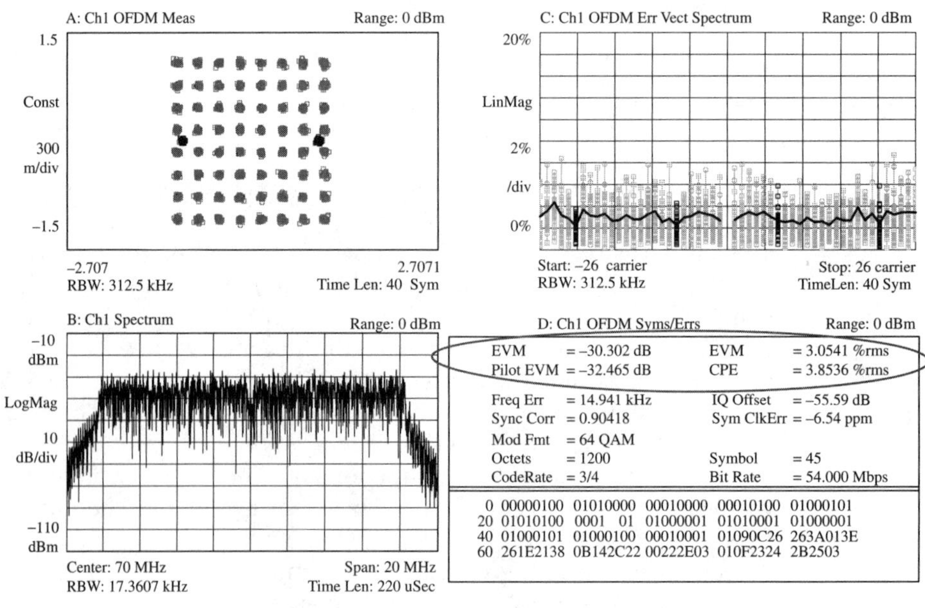

(b) A test suite for an 802.11a.

Figure 16.10 Testing using the error vector.

modes of operation for the IEEE 802.11a standard. Recall as discussed earlier 802.11a is an OFDM based wireless modem and this figure shows a scatter plot for all subcarrier matched filter outputs, the received signal spectrum, and computations on the error vector for each subcarrier. This test suite is automatically produced by an Agilent vector signal analyzer and a great deal of information can be derived from this plot about the performance of a communication system using orthogonal modulation. For example, a very useful tutorial article on how to use EVM to troubleshoot an orthogonal modulation system is [Agi00].

16.5.2 Stream Modulations

Since historically stream modulations have seen the most utility in practice several techniques have evolved to test the performance of orthogonal stream modulations. The two most common tests in stream modulation are the vector diagram plot and the eye diagram.

Vector Diagram

A generalization of the scatter plot is the vector diagram. The vector diagram was introduced in Chapter 4 as a way to visualize the 3D characteristics of the complex envelope in a 2D graph. For orthogonal stream modulation as with any IQ modulation the vector diagram can be used to get information about the vector modulated signal. For example, Figure 16.11 shows the transmitted vector diagram for a QPSK modulated stream modulation with a spectral cosine pulse with $\alpha = 0.3$ and $T_u = 8T$. A particular important vector diagram is the output of the matched filter to the transmitted pulse shape. This vector diagram with near ideal orthogonality should go through the constellation points at each sample time. Often in a vector diagram of the matched filter output, these sample times are marked with a different marker to emphasize these sample times. For example, for a QPSK modulated stream modulation with a spectral cosine pulse with $\alpha = 0.3$ and $T_u = 8T$ with little loss

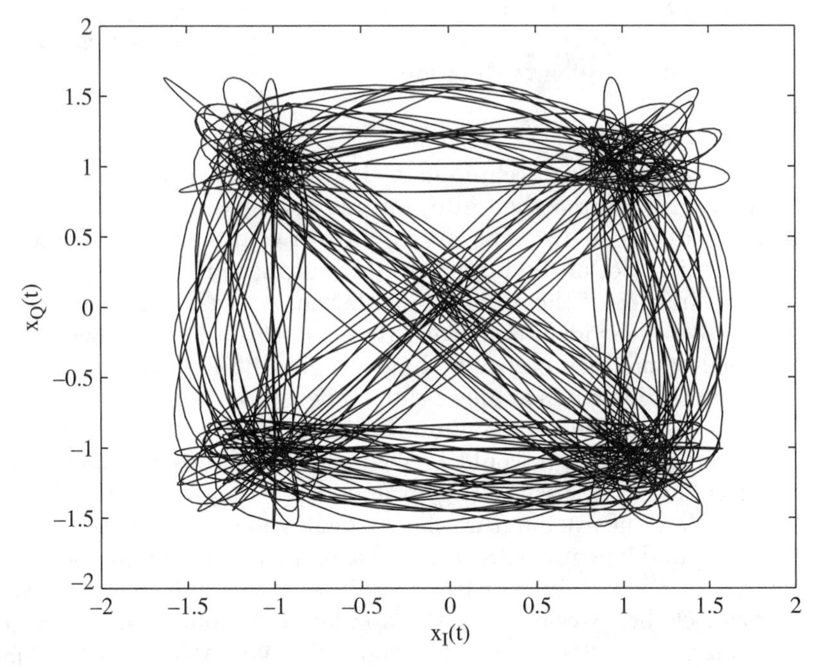

Figure 16.11 Vector diagram for a transmitted signal with QPSK, $\alpha = 0.3$ and $T_u = 8T$.

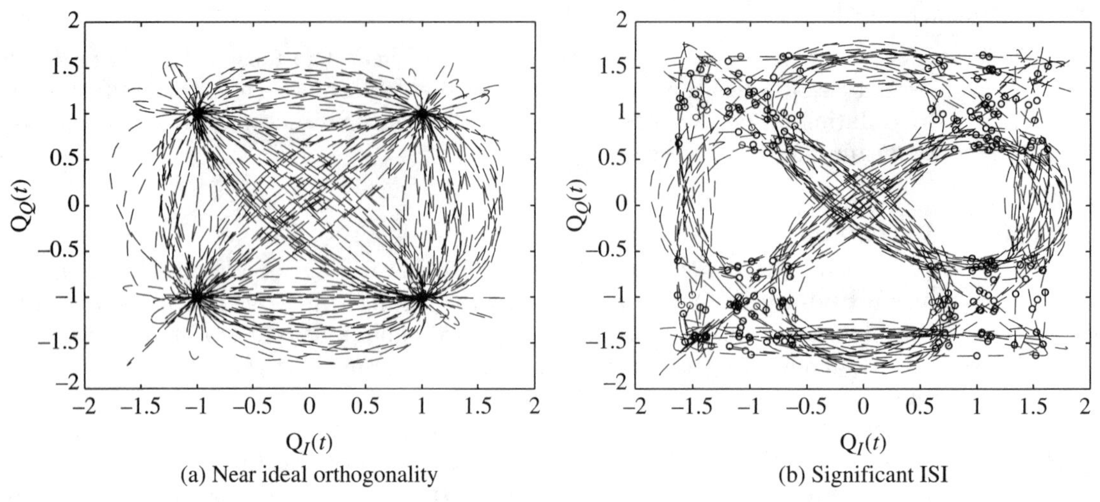

Figure 16.12 Matched filter output vector diagrams for a stream modulation with a spectral cosine pulse with $\alpha = 0.3$ and $T_u = 8T$. Sample times of the matched filter output are denoted with a circle.

in orthogonality the vector diagram of the matched filter output is shown in Figure 16.12(a). Note that the matched filter output samples taken by themselves would be the scatter plots shown in Figure 16.9. Distortion of the stream modulation signal will produce significant ISI and a vector diagram for a case of significant ISI is shown in Figure 16.12(b). Vector diagrams for stream modulations give engineers insights into the performance of practical implementations of orthogonal stream modulation.

The Eye Diagram

The eye diagram is a technique to view the time waveforms out of the matched filter. Again the matched filter output is a 3D signal and the eye diagram is a method to visualize this 3D signal in two dimensions. The eye diagram is a repetitive plot of the matched filter output (either I or Q channel) over one symbol in time of the stream modulation. An example of the eye diagram for the case of near ideal orthogonality is shown in Figure 16.13(a). The original method of producing an eye diagram goes back to the early days of digital communication when analog oscilloscopes were the primary time domain analysis tool available to a communication engineer. The I or Q channel of the matched filter output could be connected to the oscilloscope and the symbol clock could be used as the trigger and an eye diagram would then be visible on the display. The eye diagram shows the possible transitions from one symbol to the next and how impacted the orthogonality is by any distortion. An example of the eye diagram for the case of significant ISI is shown in Figure 16.13(b). The ISI causes the space between the worst case sample points of the matched filter outputs to move closer to the decision threshold. In the case of BPSK modulation as shown in Figure 16.13 the threshold will be zero. This effect of the ISI moving the optimum sample point closer to the decision boundary is often referred to as

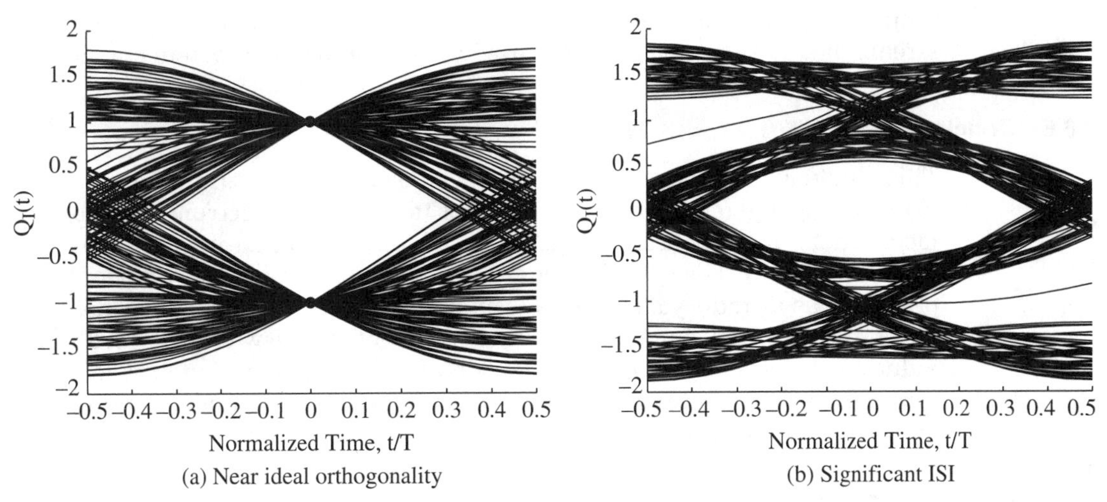

Figure 16.13 Matched filter output eye diagrams for a spectral cosine pulse with $\alpha = 0.3$ and $T_u = 8T$. BPSK modulation.

the "closing" of the eye. A communication system is often referred to as having a closed eye if there are some ISI patterns which would cause a deterministic error to occur in the simple threshold test demodulator of orthogonal demodulators. Most communications system analysis tools have the capability to produce eye diagrams and vector plots since they prove useful to communication engineers. For example, Matlab has an `eyediagram` command. The Agilent vector analyzer can produce eye diagrams for most standard stream modulations and an example of both an eye diagram and a vector diagram for a 16QAM stream modulation is shown in Figure 16.14. The tools of the eye diagram and the

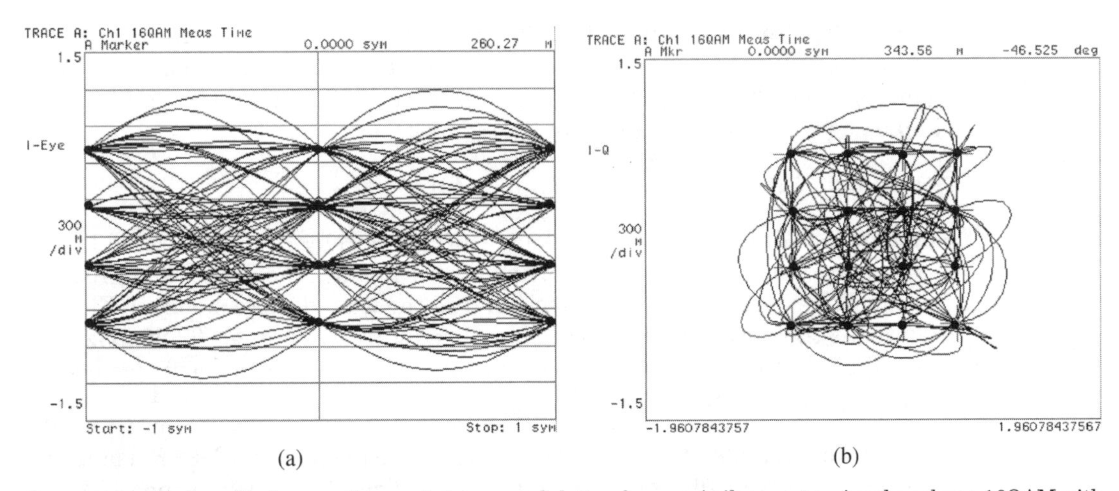

Figure 16.14 Test results for an orthogonal stream modulation from an Agilent vector signal analyzer. 16QAM with $\alpha = 0.25$.

vector plot produce a great deal of insight into the implementation of orthogonal stream modulations and are often used by communication system engineers.

16.6 Conclusions

This chapter has looked at some practical aspects of digital communications using orthogonal modulations. Techniques to shape the spectrum of orthogonal modualtions in response to requirements often mandated in real implementations were overviewed. The concept of the cosine pulse that is popular in practice was introduced. How this cosine pulse can be used in orthogonal modulations was briefly summarized. Finally, a brief overview of tools used in the industry to characterize orthogonal modulations was given. Understanding the concepts in this chapter will help the student get closer to understanding how digital communications is implemented in modern systems.

16.7 Homework Problems

Problem 16.1. Prove

$$V_u(mT) = \begin{cases} E_u & m = 0 \\ 0 & \text{elsewhere} \end{cases} \tag{16.29}$$

if and only if

$$\sum_{k=-\infty}^{k=\infty} G_u\left(f + \frac{k}{T}\right) = TE_u \tag{16.30}$$

Hint: Note the proof is based on Nyquist sampling theorem. This property indicates ways to design other pulses besides the spectral cosine pulse family to satisfy the Nyquist's criteria for zero ISI.

Problem 16.2. Prove the smallest bandwidth (100% of energy is contained within this bandwidth) that can be achieved in linear stream modulation is $B_T = 1/T$ when $D_z(k)$ can have an arbitrary constellation. If $D_z(k)$ is restricted to be real valued show the bandwidth can be $B_T = 1/(2T)$.

Problem 16.3. Prove that $u_{sc}(t) = \mathcal{F}^{-1}\{U_{sc}(f)\}$.

Problem 16.4. The company you work for, Horizon Wireless, has purchased 1 MHz of spectrum from the federal government of Elbonia. The marketing group of Horizon Wireless in Elbonia has decided that the way to make money in Elbonia is to divide the purchased spectrum into 20 equal size channels (of 50 kHz) and sell radios that use these 20 channels to the Elbonian government for use by their diplomatic corp.

(a) Give an example of a modulation that will achieve a 40 kHz transmission rate in a spectrally efficient and power efficient manner on one of these

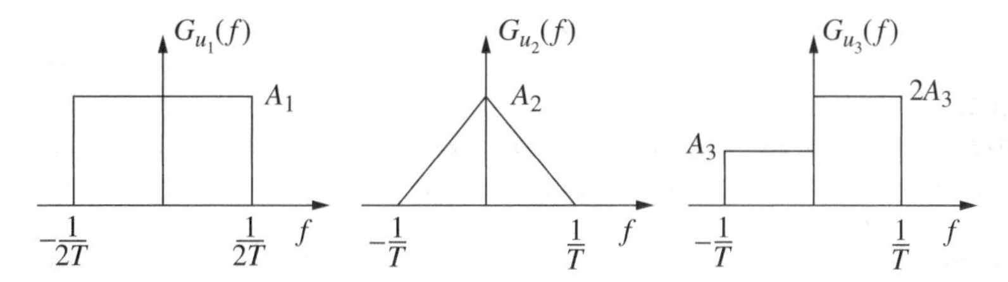

Figure 16.15 The energy spectrum of three possible pulses for linear stream modulation.

channels while sending packets of $K_b = 256$ bits and achieving a demodulation complexity that is reasonably close to $O(256)$.

(b) Give an example of a modulation that will achieve a 120 kHz transmission rate in a spectrally efficient and power efficient manner on one of these channels while sending packets of $K_b = 768$ bits and achieving a demodulation complexity that is reasonably close to $O(768)$.

(c) Assume the propagation loss in the channel is $L_p = aR^{-2}$ where R is the range of communication and a is a constant and the transmitted power remains the same for both transmission rates. If your design in part (a) can achieve a range of 1 km with a low frame error rate, what range will your design in (b) achieve and give the same frame error rate performance. (assume the noise stays constant).

Problem 16.5. The energy spectrum of three pulse shapes that can be used in a linear stream modulation is shown in Figure 16.15. Recall linear stream modulation has the form

$$X_{z,i}(t) = \sqrt{E_b} \sum_{l=1}^{K_b} D_z(l) u_i(t - (l-1)T) \tag{16.31}$$

where $u_i(t)$ corresponds to one of the three possible pulse shapes.

(a) Find the amplitude A_i for $i = 1, 2, 3$ such that $E_u = 1$.

(b) Prove that each of these three pulses can be used in a stream modulation and have simple bit demodulation result. *Hint:* These pulses have the same characteristic as the squared cosine pulse that makes the squared cosine pulse shape useful for stream modulation.

(c) Find the form of $u_3(t)$.

(d) Which pulse shape will provide the best spectral efficiency if the measure of bandwith is the 10 dB bandwidth, B_{10}?

Problem 16.6. Your boss has tasked you with designing an OFDM cable modem to send packets of $K_b = 256$ bits in less than 300 μs achieving a data rate

$W_b \geq 853.3$ kHz. She wants you to use less than 1 MHz of spectrum and cause little interference to transmissions from adjacent houses (each house is frequency multiplexed). The cable is assumed to be a frequency flat device and the goal given by your boss is to have a simple demodulator. Specify all the important parameters and pulse shapes.

Problem 16.7. The eye diagram for BPSK shown in Figure 16.13 has two levels at the optimum sample point. How many levels would the optimum sample point have in an eye diagram for

(a) QPSK modulation.

(b) 8PSK modulation.

(c) the 8-ary linear modulation introduced in Example 15.1.

(d) 16QAM introduced in Problem 15.1.

Problem 16.8. In your current job assignment you are expected to find the impact of imperfections in the hardware to be used to build your communication systems. The previous engineer who worked on the assignment had derived the impact of a amplitude imbalance in the I/Q upconverter (see Problem 4.25)

$$\tilde{x}_z(t) = Ax_I(t) + jBx_Q(t) = \sqrt{\frac{2}{1+\gamma}}x_I(t) + j\sqrt{\frac{2\gamma}{1+\gamma}}x_Q(t) \qquad (16.32)$$

where the amplitude imbalance is γ and the impact of a phase imbalance of θ in the I/Q upconverter (see Problem 4.13)

$$\tilde{x}_z(t) = x_I(t) - x_Q(t)\sin(\theta) + jx_Q(t)\cos(\theta) \qquad (16.33)$$

(a) Derive the overall response to an amplitude imbalance of γ and a phase offset of θ.

(b) Assume a linear modulation is used, $x_z(t) = D_z u(t) = (D_I + jD_Q)u(t)$ show that the error vector magnitude is given as

$$EVM(D_I, D_Q, \theta, \gamma)$$
$$= \sqrt{(1-A)^2 D_I^2 + 2(1-A)B\sin(\theta)D_I D_Q + (1 - 2B\cos(\theta) + B^2)D_Q^2}$$

Problem 16.9. The government has given the company SpyRUs 5 MHz of bandwidth to bring down video surveillence images of Odillian terrorist cells from a satellite. The video requires a data rate of 12 Mbits/s and video is divided up into packets of 1024 bits. The adjacent spectrum is being used for an ESPN broadcast of the Ohio State versus Michigan football game and cannot be disturbed (there must be at least 50 dB of attenuation in the adjacent 5 MHz channels or ESPN by contract can fine SpyRUs).

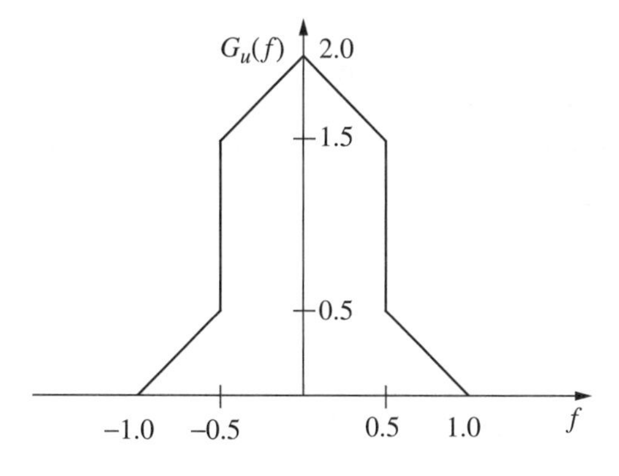

Figure 16.16 The rocket signal in frequency.

(a) The satellite channel is well modeled as frequency flat with an AWGN with a one-sided spectral density of N_0. Design a packet transmission system that will work for SpyRUs. Specify in detail how the 1024 bits will be transmitted and the precise data rate of the design.

(b) Assume the loss in the channel is L_p, in your model specify the normalization constants such the energy per bit is E_b.

(c) Estimate the error rate in demodulation for a given E_b/N_0.

Problem 16.10. **(RW)** An interesting characteristic of the Fourier transform is linearity and linearity can be used to compute the Fourier transform of complicated functions by decomposing these function into sums of simple functions.

(a) Prove if $y(t) = x_1(t) + x_2(t)$ that $Y(f) = X_1(f) + X_2(f)$.

(b) For the pulse shape spectrum given in Figure 16.16 find a symbol rate at which the Nyquist's criterion will be satisfied.

Problem 16.11. Show that the rectangular pulse, i.e.,

$$u_r(t) = \begin{cases} \sqrt{\dfrac{1}{T}} & 0 \leq t \leq T \\ 0 & \text{elsewhere} \end{cases} \tag{16.34}$$

satisfies the condition given in Eq. (16.30).

Problem 16.12. **(Design Problem)** Specify and identify all components of the transmitted waveform for a stream modulation that has a transmission rate of $W_b = 100$ kbits/s, uses BPSK modulation, achieves a spectral efficiency of $\eta_B = 0.75$ where B_T is the 30 dB bandwidth of the transmitted signal, and has an optimum demodulator with a complexity that scales linearly with the number of bits transmitted.

Problem 16.13. Often in communications a system requirement is that a certain percentage of the total power or energy must lie within a certain bandwidth for the system to be acceptable. For stream modulation using ideal an spectral cosine pulse, plot the 95% energy per bit bandwidth as a function of α. Using these results propose a bit rate for a binary stream modulation that can be achieved where 95% of the energy per bit lies within 1 MHz of bandwidth.

16.8 Example Solutions

Problem 16.2. Nyquist's criterion for stream modulation is

$$0 = \Re[D_z(l)D_z^*(k)V_u((k-l)T)] \tag{16.35}$$

so for a complex constellation this becomes $V_u(mT) = 0$ for $m \neq 0$. Using Problem 16.1 it is clear that the pulse used must satisfy

$$\sum_{k=-\infty}^{k=\infty} G_u\left(f + \frac{k}{T}\right) = TE_u \tag{16.36}$$

Hence if the pulse spectrum is

$$G_u(f) = \begin{cases} TE_u & |f| \leq \frac{1}{2T} \\ 0 & \text{elsewhere} \end{cases} \tag{16.37}$$

the condition is satisfied and $B_T = 1/T$. If the bandwidth becomes smaller than $B_T = 1/T$ then the condition in Eq. (16.36) will not hold.

For a real-valued constellation with $V_u(mT) = V_{uI}(mT) + jV_{uQ}(mT)$ the requirement becomes $V_{uI}(mT) = 0$. Note from Chapter 4 that

$$G_u(f) = G_{uI}(f) + jG_{uQ}(f) \qquad G_{uI} = \frac{G_u(f) + G_u^*(-f)}{2} \tag{16.38}$$

Using the results from Problem 16.1 the condition that must hold is

$$\sum_{k=-\infty}^{k=\infty} G_{uI}\left(f + \frac{k}{T}\right) = TE_u \tag{16.39}$$

A pulse spectrum of

$$G_u(f) = \begin{cases} 2TE_u & 0 \leq f \leq \frac{1}{2T} \\ 0 & \text{elsewhere} \end{cases} \tag{16.40}$$

will satisfy Eq. (16.39) and provide a bandwidth of $B_T = \frac{1}{2T}$.

Problem 16.4.

(a) There are a multitude of modulations that can achieve this specification of 0.8 bits/s/Hz. For example, you could use stream modulation with BPSK modulation, a symbol rate of 40 kHz ($T = 25\ \mu$s) and a spectral cosine pulse with $\alpha < 0.25$.

(b) There are a multitude of modulations that can achieve this specification of 2.4 bits/s/Hz. For example, you could use stream modulation with 8PSK modulation, a symbol rate of 40 kHz ($T = 25\ \mu$s) and a spectral cosine pulse with $\alpha < 0.25$.

(c) For the examples considered in (a) and (b) the word error probability is given in Figure 14.8. The difference in the required E_b/N_0 for BPSK and 8PSK at medium to high SNR is about 3.5 dB. Since the transmitted power is the same in both cases, the received signal power will be $P_{R_z} = L_p^2 P_{X_z}$. The received energy per bit in each case is

$$E_b(\textit{BPSK}) = P_{R_z}T \qquad E_b(8\textit{PSK}) = P_{R_z}T/3 \tag{16.41}$$

It is clear here that the received energy per bit is lower with the higher transmission rate and also the required E_b/N_0 for 8PSK is larger hence the second transmission scheme will be doubly penalized, i.e.,

$$P_{R_z}(8\textit{PSK}) = 3(10^{0.35})P_{R_z}(\textit{BPSK}) = 6.72P_{R_z}(\textit{BPSK}) \tag{16.42}$$

Consequently, the range of transmission for 8PSK will be

$$R(8\textit{PSK}) = (6.72)^{0.25}R(\textit{BPSK}) = 0.62R(\textit{BPSK}) \tag{16.43}$$

16.9 Miniprojects

Goal: To give exposure

- to a small scope engineering design problem in communications.
- to the dynamics of working with a team.
- to the importance of engineering communication skills (in this case oral presentations).

Presentation: The forum will be similar to a design review at a company (only much shorter). The presentation will be of 5 minutes in length with an overview of the given problem and solution. The presentation will be followed by questions from the audience (your classmates and the professor). All team members should be prepared to give the presentation.

16.9.1 Project 1

Project Goal: This project examines the design of a realistic pulse shape to be used with linear modulation that will meet a spectral emissions mask.

A spectral emissions mask is given in Figure 16.17. This spectral emissions mask is the one used for GSM handsets. So this problem is one of great practical

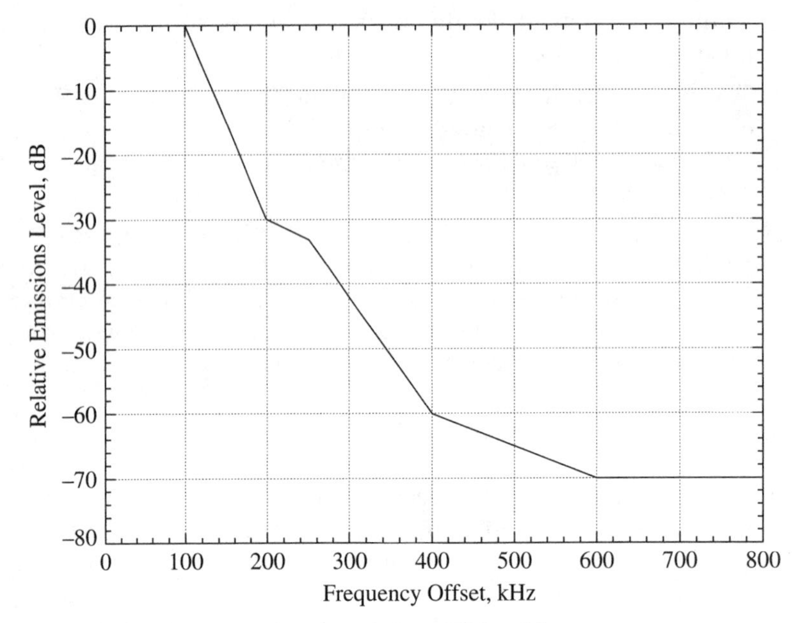

Figure 16.17 Spectral emissions mask for a GSM mobile.

interest. Consider a linear stream modulation having the form

$$X_z(t) = \sqrt{E_b} \sum_{k=1}^{K_b} D_z(k) u(t - (k-1)T) \tag{16.44}$$

The design is further constrained in that the pulse, $u(t)$, must not extend for longer than $T_u = 40\ \mu$s in time. Design a pulse shape that will satisfy Nyquist's criterion for an arbitrary constellation.

For the remainder of the project it is assumed that the pulse shape designed above is used with BPSK modulation. Any finite length pulse shape which achieves this spectral mask will have to produced some intersymbol interference (ISI) (i.e., Nyquist's criterion cannot exactly be met). Hence the matched filter demodulator given in Figure 16.18 will not be exactly optimal but can be made to be very close to optimal with a well designed pulse shape.

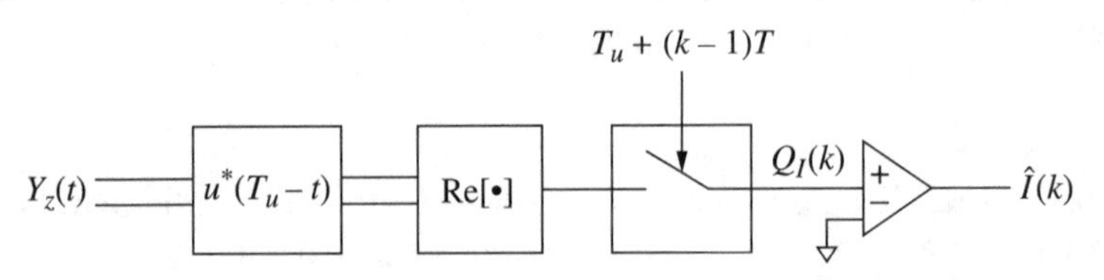

Figure 16.18 The baseband demodulator for binary pulse amplitude modulation.

(a) Assume $D_z(k) = 1$, compute the worse case degradation for an arbitrary pulse shape to the bit error probability for the demodulator given in Figure 16.18. Specifically consider the case of the kth bit where $k = K_b/2$ where K_b is large and compute the effects of ISI. Identify the values of $D_z(j),\ j \neq k$ which achieve this worse case degradation for an arbitrary pulse shape with BPSK modulation.

(b) Design a pulse shape and corresponding symbol rate such that the worse case (over all ISI patterns) degradation to the effective SNR is less than 0.25 dB and the spectral mask given in Figure 16.17 is met by the transmission. Grades will be assigned proportional to the achieved symbol rate (higher the better!). Anyone who beats the posted solution will get a bonus of 10 points for the project!

(c) Plot the transmitted complex envelope for the pulse shape that was designed in (b). Plot the real component of the matched filter output for a modulation sequence of $[1 - 111 - 11 - 1 - 1 - 11]$ for your design. Identify the symbol time sampling points.

Your grade will be penalized if a common solution to yours is found among your classmates solutions in direct proportion to the number of identical solutions. This is to encourage each of you to work independently. **Any computer code (e.g., Matlab) should be turned in with the project write-up.**

16.9.2 Project 2

Project Goal: This project examines the design of a pulse shape for a stream modulation to meet a spectral emissions mask where the constellation is limited to be real valued.

Assume a linear stream modulation where the constellation only takes real values that needs to meet the spectral emissions mask given in Figure 16.17 and be nearly optimally demodulated by a matched filter. This spectral emissions mask is the one used for GSM handsets. So this problem is one of great practical interest. The pulse shape designed for a real valued constellation can be spectrally more efficient than the pulse design for the arbitrary constellation. *Hint*: The value of $\Im\{V_u(\tau)\}$ can be of arbitrary value.

The signal will have the form

$$X_z(t) = \sqrt{E_b} \sum_{k=1}^{K_b} D_I(k) u(t - (k-1)T) \tag{16.45}$$

You are further constrained in that the pulse, $u(t)$, must not extend for longer than 40 μs in time.

For the remainder of the project it is assumed that the pulse shape designed above is used with BPSK modulation. Any finite length pulse shape which achieves this spectral mask will have produced some intersymbol interference (ISI) (i.e., Nyquist's criterion cannot exactly be met). Hence the matched filter demodulator given in Figure 16.18 will not be exactly optimal but can be made to be very close to optimal with a well-designed pulse shape.

(a) Assume $D_z(k) = 1$, compute the worse case degradation for an arbitrary pulse shape to the bit error probability for the demodulator given in Figure 16.18. Specifically consider the case of the kth bit where $k = K_b/2$ where K_b is large and compute the effects of ISI. Identify the values of $D_z(j),\ j \neq k$ which achieve this worse case degradation for an arbitrary pulse shape with BPSK modulation.

(b) If $u_1(t) = u(t)\exp[j2\pi f_o t]$, find $V_{u_1}(\tau)$. This result can help you design an infinite support pulse from the pulse shapes introduced in the text for use with real valued linear stream modulation.

(c) Design a pulse shape and corresponding symbol rate such that the worse case (over all ISI patterns) degradation to the effective SNR is less than 0.25 dB and the spectral mask given in Figure 16.17 is met by the transmission. Grades will be assigned proportional to the achieved symbol rate (higher the better!). Anyone who beats the posted solution will get a bonus of 10 points for the project!

(d) Plot the transmitted complex envelope for the pulse shape designed in (c). In addition, plot the real component of the matched filter output for a modulation sequence of $[1\ -1\ 1\ 1\ -1\ 1\ -1\ -1\ -1\ 1]$ for your design. Identify the symbol time sampling points.

Your grade will be penalized if a common solution to yours is found among your classmates solutions in direct proportion to the number of identical solutions. This is to encourage each of you to work independently. **Any computer code (e.g., Matlab) should be turned in with the project write-up.**

16.9.3 Project 3

Project Goal: To examine a typical degradation that occurs in an orthogonal modulation in a real system.

An OFDM signal is to be received that is transmitting $K_b = 16$ bits with 1 bit on each subcarrier, has a carrier frequency of $f_c = 1800$ Hz, and achieve a transmission rate of $W_b = 2000$ Hz. Unfortunately, the frequency sources at the transmitter and receiver cannot be perfectly synchronized so that

$$Y_z(t) = X_z(t)\exp[j2\pi f_o t] + W_z(t) \tag{16.46}$$

where f_o is the resulting frequency offset and

$$X_z(t) = \begin{cases} \displaystyle\sum_{l=1}^{K_b} D_z(l)\sqrt{\frac{E_b}{T_p}}\exp[j2\pi f_d(2l - K_b - 1)t] & 0 \le t \le T_p \\ 0 & \text{elsewhere} \end{cases} \tag{16.47}$$

It is desired to still use approximately the same demodulator as in the case of no frequency offset, i.e.,

$$\Re\left[H^*(k,k)\int_0^{T_p} Y_z(t)\sqrt{\frac{1}{T_p}}\exp[-j2\pi f_d(2k-K_b-1)t]dt\right]$$

$$= \Re[H^*(k,k)Q(k)] \underset{\hat{I}(k)=1}{\overset{\hat{I}(k)=0}{\gtrless}} 0 \tag{16.48}$$

This demodulator will have ISI because the orthogonality condition does not hold.

(a) Show that

$$Q(k) = \sum_{l=1}^{K_b} H(l,k)D_z(l) + N_z(k) \tag{16.49}$$

and identify $H(l,k)$.

(b) For $k = 8$ plot a scatter plot of $Q(k)$ for $f_o = 0, 1, 10$ Hz with $N_z(k) = 0$.

(c) For $k = 8$ compute the value of f_o that will result in less than a 3% RMS-EVM. Ensure that the amplitude loss of the desired signal due to the frequency offset is accounted for before computing the EVM.

16.9.4 Project 4

Project Goal: To examine a typical degradation that occurs in an orthogonal modulation in a real system.

A stream modulated signal is to be received that is transmitting $K_b = 16$ bits with a carrier frequency of $f_c = 1800$ Hz, and achieves a transmission rate of $W_b = 2000$ Hz. Unfortunately, the frequency sources at the transmitter and receiver cannot be perfectly synchronized so that

$$Y_z(t) = X_z(t)\exp[j2\pi f_o t] + W_z(t) \tag{16.50}$$

where f_o is the resulting frequency offset and

$$X_z(t) = \begin{cases} \sum_{l=1}^{K_b} D_z(l)u(t-(l-1)T) & 0 \le t \le T_p \\ 0 & \text{elsewhere} \end{cases} \tag{16.51}$$

It is desired to still use approximately the same demodulator as in the case of no frequency offset, i.e.,

$$\Re\left[H^*(k,k)\int_0^{T_p} Y_z(t)u*(t-(k-1)T)dt\right] = \Re[H^*(k,k)Q(k)] \underset{\hat{I}(k)=1}{\overset{\hat{I}(k)=0}{\gtrless}} 0 \tag{16.52}$$

This demodulator is suboptimal and will have ISI because the orthogonality condition does not hold in the matched filter outputs due to the frequency offset.

(a) Show that

$$Q(k) = \sum_{l=1}^{K_b} H(l,k)D_z(l) + N_z(k) \tag{16.53}$$

and identify $H(l,k)$.

(b) For $k = 8$ plot a scatter plot of $Q(k)$ for $f_o = 0, 1, 10$ Hz with $N_z(k) = 0$.

(c) For $k = 8$ compute the value of f_o that will result in less than a 3% RMS-EVM. Ensure that the amplitude loss of the desired signal due to the frequency offset is accounted for before computing the EVM.

Chapter

17

Orthogonal Modulations with Memory

17.1 Canonical Problems

Up to this point in this text we have introduced two general methods to communicate K_b bits.

1. General M-ary modulations
   - The advantage of a general M-ary modulation is that it can achieve very good fidelity and arbitrary spectral efficiency.
   - The disadvantage is that without more structure the optimal demodulator has complexity $O(2^{K_b})$.
2. Orthogonal memoryless modulations (including stream modulations, OFDM, OCDM)
   - The advantage of orthogonal modulation is that the optimum demodulator has complexity $O(K_b)$ and a desired spectral efficiency can be achieved with a proper design of the modulation signals.
   - The disadvantage is that the fidelity of message reconstruction is limited to that of binary modulations and this performance is far from Shannon's limit.

The goals for the final chapter of the text will be to explore the remaining areas of trade-off in fidelity, complexity, and spectral efficiency, i.e.,

- Improving the demodulation fidelity or spectral characteristics of orthogonal modulations with a goal of maintaining a demodulation complexity that is $O(K_b)$.

Shannon has given us an upperbound on the performance that can be achieved for a given spectral efficiency (see Figure 12.5) and there are a variety of ways to approach that performance with a demodulation complexity that is $O(K_b)$. The concept of orthogonal modulations with memory are what have allowed communication theory to approach the bounds provided by Claude Shannon in his ground breaking work on information theory [Sha48]. This chapter will be a brief introduction into the topic and an examination of the implications of adding memory to the performance and the spectral characteristics.

17.2 Orthogonal Modulations with Memory

This section addresses a class of communication problems where an orthogonal modulation is implemented with the modulation symbols having memory. The memory is normally included in the modulation to either improve the fidelity of demodulation or change the spectral characteristics of a memoryless orthogonal modulation. This type of modulation is referred to as modulation with memory (MWM) and incorporates most error control coding schemes [Wic95, LC04, BDMS91]. Our goal in this text is not to explore how to design these MWM but to understand the communication theory behind the fidelity, spectral efficiency, and the demodulation complexity. Design of MWM is often addressed in a course on error control coding. Suffice it to say here that a MWM adds memory to the modulation process with a goal of either changing the spectral characteristics of the transmitted signal or improving the resulting squared Euclidean distance spectrum.

This chapter first considers the special case of an orthogonal modulation with memory (OMWM) where 1 bit is sent with each symbol or orthogonal dimension. For simplicity of discussion this section will again assume that the bits to be transmitted are equally likely and independent. The generalizations for correlated bits is possible but the added notational complexity is not worth the gain in generality. The resulting data symbols can then be transmitted using an orthogonal modulation. The data modulation symbols, $\tilde{D}_z(l)$, are due to the stream of information bits, $I(l), l = 1, \ldots, K_b$. The tilde notation will be used to differentiate between modulations that have memory (tilde) and memoryless modulations (not tilde). For continuity with the previous discussions on orthogonal modulations this initial discussion considers exclusively modulations where the transmission bit rate is $R = 1$ bit per symbol. To keep a consistent normalization the mapping, $\tilde{D}_z(l) = a(J(l))$, is selected such that $E[|\tilde{D}_z(l)|^2] = 1$. The transmitted signal will have the form

$$X_z(t) = \sum_{l=1}^{N_f} \tilde{D}_z(l) s_l(t) \tag{17.1}$$

where N_f is the length of the transmitted frame in symbols and $s_l(t)$ are the orthogonal basis functions (spreading waveforms) used to transmit the symbols.

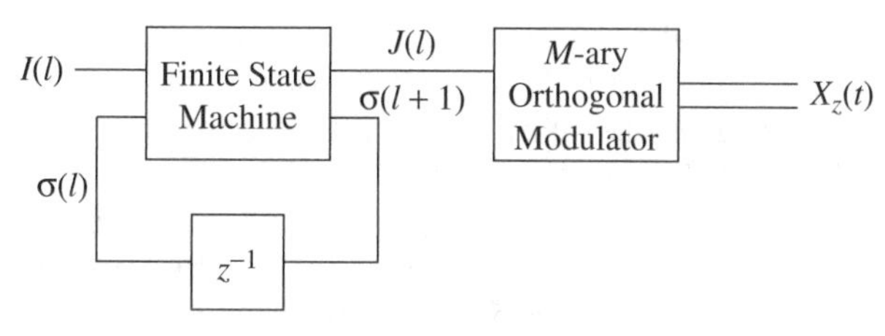

Figure 17.1 The block diagram for a modulation with memory with a transmission rate of 1 bit per symbol.

Note, this formulation is equally applicable to being used in OCDM, OFDM, or a stream modulation. A more general formulation for orthogonal modulations where the rate is not constrained to be 1 bit per symbol is considered later in this chapter.

An orthogonal modulation with memory, as considered in this section, consists of a finite state machine operating at the symbol rate where the output of the finite state machine is used as an input to an M_s-ary orthogonal modulation. Figure 17.1 shows the block diagram for a linear orthogonal modulation with memory. At each symbol time a new bit, $I(l)$ is input into the system and this produces a new constellation label, $J(l)$, and a new modulation state, $\sigma(l+1)$. Let N_s denote the number of states in the modulation and the nonlinear equations governing the finite state machine are given as

$$\sigma(l+1) = g_1(\sigma(l), I(l)) \tag{17.2}$$

$$J(l) = g_2(\sigma(l), I(l)) \tag{17.3}$$

For notational purposes denote the set of all possible state values as Ω_σ. In general, it is desirable to have ν_c extra symbols transmitted to return the modulation to a common final state at the end of the transmission frame. Again the total length of the frame for the orthogonal modulation is denoted N_f, hence, in this case when $R = 1$ bit/symbol $N_f = K_b + \nu_c$, where ν_c is a modulation dependent constant. This return to a final common state is most often known as termination in the literature of MWM [Wic95, LC04, BDMS91]. Note that the effective rate is $R_{eff} = \frac{K_b}{K_b+\nu_c} < 1$ but if a large number of bits are transmitted then the loss in efficiency by including the termination becomes small.

A couple comments about OMWM are appropriate at this point. An OMWM can only provide an improvement in demodulation fidelity or a change in spectral characteristics compared to a memoryless modulation if there is some redundancy to be exploited. An orthogonal modulation with memory has 2^{K_b} possible bit sequences as an input. The modulation symbols being mapped onto the orthogonal modulation have $M_s^{K_b+\nu_c}$ possible realizations. Hence if there is to be redundancy that the OMWM can exploit for large K_b then $M_s > 2$. The improved fidelity of demodulation or spectral efficiency is achieved by picking the best $M = 2^{K_b}$ possible transmitted symbol sequences out of the $M_s^{K_b+\nu_c}$ total possible transmitted symbol sequences. Secondly, the more states an OMWM has the more memory it contains. In general, more memory allows a designer to make better choices in the transmitted symbol sequences but at a cost of a higher complexity demodulation. Finally, an OMWM is often represented with a directed graph [Wes01]. The vertices of the directed graph represent the state of the modulation at a particular time and the edges of the directed graph represent the allowable transitions between states. In the communications literature the directed graph is often referred to as a trellis. Because of this trellis representation OMWM are often referred to as trellis coded modulation (TCM). This trellis representation of an OMWM is explored in the example considered in the next section.

17.2.1 MLWD for Orthogonal Modulations with Memory

Recall OMWM can be characterized with a finite state machine defined with

$$\sigma(l+1) = g_1(\sigma(l), I(l)) \tag{17.4}$$

$$J(l) = g_2(\sigma(l), I(l)) \tag{17.5}$$

The equivalent modulation symbol is $\tilde{D}_z(l) = a(J(l))$ with $a(\bullet)$ being the constellation mapping. Here we examine only the optimum word demodulation (MLWD). The optimum word demodulator is still the MLWD for an orthogonal modulation except now the memory of the modulation must be accounted for when computing the maximum likelihood metric. Due to the orthogonal modulation the MLWD has

$$\begin{aligned}
\hat{\boldsymbol{I}} &= \arg \max_{i=0,\ldots,M-1} T_i \\
&= \arg \max_{i=0,\ldots,M-1} \sqrt{E_b} \sum_{k=1}^{N_f} \Re[\tilde{d}_i^*(k) Q(k)] - \frac{E_b}{2} \sum_{k=1}^{N_f} |\tilde{d}_i(k)|^2 \\
&= \arg \min_{i=0,\ldots,M-1} \sum_{k=1}^{N_f} |Q(k) - \sqrt{E_b} \tilde{d}_i(k)|^2
\end{aligned} \tag{17.6}$$

where $Q(k)$ is the matched filter ouptut sample. Consequently, the output of the MLWD can be thought of as the possible transmitted modulation symbols (constrained by the modulation with memory) that is closest in Euclidean distance to the matched filter outputs that are observed ($Q(k)$) over the entire frame ($k = 1, N_f$). This minimum squared Euclidean distance decoder provides a nice analogy to the demodulation of memoryless orthogonal modulations.

As with a general M-ary modulation, the achieved fidelity in demodulation of the MLWD for an OMWM is determined by the squared Euclidean distance spectrum. Due to the use of an orthogonal modulation the squared Euclidean distance between any two words is given as

$$\Delta_E(i,j) = \int_{-\infty}^{\infty} |x_i(t) - x_j(t)|^2 dt = E_b \int_{-\infty}^{\infty} \left| \sum_{l=1}^{N_f} s_l(t)[\tilde{d}_i(l) - \tilde{d}_j(l)] \right|^2 dt$$

$$= E_b \sum_{l=1}^{N_f} \sum_{k=1}^{N_f} (\tilde{d}_i(l) - \tilde{d}_j(l))(\tilde{d}_i(k) - \tilde{d}_j(k))^* \int_{-\infty}^{\infty} s_l(t) s_k^*(t) dt \quad (17.7)$$

$$= E_b \sum_{l=1}^{N_f} |\tilde{d}_i(l) - \tilde{d}_j(l)|^2 \quad (17.8)$$

Consequently, the important characteristics in terms of fidelity of demodulation of an OMWM is the squared Euclidean distance between the code sequences produced by the modulation with memory. Consequently, the performance of the OMWM is only a function of the finite state machine that characterizes the modulation. For a $R = 1$, an improved demodulation fidelity will be achieved compared to a memoryless orthogonal modulation if the minimum Euclidean squared distance of the modulation with memory is greater than the best Euclidean squared distance for a $R = 1$ memoryless modulation. The best $R = 1$ memoryless modulation (BPSK) achieves a Euclidean squared distance of $\Delta_{E,min} = 4E_b$ so this value will be our benchmark for comparison.

17.2.2 An Example OMWM Providing Better Fidelity

Consider an OMWM consisting of a four state trellis code ($N_s = 4$) using 4PAM modulation ($M_s = 4$) as proposed by Ho, Cavers, and Varaldi [HCV93]. It should be noted here that this OMWM is a simple construction using a general combined modulation and coding technique proposed by Ungerboeck in [Ung82]. This example OMWM illustrates some important properties that will become apparent in the sequel. The modulation updates are given in Table 17.1. Here we have $\Omega_\sigma = \{1, 2, 3, 4\}$. Note that the modulation is normalized to have an average energy of unity. The trellis for this modulation is shown in Figure 17.2. As mentioned before, the vertices of the trellis diagram represent the states at each time interval (hence for this code there are $N_s = 4$ vertices at each point in time) and the edges are the possible transitions (note for instance $\sigma(l) = 1$

TABLE 17.1 The finite state machine description of an example trellis code

$\sigma(l+1) = g_1(I(l), \sigma(l))$			$J(l) = g_2(I(l), \sigma(l))$			$\tilde{D}_z(l) = a(J(l))$	
	$I(l)$			$I(l)$		$J(l)$	$\tilde{D}_z(l)$
State, $\sigma(l)$	0	1	State, $\sigma(l)$	0	1	0	$-3/\sqrt{5}$
1	1	2	1	0	2	1	$-1/\sqrt{5}$
2	3	4	2	3	1	2	$1/\sqrt{5}$
3	1	2	3	2	0	3	$3/\sqrt{5}$
4	3	4	4	1	3		

$I(k) = 0$	$I(k) = 1$	$\sigma(l)$	$\sigma(l+1)$
$-3/\sqrt{5}$	$1/\sqrt{5}$	$\sigma(l) = 1$	$\sigma(l+1) = 1$
$3/\sqrt{5}$	$-1/\sqrt{5}$	$\sigma(l) = 2$	$\sigma(l+1) = 2$
$1/\sqrt{5}$	$-3/\sqrt{5}$	$\sigma(l) = 3$	$\sigma(l+1) = 3$
$-1/\sqrt{5}$	$3/\sqrt{5}$	$\sigma(l) = 4$	$\sigma(l+1) = 4$

Figure 17.2 The trellis diagram of the Ho, Cavers, and Varaldi trellis code.

can only transition to $\sigma(l+1) = 1$ and $\sigma(l+1) = 2$). There are a total of $2N_s = 8$ edges in this example. It should be noted that the trellis labels in this text will be put to the side as in Figure 17.2. The edge labels indicate the input, $I(k)$, that causes the transition and the modulation symbol that is output for that input and state. Unless noted all the upper transitions in a trellis will correspond to the input $I(k) = 0$. As is the case for orthogonal modulation the normalization is set as

$$E\left[|\tilde{D}_z(l)|^2\right] = R = 1 \tag{17.9}$$

For instance, when $\sigma(l) = 1$ and $I(l) = 0$ the output symbol is $\tilde{D}_z(l) = -3/\sqrt{5}$. This particular code requires at most two transitions to allow the modulation to transition from any state to any other state so this code has $\nu_c = 2$. For instance, if $\sigma(l) = 2$ then to get to $\sigma(l+2) = 1$ one would have to transition to $\sigma(l+1) = 3$ first. Alternatively if $\sigma(l) = 1$ then we could immediately transition to $\sigma(l+1) = 1$.

N_f trellis sections can be combined together to get a complete description of an OMWM. For instance for $K_b = 4$, Figure 17.3 shows a trellis description for a modulation that has $\sigma(1) = 1$ and returns to $\sigma(7) = 1$. One can verify that there are 2^{K_b} paths through this trellis. To understand the fidelity of demodulation

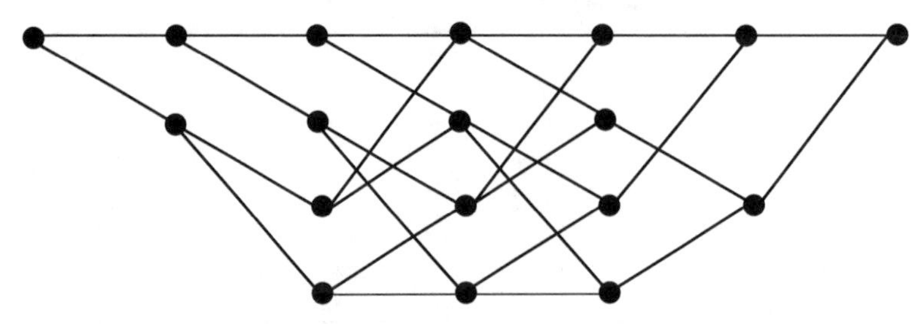

Figure 17.3 The trellis diagram of the Ho, Cavers Varaldi trellis code. $K_b = 4$.

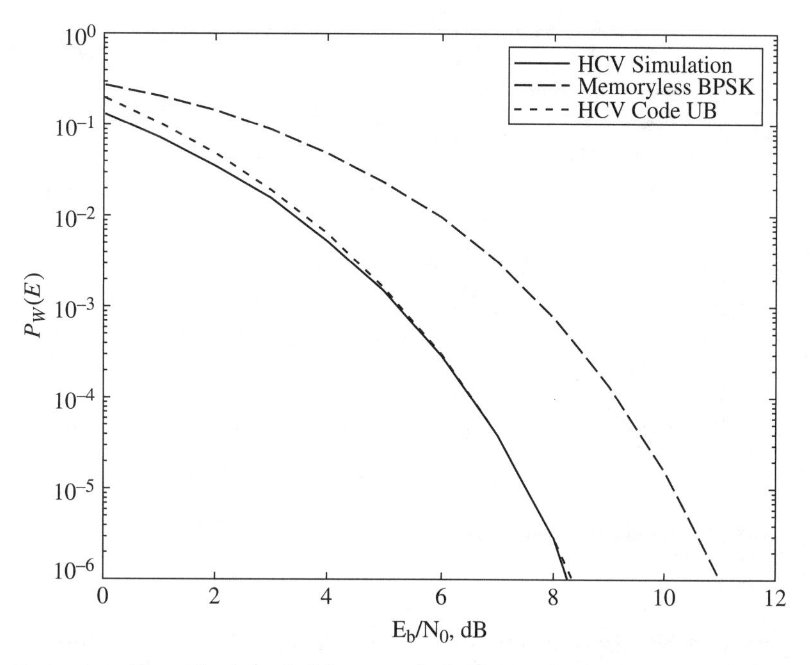

Figure 17.4 The union bound of the example OMWM and frame error rate of $K_b = 4$ memoryless orthogonally modulated bits using BPSK.

of this particular OMWM all of the $(16 \times 15)/2 = 120$ terms in the squared Euclidean distance spectrum should be computed. Consider two of the paths in the trellis corresponding to the words $\vec{I} = 0 = [0\ 0\ 0\ 0]$ and $\vec{I} = 1 = [1\ 0\ 0\ 0]$. These two words will produce modulation sequences of

$$\vec{d}_0 = [-3/\sqrt{5}\ -3/\sqrt{5}\ -3/\sqrt{5}\ -3/\sqrt{5}\ -3/\sqrt{5}\ -3/\sqrt{5}]$$

and

$$\vec{d}_1 = [1/\sqrt{5}\ 3/\sqrt{5}\ 1/\sqrt{5}\ -3/\sqrt{5}\ -3/\sqrt{5}\ -3/\sqrt{5}]$$

This results in a squared Euclidean distance of

$$\Delta_E(1,0) = E_b(4/\sqrt{5})^2 + E_b(6/\sqrt{5})^2 + E_b(4/\sqrt{5})^2 = 68E_b/5 \qquad (17.10)$$

Hence at least these two paths produce a squared Euclidean distance better than the best memoryless modulation of the same rate, i.e., BPSK has $\Delta_E(1,0) = 4E_b$. The entire distance spectrum of this modulation is explored in the homework problems (see Problem 17.1) and the union bound to the probability of word error is plotted in Figure 17.4.

17.2.3 Discussion

This section introduced one example modulation to transmit 1 bit of information per orthogonal dimension. As a final point it is worth comparing the spectral efficiency performance of this example modulation with the upperbound

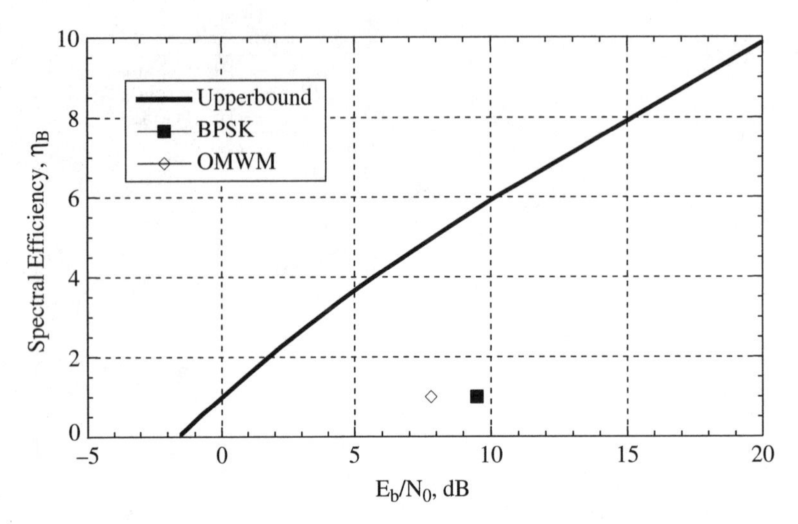

Figure 17.5 A comparison of performance of the example OMWM and Shannon's upperbound.

provided by information theory (see Section 12.3). As before we will denote reliable communication as being an error rate of 10^{-5}. This OMWM achieves reliable communications at $E_b/N_0 = 7.8$ dB. The performance of this modulation versus the Shannon capacity is plotted in Figure 17.5 and compared to the best orthogonal memoryless modulation with the same rate ($R = 1$ which is BPSK). It should be noted that by considering a very simple modulation with memory (4 states) we have moved over 2 dB closer to the achievable reliability predicted by Shannon. Using more states and a proper design of the modulation can improve the performance further. In fact, the performance can be made arbitrarily close to the Shannon bound by increasing the complexity of the OMWM.

The remaining items to be explored are the spectral characteristics of OMWM, a generalization of OMWM to an arbitrary rate, and the demodulation algorithm form and complexity. Note if we can find a demodulator that has a complexity that is $O(K_b)$ then we have succeeded in obtaining the desired characteristics (i.e., improved performance compared to memoryless orthogonal modulation while maintaining a complexity $O(K_b)$). It turns out there is a wide variety of ways to achieve linear complexity in the demodulation of OMWM [BB99, Wic95, LC04, BDMS91, CAC01]. For those of you who pursue a career as a communication engineer, algorithms like the Viterbi algorithm and the BCJR algorithm will become important tools as they provide excellent performance with a complexity of $O(K_b)$. Derivations of the algorithms, while critical in reaching Shannon's performance limits, will be left for graduate school. The remainder of this chapter will explore the spectral characteristics of stream modulation and generalize OMWM to an arbitrary transmission rate.

TABLE 17.2 The finite state machine description of the AMI trellis code

$\sigma(l+1) = g_1(I(l), \sigma(l))$			$J(l) = g_2(I(l), \sigma(l))$			$\tilde{D}_z(l) = a(J(l))$	
	$I(l)$			$I(l)$		$J(l)$	$\tilde{D}_z(l)$
State, $\sigma(l)$	0	1	State, $\sigma(l)$	0	1	0	$-\sqrt{2}$
1	1	2	1	1	0	1	0
2	1	2	2	2	1	2	$\sqrt{2}$

17.3 Spectral Characteristics of Stream OMWM

Orthogonal modulations with memory can both be used to achieve better fidelity of demodulation or a modified transmitted spectrum compared to memoryless modulation. In general, it is assumed in communications engineering that the information bits, $I(k)$, will be an independent (white), identically distributed sequence. In a stream modulation putting this white sequence into an OMWM will produce a modulation sequence that has memory. The memory in the modulation will produce a correlated time series $J(l)$, $l = 1, N_f$ and this correlation will change the average transmitted signal energy spectrum per bit, $D_{x_z}(f)$. This section will explore this effect and how to characterize the transmitted energy spectrum from the modulation description.

17.3.1 An Example OMWM with a Modified Spectrum

This section considers an example of Alternate Mark Inversion (AMI) modulation. AMI modulation is an OMWM that is defined in Table 17.2. AMI codes and related codes are used in many telecommunications applications (e.g., T1 lines on coaxial cables) for reasons that will be apparent in the sequel. Consider a transmission with $K_b = 4$ ($N_f = 5$). This transmission has a trellis diagram shown in Figure 17.6. In examining each of the paths of the trellis it is apparent that

$$\sum_{l=1}^{N_f} \tilde{d}_i(l) = 0 \tag{17.11}$$

This implies that if AMI modulation is used in a stream modulation then each transmission will not have a DC offset and that there will be a notch in the spectrum at $f = 0$. Precisely, if the data bits driving the trellis modulation are

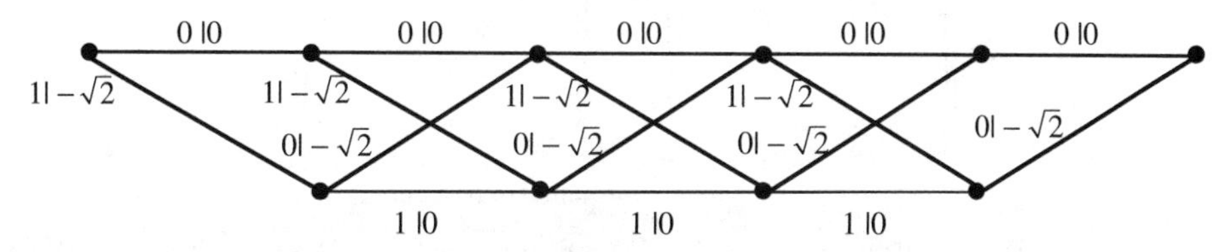

Figure 17.6 The trellis diagram for AMI with $K_b = 4$.

equally likely and independent then spectrum of the stream modulation will be

$$
\begin{aligned}
D_{x_z}(f) &= \frac{E\left[G_{X_z}(f)\right]}{K_b} \\
&= \frac{E\left[|\sqrt{E_b}\sum_l \tilde{D}_z(l)U(f)\exp[-j2\pi fT\,(l-1)]|^2\right]}{K_b} \\
&= \frac{\sum_{i=0}^{2^{K_b}-1}|\sqrt{E_b}\sum_l \tilde{d}_i(l)U(f)\exp[-j2\pi fT\,(l-1)]|^2}{2^{K_b}K_b}
\end{aligned}
\tag{17.12}
$$

For the example of AMI with $K_b = 4$ and stream modulation with

$$
u_r(t) = \begin{cases} \sqrt{\dfrac{1}{T}} & 0 \leq t \leq T \\ 0 & \text{elsewhere} \end{cases}
\tag{17.13}
$$

the average energy spectrum per bit is plotted in Figure 17.7. This figure clearly shows the spectrum of an OMWM can be shaped (compare to Figure 15.21) and this shaping ability is often why a modulation with memory is used in practice. The interesting aspect of AMI modulation is that it achieves about the same fidelity of demodulation as BPSK modulation in addition to the spectral shaping

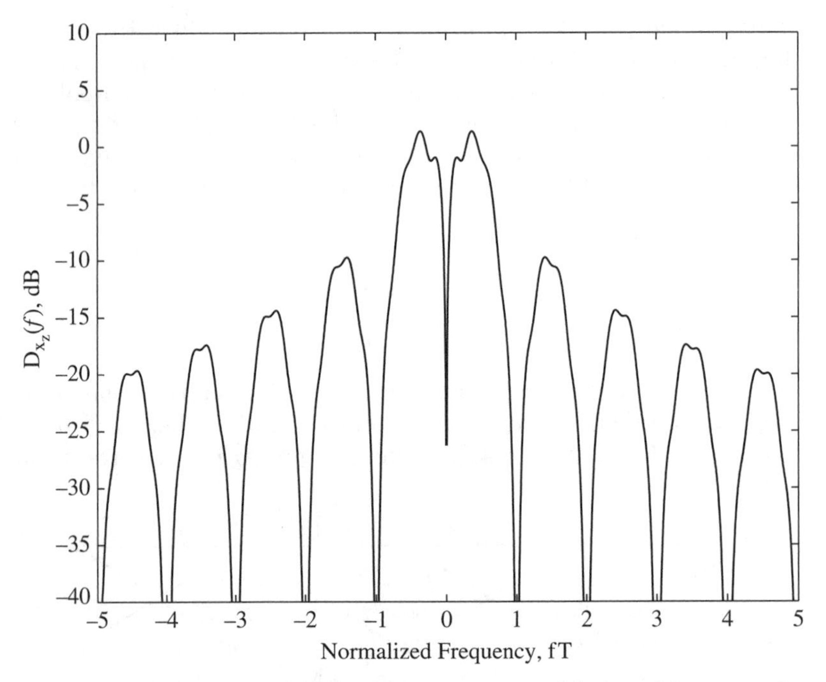

Figure 17.7 The spectrum of AMI used in a stream modulation with rectangular pulses. $K_b = 4$.

characteristics (see Problem 17.14). This combination of spectral shaping without much loss in fidelity of demodulation is the reason for AMI's adoption in many telecommunication applications.

17.3.2 Spectrum of OMWM for Large K_b

It is often of interest to evaluate the spectrum resulting from an OMWM in a stream modulation from a long frame of data (K_b large). It should be noted that the complexity of evaluating the average in Eq. (17.12) grows proportional to 2^{K_b}. Clearly for large frames this computation is increasingly impractical. Since there is a regular trellis structure to the modulation this structure can be exploited to compute the desired average energy spectrum per bit. To understand how this structure can be exploited first expand

$$K_b D_{x_z}(f) = E\left[\left|\sqrt{E_b}\sum_{l=1}^{N_f}\tilde{D}_z(l)U(f)\exp[-j2\pi fT(l-1)]\right|^2\right]$$

$$= E_b|U(f)|^2\sum_{l_1=1}^{N_f}\sum_{l_2=1}^{N_f}E[\tilde{D}_z(l_1)\tilde{D}_z^*(l_2)]\exp[-j2\pi f(l_1-l_2)T] \quad (17.14)$$

Since the trellis is the same every symbol in the frame and we assume the information bits are a white sequence, it is reasonable to assume that $\tilde{D}_z(l)$ is a stationary process. Since an OMWM typically starts from a fixed state and terminates to a fixed state this is not a true model as the distribution of the modulation at the frame boundaries will be different than the middle of the frame. If the frame is very long it is a good engineering approximation to ignore this nonstationarity at the frame boundaries. Also it is possible with certain OMWM that sequences can be produced in the middle of the frame that are not stationary time series but in general that will be the exception and not the rule. Consequently, for a presentation that captures many of the important engineering applications the discussion will assume stationarity. The stationarity assumption implies that

$$E[\tilde{D}_z(l_1)\tilde{D}_z^*(l_2)] = R_{\tilde{D}}(l_1-l_2) \quad (17.15)$$

Since all terms with the same time difference l_1-l_2 will be constant the double sum can be rearranged to give

$$K_b D_{x_z}(f) = E_b|U(f)|^2\sum_{m=-N_f+1}^{N_f-1}(N_f-|m|)R_{\tilde{D}}(m)\exp[-j2\pi fTm] \quad (17.16)$$

Looking at the limit of a long frame size and noting that

$$\lim_{K_b\to\infty}\frac{N_f}{K_b} = 1 \quad (17.17)$$

gives[1]

$$\lim_{K_b\to\infty} D_{x_z}(f) = E_b|U(f)|^2 \sum_{m=-\infty}^{\infty} R_{\tilde{D}}(m)\exp[-j2\pi f T m] = E_b|U(f)|^2 S_{\tilde{D}}(e^{j2\pi f T}) \tag{17.18}$$

where $S_{\tilde{D}}(e^{j2\pi fT})$ is the power spectrum of the stationary discrete-time random process $\tilde{D}_z(l)$ evaluated at a discrete frequency fT. It should be noted that $S_{\tilde{D}}(e^{j2\pi fT})$ is a periodic function of f with a period of $1/T$. The transmitted energy spectrum per bit for a stream modulation is seen to be a product of the pulse shape energy spectrum and the power spectral density of the discrete time modulation sequence. The form of the OMWM will determine the power spectrum of the discrete modulation sequence.

To compute $S_{\tilde{D}}(e^{j2\pi f})$ the correlation function, $R_{\tilde{D}}(m)$, needs to be identified using the trellis structure of the OMWM. For a stationary random process the correlation function is given by

$$R_{\tilde{D}}(m) = \sum_{d_i\in\Omega_{\tilde{D}}} \sum_{d_j\in\Omega_{\tilde{D}}} d_i d_j^* P_{\tilde{D}_z(l)\tilde{D}_z(l-m)}(d_i, d_j) \tag{17.19}$$

where $P_{\tilde{D}_z(l)\tilde{D}_z(l-m)}(d_i, d_j)$ is the joint probability mass function of the event $\{\tilde{D}_z(l) = d_i, \tilde{D}_z(l-m) = d_j\}$. The key point in finding the correlation function is identifying an algorithm to compute the joint PMF of the modulation symbols at various time offsets from the trellis representation of the modulation. The modulation symbol at any point in time, $\tilde{D}_z(l)$, is determined entirely by specifying the edge of the trellis. The edges of the trellis can be enumerated by considering the modulation state at the same time, $\sigma(l)$, and the modulation state at the next time, $\sigma(l+1)$.

Definition 17.1 An edge of the trellis at time l, denoted $S(l)$, is a pair of states $\{\sigma(l) = i, \sigma(l+1) = j\}$ such that $P(\sigma(l+1) = i|\sigma(l) = j) \neq 0$.

For notational purposes we define the possible values that $S(l)$ can take as being the set Ω_s with cardinality N_E. The number of edges, N_E, for a particular time near the middle of the frame for the $R = 1$ OMWM is $N_E = 2N_s$. In general, the number of edges corresponding to a trellis transition is less than the total number of pairs $\{\sigma(l+1) = i, \sigma(l) = j\}$, $\|\Omega_\sigma\|^2 = N_s^2$.

Property 17.1 There is a functional mapping from $S(l)$ to $\tilde{D}_z(l)$.

This functional mapping is due to the finite state machine characteristics of OMWM.

Consequently, the needed symbol joint probability can be computed by deriving joint edge probabilities. The goal is now identifying $P_{S(l)S(l-m)}(i, j)$ which

[1]The readers will have to excuse the slightly inconsistent notation in this equation. This is the only place in the text where a discrete frequency variable and a continuous frequency variable both denoted with f are required in the same equation.

will completely specify $P_{\tilde{D}_z(l)\tilde{D}_z(l-m)}(d_i, d_j)$ and

$$R_{\tilde{D}}(m) = \sum_{i=1}^{N_E}\sum_{j=1}^{N_E} d_s(i)d_s(j)^* P_{S(l)S(l-m)}(i,j) \tag{17.20}$$

To reflect this functional mapping we introduce a vector of modulation symbols corresponding to each edge of the trellis, i.e.,

$$\vec{d}_s = [d_s(1)\ldots d_s(N_E)] \tag{17.21}$$

where $d_s(k)$ is that value that $\tilde{D}_z(l)$ takes when $S(l) = k$. Using this notation the double sum in Eq. (17.20) can be expressed in a quadratic form as

$$R_{\tilde{D}}(m) = \vec{d}_s \mathbf{S}_E(m)\vec{d}_s^{\,H} \tag{17.22}$$

where $[\mathbf{S}_E(m)]_{i,j} = P_{S(l)S(l-m)}(i,j)$.

EXAMPLE 17.1

Consider AMI modulation again. There are $N_E = 4$ edges for AMI. The edges and the functional mapping to the modulation symbols for AMI are enumerated in Table 17.3. This results in

$$\vec{d}_s = [0 \quad \sqrt{2} \quad -\sqrt{2} \quad 0] \tag{17.23}$$

The needed joint probability of the edges, $P_{S(l)S(l-m)}(i,j)$, can be computed by forming some vectors and using linear algebra. First, the definition of conditional probability gives

$$P_{S(l)S(l-m)}(i,j) = P_{S(l)|S(l-m)}(i|j)P_{S(l-m)}(j) \tag{17.24}$$

Clearly from Eq. (17.24) the probabilities $P_{S(l-m)}(j)$ and $P_{S(l)|S(l-m)}(i|j)$ will be important quantities to evaluate the spectrum of the OMWM. To have a concise notation, define a vector (size $1 \times N_E$) of the edge probabilities at time l

$$\vec{P}_{S(l)} = [P_{S(l)}(1)\ P_{S(l)}(2)\cdots P_{S(l)}(N_E)] \tag{17.25}$$

TABLE 17.3 The edge enumeration for the AMI trellis code

$s(l)$	$\sigma(l)$	$\sigma(l+1)$	$\tilde{D}_z(l)$
1	1	1	0
2	2	1	$-\sqrt{2}$
3	1	2	$\sqrt{2}$
4	2	2	0

A one-step recursion of this probability can be obtained by using total probability and Eq. (17.24)

$$P_{S(l)}(i) = \sum_{j=1}^{N_E} P_{S(l)S(l-1)}(i, j) = \sum_{j=1}^{N_E} P_{S(l)|S(l_1)}(i|j)P_{S(l-1)}(j) \tag{17.26}$$

This summation can be represented by a matrix equation

$$\vec{P}_s(l) = \vec{P}_s(l-1)\mathbf{S}_T \tag{17.27}$$

where $\mathbf{S}_T$ is often denoted a state transition matrix [Gal01] where

$$[\mathbf{S}_T]_{ij} = P_{S(l)|S(l-1)}(j|i) \tag{17.28}$$

The state transition matrix of an OMWM is easily identified by examining $g_1(I(k), \sigma(k))$. Examining $\vec{P}_s(l+1)$ it is clear that

$$\vec{P}_s(l+1) = \vec{P}_s(l)\mathbf{S}_T = \vec{P}_s(l-1)\mathbf{S}_T^2 \tag{17.29}$$

and by induction for any nonnegative m

$$\vec{P}_s(l) = \vec{P}_s(l-m)\mathbf{S}_T^m \tag{17.30}$$

Examining Eq. (17.28) it is clear that

$$\left[\mathbf{S}_T^m\right]_{ij} = P_{S(l)|S(l-m)}(j|i) \qquad [\mathbf{S}_E(m)]_{i,j} = \left[\mathbf{S}_T^m\right]_{ij} P_{S(l-m)}(j) \tag{17.31}$$

Consequently, the characterization necessary for the PMF in Eq. (17.24) is given by Eq. (17.31) and only a form for $P_{S(l-m)}(j)$ is needed for a complete evaluation of $R_{\tilde{D}}(m)$.

To simplify the analysis the discrete random process, $S(l)$, will be assumed to be stationary. Stationary means that the value of $\vec{P}_{S(l)}$ will be constant over time. This is clearly an approximation as the OMWM starts in a known state and terminates into a known state but for a vast majority of the frame the stationary distribution is valid. Most often the stationary probability of each edge is equally likely, i.e.,

$$\vec{P}_{S(l)} = \left[\frac{1}{N_E} \quad \frac{1}{N_E} \cdots \frac{1}{N_E}\right] \tag{17.32}$$

but in the cases that an unusual chain is encountered the theory of Markov chains [Gal01] can be used to solve for the stationary probability distribution of the edges. This equally likely and stationary assumption leads to the compact expression of

$$\mathbf{S}_E(m) = \frac{\mathbf{S}_T^m}{N_E} \qquad R_{\tilde{D}}(m) = \vec{d}_s \frac{\mathbf{S}_T^m}{N_E} \vec{d}_s^H \quad \forall m \geq 0 \tag{17.33}$$

For $m < 0$ the desired correlation function is evaluated by noting that

$$R_{\tilde{D}}(-m) = (R_{\tilde{D}}(m))^* \tag{17.34}$$

Since

$$S_{\tilde{D}}(e^{j2\pi f}) = \sum_{m=-\infty}^{\infty} R_{\tilde{D}}(m) \exp[-j2\pi f m] \tag{17.35}$$

a very simple method to evaluate the spectrum of an OMWM is available.

EXAMPLE 17.2

Consider again the AMI modulation. The edge state transition matrix is

$$\mathbf{S}_T = \begin{bmatrix} 0.5 & 0 & 0.5 & 0 \\ 0.5 & 0 & 0.5 & 0 \\ 0 & 0.5 & 0 & 0.5 \\ 0 & 0.5 & 0 & 0.5 \end{bmatrix} \tag{17.36}$$

and using a uniform edge probability, $P_{S(l-m)}(i) = 1/4,\ i = 1, 4$, gives

$$\mathbf{S}_E(1) = \begin{bmatrix} 0.125 & 0 & 0.125 & 0 \\ 0.125 & 0 & 0.125 & 0 \\ 0 & 0.125 & 0 & 0.125 \\ 0 & 0.125 & 0 & 0.125 \end{bmatrix}. \tag{17.37}$$

Note that, for example,

$$R_{\tilde{D}}(1) = [0 \quad \sqrt{2} \quad -\sqrt{2} \quad 0] \begin{bmatrix} 0.125 & 0 & 0.125 & 0 \\ 0.125 & 0 & 0.125 & 0 \\ 0 & 0.125 & 0 & 0.125 \\ 0 & 0.125 & 0 & 0.125 \end{bmatrix} \begin{bmatrix} 0 \\ \sqrt{2} \\ -\sqrt{2} \\ 0 \end{bmatrix}$$

$$= -0.5. \tag{17.38}$$

Evaluation the correlation function for all value of m results in a spectrum of

$$S_{\tilde{D}}(e^{j2\pi f}) = -0.5e^{-j2\pi f} + 1 - 0.5e^{j2\pi f} = 1 - \cos(2\pi f) \tag{17.39}$$

Equation (17.39) shows that AMI used as an OMWM in a stream modulation will always produce a notch at DC. This condition is evident in Figure 17.8 which contains a plot of $G_u(f)$, $S_{\tilde{D}}(e^{j2\pi fT})$ and the resulting $D_{x_z}(f)$. Note the close match between the plots in Figure 17.8 and Figure 17.7 where a short frame was considered ($K_b = 4$). This notch at DC characteristic is advantageous in telecommunications systems as it prevents a large DC current from being driven over the coaxial or twisted pair cables prevalent in the telecommunications network.

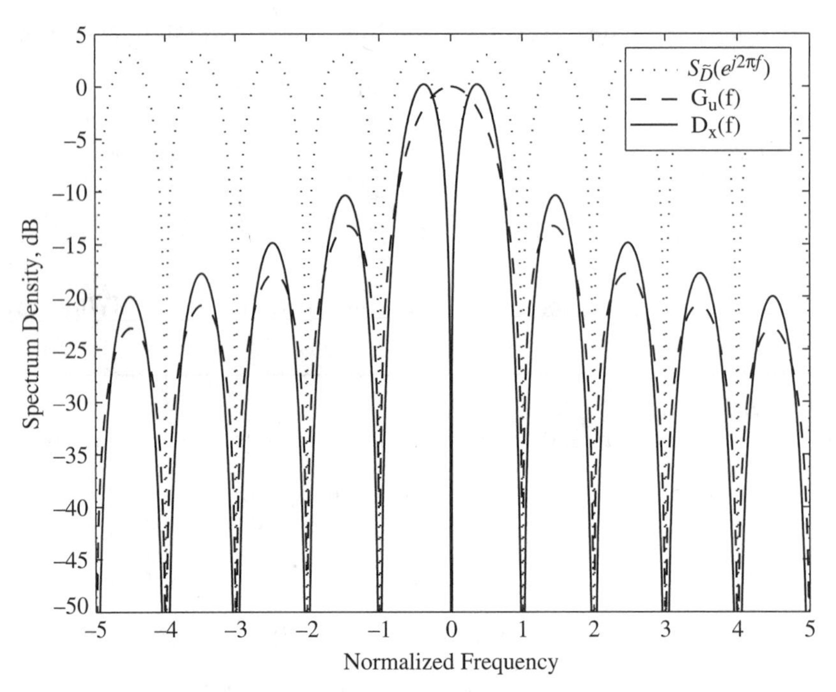

Figure 17.8 The important spectra for AMI modulation.

17.4 Varying Transmission Rates with OMWM

The OMWM that has been presented so far has limited utility as it only allows for a rate of one information bit per modulated symbol. This is usually denoted 1 bit per channel utilization by communications engineers. OMWM of this type only allows a communication engineer to design a modulation that would move horizontally on the rate performance curve of Figure 17.5. In certain situations it might be desired to lower the bit rate and achieve higher fidelity at a lower E_b/N_0 (e.g., deep space communications). Alternately it might be useful to achieve more than 1 bit per channel utilization. This is often the case when bandwidth is a scarce resource and the signal to noise ratio is relatively high (e.g., cable modems). This section will show how to generalize the OMWM to achieve these varying information transmission rates.

The key to this generalization is to enable multiple bit inputs and multiple symbol outputs in the OMWM. The general orthogonal modulation with memory, as considered in this section, consists of a finite state machine operating at an integer fraction of the symbol rate, $1/N_m T$. The K_b bits to be transmitted are broken up into blocks of K_m in length (a total of $N_b = K_b/K_m$ blocks per frame). At each symbol time a new set of K_m bits, $\vec{I}(m)$, is input into a finite state machine and this produces a new constellation label, $\vec{J}(m)$, and a new modulation state, $\sigma(m+1)$. The vector constellation label (length N_m) at the finite state machine output is used as an input to a modulator that produces

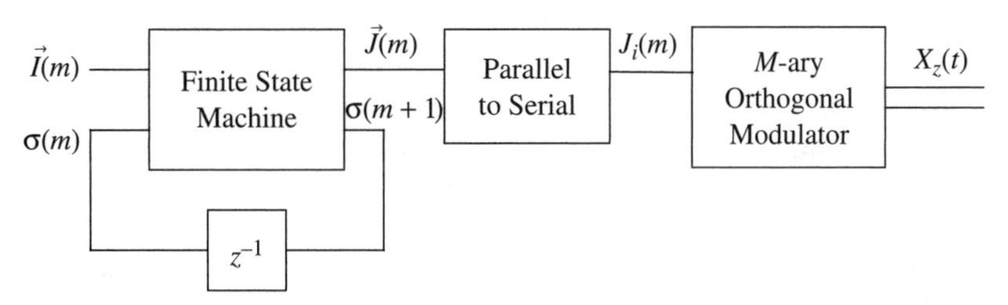

Figure 17.9 The block diagram for a general modulation with memory.

an N_m symbol block of M_s-ary modulation symbols. The output modulation symbols are

$$\tilde{D}_z((m-1)N_m + i) = a(J_i(m)) \tag{17.40}$$

where $a(\bullet)$ is the constellation mapping and $J_i(m)$ is the ith component of $\vec{J}(m)$. The output of the orthogonal modulation has the form

$$X_z(t) = \sum_{l=1} \tilde{D}_z(l)s_l(t) \tag{17.41}$$

Again to keep a consistent normalization the mapping, $\tilde{D}_z(l) = a(J_i(m))$, is selected such that $E[|\tilde{D}_z(l)|^2] = R$. A total of N_b trellis transitions are needed to communicate the K_b bits. Figure 17.9 shows the block diagram for a general OMWM. N_s again denotes the number of states in the modulation and the nonlinear equations governing the updates are

$$\sigma(m+1) = g_1(\sigma(m), \vec{I}(m)) \tag{17.42}$$

$$\vec{J}(m) = g_2(\sigma(m), \vec{I}(m)) \tag{17.43}$$

Note $\vec{J}(m) = 0, \ldots, M_s^{N_m} - 1$ where each component of $\vec{J}(m)$ only takes values $J_i(m) = 0, \ldots, M_s - 1$, $i = 1, \ldots, N_m$. The constellation label at time m and in position i will generate the modulation symbol at time $(m-1)N_m + i$ for the orthogonal modulation. In general, it is again usually desirable to have ν_c extra symbols transmitted to return the modulation to a common final state at the end of the transmission frame. The total length of the frame for the OMWM is still denoted N_f, hence, $N_f = N_b N_m + \nu_c$, where ν_c is a code dependent constant. Due to termination the effective rate is $R_{eff} = \frac{K_b}{N_b N_m + \nu_c} = \frac{K_b K_m}{K_b N_m + \nu_c}$ but if a large number of bits are transmitted then the rate becomes approximately $R = K_m/N_m$. If a rate less than one is desired then a communication engineer would choose $N_m > K_m$. If a higher transmission rate is desired then the communication engineer would choose $N_m < K_m$.

17.4.1 Spectrum of Stream Modulations with $N_m > 1$

When $N_m = 1$ the spectral characteristics of stream modulation can be computed as in Section 17.3, but if $N_m > 1$ the techniques of Section 17.3 need to be generalized. If K_b is small all the waveforms can be detailed out and the average energy spectrum can be easily computed in exactly the same way as previously detailed. This section will concentrate on the evaluation of $D_{x_z}(f)$ for long frames. Recall the $D_{x_z}(f)$ is characterized as

$$K_b D_{x_z}(f) = E_b |U(f)|^2 \sum_{l_1=1}^{N_f} \sum_{l_2=1}^{N_f} E[\tilde{D}_z(l_1)\tilde{D}_z^*(l_2)] \exp[-j2\pi f(l_1 - l_2)T] \quad (17.44)$$

in the case of K_b large. As was done for $N_m = 1$, the regular trellis structure of the modulation can be exploited in computing this spectrum. Again it is assumed that the information bits are a white sequence. Since the trellis transition is driven by a white process, the states or the edges can accurately be modeled as a stationary process for large frames. Since the modulation symbols are produced N_m at a time each trellis transition, $\tilde{D}_z(l)$ is a cyclostationary process [SW02]. Cyclostationary random process have statistics that are periodic. The contrast between stationary, cyclostationary, and nonstationary processes is best illustrated by examining the parameterization of the second order statistics contained in the correlation function. For stationary random processes the correlation function has the form

$$R_N(t_1, t_2) = g(t_1 - t_2) \quad (17.45)$$

so that the second order moments are only a function of the time difference and not the absolute time. For nonstationary processes there is no simplifying structure, i.e.,

$$R_N(t_1, t_2) = g(t_1, t_2) \quad (17.46)$$

Cyclostationary random processes have a correlation function that is periodic in time, i.e.,

$$R_N(t_1, t_2) = g(t_1, t_1 - t_2) \qquad \text{and} \qquad R_N(t_1, t_2) = R_N(t_1 + T, t_2 + T) \quad (17.47)$$

where T is the period of the process. Since $\tilde{D}_z(l)$ derived from a trellis structure that updates every N_m symbols it is apparent that this process will be a discrete cyclostationary random process with period N_m and have

$$E[\tilde{D}_z(l_1)\tilde{D}_z^*(l_2)] = R_{\tilde{D}}(l_1 - l_2, l) \qquad l = 1, \ldots, N_m \quad (17.48)$$

This cyclostationarry characteristic must be accomodated in computing the spectral characteristics.

The modeling of the modulation symbols as a cyclostationary time series enables a simplified expression for the average energy spectrum per bit. The double sum in Eq. (17.44) can be rearranged in a similar way to that of Section 17.3.2 to give

$$K_b D_{x_z}(f) = E_b |U(f)|^2 \sum_{n=-N_f+1}^{N_f-1} \sum_{l=1}^{N_m} (N_b - |n| + g(n,l)) R_{\tilde{D}}(n,l) \exp[-j2\pi f T n] \quad (17.49)$$

Again looking at the limit of a long frame size gives

$$\lim_{K_b \to \infty} D_{x_z}(f) = K_m E_b |U(f)|^2 \sum_{n=-\infty}^{\infty} \sum_{l=1}^{N_m} R_{\tilde{D}}(n,l) \exp[-j2\pi f T\, n] \quad (17.50)$$

where

$$R_{\tilde{D}}(n,l) = E[\tilde{D}_z(mN_m + l)\tilde{D}_z^*(mN_m + l - n)] \quad (17.51)$$

$R_{\tilde{D}}(n,l)$ is the correlation function between the lth symbol of a trellis transition and a symbol n symbols away in time. Note that $R_{\tilde{D}}(n,l)$ is a function of l because the number of trellis transitions that occur in n symbol times is a function of l.

EXAMPLE 17.3

If $N_m = 2$ then when $n = 1$ there can be one trellis transition, e.g.,

$$R_{\tilde{D}}(1,1) = E[\tilde{D}_z(mN_m + 1)\tilde{D}_z^*((m-1)N_m + 2)] \quad (17.52)$$

or zero trellis transitions, e.g.,

$$R_{\tilde{D}}(1,2) = E[\tilde{D}_z(mN_m + 2)\tilde{D}_z^*(mN_m + 1)] \quad (17.53)$$

It is important to realize that the function

$$\sum_{l=1}^{N_m} R_{\tilde{D}}(n,l) = \bar{R}_{\tilde{D}}(n) \quad (17.54)$$

now is analogous to the $R_{\tilde{D}}(n)$ of Section 17.3.2. In the literature $\bar{R}_{\tilde{D}}(n)$ is often denoted the time average correlation function of the cyclostationary process. Once this function is identified then the desired average energy spectrum can be computed as

$$\lim_{K_b \to \infty} D_{x_z}(f) = K_m E_b |U(f)|^2 \bar{S}_{\tilde{D}}(e^{j2\pi f T}) \quad (17.55)$$

where $\bar{S}_{\tilde{D}}(e^{j2\pi f T})$ is the discrete Fourier transform of $\bar{R}_{\tilde{D}}(n)$. It should again be noted that $\bar{S}_{\tilde{D}}(e^{j2\pi f T})$ is a periodic function of f with a period of $1/T$. The transmitted energy spectrum per bit for stream modulation is again seen to be a product of the pulse shape energy spectrum and the effective power spectral density of the discrete time modulation sequence. The form of the OMWM will determine the power spectrum of the discrete modulation sequence.

To compute $\bar{S}_{\tilde{D}}(e^{j2\pi f})$ the correlation function, $R_{\tilde{D}}(n,l)$, needs to be identified using the trellis structure of the OMWM. The correlation function is given by

$$R_{\tilde{D}}(n,l) = \sum_{d_i \in \Omega_{\tilde{D}}} \sum_{d_j \in \Omega_{\tilde{D}}} d_i d_j^* P_{\tilde{D}_z((m-1)N_m+l)\tilde{D}_z((m-1)N_m+l-n)}(d_i, d_j) \tag{17.56}$$

The key point in finding the correlation function is identifying an algorithm to compute the joint PMF of the modulation symbols at various time offsets from the trellis representation of the modulation. Recall the modulation symbol at any point in time, $\tilde{D}_z((m-1)N_m + l)$, is determined entirely by specifying the edge of the trellis, $S(m)$. The edges of the trellis can be enumerated as in Section 17.3.2 and the following property still holds:

Property 17.2 There is a functional mapping from $S(m)$ to $\tilde{D}_z((m-1)N_m + l)$.

To reflect this functional mapping a vector notation to represent the data symbols is

$$\vec{d}_s(l) = [d_s(1,l) \ldots d_s(N_m, l)] \tag{17.57}$$

where $d_s(i,l)$ is the value of $\tilde{D}_z((m-1)N_m + l)$ takes when $S(m) = i$. Again the needed joint edge probability, $P_{S(l)S(l-m)}(i,j)$, can be computed using the techniques introduced in Section 17.3.2, i.e.,

$$P_{S(l)S(l-m)}(i,j) = [\mathbf{S}_E(m)]_{i,j} = \left[\frac{\mathbf{S}_T^m}{N_E}\right]_{ij} \qquad m \geq 0 \tag{17.58}$$

Since this joint probability gives the required symbol distribution the desired correlation function can be computed with

$$R_{\tilde{D}}(n,l) = \sum_{i \in \Omega_S} \sum_{j \in \Omega_S} d_i(n,l) d_j^*(n,l) P_{S(m)S(m-m_t(n,l))}(i,j) \tag{17.59}$$

where $m_t(n,l)$ is the number of trellis transitions that will occur between time $k = mN_m + l - n$ and $k = mN_m + l$. Denoting $k = mN_m + l - n = (m - m_t(n,l))N_m + l_t(n,l)$ the double sum in Eq. (17.59) is simplified to be

$$R_{\tilde{D}}(n,l) = \vec{D}_s(l)\mathbf{S}_E(m_t(n,l))\vec{D}_s(l_t(n,l))^H = \vec{D}_s(l)\frac{\mathbf{S}_T^{m_t(n,l)}}{N_E}\vec{D}_s(l_t(n,l))^H \qquad n \geq 0 \tag{17.60}$$

The major difference in the computation of $R_{\tilde{D}}(n,l)$ compared to the $N_m = 1$ case is need to identify the number of trellis transitions $(m_t(n,l))$ as a function of l and n and the need to find which "phase" of the N_m symbols produced by the modulation that should be included in the correlation computation $(l_t(n,l))$.

EXAMPLE 17.4

From the previous example it is clear for $N_m = 2$ that

$$m_t(1,1) = 1 \qquad l_t(1,1) = 2 \tag{17.61}$$

$$m_t(1,2) = 0 \qquad l_t(1,2) = 1 \tag{17.62}$$

In words, if one symbol offset is considered in calculating the correlation function there is one transition trellis transition for $l = 1$ (the first symbol of a trellis transition) and the associated signal for the correlation calculation is the second symbol of the previous trellis update. Likewise if the second symbol is considered then no transition need be considered and the associated signal for the correlation calculation is the first symbol of the current trellis update. These computations are no more difficult than the $N_m = 1$ but are a bit more tedious.

Once the details of the computation of $\bar{R}(n)$ are understood then the mechanics of computing the average energy spectrum is straightforward. Specifically once the number of trellis transitions are identified for a given (n, l) then $R_{\tilde{D}}(n, l)$ can be computed using similar techniques as to Section 17.3.2. Once $R_{\tilde{D}}(n, l)$ is computed then $\bar{R}_{\tilde{D}}(n)$ can be evaluated. A discrete Fourier transform leads to the complete characterization of the spectrum of the OMWM. A detailed example will be considered in the sequel.

17.4.2 Example $R < 1$: Convolutional Codes

Consider an OMWM consisting of an eight-state convolutional code [LC04] ($N_s = 8$) using BPSK modulation ($M_s = 2$) that has $R = 1/2$ ($K_s = 1$ and $N_m = 2$). The modulation updates are given in Table 17.4. Since this code only sends 1 bit every two symbols the spectral efficiency of the modulation ($\eta_B \approx 1/2$ for this OMWM) will be reduced compared to BPSK, for example. Note that the modulation is normalized to have an average energy of $R = 1/2$

TABLE 17.4 The finite state machine description of an example convolutional code

$\sigma(m+1) = g_1(I(m), \sigma(m))$			$\vec{J}(m) = g_2(I(m), \sigma(m))$		
	$I(m)$			$I(m)$	
State, $\sigma(m)$	0	1	State, $\sigma(m)$	0	1
1	1	2	1	[0 0]	[1 1]
2	3	4	2	[0 1]	[1 0]
3	5	6	3	[1 1]	[0 0]
4	7	8	4	[1 0]	[0 1]
5	1	2	5	[1 1]	[0 0]
6	3	4	6	[1 0]	[0 1]
7	5	6	7	[0 0]	[1 1]
8	7	8	8	[0 1]	[1 0]

$I(k) = 0$	$I(k) = 1$
[0 0]	[1 1]
[0 1]	[1 0]
[1 1]	[0 0]
[1 0]	[0 1]
[1 1]	[0 0]
[1 0]	[0 1]
[0 0]	[1 1]
[0 1]	[0 1]

$\sigma(l)$ $\sigma(l+1)$

Figure 17.10 The trellis diagram of the example convolutional code.

per transmitted symbol, i.e.,

$$\tilde{D}_z((m-1)N_m + i) = \frac{1}{\sqrt{2}}(-1)^{J_i(m)} \tag{17.63}$$

The trellis for this modulation is shown in Figure 17.10. There are a total of $2N_s = 16$ edges in this example. The edge labels indicate the input that causes the transition and the modulation symbol that is output for that input and state. This particular code requires at most three transitions to allow the modulation to transition from any state to any other state so this code has $\nu_c = 6$. Again N_f trellis sections can be combined together to get a complete description of an OMWM. For instance for $K_b = 4$, Figure 17.11 shows a trellis description for a modulation that has $\sigma(1) = 1$ and returns to $\sigma(8) = 1$. One can verify that there are $2^{K_b} = 16$ paths through this trellis. To understand the fidelity of demodulation of this particular OMWM squared Euclidean distance spectrum should be computed. The entire distance spectrum of this modulation

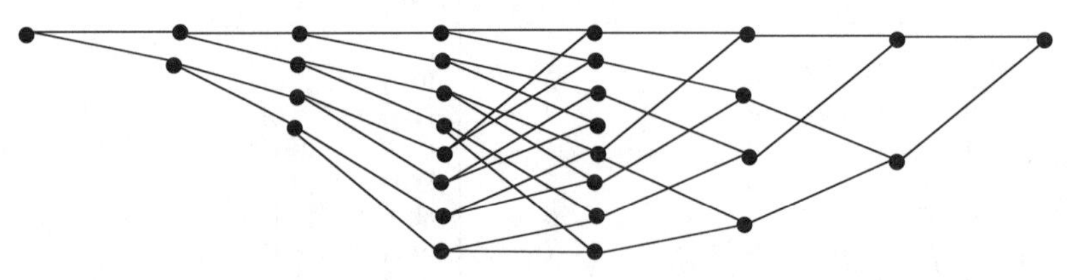

Figure 17.11 The trellis diagram of the example convolutional code. $K_b = 4$.

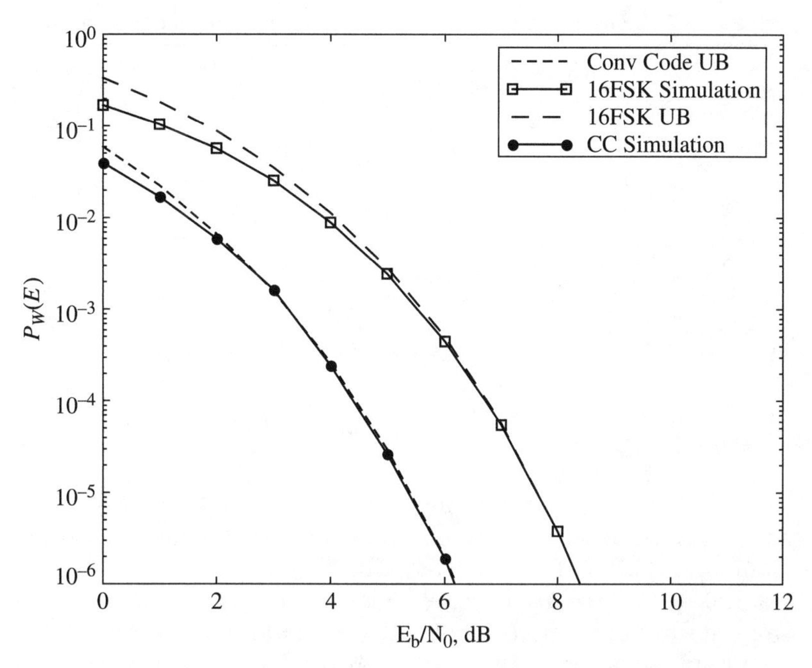

Figure 17.12 The union bound and frame error rate of $K_b = 4$ memoryless orthogonally modulated bits using 16FSK and the example convolutional code OMWM.

is explored in the homework problems (see Problem 17.16) and the union bound to the probability of word error is plotted in Figure 17.12. The important thing to notice is that the fidelity of demodulation is markedly improved compared to a modulation that does not use memory. For instance, if we compare fidelity of demodulation at $P_W(E) = 10^{-5}$ then this example convolutional code is about 2 dB better that the 16FSK, which is the memoryless modulation we considered in Chapter 14 with roughly the same spectral efficiency.

Forward error correction codes like this example convolutional code are used in many kinds of communication systems (wireless, wireline, fiber optic network, etc.) to provide significant gains in the fidelity of demodulation at rates less than 1 bit per orthogonal dimension. Forward error correction codes like turbo codes, low density parity check codes, conventional convolutional codes and Reed-Solomon codes have allowed modern communication systems to achieve performance very close to Shannon's predicted bound.

17.4.3 Example $R > 1$: Trellis Codes

Consider an OMWM consisting of a four-state trellis coded modulation [Ung82] ($N_s = 4$) using 8PSK modulation ($M_s = 8$) that has $R = 2$ ($K_m = 2$ and $N_m = 1$). The modulation updates are given in Table 17.5. Note that the modulation is

TABLE 17.5 The finite state machine description of an example trellis coded modulation

$\sigma(m+1) = g_1(\vec{I}(m), \sigma(m))$					$J(m) = g_2(\vec{I}(m), \sigma(m))$				
	$\vec{I}(m)$					$\vec{I}(m)$			
State, $\sigma(m)$	0 0	1 0	0 1	1 1	State, $\sigma(m)$	0 0	1 0	0 1	1 1
1	1	1	2	2	1	0	4	2	6
2	3	3	4	4	2	1	5	3	7
3	1	1	2	2	3	2	6	0	4
4	3	3	4	4	4	3	7	1	5

normalized to have an average energy of $R = 2$ per transmitted symbol, i.e.,

$$\tilde{D}_z(m) = \sqrt{2} \exp\left[\frac{j\pi(2J(m)+1)}{8}\right] \tag{17.64}$$

The trellis for this modulation is shown in Figure 17.13. Again in this trellis the edge labels to the left of the trellis correspond to the labels associated with edges from top to bottom for each state. There are a total of $2^{K_m} N_s = 16$ edges in this example and there are parallel edges in the trellis. Parallel edges connect the same states but are associated with different input information bits and output modulations symbols. This particular code requires at most two transitions to allow the modulation to transition from any state to any other state so this code has $\nu_c = 2$.

Again N_f trellis sections can be combined together to get a complete description of an OMWM. For instance for $K_b = 4$, Figure 17.14 shows a trellis description for a modulation that has $\sigma(1) = 1$ and returns to $\sigma(5) = 1$. One can verify that there are 2^{K_b} paths through this trellis. It is interesting to note that due to the parallel transitions there is more than one way to terminate the trellis for each of the states (4 possibilities in this example). These extra paths during termination would allow the transmission of $K_b = 6$ bits in the

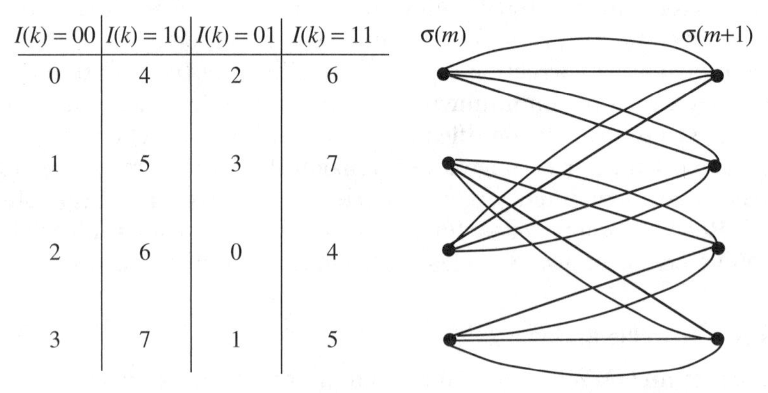

Figure 17.13 The trellis diagram of the example trellis coded modulation.

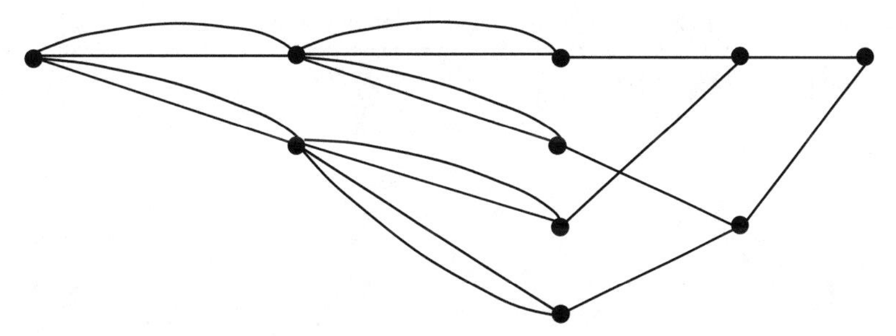

Figure 17.14 The trellis diagram of the example trellis coded modualtion. $K_b = 4$.

same time ($N_f = 4$) as $K_b = 4$ bits without much loss in fidelity of demodulation. To understand the fidelity of demodulation of this particular OMWM, the squared Euclidean distance spectrum should be computed. The entire distance spectrum of this modulation is explored in the homework problems (see Problem 17.17) and the union bound to the probability of word error is plotted in Figure 17.15. The important thing to notice is that again the fidelity of demodulation is markedly improved compared to a modulation that does not use memory. For instance, if we compare performance at $P_W(E) = 10^{-5}$ then this

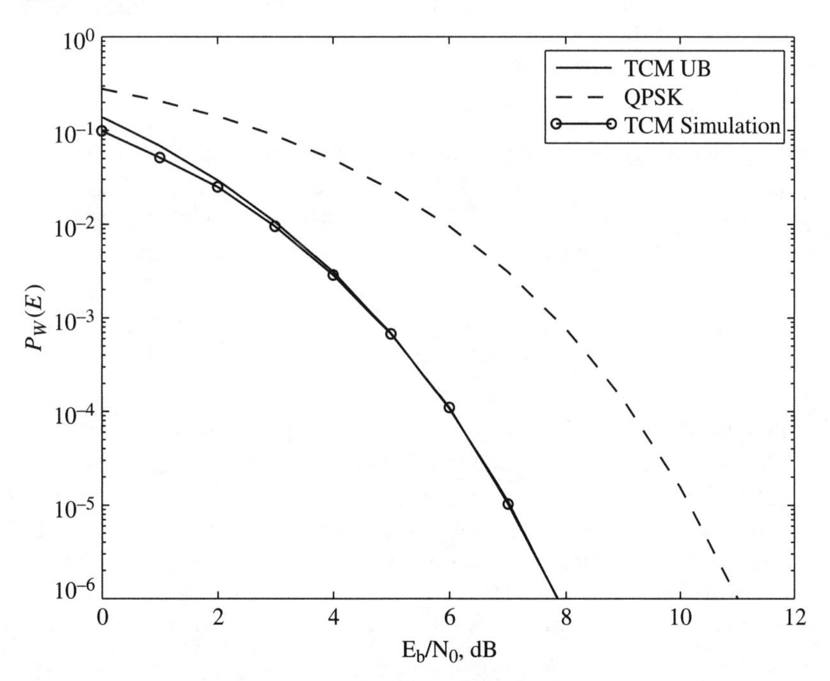

Figure 17.15 The union bound and simulated fidelity of demodulation of the example $R = 2$ OMWM and frame error rate of QPSK. $K_b = 4$.

TABLE 17.6 The finite state machine description of the Miller code

$\sigma(m+1) = g_1(I(m), \sigma(m))$		
	$I(m)$	
State, $\sigma(m)$	0	1
1	2	3
2	1	4
3	2	4
4	1	3

$\vec{J}(m) = g_2(I(m), \sigma(m))$		
	$I(m)$	
State, $\sigma(m)$	0	1
1	[0 1]	[0 0]
2	[1 0]	[1 1]
3	[0 1]	[1 1]
4	[1 0]	[0 0]

example trellis coded modulation is about 3 dB better than QPSK, which is the memoryless modulation we considered in Chapter 14 with the same spectral efficiency.

Forward error correction codes like this example trellis code can also be used in all kinds of communication systems to provide significant gains in fidelity of demodulation at rates greater than or equal to 1 bit per orthogonal dimension. A similar dizzying array of coding options exist in this case that use turbo codes, low density parity check codes, conventional convolutional codes and Reed-Solomon codes combined with higher order constellations and allow modern communication systems to achieve performance very close to Shannon's predicted bound. In general, modern communication systems are not quite as close to Shannon's bounds at high rates so there is some room for further research to fill in the remaining gap.

17.4.4 Example for Spectral Shaping: Miller Code

Consider an OMWM consisting of a four-state coded modulation [Mil63] ($N_s = 4$) using BPSK modulation ($M_s = 2$) that has $R = 1/2$ ($K_m = 1$ and $N_m = 2$) known as the Miller code. The modulation updates are given in Table 17.6. Again the modulation is normalized to have an average energy of $R = 1/2$ per transmitted symbol, i.e.,

$$\tilde{D}_z((m-1)N_m + i) = \frac{1}{\sqrt{2}}(-1)^{J_i(m)} \tag{17.65}$$

The trellis for this modulation is shown in Figure 17.16. There are a total of $2N_s = 8$ edges in this example. The edge labels indicate the input that causes the transition and the modulation symbol that is output for that input and state. A Miller code is typically not terminated as it is not used to achieve a desired fidelity of demodulation (squared Euclidean distance) and it is not possible to return to a common state in an identical number of transitions from each state. To understand the fidelity of demodulation of this particular OMWM, the squared Euclidean distance spectrum again should be computed and this distance spectrum is explored in the homework problems (see Problem 17.19).

Since the Miller code was not conceived to improve the distance properties but to change the time and spectral characteristics of an orthogonal modulation, it

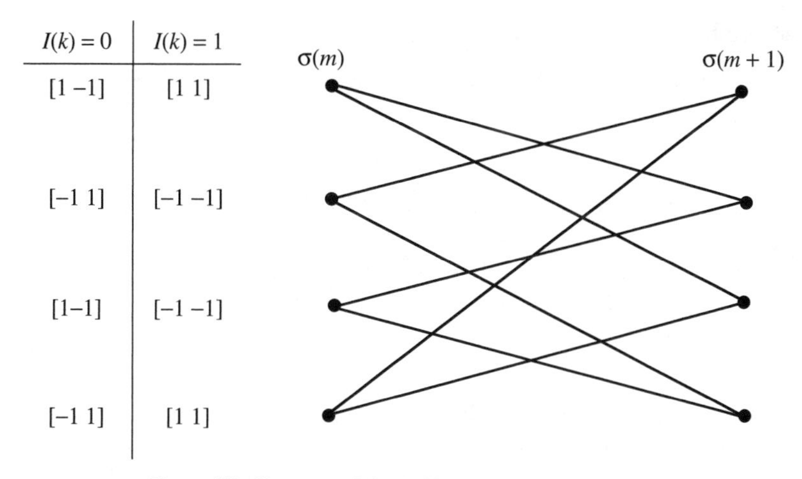

Figure 17.16 The trellis diagram of the Miller code.

is of interest to explore these properties. An interesting property of the Miller code is that at most there are 4 symbols in a row that are the same polarity. Codes of this type that limit the number of symbols between a transition are often referred to as run length limited (RLL) codes. This type of code is important in magnetic recording due to the physics of magnetic recording. An important characteristic of the Miller code is that the memory of the code shapes the transmitted spectrum. This shaped spectrum is what is advantageous in magnetic recording and playback. This spectrum can be explored using the results of Section 17.4.1.

Finding the joint probability distribution of the edges is the important step in finding the transmitted spectrum. The edge enumeration of the Miller code is given in Table 17.7. Recall the edge probability update to compute the needed modulation symbol correlation is given as

$$\vec{P}_s(m) = \vec{P}_s(m-1)\mathbf{S}_T \qquad \text{and} \qquad \vec{P}_s(m) = \vec{P}_s(m-m_t)\mathbf{S}_T^{m_t} \tag{17.66}$$

TABLE 17.7 The edge enumeration for the Miller code

$S(m)$	$\sigma(m)$	$\sigma(m+1)$	$[\tilde{D}_z(2(m-1)+1)\tilde{D}_z(2(m-1)+2)]$
1	2	1	[1 0]
2	4	1	[1 0]
3	1	2	[0 1]
4	3	2	[0 1]
5	1	3	[0 0]
6	4	3	[0 0]
7	2	4	[1 1]
8	3	4	[1 1]

where by examining Table 17.7 we have

$$\mathbf{S}_T = \begin{bmatrix} 0 & 0 & 0.5 & 0.5 & 0 & 0 & 0 & 0 \\ 0 & 0 & 0 & 0 & 0 & 0 & 0.5 & 0.5 \\ 0.5 & 0.5 & 0 & 0 & 0 & 0 & 0 & 0 \\ 0 & 0 & 0 & 0 & 0.5 & 0.5 & 0 & 0 \\ 0.5 & 0.5 & 0 & 0 & 0 & 0 & 0 & 0 \\ 0 & 0 & 0 & 0 & 0 & 0 & 0.5 & 0.5 \\ 0 & 0 & 0.5 & 0.5 & 0 & 0 & 0 & 0 \\ 0 & 0 & 0 & 0 & 0.5 & 0.5 & 0 & 0 \end{bmatrix} \quad \text{and} \quad \mathbf{S}_E(m_t) = \frac{\mathbf{S}_T^{m_t}}{8} \tag{17.67}$$

This edge transition probability can be used to compute $R_{\vec{D}}(n, l)$. Recall that when $N_m = 2$ that

$$m_t(1,1) = 1 \qquad l_t(1,1) = 2 \tag{17.68}$$

$$m_t(1,2) = 0 \qquad l_t(1,2) = 1 \tag{17.69}$$

This leads to

$$R_{\vec{D}}(1,1) = \vec{D}_s(1)\mathbf{S}_E(1)\vec{D}_s(2)^H = 0.25 \tag{17.70}$$

and

$$R_{\vec{D}}(1,2) = \vec{d}_s(2)\mathbf{S}_E(0)\vec{d}_s(1)^H = 0 \tag{17.71}$$

Summing these two results gives $\bar{R}_{\vec{D}}(1) = 0.25$. The average correlation, $\bar{R}_{\vec{D}}(n)$, can be computed numerically using results like Eqs. (17.70) and (17.71) for all n and the DFT can be taken of $\bar{R}_{\vec{D}}(n)$ to give the final result for the modulation average spectrum, $\bar{S}_{\vec{D}}(e^{j2\pi fT})$. The plot of modulation symbol average power spectrum for a Miller code is shown in Figure 17.17. Having a spectrum of the modulation concentrated at higher frequencies in magnetic recording is important since lower frequencies of the modulation tend to interfere with the servo mechanism of the magnetic read/write head. The Miller code is another example of how the OMWM can be designed to achieve a desired spectral characteristic.

17.5 Conclusions

This chapter introduced the idea of orthogonal modulations with memory (OMWM). OMWM are used in practice to get better fidelity of demodulation than orthogonal modulations or to change the spectral characteristics of orthogonal modulations while still maintaining a demodulation complexity that is $O(K_b)$. The demodulation complexity of OMWM was not addressed but interested readers are referred to [BB99, Wic95, LC04, BDMS91, CAC01]. OMWM can be implemented in a wide variety of rates and complexities. This variety of rates and complexity was illustrated with a handful of simple examples. For example, Figure 17.18 is a plot of the achieved spectral efficiency for the

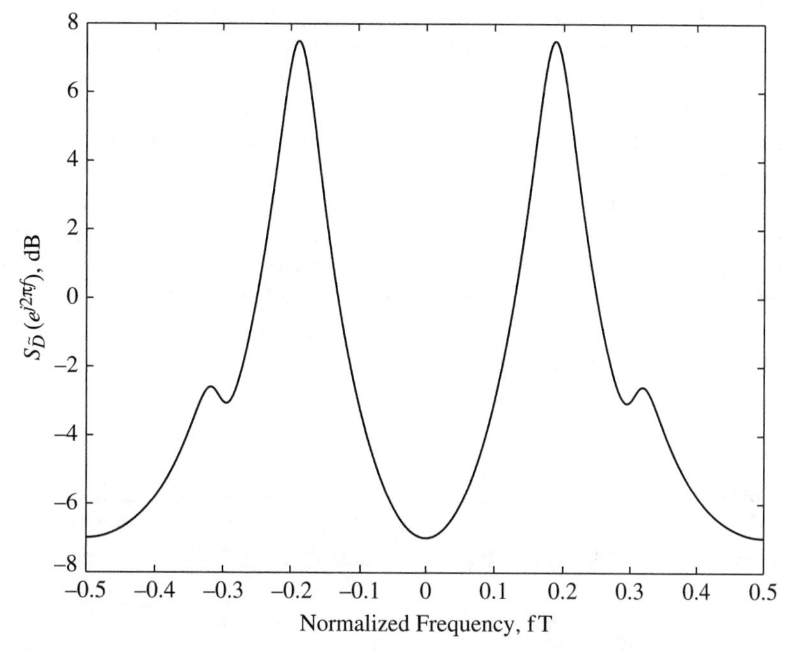

Figure 17.17 The effective power spectrum of the modulation symbols for Miller coded modulation.

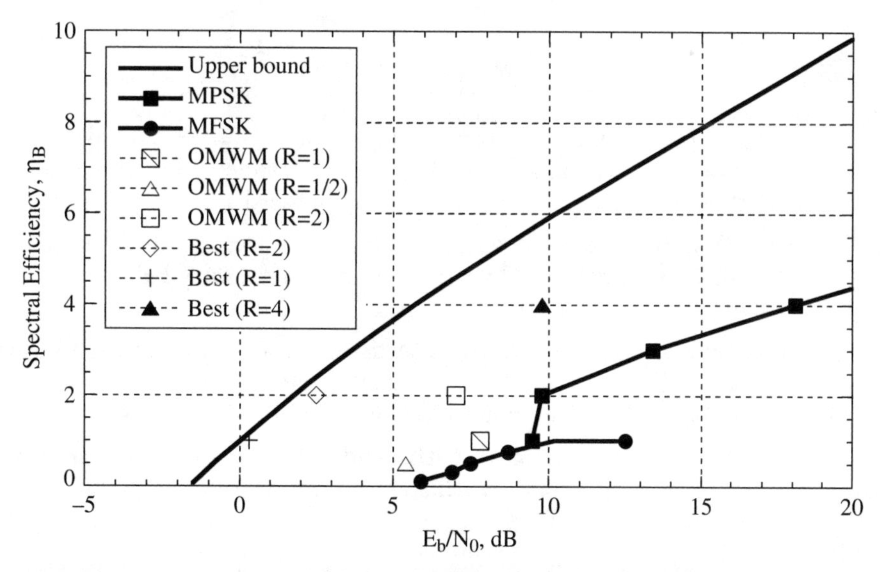

Figure 17.18 A comparison between the modulations considered in this book.

memoryless modulations and the OMWM that were considered in this book. Clearly these simple examples of OMWM have moved the achieved performance closer to that predicted by Shannon than were achieved by memoryless orthogonal modulations. These simple examples, while illustrative, do not really reflect the sophistication that is currently the state of the art in digital communications. Also included in Figure 17.18 are the operating points of several of the best OMWM that have appeared in the literature. The discussion of these modulations and the methods of demodulation requires a level of sophistication that is not appropriate for where this text is at in the development of digital communication theory, so we will not go into more details. Interested readers should take a course or read a book on modern error correcting codes [Wic95, LC04, BDMS91]. The important point is that Shannon's upperbound is one that is achievable in modern communications.

17.6 Homework Problems

Problem 17.1. Consider a $R = 1$ OMWM consisting of a four-state trellis code ($N_s = 4$) using 4PAM modulation ($M_s = 4$) as proposed by Ho, Cavers, and Varaldi [HCV93] and detailed in Section 17.2.2. The modulation is defined as

$\sigma(l+1) = g_1(I(l), \sigma(l))$

	$I(l)$	
State, $\sigma(l)$	0	1
1	1	2
2	3	4
3	1	2
4	3	4

$J(l) = g_2(I(l), \sigma(l))$

	$I(l)$	
State, $\sigma(l)$	0	1
1	0	2
2	3	1
3	2	0
4	1	3

$\tilde{D}_z(l) = a(J(l))$

$J(l)$	$D_z(l)$
0	$-3/\sqrt{5}$
1	$-1/\sqrt{5}$
2	$1/\sqrt{5}$
3	$3/\sqrt{5}$

(a) For $K_b = 4$ and the termination back to $\sigma(7) = 1$ as discussed in the text ($N_f = 6$) find the squared Euclidean distance spectrum of the code and plot the union bound to the probability of word error.

(b) Draw two transmitted paths and the corresponding error paths through the trellis associated with the minimum squared Euclidean distance error event.

(c) For $K_b = 4$ and no trellis termination (i.e., no return to a common final state, $N_f = K_b$) find the squared Euclidean distance spectrum of the code and plot the union bound to the probability of word error.

(d) Draw two transmitted paths and the corresponding error paths through the trellis associated with the minimum squared Euclidean distance error event.

Problem 17.2. The following is a commonly used $R = 1$ OMWM with four states ($N_s = 4$) and QPSK modulation ($M_s = 4$):

$\sigma(l+1) = g_1(I(l), \sigma(l))$

	$I(l)$	
State, $\sigma(l)$	0	1
1	1	2
2	3	4
3	1	2
4	3	4

$\tilde{D}_z(l) = g_2(I(l), \sigma(l))$

	$I(l)$	
State, $\sigma(l)$	0	1
1	d_0	d_2
2	d_1	d_3
3	d_2	d_0
4	d_3	d_1

where $d_0 = 1, d_1 = j, d_2 = -1$, and $d_3 = -j$.

(a) Starting in $\sigma(1) = 1$ with $\vec{I} = [1\ 1\ 0\ 1]$ give the values of $I(5)$ and $I(6)$ such that $\sigma(7) = 1$ (for $N_f = 6$). Draw the trellis and show the path through the trellis that corresponds to this transmitted word.

(b) Find the squared Euclidean distance spectrum of this code for $K_b = 4$ with termination to $\sigma(7) = 1$.

(c) With a transmitted waveform produced by $\vec{I} = [1\ 1\ 0\ 1]$ enumerate out the three unique smallest nonzero squared Euclidean distances to the transmitted waveform in the conditional union bound and show a corresponding path through the trellis.

(d) Is this code better or worse in terms of resulting performance than the HCV code used as an example in the text?

Problem 17.3. Consider the $R = 1$ OMWM given by

$\sigma(l+1) = g_1(I(l), \sigma(l))$

	$I(l)$	
State, $\sigma(l)$	0	1
1	1	2
2	3	4
3	1	2
4	3	4

$\tilde{D}_z(l) = g_2(I(l), \sigma(l))$

	$I(l)$	
State, $\sigma(l)$	0	1
1	d_0	d_2
2	d_1	d_3
3	d_2	d_0
4	d_3	d_1

Design the symbol mapping $\tilde{D}_z(l)$ (i.e., give values of d_0, d_1, d_2, d_3) to give better fidelity of demodulation than the code presented in the previous problem while leaving all constellation points on the unit circle. Assume the code is terminated back to state 1. One example is sufficient to demonstrate you understand the concepts and there is no need to find the optimum mapping to get full credit.

Problem 17.4. A concept that is useful in the analysis of coded modulations is geometric uniformity [For91]. A code is said to be geometrically uniform if the conditional squared Euclidean distance spectrum for each of the possible transmitted codes words is identical. Since the conditional squared Euclidean distance spectrum of a code is identical the union bound can be computed by only considering the conditional squared Euclidean distance spectrum for one possible path in the trellis. The concept of geometric uniformity allows one to reduce the terms needed to be computed for the union bound from $(M-1)M$ to $M-1$ where again $M = 2^{K_b}$. Consider the following two trellis codes

$\sigma(l+1) = g_1(I(l), \sigma(l))$

State, $\sigma(l)$	$I(l)$ = 0	$I(l)$ = 1
1	1	2
2	3	4
3	1	2
4	3	4

$J(l) = g_2(I(l), \sigma(l))$

State, $\sigma(l)$	$I(l)$ = 0	$I(l)$ = 1
1	0	2
2	3	1
3	2	0
4	1	3

$\tilde{D}_z(l) = a(J(l))$

$J(l)$	$\tilde{D}_z^{(1)}(l)$	$\tilde{D}_z^{(2)}(l)$
0	$-3/\sqrt{5}$	1
1	$-1/\sqrt{5}$	j
2	$1/\sqrt{5}$	-1
3	$3/\sqrt{5}$	$-j$

Consider the case of $K_b = 4$ with termination back to $\sigma(7) = 1$, are either of these codes geometrically uniform?

Problem 17.5. Consider the following $R = 1$ OMWM with seven states ($N_s = 7$) and 4-PAM modulation ($M_s = 4$):

$\sigma(l+1) = g_1(I(l), \sigma(l))$

State, $\sigma(l)$	$I(l)$ = 0	$I(l)$ = 1
1	2	4
2	3	5
3	4	6
4	1	7
5	2	4
6	3	5
7	4	6

$J(l) = g_2(I(l), \sigma(l))$

State, $\sigma(l)$	$I(l)$ = 0	$I(l)$ = 1
1	1	0
2	1	0
3	1	0
4	3	0
5	3	2
6	3	2
7	3	2

$\tilde{D}_z(l) = a(J(l))$

$J(l)$	$\tilde{D}_z(l)$
0	$-3/\sqrt{5}$
1	$-1/\sqrt{5}$
2	$1/\sqrt{5}$
3	$3/\sqrt{5}$

(a) Starting in $\sigma(1) = 4$ and that we want to return to $\sigma = 4$, show that if K_b is odd then this is possible in one transition and if K_b is even then this is possible in two transitions.

(b) For $K_b = 3$ find the squared Euclidean distance spectrum and plot the union bound. Will this modulation perform better or worse than BPSK?

(c) For $K_b = 3$ when the modulation is used in linear stream modulation, show that

$$\int_{-\infty}^{\infty} x_i(t)dt = 0 \tag{17.72}$$

(d) For $K_b = 3$ when the modulation is used in linear stream modulation, find and plot $D_{x_z}(f)$? How do you see the characteristic of (c) manifested in this plot?

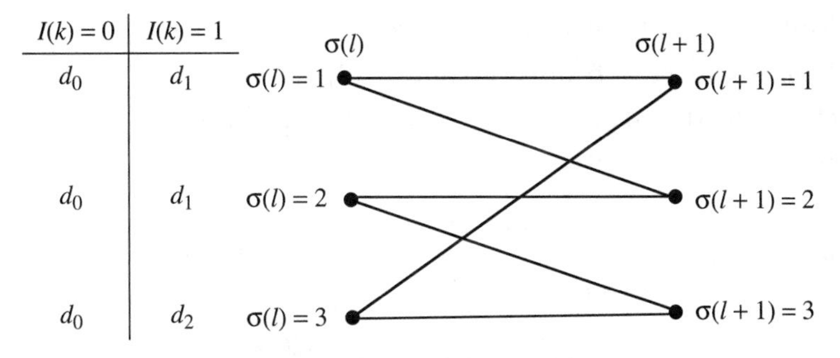

Figure 17.19 A trellis diagram for a modulation with memory.

Problem 17.6. An OMWM is defined by the trellis in Figure 17.19. Assume that $d_0 = 1$, $d_1 = \exp[j2\pi/3]$, $d_2 = \exp[-j2\pi/3]$ and $\sigma(1) = 1$.

(a) For an arbitrary $K_b = 3$ bit sequence and if the modulation has to stay as defined by the trellis how many symbols will be needed to guarantee a return to $\sigma = 1$.

(b) Will this OMWM give better fidelity of demodulation than memoryless BPSK modulation when $K_b = 3$ when terminated as in part a)? Plot the union bound.

(c) Is this OMWM geometrically uniform (see Problem 17.4 for the definition of geometrically uniform)?

(d) It is possible to define a one symbol termination (i.e., $N_f = K_b + 1$), which might not be as defined by the trellis, that would result in a minimum distance that is the same as if the modulation had stayed on the trellis and had $N_f > K_b + 1$. Give what the final symbol should be as a function of $\sigma(K_b + 1)$ to achieve this goal. Plot a union bound and compare it with the fidelity of demodulation when $N_f > K_b + 1$ and the modulation stays on the trellis.

Problem 17.7. If AMI modulation is used as an OMWM in an OFDM system, what characteristic would it produce? Would you consider AMI useful in any way as an OMWM for an OFDM system. *Hint:* AMI puts a DC notch in a stream modulation's spectrum.

Problem 17.8. Consider an OMWM consisting of a four-state trellis code ($N_s = 4$) using 4PAM modulation ($M_s = 4$) as proposed by Ho, Cavers, and Varaldi [HCV93] and detailed in Section 17.2.2. Assume this modulation is used in a stream modulation with K_b large and

$$u_r(t) = \begin{cases} \sqrt{\dfrac{1}{T}} & 0 \le t \le T \\ 0 & \text{elsewhere} \end{cases} \tag{17.73}$$

(a) Enumerate out the edges, $s(l)$, and the mappings to $\tilde{D}_z(l)$.

(b) Identify the edge transition matrix, S_T.

(c) Compute the average energy spectrum per bit, $D_{X_z}(f)$, of the resultant stream modulation.

Problem 17.9. Consider an OMWM consisting of a two-state trellis code ($N_s = 2$) using 3PAM modulation ($M_s = 3$) denoted as AMI modulation in the text. The modulation is defined as

$\sigma(l+1) = g_1(I(l), \sigma(l))$

	$I(l)$	
State, $\sigma(l)$	0	1
1	1	2
2	1	2

$J(l) = g_2(I(l), \sigma(l))$

	$I(l)$	
State, $\sigma(l)$	0	1
1	1	0
2	2	1

$\tilde{D}_z(l) = a(J(l))$

$J(l)$	$\tilde{D}_z(l)$
0	−2
1	0
2	2

(a) For $K_b = 4$ and the termination back to $\sigma(6) = 1$ as discussed in the text ($N_f = 5$) find the squared Euclidean distance spectrum of the code and plot the union bound to the probability of word error.

(b) Draw two pairs of transmitted paths and the corresponding error paths through the trellis associated with the minimum squared Euclidean distance error event.

(c) For $K_b = 4$ and no trellis termination (i.e., no return to a common final state, $N_f = K_b$) find the squared Euclidean distance spectrum of the code and plot the union bound to the probability of word error.

(d) Draw two pairs of transmitted paths and the corresponding error paths through the trellis associated with the minimum squared Euclidean distance error event.

Problem 17.10. In Problem 17.5 an OMWM is presented that, at least for small K_b, was shown to produce a notch at DC in the transmitted spectrum. Using the results of Section 17.3.2 find the average energy spectrum per bit for K_b large.

Problem 17.11. Consider an OMWM consisting of an eight-state convolutional code ($N_s = 8$) using BPSK modulation ($M_s = 2$) and detailed in Section 17.4.2. The modulation is defined in Table 17.4.

(a) For $K_b = 4$ and the termination back to $\sigma(8) = 1$ as discussed in the text ($N_f = 14$) find the squared Euclidean distance spectrum of the code and plot the union bound to the probability of word error.

(b) Draw two transmitted paths and the corresponding error paths through the trellis associated with the minimum squared Euclidean distance error event.

(c) Is this modulation geometrically uniform? (See Problem 17.4)

Problem 17.12. Consider an OMWM consisting of an eight-state convolutional code as detailed in Table 17.4 and discussed in Section 17.4.2. Assume this modulation is used in a stream modulation with K_b large and

$$u_r(t) = \begin{cases} \sqrt{\dfrac{1}{T}} & 0 \leq t \leq T \\ 0 & \text{elsewhere} \end{cases} \tag{17.74}$$

(a) Enumerate out the edges, $s(l)$, and the mappings to $\tilde{D}_z(l)$.

(b) Identify the edge transition matrix, S_T.

(c) Compute the average energy spectrum per bit, $D_{X_z}(f)$, of the resultant stream modulation.

Problem 17.13. Consider an OMWM consisting of a four-state trellis coded modulation ($N_s = 4$) using 8PSK modulation ($M_s = 8$) and detailed in Section 17.4.3. The modulation is defined in Table 17.5.

(a) For $K_b = 4$ and the termination back to $\sigma(7) = 1$ as discussed in the text ($N_f = 6$) find the squared Euclidean distance spectrum of the code and plot the union bound to the probability of word error.

(b) Draw two transmitted paths and the corresponding error paths through the trellis associated with the minimum squared Euclidean distance error event.

(c) Is this modulation geometrically uniform? (See Problem 17.4.)

Problem 17.14. When the information bits are all equally likely and independent, prove that if

$$E\left[D_z((m-1)N_m + j)|\sigma(m) = i\right] = 0 \quad j \in \{1, \ldots, N_s\}, i \in \{1, \ldots, N_s\} \tag{17.75}$$

then the transmitted energy spectrum per bit of an OMWM using stream modulation is

$$D_{x_z}(f) = E_b|U(f)|^2 \tag{17.76}$$

In other words, the spectrum is unchanged by the OMWM compared to a standard linear stream modulation [BDMS91, Big86]. Which of the example OMWM in this chapter satisfy this condition?

Problem 17.15. Consider an OMWM consisting of a four state ($N_s = 4$) OMWM using BPSK modulation ($M_s = 2$) and detailed in Section 17.4.4. The modulation is defined in Table 17.6.

(a) For $K_b = 4$ and no termination ($N_f = 8$) find the squared Euclidean distance spectrum of the code and plot the union bound to the probability of word error.

(b) Draw two transmitted paths and the corresponding error paths through the trellis associated with the minimum squared Euclidean distance error event.

(c) Is this modulation geometrically uniform? (See Problem 17.4.)

Problem 17.16. In the example that considered an $R = 2$ Ungerboeck trellis code in Section 17.4.3 it was stated that an extra 2 bits can be transmitted during the termination stages.

(a) For the example detailed in Section 17.4.3 show the $N_f = 4$ trellis where $K_b = 6$ and $\sigma(5) = 1$.

(b) Has the minimum squared Euclidean distance changed by adding these two extra bits?

(c) Compute the union bound of this new case. Plot this union bound along with the union bound of the $K_b = 4$ case. Justify the difference.

Problem 17.17. Consider an OMWM consisting of a two-state trellis code ($N_s = 2$) using QPSK modulation ($M_s = 4$) denoted as $\pi/2$-BPSK modulation. This OMWM is used in land mobile radio to control the peak-to-average-power ration (PAPR) of the transmitted signal. The modulation is defined as

$\sigma(l+1) = g_1(I(l), \sigma(l))$		
	$I(l)$	
State, $\sigma(l)$	0	1
1	2	2
2	1	1

$J(l) = g_2(I(l), \sigma(l))$		
	$I(l)$	
State, $\sigma(l)$	0	1
1	0	1
2	2	3

$\tilde{D}_z(l) = a(J(l))$	
$J(l)$	$\tilde{D}_z(l)$
0	1
1	-1
2	j
3	$-j$

(a) Compute the minimum Euclidean square distance for $K_b = 4$.

(b) If this OMWM is used in a stream modulation with a rectangular pulse shape, plot an example vector diagram for $K_b = 16$ random data bits.

(c) Compute the PAPR of this modulation.

17.7 Example Solutions

Problem 17.2.

(a) Since the code is terminated back to $\sigma = 1$ this requires $I(5) = 0$ and $I(6) = 0$. The path through the trellis is shown in Figure 17.20.

(b) Going to Matlab gives the results in Table 17.8. The plot of the performance is given in Figure 17.21.

(c) Examining the table above it is clear that the three smallest distances are $\Delta_E(1) = 10$, $\Delta_E(2) = 12$, $\Delta_E(3) = 14$. Error events corresponding to these distances are plotted in Figure 17.22.

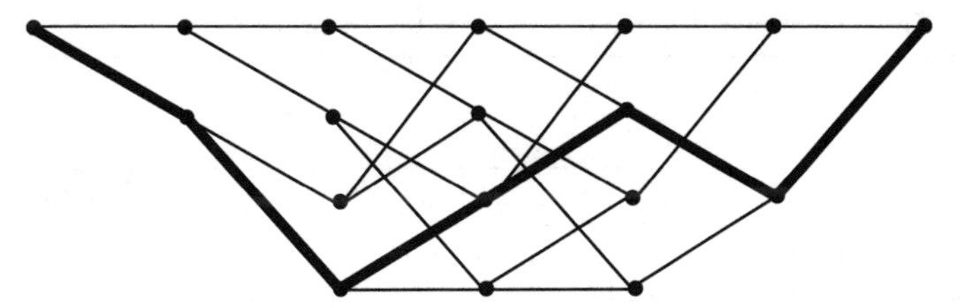

Figure 17.20 The path through the trellis.

(d) By comparing the union bound it is clear the performance is better. Comparing the minimum distances the code here has $\Delta_E(\min) = 10$ while the HCV codes has $\Delta_E(\min) = 7.2$.

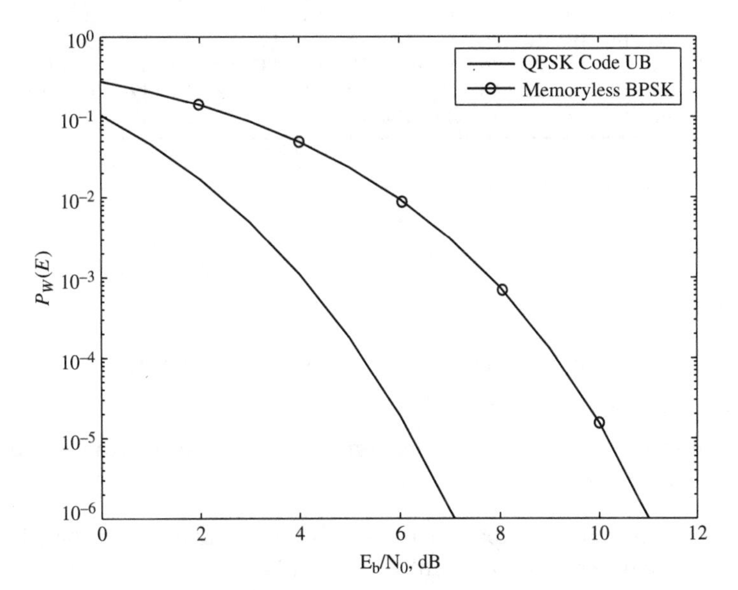

Figure 17.21 The union bound for the OMWM.

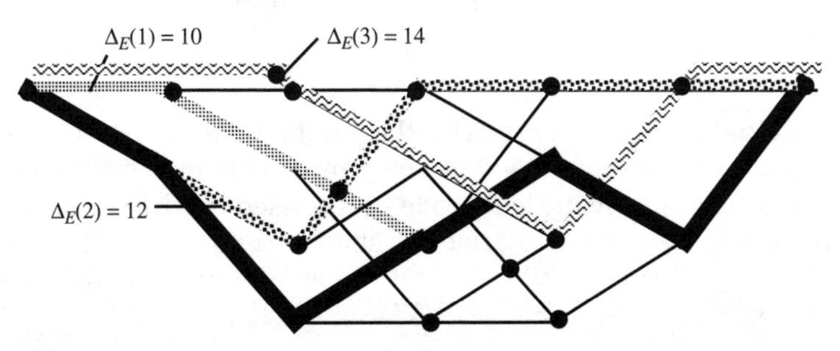

Figure 17.22 The true path and the error events.

TABLE 17.8 The tabulated Euclidean distance for the QPSK OMWM

	$\Delta_E(i, j)$															
	$\vec{I} = i$															
$\vec{I} = j$	0	1	2	3	4	5	6	7	8	9	10	11	12	13	14	15
0	0	10	10	12	10	12	12	14	10	20	12	14	12	14	14	16
1	10	0	12	10	12	10	14	12	20	10	14	12	14	12	16	14
2	10	12	0	10	12	14	10	12	12	14	10	20	14	16	12	14
3	12	10	10	0	14	12	12	10	14	12	20	10	16	14	14	12
4	10	12	12	14	0	10	10	12	12	14	14	16	10	20	12	14
5	12	10	14	12	10	0	12	10	14	12	16	14	20	10	14	12
6	12	14	10	12	10	12	0	10	14	16	12	14	12	14	10	20
7	14	12	12	10	12	10	10	0	16	14	14	12	14	12	20	10
8	10	20	12	14	12	14	14	16	0	10	10	12	10	12	12	14
9	20	10	14	12	14	12	16	14	10	0	12	10	12	10	14	12
10	12	14	10	20	14	16	12	14	10	12	0	10	12	14	10	12
11	14	12	20	10	16	14	14	12	12	10	10	0	14	12	12	10
12	12	14	14	16	10	20	12	14	10	12	12	14	0	10	10	12
13	14	12	16	14	20	10	14	12	12	10	14	12	10	0	12	10
14	14	16	12	14	12	14	10	20	12	14	10	12	10	12	0	10
15	16	14	14	12	14	12	20	10	14	12	12	10	12	10	10	0

17.8 Miniprojects

Goal: To give exposure

- to a small scope engineering design problem in communications.
- to the dynamics of working with a team.
- to the importance of engineering communication skills (in this case oral presentations).

Presentation: The forum will be similar to a design review at a company (only much shorter). The presentation will be of 5 minutes in length with an overview of the given problem and solution. The presentation will be followed by questions from the audience (your classmates and the professor). Each team member should be prepared to give the presentation.

17.8.1 Project 1

Project Goals: Build a practical orthogonal modulation with memory.

Convolutional codes are used in many practical applications. For example, a block diagram of a convolutional code that is used in wireless LAN applications and in deep space communications is shown in Figure 17.23. The summation operation is performed modulo-2 and the orthogonal modulation symbols are given as $\tilde{D}_z(2(m-1)+1) = (-1)^{J_1(m)}$ and $\tilde{D}_z(2(m-1)+2) = (-1)^{J_2(m)}$.

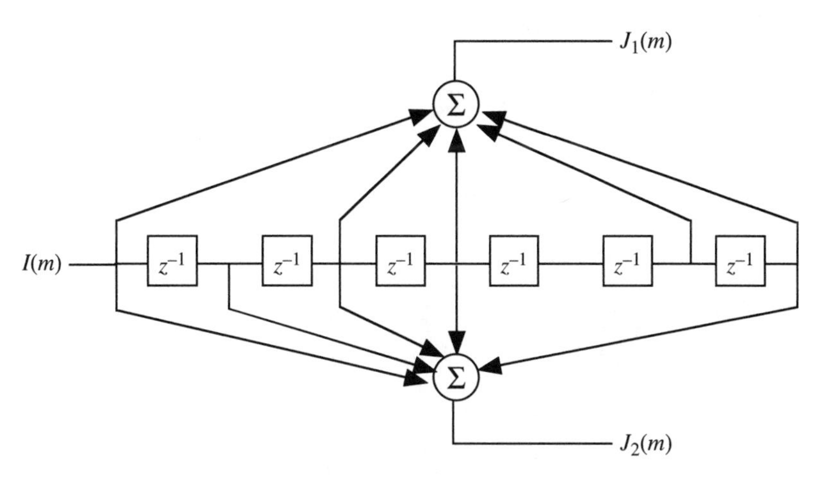

Figure 17.23 A powerful convolutional code.

(a) Characterize the OMWM by identifying the states of this modulation at time m, $\sigma(m)$ and how it relates to the sequence $I(m)$. *Hint:* There are 64 states.

(b) Enumerate the state transitions, $g_1(\sigma(m), I(m))$ for states $\sigma(m) = 1, \ldots, 8$.

(c) Assume the modulation starts in $\sigma(1) = 1$ and ends in $\sigma(N_t + 1) = 1$ where N_t is the number of trellis transitions in the frame and write software that is capable of generating the OMWM output for $K_b = 100$. A test sequence of 100 bits will be sent to the class via e-mail.

(d) Provide an estimate of the minimum Euclidean squared distance of this OMWM.

Appendix

Useful Formulas

Trigonometric Identities

$$\exp(j\theta) = \cos(\theta) + j\,\sin(\theta)$$

$$\cos(\theta) = \frac{1}{2}[\exp[j\theta] + \exp[-j\theta]]$$

$$\sin(\theta) = \frac{1}{2j}[\exp[j\theta] - \exp[-j\theta]]$$

$$\sin(a \pm b) = \sin(a)\cos(b) \pm \cos(a)\sin(b)$$

$$\cos(a \pm b) = \cos(a)\cos(b) \mp \sin(a)\sin(b)$$

$$\cos(a)\cos(b) = \frac{1}{2}[\cos(a-b) + \cos(a+b)]$$

$$\sin(a)\sin(b) = \frac{1}{2}[\cos(a-b) - \cos(a+b)]$$

$$\sin(a)\cos(b) = \frac{1}{2}[\sin(a-b) + \sin(a+b)]$$

Transcendental Functions

$$\mathrm{erf}(x) = \frac{2}{\sqrt{\pi}}\int_0^x \exp(-t^2)dt$$

$$\operatorname{erf}(x) = -\operatorname{erf}(-x) \qquad \operatorname{erfc}(x) = 1 - \operatorname{erf}(x)$$

$$J_n(x) = \frac{1}{2\pi} \int_{-\pi}^{\pi} \exp[-j(n\theta - x\sin(\theta))]d\theta$$

$$\operatorname{sinc}(x) = \frac{\sin(\pi x)}{\pi x}$$

Statistics Formulas

Gaussian density function

$$f_X(x) = \frac{1}{\sqrt{2\pi\sigma_x^2}} \exp\left(-\frac{(x-m_x)^2}{2\sigma_x^2}\right) \qquad m_x = E[X] \qquad \sigma_x^2 = E\left[(X-m_x)^2\right]$$

Bivariate Gaussian density function

$$f_{X_1,X_2}(x_1,x_2) = \frac{1}{2\pi\sigma_1\sigma_2\sqrt{1-\rho^2}} \exp\left(-\frac{1}{2(1-\rho^2)} \times \left(\frac{(x_1-m_1)^2}{\sigma_1^2} - \frac{2\rho(x_1-m_1)(x_2-m_2)}{\sigma_1\sigma_2} + \frac{(x_2-m_2)^2}{\sigma_2^2}\right)\right)$$

$$m_i = E[X_i], \quad \sigma_i^2 = E\left[(X_i-m_i)^2\right], \quad i = 1,2 \qquad \rho = E[(X_1-m_1)(X_2-m_2)]$$

Random Processes Formulas

Stationary Processes

$$R_N(\tau) = E[N(t)N(t-\tau)] \qquad S_N(f) = \mathcal{F}\{R_N(\tau)\}$$

Linear Systems and Stationary Processes

$$S_N(f) = |H_R(f)|^2 S_W(f)$$

Stationary Bandpass Noise

$$S_{N_c}(f) = \frac{1}{2}S_{N_z}(f-f_c) + \frac{1}{2}S_{N_z}(-f-f_c)$$

$$S_{N_I}(f) = \frac{S_{N_z}(f) + S_{N_z}(-f)}{4} \qquad S_{N_I N_Q}(f) = \frac{S_{N_z}(-f) - S_{N_z}(f)}{j4}$$

Fourier Transforms

TABLE 1.1 Fourier transform table

Time Function	Transform
$\exp\left[\frac{2T_1}{N_0}\right] \underset{i=0}{\overset{i=1}{\gtrless}} \exp\left[\frac{2T_0}{N_0}\right]$	$X(f) = T\,\frac{\sin(\pi f T)}{\pi f T}$
$x(t) = \begin{cases} 1 - \frac{\lvert t\rvert}{T} & -T \le t \le T \\ 0 & \text{elsewhere} \end{cases}$	$X(f) = T\left(\frac{\sin(\pi f T)}{\pi f T}\right)^2$
$x(t) = 2W\,\frac{\sin(2\pi W t)}{2\pi W t}$	$X(f) = \begin{cases} 1 & -W \le f \le W \\ 0 & \text{elsewhere} \end{cases}$
$x(t) = \begin{cases} \exp(-at) & t \le 0 \\ 0 & \text{elsewhere} \end{cases}$	$X(f) = \frac{1}{j2\pi f + a}$
$x(t) = A\exp\left(j2\pi f_0 t\right)$	$X(f) = A\delta(f - f_0)$

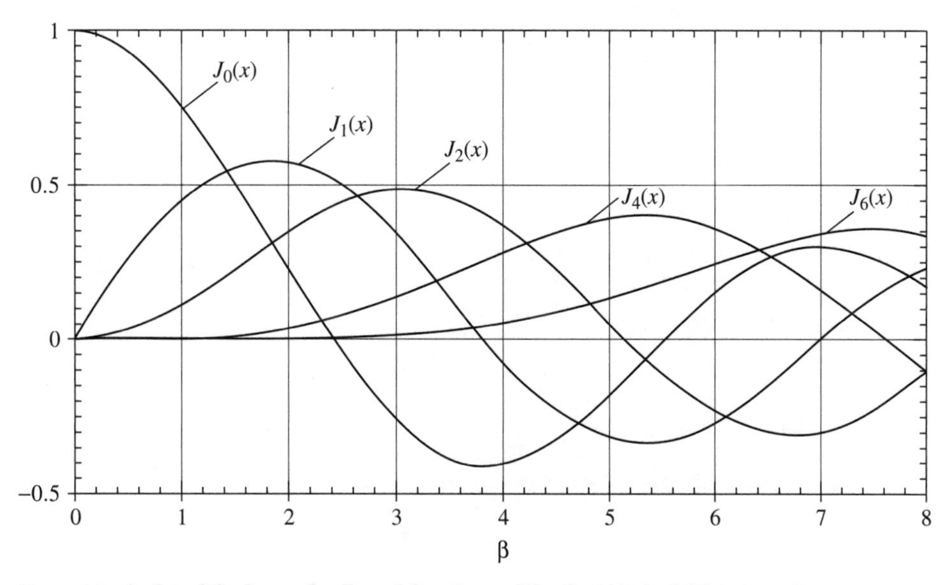

Figure 1.1 A plot of the low order Bessel functions of the first kind, $J_n(\beta)$ versus β.

Appendix

B

Notation

Signals and Systems

$x(t)$ = A signal
$\Re(\bullet)$ = Real part of a complex signal
$\Im(\bullet)$ = Imaginary part of a complex signal
x_n = Fourier series coefficients of a periodic $x(t)$
$X(f)$ = The Fourier transform of $x(t)$
$\mathcal{F}(\bullet)$ = The Fourier transform operator
P_x = The power of $x(t)$
E_x = The energy of $x(t)$
$V_x(\tau)$ = The correlation function of $x(t)$
$G_x(f)$ = The energy spectrum of $x(t)$
$h(t)$ = A filter impulse response
T_m = Measurement time interval
$S_x(f, T_m)$ = A sampled power spectral density
$\mathcal{L}(\bullet)$ = The Laplace transform operator
$H(f)$ = A filter transfer function

Random Variables

$P[A]$ = Probability of event A
$F_X(x)$ = Cumulative distribution function of a random variable X
$f_X(x)$ = Probability density function of a random variable X
σ_x = Standard deviation of a random variable X
m_X = Mean of a random variable X
$E[\bullet]$ = Expected value
ρ_{XY} = Correlation coefficient between random variable X and random variable Y
erf($\bullet$) = Erfc function
erfc($\bullet$) = Complimentaty erfc function

Bandpass Signals

$x_c(t)$ = Bandpass signal
$x_I(t)$ = In-phase signal
$x_Q(t)$ = Quadrature signal
$x_z(t)$ = Complex envelope signal
$x_A(t)$ = Amplitude signal
$x_P(t)$ = Phase signal
B_T = Transmission bandwidth
f_c = Carrier frequency

Analog Modulations

$m(t)$ = Analog message signal
W = Message signal bandwidth
τ_p = Propagation time delay
L_p = Propagation loss
ϕ_p = Propagation phase shift
E_T = Transmission efficiency
E_B = Bandwidth efficiency
A_c = Transmission amplitude
a = Modulation coefficient
$f_i(t)$ = Instantaneous frequency
$f_d(t)$ = Instantaneous frequency deviation
k_p = Phase deviation constant
k_f = Frequency deviation constant (radians/s/V)
f_k = Frequency deviation constant (Hz/V)
f_p = Peak frequency deviation
D = Bandwidth expansion factor
E_M = Multiplexing efficiency

Random Processes

$R_N(\tau), R_N(t_1, t_2)$ = Correlation function of the random process $N(t)$
$\rho_N(\tau), \rho_N(t_1, t_2)$ = Correlation coefficient of the process $N(t)$
$S_N(f)$ = Power spectrum of $N(t)$
$W(t)$ = White noise
N_0 = White noise power spectral density
B_N = Noise equivalent bandwidth
$N_z(t)$ = Complex envelope of a bandpass noise
$\tilde{N}_z(t)$ = Derotated complex envelope
SNR_b = Baseband maximum SNR
SNR_o = Demodulator output SNR

Digital Communications Basics

C = Channel capacity
E_b = Energy per bit
T = Symbol time
W_b = Bit rate
T_p = Transmission duration
$D_{x_z}(f)$ = Average energy spectrum per bit
η_B = Spectral efficiency (bits/s/Hz)
$\vec{I}$, I = Information bits
$\hat{\vec{I}}$, $\hat{I}$ = Information bit estimates
T_i = Maximum likelihood metric for $\vec{I} = i$
$T_{i|j}$ = Maximum likelihood metric for $\vec{I} = i$ when the actual transmitted signal is $\vec{I} = j$
$\Delta_E(\alpha, \beta)$ = Square Euclidean distance between $x_\alpha(t)$ and $x_\beta(t)$
$\{A_d(k), \Delta_E(k)\}$ = Squared Euclidean distance spectrum of a modulation
K_b = Number of bits to be transmitted per frame
π_i = A prior probability of $\vec{I} = i$

Linear Modulations

$u(t)$ = Modulation pulse shape
$D_i(k)$ = Modulation symbol
$a(\bullet)$ = Constellation mapping, i.e., $d_i = a(i)$, $i = 0, M - 1$
Ω_d = The constellation for a linear modulation
Q = Output of filter matched to $u(t)$

Orthogonal Modulations

$I(l)$ = Information bit to be transmitted at step l
$D_z(l)$ = Modulation symbol for orthogonal modulation
$s_l(t)$ = The spreading waveform for the lth bit for an orthogonal modulation
$u_r(t)$ = The rectangular pulse
T_u = Pulse shape duration for stream modulation
$Q(k)$ = The output for the matched filter to the kth spreading for an orthogonal modulation

Orthogonal Modulations with Memory

$\sigma(l)$ = Modulation state at time l
N_s = Number of states in a OMWM
$J(l)$ = Constellation label sequence, i.e., $D_z(l) = a(J(l))$
M_s = Number of points in the constellation, i.e., $J(l) = i$, $i = 0, M_s - 1$
$g_1(\bullet, \bullet)$ = The modulation finite state machine, i.e., $\sigma(l+1) = g_1\left(I(l), \sigma(l)\right)$

$g_2(\bullet, \bullet)$ = The output symbol selector mapping, i.e., $J(l) = g_2\left(I(l), \sigma(l)\right)$
$\vec{I}(l),\ I(l)$ = Information bits to be transmitted at symbol time l
N_f = Number of symbols transmitted
K_m = Number of bits per trellis transition
M_i = Number of branches per trellis transition, i.e., $M_i = 2^{K_s}$
N_m = Number of output symbols per trellis transition
R = Number of bits transmitted per symbol, i.e., $R = K_m/N_m$

Appendix

C

Acronyms

Signals and Systems

ADC = Analog to digital converter
BPF = Bandpass filter
DC = Direct current, often refers to zero frequency
DTFT = Discrete time Fourier transform
I = In-phase
LPF = Lowpass filter
LTI = Linear Time Invariant
Q = Quadrature
SNR = Signal to noise ratio

Random Variables and Processes

AWGN = Additive white Gaussian noise
CDF = Cumulative distribution function
ChF = Characteristic function
CLT = Central limit theorem
PDF = Probability density function
PMF = Probability mass function
PSD = Power spectral density
RV = Random variable
WSS = Wide sense stationary

Analog Communication

AM = Amplitude modulation
DSB-AM = Double sideband AM
FDM = Frequency division multiplexing
FM = Frequency modulation

LC-AM = Large carrier AM
MCPR = Message–to–Carrier Power Ratio
PAPR = Peak–to–Average Power Ratio
PLL = Phase–locked loop
PM = Phase modulation
QCM = Quadrature carrier multiplexing
SSB-AM = Single sideband AM
TV = Television
VCO = Voltage controlled oscillator
VSB-AM = Vestigial sideband AM

Digital Communication

AMI = Alternate mark inversion
APP = A posteriori probability
BCJR = Bahl, Cocke, Jelinek, and Raviv—an algorithm named after the inventors
BEP = Bit error probability
BFSK = Binary frequency shift keying
BPSK = Binary phase shift keying
DTMF = Discrete tone multiple frequency
EVM = Error vector magnitude
GSM = Global system for mobile communications
IEEE = Institute of electrical and electronic engineers
ISI = Intersymbol interference
MAP = Maximum a posteriori
MAPBD = Maximum a posteriori bit demodulation
MLBD = Maximum likelihood bit demodulation
MLWD = Maximum likelihood word demodulation
MFSK = M-ary frequency shift keying
MPSK = M-ary phase shift keying
MSK = Minimum shift keying
OCDM = Orthogonal code division multiplexing
OFDM = Orthogonal frequency division multiplexing
OMWM = Orthogonal modulation with memory
PPM = Pulse position modulation
PWM = Pulse width modulation
PWEP = Pairwise error probability
QPSK = Quadri-phase shift keying
RLL = Run length limited
VPSK = Variable phase shift keying

Appendix

D

Fourier Transforms: *f* versus ω

This course and communications engineers in general use the variable f as opposed to $\omega = 2\pi f$ when examining the frequency domain representations of signals and systems. The reason is that units of Hertz have more of an intuitive feel than the units of radians per second. Often undergraduate courses in signals and systems develop the theory with the notation ω. This appendix is provided to show the differences in the results that arise from these two views of frequency domain analysis. First we shall define the Fourier transform for the two variables

$$X(f) = \int_{-\infty}^{\infty} x(t)e^{-j2\pi f t} dt = \mathcal{F}\{x(t)\} \qquad X(\omega) = \int_{-\infty}^{\infty} x(t)e^{-j\omega t} dt \tag{4.1}$$

The inverse Fourier transform is given as

$$x(t) = \int_{-\infty}^{\infty} X(f)e^{j2\pi f t} df = \mathcal{F}^{-1}\{X(f)\} \qquad x(t) = \frac{1}{2\pi}\int_{-\infty}^{\infty} X(\omega)e^{j\omega t} d\omega \tag{4.2}$$

The other convenient thing about working with the variable f is that the term $(2\pi)^{-1}$ does not need to be included in the inverse transform. A list of Fourier transform properties of different mathematical operations are listed in Table 4.1. Additionally Fourier transforms of some commonly used signals are listed in Table 4.2.

TABLE 4.1 Fourier transforms for some common mathematical operations

Operation	Time Function, $x(t)$	Transform, $X(f)$	Transform, $X(\omega)$
Reversal	$x(-t)$	$X(-f)$	$X(-\omega)$
Symmetry	$X(t)$	$x(-f)$	$2\pi x(-\omega)$
Scaling	$x(at)$	$\frac{1}{\lvert a \rvert} X\left(\frac{f}{a}\right)$	$\frac{1}{\lvert a \rvert} X\left(\frac{\omega}{a}\right)$
Time Delay	$x(t - t_0)$	$X(f)e^{-j2\pi f t_0}$	$X(\omega)e^{-j\omega t_0}$
Time Differentiation	$\frac{d^n}{dt^n} x(t)$	$(j2\pi f)^n X(f)$	$(j\omega)^n X(\omega)$
Energy	$E_x = \int_{-\infty}^{\infty} \lvert x(t) \rvert^2 dt$	$E_x = \int_{-\infty}^{\infty} \lvert X(f) \rvert^2 dt$	$E_x = \frac{1}{2\pi} \int_{-\infty}^{\infty} \lvert X(\omega) \rvert^2 dt$
Frequency Translation	$x(t)e^{j2\pi f_c t} = x(t)e^{j\omega_c t}$	$X(f - f_c)$	$X(\omega - \omega_c)$
Convolution	$x(t) * h(t)$	$X(f)H(f)$	$X(\omega)H(\omega)$
Multiplication	$x(t)y(t)$	$X(f) * Y(f)$	$\frac{1}{2\pi} X(\omega) * Y(\omega)$

TABLE 4.2 Fourier transforms for some common signals

Time Function, $x(t)$	Transform, $X(f)$	Transform, $X(\omega)$
$x(t) = \begin{cases} 1 & -T_p/2 \le t \le T_p/2 \\ 0 & \text{elsewhere} \end{cases}$	$T_p \frac{\sin(\pi f T_p)}{\pi f T_p}$	$T \frac{\sin(\omega T_p/2)}{\omega T_p/2}$
$x(t) = \begin{cases} \exp(-at) & 0 \le t \\ 0 & \text{elsewhere} \end{cases}$	$\frac{1}{j2\pi f + a}$	$\frac{1}{j\omega + a}$
$x(t) = A$	$A\delta(f)$	$2\pi A\delta(\omega)$
$x(t) = Ae^{j\omega_0 t} = Ae^{j2\pi f_0 t}$	$A\delta(f - f_0)$	$2\pi A\delta(\omega - \omega_0)$

Appendix

Further Reading and Bibliography

[Ae72] M. Abramowitz and I. E. Stegun (eds). *Handbook of Mathematical Functions*. U. S. Department of Commerce, Washington, DC, 1972.

[Agi00] Engineers of Agilent. Using error vector magnitude measurements to analyze and troubleshoot vector-modulated signals. Agilent product note 89400-14, Agilent Technologies, 2000.

[Ala98] S. M. Alamouti. A simple transmit diversity technique for wireless communications. *IEEE Trans. on Select Areas in Comm.*, 16(8):1451–1458, October 1998.

[Arm35] Edwin A. Armstrong. A method of reducing disturbances in radio signaling by a system of frequency modulation. In *IRE Conference*, New York, November 1935.

[BB99] S. Benedetto and E. Biglieri. *Principles of Digital Transmission*. Klewer Academic Press, New York, 1999.

[BDMS91] E. Biglieri, D. Divaslar, P. J. McLane, and M. K. Simon. *Introduction to Trellis-Coded Modulations with Applications*. Macmillan, New York, 1991.

[Bed62] E. Bedrosian. The analytical signal representation of modulated waveforms. *Proceedings of the IRE*, pp. 2071–2076, October 1962.

[Big86] E. Biglieri. Ungerboeck codes do not shape the signal power spectrum. *IEEE Info. Theory*, IT-32:595–596, July 1986.

[Bra95] John Bray. *The Communications Miracle*. Plenum, New York, 1995.

[Bre59] D. G. Brennan. Linear diversity combining techniques. *Proceedings of the IRE*, 47:1075–1102, 1959.

[Bur04] Russel W. Burns. *Communications: A International History of the Formative Years*. Institute of Electrical Engineers, Herts, UK, 2004.

[CAC01] K. Chugg, A. Anastasopoulos, and X. Chen. *Iterative Detection*. Kluwer Academic Press, Boston, 2001.

[Car22] J. R. Carson. Notes on the theory of modulation. *Proc. IRE*, vol. 10:57–64, February 1922.

[Car23] J. R. Carson. Method and means for signaling with high frequency waves. Technical Report Patent 1449382, U.S. Patent Office, 1923.

[Car26] J. R. Carson. *Electric Circuit Theory and Operational Calculus*. McGraw-Hill, New York, 1926.

[CF37] J. R. Carson and T. C. Fry. Variable frequency electric circuit theory with application to the theory of frequency modulation. *Bell Systems Technical Journal*, vol. 16:513–540, 1937.

[Cha66] R. W. Chang. Synthesis of band-limited orthogonal signal for multichannel data transmission. *Bell Systems Technical Journal*, vol. 45:1775–1796, December 1966.

[CM86] G. R. Cooper and C. D. McGillem. *Probabilistic Methods of Signals and Systems Analysis*. Holt, Rinehart and Winston, New York, 1986.

[Com04] Federal Communication Commission. Telecommunication rules and regulations. Technical Report Title 47 of the Code of Federal Regulations, Government Printing Office, `http://wireless.fcc.gov/rules.html`, 2004.

[Cos56] John P. Costas. Synchronous communications. *Proceedings of the IRE*, pp. 1713–1718, December 1956.

[Cou93] L. W. Couch. *Digital and Analog Communication Systems*. Macmillan, New York, 1993.

[Cra91] J. W. Craig. A new simple and exact result for caluclating the probability of error for two dimensional signal constellations. In *IEEE Military Communication Conference*, pp. 571–575, November 1991.

[CT92] T. M. Cover and J. A. Thomas. *Elements of Information Theory*. Wiley Interscience, New York, 1992.

[Dev00] Jay L. Devore. *Probability and Statistics for Engineering and the Sciences 5*. Duxbury, 511 Forest Lodge Rd., Pacific Grove, CA 93950, 2000.

[DR87] W. B. Davenport and W. L. Root. *An Introduction to the Theory of Random Signals and Noise*. IEEE Press, New York, 1987.

[FGL+84] G. D. Forney, R. G. Gallager, G. R. Lang, F. M. Longstaff, and S. U. Qureshi. Efficient modulation for band-limited channels. *IEEE JSAC*, vol. SAC-2:632–647, 1984.

[For91] G. D. Forney. Geometrically uniform codes. *IEEE Info. Theory*, 37:1241–1960, September 1991.

[Gal01] R. Gallagher. *Discrete Stochastic Processes*. Klewer, 2001.

[Gra53] F. Gray. Pulse code communications. Technical Report Patent 2632058, U.S. Patent Office, 1953.

[Har28] R. V. L. Hartley. Transmission of information. *Bell Systems Technical Journal*, p. 535, July 1928.

[Haw05] Stephen Hawking. *God Created the Integers*. Running Press, Philadelphia, PA, 2005.

[Hay83] S. Haykin. *Communications Systems*. John Wiley and Sons, New York, 1983.

[HCV93] P. Ho, J. Cavers, and J. Varaldi. The effect of constellation density on trellis coded modulation in fading channels. *IEEE Trans. Veh. Technol.*, vol. VT-42:318–325, 1993.

[Hel91] C. W. Helstrom. *Probability and Stochastic Processes for Engineers*. Macmillan, New York, 1991.

[Huu03] Anton A. Huurdeman. *The Worldwide History of Telecommunications*. Wiley-Interscience, Hoboken, NJ, 2003.

[Jak74] W. C. Jakes. *Microwave Mobile Communications*. Wiley, New York, 1974.

[Joh28] J. B. Johnson. Thermal agitation of electricity in conductors. *Physical Review*, vol. 32:97–109, July 1928.

[KH97] E. W. Kamen and B. S. Heck. *Fundamentals of Signals and Systems Using Matlab*. Prentice-Hall, Englewood Cliffs, NJ, 1997.

[Kot60] V. A. Kotel'nikov. *Theory of Optimum Noise Immunity*. McGraw-Hill Book Company, New York, 1960.

[KR04] James F. Kurose and Keith W. Ross. *Computer Networking: A Top-Down Approach Featuring the Internet*. Addison Wesley, 2004.

[Kur76] Carl F. Kurth. Generation of single-sideband signals in multiplex communication systems. *IEEE Trans. on Circuits and Systems*, CAS-23:1–17, January 1976.

[LC04] S. Lin and D. Costello. *Error Control Coding*. Prentice-Hall, Englewood Cliffs, NJ, 2004.

[Lee82] W. C. Y. Lee. *Mobile Communications Engineering*. McGraw-Hill, New York, 1982.

[Leh86] E. L. Lehmann. *Testing Statistical Hypothesis*. Wiley Interscience, New York, 1986.

[Les69] Lawrence Lessing. *Man of High Fidelity*. Bantam, New York, 1969.

[LG89] A. Leon-Garcia. *Probability and Random Processes for Electrical Enginering*. Addison-Wesley, New York, 1989.

[LGW00] A Leon-Garcia and I. Widjaja. *Communication Networks*. McGraw-Hill, Boston, 2000.

[May84] Robert J. Mayhan. *Discrete-Time and Continuous-Time Linear Systems*. Addison-Wesley, New York, 1984.

[Mil63] A. Miller. Transmission system. Technical report, U.S. Patent Office, Patent 3,108,261, 1963.

[Mit98] S. K. Mitra. *Digital Signal Processing*. McGraw-Hill, 1998.

[MO98] S. Miller and R. O'Dea. Peak power and bandwidth efficient linear modulation. *IEEE Trans. Commun.*, 46:1639–1648, December 1998.

[Nor43] D. O. North. Analysis of factors which determine signal-noise discrimination in pulsed carrier systems. Technical Report PTR-6C, RCA Technical Report, June 1943. Reprinted in Proc. IEEE pp. 1015–1027, July 1963.

[Nyq28a] H. Nyquist. Certan topics in telegraph transmission theory. *AIEE Transactions*, vol. 47:617–644, 1928.

[Nyq28b] H. Nyquist. Thermal agitation of lectric charge in conductors. *Physical Review*, vol. 32:110–113, July 1928.

[OS99] A. V. Oppenheim and R. W. Schafer. *Digital Signal Processing*. Prentice-Hall, Englewood Cliffs, NJ, 1999.

[Osw56] A. A. Oswald. Early history of single side band. *Proc. of IRE*, vol. 44:1676–1679, December 1956.

[OW97] A. V. Oppenheim and A. S. Willsky. *Signals and Systems*. Prentice-Hall, Upper Saddle River, NJ, 1997.

[Pap84] A. Papoulis. *Probability, Random Variables and Stochastic Processes*. McGraw-Hill, New York, 1984.

[Pet89] Blake Peterson. Spectrum analysis basics. Application note 150, Hewlett Packard, 1989.

[PM88] J. G. Proakis and Manolakis. *Introduction to Digital Signal Processing*. Macmillan, 1988.

[Poo88] H. V. Poor. *An Introduction to Signal Detection and Estimation*. Springer-Verlag, New York, 1988.

[Por97] Boaz Porat. *A Course in Digital Signal Processing*. Wiley, 1997.

[Pro89] J. G. Proakis. *Digital Communications*. McGraw-Hill, New York, 1989.

[PS94] J. G. Proakis and Masoud Salehi. *Communications System Engineering*. Prentice-Hall, Englewood Cliffs, NJ, 1994.

[Ric45] S. O. Rice. Mathematical analysis of random noise. *Bell System Technical Journal*, 24:96–157, 1945.

[Ric48] S. O. Rice. Statistical properties of a sine-wave plus random noise. *Bell System Technical Journal*, vol. 27:109–157, 1948.

[SB89] D. Saha and T. Birdsall. Quadrature-quadrature phase-shift keying. *IEEE Trans. Commun.*, COM-37:437–448, May 1989.

[SBS66] M. Schwartz, W. R. Bennett, and S. Stein. *Communication Systems and Techniques*. McGraw-Hill, New York, 1966.

[Sha48] Claude Shannon. A mathematical theory of communication. *Bell System Technical Journal*, 1948.

[Skl88] B. Sklar. *Digital Communications*. Prentice-Hall, Englewood Cliffs, NJ, 1988.

[SS87] W. E. Saban and E. Schoenike. *Single Sideband Systems and Circuits*. McGraw-Hill, New York, 1987.

[Sti99] David Stirzaker. *Probability and Random Variables: A Beginner's Guide*. Cambridge University Press, 40 West 20th St. New York, NY 10011-4211, 1999.

[SW49] Claude E. Shannon and Warren Weaver. *The Mathematical Theory of Communication*. University of Illinois Press, Urbana, IL, 1949.

[SW02] Henry Stark and John W. Woods. *Probability and Random Processes with Applications to Signal Processing*. Prentice-Hall, Upper Saddle River, NJ, 2002.

[Tan02] Andrew S. Tanenbaum. *Computer Networks, Fourth Edition*. Prentice-Hall, 2002.

[Ung82] Gottfried Ungerboeck. Channel coding with multilevel/phase signals. *IEEE Info. Theory*, IT-28:55–67, January 1982.

[VO79] A. J. Viterbi and J. K. Omura. *Principles of Digital Communication and Coding*. McGraw-Hill, New York, 1979.

[Wal97] H. R. Walker. VPSK and VMSK modulation transmit digital audio and video at 15 bit/s/Hz. *IEEE Trans. on Broadcasting*, 43:96–103, 1997.

[WB00] J.C. Whitaker and K. B. Benson. *Standard Handbook of Video and Television Engineering*. McGraw-Hill, New York, 2000.

[Web87] C. L. Weber. *Elements of Detection and Signal Design*. Springer Verlag, New York, 1987. Reprint of a 1967 book.

[Wes01] Douglas B. West. *Introduction to Graph Theory*. Prentice-Hall, 2001.

[Wic95] S. B. Wicker. *Error Control Systems for Digital Communications and Storage*. Prentice-Hall, Upper Saddle River, NJ, 1995.

[Wie49] Norbert Wiener. *The Extrapolation, Interpolation, and Smoothing of Stationary Time Series with Engineering Applications*. Wiley, New York, 1949.

[WJ65] J. M. Wozencraft and I. M. Jacobs. *Principles of Communications Engineering*. John Wiley and Sons, New York, 1965.

[YG98] Roy D. Yates and David J. Goodman. *Probability and Stochastic Processes: A Friendly Introduction for Electrical and Computer Engineers*. Wiley, New York, 1998.

[Zie86] A. Van Der Ziel. *Noise in Solid State Devices*. Wiley-Interscience, New York, 1986.

[ZTF89] R. E. Ziemer, W. H. Trantor, and D. R. Fannin. *Signals and Systems: Continuous and Discrete*. Macmillan, New York, 1989.

Index